Power Systems Handbook

Short-Circuits in AC and DC Systems
ANSI/IEEE and IEC Standards
Volume 1

Power Systems Handbook

Series Author
J.C. Das
Power System Studies, Inc., Snellville, Georgia, USA

Volume 1: Short-Circuits in AC and DC Systems:
ANSI, IEEE, and IEC Standards

Volume 2: Load Flow Optimization and Optimal Power Flow

Volume 3: Harmonic Generation Effects Propagation and Control

Volume 4: Power System Protective Relaying

Short-Circuits in AC and DC Systems
ANSI, IEEE, and IEC Standards
Volume 1

J.C. Das

CRC Press is an imprint of the
Taylor & Francis Group, an **informa** business

CRC Press
Taylor & Francis Group
6000 Broken Sound Parkway NW, Suite 300
Boca Raton, FL 33487-2742

© 2018 by Taylor & Francis Group, LLC
CRC Press is an imprint of Taylor & Francis Group, an Informa business

No claim to original U.S. Government works

Printed on acid-free paper

International Standard Book Number-13: 978-1-4987-4541-3

This book contains information obtained from authentic and highly regarded sources. Reasonable efforts have been made to publish reliable data and information, but the author and publisher cannot assume responsibility for the validity of all materials or the consequences of their use. The authors and publishers have attempted to trace the copyright holders of all material reproduced in this publication and apologize to copyright holders if permission to publish in this form has not been obtained. If any copyright material has not been acknowledged please write and let us know so we may rectify in any future reprint.

Except as permitted under U.S. Copyright Law, no part of this book may be reprinted, reproduced, transmitted, or utilized in any form by any electronic, mechanical, or other means, now known or hereafter invented, including photocopying, microfilming, and recording, or in any information storage or retrieval system, without written permission from the publishers.

For permission to photocopy or use material electronically from this work, please access www.copyright.com (http://www.copyright.com/) or contact the Copyright Clearance Center, Inc. (CCC), 222 Rosewood Drive, Danvers, MA 01923, 978-750-8400. CCC is a not-for-profit organization that provides licenses and registration for a variety of users. For organizations that have been granted a photocopy license by the CCC, a separate system of payment has been arranged.

Trademark Notice: Product or corporate names may be trademarks or registered trademarks, and are used only for identification and explanation without intent to infringe.

Visit the Taylor & Francis Web site at
http://www.taylorandfrancis.com

and the CRC Press Web site at
http://www.crcpress.com

Contents

Series Preface xv
Preface to Volume 1: Short-Circuits in AC and DC Systems xvii
Author xix

1. Design and Analyses Concepts of Power Systems 1
   1.1   Static and Dynamic Systems 2
   1.2   State Variables 3
   1.3   Linear and Nonlinear Systems 5
           1.3.1   Property of Decomposition 6
   1.4   Linearizing a Nonlinear System 6
   1.5   Time-Invariant Systems 9
   1.6   Lumped and Distributed Parameters 11
   1.7   Optimization 12
   1.8   Planning and Design of Electrical Power Systems 12
   1.9   Electrical Standards and Codes 14
   1.10 Reliability Analyses 15
           1.10.1 Availability 16
                  1.10.1.1 Exponential Distribution 17
           1.10.2 Data for Reliability Evaluations 18
           1.10.3 Methods of Evaluation 18
           1.10.4 Reliability and Safety 22
   1.11 Extent of System Modeling 25
           1.11.1 Short-Circuit Calculations 26
           1.11.2 Load Flow Calculations 28
           1.11.3 Harmonic Analysis 28
   1.12 Power System Studies 28
   1.13 Power System Studies Software 29
   1.14 System of Units 30
   Problems 30
   References 32

2. Modern Electrical Power Systems 35
   2.1   Classification 35
           2.1.1   Utility Companies in the USA 36
           2.1.2   North American Power System Interconnections 37
   2.2   Deregulation of Power Industry 38
           2.2.1   Generation Company (GENCO) 38
           2.2.2   Transmission Company (TRANSCO) 39
           2.2.3   Distribution Company (DISTCO) 39
   2.3   The New Energy Platform 39
           2.3.1   Sustainable, Renewable, and Green Energy 40
           2.3.2   Green Energy 41
           2.3.3   Hydroelectric Plants 41
           2.3.4   Pumped Storage Hydroelectric Plants 43

v

		2.3.5	Nuclear Power	43
			2.3.5.1 Breeder Reactors	47
			2.3.5.2 Nuclear Fusion	47
			2.3.5.3 Nuclear Power around the Globe	48
			2.3.5.4 Is Nuclear Power Green Energy?	48

2.3.6 Geothermal Plants 49
2.3.7 Solar and Wind Energy 50
2.3.8 Biofuels and Carbon-Neutral Fuels 50
2.3.9 Local Green Energy Systems 51
2.3.10 Fuel Cells 51
2.3.11 Reducing Caron Emissions 52

2.4 Large Power Stations of the World 53
2.5 Smart Grid 57
    2.5.1 Legislative Measures 58
    2.5.2 Technologies Driving Smart Grid 58
2.6 Microgrids and Distributed Generation 59
2.7 Energy Storage 63
    2.7.1 Flywheel Storage 64
    2.7.2 Superconductivity 67
        2.7.2.1 Applications in Electrical Systems 68
2.8 Transmission Systems 68
2.9 Industrial Systems 69
2.10 Distribution Systems 71
    2.10.1 The Radial System 72
    2.10.2 The Parallel or Loop System 73
    2.10.3 Network or Grid System 73
    2.10.4 Primary Distribution System 75
2.11 Future Load Growth 77
2.12 Underground versus OH Systems 77
    2.12.1 Spot Network 78
2.13 HVDC Transmission 80
    2.13.1 HVDC Light 80
    2.13.2 HVDC Configurations and Operating Modes 80
Problems 82
Bibliography 82
IEEE Color Books 84

3. Wind and Solar Power Generation and Interconnections with Utility 85

3.1 Prospective of Wind Generation in the USA 85
3.2 Characteristics of Wind Power Generation 87
    3.2.1 Maximum Transfer Capability 91
    3.2.2 Power Reserves and Regulation 92
    3.2.3 Congestion Management 93
3.3 Wind Energy Conversion 93
    3.3.1 Drive Train 94
    3.3.2 Towers 96
    3.3.3 Rotor Blades 96
3.4 The Cube Law 97
3.5 Operation 99

Contents vii

	3.5.1	Speed Control	101
	3.5.2	Behavior under Faults and Low-Voltage Ride Through	102
3.6		Wind Generators	103
	3.6.1	Induction Generators	103
	3.6.2	Direct Coupled Induction Generator	105
	3.6.3	Induction Generator Connected to Grid through Full Size Converter	105
	3.6.4	Doubly Fed Induction Generator	106
	3.6.5	Synchronous Generators	107
3.7		Reactive Power and Wind Turbine Controls	107
3.8		Power Electronics and Harmonics	111
	3.8.1	Power Electronics	111
	3.8.2	Harmonics	112
3.9		Computer Modeling	113
	3.9.1	A Wind Turbine Controller	113
3.10		Solar Power	115
3.11		CSP Plants	116
	3.11.1	Solar Energy Collectors	116
		3.11.1.1 Parabolic Dish Concentrators	117
		3.11.1.2 Solar Tower	118
	3.11.2	Trackers	118
		3.11.2.1 Photovoltaic Trackers	119
3.12		Direct Conversion of Solar Energy through PV Cells	120
	3.12.1	Cells, Modules, Panels, and Systems	120
		3.12.1.1 PV Module	120
		3.12.1.2 PV Panel	121
		3.12.1.3 PV Array	121
		3.12.1.4 PV Array Subfield	121
3.13		Classification of Solar Cells	121
3.14		Utility Connections of Distributed Resources	123
	3.14.1	Voltage Control	123
	3.14.2	Grounding	123
	3.14.3	Synchronizing	123
	3.14.4	Distribution Secondary Spot Networks	124
	3.14.5	Inadvertent Energization	124
	3.14.6	Metering	124
	3.14.7	Isolation Device	124
	3.14.8	EMI Interference	124
	3.14.9	Surge Withstand	125
	3.14.10	Paralleling Device	125
	3.14.11	Area Faults	125
	3.14.12	Abnormal Frequencies	125
	3.14.13	Reconnection	125
	3.14.14	Harmonics	125
Problems			126
References			127

viii *Contents*

4. Short-Circuit Currents and Symmetrical Components 131
  4.1   Nature of Short-Circuit Currents 132
  4.2   Symmetrical Components 135
  4.3   Eigenvalues and Eigenvectors 138
  4.4   Symmetrical Component Transformation 139
      4.4.1   Similarity Transformation 139
      4.4.2   Decoupling a Three-Phase Symmetrical System 141
      4.4.3   Decoupling a Three-Phase Unsymmetrical System 145
      4.4.4   Power Invariance in Symmetrical Component
              Transformation 146
  4.5   Clarke Component Transformation 146
  4.6   Characteristics of Symmetrical Components 150
  4.7   Sequence Impedance of Network Components 153
      4.7.1   Construction of Sequence Networks 153
      4.7.2   Transformers 155
             4.7.2.1   Delta–Wye or Wye–Delta Transformer 155
             4.7.2.2   Wye–Wye Transformer 157
             4.7.2.3   Delta–Delta Transformer 158
             4.7.2.4   Zigzag Transformer 158
             4.7.2.5   Three-Winding Transformers 159
      4.7.3   Static Load 163
      4.7.4   Synchronous Machines 163
  4.8   Computer Models of Sequence Networks 168
  Problems 170
  Bibliography 171

5. Unsymmetrical Fault Calculations 173
  5.1   Line-to-Ground Fault 173
  5.2   Line-to-Line Fault 175
  5.3   Double Line-to-Ground Fault 177
  5.4   Three-Phase Fault 179
  5.5   Phase Shift in Three-Phase
      Transformers 180
      5.5.1   Transformer Connections 180
      5.5.2   Phase Shifts in Winding Connections 180
      5.5.3   Phase Shift for Negative Sequence
              Components 183
  5.6   Unsymmetrical Fault Calculations 186
  5.7   System Grounding 193
      5.7.1   Solidly Grounded Systems 195
      5.7.2   Resistance Grounding 196
             5.7.2.1   High-Resistance Grounded Systems 197
             5.7.2.2   Coefficient of Grounding 203
  5.8   Open Conductor Faults 204
      5.8.1   Two-Conductor Open Fault 204
      5.8.2   One-Conductor Open 204
  Problems 209
  Bibliography 211
  References 211

Contents

ix

6. Matrix Methods for Network Solutions 213
  6.1 Network Models 213
  6.2 Bus Admittance Matrix 214
  6.3 Bus Impedance Matrix 219
    6.3.1 Bus Impedance Matrix from Open-Circuit Testing 220
  6.4 Loop Admittance and Impedance Matrices 221
    6.4.1 Selection of Loop Equations 223
  6.5 Graph Theory 223
  6.6 Bus Admittance and Impedance Matrices by Graph
      Approach 226
    6.6.1 Primitive Network 226
    6.6.2 Incidence Matrix from Graph Concepts 228
    6.6.3 Node Elimination in Y-Matrix 232
  6.7 Algorithms for Construction of Bus Impedance Matrix 233
    6.7.1 Adding a Tree Branch to an Existing Node 234
    6.7.2 Adding a Link 236
    6.7.3 Removal of an Uncoupled Branch 238
    6.7.4 Changing Impedance of an Uncoupled Branch 238
    6.7.5 Removal of a Coupled Branch 238
  6.8 Short-Circuit Calculations with Bus Impedance Matrix 246
    6.8.1 Line-to-Ground Fault 246
    6.8.2 Line-to-Line Fault 246
    6.8.3 Double Line-to-Ground Fault 247
  6.9 Solution of Large Network Equations 256
  Problems 257
  Bibliography 258

7. Current Interruptions in AC Networks 259
  7.1 Rheostatic Breaker 259
  7.2 AC Arc Interruption 261
    7.2.1 Arc Interruption Theories 261
      7.2.1.1 Cassie's Theory 261
      7.2.1.2 Mayr's Theory 262
      7.2.1.3 Cassie-Mayr Theory 262
  7.3 Current-Zero Breaker 263
  7.4 Transient Recovery Voltage 264
    7.4.1 First Pole to Clear Factor 266
  7.5 The Terminal Fault 269
    7.5.1 Four-Parameter Method 269
    7.5.2 Two-Parameter Representation 270
  7.6 The Short-Line Fault 271
  7.7 Interruption of Low Inductive Currents 273
    7.7.1 Virtual Current Chopping 275
  7.8 Interruption of Capacitance Currents 276
  7.9 TRV in Capacitive and Inductive Circuits 278
  7.10 Prestrikes in Circuit Breakers 279
  7.11 Overvoltages on Energizing HV Lines 280
    7.11.1 Overvoltage Control 282
    7.11.2 Synchronous Operation 283

x   *Contents*

	7.11.3	Synchronous Capacitor Switching 283
	7.11.4	Shunt Reactors...284
		7.11.4.1 Oscillation Modes 287
7.12	Out-of-Phase Closing...288
7.13	Resistance Switching 289
7.14	Failure Modes of Circuit Breakers..293
7.15	Stresses in Circuit Breakers 295
7.16	Classification of Circuit Breakers according to Interrupting Medium 295
	7.16.1	SF_6 Circuit Breakers 296
		7.16.1.1 Electronegativity of SF_6 297
	7.16.2	Operating Mechanisms...299
	7.16.3	Vacuum Interruption..300
		7.16.3.1 Current Chopping and Multiple Ignitions 301
		7.16.3.2 Switching of Unloaded Dry-Type Transformers.......303
7.17	Part Winding Resonance in Transformers 304
	7.17.1	Snubber Circuits..306
7.18	Solid-State Circuit Breakers..306
Problems...308
Bibliography 309
References 310

8. Application and Ratings of Circuit Breakers and Fuses according to ANSI Standards...313

8.1	Total and Symmetrical Current Basis 314
8.2	Asymmetrical Ratings...316
	8.2.1	Contact Parting Time 316
8.3	Voltage Range Factor K...317
8.4	Circuit Breaker Timing Diagram...320
8.5	Maximum Peak Current 321
8.6	Permissible Tripping Delay 322
8.7	Service Capability Duty Requirements and Reclosing Capability 322
	8.7.1	Transient Stability on Fast Reclosing 323
8.8	Shunt Capacitance Switching..326
	8.8.1	Switching of Cables 331
8.9	Line Closing Switching Surge Factor 335
	8.9.1	Switching of Transformers 336
8.10	Out-of-Phase Switching Current Rating..337
8.11	Transient Recovery Voltage 337
	8.11.1	Circuit Breakers Rated Below 100 kV..338
	8.11.2	Circuit Breakers Rated 100 kV and Above.....................................338
	8.11.3	Short-Line Faults 342
	8.11.4	Oscillatory TRV...344
		8.11.4.1 Exponential (Overdamped) TRV 344
	8.11.5	Initial TRV..345
	8.11.6	Adopting IEC TRV Profiles in IEEE Standards 345
	8.11.7	Definite-Purpose TRV Breakers...350
8.11.8 TRV Calculation Techniques 350

Contents   xi

8.12	Generator Circuit Breakers	353
8.13	Specifications of High-Voltage Circuit Breakers	358
8.14	Low-Voltage Circuit Breakers	358
	8.14.1 Molded Case Circuit Breakers	358
	8.14.2 Insulated Case Circuit Breakers (ICCBs)	359
	8.14.3 Low-Voltage Power Circuit Breakers (LVPCBs)	359
	8.14.3.1 Single-Pole Interrupting Capability	361
	8.14.3.2 Short-Time Ratings	361
	8.14.3.3 Series Connected Ratings	362
8.15	Fuses	363
	8.15.1 Current-Limiting Fuses	364
	8.15.2 Low-Voltage Fuses	365
	8.15.3 High-Voltage Fuses	365
	8.15.4 Interrupting Ratings	366
Problems		367
References		368

9. Short Circuit of Synchronous and Induction Machines and Converters 371

9.1	Reactances of a Synchronous Machine	372
	9.1.1 Leakage Reactance X_l	372
	9.1.2 Subtransient Reactance X_d''	372
	9.1.3 Transient Reactance X_d'	372
	9.1.4 Synchronous Reactance X_d	372
	9.1.5 Quadrature Axis Reactances X_q'', X_q', and X_q	373
	9.1.6 Negative Sequence Reactance X_2	374
	9.1.7 Zero Sequence Reactance X_0	374
	9.1.8 Potier Reactance X_p	374
9.2	Saturation of Reactances	375
9.3	Time Constants of Synchronous Machines	375
	9.3.1 Open-Circuit Time Constant T_{do}'	375
	9.3.2 Subtransient Short-Circuit Time Constant T_d''	375
	9.3.3 Transient Short-Circuit Time Constant T_d'	375
	9.3.4 Armature Time Constant T_a	375
9.4	Synchronous Machine Behavior on Short Circuit	375
	9.4.1 Equivalent Circuits during Fault	380
	9.4.2 Fault Decrement Curve	383
9.5	Circuit Equations of Unit Machines	386
9.6	Park's Transformation	390
	9.6.1 Reactance Matrix of a Synchronous Machine	390
	9.6.2 Transformation of Reactance Matrix	393
9.7	Park's Voltage Equation	395
9.8	Circuit Model of Synchronous Machines	397
9.9	Calculation Procedure and Examples	399
	9.9.1 Manufacturer's Data	406
9.10	Short Circuit of Synchronous Motors and Condensers	408
9.11	Induction Motors	409
9.12	Capacitor Contribution to the Short-Circuit Currents	413
9.13	Static Converters Contribution to the Short-Circuit Currents	414

9.14	Practical Short-Circuit Calculations	417
Problems		418
References		419
Bibliography		419

10. Short-Circuit Calculations according to ANSI Standards 421

10.1	Types of Calculations	421
	10.1.1 Assomptions	422
	10.1.2 Maximum Peak Current	422
10.2	Accounting for Short-Circuit Current Decay	423
	10.2.1 Low-Voltage Motors	424
10.3	Rotating Machine Model	425
10.4	Type and Severity of System Short Circuits	426
10.5	Calculation Methods	427
	10.5.1 Simplified Method $X/R \leq 17$	427
	10.5.2 Simplified Method $X/R > 17$	427
	10.5.3 E/X Method for AC and DC Decrement Adjustments	427
	10.5.4 Fault Fed from Remote Sources	428
	10.5.5 Fault Fed from Local Sources	430
	10.5.6 Weighted Multiplying Factors	435
10.6	Network Reduction	435
	10.6.1 E/X or E/Z Calculation	436
10.7	Breaker Duty Calculations	437
10.8	Generator Source Asymmetry	437
10.9	Calculation Procedure	439
	10.9.1 Necessity of Gathering Accurate Data	439
	10.9.2 Calculation Procedure	440
	10.9.3 Analytical Calculation Procedure	441
	10.9.4 Hand Calculations	441
	10.9.5 Dynamic Simulation	441
	10.9.6 Circuit Breakers with Sources on Either Side	441
	10.9.7 Switching Devices without Short-Circuit Interruption Ratings	443
	10.9.8 Adjustments for Transformer Taps and Ratios	443
10.10	Examples of Calculations	444
	10.10.1 Calculation of Short-Circuit Duties	444
	10.10.2 K-Rated 15 kV Breakers	448
	10.10.3 4.16 kV Circuit Breakers and Motor Starters	452
	10.10.4 Transformer Primary Switches and Fused Switches	452
	10.10.5 Low-Voltage Circuit Breakers	452
	10.10.6 Bus Bracings	452
	10.10.7 Power Cables	455
	10.10.8 Overhead Line Conductors	456
	10.10.9 Generator Source Symmetrical Short-Circuit Current	460
	10.10.10 Generator Source Asymmetrical Current	461
	10.10.11 System Source Symmetrical Short-Circuit Current	461
	10.10.12 System Source Asymmetrical Short-Circuit Current	462
	10.10.13 Required Closing Latching Capabilities	462
	10.10.14 Selection of the Generator Breaker	463

Contents xiii

10.11 Deriving an Equivalent Impedance 464
10.12 Thirty-Cycle Short-Circuit Currents 469
10.13 Fault Current Limiters..470
      10.13.1 Superconducting Fault Current Limiters 473
Problems...474
References 477

11. Short-Circuit Calculations according to IEC Standards............................479
11.1 Conceptual and Analytical Differences..479
      11.1.1 Breaking Capability...479
      11.1.2 Rated Restriking Voltage 480
      11.1.3 Rated Making Capacity 480
      11.1.4 Rated Opening Time and Break Time 480
      11.1.5 Initial Symmetrical Short-Circuit Current......................480
      11.1.6 Peak Making Current 481
      11.1.7 Breaking Current 481
      11.1.8 Steady-State Current 481
      11.1.9 Highest Short-Circuit Currents..482
11.2 Prefault Voltage 483
11.3 Far-From Generator Faults...483
      11.3.1 Nonmeshed Sources..485
      11.3.2 Meshed Networks 487
          11.3.2.1 Method A: Uniform Ratio R/X or X/R
                    Ratio Method 487
          11.3.2.2 Ratio R/X or X/R at the Short-Circuit Location 487
          11.3.2.3 Method C: Equivalent Frequency Method 488
11.4 Near-to-Generator Faults 489
      11.4.1 Generators Directly Connected to Systems 489
      11.4.2 Generators and Unit Transformers of Power
          Station Units 490
      11.4.3 Motors..491
      11.4.4 Short-Circuit Currents Fed from One Generator 491
          11.4.4.1 Breaking Current...491
          11.4.4.2 Steady-State Current 492
      11.4.5 Short-Circuit Currents in Nonmeshed Networks...........493
      11.4.6 Short-Circuit Currents in Meshed Networks 494
11.5 Influence of Motors..495
      11.5.1 Low-Voltage Motor Groups 496
      11.5.2 Calculations of Breaking Currents of
          Asynchronous Motors...496
      11.5.3 Static Converter Fed Drives..497
11.6 Comparison with ANSI/IEE Calculation Procedures.................497
11.7 Examples of Calculations and Comparison with
     ANSI Methods 499
11.8 Electromagnetic Transients Program Simulation of a
     Generator Terminal Short Circuit..513
      11.8.1 The Effect of PF 513
Problems...517
References 519

xiv   *Contents*

12. Calculations of Short-Circuit Currents in Direct Current Systems................521

    12.1   DC Short-Circuit Current Sources...521

    12.2   Calculation Procedures 523

            12.2.1 IEC Calculation Procedure 523

            12.2.2 Matrix Methods...525

    12.3   Short-Circuit of a Lead Acid Battery 525

    12.4   Short-Circuit of DC Motors and Generators 531

    12.5   Short-Circuit of a Rectifier 537

    12.6   Short-Circuit of a Charged Capacitor...543

    12.7   Total Short-Circuit Current..544

    12.8   DC Circuit Breakers 545

    12.9   DC Rated Fuses 548

    12.10 Protection of the Semi-Conductor Devices 548

    12.11 High-Voltage DC Circuit Breakers...550

    Problems..553

    References 553

Appendix A: Matrix Methods...555

Appendix B: Sparsity and Optimal Ordering 587

Appendix C: Transformers and Reactors..595

Appendix D: Solution to the Problems 629

Index..709

Series Preface

This handbook on power systems consists of four volumes. These are carefully planned and designed to provide state-of-the-art material on the major aspects of electrical power systems, short-circuit currents, load flow, harmonics, and protective relaying.

An effort has been made to provide a comprehensive coverage, with practical applications, case studies, examples, problems, extensive references, and bibliography.

The material is organized with sound theoretical base and its practical applications. The objective of creating this series is to provide the reader with a comprehensive treatise that could serve as a reference and day-to-day application guide for solving the real-world problem. It is written for plasticizing engineers and academia at the level of upper-undergraduate and graduate degrees.

Though there are published texts on similar subjects, this series provides a unique approach to the practical problems that an application engineer or consultant may face in conducting system studies and applying it to varied system problems.

Some parts of the work are fairly advanced on a postgraduate level and get into higher mathematics. Yet the continuity of the thought process and basic conceptual base are maintained. A beginner and advanced reader will equally benefit from the material covered. An underground level of education is assumed, with a fundamental knowledge of electrical circuit theory, rotating machines, and matrices.

Currently, power systems, large or small, are analyzed on digital computers with appropriate software. However, it is necessary to understand the theory and basis of these calculations to debug and decipher the results.

A reader may be interested only in one aspect of power systems and may choose to purchase only one of the volumes. Many aspects of power systems are transparent between different types of studies and analyses—for example, knowledge of short-circuit currents and symmetrical component is required for protective relaying and fundamental frequency load flow is required for harmonic analysis. Though appropriate references are provided, the material is not repeated from one volume to another.

The series is a culmination of the vast experience of the author in solving real-world problems in the industrial and utility power systems for more than 40 years.

Another key point is that the solutions to the problems are provided in Appendix D. Readers should be able to independently solve these problems after perusing the contents of a chapter and then look back to the solutions provided as a secondary help. The problems are organized so these can be solved with manual manipulations, without the help of any digital computer power system software.

It is hoped the series will be a welcome addition to the current technical literature.

The author thanks CRC Press editor Nora Konopka for her help and cooperation throughout the publication effort.

—**J.C. Das**

Preface to Volume 1: Short-Circuits in AC and DC Systems

The first three chapters of this volume—design and analyses concepts of power systems, modern electrical power systems, and wind and solar power generation—are of a general nature applicable to this series.

Short-circuit studies are the very first studies that are conducted for a power system. It is important to understand the nature of short-circuit currents, the symmetrical components for unsymmetrical faults, and matrix methods of solutions that are invariably used on digital computers. This material is covered in Chapters 4, 5, and 6. A long-hand calculation of unsymmetrical fault current even in a simple system in Chapter 5 shows the complexity of such calculations and paves the way for discussions of matrix methods in Chapter 6.

A description of the AC current interruption process is provided in Chapter 7. An understanding is important to appreciate the interruption of various types of short-circuit currents, overvoltages, TRV, restrikes, and multiple reignitions in circuit breakers. This is followed by Chapter 8, which details the rating structures of circuit breakers and fuses according to ANSI/IEEE standards. The major short-circuit contributing sources are synchronous generators and rotating motors. Their models are developed using Park's transformation and circuits of a unit machine followed by their behavior on short circuit and the resulting decaying transients.

Chapters 10 and 11 detail the methodology of short-circuit calculations using ANSI/IEEE and IEC standards. There are conceptual and analytical differences in the calculations illustrated with real-world examples. Chapter 12 is devoted to the short-circuit currents in DC systems.

This book, therefore, provides an understanding of the nature of short-circuit currents, current interruption theories, circuit breaker types, calculations according to ANSI/IEEE and IEC standards, theoretical and practical basis of short-circuit current sources, the rating structure of switching devices, and the short-circuit currents in DC systems. This volume covers a wide base of short-circuit current in the power systems and can be considered a treatize on this subject of practical and academic value.

—J.C. Das

Author

J.C. Das is an independent consultant, Power System Studies, Inc. Snellville, Georgia. Earlier, he headed the electrical power systems department at AMEC Foster Wheeler for 30 years. He has varied experience in the utility industry, industrial establishments, hydroelectric generation, and atomic energy. He is responsible for power system studies, including short circuit, load flow, harmonics, stability, arc flash hazard, grounding, switching transients, and protective relaying. He conducts courses for continuing education in power systems and is the author or coauthor of about 70 technical publications nationally and internationally. He is the author of the following books:

- *Arc Flash Hazard Analysis and Mitigation*, IEEE Press, 2012.
- *Power System Harmonics and Passive Filter Designs*, IEEE Press, 2015.
- *Transients in Electrical Systems: Analysis Recognition and Mitigation*, McGraw-Hill, 2010.
- *Power System Analysis: Short-Circuit Load Flow and Harmonics*, Second Edition, CRC Press 2011.
- *Understanding Symmetrical Components for Power System Modeling*, IEEE Press, 2017.

These books provide extensive converge, running into more than 3000 pages, and are well received in the technical circles. His interests include power system transients, EMTP simulations, harmonics, passive filter designs, power quality, protection, and relaying. He has published more than 200 electrical power system study reports for his clients.

He has published more than 200 study reports of power systems analysis addressing one problem or the other.

Das is a Life Fellow of the Institute of Electrical and Electronics Engineers, IEEE, (USA), Member of the IEEE Industry Applications and IEEE Power Engineering societies, a Fellow of the Institution of Engineering Technology (UK), a Life Fellow of the Institution of Engineers (India), a Member of the Federation of European Engineers (France), a Member of CIGRE (France), etc. He is registered Professional Engineer in the states of Georgia and Oklahoma, a Chartered Engineer (CEng) in the UK, and a European Engineer (EurIng) in Europe. He received a meritorious award in engineering, IEEE Pulp and Paper Industry in 2005.

He earned a PhD in electrical engineering at Atlantic International University, Honolulu, an MSEE at Tulsa University, Tulsa, Oklahoma, and a BA in advanced mathematics and a BEE at Panjab University, India.

1

Design and Analyses Concepts of Power Systems

Electrical power systems are the most complex man-made nonlinear systems on earth. Also these are highly dynamic in nature. Consider that the circuit breakers are closing and opening, the generation is varying according to load demand, and the power systems are subjected to disturbances of atmospheric origin and switching operations. The energy state of the power systems is constantly changing and the energy is being redistributed into electromagnetic and mechanical systems.

Yet, we study electrical power systems in the steady state; though, it may last for a short duration. The transition from one steady state to another does not take place instantaneously and transition to each steady state should be acceptable and stable.

The study of power systems in steady-state amounts to taking a still picture of continuously varying natural phenomena like sea waves. The study of transients that lead from one steady state to another requires transient analysis. The four volumes of this series present steady-state analysis of power systems for the material covered in these volumes. However, the transient analysis is presented as required. There cannot be a water tight compartment between the steady state and the transient state—one may lead to another. For example, short circuits in AC systems are decaying transients and subject the power system to severe stresses and stability problems, yet empirical calculations as per standards are applied for their calculations.

However, this series does not provide an insight into the transient behavior of the power systems and confines the analyses *mostly* to steady state. Some sporadic reference to transients may be seen in some chapters and appendices of the book. A section on effect of protective relaying on stability and fundamental concepts of power system stability are included in Volume 4.

The time duration (frequency) of the transient phenomena in the power system varies. The International Council for Large Electrical Power Systems (CIGRE) classifies the transients with respect to frequency in four groups. These groups are low-frequency oscillations (0.1 Hz–3 kHz), slow front surges (50/60 Hz–20 kHz), fast front surges (10 kHz–3 MHz), and very fast front surges (100 kHz–50 MHz). It is very difficult to develop a power system component (say transmission lines) model, which is accurate from low to very high frequencies. Thus, models are good for a certain frequency range. A model must reproduce the frequency variations, saturation, nonlinearity, surge arrester characteristics, power fuse, and circuit breaker operations accurately for the frequency range considered. The models in this book are mostly steady-state models. For harmonic analysis, the impacts of higher frequencies involved on some component models are discussed in Volume 3.

1.1 Static and Dynamic Systems

A time-invariant resistor connected to a sinusoidal source takes a current depending on the value of the resistor and the applied voltage. The output, the voltage across the resistor, is solely dependent on the input at that instant. Such a system is *memoryless* and is a static system. The energy is dissipated as heat.

On the other hand, a capacitor or an inductor is not memoryless. Figure 1.1 shows a resistor and a capacitor connected to a voltage source. For the resistor at any time: $e(t) = R \times i(t)$. For the capacitor

$$e(t) = e(t_0) + \int_{t_0}^{t} i(t)\,dt \tag{1.1}$$

The output depends not only on the input from t_0 to t but also on the capacitor voltage at t_0 due to the past current flow.

The state of the system with memory is described by *state variables* that vary with time (see Section 1.2). The state transition from $\mathbf{x}(t_0)$ at time t_0 to $\mathbf{x}(t)$ at time $t > t_0$ is a dynamic process that can be described by differential equations or difference equations:

$$\frac{dx}{dt} = \frac{1}{C}r(t) \tag{1.2}$$

$$r(t) = i(t)$$

We defined a dynamic system as the one whose behavior changes with time. Neglecting wave propagation and assuming that the voltages and currents external to the system are bounded functions of time; a broad definition can be that a dynamic circuit is a network of n-terminal resistors, capacitors and inductors, which can be nonlinear. We call a capacitor and an inductor as dynamic elements. We can define the nonlinear resistors, inductors, and capacitors using the following equations:

$$R(v, i) = 0$$
$$C(q, v) = 0 \tag{1.3}$$
$$L(\phi, i) = 0$$

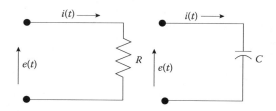

FIGURE 1.1
A resistor and capacitor connected to a voltage source—to illustrate memoryless and systems with memory.

Design and Analyses Concepts of Power Systems 3

These relations are *constituency relations.* A nonlinear resistor is described by the current and voltage and a nonlinear capacitor by the charge q and voltage v across it. If an element is time varying, we write

$$R(v,i,t)=0$$

$$C(q,v,t)=0 \tag{1.4}$$

$$L(\phi,i,t)=0$$

A resistor is voltage controlled, if the current i is a function of the voltage across it. The constituency relation written as $R(i,v)=0$ is expressed as follows:

$$i = \hat{i}v \tag{1.5}$$

If it is current controlled, we write as follows:

$$v = \hat{v}i \tag{1.6}$$

Same notations apply to capacitors and inductors. For a capacitor, we write as follows:

$$C(q,v)=0 \quad \text{or} \quad q=\hat{q}v \text{ (voltage controlled).}$$
$$v = \hat{v}q \text{ (charge controlled)} \tag{1.7}$$

And for an inductor

$$i = \hat{i}(\phi) \quad \phi = \hat{\phi}(i) \tag{1.8}$$

A nonlinear resistor with constituency relationship $i=g(v)$ is passive, if $vg(v) \geq 0$ for all v. It is strictly passive if $vg(v) > 0$ for all $v \neq 0$. It is eventually passive if there exists a $k > 0$ such that $vg(v) \geq 0, |v| > k$. It is strictly passive if $vg(v) > 0, |v| > k$.

$$vg(v) \geq 0, |v| > k \tag{1.9}$$

1.2 State Variables

When we are interested in terminal behavior of a system, i.e., for certain inputs, there are certain outputs, the system can be represented by a simple box. There can be multiple inputs and outputs, Figure 1.2. The mathematical model specifies the relations between the inputs and outputs:

$$r_i(t), t = 1,2,3,....m$$
$$y_i(t), t = 1,2,3,....p \tag{1.10}$$

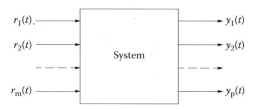

FIGURE 1.2
A system with multiple inputs and outputs.

The response $y_i(t)$ will be a function of initial state vector $x(t_0)$ and the inputs $r_i(t)$ for $t > t_0$. In general, the state of the system at any time is given by $\mathbf{x}(t)$. The components of this vector $x_1(t)$, $x_2(t)$, ..., $x_n(t)$ are called the *state variables*.

The input-output-state relations for the system can be represented by the following:

$$
\begin{aligned}
y_1(t) &= \varphi_1\left[x_1(t_0),\ldots,x_n(t_0); r_1(t_0,t),\ldots,r_m(t_0,t)\right]; t \geq t_0 \\
&\vdots \\
y_p(t) &= \varphi_p\left[x_1(t_0),\ldots,x_n(t_0); r_1(t_0,t),\ldots,r_m(t_0,t)\right]; t \geq t_0
\end{aligned}
\tag{1.11}
$$

The dependence of $\mathbf{x}(t)$ on $\mathbf{x}_0(t)$ and $r(t_0, t)$ is

$$
\begin{aligned}
x_1(t) &= \phi_1\left[x_1(t_0),\ldots,x_n(t_0); r_1(t_0,t),\ldots,r_m(t_0,t)\right]; t \geq t_0 \\
&\vdots \\
x_n(t) &= \phi_n\left[x_1(t_0),\ldots,x_n(t_0); r_1(t_0,t),\ldots,r_m(t_0,t)\right]; t \geq t_0
\end{aligned}
\tag{1.12}
$$

In matrix form

$$
\mathbf{x}(t) = \begin{vmatrix} x_1(t) \\ x_2(t) \\ \cdot \\ x_n(t) \end{vmatrix} \quad \phi(.) = \begin{vmatrix} \phi_1(.) \\ \phi_2(.) \\ \cdot \\ \phi_n(.) \end{vmatrix} \quad \mathbf{y}(t) = \begin{vmatrix} y_1 \\ y_2 \\ \cdot \\ y_p \end{vmatrix} \quad \varphi(.) = \begin{vmatrix} \varphi_1(.) \\ \varphi_2(.) \\ \cdot \\ \varphi_p(.) \end{vmatrix}
$$

$$
\mathbf{x}(t_0) = \begin{vmatrix} x_1(t_0) \\ x_2(t_0) \\ \cdot \\ x_n(t_0) \end{vmatrix} \quad \mathbf{r}(t_0,t) = \begin{vmatrix} r_1(t_0,t) \\ r_2(t_0,t) \\ \cdot \\ r_m(t_0,t) \end{vmatrix}
\tag{1.13}
$$

In abbreviated form

$$
\begin{aligned}
\mathbf{x}(t) &= \phi[\mathbf{x}(t_0), \mathbf{r}(t_0,t)]; \quad t \geq t_0 \\
\mathbf{y}(t) &= \varphi[\mathbf{x}(t_0), \mathbf{r}(t_0,t)]; \quad t \geq t_0
\end{aligned}
\tag{1.14}
$$

Figure 1.3 shows the concept of state variables in a general system.

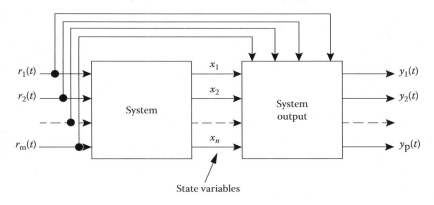

FIGURE 1.3
A general system to show state variables.

1.3 Linear and Nonlinear Systems

It is pertinent to define mathematically linearity and nonlinearity, static, and dynamic systems.

Linearity implies two conditions:

1. Homogeneity
2. Superimposition

Consider the state of a system defined by the following:

$$\dot{x} = f[x(t), r(t), t] \qquad (1.15)$$

If $x(t)$ is the solution to this differential equation with initial conditions $x(t_0)$ at $t = t_0$ and input $r(t)$, $t > t_0$:

$$x(t) = \phi[x(t_0), r(t)] \qquad (1.16)$$

Then, homogeneity implies that

$$\phi[x(t_0), \alpha r(t)] = \alpha \phi[x(t_0), r(t)] \qquad (1.17)$$

where α is a scalar constant. This means that $x(t)$ with input $\alpha r(t)$ is equal to α times $x(t)$ with input $r(t)$ for any scalar α.

Superposition implies that

$$\phi[x(t_0), r_1(t) + r_2(t)] = \phi[x(t_0), r_1(t)] + \phi[x(t_0), r_2(t)] \qquad (1.18)$$

That is, $\mathbf{x}(t)$ with inputs $\mathbf{r}_1(t) + \mathbf{r}_2(t)$ is = sum of $\mathbf{x}(t)$ with input $\mathbf{r}_1(t)$ and $\mathbf{x}(t)$ with input $\mathbf{r}_2(t)$. Thus, linearity is superimposition plus homogeneity.

1.3.1 Property of Decomposition

A system is said to be linear if it satisfies the decomposition property and the decomposed components are linear.

If $x'(t)$ is the solution of Equation 1.14 when system is in zero state for all inputs $\mathbf{r}(t)$, i.e.,

$$x'(t) = \phi\big(0, \mathbf{r}(t)\big) \tag{1.19}$$

And $\mathbf{x}''(t)$ is the solution when all states $x(t_0)$, the input $\mathbf{r}(t)$ is zero, i.e.,

$$x''(t) = \phi\big(x(t_0), 0\big) \tag{1.20}$$

Then, the system is said to have decomposition property, if

$$\mathbf{x}(t) = \mathbf{x}'(t) + \mathbf{x}''(t) \tag{1.21}$$

The zero-input response and zero-state response satisfy the property of homogeneity and superimposition with respect to initial states and initial inputs, respectively. If this is not true then the system is nonlinear.

For nonlinear systems, general methods of solutions are not available and each system must be studied specifically. Yet, we apply linear techniques of solution to nonlinear systems over a certain time interval. Perhaps the system is not changing so fast, and for certain range of applications, linearity can be applied. Thus, the linear system analysis forms the very fundamental aspect of the study.

For the short-circuit calculations, we assume that the system components are linear. In load flow, the voltage drop across a reactor on flow of reactive current is nonlinear, and iterative techniques are applied. For the harmonic penetration, the system components have nonlinear relation with respect to frequency.

1.4 Linearizing a Nonlinear System

In electrical power systems, to an extent all system components can be considered nonlinear. Saturation, fringing, eddy current and proximity effects, thermal effects, especially at high frequency, cannot be ignored. This becomes of special importance for switching transient studies. Handling nonlinearity requires knowledge of

- Piecewise linearization, one example is saturation in transformers
- Exponential segments method is used in arrester models
- One-time step delay methods are used in pseudo-nonlinear devices
- Iterative Newton methods are discussed in Volume 2

Design and Analyses Concepts of Power Systems

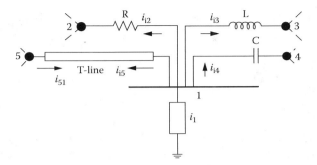

FIGURE 1.4
A node (Bus1) with multiple connections and current injections—to illustrate nonlinearity models.

- Trapezoidal rule of integration
- Runge–Kutta methods
- Taylor series and approximations

These techniques are not discussed. Consider a network of connections as shown in Figure 1.4. For the inductance branch, nodes 1 and 3, we can write as follows:

$$v = L \frac{di}{dt} \quad (1.22)$$

In terms of difference equation

$$\frac{v(t) + v(t - \Delta t)}{2} = L \frac{i(t) - i(t - \Delta t)}{\Delta t} \quad (1.23)$$

This can be written as follows:

$$i_{13}(t) = \frac{\Delta t}{2L}(v_1(t) - v_3(t)) + hist_{13}(t - \Delta t) \quad (1.24)$$

where $hist_{13}$ term is known from the preceding time step.

$$hist_{13}(t - \Delta t) = i_{13}(t - \Delta t) + \frac{\Delta t}{2L}(v_1(t) - v_3(t)) \quad (1.25)$$

For the capacitance circuit, we can similarly write as follows:

$$hist_{14}(t - \Delta t) = -i_{14}(t - \Delta t) - \frac{2C}{\Delta t}(v_1(t - \Delta t) - v_4(t - \Delta t)) \quad (1.26)$$

For the transmission line, ignoring losses

$$i_{15}(t) = \frac{1}{Z}\,\upsilon_1(t) + hist_{15}(t-\tau)$$

$$hist_{15}(t-\tau) = -\frac{1}{Z}\,\upsilon_5(t-\tau) - i_{51}(t-\tau)$$

(1.27)

where Z = surge impedance and τ = line length/velocity of propagation. Therefore, for node 1, we can write as follows:

$$\left(\frac{1}{R} + \frac{\Delta t}{2L} + \frac{2C}{\Delta t} + \frac{1}{Z}\right)\upsilon_1(t) - \frac{1}{R}\,\upsilon_2(t) - \frac{\Delta t}{2L}\,\upsilon_3(t) - \frac{2C}{\Delta t}\,\upsilon_4(t)$$

$$= i_1(t) - hist_{13}(t-\Delta t) - hist_{14}(t-\Delta t) - hist_{15}(t-\tau)$$

(1.28)

For any type of network with n nodes, we can write the general equation as follows:

$$\overline{G}\overline{v}_t = \overline{i}_t - \overline{hist}$$

(1.29)

where
$\overline{G}$ = nxn symmetrical nodal conductance matrix,
$\overline{v}_t$ = vector of n node voltages,
$\overline{i}_t$ = vector of current sources,
$\overline{hist}$ = vector of n known history terms.

Some nodes will have known voltages or may be grounded. Equation 1.29 can be partitioned into set of nodes A with known voltages and a set B with unknown voltages. The unknown voltages can be found by solving for $\overline{v}_{At}$

$$\overline{G}_{AA}\overline{v}_{At} = \overline{i}_{At} - \overline{hist}_A - \overline{G}_{AB}\overline{v}_{Bt}$$

(1.30)

Example 1.1

Consider a function as depicted in Figure 1.5. The relation is nonlinear but continuous. At point P, a tangent can be drawn and the small strip around P can be considered a linear change in the system. Let the function be represented by the following:

$$y = \varphi(r)$$

(1.31)

Expand according to Taylor's series:

$$y = \varphi(r) = \varphi(r_0) + \left(\frac{d\varphi}{dr}\right)_0 \frac{(r-r_0)}{1!} + \left(\frac{d^2\varphi}{dr^2}\right)_0 \frac{(r-r_0)^2}{2!} + \cdots$$

(1.32)

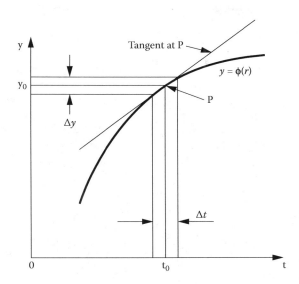

FIGURE 1.5
Linearizing a nonlinear system with Taylor's series.

As the variation at point P is small, neglect the higher order terms in Taylor's series:

$$y = \varphi(r) = \varphi(r_0) + \left(\frac{d\varphi}{dr}\right)_0 \frac{(r-r_0)}{1!} = y_0 + m(r - r_0) \qquad (1.33)$$

where m is the slope at the operating point.
This forms the basic concept of load flow with Newton Raphson method, Volume 2.

1.5 Time-Invariant Systems

When the characteristics of a system do not change with time it is called a time-invariant system. If these change with time the system is time-variant.

Figure 1.6a represents a time-invariant system whereas Figure 1.6b does not represent a time-invariant system. Mathematically, the operator $Z^{-\tau}$ acting on the system results in an advancement and retardation without impacting the shape of the signal:

$$Z^{\tau}[f(t)] = f(t + \tau)$$
$$Z^{-\tau}[f(t)] = f(t - \tau) \qquad (1.34)$$

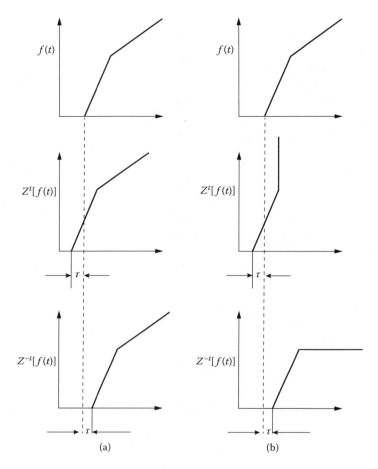

FIGURE 1.6
The effect of translation operator on a (a) time-invariant system and (b) nontime-invariant system.

The state variable model of a time-invariant system can be written as follows:

$$\begin{vmatrix} \dot{x}_1 \\ \dot{x}_2 \\ \cdot \\ \dot{x}_n \end{vmatrix} = \begin{vmatrix} a_{11} & a_{12} & \cdot & a_{1n} \\ a_{21} & a_{22} & \cdot & a_{2n} \\ \cdot & \cdot & \cdot & \cdot \\ a_{n1} & a_{n2} & \cdot & a_{nn} \end{vmatrix} \begin{vmatrix} x_1 \\ x_2 \\ \cdot \\ x_n \end{vmatrix} + \begin{vmatrix} b_{11} & b_{12} & \cdot & b_{1m} \\ b_{21} & b_{22} & \cdot & a_{2m} \\ \cdot & \cdot & \cdot & \cdot \\ b_{m1} & b_{m2} & \cdot & a_{mm} \end{vmatrix} \begin{vmatrix} r_1 \\ r_2 \\ \cdot \\ r_m \end{vmatrix} \qquad (1.35)$$

and

$$\begin{vmatrix} y_1 \\ y_2 \\ \cdot \\ y_p \end{vmatrix} = \begin{vmatrix} c_{11} & c_{12} & \cdot & c_{1n} \\ c_{21} & c_{22} & \cdot & c_{2n} \\ \cdot & \cdot & \cdot & \cdot \\ c_{p1} & c_{p2} & \cdot & c_{pn} \end{vmatrix} \begin{vmatrix} x_1 \\ x_2 \\ \cdot \\ x_n \end{vmatrix} + \begin{vmatrix} d_{11} & d_{12} & \cdot & d_{1m} \\ d_{21} & d_{22} & \cdot & d_{2m} \\ \cdot & \cdot & \cdot & \cdot \\ d_{p1} & d_{p2} & \cdot & d_{pm} \end{vmatrix} \begin{vmatrix} r_1 \\ r_2 \\ \cdot \\ r_m \end{vmatrix} \qquad (1.36)$$

Design and Analyses Concepts of Power Systems

The state variables are a linear combination of system states and inputs and similarly output is a linear combination of states and inputs. Or in the abbreviated form

$$\dot{x} = Ax + Br \quad \text{state equation}$$
$$y = Cx + Dr \quad \text{output equation} \tag{1.37}$$

Here, $x = n$-dimensional state vector, $r = m$-dimensional input vector, $y = p$-dimensional output vector, $A = nxn$ square matrix, $B = nxm$ matrix, $C = pxn$ matrix, and $D = pxm$ matrix.

1.6 Lumped and Distributed Parameters

Consider a system of circuit elements like resistors, capacitors, and inductors connected in a certain manner, and let us call it a system. A system is called a lumped parameter system if a disturbance or input applied to any point of the system propagates *instantaneously* to all other parts of the system. This is a valid assumption if the physical dimension of the system is small compared to the wavelength of the highest significant frequency. For example, the transmission lines [less than approximately 80 km (50 miles)] can be modeled with lumped parameters without much error. For transient analysis, these systems can be modeled with ordinary differential equations. The energy is dissipated or stored in isolated components like resistors, inductors, and capacitors.

In a distributed parameter system, it *takes a finite time for an input to travel to another point in the system*. We need to consider the space variable in addition to time variable. The equations describing such systems are partial differential equations.

In power systems, all systems are to an extent distributed systems. A rigorous model of a motor winding will consist of a resistance, inductance, and capacitance and capacitance to ground of each turn of the winding. This model may be required to study the surge phenomena in the motor windings say on impact of a switching surge or lightning impulse.

A long transmission line has resistance, inductance, and capacitance distributed along its length. Wave propagation occurs over long lines even at power frequencies. Figure 1.7 shows

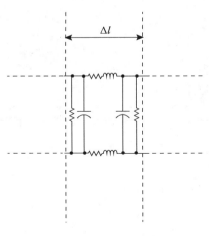

FIGURE 1.7
Lumped equivalent model of a system with distributed parameters.

a small section of a long line, Δl, where the series resistance and inductance and shunt admittance and susceptance are shown lumped. This assumption of lumping is acceptable as the line section consists of a small section Δl, the entire line length being represented with series connections of infinite number of such small sections.

1.7 Optimization

The problem of optimization occurs in power system studies quite often. A solution must optimize between various conditions and constraints. See Volume 2 for optimal power flow and various techniques to reach an optimal solution. We can describe optimization, in general, as a mathematical attempt to find the best solution with interplay of design, available resources, constraints, and costs that is spread over a period of time from a couple of seconds to years. Automatic Generation Control (AGC) must respond fast for system integrity within seconds while optimizing generation planning and minimizing investment, which evolves over the course of years.

1.8 Planning and Design of Electrical Power Systems

The electrical power systems are like a *chain with many complex links*. A weakness in any *chain-link* can jeopardize the integrity of the complete system. In general, the electrical power systems should be

- Secure
- Safe
- Expandable
- Maintainable
- Reliable

A number of criteria apply to the planning and design of electrical systems, namely,

1. All switching devices must be selected to interrupt and withstand system short-circuit currents, Volume 1.
2. All interlocks must guarantee that human errors of operation are eliminated, for example, when switching between sources that may be out of phase, when paralleling two different sources of power, and when bringing another standby source in service.
3. The protective systems should isolate the faulty section without escalating the fault to nonfaulted areas, Volume 4.
4. All system loads must be served without excessive voltage dips and voltage instability under various switching conditions, Volume 2.
5. The steady state and transient stability of the systems must be ensured under the studied upset conditions, Volume 4.

Design and Analyses Concepts of Power Systems

6. The starting of large motors should not create excessive voltage dips and instability.

7. The system and equipment grounding should receive adequate considerations and protections. Proper grounding of equipment (safety grounding) and system neutrals (system grounding) is essential for human safety and ground fault relaying, Volume 4.

8. Proper mitigation of harmonics from nonlinear loads should be provided to meet the required standards and protect electrical equipment from deleterious effects of harmonics, Volume 3.

9. In industrial environment, all electrical equipment should be heavy duty to endure more frequent usage and for reliability. The listed equipment should be provided.

10. A reliability analysis of the distribution system in the design stage should be conducted.

11. Special requirements are applicable for standby and emergency systems, continuity of essential loads on normal power interruption, standby generators and the like.

12. Design of electrical systems for special facilities like hospitals, nuclear plants, generating plants, aircraft hangers, railway electrifications, renewable power sources, interconnections with grid demand special precautions, and analyses. Solar and wind generation plants require specific knowledge of these facilities and electrical and equipment characteristics. Separate set of standards and equipment specifications are applicable to nuclear facilities. *To an extent all power distribution systems even for the same type of facility differ, to meet the specific requirements* and generalizations cannot be applied. Generalization can be a guideline only; and the experience of the Engineer and Planner becomes the prime factor.

This list of tasks for effective planning and development and for meeting the specific performance requirements of electrical power systems can be long—depending upon the electrical power system under consideration.

The integrity of the electrical equipment should be maintained with emphasis on the type of enclosures, insulation coordination, operating mechanisms, grounding and protective relaying. Yet, the power system designs may fall short from engineering design and safety considerations. Even the functionality for which these systems are designed for adequate performance may be compromised.

Though there is a spate of guidelines and standards, it is not unusual to see inadequately designed systems, lacking in some respect or the other. Competition and economical constraints can make even expert designers and planners to cut corners, which may ultimately result in spending more funds for the short-term fixes and long-term upgrades.

The power system studies can be an effective tool in the design of power systems. These can also identify weak spots, limitations of the current systems, and point to remedial measures. For example,

- The switching devices may be underrated from short-circuit conditions.
- The load flow may indicate problems of voltage drops or voltage instability under certain operating and switching conditions.
- The protective relaying may be inadequate; lack of coordination may result in nuisance trips and shutdowns.

- The harmonic pollutions may overload the system components and application of capacitors may result in harmonic resonance problems. The harmonics can seriously derate the cables and transformers and overload the rotating equipment.
- Adequate surge protection may not be provided.
- Considerations may not be applied for transient stability for disturbances and faults in the system.
- The equipment may be underrated to carry the system load currents in one condition or the other.

This series attempts to analyze these problems. The surge protection is not discussed in this series.

1.9 Electrical Standards and Codes

There are a number of current standards and guidelines for planning and designing of electrical power systems for commercial, industrial, and utility applications. IEEE Industry Applications Society and Power Engineering Society have immense database of publications exploring the new technology, drafting new standards, revising the existing standards, and providing guidelines for specifying, engineering, maintenance, equipment selection, and applications. The IEEE website [1] compiles more than three million documents that a user can access,—which includes standards, technical papers in conferences, transactions, and journals. Reference [2] provides titles of IEEE "color books." The electrical requirements of protection, wiring, grounding, control and communication systems, contained in [3] National Electrical Code (NEC) are important for electrical engineers. In most instances, the NEC is adopted by local ordnance as a part of building code. Then, there are numbers of NFPA standards containing requirements on electrical equipment and systems [4], which include NFPA 70E-2015.

Legislation by the US federal government has the effect of giving certain ANSI standards the impact of law. Not all standards are ANSI approved. Occupational safety and Health Administration (OSHA) requirements for electrical systems are contained in 29 CFR Part 1910 of the Federal Register [5].

The US National Institute of Occupational Safety and Health (NIOSH) publish Electrical Alerts to warn unsafe practices of hazardous electrical equipment, [6]. The US Department of Energy has advanced energy conservation standards. These include ASHRA/IES legislation embodying various energy conservation standards, such as ASHRAE/IES 90.1P. These standards impact architectural, mechanical, and electrical designs.

The Underwriters Laboratories (ULs) and other independent testing laboratories may be approved by an appropriate jurisdictional authority like OSHA. A product may be UL labeled or listed. Generally, the designers insist for UL labels on the electrical components in a design and planning process and specifications for the electrical equipment. The UL publishes an Electrical Construction Materials Directory, an Electrical Appliance and Utilization Directory, and other standards. The Electrification Council (TEC) is representative of investor-owned utilities and publishes several informative handbooks [7].

The National Electrical Manufacturer's Association (NEMA) [8] represents the equipment manufacturers and their publications that standardize certain design features of

Design and Analyses Concepts of Power Systems 15

electrical equipment and provide testing and operating standards. NEMA publishes a number of standards on electrical equipment, which contain important application and selection guidelines.

Further, there are many handbooks, which, over the course of years have established reputations in the electrical field; some of these are cited in [9].

Safety for operating personnel is achieved through proper design. The National Electric Safety Code (NESC) [10] covers basic provisions for safeguarding from hazards arising out of conductors in electric supply substations, overhead and underground electric supply, and communication lines. It also covers work rules and safe clearances from live parts to ground and between phases. The Electrical Generating System Association [11] publishes performance standards for emergency, standby, and cogeneration equipment.

System protection is a fundamental requirement for all electrical systems. All switching devices must be applied safely within their interrupting ratings; faulted circuits must be isolated and relay protection should be properly designed and coordinated. Physical protection of equipment from tempering, damage, and environment must be provided. The operating personnel must be trained for the specific jobs. Essentially, all states in the US require that the system designs be performed under the seal of a licensed Professional Engineer, registered in that particular state.

In spite of this spate of standards, codes, and guidelines, the electrical power and distribution systems as designed and implemented currently may be inadequate with respect to personnel safety. Prevention through Design (PtD) is a new initiative sponsored by the NIOSH. This initiative was launched in a July 2007 workshop held in Washington, DC, to create a *national strategy* for PtD [12]. It may be defined as: "PtD involves addressing the occupational safety and health needs and redesign processes to prevent or minimize work-related hazards and risks associated with the construction; use; maintenance; and disposal of facilities, materials, and equipment."

This definition has a much wider base than merely the system designs of electrical installations. References [13–27] provide further reading.

1.10 Reliability Analyses

Reliability analyses are not covered in this series; however, some fundamental aspects are discussed in this chapter. Reliability is the probability of successful operation and is time dependent. For a system that has components with relatively constant failure rates, reliability is an exponentially decaying function with time; longer the time interval the longer is the reliability irrespective of system design. The reliability curve will be flatter for a well-designed system as compared to a poorly designed system.

System reliability assessment and evaluation methods are based upon probability theory that allows reliability of a proposed system to be assessed quantitatively. This is finding a wide application. Alternative system designs, redundancy, impact on cost of changes, service reliability, protection and switching, and system maintenance policy can be quantitatively studied. Using reliability evaluation methods, system reliability indexes can be computed. The two basic system reliability indexes *are the load interruption frequency and expected duration of load interruption events*. These can be used to compute other indexes, i.e., total expected average interruption time per year, system availability or unavailability at the load supply point, expected energy demanded, but unsupplied per year. Reliability also addresses emergency and standby power systems. Preventive maintenance has a large impact not only on the availability of systems but also on arc flash hazards—a poorly maintained system is more prone to failures including personal hazards.

If the time t over which a system must operate and the underlying distributions of failures of its constituent elements are known, then the system reliability can be calculated by taking the integral, essentially the area under the curve defined by the probability density function (PDF) from t to infinity:

$$R(t) = \int_{t}^{\infty} f(t)\,dt \tag{1.38}$$

1.10.1 Availability

Availability can be defined as the percent of time a system is immediately ready for use, or an instant probability of a system being immediately ready for use. We speak of inherent availability (Ai) and operational availability (Ao). Ai consists of component failure rates and average repair time; Ao goes beyond Ai, in the sense, that maintenance downtime (Mdt), parts procurement time, logistics, etc. are included.

For Ai, we can write as follows:

$$Ai = \frac{MTBF}{MTBF + MTTR} \tag{1.39}$$

where,
  MTBF: mean time between failures
  MTTR: mean time to repair
  Each probability distribution has unique PDF with notation $f(t)$. The area under that curve shows the relative probability of a failure occurring before time t, Figure 1.8a. Cumulative distribution function (CDF) can be calculated by the integral in the following equation:

$$F(t) = \int_{0}^{t} f(t)\,dt \tag{1.40}$$

where $F(t)$ is the probability occurring before time t. Plotting $F(t)$ gives CDF, Figure 1.8b. Finally, the reliability fuction $R(t)$ is the probability of a component not failing by time t:

$$R(t) = 1 - F(t) \tag{1.41}$$

The hazard rate or hazard function is defined for the remainder of the time:

$$H(t) = \frac{f(t)}{R(t)} \tag{1.42}$$

Design and Analyses Concepts of Power Systems

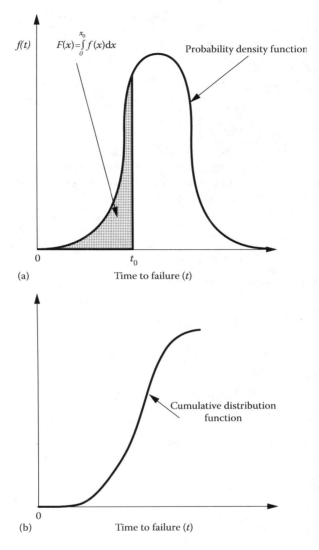

FIGURE 1.8
Probability of failure represented by the area under the curve of (a) PDF and (b) the cumulative distribution.

1.10.1.1 Exponential Distribution

The PDF for exponential distribution is given by the following:

$$f(t) = \lambda e^{-\lambda t} \tag{1.43}$$

The CDF is

$$f(t) = 1 - e^{-\lambda t} \tag{1.44}$$

And probability function is

$$R(t) = e^{-\lambda t} \tag{1.45}$$

The hazard function is therefore $= \lambda$.

1.10.2 Data for Reliability Evaluations

IEEE 493, Reference [28], contains the reliability data collected from equipment reliability surveys and a data collection program over a period of 35 years or more. Data needed for a reliability analysis will depend upon nature of the system being studied and the details of the study. Usually, data on individual system components and the time required to do various switching operations will be needed. System component data required are as follows:

1. Failure rates or forced outage rates associated with various modes of component failure
2. Expected average time to repair or replace a component
3. Scheduled maintenance outage rate of the component
4. Expected duration of a scheduled outage event

Switching the time data needed includes expected time to open and close a circuit breaker, disconnect or throw-over switch, replace a fuse link, and perform emergency operations such as installing jumpers.

The service reliability requirements of the loads and processes supplied are assessed to decide a proper definition of service interruption. It is not only the total collapse of voltage, but voltage sag or swell may also cause a shutdown. A *failure modes and effects* analysis is carried out. This means listing of all component outage events or combinations of component outages that result in an interruption of service at the load point being studied. Component outages are categorized as follows:

1. Forced outages and failures
2. Scheduled or maintenance outages
3. Overload outages

Component failure can be categorized by physical mode or type of failure.

1.10.3 Methods of Evaluation

- Cut-set method
- Network reduction
- Go algorithms
- State space
- Monte-Carlo simulations

The details of these analyses are beyond the scope of this series. In the cut-set method, computation of the quantitative reliability indexes can proceed once the minimal cut sets

Design and Analyses Concepts of Power Systems 19

of the system have been found. The first step is to compute the frequency, expected duration and expected downtime of each minimal cut set. Statistical methods and expressions for frequency and expected duration of the most commonly considered interruption events associated with first-, second-, and third-order cut sets are then applied to assess overall reliability. Digital computers are used for the reliability analysis as the statistical manipulations become complex, even for small systems.

The minimum cut-set method is well suited for industrial power systems and commercial distribution systems. It lends itself to computer simulations and many commercial programs that are available. The procedure is

- Access the service reliability of equipment, loads, and processes.
- Perform a failure mode and effect analysis (FMEA), which amounts to listing of component outages or listing of component outages that result in interruption of services at the load point.
- Different types of modes and outage of components exist, which may be classified as follows:
  1. Forced outages or failures,
  2. Scheduled outages or maintenance outages,
  3. Overload outages.
- Compute the interruption frequency distribution, the expected interruption duration, and probability of each of the minimum cut sets.
- Combine the results to produce system reliability indexes.

The FMEA and determination of minimal cut sets is conducted by considering first the effect of outage of a single component and then the effect of overlapping outages of increasing number of components.

Define

f_s = interruption frequency = f_{csi}
r_s = expected interruption duration

$$r_s = \sum_{\text{min cut-set}} \frac{f_{csi} r_{csi}}{f_s} \tag{1.46}$$

$$f_s r_s = \text{total interruption time per time period.} \tag{1.47}$$

The first-, second-, and third-order cut sets are shown in Table 1.1

TABLE 1.1

Frequency and Expected Duration Expressions for Interruption Associated with Minimal Cut Sets

	Forced Outages	
First-Order Minimal Cut Set	**Second-Order Minimal Cut Set**	**Third-Order Minimal Cut Set**
$f_{cs} = \lambda_i$	$f_{cs} = \lambda_i \lambda_j (r_i + r_j)$	$f_{cs} = \lambda_i \lambda_j \lambda_h (r_i + r_j + r_i r_h + r_j r_h)$
$r_{cs} = r_i$	$r_{cs} = \dfrac{r_j r_i}{r_i + r_j}$	$r_{cs} = \dfrac{r_j r_i r_h}{r_i + r_j + r_i r_h + r_j r_h}$

Forced outage overlapping schedule outage is given by the following expressions:
Second-order minimum cut set

$$r_{cs} = \left[\lambda'_i \lambda'_j \left[\frac{r'_i\, r'_j}{r'_i + r'_j} \right] + \frac{\lambda'_j \lambda'_i\, r'_j \left[\dfrac{r'_j\, r'_i}{r'_j + r'_i} \right]}{f_{cs}} \right] \quad (1.48)$$

Define

f_{cs} = frequency of cut-set event
r_{cs} = expected duration of cut-set event
λ_i = forced outage rate of ith component
λ'_i = scheduled outage rate of ith component
r_i = expected repair or replacement time of ith component
r'_i = expected schedule outage duration of ith component.

This standard provides a summary of all-industry failure rate and equipment outage duration data. Table 2.1 of this standard lists inherent availability and reliability data,

TABLE 1.2

Definition Summary for the Reliability Analysis

Parameter	Equation	Explanation of Symbols
Ai, inherent availability	MTBF/(MTBF + MTTR)	
Ao, operational availability	MTBM/(MTBM + MDT)	
Λ, failure rate (f/h)	T_f/T_p	T_f, total number of failures during a given T_p
MDT, mean down time	$(R_{dt} + R_{lt} + M_{dt})/T_{de}$	T_{de}, total downtime events, R_{dt}, total downtime for unscheduled maintenance, R_{lt}, total logistic time for unscheduled maintenance and M_{dt}, maintenance down time
MTBF, mean time between failure (h)	T_p/T_f	
MTBM, mean time between maintenance (h)	T_p/T_{de}	
MTTM, mean time to maintain (h)	M_{dt}/T_{ma}	T_{ma}, total number of scheduled maintenance in T_p
MTTR, mean time to repair (h)	R_{dt}/T_f	
R(t) reliability	$e^{-\lambda t}$	

Note: Based on IEEE 493 [2].

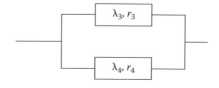

FIGURE 1.9
Two systems in parallel with different λ and r.

Design and Analyses Concepts of Power Systems 21

which is fundamental to the reliability analysis. Table 1.2 shows the definition summary from this standard.

When two repairable components are connected in parallel, as shown in Figure 1.9, the following expressions apply:

$$f_P = \frac{\lambda_3 \lambda_4 (r_3 + r_4)}{8760}$$

$$f_P r_P = \frac{\lambda_3 r_3 \lambda_4 r_4}{8760} \qquad (1.49)$$

$$r_P = \frac{r_3 r_4}{r_3 + r_4}$$

Equations 1.49 are approximate and should be used when $\lambda_3 r_3/8760$ and $\lambda_4 r_4/8760$ are less than 0.01.

Table 1.3 shows the results of IEEE survey of reliability of electrical utility power supplies for industrial plants. This shows importance of utility supply source to which the distribution system is connected. Except for small distribution plants where the process interruptions can be tolerated without much loss of revenue, single source utility sources are not recommended. Consider that in a process plant with production rate of thousand of tonnes of product per day, a supply source interruption can amount to hourly loss of thousands of dollars. Then, there are processes where the loss of power can damage or jam the equipment and it takes days to clean up and restart the processes. For critical electronic equipment manufacture, like a power supply to Intel manufacturing facilities, the complete power supply is conditioned to be immune to voltage sags and swells, other power quality problems using solid-state devices that are discussed in Volume 2.

Table 1.4 shows failure modes of circuit breakers. Note that 42% of failures occur because of nuisance trips; the circuit breaker opens when it should not.

Example 1.2

Calculations of reliability of a simple radial system, shown in Figure 1.10 are from IEEE standard 493–2007. The results of calculations for the forced hour down time are shown in Table 1.5. The single utility power source and the single transformer are the greatest contributors to the failure rate. If a spare transformer is at hand or with cogeneration,

TABLE 1.3

IEEE Survey of Reliability of Utility Power Supplies to Industrial Plants

Number of Circuits, all Voltages	λ	r	λ_r
Single circuit	1.956	1.32	2.582
Double circuit Loss of both circuits[a]	0.312	0.52	0.1622
Double circuit-calculated value for loss of source 1 while source 2 is okay	1.644	0.15[b]	0.2466
Calculated two utility power sources at 13.8 kV that are assumed to be completely independent	0.00115[c]	0.66[c]	0.00076

[a] Data for double circuit that had all circuit breakers closed.
[b] Manual switchover time of 9 min to source 2.
[c] Calculated using single circuit utility power supply data and equations for parallel reliability.

TABLE 1.4

Failure Mode of Circuit Breakers–Percentage of Total Failure in Each Mode

Percentage of T_f (All Voltages)	Failure Characteristics
9	Backup protective equipment required, failed while opening
Other Circuit Breaker Failures	
7	Damaged while successfully opening
32	Failed in service, not while opening or closing
5	Failed to close when it should
2	Damaged while closing
42	Opened when it should not
1	Failed during testing or maintenance
1	Damage discovered during testing or maintenance
1	Other
100%	Total percentage

the λ_r is reduced. The results for some primary selective and secondary selective systems from this standard are shown in Table 1.6. Figure 1.11a through c applies to the primary selective systems and Figure 1.12 to the secondary selective system. It is seen that the primary selective systems (a) and (b) are almost identical with respect to λ_r, thus the additional cost of a 13.8 kV circuit breaker and cable connections can be avoided.

1.10.4 Reliability and Safety

A system designed for high MTBF is not necessarily the safest from the point of view of injuries and hazards to the worker. For example, an industrial distribution system served from a single source of power; one transformer connected to the utility source will not be very reliable. In such a system, any failure of a component connected in radial circuit can result in total failure of the served loads, i.e., failure of main transformer, circuit breaker, or interconnecting cables. This will call for more human intervention to attend to the failures. Yet, this simple radial system can be designed safe from arc flash hazard and worker safety considerations. A system with redundant parallel running transformers, alternate sources of power, standby generation and UPS systems, auto-transfer switches of power between sources will be more reliable with respect to reliability indices, but may not have the required worker safety features; for example, a proper interlocking of the sources may be missing or improperly designed. *It should not be construed that complex systems designed with higher reliability cannot be made equally or even safer than simpler systems.* For properly designed automated systems, lesser interruptions and human interventions will be needed. This will improve both—the reliability as well as the safety.

The maintenance aspects will be common for reliability and safety. Poorly maintained systems are neither safe nor reliable. The aspects of reliability and safety should converge for the effective system design.

To avoid man-machine interface (MMI) errors, the emphasis is upon *self-monitoring* and *self-correcting* systems with least amount of manual interference.

The IEC 61511 [24] has been developed as process sector implementation of the IEC 61508 [25]: "Functional Safety of Electrical/Electronic/Programmable Electronic Safety Related Systems." It has two basic concepts: The safety lifecycle and safety integrity levels (SILs). The safety lifecycle forms the central framework, and it is a good engineering procedure for safety instrumented systems (SIS) design in process industry (sensors, logic solvers,

Design and Analyses Concepts of Power Systems

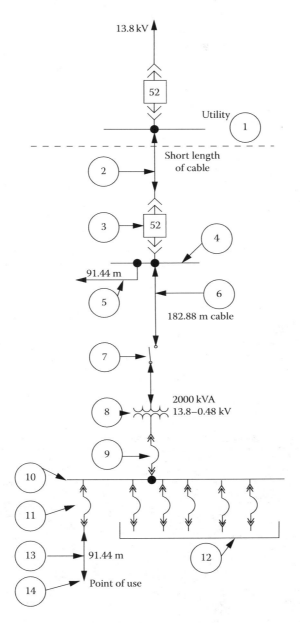

FIGURE 1.10
A radial distribution system, with one utility source for reliability evaluation using cut-set method.

and final elements are a part of SIS). In the safety lifecycle process, risks are evaluated and SIS performance requirements are established. Layers of protection are designed and analyzed. Then, an SIS is optimally designed to meet the particular process risk. SIL levels indicate order of magnitude of risk reduction. Table 1.7 shows safety integrity levels. SIL 1 has the lowest level of risk reduction and SIL 4 has the highest level of risk reduction. The standard suggests that applications that require the use of SIS of SIL4 are rare in the process industry; an exception can be nuclear plants. The standard mainly deals SIS and

TABLE 1.5

Reliability Analysis of Simple Radial System

Component Number	Component	λ	λ$_r$	Ai
1	13.8 kV utility power source	1.956000	2.582000	0.999705338
2	Primary protection and controls	0.000600	0.003000	0.999999658
3	13.8 kV metal clad breaker	0.001850	0.000925	0.999999894
4	13.8 kV switchgear bus insulated	0.004100	0.153053	0.999982529
5	Cable (13.8 kV) 274.32 m (900 ft)	0.002124	0.033347	0.999996193
6	Cable terminations	0.002960	0.002220	0.999999747
7	Disconnect switch	0.001740	0.001740	0.999999801
8	Transformer	0.010800	1.430244	0.999836757
9	480 V metal-clad circuit breaker	0.000210	0.001260	0.99999856
10	480 V switchgear bus	0.009490	0.069182	0.999992103
11	480 V metal-clad circuit breaker	0.000210	0.001260	0.999999856
12	480 V metal-clad circuit breakers (5) failed while opening	0.000095	0.000378	0.9999999957
13	Cable 480 V, 91.44 m (300 ft)	0.000021	0.000168	0.9999999981
14	Cable terminations (2) at 480 V	0.000740	0.000555	0.9999999937
	Total at 480 V point of use	1.990940	4.279332	0.999511730

Source: IEEE standard 493.
Note: The data for hours of downtime per failure are based upon repairing the failed unit.

TABLE 1.6

Reliability Comparison, Different System Configurations

Distribution System	Switchover in Less Than 5 s		Switchover in 9 min			
	λ	λ$_r$	λ	λ$_r$	λ	λ$_r$
Simple radial					1.990940	4.279332
Simple radial with a spare transformer					1.990940	3.367488
Simple radial with cogeneration					0.053069	1.741527
Primary selective to 13.8 kV utility supply	0.344490	1.855647	1.990940	2.102614		
Primary selective to load side of 13.8 kV circuit breaker	0.345938	1.867318	1.992388	2.114285		
Primary selective to primary of the transformer	0.333566	1.665287	1.992018	1.914055		
Secondary selective	0.322499	0.233556	1.990883	0.483814		

interface between SIS and other safety systems in requiring that a process hazard and risk assessment be carried out.

The difference between SIL related to *a single piece of equipment* and *a set of processes* must be clearly distinguished. In the USA, the onus of specifying a particular equipment meeting the required safety levels (i.e., OSHA, UL, ANSI/IEEE standard requirements) lies

Design and Analyses Concepts of Power Systems 25

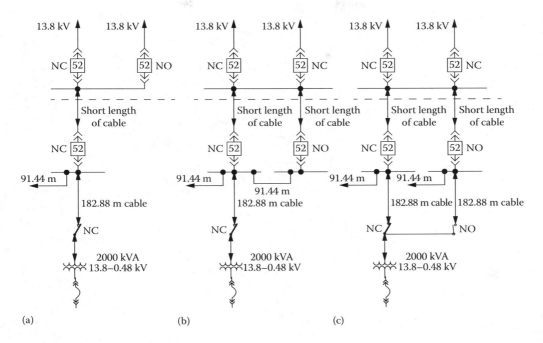

FIGURE 1.11
The system with duplicate utility sources: (a) primary selective; (b) primary selective system to the load side of the 13.8 kV circuit breaker; and (c) primary selective system to the primary of the transformer.

with the user. In Europe, the machine manufacturers must meet the specified SIL level. Relevant references are Standards: IEC 62061 also ISO 13849-1, [26,27].

IEC 61511 Part 3 details the guidelines for selecting SIL in hazards and risk analysis. The information is intended to provide a broad range of global methods used to do such analysis. There are several informative annexes, of which Annex A covers As Low As Reasonably Practical (ALARP) principle and tolerable risk concepts, Figure 1.13.

1.11 Extent of System Modeling

The system to be studied is a small part of an overall grid system, Figure 1.14. It is important that the impact of the larger system should be carefully accounted for the accuracy of the type of study being considered. This requires that the boundary conditions should be clearly established. Consider a large system consisting of many generators, transmission lines, transformers, and cables. According to Thévenin theorem, any node of interest can be pulled out from a larger system and the system impact modeled by equivalent Thévenin impedance in series with an equivalent voltage source at the point of interconnection. It is, however, obvious that this equivalent Thévenin impedance cannot be a fixed number and will vary with the changes in the operation of the system; for example, some of the generators may be off-line and transmission lines may be out of service.

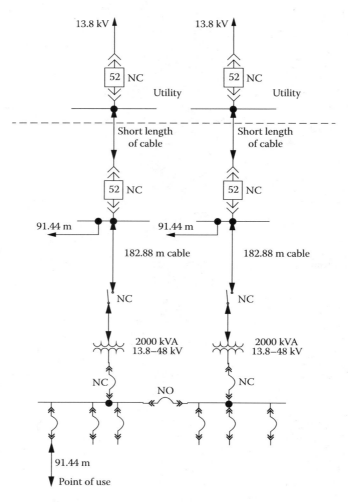

FIGURE 1.12
A secondary selective system.

1.11.1 Short-Circuit Calculations

For the short-circuit conditions, the equivalence is established at the point of interconnection with one single number *representing the short-circuit levels in symmetrical kA*.

Three-phase short level in symmetrical kA and its X/R ratio, maximum and minimum values

Single line to ground fault in symmetrical kA and its X/R ratio, maximum and minimum values

Based on this data, the maximum and minimum values of positive, negative, and zero sequence impedances can be calculated for the short-circuit study.

According to IEEE standards for the short-circuit calculations, this utility source can be represented by an *invariant impedance*. Care has to be exercised in making this assumption. Consider that a large generating station is in the vicinity of the

Design and Analyses Concepts of Power Systems

TABLE 1.7

SIL for Safety Functions Operating on Demand or in a Continuous Demand Mode

Safety Integrity Level	Demand Mode of Operation (Average Probability of Failure to Perform Its Design Function on Demand-Pdf)	Continuous/High Demand Mode of Operation (Probability of a Dangerous Failure per Hour)
4	$\geq 10^{-5}$ to $<10^{-4}$	$\geq 10^{-9}$ to 10^{-8}
3	$\geq 10^{-4}$ to 10^{-3}	$\geq 10^{-8}$ to 10^{-7}
2	$\geq 10^{-3}$ to 10^{-2}	$\geq 10^{-7}$ to 10^{-6}
1	$\geq 10^{-2}$ to 10^{-1}	$\geq 10^{-6}$ to 10^{-5}

Source: IEC 61508-1, Tables 2 and 3.

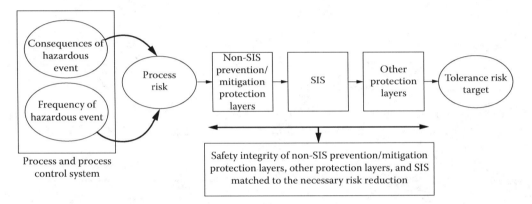

FIGURE 1.13
Risk and safety integrated systems, see text.

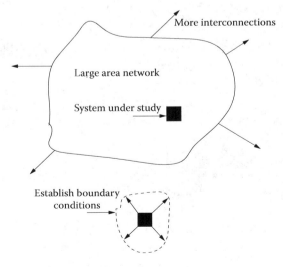

FIGURE 1.14
The extent of system modeling for power system studies.

industrial plant for which the studies are to be undertaken. It will be erroneous to represent the source impedance with an invariant single number representative of symmetrical and asymmetrical fault current situation—this will give erroneous results with respect to momentary or first cycle short-circuit calculations. It will be necessary to extend the boundary of modeling to include the generating station, and its interconnections.

1.11.2 Load Flow Calculations

The concept of the swing bus is defined in Volume 2. It may be considered an ideal Thévenin source as well an ideal Norton equivalent. Any amount of current taken from this source will not alter the voltage a bit. A stiff utility source may approximate it and may be represented as swing bus, but there are lots of electrical distribution systems connected to weak utility sources. On a load flow, declaring the utility source as a swing bus will give erroneous results as it will ignore the voltage dip occurring in the impedance of the utility source, which is not zero. A fictitious impedance representative of the source can be calculated and provided in series with a large stiff source, see Volume 2.

1.11.3 Harmonic Analysis

In power system harmonic analysis, inadequate representation of the boundary limits may give rise to totally invalid results, see Volume 3 for further discussions.

1.12 Power System Studies

Power system analysis is a vast subject. Consider the following broad categories:

- Short-circuit studies, in AC and DC systems Volume 1
- Load flow studies, Volume 2
- Stability studies; large rotor angle, small disturbances, and voltage instability—a brief introduction is provided in Volume 4
- Motor starting studies, Volume 2
- Optimal load flow, contingency, and security analysis studies, Volume 2
- Harmonic analysis studies, harmonic mitigation, flicker mitigation, and harmonic filter designs studies, Volume 3
- Application of FACTS and power electronics studies, Volume 2
- Transmission and distribution lines planning and design studies, underground transmission, and distribution studies
- Insulation coordination and application of surge protection studies
- Switching transient and transient analysis studies
- Power system reliability studies
- Cable ampacity calculation studies in UG duct banks, submarine cables, oil filled cables, and the like

Design and Analyses Concepts of Power Systems 29

- Torsional dynamics studies
- HVDC transmission studies, short-circuits, and load flow in DC systems studies—a brief description is provided in Volumes 1 and 2
- Renewable generation interconnections with utility grid—brief description is included in Chapter 3
- Wind power and renewable energy sources, feasibility studies, and their integration in the grid studies
- Power quality for sensitive and electronic equipment studies
- Ground mat (grid) design for safety studies, system grounding studies, grounding for electronic equipment studies
- Protective relaying and relay coordination studies, Volume 4
- Arc flash hazard analysis studies—a brief description is provided in Volume 4
- Energy conservation studies
- Studies for design and applications of standby power systems

In addition, specific studies may be required for a specific task, for example, transmission line designs or generating stations, transmission substations, and consumer load substations.

In power system studies, there is some data that are transparent between the various types of studies. For example, correct impedance data model is required for short-circuit, load flow, and harmonic analyses. The load flow algorithms are modified for harmonic penetration and harmonic power flow, see Volume 3. Also short-circuit calculations are a prior requirement for harmonic analyses. For the protective relaying studies, a prior knowledge of symmetrical and unsymmetrical fault current calculations and symmetrical components is required. *The repetitions from one volume to another are avoided without loss of continuity. A reader may not be interested in all aspects of the power system studies, and this limitation should be noted. Appropriate references are provided as required.*

1.13 Power System Studies Software

Around 25 years back, there was no power system studies software for PC use. Today, the market is flooded with many options available for the choice of softwares for a particular study. However, all the software programs available may not have the same or identical capabilities. This puts an onerous of selection of the right software not only for the current system to be studied but also for the future studies. Further, every vendor claims that the software meets IEEE and IEC requirements. However, there is no independent body to verify this claim and some variations are noted in the end results for the same system configuration. No specifications have been written to verify the claims of the vendors. There are no standards established on the input data and output formats.

The databases vary in their capabilities. User-defined models are possible in some cases. The databases are, generally, not transparent between various softwares. This means that if a different software is to be used, other than that on which the system was modeled, all data will need to be reentered, which can be very time consuming.

The desirable capabilities of softwares are discussed in appropriate sections in this series.

1.14 System of Units

The SI units are based on meter-kilogram-second-ampere (mksa) system. These have been adopted by standardization bodies of the world, including IEC, ANSI, and IEEE. The USA is the only industrialized nation in the world that does not mandate the use of SI units. Even in many IEEE technical papers in conferences and IEEE Transactions, SI units are not strictly followed, though IEEE insists that SI units are used. The engineering work in the USA still holds foot-pound-second (FPS) system. The US congress has the constitutional right to establish measuring units; and has not enforced any system. The metric system (now SI) was legalized in 1866, and is the only legal measuring system, but other non-SI units are legal as well.

Other countries adopted SI units in 1960–1970. Some denounced as branding it "un-American." Progressive businesses and educational institutes asked Congress to mandate it. As a result, in 1988 Omnibus Trade and Competitiveness Act, Congress established SI as the *preferred* system for the US trade and commerce and urged all federal agencies to adopt it by 1992 or as early as possible. SI remains voluntary for the private US businesses.

This series uses both the units, SI and FPS units.

Problems

1.1 Figure P1.1 shows a parallel RLC circuit A differential equation for the total current drawn by the circuit that can be written as follows:

$$i(t) = \frac{e_C}{R} + C\frac{de_C}{dt} + \frac{1}{L}\int_{-\infty}^{t} e_C(t)\,dt$$

Write equations for the state model.

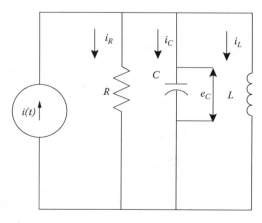

FIGURE P1.1
A parallel RLC circuit excited by a current source.

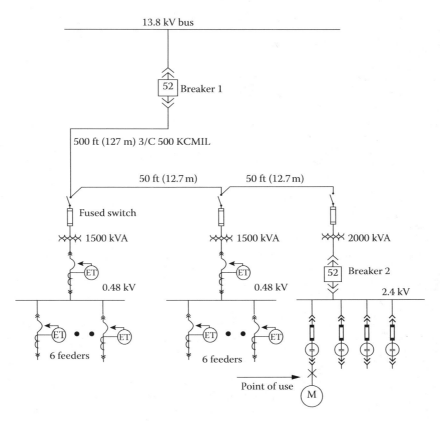

FIGURE P1.2
A simple distribution system for calculation of reliability indices.

1.2 Consider the following expressions:

$$y(t) = ax(0) + br(t)$$

$$y(t) = [r(t)]^2$$

Do they represent a linear or nonlinear systems?

1.3 Calculate reliability indices similar to Table 1.5 for the configuration shown in Figure P1.2 at the point of use.

1.4 Consider that two systems continuously operate in parallel and have the following parameters, pertaining to each subsystem:

$$\lambda_1 = 1.3, \lambda_2 = 1.5$$

$$r_1 = 0.67, r_2 = 0.34$$

Calculate overall λ_r

1.5 Describe different maintenance strategies for electrical power systems (not discussed in the text)

References

1. www.ieee.xplore, Digital Library 2017.
2. IEEE color books

141: IEEE Recommended Practice for Electric Power Distribution for Industrial Plants (IEEE Red Book), 1993.

142: IEEE Recommended Practice for Grounding of Industrial and Commercial Power Systems (IEEE Green Book), 1991.

241: IEEE Recommended Practice for Electric Power Systems in Commercial Buildings (IEEE Gray Book), 1997.

242: IEEE Recommended Practice for Protection and Coordination of Industrial and Commercial Power Systems (IEEE Buff Book), 2001.

399: IEEE Recommended Practice for Industrial and Commercial Power Systems Analysis (IEEE Brown Book), 1997.

446: IEEE Recommended Practice for Emergency and Standby Power Systems for Industrial and Commercial Applications (IEEE Orange Book), 2000.

493: IEEE Recommended Practice for the Design of Reliable Industrial and Commercial Power Systems (IEEE Gold Book), 2007.

602: IEEE Recommended Practice for Electric Systems in Health Care Facilities (IEEE White Book), 1996.

739: IEEE Recommended Practice for Energy Management in Commercial and Industrial Facilities (IEEE Bronze Book), 1995.

902: IEEE Guide for Maintenance, Operation and Safety of Industrial and Commercial Power Systems (IEEE Yellow Book), 1998.

1015: IEEE Recommended Practice for Applying Low-Voltage Circuit Breakers Used in Industrial and Commercial Power Systems (IEEE Blue Book).

1100: IEEE Recommended Practice for Powering and Grounding Electronic Equipment (IEEE Emerald Book), 1999.

551: Recommended Practice for Calculating Short-Circuit Currents in Industrial and Commercial Power Systems (IEEE Violet Book), 2006.
3. National Electrical Code (NEC) NFPA-70-2014.
4. NFPA Publications:

a. HFPE and Society of Fire Protection Engineers, SFPE Handbook of Fire Protection Engineering.

b. 101H, Life Safety Code Handbook.

c. 20. Centrifugal Fire Pumps.

d. 70B. Electrical Equipment Maintenance.

e. 70E. Electrical Safety Requirements for Employee Workplaces, 2015.

f. 72. National Fire Alarm Code.

g. 75. Protection of Electronic Computer/Data Processing Equipment.

h. 77. Static Electricity.

i. 78. Lightning Protection Code.

j. 79. Electrical Standards for Industrial Machinery.

k. 92A. Smoke Control Systems.

l. 99. Health Care Facilities.

m. 110. Emergency and Standby Power Systems.

n. 130. Fixed Guide-Way Transit System.
5. Federal Register. Superintendent of Documents, U.S. Government Printing Office, Washington DC 2040.
6. NIOSH. Publications Dissemination, 4676 Columbia Parkway, Cincinnati, OH 45226.
7. TEC. 1111, 19th Street, Washington DC 20036.

Design and Analyses Concepts of Power Systems

8. NEMA 2101 L Street, NW, Suite 300, Washington DC 20037.
9. Popular Handbooks.

   DG Fink, HW Beaty. *Standard Handbook for Electrical Engineers*, 15th Edition, McGraw-Hill, New York.

   T Croft, CC Carr, JH Watt. *American Electricians Handbook*, 12th Edition, McGraw-Hill, New York.

   JG Webster, Ed. *Wiley Encyclopedia of Electrical and Electronics Engineering*, 21 Volumes, John Wiley & Sons, New York, 1999.

   Illuminating Engineering Society (IES) Handbook, Vols. 1 and 2. 1221, Avenue of the Americas, New York, 10020.

   Electrical Transmission and Distribution Reference Book, 4th Edition, East Pittsburg, PA, Westinghouse Electric Corporation, 1964. *Applied Protective Relaying*, Westinghouse Electric Corporation, Coral Springs, Florida, 1982.

   RS Smeaton, Ed. *Motor Applications and Maintenance Handbook*, McGraw-Hill, New York, 1987.

   DL Beeman. *Industrial Power Systems Handbook*, McGraw-Hill, New York, 1955.

   Edison Electrical Institute. *Underground Systems Handbook*, 1957.

   RS Smeaton, Ed. *Switchgear and Control Handbook*, McGraw-Hill, 1987.

   JM McPartland. *Handbook of Practical Electrical Design*, McGraw-Hill, New York, 1984.
10. NESC, C-2, National Electrical Safety Code, C-2.
11. EGSA. P.O. Box 9257, Corel Springs, FL 33065.
12. E Manuele. Prevention through design: Addressing occupational risks in the design and redesign processes, In special issue of By Design, Engineering Practice Specialty of American Society of Safety Engineers, pp. 1–13, October 2007.
13. ANSI Z10, Occupational Safety and Health Management Systems, 2004.
14. HL Floyd. The NIOSH prevention through design initiative, in *Conf. record IEEE IAS Petroleum and Chemical Industry Committee Tech. Conference*, pp. 363–369, September 1999.
15. JC Cawley, GT Homce. Trends in the electrical injuries in the U.S. 1992–2002, *IEEE Trans Ind Appl Mag*, 44(4), 962–972, 2008.
16. JA Gambatese, J Hinze, M Behm, Investigations of the viability of designing for safety, The Center to Protect Workers' Rights, Silver Spring, MD, Tech. Rep., May 2005.
17. Council; Directive 92/57/EEC. The implementation of minimum safety and health requirements at temporary or mobile construction sites, European Union Regulation, Official Journal L245, pp. 6–22, June 1992.
18. Construction Design and Management Regulations, UK Health and Safety Executive, Statutory Instrument No. 320, pp. 1–9, 2007.
19. Commonwealth of Australia, National OHS Strategy 2002–2012, Tech. Rep., pp. 1–9, 2002.
20. S Jamil, A Golding, HL Floyd, M Capelli-Schellpfeffer, Human factors in electrical safety, in Proc. *IEEE IAS Petroleum and Chemical Industry Tech. Conf.* pp. 349–356, September 2007.
21. WC Christensen, Safety through design: Helping design engineers: 10 key questions, American Society of Safety Engineers Prof Saf pp. 32–39, March 2003.
22. ASSE: TR-Z790.001, Prevention through design guidelines for addressing occupational risks in design and redesign processes, American Society of Safety Engineers, Tech. Rep., pp. 1–28, October 2009.
23. NETA. International Electric Testing Association. MTS1993. Maintenance Testing Specifications.
24. IEC 61511-SER Ed. 1.0, Functional Safety-Safety Instrumentation Systems for Process Industry Sector—All Parts, 2004.
25. IEC 61508, Ed. 2.0 Functional Safety of Electrical/Electronic/Programmable Electronic Safety—Related Systems—All Parts, 2010.
26. IEC 62061-1, Ed. 1.0. Guidance for Application of ISO 13849-1 and IEC 62601 in the Design of Safety Related Control Systems for Machinery, 2010.
27. IEC 62061. Safety of Machinery—Functional Safety of Safety in Related Electrical Electronic and Programmable Electronic Control Systems, 2005.
28. IEEE Standard 493. Design of Reliable Industrial and Commercial Power Systems.

2

Modern Electrical Power Systems

2.1 Classification

Electrical power systems can be broadly classified into generation, transmission, subtransmission, and distribution systems. Individual power systems are organized in the form of electrically connected areas of *regional* grids, which are interconnected to form *national* grids and also international grids. Each area is interconnected to another area in terms of some contracted parameters like generation and scheduling, tie-line power flow, and contingency operations. The business environment of power industry has entirely changed due to deregulation and decentralization with more emphasis on economics and cost centers.

Figure 2.1 shows a general concept of the electrical power systems. This shows generation, transmission, and distribution, with distributed generation, renewable sources like wind and solar power and also HVDC links in a schematic manner.

Figure 2.2 may be considered a small section of a large grid system, somewhat hypothetical for illustration purposes. It is only a 24-bus system—note that even a large industrial distribution system may comprise more than 800 buses. Figure 2.2 shows, generation, transmission, subtransmission, and distribution; an industrial plant in a cogeneration mode with a 30 MVA generator, HVDC link, Static Synchronous Compensator (STATCOM, see Volume 2) for plus minus reactive power compensation. For transmission of electrical power over distances of more than 500 km, HVDC links are preferable. Converter stations are required at each end; however, only a maximum of two lines are required. DC lines do not have reactance and are asynchronous links, eliminating the instability problems inherent in AC transmission and can transmit more electrical power than AC lines. At a certain line length, the cost of the converter equipment at either end is offset by transmission of higher power and reduction in the cost of the line. These do generate harmonics and AC and DC filters are required. HVDC systems are discussed in Volume 2 and the harmonic impacts in Volume 3.

Single shaft steam units of 1500 MW are in operation, and superconducting single units of 5000 MW or more are a possibility. On the other hand, dispersed generating units, integrated with grid may produce only a few kilowatts of power.

The generation voltage is low 13.8–25 kV because of problems of inter-turn and winding insulation at higher voltages in the generator stator slots, though a 110 kV Russian generator is in operation. The transmission voltages have increased to 765 kV or higher and many HVDC links around the world are in operation. Maintaining acceptable voltage profile and load frequency control are major issues. Synchronous condensers, shunt capacitors, static VAR compensators, and flexible AC transmission system (FACTS, see Volume 2) are employed to improve power system stability and enhance power handling capability of

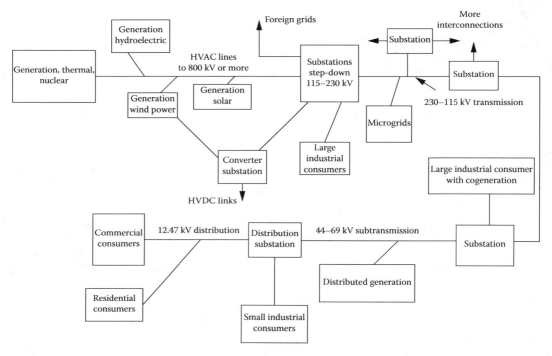

FIGURE 2.1
Electrical power systems, generation, transmission and distribution, conceptual configuration.

transmission lines. Automatic generation control (AGC) and automatic frequency control (AFC) are discussed in Volume 2. The subtransmission voltage levels are 23 kV to approximately 69 kV, though for large industrial consumers voltages of 230 and 138 kV are in use. Subtransmission systems connect high voltage substations to local distribution substations. The voltage is further reduced to 12.47 kV and several distribution lines and cables emanate from distribution stations. At the consumer level, the variety of load types, their modeling, and different characteristics present a myriad of complexities. A pulp mill may use single synchronous motors of 30,000 hp or more and ship propulsion can use even higher ratings.

If an aerial map is taken of the electrical power lines and interconnections in an area, it will be denser than all the routes of highways, subhighways lanes, and by-lanes combined together. There are approximately 200,000 miles of transmission lines in the USA.

2.1.1 Utility Companies in the USA

The utilities in the USA may be classed as follows:

- Publicly owned utilities, which are nonprofit state or local agencies and include Municipals, Public Power Districts, and Irrigation Districts, accounting for 10.7%.
- Federally owned utilities such as Tennessee Valley Authority and Bonneville Power Administration, and the US Navy corps of Engineers accounting for 8.2%.
- Cooperatively owned utilities, owned by farmers and communities, mostly provide power to the members, account for 3.2%.

Modern Electrical Power Systems

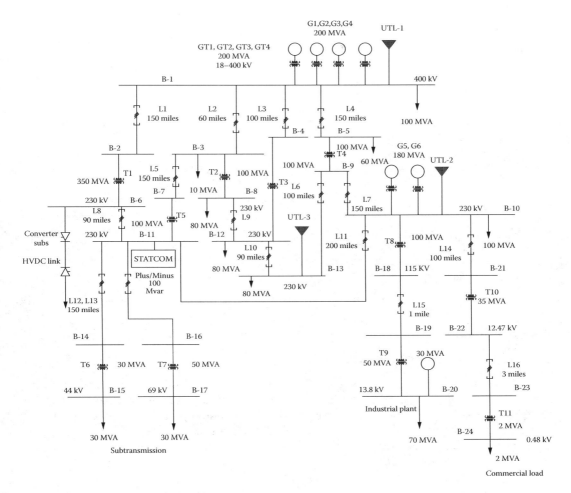

FIGURE 2.2
A 24-bus system, showing generation, transmission, subtransmission, industrial, and commercial loads.

- Nonutilities, which generate power for their own use and/or sale in wholesale power markets, account for 11.9%.

Today, investor owned utilities produce most of the US electrical power, which is changing as independent power producers are increasing.

2.1.2 North American Power System Interconnections

The transmission systems of the USA and Canada are interconnected into a large power grid known as North American Power Systems Interconnections. It has evolved into three major areas:

1. The Eastern Interconnection that includes all eastern and central states. This is tied to other two major regions through HVDC.

2. The Western Interconnection comprised of Western North America, from Rocky Mountains to Pacific coast. It is interconnected to Eastern region at six points, and also to systems in Northern Canada and Northwest Mexico.

3. The Texas Interconnection that includes most of the state of Texas. It is tied to Eastern interconnection at two points and also has ties to systems in Mexico.

Each area consists of several neighboring utilities, which operate jointly to schedule generation in a cost-effective manner.

After the November 1965 great blackout, the National Reliability Council was created in 1968 and later, changed its name to North American Reliability Council (NERC). It lays down guidelines and rules for utility companies to follow and adhere to. Yet, it seems that adequate dynamic system studies are not carried out during the planning stage. This is attributed to lack of resources, data, or expertise (lack of models for dispersed generation, verified data and load models, models for wind generation). In a modern complex power system, the potential problem area may lay hidden, and may not be possible to identify intuitively or even with some power system studies. It is said that cascading type blackouts can be minimized and time frame prolonged but these cannot be entirely eliminated.

2.2 Deregulation of Power Industry

Prior to 1992, within a certain geographical area, a single company owned and operated all subsystems, like generation, transmission, distribution, and power delivery. It was a vertically integrated monopoly that was regulated by state utility commission or another appropriate regulatory authority. It served all consumers who had no choices. The tariff rates of electricity included generation, transmission, distribution, maintenance, operating, and fuel costs plus profit margins.

In 1992, the US Senate passed a comprehensive National Energy Policy Act (NEPA) to promote competition and growth of a free market. The act includes 30 Titles; Title 1 of the Act encourages integrated Resource Planning (IRP) and demand side management (DSM). The emphasis is on reducing the costs for efficiency improvements of generation, transmission, and distribution facilities.

In Title VII, exempt wholesale generators (EWGs) are defined as any company owning or operating all or part of an eligible facility and selling electricity at wholesale costs. Utilities are permitted to purchase from EWGs under jurisdiction of state commission. The Federal Energy Regulatory Commission (FERC) is given authority to mandate contract signing as long as these contracts are in the public interest. This means acquiring open access to transmission facilities. The focus is competition and improvement of efficiencies instead of regulations.

In 1993, the FERC notice of proposed rulemaking (NOPR) was a major step in restructuring:

2.2.1 Generation Company (GENCO)

The GENCOs are responsible for selling electrical generation according to contracts together with needed ancillary services, which enhance reliability and quality of power delivered to the consumer.

Modern Electrical Power Systems 39

2.2.2 Transmission Company (TRANSCO)

The TRANSCOs are responsible for maintaining all transmission equipment to facilitate transport of electricity through grids according to prearranged schedules together with any ancillary services.

2.2.3 Distribution Company (DISTCO)

The DISTCOs are responsible for providing the distribution systems to connect the end users to the transmission grids.

It is envisaged that these three companies will have a horizontally integrated environment and will be the main players in the new deregulated market. They will be coordinated by a central coordinator or an independent administrator (ICA) who will act as an agent between the producers and the consumers. He will handle monetary transactions for sale and purchase of energy and will earn certain percentage by the profits made from sale and purchase of energy.

In March 1995, the FERC issued the ancillary services NOPR. The ancillary or support services could be provided by a third party, referred to as ancillary services company (ANCILCO). The provider of ancillary services will be regulated and the cost of services will be set by state utility commission or other regulatory authority. Energy Service Companies (ESCOs) are emerging players in this scenario.

Thus, the deregulated structure may be simplified as five different but coordinated entities (GENCO, TRANSCO, DISTCO, ANCILCO, and ESCO).

2.3 The New Energy Platform

At the turn of the century, the global energy forecast seemed bleak, as the demand overpaced the production sending cost of oil to $150.00 per barrel. This put a downward pressure on global economic growth. In 2006, the USA—the world largest consumer—imported 56% of its total oil consumption.

The energy landscape is much different today. "The New Oil Order" reflects three major trends:

- In North America, decades of investments in Shale technologies have resulted in soaring oil and gas production
- Clean energy sources like solar and wind have rapidly advanced
- Improvements in energy efficiency are becoming a major factor
- A cleaner and more efficient energy footprint is good for environment and economy

The USA has become one of the largest producers of crude oil. By 2014, oil production reached 12 million barrels a day, *nearly 70% jump over 2008,* surpassing every OPEC country, including Saudi Arabia. To explore, the shale revolution requires long-term demand side investments, cleaner and more efficient extraction technologies, new refining capabilities, and pipelines.

2.3.1 Sustainable, Renewable, and Green Energy

We can define sustainable energy as "The form of energy obtained from nonexhaustible sources, such that the provision of this form of energy serves the need of the present *without compromising the ability of future generations to meet their needs.*"

The irreplaceable sources of power generation are petroleum, natural gas, oil, and nuclear fuels. The fission of heavy atomic weight elements like uranium and thorium and fusion of lightweight elements, i.e., deuterium offer almost limitless reserves. Replaceable sources are elevated water, pumped storage systems, solar, geothermal, wind, and fuel cells, which in recent times have received much attention driven by strategic planning for independence from foreign oil imports. Technologies promoting sustainable energy include renewable sources such as

- Hydro and pumped storage
- Solar
- Wind energy
- Geothermal
- Bioenergy
- Tidal power
- Technologies designed to improve energy efficiency

The costs have fallen dramatically and effective government policies support investor confidence. Considerable progress is made in energy transition from fossil fuels to ecologically sustainable systems and some studies support 100% renewable energy. To distinguish sustainable energy from other renewable energy terminologies like alternate energy—we say sustainable energy source continues to provide energy. It is also distinct from low-carbon energy, which implies that it does not add to the carbon dioxide in the atmosphere.

Figure 2.3 shows the World's electricity generation growth, and the type of generation source. It predicts a 54% growth in 2015 based on 1995 levels; though many efforts are directed toward energy conservation.

In 2012, the world relied on renewable sources for around 13.2% of its total energy supply, and in 2013, renewable accounted for almost 22% of global energy generation, a 5%

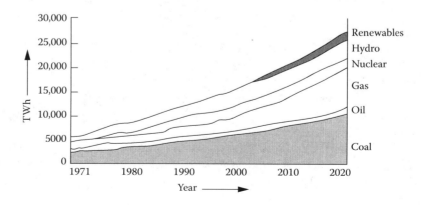

FIGURE 2.3
The world's projected electrical demand, and sources of energy production.

Modern Electrical Power Systems

TABLE 2.1

Total World Energy Production—Renewables-2010 Data

Renewable	Percentage
Biomass heat	11.44
Solar hot water	0.17
Geothermal heat	0.12
Hydropower	3.34
Ethanol	0.50
Biodiesel	0.17
Biomass electricity	0.28
Wind power	0.51
Geothermal electricity	0.07
Solar PV power	0.06
Solar CSP	0.002
Ocean power	0.001

Renewables = 17% in 2010.

increase. (This includes hydro generation.) Table 2.1 shows the world's renewable energy generation, which amounts to 17% of the total energy generation in 2010.

2.3.2 Green Energy

Green energy is the energy that can be extracted, generated, or consumed without significant impact on the environment. Earth has the capability to recover from some pollution and the pollution that does not go beyond that capability can still be termed green.

Green power is a subset of renewable energy. The US Environmental Protection Agency defines green power as the power produced from solar, wind, geothermal, biogas, biomass, and low-impact small hydroelectric sources.

We can talk of three generation technologies:

1. First generation technologies evolved at the end of 19th century, and include hydropower, biomass combustion, geothermal power, and heat.

2. Second generation includes wind power, solar heating and cooling, modern forms of bioenergy, and solar photovoltaic.

3. Third generation technologies are still under development and include biomass gasification, biorefinery technologies, concentrating solar thermal power, hot dry rock geothermal energy, and ocean energy.

2.3.3 Hydroelectric Plants

Hydropower is one of the oldest sources of power on the planet. Hydroelectric plants can be constructed in large capacity and have the advantage of long life among the sources of renewable energy—many existing plants have operated for more than 100 years. The largest electric power generating facility ever built is the Three Gorges Dam hydroelectric plant in China. The facility generates power by utilizing 32 Francis turbines each of capacity 700 MW and two 50 MW turbines, totaling an installed capacity of 22,500 MW, largest in the world.

Criticism has been levied that significant amount of CO_2 is produced during construction and flooding of reservoirs. However, it has been found that high emissions are associated only with shallow reservoirs in tropical locales, and recent innovations in hydropower turbine technologies are enabling efficient development of low-impact run-of-the-river hydroelectricity projects. Generally speaking, hydroelectric plants produce much lower life-cycle emissions than other types of generation. Hydroelectric power is experiencing resurgence in many countries.

- Often we think in terms of large hydropower plants but these can be small too, taking advantage of water flows in municipal facilities and irrigation ditches. These can be damless with diversion of the run of the river.
- Some states in the USA use lots of hydropower. About 66% of electricity in Washington State comes from hydropower.
- America's hydropower industry has more than 100 GW of hydropower capacity and employs 20,000 to 30,000 people.
- Hydropower costs less than most energy sources.
- Dams are built for a number of reasons—irrigation, shipping and navigation, and flood control. Only 3% of the nation's dams generate electricity. An energy department funded study found that 12 GW of hydrogenerating capacity could be added to existing power generation.

A hydroelectric plant converts the inherent energy of water under pressure (potential energy, because of high-level storage) in to electric energy. Its main elements are as follows:

- An upper high-level reservoir, usually built by constructing a diversion dam on a stream or a river.
- An intake consisting of a canal or concrete passageway to carry the water directly to low-head turbines or the pressure conduit used for medium or high-head turbines.
- A pressure conduit, consisting of a tunnel, pipeline, or penstock to carry water to medium or high-head turbines.
- A surge tank to prevent sudden pressure rise and drops during load change, trash racks, penstock shut-off valves.
- A hydraulic turbine, consisting of a runner and shaft coupled to a synchronous generator.
- The draft tube to carry the water away from the turbine and tailrace.

The difference in the water level of the reservoir and the tailrace called the "head" determines the type of hydraulic turbine—reaction or impulse. The percent efficiency is high, approximately 90% at full load.

The power produced is given by the following:

$$P = g\rho WH \tag{2.1}$$

Modern Electrical Power Systems 43

where P is the power produced in watts, W = discharge (m³/s) through the turbine, ρ = density (1000 kg/m³), H = head in m, and g = gravitational constant = 9.81 m/s².

The kind of water turbines that are employed are as follows:

- Pelton: for heads 184–1840 m and consists of a bucket wheel with adjustable flow nozzles
- Francis: for heads 37–490 m and is of mixed flow type
- Kaplan: for heads of up to 61 m has axial flow rotor with variable pitch blades

Water hammer is defined as the change in pressure above or below the normal pressure caused by the sudden change in the rate of flow. For long penstocks, water hammer may be great and surge tanks are provided to counteract this.

Whenever pressure in any turbine drops below the evaporation pressure, it will result in vapor formation. This vapor is trapped in liquid in the form of bubbles that may be carried to high pressure regions where they collapse with great force and cause pitting or cavitations.

$$\sigma = \frac{H_b - H_s}{H} \tag{2.2}$$

where

H_b = barometric pressure,

H_s = suction pressure or height of runner outlet above tail race,

σ should have some specific value to avoid cavitations.

The advantage of hydroelectric power plants is lower running costs, because a major component of the cost—fuel cost is eliminated; however, there are other costs like costs of dams, area flooded for pondage, intake works, discharge arrangements, etc. However, the generation is dependent upon renewable water sources and one major reason for recent power crises in California was inadequate power generation from the hydro-plants.

Figures 2.4 and 2.5 show typical vertical and axial flow turbine unit arrangements, reference IEEE Standard 1020.

2.3.4 Pumped Storage Hydroelectric Plants

The pumped storage systems utilize the off-peak power to pump water from the tailrace into an upstream reservoir and utilize it to generate power during peak load hours.

2.3.5 Nuclear Power

The basic heat producing process is the fission of radioactive materials. Enriched uranium (isotopes 233, 235, and 238 and 239, isotope 235 occurs naturally) is the common fission material. The system consists of a controlled fission heat source, a coolant system to remove and transfer the heat produced, and equipment to convert the thermal energy in the hot coolant to electric power. Regardless of the type of fission heat source used, the basic process is fission of nuclear fuel to produce thermal energy. This thermal energy is removed from the heat source (reactor core) by a coolant, which can be used directly as the working

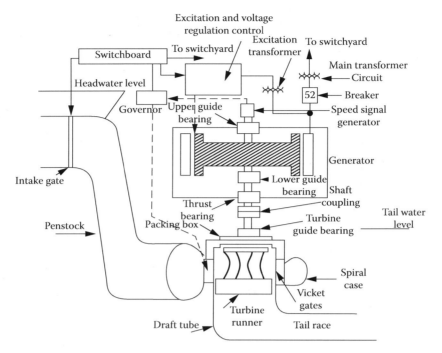

FIGURE 2.4
Vertical Francis unit. (IEEE Standard 1020, 1994.)

fluid in the power conversion cycle, or sometimes a secondary cooling loop is added. Some features of the nuclear plants are as follows:

- Nuclear fuel has a relatively long life, usually measured in months or years as compared to fossil fueled plants. But the burned nuclear fuel is radioactive, and has thousands of years of half-life. U-238 has a half-life of 4.5×10^{10} years. Disposal of the nuclear waste requires special processing and disposal. Due to extremely long life to which the material can remain radioactive, its safe disposal is the major public and safety concern. Nobody wants it in his backyard. Yet, countries like France and Japan produce a considerable amount of nuclear power. The recent power crises has renewed interest in nuclear. Low-level wastes (LLW) and high-level wastes (HLW) are both produced. The Nuclear Regulatory Commission (NRC) regulation for LLW is 10 CFR 61. It defines three classes of nuclear wastes A, B, and C depending upon isotope half-life. For HLW, the NRC regulation is 10 CFR 60, descriptions are not provided here.

- The major portions of the nuclear plant are radioactive during operation and require special precautions for maintenance. Also special measures are required to prevent radioactivity release during normal operation and due to an accident.

- Control and instrumentation is strongly influenced by safety requirements and are related to reactor stability, and the capability of the reactor to respond to the load demands. Nuclear fuel is a highly processed material, and its use does not require combustion air.

Modern Electrical Power Systems 45

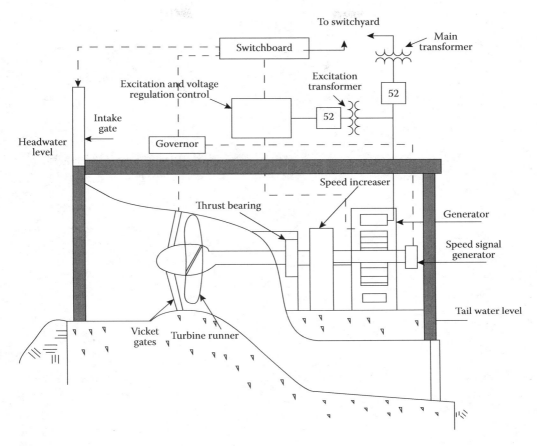

FIGURE 2.5
Horizontal axial flow arrangement. (IEEE Standard 1020, 1994.)

- The need for containment shells, shielding, instrumentation, safety measures and comparatively low steam pressures and temperatures, and consequent lower thermal efficiency requires large turbines and piping as compared to modern thermal power plants and make the nuclear power plants capital intensive. These are not considered in any size less than 500 MW.

The basic process of fission is splitting of the heavier nucleus. A neutron can be captured by a nucleus of a large atomic weight molecule to form a new nucleus, which may be unstable. The split may occur into two approximately equal nuclei, with emission of several neutrons and considerable energy. Since more than one neutron is emitted for each absorbed atom, it leads to a chain reaction. When this chain reaction becomes uncontrolled, immense amount of heat and energy is released (atomic bomb). One ton of uranium is equivalent of 10,000 tons of coal and if all the end products of reaction can be recycled it is equivalent to 100,000 tons of coal.

Figure 2.6 shows a symbolic picture of the chain reaction. The prompt neutron production is not in geometric progression. Neutron kinetics is complex. A percentage of absorbed

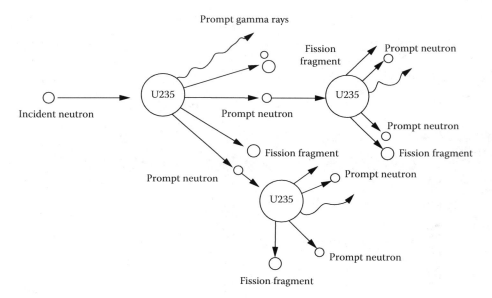

FIGURE 2.6
Fission reaction-schematic diagram.

neutrons does not cause fission. The number of neutrons released per capture of a neutron is given by the following:

$$\eta = v \frac{\sum f}{\sum u} \quad (2.3)$$

where,
 η = number of neutrons released per capture
 v = number of neutrons released per fission
 $\sum f$ = cross section of the fuel for fission
 $\sum u$ = absorption cross section (fission and nonfission) in the fuel

The various types of reactors are not discussed. In a boiling water reactor, the reactor vessel also contains steam separation apparatus since the coolant is converted to steam in the core. The fission process is regulated by absorption in a controlled manner. This controlled absorption may be provided by control elements or rods, which are mechanically inserted into or withdrawn from the core. It can also be provided by an absorber material within the reactor core.

The fission fragments that result from the fission process are radioactive and decay by emission of beta particles, gamma rays, and to a lesser extent alpha particles and neutrons. The neutrons that are emitted after fission are called delayed neutrons, and permit control of fission chain reaction.

The greatest concern is release of radioactivity. The barriers are the fuel matrix, the fuel jacket, or cladding, the primary system, and the containment structure. Protective systems may be designed with triple redundancy and shutdown the reactor in a situation, which

Modern Electrical Power Systems 47

will lead to excessive release of radioactivity past a certain boundary. A guiding principle applied throughout the industry is known as As Low an exposure As is Reasonably Achievable (ALARA). Title 10 of Code of Federal Regulations (10CFR) has many parts that lay down standards for protection against radiation.

Electrical design of nuclear plants requires special consideration; many system components are labeled "Nuclear Grade" double or triple redundancy in power distribution, control, and protective relaying. Large standby units for critical power supply are some essential features that are not elaborated in this volume.

2.3.5.1 Breeder Reactors

Breeder reactors are the technology of the future. It produces more fissionable material than it consumes. It is designed to create more nuclear reactions created by neutrons in the high-energy region than by energy in the thermal-energy region as used by light water reactors. Much research concentrated in Germany, Japan, France, India, and the USA. The capacity of various Fast breeder reactors (FBRs) will increase from 23 GW in 2010 to 600 GW in 2050.

2.3.5.2 Nuclear Fusion

Fission is the process of splitting an atom of heavy element into two elements of lighter elements. The resulting mass is less than that of the fissioning atoms. In *fusion*, two nuclei of lighter element are combined into one nucleus of heavier element. It creates energy as resulting mass is less than that of fusing nuclei. Similarly, in fission, the splitting mass is less than the base mass.

This has an immense potentiality as an isotope of deuterium exists as one part in 6500 parts of an isotope of hydrogen in natural hydrogen. *The deuterium available from the ocean could fuel the power requirements of the world for billions of years.* The fusion temperatures are of the order of 100 million degree Kelvin, generated by plasma in a toroidal field coil. A large number of fusion research facilities are in operation worldwide, yet there are no commercial power generating units. The only man-made fusion device is the hydrogen bomb, the detonation of the first device occurred in 1952.

The first generation fusion plants have so-called D–T–Li cycle, which involves two types of reactions. At first deuterium and tritium react to produce an alpha particle (helium 4 nucleus) and a high energy neutron:

$$D + T \rightarrow {}^4He + n \tag{2.4}$$

The helium-4 ion emerges with an energy of 3.5 MeV, and neutron with 14.1 MeV. A mass change in the reaction appears as kinetic energy of products in agreement with kinetic $E = \Delta mc^2$, where Δm is the change in rest mass of particles and c is the speed of light = 300,000 m/s (186,000 miles/s).

The second reaction involves production of tritium by lithium, which is contained in a blanket surrounding the plasma:

$$
\begin{aligned}
{}^6Li + n &\rightarrow T + {}^4He \\
{}^7Li + n &\rightarrow T + {}^4He + n
\end{aligned}
\tag{2.5}
$$

The tritium used in the first reaction is regenerated by lithium reactions in the blanket.

The sun is the main-sequence star and generates its energy by nuclear fusion of hydrogen nuclei into helium. In its core it fuses 620 million metric tons of hydrogen every second.

2.3.5.3 Nuclear Power around the Globe

The first nuclear power station started operation in the 1950s. Now there are 440 commercial power reactors operating in 30 countries with 377,000 MWe of total capacity. These provide 15% of the world's electricity. Fifty-six countries operate a total of 250 research reactors and a further 180 nuclear reactors power some 140 ships and submarines. Only eight countries are known to have nuclear weapon capability. Over 60 further nuclear reactors are under construction. Figure 2.7 shows the world's nuclear energy production, country wise. Note that France produces approximately 77% of its total energy requirements from nuclear sources. The bar width shows the amount of electricity in each country.

2.3.5.4 Is Nuclear Power Green Energy?

Some personalities like Bill Gates have classified nuclear power as green energy. However, this is controversial, as some disagree, claiming that problems associated with radioactive waste and risk of nuclear accidents such as the Chernobyl disaster pose an unacceptable risk to the environment. However, in some integral fast reactors, the reactor designs are capable of utilizing "nuclear waste" until it is dramatically less dangerous.

Recently, in a nuclear letdown, the US nuclear fleet dipped below 100 for the first time. The Vermont Yankee 604 MW plant was shut down, due to many safety issues and its

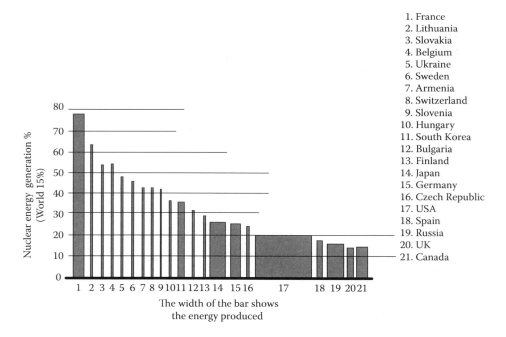

FIGURE 2.7
World-wide nuclear energy production.

Modern Electrical Power Systems

owner Entergy, just did not have enough money to make upgrades. There are several reactors including a handful in Illinois and New York on the verge of shutdown. The International Energy Agency's most recent blueprint for holding global warming temperature to 2°C requires an expansion of nuclear power in every country of the world. Without nuclear power, it will be hard and expensive to meet the projected carbon dioxide emission from all power plants.

2.3.6 Geothermal Plants

Geothermal plants can operate continuously and serve as base load plants. The potential capacity for geothermal plants is estimated at 85 GW over the next 30 years. However, geothermal power is accessible in limited areas of the world, which includes the USA, Central America, East Africa, Iceland, Indonesia, and the Philippines.

Geothermal energy is derived from heat within the earth in the form of underground hot water or steam. "Magma"—molten rocks are found beneath the earth's surface. This Magma heats the surrounding rocks and subterranean water. Hot springs originate because some of this water can come to earth's surface through cracks and fissures. When this hot water and steam are trapped in permeable rocks under a bed of impermeable rocks, geothermal reservoirs are formed. The thermal energy can be taken from these reservoirs through wells to depths of hundreds or thousands of feet.

Geothermal power is cost-effective, environmentally friendly, and unlike solar or wind power does not depend upon weather.

The geothermal plants can be classified as follows:

- Dry steam plant
- Flash steam plant
- Binary cycle plant

Dry steam plants use dry steam as it comes out of production wells. This steam directly drives a turbine generator. The steam is condensed and pumped down in another well called injection well. The largest plant of this nature is *geysers* north of San Francisco, a total of 15 plants producing 727 MW.

The flash steam plants utilize hot water above 350°F (176°C) from the production well. The hot water is depressurized or "flashed" into steam, which runs a steam-turbine generator. The exhaust steam is condensed in a condenser cooled by cold water from a cooling tower and provides make up water for the cooling tower. Hot water not flashed into steam is returned through injection well. This is the most common type of geothermal generating plant.

In the binary cycle plant, when the geothermal water temperature is not high enough, the hot water is passed through a heat exchanger. The heat is recovered by a *secondary fluid* with a boiling point temperature much below that of water. The secondary fluid flashes to vapor, which drives the turbine. The vapor is then condensed and circulated back to the heat exchanger. The cooled water is returned to the geothermal reservoir. The binary cycle plants are the fastest growing geothermal plants. A schematic of a binary cycle geothermal plant is illustrated in Figure 2.8.

The USA leads the geothermal power production. The plants are mostly located in the Western states, Alaska, and Hawaii. According to Geothermal Energy Association, the USA has a total installed capacity of around 3.0 GW.

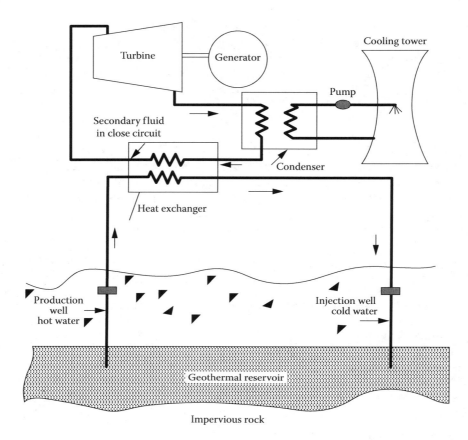

FIGURE 2.8
Schematic of a binary cycle geothermal plant.

Every technology has some limitations. A Geothermal plant in California with millions of dollars of funding was halted in 2009—the concerns were that it is causing increase in regional earthquake activity.

2.3.7 Solar and Wind Energy

These are the most important components of renewable and green energy sources. Entire Chapter 3 is devoted to these.

2.3.8 Biofuels and Carbon-Neutral Fuels

Biofuels may be described as renewable, but these are not sustainable due to soil degradation. As of 2012, 40% of the corn produced goes for production of ethanol. It is still debatable whether ethanol should be considered clean energy.

According to the International Energy Agency, new bioenergy technologies being developed today, notably cellulosic ethanol biorefinery could pave way for a much bigger role in the future. Cellulosic ethanol can be produced from plant matter composed primarily of inedible cellulose fibers, from stems and branches of most plants. Corp residues, such

Modern Electrical Power Systems 51

as corn stalks, wheat straw, and wood waste, a solid waste, are the potential sources of cellulosic biomass.

Carbon-neutral fuels are synthetic fuels including methane, gasoline, diesel fuel, jet fuel, or ammonia produced by hydrogenating waste carbon dioxide from power plant flue-emissions; recovered from automotive exhaust gas or derived from carbonic acid in sea water. Many commercial ventures claim that they can economically produce synthetic fuels when cost per barrel of oil is greater than $55. Renewable methanol (RM) is a fuel produced from hydrogen and carbon dioxide by catalytic hydrogenation—the hydrogen obtained from electrolysis of water.

From 1978 to 1996, the National Renewable Energy Laboratory experimented with producing algae fuel. Algae has a natural oil content greater than 50%, which can be grown on algae ponds at waste water treatment plants. This oil rich algae can then be extracted from the system and processed into biofuels. This is only in feasibility studies and there are no commercial productions.

2.3.9 Local Green Energy Systems

Public can install their own green energy based systems-like solar or wind or even micro-hydroelectric. Geothermal heat pumps tap the high temperatures below soils, which are 7°C–15°C a few feet underground and increases dramatically at greater depths. In the USA, many states offer the incentives to offset the installation of renewable energy systems. In states like California, MA, a new approach to community energy supply called Community Choice Aggregation has provided communities options to solicit a competitive electricity supplier and use municipal bonds to finance development of local green energy sources.

More than $1 billion has been spent on research and development of hydrogen fuel in the USA. Hydrogen is suitable for energy storage and for use in airplanes.

2.3.10 Fuel Cells

The fuel cells convert chemical energy of a fuel directly into electrical energy. The fuel reactant is supplied from an external source. The fuel cells have higher efficiency (36%–40%) as these produce electricity directly, compared to the energy generated by electromechanical process.

A typical fuel cell has a DC voltage of 0.7 V, and many fuel cells are grouped in series and parallel to obtain the desired voltage. These require power conditioning units to convert unregulated DC-to-DC to high voltage DC or DC-to-AC inverters for connection to the grid. The fuel cells can be

- Polymer electrolyte fuel cell (PEFC)
- Molten carbon fuel cell (MCFC)
- Solid oxide fuel cell (SOFC)
- Phosphoric acid fuel cell (PAFC)

The PAFC is depicted in Figure 2.9. The electrolyte is liquid phosphoric acid, which is stored in a Teflon-bonded silicon carbide matrix, and uses porous carbon electrodes containing a platinum catalyst. The hydrogen is fed into the anode, and the oxygen enters the fuel cell through the cathode. Hydrogen atoms are split into electrons and

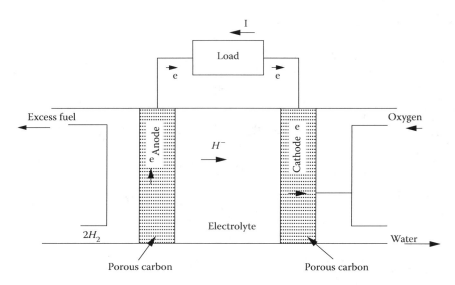

FIGURE 2.9
Schematic of a phosphoric acid fuel cell.

protons. The protons pass through the electrolyte membrane. Electrons flow through an external circuit before returning to cathode. The oxygen entering the cathode combines with electrons and protons to form water—water vapor and heat are the byproducts of the reaction. Approximately 300 units have been installed in the USA, Japan, and Europe. When used in combination with space and water heating, the efficiency amounts to 85%.

An article published, as early as 2002, in *Scientific American* talks about fuel cell technology for automotive transportation versus internal-combustion engine. Quoting a General Motor venture autonomy, it describes far simpler designs and twice the efficiency of the internal combustion engine. By 2020, approximately 15% of the world population will own a vehicle as compared to 9% in 2002. Though manufacturers have prototype designs for hydrogen powered vehicles concerns have been expressed on the safety of on-board hydrogen and cost deterrent.

2.3.11 Reducing Carbon Emissions

The coal-based plants produce SO_2, nitrogen oxides, CO, CO_2, hydrocarbons, and particulates. Coal is the biggest offender and is accountable for 80% of power plant carbon emissions. Approximately 70% of the electricity is produced by coal and natural gas.

Transportation of fossil fuels also requires a large amount of energy. While coal is transported by trains; oil and natural gas are moved around in ships and pipelines. About 30% of greenhouse gas emissions are attributed to transportation.

Coal, natural gas, and nuclear plants require tons of water to cool, taking another precious source from the planet.

Figure 2.10 shows the USA initiative (with China) to reduce carbon pollutions to 40% by 2025. This amounts to 21 million metric tons of emission reductions—which amount to same as taking 4.2 million cars off the road. Achieving this goal will save taxpayers up to $18 billion in avoided energy costs.

Modern Electrical Power Systems 53

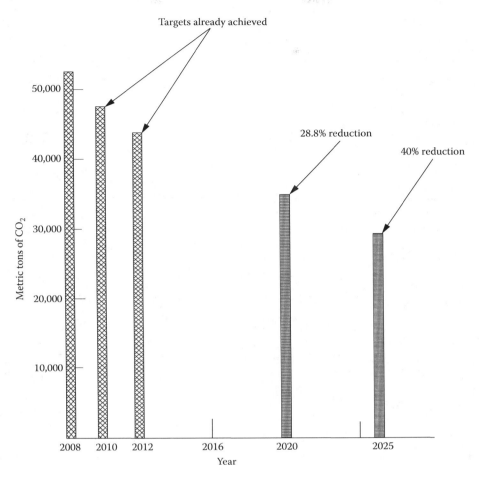

FIGURE 2.10
USA initiative (with China) to reduce carbon pollutions by 40% by 2025.

Department of energy takes energy management and greenhouse gas (GHG) reductions seriously. It has set a goal of reducing GHG to 28% by 2029 and as of 2013, the department has exceeded this goal. Other agencies involved are Environment Protection Agency (EPA) and General Service Administration (GSA). Working together, the federal government and private sector can have a profound impact in the fight against climate change.

2.4 Large Power Stations of the World

The world's large generating stations are shown in Tables 2.2 through 2.10.

- Table 2.2: Twenty largest generating stations in the world. Note that the utilization of the installed capacity varies, and is not necessarily in relation to the installed MW.
- Table 2.3: Ten largest nuclear plants.

TABLE 2.2
Twenty Largest Generating Stations in the World

Rank	Station	Country	Capacity (MW)	Annual Generation (TWh)	Type
1	Three Gorges Dam	China	22,500	98.8	Hydro
2	Itaipu Dam	Brazil	14,000	98.63	Hydro
3	Xiluodu	China	13,860	57.1	Hydro
4	Guri	Venezuela	10,235	47	Hydro
5	Tucurui	Brazil	8,370	21.4	Hydro
6	Kashiwazaki-Kariwa	Japan	7,965	60.3	Nuclear
7	Grand Coulee	USA	6,809	21	Hydro
8	Longtan	China	6,426	18.7	Hydro
9	Sayano-Shushenskaya	Russia	6,400	23.5	Hydro
10	Bruce	Canada	6,300	45	Nuclear
11	Krasnoyarsk	Russia	6,000	23	Hydro
12	Hanul	South Korea	5,881	48.16	Nuclear
13	Hanbit	South Korea	5,875		Nuclear
14	Zaporizhia	Ukraine	5,700	48.16	Nuclear
15	Robert-Bourassa	Canada	5,616		Hydro
16	Shoaiba	Saudi Arabia	5,600		Fuel oil
17	Surgut-2	Russia	5,597	39.85	Natural gas
18	Taichung	Taiwan	5,500	42	Coal
19	Gravelines	France	5,460	38.5	Nuclear
20	Churchill Falls	Canada	5,428	35	Hydro

TABLE 2.3
Largest Nuclear Stations

Rank	Station	Country	Capacity (MW)
1	Kashiwazaki-Kariwa	Japan	7,965
2	Bruce	Canada	6,300
3	Hanul	South Korea	5,881
4	Hanbit	South Korea	5,875
5	Zaporizhia	Ukraine	5,700
6	Gravelines	France	5,460
7	Paluel	France	5,528
8	Cattlenom	France	5,448
9	Ōi	Japan	4,710
10	Fukushima Daini	Japan	4,400

TABLE 2.4

Solar Power Generation-Photovoltaic-Flat Panels

Rank	Station	Country	Capacity (MW)
1	Topaz	USA	550
2	Desert Sunlight	USA	550
3	Longyangxia Dam Solar Park	China	320
4	Solar Star I and II	USA	309
5	California Valley Solar Ranch	USA	292
6	Auga Caliente	USA	292

TABLE 2.5

Solar Power Generation-Concentrated Photovoltaic

Rank	Station	Country	Capacity (MW)
1	Alamosa	USA	3.7
2	Navarra	Spain	7.8
3	Hatch	USA	5
4	Casaquemada	Spain	1.9
5	Sevilla	Spain	1.2
6	Victor Valley College CPV	USA	1
7	Questa	USA	1

TABLE 2.6

Solar Power Generation-Concentrated Solar Thermal

Rank	Station	Country	Capacity (MW)
1	Ivanpath	USA	377
2	SEGS	USA	354
3	Solana	USA	280
4	Genesis	USA	250
5	Solaben	Spain	200
6	Solanova	Spain	150
7	Andasol	Spain	150
8	Extresol	Spain	150
9	Palma del Rio	Spain	100
10	Manchasol	Spain	100
11	Valley	Spain	100

TABLE 2.7

Wind Power Generation-Off-shore

Rank	Station	Country	Capacity (MW)
1	London Array	UK	630
2	Greater Gabbard	UK	504
3	Anholt	Denmark	400
4	Bard off-shore-1	Germany	400
5	Walney	UK	367
6	Thorntonbank	Belgium	325
7	Sheringham Shoal	UK	315
8	Thanet	UK	300
9	Lines	UK	270
10	Horns Rev-2	Denmark	209

TABLE 2.8

Pumped Storage and Run of the River

Type	Rank	Station	Country	Capacity (MW)
Pumped storage	1	Bath County	USA	3003
	2	Huizhou	China	2448
	3	Guangdong	China	2400
	4	Okutataragi	Japan	1932
	5	Ludington	USA	1872
Run of the river	1	Chief Joseph	USA	2620
	2	John Day	USA	2160
	3	Beauharnois	Canada	1903
	4	The Dalles	USA	1780
	5	Nathpa Jhakri	India	1500

TABLE 2.9

Geothermal Plants

Rank	Station	Country	Capacity (MW)
1	The Geysers	USA	1808
2	Cerro Prieto	Mexico	958
3	Hellisheioi	Iceland	303
4	Olkaria	Kenya	260
5	Darajat	Indonesia	255
6	Malitbog	Philippines	233
7	Wayang Windu	Indonesia	227
8	Kamojang	Indonesia	203
9	Navy	USA	240

Modern Electrical Power Systems

TABLE 2.10

Biomass

Rank	Station	Country	Capacity (MW)
1	Tilbury	UK	750
2	Drax	UK	660
3	Ironbridge	UK	600
4	Alholmens Kraft	Finland	265
5	Maasvlakte 3	Netherlands	220
6	Polaniec	Poland	205
7	Atikokan	Canada	205
8	Rodenhuize	Belgium	180
9	Ashdown Paper Mill	USA	157
10	Wisapower	Finland	150

- Tables 2.4 through 2.6: Solar generation.
- Table 2.7: Off-shore wind power generation.
- Table 2.8: Pumped storage and run-of-the river plants.
- Table 2.9: Geothermal generation.
- Table 2.10: Biomass generation.

2.5 Smart Grid

While we may define electrical systems as entity of generation, transmission and distribution, storage and end use, the grid may be defined as the infrastructure that lies between generation sources and the consumer, i.e., transmission, distribution, and electricity delivery.

By smart grid is implied a class of technology that brings the grid systems into the 21st century by using computer-based remote control and automation. These systems are made possible by two-way communication technologies and computer processing, and are beginning to be used on electric networks from the power plants and wind farms all the way to the consumers. A smart meter at a consumer premises is one example—rather than a crew taking the meter readings, a smart meter can communicate, reset, and signal power quality data, with time stamp.

In the years to come the power industry will undergo profound changes, need based—environmental compatibility, reliability, improved operational efficiencies, integration of renewable energy sources, and dispersed generation. The dynamic state of the system will be known at all times and under any disturbance.

Consumers want more choice and control over their energy sources, and new unregulated entities are entering the market with new choices and incentives. It is imperative that grid and grid operations evolve.

In the vision of the future grid, it is necessary to look at the grid in the context of entire value chain of electrical system. The electrical system of the future will include both central and distributed generation sources with a mix of controllable and noncontrollable sources.

- Energy storage will be a key component but it will not replace dispatched sources. Consumers may generate their own electricity. Multi-consumer and single-consumer microgrids will complement the future grid.
- There will be a mix of regulated and competitive services. With third party providers entering the market to offer consumers new services that are not a part of the regulated pricing, a mix of regulated and competitive services will emerge. There will be a retail market for services.
- While the grid will be more complex, it will be more flexible, adaptable, and responsive.
- Balancing supply and demand will remain a key function of the operator. Currently, the grids are designed for one-way power flow from the grid to the consumer. The future grids should be able to accept two-way or multiple way of power flow. (Even now large cogeneration facilities exist and operate successfully in synchronism with the grid, pumping excess power into the grid.)
- Grid operators should be able to predict conditions in real time with sophisticated modeling and state estimation capabilities. New tools will be available to manage new dynamic system and continued balance of supply-and-demand. Also, see Volume 2 for basic concepts.
- Interdependence and interconnections between transmission and distribution systems operations will grow. A high-bandwidth, low-latency, cost-effective communication system will be needed to overlay the entire grid.
- The grids shall be self-learning and self-healing in all aspects of the system, down to end-use devices, without sacrificing security and privacy.

2.5.1 Legislative Measures

In 2007, Title XIII of the Energy Independence and Security Act (EISA) was passed. This provided legislative support for Department of Energy (DOE's) smart grid activities and reinforced its role in leading and coordinating national Grid modernization efforts. Key provisions of Title XIII are as follows:

- Section 1303 establishes DOE the Smart Grid Advisory Committee and Federal Smart Grid Task Force.
- Section 1304 authorizes DOE to develop a "Smart Grid Regional Demonstration Initiative."
- Section 1305 directs the National Institute of Standards and Technology (NIST), with DOE and others to develop Smart Grid Interoperability Framework.
- Section 1306 authorizes DOE to develop a "Federal Matching Fund for Smart Grid Investment Costs."

2.5.2 Technologies Driving Smart Grid

The technologies driving smart grids are

- RAS (remedial action scheme)
- SIPS (system integrated protection system)

Modern Electrical Power Systems 59

- WAMS (wide area measurement system)
- FACTS (flexible AC transmission system), see Volume 2
- EMS (energy management system)
- PMUs (phasor measurement units)
- Overlay of 750 kV lines and HVDC links
- DCA (dynamic contingency analysis) all somewhat related

It is amply clear that stability of a system is not a fixed identity and varies with the operating and switching conditions. Some, not so common contingencies, in a system can cascade and bring about a shutdown of a vital section. Historically, the Great North East Blackout of November 9–10, 1965, and more recently, 2003, East Coast Blackout can be mentioned.

Volume 4 of this series provides a description of

- Wide area network measurements
- Phasor measurements
- SIPS
- Adaptive relaying

Other smart grid issues can be itemized as follows:

- Requirements for renewable portfolio standards (RPS), limits on greenhouse gases (GHGs) and demand response (DR).
- Advanced metering structures at consumer loads
- Integration of solar, wind, nuclear, geothermal facilities that poses their own challenges. For example, the large scale solar plants or wind generation may be located in areas distant from existing transmission facilities. New protection and control strategies, interconnection standards, for example, low-voltage ride through (LVRT) capabilities, forecasting, and scheduling are required.
- Circuit congestion, managing distribution system overloads.
- Role of information and automation technologies.

2.6 Microgrids and Distributed Generation

Distributed generation encompasses a wide range of technologies such as gas turbines, microturbines, photovoltaics, fuel cells, solar, and wind power. These have an inverter to interface with the electrical distribution systems. These have lower emissions. The applications include power support at the substations. Penetration of distributed generation in the USA has not reached a significant level; however, this situation is changing rapidly.

Microgrids take a system approach to the generation and associated loads and local control, thereby reducing the need for central dispatch. During the disturbances defined in IEEE, the loads and generation are separated from the distribution system to maintain a high level of service. This islanding from the main transmission grid has the potentiality of providing higher reliability. During islanding, microgrid implementations try to use the available waste heat. These solutions rely upon complex communications and controls. Intelligent power electronics interfaces, a single smart grid disconnection switch and resynchronization specifically lower the costs and improve reliability.

Microturbines are a new emerging technology. These are single shaft, mechanically simple devices with no bearings or lubricants. The generator is usually of permanent magnet type. Combustion systems with low turbine temperatures and lean fuel to air mixtures result in inherently low carbon emissions.

DG has the potentiality to increase system reliability and power quality due to decentralization of the electrical supply. Figure 2.11 shows a microgrid concept. The static switch has the ability to autonomously island the microgrid from disturbances. After islanding the reconnection is achieved automatically when the tripping event is no longer present. This synchronization is achieved by using frequency difference between the islanded microgrid and the utility grid, ensuring a transient-free operation. Each micro source can seamlessly balance the power on the islanded microgrid using a power versus frequency droop controller. Figure 2.11 shows three feeders, with micro sources that control the operation using local voltage and current measurements. When there is a problem with the utility the static switch opens. Nonsensitive loads connected to feeder D ride through the event. It is assumed that the micro sources provide sufficient generation on feeders to which these are connected to serve the feeder loads.

The micro-source controls must ensure that the system is expandable and more sources can be added, without modifications to the system and set points, the microgrid can island and reconnect, and the active and reactive power load demand can be served.

When regulating the output power, each source has a constant negative slope with respect to frequency versus power, Figure 2.12, under steady state. The dashed line shows the slope as power varies from zero to maximum. The operating points for the

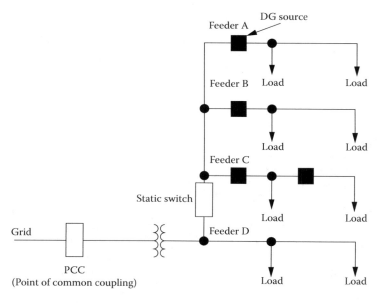

FIGURE 2.11
Microgrid concept, see text.

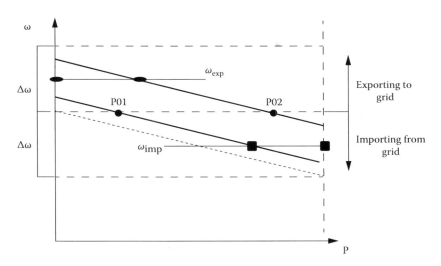

FIGURE 2.12
Regulation of output power versus frequency in a microgrid, Figure 2.11.

two micro units, P01 and P02, are also shown, which represents the power thrown into the grid when the units are connected to the grid, at system frequency. If the system islanded, when importing power from the grid, then the generation needs to increase for the power balance. The new operating points will be at a frequency lower than the nominal value. In this case, both units have increased their power, with the Unit 2 at its maximum output. If the system is islanded when exporting power to the grid, the new frequency will be higher.

A power management strategy is required, which should meet the following main criteria:

- Load sharing between units while minimizing the total system loss.
- Considerations of specific limits of each unit, including type of DG unit, cost of generation, time dependency of the prime source.
- Maintaining power quality, voltage, voltage fluctuations, and limiting the harmonics.
- Improving the dynamic response, maintaining stability margin, and voltage/frequency restoration of the system during and after transients.

We can talk of static and dynamic grid codes, as shown in Table 2.11. These define the performance expectations of the plants for small rotating generators. Table 2.11 is based on

TABLE 2.11

Grid Codes-Germany

Grid Code	Parameter	Expectation
Static	Voltage variations	Plus/minus 10%
	Power factor range	0.95 lead to 0.95 lag
Dynamic	Voltage and fault ride through time	30% for 150 ms

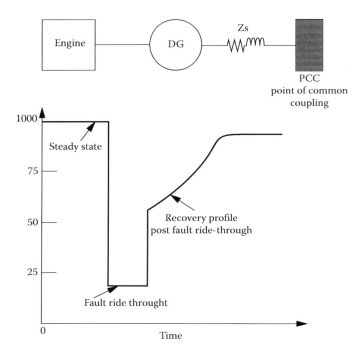

FIGURE 2.13
Fault ride through and recovery event of a DG.

a grid code document in Germany—voltage, power factor during steady-state conditions, and connection period during the fault ride through.

Figure 2.13 shows a typical ride through event. The voltage at the point of common coupling drops to a low value and the active load on the generating set also drops. The kinetic energy in the shaft (which is low for small machines) accelerates the rotor (typical phenomena as experienced in larger rotating machines), which exposes the rotating generator to thermal and mechanical stresses. As the grid starts to recover, the generator is now trying to reconnect to the grid, with a different frequency and phase angle as compared to the grid voltage, which mechanically stresses the generator. The small generators for microgrid applications have special designs. The transient effects can be studied with appropriate modeling.

Producers of large renewable power must condition the power to interconnect with the utility. A STATCOM and DVR can serve this need, act as a reactive power compensator with a high transient capability of 250% or more. See Volume 2 for FACT devices.

These devices can seamlessly integrate with low cost shunt capacitors, reactors and provide a highly effective cost-effective solution for steady-state voltage regulation, power factor correction and LVRT capability of entire renewable generation. It can manage the switching of the capacitors and reactors to provide steady state or dynamic compensation as needed with a response time of no more than **half** cycle to one cycle on a 60 Hz basis. Voltage sags and swells originating from the transmission grid can be mitigated. This enhances the capability of renewables to stay on line and prevent nuisance tripping of the generators and enhance stability. Step changes due to local or remote capacitor switching can be mitigated. The wind farms can maintain a smoother voltage profile.

Modern Electrical Power Systems

2.7 Energy Storage

The storage of energy is an important link and much research work is being undertaken in this direction. The energy is required to be stored in some medium for use on-demand and in stand-alone systems, not connected to the grid. The electrical generation is relatively fixed, but the demand for electricity fluctuates throughout the day. The load profile varies depending upon the consumers—say continuous—process plant will have a much higher load factor compared to residential consumers. Developing technologies to store energy, so that it can be made available on-demand will be a major breakthrough in power distribution. The storage devices can supply power on peak demand and result in demand shaving. Solar and wind energy are highly variable, depending on the sunshine or weather and storage can flatten the daily usage. These can balance micro grids to achieve a good match between generation and load.

At present, the USA has about 24.6 GW (approximately 2.3% of the electrical production capability) of grid storage. Ninety-five percent of this storage comes from pumped storage hydro; Europe and Japan have noticeably higher fractions of grid energy storage.

Energy storage technologies such as pumped hydro, compressed air energy storage, various types of batteries, flywheels, capacitors, etc. provide for multiple applications, namely, energy management, backup power, load leveling, frequency regulation, and voltage support.

DOE is addressing these challenges as shown in Table 2.12. The maturity of energy storage technologies is shown in Figure 2.14. Table 2.13 shows the primary applications, challenges, and current utilization of the technology.

TABLE 2.12

DOE Energy Storage Strategy

Goal	Strategy Summary
Cost competitive energy storage technology	• Targeted scientific investigations of fundamental materials, transport processes, and phenomena enabling discovery of new or enhanced storage technologies with increased performance • Materials and system engineering research to resolve key technology costs and performance challenges of known emerging storage technologies, including manufacturing • Seeded technology innovation for new storage concepts • Development of storage technology cost models to guide R&D and assist innovators • Resolution of grid benefits of energy storage to guide technology development and facilitate market penetration
Validated reliability and safety	• R&D programs focused on deregulation and failure mechanisms and their mitigation, and accelerated life testing • Development of standard testing protocols and independent testing of storage devices under accepted utility use cases • Track, document, and make available performance of installed storage systems
Equitable regulatory environment	• Collaborate public and private sector characteristics and evolution of grid benefits of storage • Exploration of technology-neutral mechanisms for monetizing grid services provided by storage • Development of industry and regulatory agency-accepted standards for siting, grid integration, procurement, and performance evaluation
Industry acceptance	Collaborative, co-funded field trials and demonstrations enabling accumulation of experience and evaluation of performance—especially for facilitating renewable integration and enhanced grid resilience Adaptation of industry-accepted planning and operational tools to accumulate energy storage Development of storage system design tools for multiple grid services

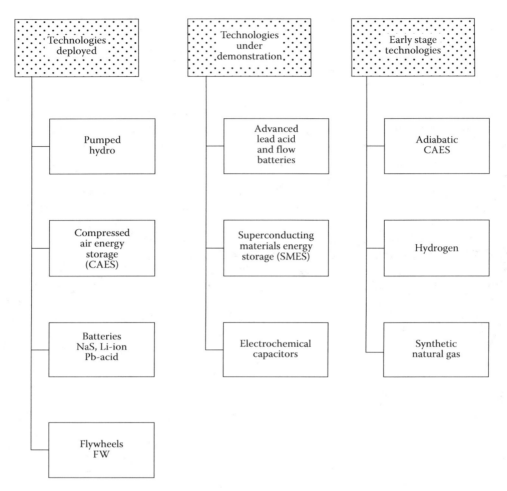

FIGURE 2.14
The maturity of energy storage technologies (DOE).

2.7.1 Flywheel Storage

Figure 2.15 shows a schematic of the flywheel storage system. It consists of a rotor suspended inside a vacuum chamber to reduce friction, connected to a combination of electrical motor/generator. The earlier large steel flywheels and mechanical bearings have been replaced with carbon-fiber composite rotors that have a higher tensile strength than steel and less heavy.

Compared to other energy storage systems, flywheel energy storage systems (FESs) have longer life time and less maintenance, in excess of 10^5–10^7 full cycles of use. The maximum energy density is given by

$$\frac{E}{m} = K\left(\frac{\sigma}{\rho}\right) \tag{2.6}$$

TABLE 2.13

Energy Storage Technologies

Technology	Primary Application	What Is Currently Known?	Challenges	
Compressed air energy storage (CAES)	Energy management Backup and seasonal reserves Renewable integration	Better ramp rated than gas turbine plants Established technology in operation since 1970	Geographically limited Lower efficiency due to round trip conversion Slower response time than flywheels or batteries Environmental impact	First commercial plant was built in Germany in 1978 and had 280 MW capacity. An additional plant with 110 MW capacity was built in Alabama in 1991. Advanced Adiabatic CAES systems are under development
Pumped hydro	Energy management Backup and seasonal reserves Regulation service also available through variable speed pumps	Developed and mature technology Very high ramp rate Currently most effective form of storage	Geographically limits Plant sites Environmental impacts High overall project cost	It represents 3% of global generation capacity—more than 90GW of PH storage has been installed worldwide
Fly-wheels	Load leveling Frequency regulation Peak shaving and off peak storage Transient stability	Modular technology Proven growth potential to utility scale Long cycle life High peak power without overheating concerns Rapid response High round trip efficiency	Rotor tensile strength limitations Limited energy storage time due to frictional losses	Currently FW are used in many uninterruptable power supply systems and aerospace applications. FW farms are being built to store megawatts of electricity for short-duration regulation services
Advanced lead acid batteries with carbon enhanced electrodes (ALA-CEE)	Load levelling and regulation Grid stabilization	Mature battery technology Low cost High recycled content Good battery life	Limited depth of discharge Low energy density Large footprint Electrode corrosion limits useful life	ALA-CEE was developed as an inexpensive battery for hybrid electrical vehicles. The largest traditional lead acid battery in service is 10 MW/40 MWh capacity

(Continued)

TABLE 2.13 (*Continued*)

Energy Storage Technologies

Technology	Primary Application	What Is Currently Known?	Challenges	
Sodium sulphide batteries (NaS)	Power quality Congestion relief Renewable source integration	High energy density Long discharge cycles Fast response Long life Good scaling potential	Operating temperatures required between 250°C and 300°C Liquid containment issues, corrosion and brittle glass seals	The battery has been demonstrated at more than 190 sites in Japan, totally more than 270 MW capacities US utilities have installed 9 MW for peak shaving, firming wind power and other applications. Further developments in progress
Lithium-ion batteries (Li-ION)	Power quality Frequency regulation	High energy densities Good cycle life High charge/discharge efficiency	High production cost Extremely sensitive to overcharge, temperature, and internal pressure built up	These dominate the consumer electronic market. Manufacturers are working to reduce system costs, increase safety, enabling these to be used in large scale markets

Modern Electrical Power Systems

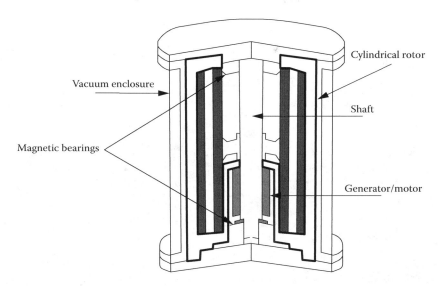

FIGURE 2.15
Schematic of a flywheel energy storage system (FES).

where,
E = kinetic energy of the rotor (J)
m = mass of the rotor (kg)
K = rotor geometric shape factor
σ = tensile strength of the material (Pa)
ρ = material density (kg/m^2)

FES finds applications in automotive, rail vehicles, uninterruptible power supply system (UPS), aircraft launchers, pulse power, motor sports wind turbines, and grid energy storage. Beacon Power opened a 5 MWh (20 MW over 15 min) energy storage system in Stephentown, New York, in 2011. Lower carbon emissions and faster response times and ability to buy power at off-peak hours are among some advantages. Flywheels are not as adversely affected by temperature changes, can operate at much wider temperature ranges and are not subject to many failures inherent in chemical batteries.

2.7.2 Superconductivity

Superconductivity has a wide field of applications not only in electrical power, but in many other fields. It can revolutionize the future technologies. The superconductivity was discovered by Dutch physicist, H.K. Onnes, in 1911. Since 1960, a Niobium–Titanium (Ni–Ti) alloy has been the material of choice for commercial superconducting magnets. More recently, a brittle Niobium–Tin inter-metallic material has emerged. In 1986, an oxide-based ceramic material demonstrated superconducting properties above 35 K. A superconducting material with critical temperature above 23.3 K is called high-temperature superconductor (HTS).

The unique properties of superconductors are as follows:

- Zero resistance to direct current
- Extremely high current carrying density

- Extremely low resistance at high frequencies
- Extremely low signal dispersion
- High sensitivity to magnetic field
- Exclusion of externally applied magnetic field
- Rapid single flux quantum transfer
- Close to speed of light signal transmission

Superconducting-based products are extremely environmentally friendly, and generate no GHGs. These are cooled by noninflammable liquid nitrogen as opposed to conventional oil coolants that are both inflammable and toxic. Typically, they are 50% smaller and lighter than equivalent conventional units.

2.7.2.1 Applications in Electrical Systems

Superconductivity finds applications in generators, transformers, underground cables, synchronous condensers, fault current limiters, and energy storage.

The relieving of overburdened grid—where large chunk of power from their generation sources have to be transmitted to the end users can be revolutionized with HTS technologies.

HTS cables can carry three to five times more power than conventional copper cables. Significant progress toward commercialization of HTS cable is underway. Three major in-grid demonstrations have been completed in the USA, including the world's first HTS transmission cable system in commercial grid capable of transmitting up to 574 MW of power. New demonstrations are in planning stage in the USA and around the world.

HTS fault current limiters (FCLs) can interrupt large short-circuit currents quickly to limit the short-circuit duties on switching devices. The superconductor changes from a path of zero resistance to high resistance instantaneously.

HTS transformers are cooled by liquid nitrogen and have twice the overload capability. Direct drive wind generators are utilizing a new high efficiency stator design and replacing copper with HTS wire on the rotor. A 10 MW drive utilizing HTS will weigh approximately one-third the weight of a conventional generator.

Small-scale superconducting magnetic energy storage (SEMS) systems using low-temperature superconductor have been in use for many years. FES based on frictionless superconducting bearings is in operation.

The other applications of superconductivity include transportation, marine propulsion-using HTS motors and generators, magnetically levitated trains, medical imaging and diagnosis (radiation free imaging—ultra-lows field resonance imaging—ULF-MRI), industrial processing, etc.

2.8 Transmission Systems

Figure 2.16 depicts progressive increase in AC transmission voltages. There are many transmission lines operating at 800 kV around the globe. UHVs (ultra-high voltages) to 1200 kV are in testing or development stage. A double circuit 1000 kV transmission line tower rises to approximately 108 m from the ground, the lowest conductor being at a clearance of 50 m from the ground, with conductor to conductor spacing of approximately 21 m.

Modern Electrical Power Systems

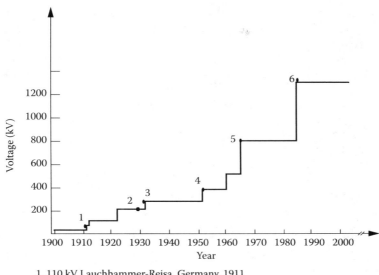

1. 110 kV Lauchhammer-Reisa, Germany, 1911
2. 220 kV Brauweiler-Hoheneck, Germany, 1929
3. 287 kV Boulder Dam, Los Angeles, USA 1932
4. 380 kV Harspranget-Halsberg, Sweden, 1952
5. 735 kV Montreal-Manicouagan, Canada, 1965
6. 1200 kV Ekibastuz-Kokchetave, USSR 1985

FIGURE 2.16
Developments in AC transmission voltages over the course of years.

A high-voltage grid system may incorporate hundreds of generators, transmission lines, and transformers. Thus, the extent to which a system should be modeled has to be decided. The system to be retained and deriving an equivalent of the rest of the system can be addressed properly only through a sensitivity study. Transmission systems have higher X/R ratios and lower impedances and the harmonics can be propagated over much longer distances. The operating configuration range of a transmission system is much wider than that of a distribution system. A study may begin by identifying a local area, which must be modeled in detail. The distant portions of the system are represented as lumped equivalents, Chapter 1. The transmission systems use SVCs TCRs, STATCOMs, TCSC, and other FACTS controllers (Volume 2), which need to be properly modeled.

Volume 2 discusses the transmission line modeling, their compensation, steady-state and transient stability characteristics, traveling wave phenomena, voltage, and reactive power control.

2.9 Industrial Systems

Industrial systems vary in size and complexity. While a saw mill with a load of 1000 kVA served from 480-V pole mounted transformer may be called an industrial system, the largest utility substation with an installed capacity of 260 MW, in the state of Georgia, USA serves a newsprint mill at 230 kV. Some industrial plants may generate their own power and may operate in cogeneration mode with the utility systems at high voltages. The utility service may be at voltages of 115, 138, and 230 kV and the load demand at these voltages can be upward of 100–200 MW. It is usual to represent the utility source by its short-circuit impedance, however, nearby harmonic loads should be considered. The large rolling mill or arc furnace loads can impact an adjacent industrial system that does not have any harmonic producing load, see Volume 3. Consider an example of a large distribution system:

- Two utility interconnecting transformers of 50 MVA at 230 kV.
- Four generators totaling 120 MW.
- Load demand of 100 MW, at certain times the excess generated power can be supplied into the utility system.
- Seventy-six secondary unit substations (0.5–2.5 MVA) at low voltage, some of them double ended.
- Two hundred and eighty low-voltage MCCs (motor control centers).
- Two-thousand one hundred low-voltage motors, ranging from 5 to 250 hp.
- A number of emergency tie connections between buses for alternate supply of power.
- Six redundant storage battery systems, with duplicate charges. Uninterruptible power supply systems for critical loads.
- Auto-switching and transfer of power for critical process loads and generation auxiliary loads.
- Twenty-six primary unit substations at 2.4 or 4.16 kV (2.5–7.5 MVA) serving medium voltage motor loads, induction and synchronous, 5000-hp. (Some pulping mill operations use synchronous refiner motors of 45,000-hp).
- 10 miles of cable interconnections at medium and low voltage.
- 5 miles of 13.8 kV overhead lines.
- Six current limiting reactors to control short-circuit levels to acceptable limits within ratings of the circuit breakers.
- 25% of operating load consisting of drive systems, 6–18 pulse.
- Three harmonic filters totaling 30 MVAr for voltage support and harmonic mitigation.

The location of an industrial plant impacts the modeling strategy. Many a times, the utility high-voltage substations, 230–66 kV (sometimes operated by the industrial plant personnel) are located within the industrial plant facilities. An industrial plant may be located close to a large generating station or there may be nearby harmonic sources or capacitors, which will impact the extent of external system modeling. Generators and large rotating loads may be modeled individually, while an equivalent AC motor model can be derived connected through fictitious transformer impedance representing a number of transformers to which the motors are connected. This aggregating of loads is fairly accurate, provided that harmonic source buses and the buses having capacitor compensations are modeled in detail.

The major sources of harmonics in industrial systems are adjustable speed drives (ASDs) and the harmonic emission can vary over wide limits depending upon topology. The sources of nonlinear loads vary depending upon processes. Electrolyzing and metallurgical plants will have large chunks of rectifier loads. Cement plants have gearless load-commutated inverter (LCI) fed large synchronous motors. Steel rolling mills may have large electronically fed DC motors. See Volume 3.

2.10 Distribution Systems

The primary distribution system voltage levels are 4–44 kV, while the secondary distribution systems are of low voltage (<600 V). Distribution embraces all electrical utility system between the bulk power supply source and the consumer disconnect switch/meter. Figure 2.17 shows the fundamental configuration of distribution systems. A distribution system consists of

- Subtransmission circuits operating from 12.47 to 345 kV, which deliver energy to distribution substations.
- Distribution substations, which convert energy to lower voltages for local distribution and regulate the voltage delivered to load centers, i.e., use of induction voltage regulators, shunt power capacitors.
- Primary circuits for feeders, which may operate between 2.4 and 34.5 kV and supply loads to certain areas.

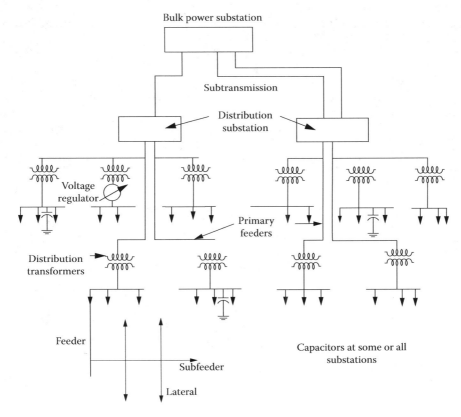

FIGURE 2.17
A distribution system configuration.

- Pole or pad-mounted distribution transformers or transformers, rated 10–2500 KVA located in underground vaults, close to consumer loads, which convert energy to the voltage of utilization.
- Secondary circuits at utilization voltage, which deliver the energy along the street or alleys to within short distance of users.
- Finally, the service drops or underground circuits, which bring the energy from the secondary mains to the consumer service switches.

The subtransmission circuits can be

- Simple radial circuits
- Parallel or loop circuits
- Interconnected circuits forming a grid or network

2.10.1 The Radial System

The radial form is the simplest, costless, but is not used when power is transmitted over one radial circuit, Figure 2.18. The fault on the single circuit feeder will result in interrupting a large block of load, and the electrical service to many customers will be interrupted which is not acceptable.

A modified form of radial system with duplicate automatic and manual throw over switches/circuit breakers is shown in Figure 2.19. In the event of failure of one circuit, the faulty circuit is quickly relayed open, and the alternate normally open breaker can be

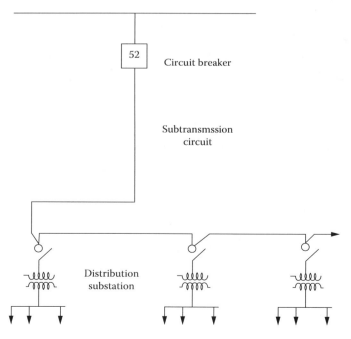

FIGURE 2.18
A radial distribution system.

Modern Electrical Power Systems

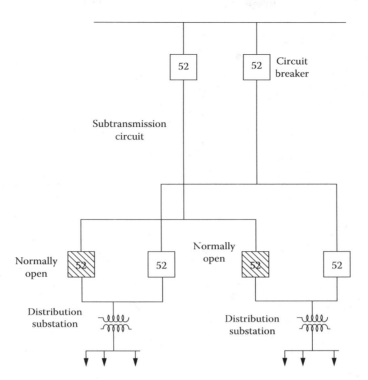

FIGURE 2.19
A radial system with duplicate automatic and manual throw over switches/circuit breakers.

manually or automatically closed. This arrangement does not preclude large-scale short-time interruption, but the power supply can be quickly restored.

2.10.2 The Parallel or Loop System

A parallel or loop system is shown in Figure 2.20. There are duplicates or parallel feeders to each of the distribution substations. All the circuit breakers shown in this figure are normally closed, and therefore the load has two routes of power flow. Note that the circuit to substations 1–3 forms a loop. Now consider a fault, say at location F1 between substations 1 and 2. By appropriate discriminative relaying, breakers B2 and C2 will be opened, while all other breakers remain closed. The power supply to any of the substations is not interrupted, though the fault clearance may cause a voltage dip, which will be experienced by all the consumers, depending upon their location and impedance to fault point.

2.10.3 Network or Grid System

A grid or network system is shown in Figure 2.21. This system provides the best reliability of the power supply. Generally more than one bulk power supply source buses will be tied together. In this system, the power can flow from any source to any substation. Network construction permits adding new substations without much capital expenditure or construction. However, it has the disadvantage that there are many circuit breakers, and the selective, protective relaying is complex. The reactive power compensation through capacitors may be spread throughout the system, locations decided by power system studies.

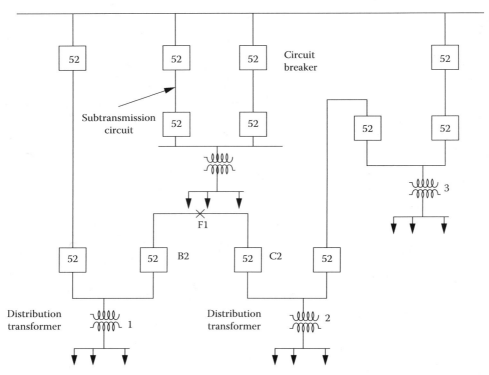

FIGURE 2.20
A parallel or loop system of distribution.

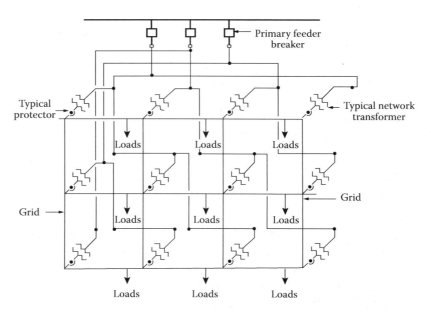

FIGURE 2.21
A grid or network system of distribution.

2.10.4 Primary Distribution System

The primary distribution system takes energy from the low-voltage buses of the distribution substations (Figures 2.18 through 2.21) and delivers it to the primaries of the distribution transformers.

Most primary distribution circuits are radial. These consist of primary feeders that take the energy to a load area, subfeeders that distribute the energy to laterals, and finally the laterals that connect to the individual transformers.

Figure 2.22 illustrates a tree type of radial circuit. The current tapers off from the substation to the distal radials and the conductor size can be reduced as one proceeds from the trunk of the tree to the branches. It is an economical and simple circuit, but the voltages drops may increase toward the remote laterals from the substation. The maximum voltage drop is limited to 5% or even less.

Figure 2.23 shows another typical distribution system. Note the recloser and switched power capacitors. Many faults are transient in nature, for example, an insulator on overhead (OH) line may temporarily flash over to ground, creating a temporary line-to-ground fault. A recloser will open on such a fault and reclose after a short delay. If the fault is of a temporary nature, the service will be restored quickly; however, if the fault is of permanent nature, it will again trip. As many as three to four attempts may be made to restore the service, before the recloser will lockout. The power capacitors are used to support the voltage on heavy loads and automatic voltage dependent switching is resorted to.

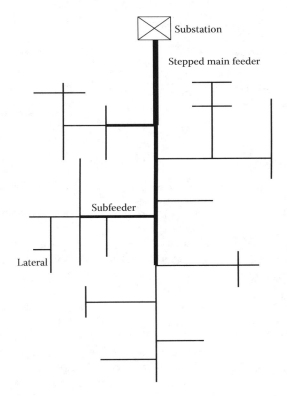

FIGURE 2.22
A tree type of radial circuit.

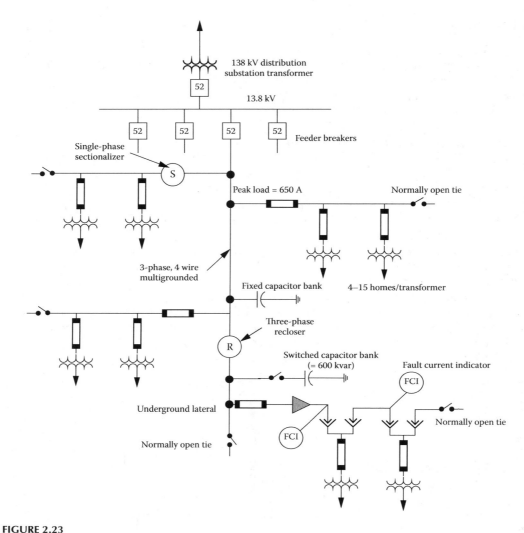

FIGURE 2.23
A typical distribution system with sectionalizer, reclosure, and switched and fixed capacitor banks.

Alternate circuit connections can restore power in case of a fault in one section. This figure also shows fuses at single phase tap points and transformers with internal or external fusing. The voltage profile is maintained for about 50 km from the nearest to the farthest customer by application of line regulators or shunt capacitors, and in some instances both these devices are used.

When a failure occurs on the radial system, action must be taken by utility employees to restore power, which may not occur automatically, prolonging the period of shutdown. In some densely populated areas and large cities, utilities have installed secondary networks, consisting of multiple distribution feeders; each serving an underground transformer installed in vaults. The low-voltage secondary of these transformers are interconnected and the service from these secondaries provide high reliability. *Network protectors* are installed due to high short-circuit current and current limiters on the secondaries.

Investments in distribution systems constitute 50% of the capital investment of a typical utility system.

2.11 Future Load Growth

The distribution system must be designed so that anticipated load growth can be served at minimum expense. This flexibility is needed to handle load growth in existing areas as well as load growth in new areas of development. It is far more cost-effective to design a system initially with prospective load growth based upon statistics, development plans, and the location, than to later on augment the systems. The various load types, their characteristics, and load windows are discussed in Volume 2, Noteworthy points are

- Electrical generation, transmission, and distribution facilities cannot be added overnight and takes many years of planning and design engineering efforts.
- In industrial plant distribution systems, it is far easier and economical to add additional expansion capacity in the initial planning stage rather than to make subsequent modifications to the system, which are expensive, may result in partial shutdown of the facility and loss of vital production. The experience shows that most industrial facilities grow in the requirements of power demand.
- Energy conservation strategies should be considered and implemented in the planning stage itself.
- Load management systems. An ineffective load management and load dispatch program can offset the higher capital layout and provide better use of plant, equipment, and resources.

Thus, load forecasting is very important in order that a plant system and apparatus of the most economical size be constructed at the correct place and the right time to achieve the maximum utilization.

For industrial plants, the load forecasting is relatively easy. The data are mostly available from the similar operating plants. The vendors have to guarantee utilities within narrow parameters with respect to the plant capacity.

Something similar can be said about commercial and residential loads. Depending upon the size of the building and occupancy, the loads can be estimated.

Though these components will form utility system loading, the growth is sometimes unpredictable. The political climate and the migration of population are not easy to forecast. The sudden spurs of industrial activity in a particular area may upset the past load trends and forecasts.

Load forecasting is complex, statistical, and econometric models and regression analysis is used, which is discussed in Volume 2.

2.12 Underground versus OH Systems

The OH and underground systems are both used. Under severe weather and storms, the OH systems are more prone to failures. Given the essential nature of the electrical power utilization and demand, consumers expect a reliable continuous power supply. Utilities are placing a significant number of lines underground. However, it will be hasty to jump to the conclusions that undergrounded systems always improves the reliability and continuity of power.

- It is estimated that in undergrounded systems the outage rate is about one-third compared to that of the OH systems. Conversely, it takes much longer time to rectify and locate an underground fault and that is where the undergrounded systems lose much advantage.
- It is relatively easy to locate a fault on OH systems and repair it. Underground services require specialized crews and equipment to locate a fault.
- In urban areas underground lines are four times more expensive as compared to OH facilities.
- Two Maryland utilities have gone in the other direction and replaced underground distribution systems with OH systems to improve reliability.
- In the initial phase of underground services, the dig-ins may result in more interruptions. In OH systems, motor-vehicle accidents caused by collision with poles can impair the service.
- An underground service presents an esthetic look.
- Toward the end of their life, after 25–35 years, underground cables are likely to fail more often, and localization and repairs become time consuming and difficult.
- During hurricanes, water and moisture penetration can cause damage to the underground services.

2.12.1 Spot Network

Figure 2.24 shows a spot network. Two or more transformers are supplied from a separate primary feeder. The secondaries are connected in a parallel through a special device called a network protector, to a common secondary bus. Radial secondary feeders are tapped from this bus to supply the loads.

If the primary feeder fails or a fault occurs in one of the transformers or primary side of the network protector, the fault current will flow as shown in Figure 2.24. The parallel transformers all feed into the fault. This causes the associated network protector to open fast and disconnect the supply circuit from the secondary bus. The network protector operates so fast that there is a minimal exposure of the secondary connected equipment to the voltage dip.

The power interruption can only take place if there is simultaneous failure of all the feeders serving the secondary bus or if there is a fault is on the secondary bus itself.

These networks are expensive because of duplication of the transformer installed capacity and of the network protectors, which are required, if one or two transformers go out of service; also the full loads should be capable of being supplied and the cost of network protectors is a consideration. Also the short-circuit ratings on the secondary bus increases due to a number of transformers operating in parallel. The scheme is used only in low-voltage applications serving high-density loads.

To ensure high reliability, the secondary bus is specially designed and insulated to limit the possibility of a fault on the bus itself.

Any feeder at voltage higher than 34.5 kV should not be considered for spot network service because the cable charging current begin to impact the network relay response. In Figure 2.25, the three positions of the primary switch are shown as open, close, and ground. These switches have interlocks, coils 1 and 2, which prevent unsafe operation. The network

Modern Electrical Power Systems

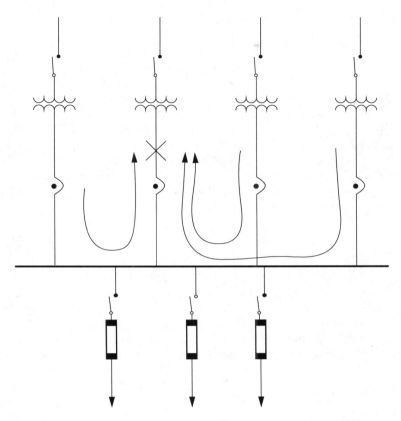

FIGURE 2.24
Configuration of a spot network, showing current flows on a transformer secondary short circuit.

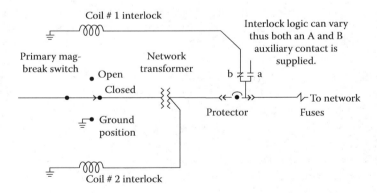

FIGURE 2.25
Three-position switch interlock for network transformers.

transformers are manufactured with the three-position magnetic-break switch integral to transformers. Primary fuse protection is sometimes added. Network protector can be provided in NEMA 4 weatherproof enclosure, submersible, or for mounting in the low-voltage switchgear. Network protectors are special air power break units having a full complement of current potential and control transformers and relay functions.

2.13 HVDC Transmission

The cumulative megawatts of HVDC systems around the world approach 100 GW. HVDC has long history, but a transition point occurred when thyristor valves took over mercury arc rectifiers in late 1970. The major technology leap was in Brazil with the 3150 MW ± 600 kV Itaipu project commissioned from 1984 to 1987. The OH line is 800 km long and each 12-pulse converter is rated 790 MW, 300 kV. HVDC is finding major applications in countries like India and China, and a large number of thyristor valve-based systems are planned—the power levels and distances are such that ±800 kV may be needed. HVDC project list worldwide can be seen on web site. Another web site of interest is CIGRE Study Committee B4, HVDC and Power Electronic Equipment.

2.13.1 HVDC Light

The IGBTs (Insulated gate bipolar transistor) for motor drives have begun to find applications in HVDC systems at the lower end of power usage. These operate using PWM techniques; there is practically no or little need for reactive power compensation as the converters can generate active and reactive power. The systems have found applications in off-shore wind farms and short-distance XLPE type cable systems. The largest system to date is the 330-MW Cross-Sound DC link between Connecticut and Long Island.

2.13.2 HVDC Configurations and Operating Modes

Figure 2.26 shows common DC transmission system configurations, which are only partially explained as follows; referring to configurations marked (a)–(f) in this figure:

a. Monopolar systems are the simplest and most economical for moderate power transfer. Only two converters and one high-voltage connection is required. These have been used with low-voltage electrode lines and sea electrodes to carry return currents.

b. In congested areas, or soils of high resistivity conditions may not be conductive to monopolar systems. In such cases, a low-voltage cable is used for the return path and DC circuit uses local ground connection for potential reference.

c. An alternative of monopolar systems with metallic return is that the midpoint of a 12-pulse converter can be connected to earth and two-half voltage cables or line conductors can be used. The converter is operated only in 12-pulse mode, so that there are no stray currents.

d. Back-to-back systems are used for interconnection of asynchronous networks and use AC lines to connect on either side. The power transfer characteristics are limited by the relative capabilities of adjacent AC systems. There are no DC lines. The purpose is to provide bidirectional exchange of power, easily and quickly. An AC link will have limitations in control over direction and amount of power flow. Twelve-pulse bridges are used. It is preferable to connect two back-to-back systems in parallel between the same AC buses.

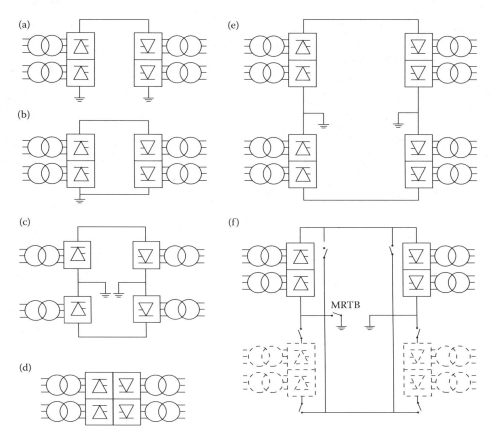

FIGURE 2.26
Various system configurations for HVDC, see text.

e. The most common configuration is 12-pulse bipolar converter for each pole at the terminal. This gives two independent circuits each of 50% capacity. For normal balanced operation, there is no earth current. Monopolar earth return operation can be used during outage of the opposite pole.

f. The earth return option can be minimized during monopolar operation by using opposite pole line for metallic return through pole/converter bypass switches at each end. This requires a metallic return transfer breaker in the ground electrode line at one of the DC terminals to commutate the current from relatively low resistance of earth into that of DC line conductor. This metallic return facility is provided for most DC transmission systems.

For voltages above ±500 kV series connected converters are used to reduce energy unavailability for individual converter outage or partial line insulation failure. By using two-series connected converters per pole in a bipolar system, only 25% of the line capability is lost for a converter outage or if the line insulation is degraded and it can support only 50% of the rated line voltage.

A complete terminal layout of an HVDC station is shown in Figure 2.27. The HVDC discussions continue in Volumes 2 and 3.

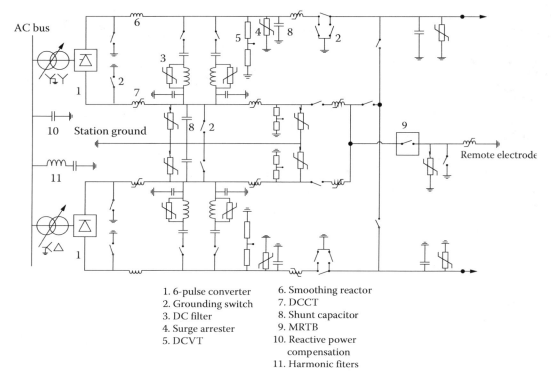

FIGURE 2.27
A typical HVDC terminal layout.

Problems

2.1 Calculate the specific speed and the power generated in a hydro-plant, given that: penstock efficiency=98%, generator efficiency=98.5%, turbine efficiency=76%., head=100 m, and average water flow=1000 m^3/s.

2.2 Draw a schematic showing structure of utility companies in the USA.

2.3 Research and write half page on compressed energy storage.

2.4 Research and draw a schematic diagram of proton–proton fusion cycle in stars.

Bibliography

40% of the Corn Goes to Ethanol, 2011. www.nationalreview.com.

Advanced Materials and Devices for Stationary Electrical Energy Storage Applications, Prepared by Nexight Group, Sponsored by US Department of Energy, 2010.

American Council for an Energy Efficient Economy. The Twin Pillars of Sustainable Energy: Synergies between Energy Efficiency and Renewable Energy Technology and Policy, 2007.

S Atzeni, J Meyer-ter-Vehn. *Nuclear Fusion Reactions. The Physics of Internal Fusion*. Oxford University Press, Oxford, 2004.

MG Bell et al. Deuterium-tritium plasmas in novel regimens in the Tokamak fusion test reactor. *Physics, Plasmas*, 4(5), 1714–1724, 1997.

Boeing Flywheel Energy Storage Technology. www.uaf.edu/files/acep/Boeing.

CIGRE study committee B4, HVDC and Power Electronic Equipment, Web site, www.cigre-b4.org.

Concepts of New Sustainable Energy Technologies. www.pitb.de/nolting/energytech.

SO Dean. Fifty years of fusion research, *Nuclear News*, pp. 34–41, July 2002.

TJ Dolan. *Fusion Research*, Pergamon Press, New York, 1982.

EHV Transmission. *IEEE Trans.*, 1966 (special issue), No. 6, PAS-85-1966, pp. 555–700.

EPRI, Grid Energy Storage: Challenges and Research Needs, Draft White Paper, July 8, 2013.

Eyer, G Corey. Energy Storage for the Electrical Grid: Benefits and Market Potential Assessment Guide. Sandia National Laboratories Report, SAND2010-0815, Albuquerque, NM.

Flybrid System KERS Using Carbon Fiber Flywheel. www.flybridsystem.com.

Goldman Sachs: Unlocking the Economic Potential of North America Energy Resources, June 2014.

Green Power Defined. www.epa.gov/greenpower/gpmarket/index.htm.

Gridwise Alliance; The Future of the Grid: Evolving to Meet America's Needs. Prepared for US Department of Energy by Energetics Incorporated. No. GS-10F-0103J, December 2014.

JS Gulliver, REA Arndt. *Hydropower Engineering Handbook*, McGraw-Hill, New York, 1991.

ND Hatziargyriou, APS Melipooulous. Distributed energy sources; Technical challenges, in *Proceeding of IEEE Power Engineering Society* Winter Meeting, vol. 2, pp. 1017–1022, New York, January 2002.

How to Manage our Oil Addiction—CESP. http://cesp.stanford.edu/news/oil-Addiction.

IEEE/PES Transmission and Distribution Committee Web site. www.ece.uidaho.edu/HVDCfacts.

MG Jog. *Hydro-Electric and Pumped Storage Plants*, Wiley and Sons, New York, 1989.

K Katiraci, MR Iravani. Power management strategies for a microgrid with multiple distributed generation units. *IEEE Trans. Power Syst.*, 21(4), 1821–1831, 2006.

Largest US Solar Photovoltaic System Begins Construction at Nellie Air Force. www.prnewswire.com.

R Lasseter. Microgrids. *IEEE PES* Winter Meeting, January 2002.

RH Lasseter. Microgrids and distributed generation. *J. Energy Eng. Am. Soc. Civil Eng.*, 133, 144–149, 2007.

JV Milanovic, TM David. Stability of distributed networks with embedded generators and induction motors, in *Proceedings of IEEE Power Engineering Society* Winter Meeting, vol. 2, pp. 1023–1028, New York, January 2002.

MS Naidu, V Kamaraju. *High Voltage Engineering*, 2nd ed., McGraw-Hill, New York, 1999.

SR Narayanan, GK Prakash, A Manohar, B Yang, S Malkhandi, A Kindler. Material challenges and technical approaches for relaizing inexpensive and robust iron-air batteries for large scale energy storage. *Solid State Ion*, 216, 105–109, 2012.

Next-Generation of Flywheel Energy Storage. www.pddnet.com/ Product Design and Development, 2009.

EJ Nuttall. Fusion as an Energy Source: Challenges and Opportunities, Institute of Physics Report, 2008. www.iop.org/publications/iop.

OECD, International Energy Agency, Renewables in Global Energy Supply: An IEA Factsheet, 2006. www.iea.org/textbase/papers/2006/renewable_factsheet.

Office of Electricity Delivery and Energy Reliability. http://energy.gov/oe/services/technology-development/energy-storage.

KR Padiyar. *HVDC Power Transmission Systems*, New Academic Science, Waltham, CA, 2011.

S Rao. *EHV-AC, HVDC Transmission and Distribution Engineering*, Khanna Publishers, New Delhi, 2004.

Renewable Energy and Efficiency Partnership, 2004. www.reep.org/sites/default/files/.

Solar, Wind, Hydropower: Home Renewable Energy Installation. http://energy.gov/articles/solar-wind-hydropower-home-renewable-energy-installations.

US Department of Energy, Grid Energy Storage, December 2013. www.energy.gov/oe/services/technology.

AM Wolsky. The status and prospects for flywheels and SEMS that incorporate HTS. *Physica C*, 372–376, 1495–1499, 2002.

Z Yang et al. Electrochemical energy storage for green grid. *Chem. Rev.*, 111(5), 3577–3613, 2011.

H Zang, M Chandorkar, G Venktaramanan. Development of static switchgear for utility interconnection in microgrid. *IEEE PES*, Palm Springs, February 2003.

IEEE Color Books

141: IEEE Recommended Practice for Electric Power Distribution for Industrial Plants (IEEE Red Book), 1993.

142: IEEE Recommended Practice for Grounding of Industrial and Commercial Power Systems (IEEE Green Book), 1991.

241: IEEE Recommended Practice for Electric Power Systems in Commercial Buildings (IEEE Gray Book) 1997.

242: IEEE Recommended Practice for Protection and Coordination of Industrial and Commercial Power Systems (IEEE Buff Book), 2001.

399: IEEE Recommended Practice for Industrial and Commercial Power Systems Analysis (IEEE Brown Book), 1997.

446: IEEE Recommended Practice for Emergency and Standby Power Systems for Industrial and Commercial Applications (IEEE Orange Book), 2000.

493: IEEE Recommended Practice for the Design of Reliable Industrial and Commercial Power Systems (IEEE Gold Book), 2007.

602: IEEE Recommended Practice for Electric Systems in Health Care Facilities (IEEE White Book), 1996.

739: IEEE Recommended Practice for Energy Management in Commercial and Industrial Facilities (IEEE Bronze Book), 1995.

902: IEEE Guide for Maintenance, Operation and Safety of Industrial and Commercial Power Systems (IEEE Yellow Book), 1998.

1015: IEEE Recommended Practice for Applying Low-Voltage Circuit Breakers Used in Industrial and Commercial Power Systems (IEEE Blue Book).

1100: IEEE Recommended Practice for Powering and Grounding Electronic Equipment (IEEE Emerald Book), 1999.

551: Recommended Practice for Calculating Short-Circuit Currents in Industrial and Commercial Power Systems (IEEE Violet Book), 2006.

3

Wind and Solar Power Generation and Interconnections with Utility

3.1 Prospective of Wind Generation in the USA

See Chapter 2 for discussions on renewable energy sources (RESs). This chapter is solely devoted to wind and solar power due to their gaining importance in the energy mix of renewable sources. More than 94 GW of wind power generation has been added worldwide by the end of 2007, out of which over 12 GW is in the USA and 22 GW in Germany alone. Looking at energy penetration levels (ratio of wind power delivered by total energy delivered), Denmark leads by reaching a level of 30% or more which is followed by Spain. In some hours of the year in Denmark, the wind energy penetration exceeds 100%, with excess sold to Germany and Nord Pool. Nineteen off-shore projects operate in Europe producing 900 MW. The US off-shore wind energy resources are abundant.

The USA now ranks second in the world for installed wind capacity, equal to approximately 4.5% of the total electrical demand. Between 2003 and 2013, a total of 842 MW of wind turbines were installed in distributed applications, reflecting nearly 72,000 units, see Figure 3.1. Note the increasing addition of turbines of less than 100 kW. DOE Wind Program encourages anyone interested in purchasing small- or medium-sized turbines to consider using a certified product, and public funds can only be expanded on certified machines. In 2013, the wind power additions have been comparatively low, but in future, larger installed capacity is expected. An interesting trend is that, in 2012, the exports stood at 8 MW and these increased to 13.6 MW in 2013. Growth after 2015 remains uncertain, dictated in part by future natural gas prices, fossil plant retirements, and policy decisions. The total US wind power installed capacity as in 2013 is approximately 65 GW (DOE, 2013 report).

Figure 3.2 shows the bar chart of estimated wind production as a percent of electricity consumption.

All countries are not included. Note that the wind power approximated to less than 5% of the world electrical power demand.

Renewable Energy Laboratory (DOE/NREL) took an investigation how 20% of energy from wind will look like in 2030, see Figure 3.3 [1]. American Electric Power (AEP), produced a white paper that included 765 kV network overlay for the US power system that will increase reliability and allow for 400 GW of wind or other generation to be added. (This project is currently shelved.) The political climate change in Washington can impact these decisions and accelerate exploitation of wind power. The modern wind power plants can be as large as 300 MW and are often located within a short distance of each other. The microgrids and independent standalone power generation can serve to off-load burdened transmission systems.

85

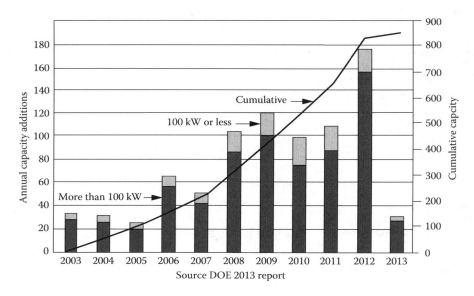

FIGURE 3.1
Wind power generation in the USA (DOE).

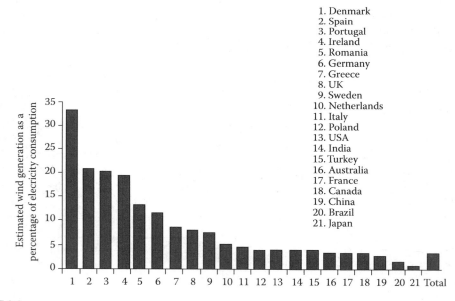

FIGURE 3.2
Estimated wind power generation as a percentage of electricity consumption, worldwide-major wind generating countries.

With respect to power system analysis, transients and stability analyses and interconnections with grid systems the wind power stations pose even larger problems, which require more thorough analysis of the interconnection and system isolation under disturbances.

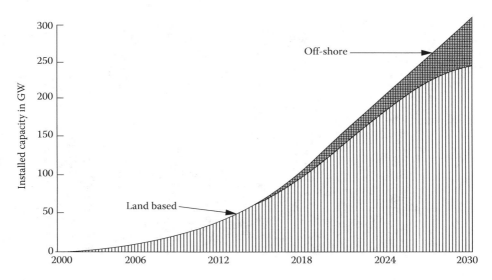

FIGURE 3.3
Twenty percent (proposed) wind power penetration of land-based and off-shore wind power installed capacity by 2030.

3.2 Characteristics of Wind Power Generation

A challenge of increasing the contributions from RES is that their outputs vary directly with the weather conditions. Figure 3.4 shows hourly wind production over 365 days and statistical nature of wind power production. The patterns will vary depending upon the location and this figure shows the wide variations in wind power that can occur. Though the tools for predicting generation capabilities from RES have considerably improved; the scheduling of these units remain difficult. The power produced in off-shore units may not be available even for days or weeks. Thus, the emphasis is upon RES having energy storage capabilities. Increasing system flexibility will be required to manage the task of system balancing. The conventional power plants must accommodate the swings in power generation of RES. The conventional thermal plants have to exercise a greater amount of control, and will frequently operate inefficiently at partial load or will be shutdown and restarted. Thus, emission and fuel costs increase, besides increasing the cost of energy produced from these units. The capacity of existing storage power plants, mainly pumped hydro is not adequate and system flexibility is needed for balancing functions.

Transmission systems and distribution grids have been planned so that large power stations are as close as possible to the load centers. A large potential of RES is available far from the load centers. This leads to a new load flow situation and bottlenecks that already occur in transmission systems with large wind power. Contingency management serves short-term goals and in the long run appropriate reinforcement of the grids is a must. In the USA, an overlay of 765 kV AC transmission network was proposed.

Expanded EHV systems are required to provide long-term transmission capacity and integration of RES. High-voltage, direct current (HVDC) can transmit more power over longer distances. In HVDC transmission, the transfer capability is influenced by the losses and not by reactive power or stability concerns as in AC transmission systems. There are

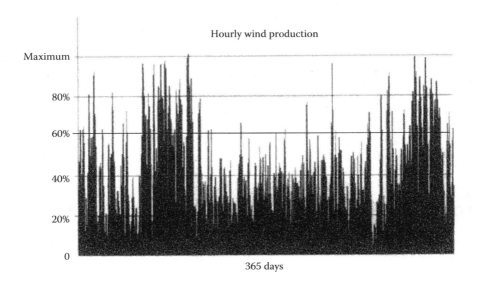

FIGURE 3.4
Hourly variations of wind power, pattern depending upon location.

two converter technologies used in the HVDC transmission: (1) current source converters (CSCs) and (2) voltage source converters (VSCs); see Volume 3. The CSC-based technology has been applied at ±600 kV and 6300 MW on double bipoles (Brazil) and will be soon applied at ±800 kV up to 6400 MW (China). VSC-based HVDC transmission is available through ±320 kV and 1100 MW for symmetrical monopole configuration and up to 2200 MW at ±640 kV in bipolar configuration. References [2–4] describe HVDC transmission system configurations, also see Volume 2.

There is an interdependence of HVDC and HVAC transmission systems. For example, AC network where the CSC-based HVDC is connected must be strong relative to their transmitted power. CSC converters draw reactive power at the point of interconnection and the reactive power demand increases with loading. Harmonic filters and other reactive power compensation methods are required, Volume 3. VSC-based systems do not require a strong AC network to operate properly and can serve to interconnect small-scale wind generation using induction generators. To support local system voltage, a VSC station can act as an SVC or STATCOM.

AC transmission provides more accessibility and flexibility. AC transmission systems can be tapped more easily to serve loads or pick up resources over moderate distance. A system that is flexible enough to integrate new resources and additional future generation is required. In some cases, HVDC may be an advantage for point-to-point transmission of bulk power as it is an asynchronous link and short-circuit power is not transferred. For underground or submarine connections, HVDC cable systems provide high capacity over significant distances without reactive power compensation required by AC systems. This is not a comparison of HVDC vis-à-vis HVAC; the two systems must coexist. A technology comparison of EHV AC and HVDC systems is shown in Table 3.1 [5].

Figure 3.5 shows the line loadability versus distance adapted from [5]. For a 200-mile transmission distance, the number of AC lines required for 6000 MW capacity will be two 765 kV lines, six 500 kV lines, and twelve 345 kV lines. These point to the reduction of ROW with 765 kV transmission lines. These numbers of lines are calculated without series compensation, Volume 2. With series compensation, the lines required for 6000 MW are

TABLE 3.1

EHV AC and HVDC Systems for 6000 MW Delivery

Voltage (kV)	No. of Lines	Circuits per Tower	Circuit SIL (MW)	Capacity (MW)	Series Comp.	Shunt React.	Shunt Caps.	Conductor Type	No. in Bundle	Area KCMIL
345 AC	4	2	400	800	Yes	No	Yes	ACSR	2	1590
500 AC	4	1	900	1800	Yes	Yes	Yes	ACSR	3	1590
765 AC	2	1	2400	3100	No	Yes	Yes	ACSR/TW	6	957
±500 DC	2	1		3000				ACSR	3	2515
±800 DC	1	1		6000				ACSR	4	2515

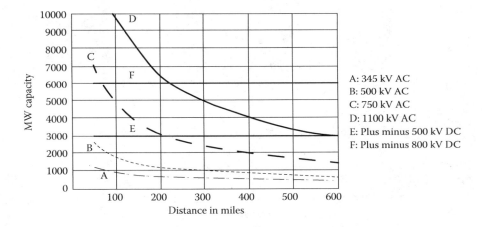

FIGURE 3.5
Line loadability versus distance. Comparative for AC and DC transmission voltages as shown.

four and eight, respectively, for 500 kV and 345 kV and one series compensated 765 kV line can carry 4800 MW.

The energy efficiency is another important criteria and Figure 3.6 compares losses at full load, including line resistive and terminal equipment losses for the 6000 MW point-to-point transmissions. It may be said that for long-distance point-to-point transmission without intermediate accessibility to line, HVDC systems are rather uncommon in much of the USA, also see References [6–11].

The following aspects become important:

1. *Power Flow Control*: Large unexpected fluctuations in wind power can cause additional loop that flows through transmission grids. When a large deviation has to be balanced by other sources, these may be located far away from the wind generation. The flow through parallel transmission paths may reverse, making it difficult to operate them. The preferred solution is to build high-voltage transmission systems.

2. *Congestion Management*: This needs to be addressed from market side view point as well as from the technical view using FACTs devices, Volume 2. It is known that installed wind power in some grid areas has already reached a level, which entails

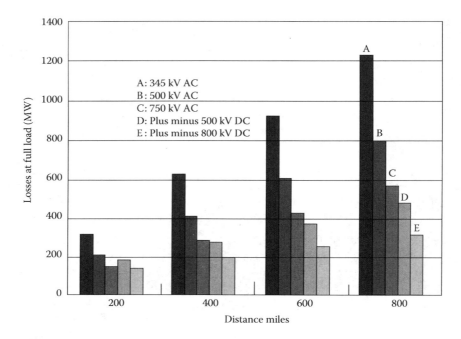

FIGURE 3.6
Comparative losses versus distance, AC and DC transmission voltages.

problems in grid management and grid control in strong wind periods caused by large wind fluctuations.

3. *Long-Term and Short-Term Voltage Stability*: By replacing large synchronous generators with small generators integrated into subtransmission systems or even distribution systems far from load centers, the amount of reactive power reserve in the system is considerably reduced. The reactive power contribution of the wind power plants is highly dependent upon the technology used, connection point, as well as additional reactive power support; SVC, STATCOM, switched capacitors, or increasing the "must run" units. There is already reduction of long-term stability limit resulting in a reduction of maximum power transfer along interconnecting lines. The short-term voltage stability limit remains above long-term voltage stability limit and is not as critical, yet it should be addressed and planned for in the grid operation.

4. *Transient Stability and Low-Voltage Ride through Capability*: This is further discussed; the stability is also a function of the generator type.

5. *Greenhouse Gas Emissions*: The greenhouse gas reduction is in the range of 350–450 kg CO_2 per MWh of power generated by wind power plants.

6. *Reserve Requirement*: The impact of wind power on tertiary control requirements can be estimated. Significant hourly fluctuations and forecast errors have to be managed. It is necessary to consider the effect of wind and solar energy together.

7. *Energy Storage*: Large-scale storage is possible in pumped-hydro power stations. In the future, compressed air energy storage is seen. The production and storage of hydrogen could be a solution, though the overall efficiency of hydrogen chain, production, storage, and combustion is low, of the order of 30%–35%. Developments in battery technology could lead to a revolution in car technology, see Chapter 2.

3.2.1 Maximum Transfer Capability

In May 1965, NERC introduced a term Transmission Transfer Capability (TTC), and later in 1996 approved a document entitled Available Transfer Capability (ATC) definition and determination. According to this document, calculation of ATC requires calculation of several related terms including TTC, which is defined as the maximum power that can be transferred in a reliable manner between a pair of defined sources and sink locations in the interconnected system while meeting all predefined sets of pre- and post-contingency system conditions [12]. A number of contingency system conditions are considered and TTC is determined at each of these conditions. This maximum transferable power is commonly known as Maximum Transfer Capability (MTC). A number of methods have been proposed for calculating it [13–17], Optimal Power Flow (OPF), Volume 2 is one such method. Consider Figure 3.7 and let the additional power be represented by scalar γ. This additional power is accounted for by modifying the active and reactive power demands at the sink bus. Assuming a constant power factor

$$P = P_0 + \gamma \cos \phi_0$$
$$Q = Q_0 + \gamma \sin \phi_0, \tag{3.1}$$

where P_0 and Q_0 are the base active and reactive powers and ϕ_0 is the power factor angle. The optimization is formulated as follows:

$$\begin{aligned}&\max \gamma \\ &s.t \\ &G(x,\gamma) = 0 \\ &H(x,\gamma) \le 0,\end{aligned} \tag{3.2}$$

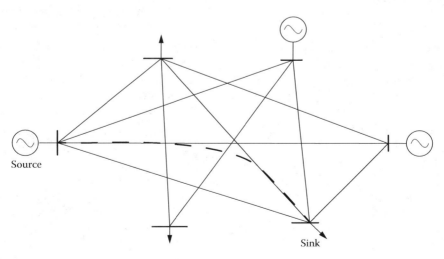

FIGURE 3.7
Representation of TCC, see text.

where x is a vector of system control and state variables, $G(x, \gamma) = 0$ are a set of equality constraints accounting the power balance equations and $H(x, \gamma) \leq 0$ are a set of inequality constraints according to system security limits. These can be

- Generation capacity
- Transient stability limit
- Voltage stability limit
- Voltage excursions limits
- Transmission line thermal capability

Then, the solution of OPF gives MTC:

$$\text{MTC} = P_0 + \gamma_{\max}\phi_0. \tag{3.3}$$

Investigations of TTC with wind power generation become important. The varying nature of wind power generation and system voltage instability and transient stability are essential parameters. Monte Carlo methods are well suited to investigate these variables and effects of system performance. A simplified description of Monte Carlo methods is given in Reference [18], also see Volume 2.

3.2.2 Power Reserves and Regulation

Different generation types behave differently with respect to regulation features.

Hydrogenation: Hydropower offers good power regulation. Approximately 40% of the capacity can be regulated in 1 min if the reservoir holds enough water. Hydro plants with pumped storage facilities offer good storage capability.

Nuclear Power Plants: These are operated at their maximum capacity throughout the year as the base load plants and do not offer much regulation capability. One minute regulation capability of boiling water nuclear reactors is generally of the order of 1%–3% or little higher.

Condensing Power Plants: Condensing plants using coal as fuel can be regulated by approximately 5% within 1 min. The regulation of peat fuelled plants is 3%. Starting time of a condensing plant is dependent upon time that elapsed from the last use. Warm plants can be started in some hours while cold plant starting time may be approximately 10 h.

Gas Turbines: Gas turbines are capable of fast and reliable power regulation, and can be started and synchronized to the grid in about 3 min. These can be regulated to approximately 40% of their capacity in 1 min.

Combined Heat and Power Plants (CHPs): These have different capabilities depending upon the operation. For example, bypassing a high pressure preheater can increase electrical power in the range 5%–10% and by bypassing the condenser in an extraction back pressure plant it is possible to increase power quickly by 10%–15%.

Wind Power: Wind power also offers upward and downward regulation capability by rotor blade controls.

Wind and Solar Power Generation and Interconnections with Utility 93

Load Shedding: Load shedding is an important tool of regulation in case of disturbances, and the loads for fast active disturbances are disconnected to maintain transient stability. HVDC systems serve as reserves between two AC systems.

3.2.3 Congestion Management

Reliable operation means that the transmission system has to operate within constraints of thermal limits and contingency analyses. For large-scale wind generation, transmission system operators (TSOs) and wind farm operators must coordinate the data in advance for operation planning, bottlenecks on the tie lines in advance and congestion forecast. We talk about short-term, middle-term, and long-term congestion forecasts. The main objective is to determine power system state in advance that fulfills the network security requirements. Network- and market-based measures are adopted to which a third set of measures can also be added. Again it can be formulated as an optimization problem with an objective function to minimize:

$$\text{Minimize}: F_t = \sum_{a=1}^{N^A} \left(W_a^{\text{R}} f_{t,a}^{\text{R}} + W_a^{\text{M}} f_{t,a}^{\text{M}} \right), \tag{3.4}$$

where F_t is the objective function in time t, which is divided into number of intervals to be analyzed. Index a describes the control area of the transmission system, f^{R} describes power system requirements of current state, f^{M} describes total cost measures to avoid congestion, W^{R} is a weighting factor of subobjective of network security requirements, and W^{M} describes weight factor for subobjective of total costs of the measures. The weights fulfill the condition:

$$W_a^{\text{R}} \gg W_a^{\text{M}}. \tag{3.5}$$

The criteria to determine a secure state can be maximum current on each line, maximum and minimum node voltages and $n - 1$ security criteria. Change of topology and rerouting of power can be considered as possible actions to avoid congestions and the short-circuit currents for the new configuration are calculated. Criteria are modeled with penalty functions that become zero for secure power system state [19–21].

Market-based measures consider redispatch with minimal total cost. The forecast of wind generation is considered and wind farms cannot deliver more active power than predicted with the forecasted wind speed. To avoid congestion, only a reduction in the wind power projected output is possible, which depends upon regulation. The generation management is resorted to when network- and market-based measures are not sufficient to avoid congestion.

3.3 Wind Energy Conversion

In the beginning of wind power development, the energy produced by small wind power turbines was expensive and subsidies were high. Today, single units of 5 MW are on the market. Today's wind turbines all over the world have three-blade rotors, diameters ranging from 70 to 80 m, and mounted atop 60–80 m or higher towers. The tower heights up

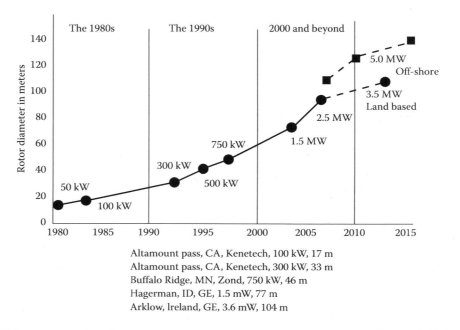

FIGURE 3.8
Development of single unit wind turbines in the USA.

to 160 m are a technical feasibility. The typical turbine installed in the USA, in 2006, can produce about 1.5 MW of power. Higher rated units and off-shore wind base plants may see a unit size of 5 MW or more by 2010. Figure 3.8 shows the developments of single unit wind power turbines in the USA.

Figure 3.9 shows a schematic representation of electrical and mechanical features of a wind converter unit, with *upwind* rotor. The rotor speed which is of the order of 8–22 rpm is the input to gear box, and on the output side, the speed is 1500–1800 rpm. The drive train dimensions are large, increasing the horizontal dimension of nacelle. The essential components are as follows:

- Mechanical components like: towers, rotor, shaft with drive train, electrical generator, and yaw mechanism such as tail vane, sensors, and control
- Numerous sensors are used to regulate mechanical and electrical parameters
- Anemometers to measure the wind speed and transmit the data to the controller
- Stall controller that starts and shuts-off the system at low and high speeds
- Power electronics
- Energy storage system
- Transmission link or grid connections

3.3.1 Drive Train

Several designs are under development to reduce the drive train weight and cost. One approach is to build direct drive permanent magnet generators that eliminate complexity of gear box. The slowly rotating generator will be larger in diameter, 4–10 m, and

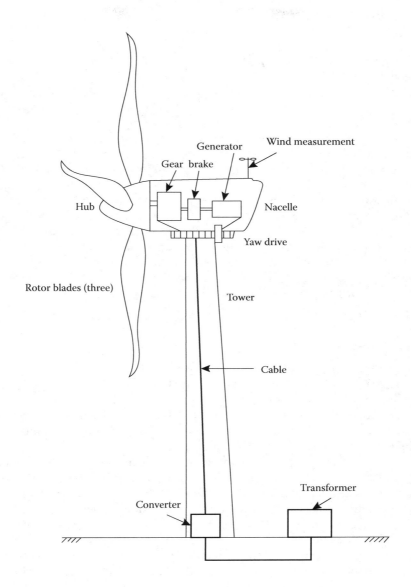

FIGURE 3.9
A schematic of wind conversion unit, showing major components.

quite heavy. The decrease in cost and availability of rare earth permanent magnets is expected to significantly affect the size and weight. The generator designs tend to be compact and lightweight and reduce electrical losses in windings; as compared to wound rotor machines. Prototypes have been built—a 1.5 MW design with 56 poles is only 4 m in diameter, versus 10 m for a wound rotor design. It is undergoing testing at National Wind Technology Center.

A hybrid of direct drive approach uses low speed generator. WindPACT drive train project has developed a single-stage planetary drive operating at a gear box ratio of 9.16:1. The gear box drives a 72 pole PM generator and reduces the diameter of a 1.5 MW generator to 2 m, [22]. Another development is the distributed drive train.

3.3.2 Towers

Depending upon the nacelle weight, the towers are constructed from steel or steel/concrete. The tower heights are up to 100 m. The towers must be at least 25–30 m high to avoid turbulence caused by trees and buildings. Wind speeds are height dependent, and lattice towers to 160 m height can be constructed. A relationship between tower mass and tower height shows sharp increase in the tower mass per meter height as turbine output increases, approximately 2000 kg/m for turbines in the output range of 1.5 MW. Large plants tend to have significantly greater mass per meter as the mast height increases. There is an ongoing effort to develop advanced tower designs that are easily transported and installed and are cost-effective.

The wind speeds are not uniform over the area of the rotor. The rotor profiles result in different wind speeds at the blades nearest to the ground level compared to the top of the blade travel. Gusts and changes in wind speed impact the rotor unequally. Influences due to tower shadow or windbreak effects cause fluctuations in power or torque.

Structural dynamics play an important role in the tower design. The vibrations, and resulting fatigue cycles under wind speed fluctuations are avoided. This requires avoidance of the resonant frequencies of the tower, nacelle, and rotor from wind fluctuations frequencies.

3.3.3 Rotor Blades

As the wind turbines increase in output so does their blades, from 8 m in 1980 to 70 m for many land-based units. Improved blade designs have kept weight growth much lower than the geometric escalation of blade lengths. Work continues in the application of lighter and stronger carbon fiber in highly stressed areas to stiffen the blades, improve fatigue resistance, and simultaneously reduce the weight. Research continues in the development of lighter blades—carbon fiber, and fiberglass.

The blades are aerofoil and utilize aerodynamic principles to capture maximum wind power. The design uses a longer upper side surface and the bottom surface somewhat remains uniform. A lift is created on the aerofoil by the difference in pressure in the wind flowing over top and bottom surfaces—a drag force is simultaneously created, which acts perpendicular to blades, impeding the lift effect and slowing blades. The design objective is to get maximum lift-to-drag ratio.

It is hard to fix a nameplate of the wind turbine rating, and there is no globally accepted standard. Most manufacturers supply the rating as the generator peak electrical capacity/rotor diameter. The specific rated capacity is defined as follows:

$$\text{SRC} = \frac{\text{MW}_{\text{generator}}}{R_{\text{sweptarea}}}. \tag{3.6}$$

Say for a 500 kW generator and 30 m diameter turbine, SRC is

$$\text{SRC} = \frac{0.5}{\pi \times 15^2} = 0.707 \times 10^{-3}.$$

Figure 3.10 shows the torque and power output of a wind turbine at two speeds, V_1 and V_2. The speed at maximum power is not the same as the torque. A desirable strategy is the

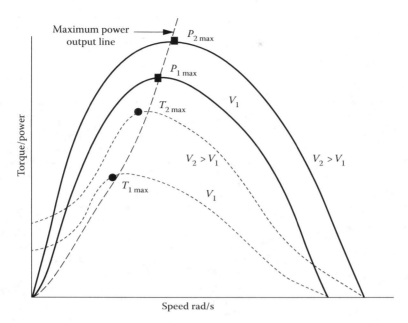

FIGURE 3.10
Torque and power output versus speed.

generate maximum power at varying speeds, which requires a variable speed operation an adjustment of speed to generate maximum power. The tip speed ratio is defined as follows:

$$\text{TSR} = \frac{\omega R}{V}, \tag{3.7}$$

where,
  ωR = linear speed of the uppermost tip of the blades, where R is the rotor radius in m
  V = free upstream wind velocity

The rotor efficiency varies with TSR, and the maximum occurs approximately at a speed that gives maximum power. Most manufacturers offer now variable-speed systems in combination with pitch regulation.

3.4 The Cube Law

Wind originates from difference in temperature and pressure in air mass. A change in these parameters alters the air density, ρ. The force exerted on volume of air V is given by the following:

$$F = Vg\Delta\rho, \tag{3.8}$$

where g is the gravitational constant. This produces kinetic energy:

$$E = \frac{1}{2}mv^2, \tag{3.9}$$

where m is the mass of the air and v, *the velocity is assumed constant*, then E = wind power. This is rather an oversimplification. If we consider an air volume of certain cross section, and a swirl free speed, upstream of the turbine, downstream of the turbine it will result in a reduction in speed, with a corresponding broadening of the cross sectional area (wake decay). However, with this simplification of constant speed, the basic tenets of turbine output are still valid.

The following equation can be written for the air mass:

$$m = \rho A v, \tag{3.10}$$

where A is the rotor area. Substituting in Equation 3.10 in Equation 3.9, the theoretical power output is as follows:

$$P = \frac{1}{2}\rho A v^3 c_P \tag{3.11}$$

where the wind speed-dependent coefficient, c_P, describes the amount of energy converted by the wind turbine. It is of the order 0.4–0.5.

All the energy in a moving stream of air cannot be captured. A block wall cannot be constructed, because some air must remain in motion after extraction. On the other hand, a device that does not slow the air will not extract any energy. The optimal blockage is called Betz limit, around 59%. The aerodynamic performance of blades has improved dramatically, and it is possible to capture about 80% of the theoretical limit. The new aerodynamic designs also minimize fouling due to dirt and bugs that accumulate at the leading edge and can reduce efficiency.

According to Betz, the maximum wind power turbine output is as follows:

$$P = \frac{16}{27} A_R \frac{\rho}{2} v_1^3, \tag{3.12}$$

where A_R is the air flow in the rotor area and v_1 is the wind velocity far upstream of the turbine.

The maximum is obtained when

$$v_2 = \frac{2}{3}v_1 \quad \text{and} \quad v_3 = \frac{1}{3}v_1, \tag{3.13}$$

where v_3 is the reduced velocity after broadening of the air stream past the rotor and v_2 is the velocity in the rotor area.

The ratio of power absorbed by turbine to that of moving air mass:

$$P_0 = A_R \frac{\rho}{2} v_1^3. \tag{3.14}$$

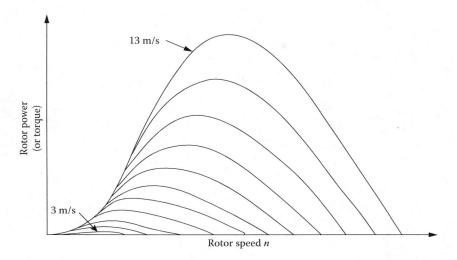

FIGURE 3.11
Rotor power of a wind turbine unit based on rotor speed and wind velocity.

And

$$c_p = \frac{P}{P_0}. \tag{3.15}$$

Another way of defining c_p is with respect to λ, which is defined as follows:

$$\lambda = \frac{v_{TS}}{v_{RP}} = \frac{2\pi n r}{v_{RP}} \tag{3.16}$$

where v_{TS} is the blade tip speed and v_{RP} is speed of rotor plane, and v_{TS} can be written as follows:

$$v_{TS} = 2\pi n r, \tag{3.17}$$

where the rotor radius r is in meters and n is the speed in s^{-1}. Typical values of λ are 8–10 and the tip speed ratio influences the power coefficient c_p. Also c_p is dependent upon wind speed. Combining these relations, Figure 3.11 depicts the power or torque versus the wind speed. (Torque is simply power divided by angular velocity ω.) Figure 3.12 shows the performance coefficient as a function of tip speed ratio with blade pitch angle, (see section below) as a parameter.

3.5 Operation

Figure 3.13 shows the power curves for typical modern turbine. The turbine output is controlled by rotating the blades about their long axis to change angle of attack called, "controlling the blade pitch." The turbine is pointed into the wind by rotating the nacelle

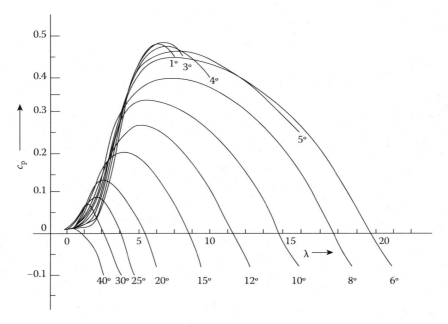

FIGURE 3.12
$c_p - \lambda$ characteristics of a wind turbine unit with blade pitch angle.

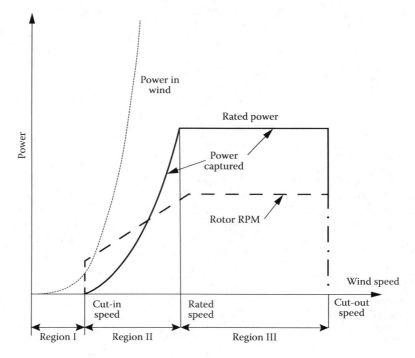

FIGURE 3.13
Typical wind generating unit operating curve, power output versus wind speed.

Wind and Solar Power Generation and Interconnections with Utility   101

about tower, which is called the "yaw control." Modern turbines operate with rotor positioned on the windward side of tower, which is referred as an "upward rotor." A turbine generally starts producing in winds of about 12 mph, reach a maximum power at about 28–30 mph and shutdown "feather the blades" at about 50 mph. The amount of energy in the wind available for extraction by turbine increases with cube of speed; thus a 10% increase in speed means a 33% increase in energy. While the output increases proportional to rotor swept area; the volume of material and, thus, the cost increases as cube of the diameter. Controllers integrate signals from dozens of sensors to control rotor speed, blade pitch angle, and generator torque and power conversion voltage, and phase angle.

Wind speed changes may occur over long periods of time or suddenly within a matter of seconds. The system has to be protected from the sudden gusts of wind. The mechanical loads are determined by dynamic forces and knowledge of dynamic wind behavior at a location is necessary for proper component ratings and the mechanical power acting on the turbine should be limited.

There are three methods to achieve it:

1. Stall
2. Active stall
3. Pitch control

The stall control exercises adjustable clutches on rotating blade tips to shut down. Under normal conditions, laminar air flow occurs on the blades. The lift values corresponding to angle of attack are achieved at low drag components. With wind speeds exceeding nominal values at which the generator rated output occurs, higher angles of attack and stalling occurs. This is achieved by properly profiling the blades. The lift force and lift coefficient are reduced, and the drag forces and coefficients are increased. Stall regulated machines are often designed with asynchronous generators of higher nominal output and rigid coupling with the grid is obtained. An active stall control is achieved by turnable rotor blades.

Variable blade pitch allows direct control of turbine. By varying the blade pitch, it is possible to control the torque of the turbine, and by further adjustments bring about a stall condition. The control and regulation system is complex. In high-power applications, pitch control is used. For the design and control of pitch adjustments, a host of mechanical moments and forces must be considered. These include the forces and moments due to blade deflection, lift forces on blades, propeller moment, because of teetering and frictional moments [23]. Figure 3.14 shows three types of controls.

3.5.1 Speed Control

The curves of fixed and variable speed generators are marked in Figure 3.15. This shows that when the turbine is driven by a synchronous generator, varying the generator frequency at a certain wind speed will give operation at $n/n_1 = 1$, where n corresponds to the grid frequency. The turbine is constrained to follow the grid frequency. Sufficient turbine torque to drive the generator is available for wind speeds above approximately 3.6 m/s for synchronous generators to about 3.8 m/s for asynchronous generators. Under variable frequency generator operation, the speed of rotation can be freely set within given limits. The turbine utilization of available wind power is optimized. A ramp control of the active power output is possible [23, 24].

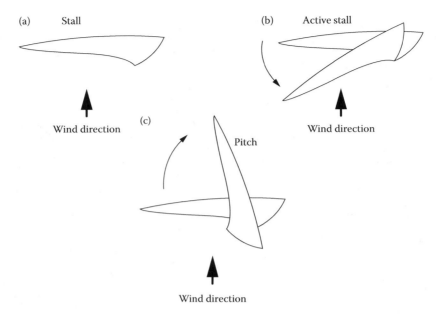

FIGURE 3.14
(a) Stall control, (b) active stall control, and (c) pitch control; see text.

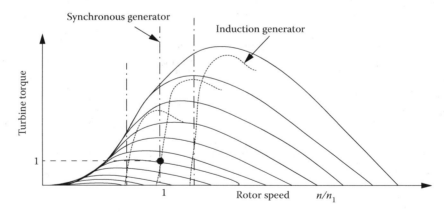

FIGURE 3.15
Wind turbine torque–speed characteristics by variation of generator frequency superimposed on wind power curves.

3.5.2 Behavior under Faults and Low-Voltage Ride Through

Figure 3.16 shows the recommendation of Western Electricity Coordinating Council (WECC) wind generation task force (WGTF) with respect to proposed voltage ride through requirements for all wind generators [25]. A three-phase fault is cleared in nine cycles, and the post fault voltage recovery dictates whether the wind power generating plant can remain on line. The requirement does not apply to faults that will occur between the wind

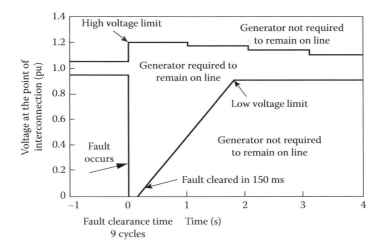

FIGURE 3.16
Proposed WECC voltage ride through requirements of wind generators.

generator terminals and the high side of generator step up unit transformer (GSU) and the wind plants connected to transmission network via a radial line will not be required to ride through the fault on that line.

Figure 3.16 also addresses the high and low voltages. Generators based upon simple induction machines, degrade power system voltage performance as these require excitation reactive power. The Federal Energy Regulatory Commission (FERC) orders 661 and 661A require new wind generators to have capability to control their reactive power within 0.95 leading to 0.95 lagging range. As this requirement can be expensive to comply, FERC requires it only if the interconnection study shows that it is needed. Modern wind generators provide this capability from power electronics that control the real power operation of the machine, also see References [26,27].

3.6 Wind Generators

3.6.1 Induction Generators

The operation and characteristics of induction motors are described in Volume 2. The characteristics of an induction machine for negative slip are depicted in Figure 3.17a. An induction motor will act as induction generator with negative slip. At $s = 0$, the induction motor torque is zero and if it is driven above its synchronous speed, the slip becomes negative and generator operation results. The maximum torque can be written as follows:

$$T_m = \frac{V_s^2}{2\left[\sqrt{\left(r_s^2 + (X_s + X_r)^2\right)} \pm r_s\right]}. \tag{3.18}$$

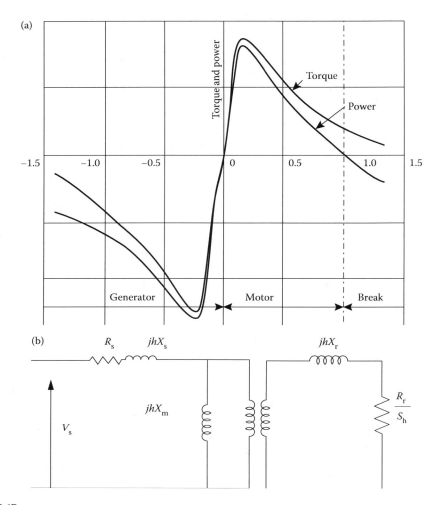

FIGURE 3.17
(a) Torque and power characteristics of an induction machine versus slip: induction motor, induction generator, and a brake (b) equivalent circuit model of an induction generator.

The negative r_s, in Equation 3.18, represents the maximum torque required to drive the machine as generator. The maximum torque is independent of the rotor resistance. For subsynchronous operation, the rotor resistance does affect the slip at which the maximum torque occurs. For maximum torque at starting, $s = 1$, and

$$r_r \approx \sqrt{r_s^2 + (X_s + X_r)^2}. \tag{3.19}$$

For supersynchronous operation (generator operation), the maximum torque is independent of r_r, same as for motor operation, but increases with the reduction of stator and rotor reactances. Therefore, we can write

$$\frac{T_{m,\text{gen}}(\text{supersyn.})}{T_{m,\text{motor}}(\text{subsyn.})} = \frac{\sqrt{r_s^2 + (X_s + X_r)^2} + r_s}{\sqrt{r_s^2 + (X_s + X_r)^2} - r_s} \approx \frac{X_s + X_r + r_s}{X_s + X_r - r_s}. \tag{3.20}$$

Wind and Solar Power Generation and Interconnections with Utility 105

The approximation holds as long as $r_s \ll X_s$. The torque–speed characteristics of the machine above synchronism are similar to that for running as an induction motor. If the prime mover develops a greater driving torque than the maximum counter torque, the speed rises into an unstable region and the slip increases. At some high value of slip, the generating effect ceases and the machine becomes a brake.

Induction generators do not need synchronizing and can run in parallel without hunting and at any frequency, the speed variations of the prime mover are relatively unimportant. Thus, these machines are applied for wind power generation.

An induction generator must draw its excitation from the supply system, which is mostly reactive power requirement. On a sudden short-circuit the excitation fails, and with it the generator output; so in a way the generator is self-protecting.

Figure 3.17b shows a simplified equivalent circuit.

As the rotor speed rises above synchronous speed, the rotor EMF becomes in phase opposition to its subsynchronous position, because the rotor conductors are moving faster than the stator rotating field. This reverses the rotor current also and the stator component reverses. The rotor current locus is a complete circle. The stator current is clearly a leading current of definite phase angle. The output cannot be made to supply a lagging load.

An induction generator can be self-excited through a capacitor bank, without external DC source, but the frequency and generated voltage will be affected by speed, load, and capacitor rating. For an inductive load, the magnetic energy circulation must be dealt with by the capacitor bank as induction generator cannot do so.

For wind power applications, generators are either squirrel cage or wound rotor induction types with rotating field windings. The coupling with the grid, directly or through inverters is of significance. Mostly induction generators are used.

The induction generator must draw its reactive power requirement from the grid source. When capacitors, SVC's, rotary phase shifters are connected, the operational capabilities can be parallel with synchronous machines, though resonance with grid inductance is a possibility. Induction generators produce harmonic and synchronous pulsating torques, akin to induction motors. A synchronous machine provides control of operating conditions, leading or lagging by excitation control. With respect to interconnection with the grid, the following schemes exist.

3.6.2 Direct Coupled Induction Generator

The direct coupled induction machine is generally of 4-pole type; a gear box transforms the rotor speed to a higher speed for generator operation above synchronous speed. It requires reactive power from grid or ancillary sources, and starting after a blackout may be a problem. Wind-dependent power surges produce voltage drops and flicker. The connection to the grid is made through thyristor switches that are bypassed after start. A wound rotor machine has the capability of adjusting the slip and torque characteristics by inserting resistors in the rotor circuit and the slip can be increased at an expense of more losses, heavier weight, Figure 3.18a. The system will not meet the current regulations of connection to grid and may be acceptable for isolated systems.

3.6.3 Induction Generator Connected to Grid through Full Size Converter

The induction generator is connected to the grid through two back-to-back VSCs. Because of full power rating of the inverter, the cost of electronics is high. The wind-dependent

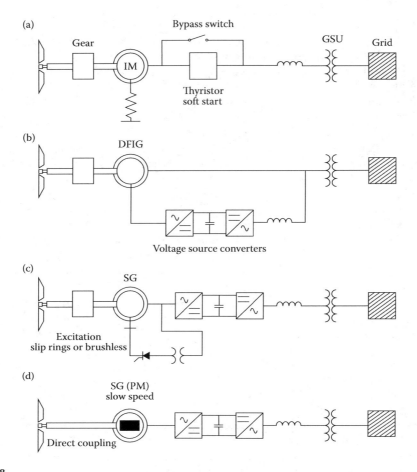

FIGURE 3.18
Grid connections of wind generators. (a) Direct connection of an induction generator, stall regulated. (b) Connection of a DFIG variable speed generator, pitch regulated. (c) Synchronous generator brush type or brushless type with voltage source converters, pitch regulated. (d) Gearless connections of a low-speed permanent magnet generator.

power spikes are damped by the DC link. The grid side inverter need not be switched in and out so frequently and harmonic pollution occurs.

3.6.4 Doubly Fed Induction Generator

The stator of the induction machine is directly connected to the grid, while the rotor is connected through VSC, Figure 3.18b. The energy flow over the converter in the rotor circuit is bidirectional. In subsynchronous mode, the energy flows to the rotor and in supersynchronous mode, it flows from rotor to the grid. The ratings of the converter are much reduced, generally one-third of the full power and depend upon the speed range of turbine. The power rating is

$$P = P_s \pm P_r, \tag{3.21}$$

where P_s and P_r are the stator and rotor powers. But the rotor has only the slip frequency induced in its windings, therefore, we can write

$$P_r = P_a \times s, \tag{3.22}$$

where s is the slip. For a speed range of ±30%, the slip is ±0.3, and a third of converter power is required. Also we can write

$$n_s = \frac{f_r \pm f}{P} 120, \tag{3.23}$$

where p are the number of pairs of poles.

3.6.5 Synchronous Generators

Synchronous generators can be brush type or brushless type of permanent magnet excitation systems. These are also connected to the grid much like asynchronous machines. The excitation power has to be drawn from the source, unless the generator is of permanent magnet type. Figure 3.18c and d shows typical connections.

3.7 Reactive Power and Wind Turbine Controls

A prior reactive power compensation study is a must. As we have previously seen, the reactive power compensation, power factor, and voltage profiles are interrelated, with the system impedance playing an important role.

With reference to Figure 3.17b, equivalent circuit, the reactive power required by an induction generator can be written as follows:

$$Q = \frac{-b}{2a} V_s^2 + \frac{\sqrt{(b^2 - 4ac)V_s^4 + 4aPV_s^2}}{2a}, \tag{3.24}$$

where

$$X_s = X_1 + X_m, \quad a = \frac{R_r X_s^2}{X_m \sin^2 \phi}, \quad b = \frac{2R_r X_s}{X_m^2} + \frac{1-s}{\tan \phi}, \quad c = \frac{R_r}{X_m^2}, \tag{3.25}$$

where P is the active power and ϕ is the power factor angle.

The wind turbine dynamic models consist of pitch angle control, active and reactive power control, the drive train model, and the generator model. These models are considered the proprietary of the turbine manufacturer and can be obtained only under confidentiality agreements. Some efforts have been directed by WECC toward generic models, which are now available in some commercial software packages.

The overall control system diagram of a wind generation using doubly fed induction generator (DFIG) is shown in Figure 3.19. It has four major control components:

- Pitch angle control model
- Vector decoupling control system of DFIG
- Grid VSC control model
- Rotor VSC control system

Wind turbines are not able to maintain the voltage level and required power factor. According to one regulation, it should be capable of supplying rated MW at any point between 0.95 power factor lagging to leading at the point of common coupling (PCC). The reactive power limits defined at rated MW at lagging power factor will apply at all active power output levels above 20% of the rated MW output. Also the reactive power limits defined at rated MW at leading power factor will apply at all active power output levels above 50% of the rated MW output, see Figure 3.20 for further details; this

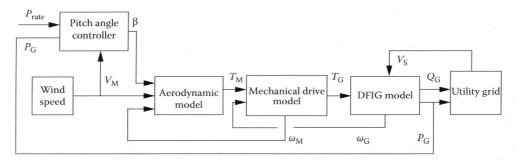

FIGURE 3.19
Components of main control circuit of wind power generation.

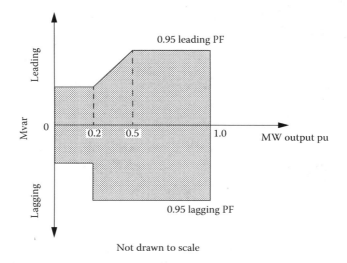

FIGURE 3.20
Operation requirements of a wind generating farm for connections at the utility source, output versus power factor.

Wind and Solar Power Generation and Interconnections with Utility 109

figure is for interconnection at the grid, PCC, *and not for individual operating units.* Thus, the reactive power compensation, fault levels and short-circuit analysis, the variations in the active power due to wind speed in the particular area over the course of a day, month-to-month, peak and lowest raw electricity that will be generated are the first set of studies performed in the planning stage. These allow fundamental equipment ratings to be selected and protection system designs. The considerations like cables versus overhead lines for connection of collector buses to grid step-up transformer also arise. Further need for dynamic studies, voltage profiles, fault clearance times, and studies documenting the grid connection requirements also arise. In fact, wind power generation and utility interconnections required extensive studies apart from harmonic considerations.

Due to the stochastic nature of wind turbine harmonics, probability concepts have been applied. Autoregressive moving average (ARMA) model is the statistical analysis of time series and provides parsimonious description of a stationary stochastic process in terms of two polynomials, one for auto regression and the other for moving average.

The mathematical model of a variable speed constant frequency generator under a *dq* synchronous rotating coordinate system is given by

$$
\begin{vmatrix} v_{sd} \\ v_{sq} \\ v_{rd} \\ v_{rq} \end{vmatrix} = \begin{vmatrix} pL_s + R_s & -\omega_1 L_s & pL_m & -\omega_1 L_m \\ \omega_1 L_s & pL_s + R_s & \omega_1 L_m & pL_m \\ pL_m & -(\omega_1 - \omega_r)L_m & pL_r + R_r & -(\omega_1 - \omega_r)L_r \\ (\omega_1 - \omega_r)L_m & pL_m & (\omega_1 - \omega_r)L_r & pL_r - R_r \end{vmatrix} \begin{vmatrix} i_{sd} \\ i_{sq} \\ i_{rd} \\ i_{rq} \end{vmatrix}. \quad (3.26)
$$

The subscripts *s* and *r* that denote stator and rotor, and *p* is the differential operator
Generator flux is given by

$$
\begin{vmatrix} \varphi_{sd} \\ \varphi_{sq} \\ \varphi_{rd} \\ \varphi_{rq} \end{vmatrix} = \begin{vmatrix} L_s & 0 & L_m & 0 \\ 0 & L_s & 0 & L_m \\ L_m & 0 & L_r & 0 \\ 0 & L_m & 0 & L_r \end{vmatrix} \begin{vmatrix} i_{sd} \\ i_{sq} \\ i_{rd} \\ i_{rq} \end{vmatrix}. \quad (3.27)
$$

If stator flux linkage is in the same direction as *d*-axis of rotating coordinate system, then $\varphi_{sd} = 0$, Then

$$
\varphi_{sd} = \varphi_s
$$
$$
\varphi_{sq} = 0. \quad (3.28)
$$

If stator coil resistance is ignored

$$
v_{sd} = 0
$$
$$
v_{sq} = |v_s| \quad (3.29)
$$

The generator side active and reactive powers are

$$P_s = v_{sd} i_{sd} + v_{sq} i_{sq} = v_{sq} i_{sq}$$
$$Q_s = v_{sq} i_{sd}.$$

(3.30)

The P_s and Q_s can be decoupled using the above equations.

The crowbar protection is specific to DFIG. The rotor side converter must be protected in case of nearby faults. When the currents exceed a certain limit, the rotor side converter is bypassed to avoid any damage.

Apart from the reactive power need of the induction machine itself, the passive connecting element consumes reactive power. These elements are transformers and cable connectors to the point of grid. Then, the reactive power is required by loads. Problems of transient low voltages can occur when the wind power generation is connected to relatively weak grid systems.

Some wind generating plants have added SVCs, DSTATCOM, (Volume 2), which control the power factor to unity at the point of interconnection. The Argonne Mesa wind plant in New Mexico has a DSTATCOM, which controls the power factor to utility at the point of interconnection at Guadalupe 345 kV station bus. Four mechanically switched capacitor banks are located in the collector substation approximately 2 miles away from the interconnect substation. The DSTATCOM controls determine the required reactive power output based upon voltage and current measurements at 345 kV collector bus.

Figure 3.21 shows the impact of high winds on the 1.5 MW generators; connected to a 230 kV transmission line. Line drop compensation algorithms are utilized to synthesize voltage at the point of interconnection, located approximately 75 km from the wind plant. The flicker index of the voltage is less than 2% at the point of interconnection. In spite of considerable variation in the wind speed, the plant output is relatively stable.

Figure 3.22 shows the field tests result on an active power regulator and power rate limiter on an operating 30 MW wind plant. Initially, the output is curtailed to 10 MW, and during the tests, the active power command is raised in four 5 MW increments. The transition between each step ramp rate is controlled to 2 MW/min.

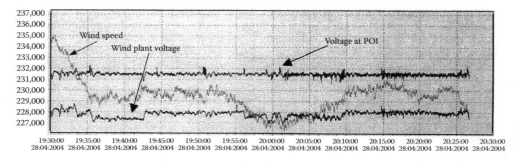

FIGURE 3.21
Wind plant voltage response and regulation at the point of connection.

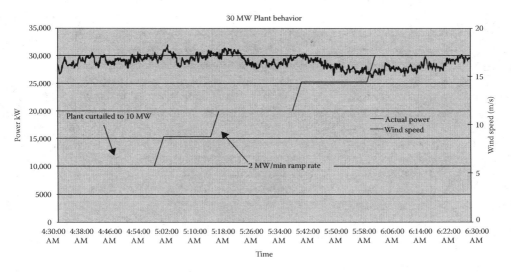

FIGURE 3.22
Active power response of a wind plant with ramp control.

3.8 Power Electronics and Harmonics

3.8.1 Power Electronics

The entire output of variable voltage, variable frequency output of a wind generator, sometimes called the wild AC is converted to direct current, which is then converted to utility quality AC power. The frequency converters condition the electrical energy from the wind generators and damp the influence on the grid connections. With respect to the topology of the electronic devices, we have

- Current-controlled rectifier
- Voltage source rectifier
- Voltage source inverters
- Current source inverters
- Newer technology of AC/DC/AC ZSI converters

The rectifiers may be (1) uncontrolled, (2) bridges with DC/DC regulators, and (3) controlled rectifiers. Again, pulse-controlled insulated gate bipolar transistor (IGBT) topology with phase multiplication to reduce harmonic generation is the preferred choice, as active and reactive control of power can be exercised. Practically, all the installed variable speed synchronous generators wind turbines that employ fully rated converters use VSI topology. As the power of the converter rises, devices for pulse width modulation (PWM) are needed to switch at higher frequencies. For such applications, as an alternative, CSI is proposed [28], however harmonic performance suffers and VAR compensation is required.

Consider a synchronous generator grid connection. The constant voltage DC link is supplied by a controlled rectifier bridge and the rectifier is current controlled so that magnitude and phase angle of the generator current is controlled by triggering of the rectifier. By phase shifting, the generator current in under excited and overexcited regions, the generator voltage can be controlled and matched to the DC link, also see Reference [29].

Short-circuit current calculations according to machines integrated in industrial systems may not be always valid. For example in a doubly fed induction motor (DFIM), the stator current for a nearby fault may be limited to nearly rated current if rotor power converter remains active. It may, however, be disabled by crowbar circuit for protection during fault, in which case the fault current will be several times the rated current for a few cycles.

Switching of devices in converters gives rise to interference emission over a wide spectrum. The electromagnetic compatibility should be insured according to relevant standards.

References [30, 31] provide further reading. Reference [32] describes its application to a wind turbine, linking with the utility. Basically, a Z-network is introduced between the front end diode rectifier and neutral point clamped inverter. This three-level diode clamped AC/DC/ACZSI has three-level output waveform, with use of minimal passive components. Before the front-end diode rectifier, passive harmonic filters are placed and the neutral potential needed by the three-level inverter circuitry that can be tapped from the wye-connected filter capacitors' common point. See Volume 3 for details of power electronics.

3.8.2 Harmonics

Generation of harmonics and mitigation is discussed in Volume 3. Even harmonics in wind generation can arise due to unsymmetrical half waves and may appear at fast load changes. Subharmonics can be produced due to periodical switching with variable frequency. Interharmonics can be generated when the frequency is not synchronized to the fundamental frequency, which may happen at low- and high-frequency switching.

The interharmonics due to back-to-back configuration of two converters can be calculated according to IEC [33]:

$$f_{n,m} = \left[(p_1 k_1) \pm 1 \right] f \pm (p_2 k_2) F, \tag{3.31}$$

where $f_{n,m}$ is the interharmonics frequency, f_1 is the input frequency, F is the output frequency, and p_1 and p_2 are the pulse numbers of the two converters. Interharmonics are also generated due to speed-dependent frequency conversion between rotor and stator of DFIM, and as sidebands of characteristic harmonics of PWM converters. Noncharacteristic harmonics can be generated due to grid unbalance.

A topology shown in Figure 3.23 is advocated in Reference [28]. Inverters 1 and 2 are series connected bridge circuits that employ fully controllable switches with bidirectional blocking capabilities. The bridges are switched at line frequency using phase-control techniques, which avoids PWM and the associated switching losses and allows use of high power, but relatively slow devices such as gate-controlled thyristors, or even silicon-controlled thyristors. Using phase control, the DC link voltage is modulated and thus DC-link current, but the phase angle of phase-controlled inverter is tied to the firing angle, Volume 3. By controlling the phase angle of the two converters so that $\phi_2 = \phi_1$, inverter DC link voltage is modulated without affecting the fundamental power factor at the mains. See Volume 3 for a case study and further analysis.

Wind and Solar Power Generation and Interconnections with Utility

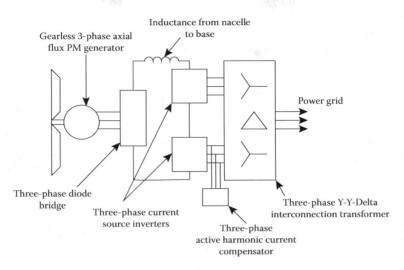

FIGURE 3.23
Current source topology for wind turbines, providing reactive power control and reduced harmonics. (Adapted from P Tenca et al. *IEEE Trans Ind Appl*, 43, 1050–1060, 2007.)

3.9 Computer Modeling

Modeling is required in the planning stage to conduct interconnection studies, grid reliability, and simulate energy capture for hybrid plants. The energy capability of the wind power generation for typical time history wind samples or wind probability density curve can be forecasted. Also simulation is required for aerodynamics, mechanical dynamics, structures—this gives ideas of static and dynamic loads, predicted power curves, vibration modes, and control system response. Electrical transients in generators and power electronics need to be simulated. Excess staring currents, behavior under short circuit, and under voltage transients, isolation from grid can be studied in a time frame varying from a few microseconds to seconds. GE PSLF/PSDS and Siemens PTI/PSSE programs are designed for study of large-scale interconnected systems; yet, there is not much sharing of the data between consultants, utilities, and power planners. Engineering design models are implemented in a three-phase simulation programs like EMTP and PSCAD [34–39].

3.9.1 A Wind Turbine Controller

Classically, the wind turbine control principles can be depicted with reference to Figure 3.24. In zone 1, the system is operated at optimal rotor speed according to rotor aerodynamics to extract maximum energy from the wind. A relation of power reference to rotor speed can be written as

$$P_{ref} = f(\Omega_{\omega t}) \tag{3.32}$$

and

$$\Omega_{\omega t\text{-ref}} = f(P_{meas}).$$

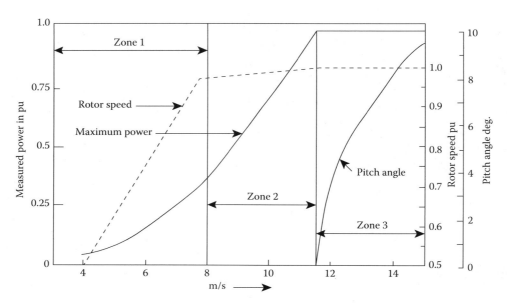

FIGURE 3.24
Wind turbine control zones.

In zone 2, the rotor speed is limited to the nominal rotor speed, $\Omega_{wt\text{-}nom}$. For controller stability, a slope may be added to the power reference to rotor speed characteristics. In zone 3, the wind speed is high enough to allow nominal power production. In this zone, the objective is to maintain the rotor speed and the measured produced power at their nominal values.

Figure 3.25 shows one-wind controller implementation strategy. The two quantities of interest are rotor speed nr and measured power P_{meas}. In order to regulate these two quantities, two control variables are available: the blade pitch angle β_{ref} and electromagnetic torque $T_{em\text{-}ref}$. The rotor speed is regulated by the pitch control angle and measured power is regulated by acting upon electromagnetic torque $T_{em\text{-}ref}$. A block circuit diagram is shown in Figure 3.25 [39]. This represents two control loops, one for the measured power P_{meas} and the other for the rotor speed Ω_{wt}. For the power control loop, P_{ref} is extracted from optimal power set point to rotor speed and an external set point $P_{setpoint}$ can be introduced in a saturation element to limit maximum value of P_{ref}. Then, the internal power set point and the measured produced power are used to compute power error. From this, the torque error Terror is calculated using generator rotor speed Ω_{em}. Then, the torque reference is used by a PI controller to provide electromagnetic torque reference $T_{em\text{-}ref}$.

Analysis of any pitch-controlled turbine reveals a nonlinearity caused by relation of wind turbine torque T_{wt} and pitch angle β. The capability to change power set point requires adjustment of the process gain depending upon wind *and* power set point also. The principle of model inversion is used to tackle this problem and is called general gain scheduling. Each element of the process is inverted in the control algorithm, third loop in Figure 3.25. The pitch angle reference β_{ref} is obtained from power coefficient $c_{p\text{-}ref}$ using the approximated tip speed ratio $\tilde{\lambda}$. The power coefficient $c_{p\text{-}ref}$ is obtained from wind turbine torque reference $T_{wt\text{-}ref}$ provided by the PI controller using wind speed, v, the approximated rotor speed, $\tilde{\Omega}_{wt}$, and the mathematical expression of the coefficient reference $c_{p\text{-}ref}$ derived from the equation of rotor torque [40,41], also see Reference [42].

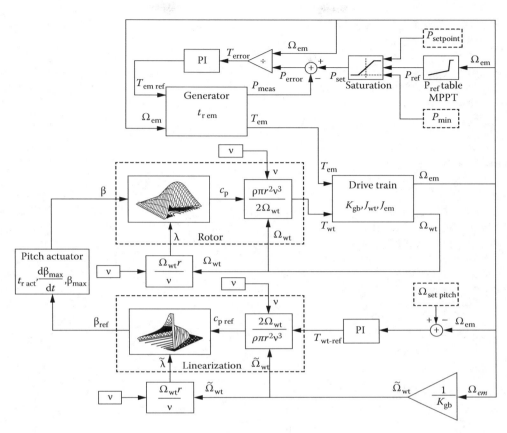

FIGURE 3.25
Block circuit diagram of a wind turbine controller. (From P Venne, X Guillaud. *CIGRE and IEEE PES Joint Symposium*, Calgary, Alberta, Canada, 2009.)

3.10 Solar Power

Earth receives 174 PW (Peta watts) of incoming solar radiation at the upper atmosphere. About 30% is reflected back to space and the rest is absorbed by clouds, ocean, and land masses. The spectrum of solar light at earth's surface is mostly spread across visible light and near infra-red ranges with a small part in ultraviolet region. Photosynthesis captures approximately 3000 EJ per year in biomass. The amount of solar energy reaching the surface of planet the is so vast that in one year it is twice as much as will ever be obtained from non-renewable sources like coal, oil, natural gas, and mined uranium combined. The solar energy can be used for water heating, cooling and ventilation, industrial process heating, homes, fans, pumps, traffic lights, hybrid PV powered homes, cooking, etc. We will be more interested in electricity production. The two technologies for converting solar energy into electricity are as follows:

- In the solar thermo-mechanical systems, the solar radiation is used to heat a working fluid, which runs the turbines. Concentrated solar power (CSP) systems use lenses or mirrors and tracking systems. CSP-Sterling (described further) is known to have highest efficiencies around 30%.

- Solar photovoltaic cells, which directly convert solar radiation to electric current. The efficiency is low around 15%.

These may be called active solar technologies. The passive solar technologies include orienting a building to the sun, selecting materials with favorable thermal mass, or light dispersing properties and designing spaces that naturally circulate air.

3.11 CSP Plants

Commercial CSP plants were first developed in 1980. Since 985, 354 MW SWGS CSP installations in Mojave Desert of California is the largest solar plant in the world. Other large CSP plants include 150 MW Solnova solar plant and 100 MW Andasol solar plant, both in Spain. A new plant called "Solana" of 280 MW in Arizona and another 550 MW in Mojave solar park in California are under planing. The 250 MW Agua Caliente Solar project in the USA and 221 MW Charanka solar park in India are the largest photovoltaic plants.

The major components of a CSP are the following:

1. Parabolic mirrors or other type of solar collectors that concentrate the solar energy and heat-transfer fluid
2. Hot fluid returns from solar field
3. The hot fluid transfers its heat to water, creating steam
4. The steam produced drives a steam turbine-generator set
5. The hot fluid also heats molten salt
6. If the sun is not shining the fluid can be heated by the molten salt
7. The fluid is sent back to the solar field for reheating when the sun is shining; it is a close circuit circulation

3.11.1 Solar Energy Collectors

The three types of solar energy collectors are as follows:

- Parabolic troughs
- Central receiver
- Parabolic dish

Figure 3.26 shows a parabolic trough; and it is the most mature technology. It concentrates the sunlight on a glass-encapsulated tube running along the focal line of the collector. The tube carries a heat-absorbing liquid, typically synthetic oil, or molten salt. The hot oil (temperature to 750°F) is pumped to heat exchangers to generate steam, which drives a conventional turbo-generator set. Figure 3.27 shows a schematic diagram of the power plant.

The molten salt is cheaper and safer than oil and is an effective storage medium. The spare solar power is used in the form of heated molten salt in storage tanks, for use when solar power is not available. The concentrators track the sun on two axes.

Wind and Solar Power Generation and Interconnections with Utility

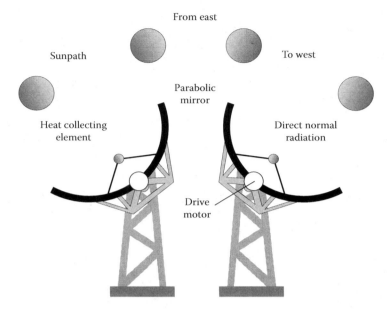

FIGURE 3.26
Parabolic trough with sun tracking.

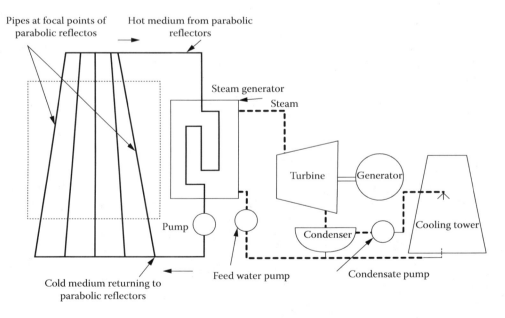

FIGURE 3.27
Schematic diagram of a CSP solar power plant.

3.11.1.1 Parabolic Dish Concentrators

This consists of a dish-shaped mirror that reflects sun radiation on to a receiver located at the focal point of the dish. The dish structure is designed to track the sun on two axes. A schematic diagram is shown in Figure 3.28. The concentration ratio is much higher than

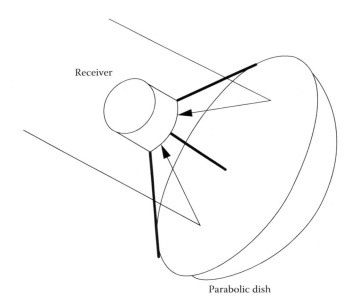

FIGURE 3.28
Schematic diagram of a parabolic dish collector.

that of parabolic trough, over 2000, and working fluid temperature of 1300°F. The receiver consists of a bank of tubes filled with a cooling fluid, usually hydrogen or helium; which is the heat transfer fluid and also the working fluid for an engine. The heat is delivered to a Sterling engine, which is attached to the receiver. The Sterling engine is attached to a generator. A commercial scale Dish Sterling system is being built near Phoenix, AZ. The 1.5 MW plant consists of sixty 25 kW units, known as sun catcher. The sun catcher dish is formed into parabolic shape using multiple arrays of curved glass mirrors. Each unit is designed to effectively track the sun for maximum concentration of radiation.

3.11.1.2 Solar Tower

The mirrors track the sun and reflect the sunlight on to a central receiver mounted on top of a tower. A working fluid, oil or molten salt circulates in the receiver, where it is heated to 1300°F. The heated fluid is pumped to heat exchangers to generate steam.

The earliest solar power towers were 10 MW Solar One and Solar Two projects in Mojave desert. In 2009, Pacific Gas and Light Company entered into a Bright Source Energy, Inc. for a total of 1310 MW of solar power, project to progress in seven phases. Each plant will consist of thousands of computer-controlled heliostats that track the sun and reflect the energy onto a water boiler located on a centralized tower, Figure 3.29. The water is heated to 1000°F creating superheated steam.

3.11.2 Trackers

Tracking systems are applicable to solar concentrators as well as PV cell arrays. These systems track the devices to point at the sun to capture the maximum radiation. For example, the effective collection area of a flat panel solar collector varies with the cosine of the misalignment of the panel with the sun.

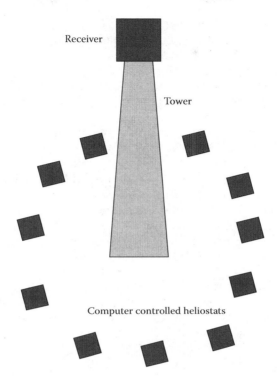

FIGURE 3.29
Schematic diagram of a solar tower.

Tracking systems may be configured as follows:

- Fixed collector/moving mirror, i.e., heliostat. Generally, it is more complex to align a moving mirror with the sun. Many collectors cannot be moved, for example, high-temperature collectors where the energy is recovered as hot liquid or steam. Other examples are direct heating of buildings. In such cases, it is necessary to employ a moving mirror, so that, regardless of where the sun is positioned in the sky, the sun's rays are redirected on to the collector. The level of precision required to correctly target the sun's rays using a heliostat mirror that generally employs a dual axis tracking system, with at least one axis mechanized. In different applications the mirrors may be flat or concave.
- Moving collector. A tracker rotating in east-west direction is known as single axis tracker. The sun also moves through 46 degrees north and south during a year. The same set of panels set at the midpoint between two local extremes will see the sun move 23 degrees on either side, causing a loss of 8.3%. A tracker system that accounts for both the daily and seasonal motions of the sun is called a dual-axis tracker.

3.11.2.1 Photovoltaic Trackers

The photovoltaic trackers can be classified into two types: (1) standard photovoltaic trackers and (2) concentrated photovoltaic trackers (CPV). Each of these types can be further characterized by the number and orientation of their axes, their actuation architecture and drive type, their intended application and their vertical supports and foundations.

In standard, nonconcentrating photovoltaic trackers the functionality is to optimize the angle of incidence between incoming light and photovoltaic panel. This increases the amount of energy gathered from the direct component of the incoming sunlight. The technology can be applied to all types of crystalline silicon panels; either mono-Si or multi-Si and all types of thin film panels, amorphous silicon, CdTE, CiGS, microcrystalline—see sections to follow.

In concentrator photovoltaic (CPV) trackers, the functionality is used to orient the optics so that the incoming light is focused to a photovoltaic collector. The CPV modules that concentrate in one dimension must be tracked normal to the sun in one axis, CPV modules that concentrate in two dimensions must be tracked normal to the sun in two axes. In typical concentrator systems, the tracking accuracy must be of the order of $\pm 0.1°$ range to deliver approximately 90% of the rated power output. The technology is used with refractive and reflective based concentrator systems—from crystalline silicon-based photovoltaic receivers to germanium based triple junction receivers.

The axis of rotation for horizontal axis tracker is horizontal with respect to ground. The single axis trackers can be horizontal single axis trackers (HSATs), horizontal single axis trackers with tilted modules (HTSATs), vertical axis trackers (VSATs), tilted single axis trackers (TSATs), and polar aligned single axis trackers (PSATs).

The dual axis trackers have two degrees of freedom that act as axes of rotation. These axes are typically normal to each other. Two common applications are tip-tilt dual axis trackers (TTDATs) and azimuth-altitude dual axis trackers (AADATs).

3.12 Direct Conversion of Solar Energy through PV Cells

The photoelectric effect was first discovered by French physicist Edmond Becquerel, in 1839, at the age of 19. Albert Einstein explained the underlying mechanism of light instigated carrier excitation–photoelectric effect in 1905, for which he got the Nobel Prize. The first practical photoelectric cell was demonstrated in April 1954 at Bell Laboratories. The inventors were Daryl Chapin, Calvin Souther Fuller, and Gerald Pearson.

PV cells were first used to power satellites. PV is a device that converts sunlight directly into electricity. The basic structure of a PV cell is shown in Figure 3.30. The incident photons cause generation of electron-hole pairs in both p and n layers. These photon generated minority carriers freely cross the junction, which increases minority flow many times. The major component is the light-generated current, when load is connected to the cell. There is also thermally generated reverse current, also called dark current as it flows even in absence of light. The light current flows in opposite direction to the forward diode current of the junction.

3.12.1 Cells, Modules, Panels, and Systems

3.12.1.1 PV Module

Multiple solar cells in an integrated group, all oriented in one plane, constitute a photoelectric module. These modules have a sheet of glass on the sun facing side, allowing the light topass, while protecting the semiconductor wafers. Solar cells are usually connected in series in the modules. A parallel connection can cause substantial power loss due to shadowing effect and reverse bias applied by their illuminated partners. This includes receivers and optics (concentrator type) and related components, such as interconnects and mounting, that accepts un-concentrated sunlight.

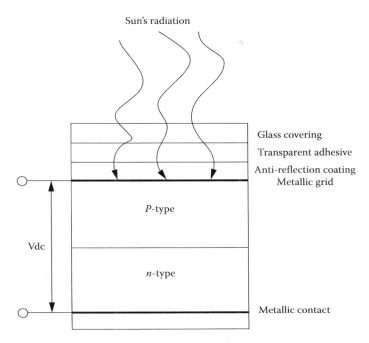

FIGURE 3.30
PV solar cell, basic constructional features.

3.12.1.2 PV Panel

One or more PV modules assembled and wired and designed to provide a field-installable unit.

3.12.1.3 PV Array

PV array is the smallest installed assembly of PV panels support structures, foundations, and other components as required, such as trackers.

3.12.1.4 PV Array Subfield

It is one or more arrays associated by a distinguishing feature, such as field geometry or electrical interconnection.

3.13 Classification of Solar Cells

A PV cell is classified as follows:

- In terms of material, i.e., noncrystalline silicon, polycrystalline silicon, amorphous silicon, gallium arsenide, cadmium telluride, cadmium sulfide, etc.
- In terms of technology, i.e., single crystal bonds, thin film, etc.

The conversion efficiency varies. Under normal temperature of 25°C and illumination level of 100 mW/cm^2, amorphous silicon has an efficiency of 5%–6% while gallium arsenide has an efficiency of 20%–25%.

Another classification is first, second, and third generation cells.

First generation cells, also called conventional, traditional, or wafer-based that cells are made of crystalline silicon. This is commercially predominant PV technology that includes materials such as polysilicon and monocrystalline silicon.

Second generation cells are thin film solar cells that include amorphous silicon, CdTe (cadmium telluride), and CIGS (copper indium gallium selenide) cells. These are prominent in utility scale photovoltaic power stations or in standalone power systems.

Third generation cells include a number of thin film emerging technologies; most of them are not yet a commercial venture. Many use organic materials, often organometallic compounds as well inorganic substances. Despite the fact that efficiencies are low and stability of absorber material is too short for commercial applications, much research is invested in these technologies, as these promise to mass produce low-cost, high-efficiency solar cells.

The capacity to produce solar cells in the USA is increased by 20%, with the official inauguration of new thin film facility, United Solar Systems, which can produce 30 MW of solar cells per year. The PV panels are installed at the top of CN tower in Toronto, the tallest such tower in the world. Most of the solar capacity of 6.5 MW in Canada alone is for

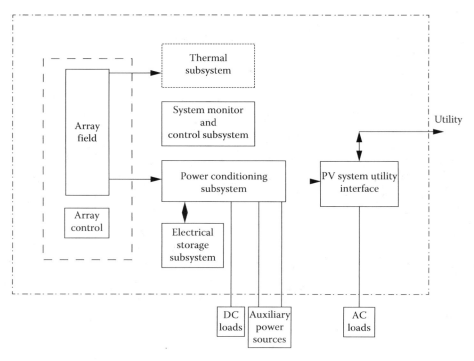

FIGURE 3.31
Block and interface diagram of photovoltaic power system.

Wind and Solar Power Generation and Interconnections with Utility   123

off-grid applications for lighthouses, homes, navigation buoys, remote telecommunication towers, etc. There are an estimated 12,000 residential solar water systems, seasonal pool heaters, and hot water units in Canada alone.

Figure 3.31 shows a block circuit diagram of photovoltaic/thermal system, [55], also see References [43–56] for solar power.

3.14 Utility Connections of Distributed Resources

Till the mid-1990s, the applications were standalone systems located where connections to utility grids were impractical. Utility interactive PV cells are classified by IEEE Standard 929. Generally, utility interactive systems do not incorporate any form of energy storage and supply power to the grid when operating. As the output is DC, it is necessary to convert it to AC before interconnection with utility. An inverter is used, called as power conditioning unit (PCU). Automatic disconnects are provided in the event of loss of utility voltage. PCUs are synchronized with the utility voltage.

The following is an extract from IEEE Std. 1547-2 [57], for all Distributed Resources (DR) technologies aggregate capacity of 10 MVA at the PCC that are interconnected with an area electrical power system (EPS) at typical primary or secondary distribution voltages, also see Reference [58].

3.14.1 Voltage Control

The voltage limits at the PCC, where the area EPS is connected with a local EPS shall be as specified in ANSI C84.1, range A [59]. This is further discussed in Volume 2 with an extract of voltage requirements.

3.14.2 Grounding

The grounding scheme of DR interconnection shall not cause overvoltages that exceed the rating of the equipment connected to area EPS and shall not disrupt the coordination of ground fault protection on the area EPS.

3.14.3 Synchronizing

The DR unit shall parallel with the area EPS without causing a voltage fluctuation at the PCC greater than ±5% of the prevailing voltage level of the area EPS at the PCC, and meet the flicker requirements of IEEE Std. 1547-2003 [60]. The synchronization parameters to be met are shown in Table 3.2. Self-excited induction generators shall be tested according to IEEE Std. 1547.

An inverter interconnection system that produces fundamental voltage before the paralleling device is closed shall be tested as per procedures for synchronous interconnection according to IEEE Std. 1547. All other inverter interconnection systems shall be tested to determine the maximum start-up current. The results shall be used to estimate starting voltage magnitude change and verify that the unit shall meet the synchronization requirements.

TABLE 3.2

Synchronization Parameter Limits for Synchronous Interconnection to an EPS or an Energized local EPS to an Energized Area EPS

Aggregate Rating of DR Units (kVA)	Frequency Difference (Δf, Hz)	Voltage Difference (ΔV, %)	Phase Angle Difference ($\Delta\phi°$)
0–500	0.3	10	20
>500–1500	0.2	5	15
>1500–10,000	0.1	3	10

3.14.4 Distribution Secondary Spot Networks

Network protectors shall not be used to separate, switch, serve as breaker failure backup or in any manner isolate a network or network primary feeder to which DR is connected from the remainder of area EPS, unless the protectors are rated and tested as per applicable standards for such an application.

Any DR installation connected to a spot network shall not cause operation or prevent reclosing of any network protectors installed on the spot network. This coordination shall be accomplished without requiring any change to prevailing network protector clearing time practices of the area EPS.

Connection of DR to area EPS is only permitted if the area EPS network bus is already energized to more than 50% of the installed network protectors.

The DR unit shall not cause any cycling of the network protectors.

DR installations on a spot network, using an automatic transfer scheme (load transferred between DR and EPS in a momentary make-before-break operation) shall meet all the above requirements.

3.14.5 Inadvertent Energization

The DR unit shall not energize area EPS when area EPS is de-energized.

3.14.6 Metering

Each DR unit of 250 kVA or more or DR aggregate of 250 kVA or more at a single PCC shall have provisions for monitoring its connection status, real power output, reactive power output, and voltage at the point of DR connection.

3.14.7 Isolation Device

Where required by EPS operating practices, a readily accessible, lockable, visible break isolation device shall be located between the area EPS and DR unit.

3.14.8 EMI Interference

The interconnection shall have the capability to withstand electromagnetic interference (EMI) environment according to IEEE Std. C37.90.2 [61]. The influence of EMI shall not result in change in state or misoperation of the interconnected system.

Wind and Solar Power Generation and Interconnections with Utility

3.14.9 Surge Withstand

The interconnection system shall have capability to withstand voltage and current surges in accordance with the environment described in IEEE Std. C62.41.2 or IEEE Std. C37.90.1 [62,63] as applicable.

3.14.10 Paralleling Device

The interconnection system paralleling device shall be capable of withstanding 220% of the interconnected system voltage.

3.14.11 Area Faults

The DR unit shall cease to energize the area EPS for faults in the EPS circuit to which it is connected. Also, the DR will cease to energize the area EPS circuit to which it is connected prior to recloser by the area EPS.

3.14.12 Abnormal Frequencies

When the frequency is in range given in Table 3.3, the DR will cease to energize the area EPS within the clearing time as indicated. For DR less than or equal to 30 kW peak, the frequency set points shall be either fixed or field adjustable. For DR greater than 30 kW, the frequency points shall be field adjustable.

3.14.13 Reconnection

After an area EPS disturbance, no DR reconnection shall take place until area EPS voltage is within Range B of ANSI 84.1, and frequency range 59.3–60.5 Hz. The DR connection shall include an adjustable delay or fixed delay of 5 min that may delay reconnection for up to 5 min after the area EPS steady state voltages and frequency are restored to ranges as defined above.

3.14.14 Harmonics

When the DR is serving balanced linear loads, the harmonic current injection into the area EPS at the PCC shall not exceed the limits in Table 3.4. The harmonic current injections shall be exclusive of any harmonic currents due to voltage distortion present in area EPS without DR connected.

TABLE 3.3

Interconnection System Response to Abnormal Frequencies

DR Size	Frequency (Hz)	Clearing Time (s)
≤30 kW	>60.5	0.16
	<59.3	0.16
>30 kW	>60.5	0.16
	<{59.8–57} (adjustable set point)	Adjustable 0.16–300
	<57.0	0.16

Note: DR ≤ 30 kW, maximum clearing times, DR > 30 KW, default clearing times.

TABLE 3.4
Maximum Harmonic Distortion in Percentage of Current $I*$

H Order, Odd Harmonics	$h < 11$	$11 < h < 17$	$17 < h < 23$	$23 < h < 35$	$35 \geq h$	Total Demand Distortion
%	4.0	2.0	1.5	0.6	0.3	5

Note: $I*$ = The greater of the local EPS maximum load current integrated demand (15 or 30 min) without the DR unit or the DR unit rated current capacity (transformed to PCC when a transformer exists between DR unit and PCC).
Even harmonics are limited to 25% of the odd harmonics shown.

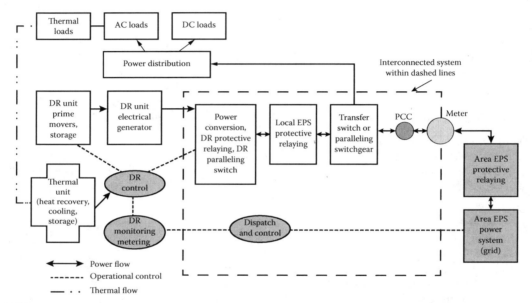

FIGURE 3.32
Interconnected system, controls and interface of solar generating plant with area EPS (Electrical Power System).

A functional diagram of the interconnected system is shown in Figure 3.32. Reference [64] is IEEE recommended practice for residential and intermediate photovoltaic systems utility interface.

Problems

3.1 A wind turbine is mounted at a height of 120 m and has a rotor diameter of 110 m. The average wind speed velocity at height of mounting = 120 m, is 11 m/s. Consider that the air density is 1.18 at a temperature of 20°C. The turbine power coefficient is 0.49. The turbine blade efficiency = 95% and the generator efficiency = 98%, coupling efficiency is 97%. What is the output power?

3.2 Draw a schematic drawing showing the formation of a solar array, starting from basic PV cell.

References

1. US Department of Energy, Fact Sheet for 20% Wind Energy Report.
2. KR Padiyar. *Power Transmission Systems*, John Wiley, New York, 1990.
3. J Arrillaga. *HVDC Transmission*, 2nd ed., IEEE Press, Piscataway, NJ, 1998.
4. JC Das. *Transients in Electrical Systems*, McGraw-Hill, New York, 2010.
5. JA Fleeman, R Gutman, M Heyeck, M Baharman, B Normark. EHV and HVDC transmission working together to integrate renewable power. *CIGRE and IEEE PES Joint Symposium*, Calgary, Canada, 2009.
6. RD Dunlop, R Gutman, PP Marchenko. Analytical development of loadability characteristics for EHV and UHV transmission lines. *IEEE Trans Power App Syst*, 98(2), 606–617, 1979.
7. AEP, 2007. Interstate Vision for Wind Integration. www.aep.com.
8. IEEE Transmission and Distribution Committee. An IEEE Survey of US and Canadian Overhead Transmission Outages at 230 kV and Above. Paper no. WM054-7PWRD, *IEEE PES Winter Meeting*, Columbus, Ohio, January/February 1993.
9. HC Barnes, TJ Nagel. AEP 765-kV system: General background relating to its development. *IEEE Trans Power App Syst*, 88(9), 1313–1319, 1969.
10. JD McDonald. The next generation grid: Energy infrastructure of the future. *IEEE Power and Energy Mag*, 7(2), 52–62, 2009.
11. MP Bahrman, BK Johnson. The ABCs of HVDC transmission technologies. *IEEE Power and Energy Mag*, 5(2), 32–44, 2007.
12. NERC. Transmission Transfer Capability Task Force. *Available Transfer Capability Definitions and Determinations*, New Jersey, June 1996.
13. GC Ejebe, JG Waight, M Saots-Nieto, WF Tinney. Fast calculation of linear available transfer capability. *IEEE Trans Power Syst*, 15, 1112–1116, 2000.
14. M Liang, A Abur. Total transfer capability computation for multi-area power systems. *IEEE Trans Plasma Sci*, 21, 1141–1147, 2006.
15. Y Ou, C Singh. Assessment of available transfer capability and margins. *IEEE Trans Plasma Sci*, 21, 2006.
16. PowerCon, R Sun, Y Fan, Y Song, Y Sun. Development and application of software for ATC calculation, *International Conference on Power Systems Technology*, pp. 463–468, 22–26 October, 2006.
17. K Audomvongseree, A Yokoyama. Consideration of an appropriate TTC by probabilistic approach. *IEEE Trans Plasma Sci*, 22, 2017–2107, 2002.
18. R Billinton, L Wenyuan. *Reliability Assessment of Electric Power Systems Using Monte Carlo Methods*. Plenum, New York, 1994.
19. ETSO. Counter Measures for Congestion Management Definition and Basic Concepts, June 2003. www.etso-net.org.
20. WESC (Sixth World Energy System Conference). AFK Kamga, JF Verstege. A Cross Border Congestion Management System Integrating AC and DC Load Flow Models. Turin, June 2006.
21. AFK Kamga, S Völler, JF Verstege. Congestion Management in Transmission Systems with Large Scale Integration of Wind Energy. *CIGRE and IEEE PES Joint Symposium*, Calgary, Alberta, Canada, July 2009.
22. www.nrel.gov/publications/ Following Reports can be obtained:
    i. DA Griffen, WindPACT turbine design scaling studies technical area 1—composite blades for 80–120 m rotor: 21 March 2000–15 March 2001. NREL Rep. SR-500-29492.
    ii. G Bywaters et al. Northern Power systems WindPact drive train alternative design study report: Period of performance: April 12, 2001 to January 31, 2005. NREL Rep. SR-500-35524, 2004.
    iii. MW LaNier. LWST phase 1 conceptual design study: Evaluation of design and construction approaches for economical hybrid steel/concrete wind turbine towers; June 28, 2002–July 31, 2004. NREL Rep. SR-500-36777.

23. S Heier. *Grid Integration of Wind Energy Conversion Systems*, 2nd ed., John Wiley, New York, 2009.
24. T Burton, D Sharpe, N Jenkins, E Bossanyi. *Wind Power Handbook*. John Wiley, New York, 2001.
25. Western Electricity Coordinating Council Disturbance Monitoring Reports. www.wecc-biz/.
26. IEEE. Guide for Interfacing Dispersed Storage and Generation Facilities with Electric Facility Systems, Standard 1001.1988.
27. IEEE. Standard for Interconnecting Distributed Resources with Electrical Power Systems. Standard 1547. 2003. (Available with subject areas: Energy generation/Power Generation Smart Grid.)
28. P Tenca, AA Rockhill, TA Lipo. Wind turbine current source converter providing reactive power control and reduced harmonics. *IEEE Trans Ind Appl*, 43, 1050–1060, 2007.
29. R Strzelecki, G Benysek, Eds. *Power Electronics in Smart Electrical Energy Networks*, Springer, London, 2008.
30. FL Luo, H Ye. *Power Electronics: Advanced Conversion Technologies*. CRC Press, Boca Raton, FL, 2010.
31. FZ Peng. Z-source inverter. *IEEE Trans Ind Appl*, 39, 504–510, 2003.
32. PC Loh, F Gao, PC Tan, F Blaabjerg. Three-level AC-DC-AC Z-source converter using reduced passive component count. *Proceedings of IEEE PESC*, pp. 2691–2697, 2007.
33. IEC 61000-2-4, Electromagnetic Compatibility, Part 2. Environmental Section 4: Compatibility Levels in Industrial Plants for Low-Frequency Conducted Disturbances.
34. Dynamic Models for Wind Farms for Power System Studies. www.energy.sintef.no/wind/iea.asp.
35. UWIG Modeling User Group, Dynamic Model Validation for the GE Wind Turbine. www.uwig.org.
36. EPA, Renewable Portfolio Standards Fact Sheet. www.epa.gov/chp/state-policy/renewable_fs.html.
37. IEC-61400-21, Wind Turbine Generator Systems. Part 21: Measurement and Assessment of Power Quality Characteristics of Grid Connected Wind Turbines, 2001.
38. EA DeMeo, W Grant, MR Milligan, MJ Schuerger. Wind power integration. *IEEE Power Energy Mag*, 3(6), 38–46, 2005.
39. GE Energy and AWS Truewind, Ontario Wind Integration Study. www.ieso.ca/imoweb/pubs/marketreports/opa-report-200610-1.pdf.
40. P Venne, X Guillaud. Impact of turbine control strategy on deloaded operation. *CIGRE and IEEE PES Joint Symposium*, Calgary, Alberta, Canada, 2009.
41. T Ackermann, Ed. *Wind Power in Power Systems*, Wiley-Interscience, New York, 2005.
42. IEEE Std. 1094. IEEE Recommended Practice for Electrical Design and Operation of Windfarm Generating Stations, 1991.
43. International Energy Agency. Solar energy perspectives: Executive summary. www.iea.org/.
44. Canadian Renewable Energy Network. Solar energy technologies and applications. www.canren.gc.ca/.
45. J Bolton. *Solar Power and Fuels*. Academic Press, New York, 1977.
46. F Daniels. *Direct use of Suns Energy*. Ballantine Books, New York, 1964.
47. D Halacy. *The Coming Age of Solar Energy*. Harper and Row, New York, 1973.
48. E Mazaria. *The Passive Solar Energy Book*. Rondale Press, Emmaus, PA, 1979.
49. J Perlin. *From Space to Earth: The Story of Solar Electricity*. Harvard University Press, Cambridge: MA, 1999.
50. EPIA. Solar photovoltaic competing in the energy sector. www.epia.org/.
51. H Mousazadeh, A Keyhani, A Javadi, H Mobli, K Abrinia, A Sharifi. A review of principle of sun-tracking methods for maximizing. *Renewable and Sustainable Energy Rev*, 13, 1800–1818, 2009.
52. P Gevorkin. *Sustainable Energy Systems Engineering: The Complete Building Design Resource*, McGraw-Hill Professional, New York, 2007.
53. J Chavas. *Introduction to Nonimaging Optics*. CRC Press, Boca Raton, FL, 2008.
54. L Fraqas, L Partain. *Solar Cells and Their Applications*, 2nd ed., John Wiley, New York, 2010.

55. ANSI/IEEE Std. 928. IEEE Recommended Criteria for Terrestrial Photovoltaic Power Systems, 1986.
56. IEEE Std. 1262. IEEE Recommended Practice for Qualifications of Photovoltaic (PV) Modules, 1995.
57. IEEE Std. 1547.2. IEEE Application Guide for IEEE Std. 1547 TM, IEEE Standard for Interconnecting Distributed Resources with Electrical Power Systems, 2008.
58. IEEE Std. 1021. IEEE Recommended Practice for Utility Interconnection of Small Wind Energy Conversion Systems, 1989.
59. ANSI C84.1. American National Standard for Electrical Power Systems and Equipment-Voltage Ratings, 2006.
60. IEEE Std. 1547. IEEE Standard for Interconnecting Distributed Resources with Electric Power Systems, 2003.
61. IEEE Std. C37.90.2. IEEE Standard Withstand Capability of Relay System's to Radiated Electromagnetic Interference from Transceivers, 2004.
62. IEEE Std. C62.41.2. IEEE Recommended Practice on Characterization of Surges in Low Voltage (1000 V and less) AC Power Circuits, 2002.
63. IEEE Std. C37.90.1. IEEE Standard Surge Withstand Capability (SWC) Tests for Protective Relays and Relay Systems Associated with Electrical Power Apparatus, 2002.
64. ANSI/IEEE Std. 929. IEEE Recommended Practice for Utility Interface of Residential and Intermediate Photovoltaic (PV) Systems, 1988.

4

Short-Circuit Currents and Symmetrical Components

Short circuits occur in well-designed power systems and cause large *decaying* transient currents, generally much above the system load currents. This results in disruptive electrodynamic and thermal stresses that are potentially damaging. Fire risks and explosions are inherent. One tries to limit short circuits to the faulty section of the electrical system by appropriate switching devices capable of operating under short-circuit conditions without causing any damage and isolating only the faulty section, so that a fault is not escalated. The faster the operation of sensing and switching devices, the lower is the fault damage, and the better is the chance of systems holding together without loss of synchronism.

Short circuits can be studied from the following angles:

1. Calculation of short-circuit currents
2. Interruption of short-circuit currents and rating structure of switching devices
3. Effects of short-circuit currents
4. Limitation of short-circuit currents, i.e., with current-limiting fuses and fault current limiters
5. Short-circuit withstand ratings of electrical equipment like transformers, reactors, cables, and conductors
6. Transient stability of interconnected systems to remain in synchronism until the faulty section of the power system is isolated, briefly discussed in volume 4

We will confine our discussions to the calculations of short-circuit currents and the basis of short-circuit ratings of switching devices, i.e., power circuit breakers and fuses. As the main purpose of short-circuit calculations is to select and apply these devices properly, it is meaningful for the calculations to be related to current interruption phenomena and the rating structures of interrupting devices. The objectives of short-circuit calculations, therefore, can be summarized as follows:

- Determination of short-circuit duties on switching devices, i.e., high-, medium-, and low-voltage circuit breakers and fuses
- Calculation of short-circuit currents required for protective relaying and coordination of protective devices
- Evaluations of adequacy of short-circuit withstand ratings of static equipment like cables, conductors, bus bars, reactors, and transformers
- Calculations of fault voltage dips and their time-dependent recovery profiles

The type of short-circuit currents required for each of these objectives may not be immediately clear but will unfold in the chapters to follow.

In a three-phase system, a fault may equally involve all three phases. A *bolted fault* means as if three phases were connected together with links of zero impedance prior to the fault, i.e.,

131

the fault impedance itself is zero and the fault is limited by the system and machine impedances only. Such a fault is called a symmetrical *three-phase bolted fault* or *solid fault*. Bolted three-phase faults are rather uncommon. Generally, such faults give the maximum short-circuit currents and form the basis of calculations for short-circuit duties on switching devices.

Faults involving one, or more than one, phase and ground are called unsymmetrical faults. Under certain conditions, the line-to-ground fault or double line-to-ground fault currents may exceed three-phase symmetrical fault currents, discussed in the chapters to follow. Unsymmetrical faults are more common as compared to three-phase faults, i.e., a support insulator on one of the phases on a transmission line may start flashing to ground, ultimately resulting in a single line-to-ground fault.

Short-circuit calculations are, thus, the primary study whenever a new power system is designed or an expansion and upgrade of an existing system are planned.

4.1 Nature of Short-Circuit Currents

The transient analysis of the short circuit of a passive impedance connected to an alternating current (AC) source gives an initial insight into the nature of the short-circuit currents. Consider a sinusoidal time-invariant single-phase 60 Hz source of power, $E_m \sin \omega t$, connected to a single-phase short distribution line, $Z = (R + j\omega L)$, where Z is the complex impedance, R and L are the resistance and inductance, E_m is the peak source voltage, and ω is the angular frequency $= 2\pi f$, f being the frequency of the AC source. For a balanced three-phase system, a single-phase model is adequate, as we will discuss further. Let a short circuit occur at the far end of the line terminals. As an ideal voltage source is considered, i.e., zero Thévenin impedance, the short-circuit current is limited only by Z, and its steady-state value is vectorially given by E_m/Z. This assumes that the impedance Z does not change with flow of the large short-circuit current. For simplification of empirical short-circuit calculations, the impedances of static components like transmission lines, cables, reactors, and transformers are assumed to be time invariant. Practically, this is not true, i.e., the flux densities and saturation characteristics of core materials in a transformer may entirely change its leakage reactance. Driven to saturation under high current flow, distorted waveforms and harmonics may be produced.

Ignoring these effects and assuming that Z is time invariant during a short circuit, the transient and steady-state currents are given by the differential equation of the R–L circuit with an applied sinusoidal voltage:

$$L\frac{di}{dt} + Ri = E_m \sin(\omega t + \theta) \tag{4.1}$$

where θ is the angle on the voltage wave, at which the fault occurs. The solution of this differential equation is given by

$$i = I_m \sin(\omega t + \theta - \phi) - I_m \sin(\theta - \phi)e^{-Rt/L} \tag{4.2}$$

where I_m is the maximum steady-state current, given by E_m/Z, and the angle

$$\phi = \tan^{-1}(\omega L)/R$$

In power systems, $\omega L \gg R$. A 100 MVA (mega volt ampere), 0.85 power factor synchronous generator may have an X/R of 110, and a transformer of the same rating, an X/R of 45. The X/R ratios in low-voltage systems are of the order of 2–8. For present discussions, assume a high X/R ratio, i.e., $\phi \approx 90°$.

If short circuit occurs at an instant $t = 0$, $\theta = 0$ (i.e., when the voltage wave is crossing through zero amplitude on the X-axis), the instantaneous value of the short-circuit current from Equation 4.2 is $2I_m$. This is sometimes called as the *doubling effect*.

If short circuit occurs at an instant when the voltage wave peaks, $t = 0$, $\theta = \pi/2$, the second term in Equation 4.2 is zero and there is no transient component.

These two situations are shown in Figure 4.1a and b. The voltage at the point of bolted fault will be zero. The voltage E shown in Figure 4.1a and b signifies that prior to and after the fault is cleared, the voltage remains constant.

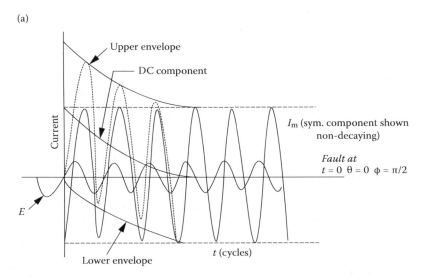

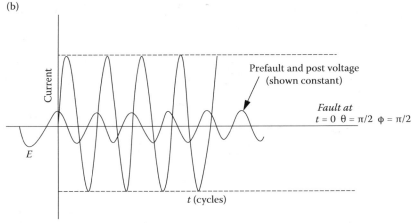

FIGURE 4.1
(a) Terminal short circuit of time varying impedance, current waveforms with maximum asymmetry and (b) current waveform with no DC component.

A simple explanation of the origin of the transient component is that in power systems, the inductive component of the impedance is high. The current in such a circuit is zero when the voltage is at peak, and for a fault at this instant, no direct current (DC) component is required to satisfy the physical law that states the current in an inductive circuit cannot change suddenly. When the fault occurs at an instant θ = 0, there has to be a transient current whose initial value is equal and opposite to the instantaneous value of the AC short-circuit current. This transient current, the second term of Equation 4.2, can be called a DC component that decays at an exponential rate. Equation 4.2 can be simply written as

$$i = I_m \sin \omega t + I_{dc} e^{-Rt/L} \qquad (4.3)$$

where the initial value of

$$I_{dc} = I_m \qquad (4.4)$$

The following inferences can be drawn from the above discussions:

1. There are two distinct components of a short-circuit current: (1) a nondecaying AC component or the steady-state component, and (2) a decaying DC component at an exponential rate, the initial magnitude of which is a maximum of the AC component and it depends on the time on the voltage wave at which the fault occurs.

2. The decrement factor of a decaying exponential current can be defined as its value any time after a short circuit expressed as a function of its initial magnitude per unit. Factor L/R can be termed the time constant. The exponential factor then becomes $e^{-t/t'}$, where $t' = L/R$. In this equation, making $t = t'$ = time constant will result in a decay of approximately 62.3% from its initial magnitude, i.e., the transitory current is reduced to a value of 0.368 per unit after an elapsed time equal to the time constant, as shown in Figure 4.2.

3. The presence of a DC component makes the fault current wave shape envelope asymmetrical about the zero line and axis of the wave. Figure 4.1a clearly shows the profile of an asymmetrical waveform. The DC component always decays to zero in a short time. Consider a modest X/R ratio of 15, say for a medium-voltage

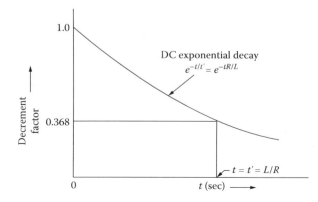

FIGURE 4.2
Time constant of DC component decay.

13.8-kV system. The DC component decays to 88% of its initial value in five cycles. The higher the X/R ratio, the slower is the decay, and the longer is the time for which the asymmetry in the total current will be sustained. The stored energy can be thought to be expanded in I^2R losses. After the decay of the DC component, only the symmetrical component of the short-circuit current remains.

4. Impedance is considered as time invariant in the above scenario. Synchronous generators and dynamic loads, i.e., synchronous and induction motors are the major sources of short-circuit currents. The trapped flux in these rotating machines at the instant of short circuit cannot change suddenly and decays, depending on the machine time constants. Thus, the assumption of constant L is not valid for rotating machines and decay in the AC component of the short-circuit current must also be considered.

5. In a three-phase system, the phases are time displaced from each other by 120 electrical degrees. If a fault occurs when the unidirectional component in phase a is zero, the phase b component is positive and the phase c component is equal in magnitude and negative. Figure 4.3 shows a three-phase fault current waveform. As the fault is symmetrical, $I_a + I_b + I_c$ is zero at any instant, where I_a, I_b, and I_c are the short-circuit currents in phases a, b, and c, respectively. For a fault close to a synchronous generator, there is a 120 Hz current also, which rapidly decays to zero. This gives rise to the characteristic nonsinusoidal shape of three-phase short-circuit currents observed in test oscillograms. The effect is insignificant and ignored in the short-circuit calculations. This is further discussed in Chapter 9.

6. The load current has been ignored. Generally, this is true for empirical short-circuit calculations, as the short-circuit current is much higher than the load current. Sometimes the load current is a considerable percentage of the short-circuit current. The load currents determine the effective voltages of the short-circuit sources, prior to fault.

The AC short-circuit current sources are synchronous machines, i.e., turbo generators and salient pole generators, asynchronous generators, and synchronous and asynchronous motors. Converter motor drives may contribute to short-circuit currents when operating in the inverter or regenerative mode. For extended duration of short-circuit currents, the control and excitation systems, generator voltage regulators, and turbine governor characteristics affect the transient short-circuit process.

The duration of a short-circuit current depends mainly on the speed of operation of protective devices and on the interrupting time of the switching devices.

4.2 Symmetrical Components

The method of symmetrical components has been widely used in the analysis of unbalanced three-phase systems, unsymmetrical short-circuit currents, and rotating electrodynamic machinery. The method was originally presented by C.L. Fortescue in 1918 and has been popular ever since.

Unbalance occurs in three-phase power systems due to faults, single-phase loads, untransposed transmission lines, or nonequilateral conductor spacings. In a three-phase

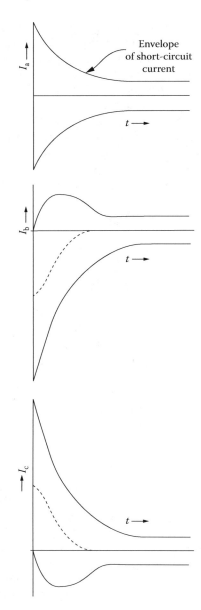

FIGURE 4.3
Asymmetries in phase currents in a three-phase short circuit.

balanced system, it is sufficient to determine the currents and voltages in one phase, and the currents and voltages in the other two phases are simply phase displaced. In an unbalanced system, the simplicity of modeling a three-phase system as a single-phase system is not valid. A convenient way of analyzing unbalanced operation is through symmetrical components. The three-phase voltages and currents that may be unbalanced are transformed into three sets of balanced voltages and currents called symmetrical components. The impedances of various power system components, i.e., transformers, generators, and

Short-Circuit Currents and Symmetrical Components

transmission lines, to symmetrical components are *decoupled* from each other, resulting in independent networks for each component. These form a balanced set. This simplifies the calculations.

Familiarity with electrical circuits and machine theory, per unit system, and matrix techniques is required. A review of the matrix techniques in power systems is included in Appendix A. The notations described in this appendix for vectors and matrices are followed throughout.

The basic theory of symmetrical components can be stated as a mathematical concept. A system of three coplanar vectors is completely defined by six parameters, and the system can be said to possess six degrees of freedom. A point in a straight line being constrained to lie on the line possesses but one degree of freedom, and by the same analogy, a point in space has three degrees of freedom. A coplanar vector is defined by its terminal and length and, therefore, possesses two degrees of freedom. A system of coplanar vectors having six degrees of freedom, i.e., a three-phase unbalanced current or voltage vectors, can be represented by three symmetrical systems of vectors each having two degrees of freedom. In general, a system of n numbers can be resolved into n sets of component numbers each having n components, i.e., a total of n^2 components. Fortescue demonstrated that an unbalanced set of n phasors can be resolved into $n-1$ balanced phase systems of different phase sequence and one zero sequence system, in which all phasors are of equal magnitude and cophasal

$$V_a = V_{a1} + V_{a2} + V_{a3} + \cdots + V_{an}$$

$$V_b = V_{b1} + V_{b2} + V_{b3} + \cdots + V_{bn} \tag{4.5}$$

$$V_n = V_{n1} + V_{n2} + V_{n3} + \cdots + V_{nn}$$

where

$V_a, V_b, \ldots, V_n$ are original n unbalanced voltage phasors
$V_{a1}, V_{b1}, \ldots, V_{n1}$ are the first set of n *balanced* phasors at an angle of $2\pi/n$ between them
$V_{a2}, V_{b2}, \ldots, V_{n2}$ are the second set of n balanced phasors at an angle $4\pi/n$
$V_{an}, V_{bn}, \ldots, V_{nn}$ is the zero sequence set of all phasors at $n(2\pi/n) = 2\pi$, i.e., cophasal

In a symmetrical three-phase balanced system, the generators produce balanced voltages, which are displaced from each other by $2\pi/3 = 120°$. These voltages can be called positive sequence voltages. If a vector operator a is defined, which rotates a unit vector through $120°$ in a counterclockwise direction, then $a = -0.5 + j0.866$, $a^2 = -0.5 - j0.866$, $a^3 = 1$, $1 + a^2 + a = 0$. Considering a three-phase system, Equation 4.5 reduces to

$$V_a = V_{a0} + V_{a1} + V_{a2}$$

$$V_b = V_{b0} + V_{b1} + V_{b2} \tag{4.6}$$

$$V_c = V_{c0} + V_{c1} + V_{c2}$$

We can define the set consisting of V_{a0}, V_{b0}, and V_{c0} as the zero sequence set, the set of V_{a1}, V_{b1}, and V_{c1} as the positive sequence set, and the set of V_{a2}, V_{b2}, and V_{c2} as the negative sequence set of voltages. The three original unbalanced voltage vectors give rise to nine voltage vectors, which must have constraints of freedom and are not totally independent. By definition of positive sequence, V_{a1}, V_{b1}, and V_{c1} in a normal balanced system should be related as follows:

$$V_{b1} = a^2 V_{a1}, V_{c1} = a V_{a1}$$

Note that V_{a1} phasor is taken as the reference vector.

The negative sequence set can be defined similarly but with the opposite phase sequence,

$$V_{b2} = a V_{a2}, V_{c2} = a^2 V_{a2}$$

Also, $V_{a0} = V_{b0} = V_{c0}$. With these relations defined, Equation 4.6 can be written as

$$\begin{vmatrix} V_a \\ V_b \\ V_c \end{vmatrix} = \begin{vmatrix} 1 & 1 & 1 \\ 1 & a^2 & a \\ 1 & a & a^2 \end{vmatrix} \begin{vmatrix} V_{a0} \\ V_{a1} \\ V_{a2} \end{vmatrix} \tag{4.7}$$

The same equation can be written in the abbreviated form as follows:

$$\bar{V}_{abc} = \bar{T}_s \bar{V}_{012} \tag{4.8}$$

where $\bar{T}_s$ is the transformation matrix. Its inverse will give the reverse transformation.

While this simple explanation may be adequate, a better insight into the symmetrical component theory can be gained through matrix concepts of similarity transformation, diagonalization, eigenvalues, and eigenvectors.

The discussions are the following:

- Eigenvectors giving rise to symmetrical component transformation are the same though the eigenvalues differ. Thus, these vectors are not unique.
- The Clarke component transformation is based on the same eigenvectors but different eigenvalues.
- The symmetrical component transformation does not decouple an initially unbalanced three-phase system. *Prima facie*, this is a contradiction of what we said earlier that the main advantage of symmetrical components lies in decoupling unbalanced systems, which could then be represented much akin to three-phase balanced systems. This would be clear as we proceed.

4.3 Eigenvalues and Eigenvectors

The concept of eigenvalues and eigenvectors is related to the derivation of symmetrical component transformation. It can be briefly stated as follows:

Consider an arbitrary square matrix $\bar{A}$. If a relation exists such that

$$\bar{A}\bar{x} = \lambda \bar{x} \tag{4.9}$$

where λ is a scalar quantity called an eigenvalue, characteristic value, or root of the matrix $\bar{A}$, and $\bar{x}$ is a vector called the eigenvector or characteristic vector of $\bar{A}$.

Short-Circuit Currents and Symmetrical Components 139

Then, there are n eigenvalues and corresponding n sets of eigenvectors associated with an arbitrary matrix $\bar{A}$ of dimension $n \times n$. The eigenvalues are not necessarily distinct, and multiple roots may occur.

Equation 4.9 can be written as

$$\lfloor \bar{A} - \lambda I \rfloor [\bar{x}] = 0 \tag{4.10}$$

where I the is identity matrix. On expanding, we get

$$\begin{vmatrix} a_{11} - \lambda & a_{12} & a_{13} & \cdots & a_1 n \\ a_{12} & a_{22} - \lambda & a_{23} & \cdots & a_2 n \\ \cdots & \cdots & \cdots & \cdots & \cdots \\ a_{n1} & a_{n2} & a_{n3} & \cdots & a_{nn} - \lambda \end{vmatrix} \begin{vmatrix} x_1 \\ x_2 \\ \cdots \\ x_n \end{vmatrix} = \begin{vmatrix} 0 \\ 0 \\ \cdots \\ 0 \end{vmatrix} \tag{4.11}$$

This represents a set of homogeneous linear equations. Determinant $|A - \lambda I|$ must be zero as $\bar{x} \neq 0$.

$$\left| \bar{A} - \lambda I \right| = 0 \tag{4.12}$$

This can be expanded to yield an nth order algebraic equation:

$$a_n \lambda^n + a_n - I \lambda^n - 1 + \cdots + a_1 \lambda + a_0 = 0, \text{ i.e.,} \tag{4.13}$$
$$(\lambda_1 - a_1)(\lambda_2 - a_2)\ldots(\lambda_n - a_n) = 0$$

Equations 4.12 and 4.13 are called the *characteristic equations* of the matrix $\bar{A}$. The roots λ_1, $\lambda_2, \lambda_3, \ldots, \lambda_n$ are the eigenvalues of matrix $\bar{A}$. The eigenvector $\bar{x}_j$ corresponding to $\bar{\lambda}_j$ is found from Equation 4.10 (see Appendix A for details and an example).

4.4 Symmetrical Component Transformation

Application of eigenvalues and eigenvectors to the decoupling of three-phase systems is useful when we define similarity transformation. This forms a diagonalization technique and decoupling through symmetrical components.

4.4.1 Similarity Transformation

Consider a system of linear equations:

$$\bar{A}\bar{x} = \bar{y} \tag{4.14}$$

A transformation matrix $\bar{C}$ can be introduced to relate the original vectors $\bar{x}$ and $\bar{y}$ to new sets of vectors $\bar{x}_n$ and $\bar{y}_n$ such that

$$\bar{x} = \bar{C}\bar{x}_n$$

$$\bar{y} = \bar{C}\bar{y}_n$$

$$\bar{A}\bar{C}\bar{x}_n = \bar{C}\bar{y}_n$$

$$\bar{C}^{-1}\bar{A}\bar{C}\bar{x}_n = \bar{C}^{-1}\bar{C}\bar{y}_n$$

$$\bar{C}^{-1}\bar{A}\bar{C}\bar{x}_n = \bar{y}_n$$

This can be written as

$$\bar{A}_n\bar{x}_n = \bar{y}_n$$

$$\bar{A}_n = \bar{C}^{-1}\bar{A}\bar{C} \tag{4.15}$$

$\bar{A}_n\bar{x}_n = \bar{y}_n$ is distinct from $\bar{A}\bar{x} = \bar{y}$. The only restriction on choosing $\bar{C}$ is that it should be nonsingular. Equation 4.15 is a set of linear equations derived from the original Equation (4.14) and yet distinct from them.

If $\bar{C}$ is a nodal matrix $\bar{M}$, corresponding to the coefficients of $\bar{A}$, then

$$\bar{C} = \bar{M} = [x_1, x_2 \ldots x_n] \tag{4.16}$$

where $\bar{x}_i$ are the eigenvectors of the matrix $\bar{A}$, then

$$\bar{C}^{-1}\bar{A}\bar{C} = \bar{C}^{-1}\bar{A}[x_1, x_2, \ldots, x_n]$$

$$\bar{C}^{-1}[\bar{A}x_1, \bar{A}x_2, \ldots, \bar{A}x_n]$$

$$= \bar{C}^{-1}[\lambda_1 x_1, \lambda_2 x_2, \ldots, \lambda_n x_n]$$

$$= C^{-1}[x_1, x_2, \ldots x_n] \begin{vmatrix} \lambda_1 & & & \\ & \lambda_2 & & \\ & & \cdot & \\ & & & \lambda_n \end{vmatrix} \tag{4.17}$$

$$= \bar{C}^{-1}\bar{C} \begin{vmatrix} \lambda_1 & & & \\ & \lambda_2 & & \\ & & \cdot & \\ & & & \lambda_n \end{vmatrix}$$

$$= \bar{\lambda}$$

Thus, $\bar{C}^{-1}\bar{A}\bar{C}$ is reduced to a diagonal matrix $\bar{\lambda}$ called a *spectral matrix*. Its diagonal elements are the eigenvalues of the original matrix $\bar{A}$. The new system of equations is an

Short-Circuit Currents and Symmetrical Components 141

uncoupled system. Equations 4.14 and 4.15 constitute a similarity transformation of matrix $\bar{A}$. The matrices $\bar{A}$ and $\bar{A}1_n$ have the same eigenvalues and are called similar matrices. The transformation matrix $\bar{C}$ is nonsingular.

4.4.2 Decoupling a Three-Phase Symmetrical System

Let us decouple a three-phase transmission line section, where each phase has a mutual coupling with respect to ground. This is shown in Figure 4.4a. An impedance matrix of the three-phase transmission line can be written as

$$\begin{vmatrix} Z_{aa} & Z_{ab} & Z_{ac} \\ Z_{ba} & Z_{bb} & Z_{bc} \\ Z_{ca} & Z_{cb} & Z_{cc} \end{vmatrix} \tag{4.18}$$

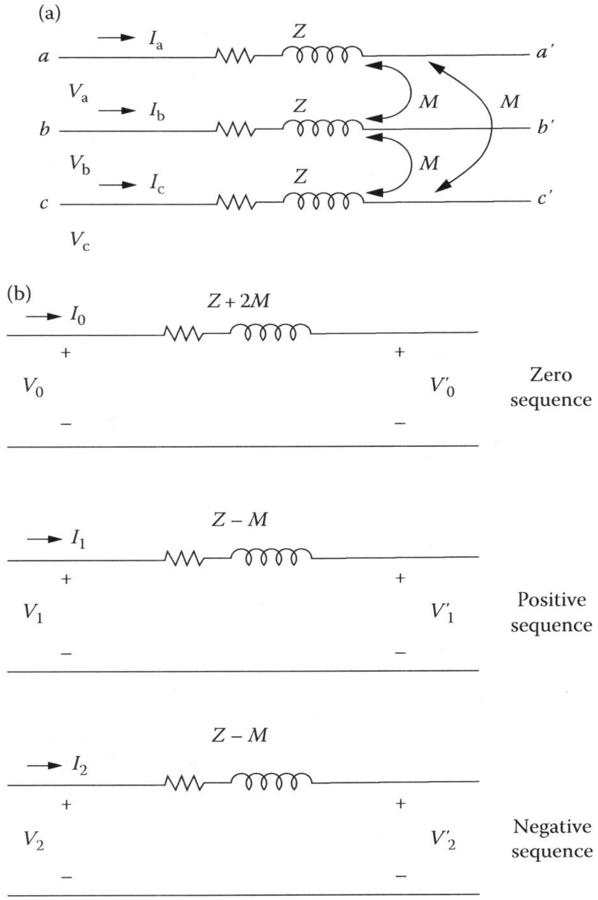

FIGURE 4.4
(a) Impedances in a three-phase transmission line with mutual coupling between phases and (b) resolution into symmetrical component network.

where Z_{aa}, Z_{bb}, and Z_{cc} are the self-impedances of the phases a, b, and c, respectively; Z_{ab} is the mutual impedance between phases a and b, and Z_{ba} is the mutual impedance between phases b and a.

Assume that the line is *perfectly symmetrical*. This means that all the mutual impedances, i.e., $Z_{ab} = Z_{ba} = M$ and all the self-impedances, i.e., $Z_{aa} = Z_{bb} = Z_{cc} = Z$ are equal. This reduces the impedance matrix to

$$\begin{vmatrix} Z & M & M \\ M & Z & M \\ M & M & Z \end{vmatrix} \tag{4.19}$$

It is required to decouple this system using symmetrical components. First, find the eigenvalues:

$$\begin{vmatrix} Z-\lambda & M & M \\ M & Z-\lambda & M \\ M & M & Z-\lambda \end{vmatrix} = 0 \tag{4.20}$$

The eigenvalues are

$$\lambda = Z + 2M$$
$$= Z - M$$
$$= Z - M$$

The eigenvectors can be found by making $\lambda = Z + 2M$ and then by $Z - M$. Substituting $\lambda = Z + 2M$ in Equation 4.20,

$$\begin{vmatrix} Z-(Z+2M) & M & M \\ M & Z-(Z+2M) & M \\ M & M & Z-(Z+2M) \end{vmatrix} \begin{vmatrix} X_1 \\ X_2 \\ X_3 \end{vmatrix} = 0 \tag{4.21}$$

This can be reduced to

$$\begin{vmatrix} -2 & 1 & 1 \\ 0 & -1 & 1 \\ 0 & 0 & 0 \end{vmatrix} \begin{vmatrix} X_1 \\ X_2 \\ X_3 \end{vmatrix} = 0 \tag{4.22}$$

Short-Circuit Currents and Symmetrical Components   143

This gives $X_1 = X_2 = X_3 =$ any arbitrary constant k. Thus, one of the eigenvectors of the impedance matrix is

$$\begin{vmatrix} k \\ k \\ k \end{vmatrix} \tag{4.23}$$

It can be called the *zero sequence* eigenvector of the symmetrical component transformation matrix and can be written as

$$\begin{vmatrix} 1 \\ 1 \\ 1 \end{vmatrix} \tag{4.24}$$

Similarly, for $\lambda = Z - M$,

$$\begin{vmatrix} Z - (Z - M) & M & M \\ M & Z - (Z - M) & M \\ M & M & Z - (Z - M) \end{vmatrix} \begin{vmatrix} X_1 \\ X_2 \\ X_3 \end{vmatrix} = 0 \tag{4.25}$$

which gives

$$\begin{vmatrix} 1 & 1 & 1 \\ 0 & 0 & 0 \\ 0 & 0 & 0 \end{vmatrix} \begin{vmatrix} X_1 \\ X_2 \\ X_3 \end{vmatrix} = 0 \tag{4.26}$$

This gives the general relation $X_1 + X_2 + X_3 = 0$. Any value of X_1, X_2, X_3 that satisfies this relation is a solution vector. Some examples are as shown below:

$$\begin{vmatrix} X_1 \\ X_2 \\ X_3 \end{vmatrix} = \begin{vmatrix} 1 \\ a^2 \\ a \end{vmatrix}, \begin{vmatrix} 1 \\ a \\ a^2 \end{vmatrix}, \begin{vmatrix} 0 \\ \dfrac{\sqrt{3}}{2} \\ -\dfrac{\sqrt{3}}{2} \end{vmatrix}, \begin{vmatrix} 1 \\ -\dfrac{1}{2} \\ -\dfrac{1}{2} \end{vmatrix} \tag{4.27}$$

where a is a unit vector operator, which rotates by $120°$ in the counterclockwise direction, as defined before.

Equation 4.27 is an important result and shows that, for perfectly symmetrical systems, the common eigenvectors are the same, although the eigenvalues are different in each system. The Clarke component transformation (described in Section 4.5) is based on this observation.

The symmetrical component transformation is given by the following solution vectors:

$$\begin{vmatrix} 1 \\ 1 \\ 1 \end{vmatrix} \begin{vmatrix} 1 \\ a \\ a^2 \end{vmatrix} \begin{vmatrix} 1 \\ a^2 \\ a \end{vmatrix} \tag{4.28}$$

A symmetrical component transformation matrix can, therefore, be written as

$$\bar{T}_s = \begin{vmatrix} 1 & 1 & 1 \\ 1 & a^2 & a \\ 1 & a & a^2 \end{vmatrix} \tag{4.29}$$

This is the same matrix as that was derived in Equation 4.8. Its inverse is

$$\bar{T}_s^{-1} = \frac{1}{3} \begin{vmatrix} 1 & 1 & 1 \\ 1 & a & a^2 \\ 1 & a^2 & a \end{vmatrix} \tag{4.30}$$

For the transformation of currents, we can write as follows:

$$\bar{I}_{abc} = \bar{T}_s \bar{I}_{012}, \tag{4.31}$$

where $\bar{I}_{abc}$, the original currents in phases a, b, and c, are transformed into zero sequence, positive sequence, and negative sequence currents, $\bar{I}_{012}$. The original phasors are subscripted as abc and the sequence components are subscripted as 012. Similarly, for transformation of voltages,

$$\bar{V}_{abc} = \bar{T}_s \bar{V}_{012} \tag{4.32}$$

Conversely,

$$\bar{I}_{012} = \bar{T}_s^{-1} \bar{I}_{abc}, \quad \bar{V}_{012} = \bar{T}_s^{-1} \bar{V}_{abc} \tag{4.33}$$

The transformation of impedance is not straightforward and is derived as follows:

$$\bar{V}_{abc} = \bar{Z}_{abc} \bar{I}_{abc}$$

$$\bar{T}_s \bar{V}_{012} = \bar{Z}_{abc} \bar{T}_s \bar{I}_{012} \tag{4.34}$$

$$\bar{V}_{012} = \bar{T}_s^{-1} \bar{Z}_{abc} \bar{T}_s \bar{I}_{012} = \bar{Z}_{012} \bar{I}_{012}$$

Short-Circuit Currents and Symmetrical Components 145

Therefore,

$$\bar{Z}_{012} = \bar{T}_s^{-1} \bar{Z}_{abc} \bar{T}_s \tag{4.35}$$

$$\bar{Z}_{abc} = \bar{T}_s \bar{Z}_{012} \bar{T}_s^{-1} \tag{4.36}$$

Applying the impedance transformation to the original impedance matrix of the three-phase symmetrical transmission line in Equation 4.19, the transformed matrix is as follows:

$$\bar{Z}_{012} = \frac{1}{3}
\begin{vmatrix} 1 & 1 & 1 \\ 1 & a & a^2 \\ 1 & a^2 & a \end{vmatrix}
\begin{vmatrix} Z & M & M \\ M & Z & M \\ M & M & Z \end{vmatrix}
\begin{vmatrix} 1 & 1 & 1 \\ 1 & a^2 & a \\ 1 & a & a^2 \end{vmatrix}$$

$$= \begin{vmatrix} Z+2M & 0 & 0 \\ 0 & Z-M & 0 \\ 0 & 0 & Z-M \end{vmatrix} \tag{4.37}$$

The original three-phase coupled system has been decoupled through symmetrical component transformation. It is diagonal, and all off-diagonal terms are zero, meaning that there is no coupling between the sequence components. Decoupled positive, negative, and zero sequence networks are shown in Figure 4.4b.

4.4.3 Decoupling a Three-Phase Unsymmetrical System

Now consider that the original three-phase system is not completely balanced. Ignoring the mutual impedances in Equation 4.18, let us assume unequal phase impedances, Z_1, Z_2, and Z_3, i.e., the impedance matrix is

$$\bar{Z}_{abc} = \begin{vmatrix} Z_1 & 0 & 0 \\ 0 & Z_2 & 0 \\ 0 & 0 & Z_3 \end{vmatrix} \tag{4.38}$$

The symmetrical component transformation is

$$\bar{Z}_{012} = \frac{1}{3}
\begin{vmatrix} 1 & 1 & 1 \\ 1 & a & a^2 \\ 1 & a^2 & a \end{vmatrix}
\begin{vmatrix} Z_1 & 0 & 0 \\ 0 & Z_2 & 0 \\ 0 & 0 & Z_3 \end{vmatrix}
\begin{vmatrix} 1 & 1 & 1 \\ 1 & a^2 & a \\ 1 & a & a^2 \end{vmatrix}$$

$$= \frac{1}{3}
\begin{vmatrix} Z_1 + Z_2 + Z_3 & Z_1 + a^2 Z_2 + a Z_3 & Z_1 + a Z_2 + Z_3 \\ Z_1 + a Z_2 + a Z_3 & Z_1 + Z_2 + Z_3 & Z_1 + a^2 Z_2 + a Z_3 \\ Z_1 + a^2 Z_2 + a Z_3 & Z_1 + a Z_2 + a Z_3 & Z_1 + Z_2 + Z_3 \end{vmatrix} \tag{4.39}$$

The resulting matrix shows that the *original unbalanced system is not decoupled.* If we start with equal self-impedances and unequal mutual impedances or vice versa, the resulting matrix is nonsymmetrical. It is a minor problem today, as nonreciprocal networks can be easily handled on digital computers. Nevertheless, the main application of symmetrical components is for the study of unsymmetrical faults. Negative sequence relaying, stability calculations, and machine modeling are some other examples. It is assumed that the system is perfectly symmetrical before an unbalance condition occurs. *The asymmetry occurs only at the fault point.* The symmetrical portion of the network is considered to be isolated, to which an unbalanced condition is applied at the fault point. In other words, the unbalanced part of the network can be assumed to be connected to the balanced system at the point of fault. Practically, the power systems are not perfectly balanced and some asymmetry always exists. However, the error introduced by ignoring this asymmetry is small. (This may not be true for highly unbalanced systems and single-phase loads.)

4.4.4 Power Invariance in Symmetrical Component Transformation

Symmetrical component transformation is power invariant. The complex power in a three-phase circuit is given by

$$S = V_a I_a^* + V_b I_b^* + V_c I_c^* = \bar{V}'_{abc} \bar{I}^*_{abc} \tag{4.40}$$

where I_a^* is the complex conjugate of I_a. This can be written as

$$S = \left[\bar{T}_s \bar{V}_{012} \right] \bar{T}_s^* \bar{I}^*_{012} = \bar{V}'_{012} \bar{T}'_s \bar{T}_s^* \bar{I}^*_{012} \tag{4.41}$$

The product $\bar{T}_s \bar{T}_s^*$ is given by (see Appendix A)

$$\bar{T}'_s \bar{T}_s^* = 3 \begin{vmatrix} 1 & 0 & 0 \\ 0 & 1 & 0 \\ 0 & 0 & 1 \end{vmatrix} \tag{4.42}$$

Thus,

$$S = 3 V_1 I_1^* + 3 V_2 I_2^* + 3 V_0 I_0^* \tag{4.43}$$

This shows that complex power can be calculated from symmetrical components and is the sum of the symmetrical component powers.

4.5 Clarke Component Transformation

It has been already shown that, for perfectly symmetrical systems, the component eigenvectors are the same, but eigenvalues can be different. The Clarke component transformation is defined as

Short-Circuit Currents and Symmetrical Components 147

$$
\begin{vmatrix} V_a \\ V_b \\ V_c \end{vmatrix} = \begin{vmatrix} 1 & 1 & 0 \\ 1 & -\dfrac{1}{2} & \dfrac{\sqrt{3}}{2} \\ 1 & -\dfrac{1}{2} & -\dfrac{\sqrt{3}}{2} \end{vmatrix} \begin{vmatrix} V_0 \\ V_\alpha \\ V_\beta \end{vmatrix} \tag{4.44}
$$

Note that the eigenvalues satisfy the relations derived in Equation 4.27, and

$$
\begin{vmatrix} V_0 \\ V_\alpha \\ V_\beta \end{vmatrix} = \begin{vmatrix} \dfrac{1}{3} & \dfrac{1}{3} & \dfrac{1}{3} \\ \dfrac{2}{3} & -\dfrac{1}{3} & -\dfrac{1}{3} \\ 0 & \dfrac{1}{\sqrt{3}} & -\dfrac{1}{\sqrt{3}} \end{vmatrix} \begin{vmatrix} V_a \\ V_b \\ V_c \end{vmatrix} \tag{4.45}
$$

Similar equations can be written for the current. Note that

$$
\begin{vmatrix} 1 \\ 1 \\ 1 \end{vmatrix}, \quad \begin{vmatrix} 1 \\ -\dfrac{1}{2} \\ -\dfrac{1}{2} \end{vmatrix}, \quad \begin{vmatrix} 0 \\ \dfrac{\sqrt{3}}{2} \\ -\dfrac{\sqrt{3}}{2} \end{vmatrix}
$$

are the eigenvectors of a perfectly symmetrical impedance.

The transformation matrices are

$$
\bar{T}_c = \begin{vmatrix} 1 & 1 & 0 \\ 1 & -\dfrac{1}{2} & \dfrac{\sqrt{3}}{2} \\ 1 & \dfrac{1}{2} & \dfrac{\sqrt{3}}{2} \end{vmatrix} \tag{4.46}
$$

$$
\bar{T}_c^{-1} = \begin{vmatrix} \dfrac{1}{3} & \dfrac{1}{3} & \dfrac{1}{3} \\ \dfrac{2}{3} & -\dfrac{1}{3} & -\dfrac{1}{3} \\ 0 & \dfrac{1}{\sqrt{3}} & -\dfrac{1}{\sqrt{3}} \end{vmatrix} \tag{4.47}
$$

148   *Short-Circuits in AC and DC Systems*

and as before,

$$\bar{Z}_{0\alpha\beta} = \bar{T}_c^{-1}\bar{Z}_{abc}\bar{T}_c \tag{4.48}$$

$$\bar{Z}_{abc} = \bar{T}_c\bar{Z}_{0\alpha\beta}\bar{T}_c^{-1} \tag{4.49}$$

The Clarke component expression for a perfectly symmetrical system is

$$\begin{vmatrix} V_0 \\ V_\alpha \\ V_\beta \end{vmatrix} = \begin{vmatrix} Z_{00} & 0 & 0 \\ 0 & Z_{\alpha\alpha} & 0 \\ 0 & 0 & Z_{\beta\beta} \end{vmatrix} \begin{vmatrix} I_0 \\ I_\alpha \\ I_\beta \end{vmatrix} \tag{4.50}$$

The same philosophy of transformation can also be applied to systems with two or more three-phase circuits in parallel. The instantaneous power theory, Volume 3 and Electromagnetic Transient Program modeling of transmission lines, Volume 2 are based upon Clarke's transformation.

Example 4.1

The symmetrical component transformation concepts outlined above demonstrate that the eigenvectors are not unique. Choose some different eigenvectors and design an entirely new transformation system.

Let us arbitrarily choose the eigenvectors as

$$\begin{vmatrix} 1 \\ 1 \\ 1 \end{vmatrix}, \begin{vmatrix} 1 \\ -\dfrac{1}{4} \\ -\dfrac{3}{4} \end{vmatrix}, \begin{vmatrix} 1 \\ -\dfrac{1}{2} \\ -\dfrac{1}{2} \end{vmatrix}$$

Then the transformation matrix is

$$\bar{T}_t = \begin{vmatrix} 1 & 1 & 1 \\ 1 & -\dfrac{1}{4} & -\dfrac{1}{2} \\ 1 & -\dfrac{3}{4} & -\dfrac{1}{2} \end{vmatrix}$$

Its inverse is

$$\bar{T}_t^{-1} = \begin{vmatrix} \dfrac{1}{3} & \dfrac{1}{3} & \dfrac{1}{3} \\ 0 & 2 & -2 \\ \dfrac{2}{3} & -2\dfrac{1}{3} & 1\dfrac{2}{3} \end{vmatrix}$$

Short-Circuit Currents and Symmetrical Components

149

Then,

$$\bar{Z}_{012} = \begin{vmatrix} \dfrac{1}{3} & \dfrac{1}{3} & \dfrac{1}{3} \\ 0 & 2 & -2 \\ \dfrac{2}{3} & -2\dfrac{1}{3} & 1\dfrac{2}{3} \end{vmatrix} \begin{Vmatrix} Z & M & M \\ M & Z & M \\ M & M & Z \end{Vmatrix} \begin{vmatrix} 1 & 1 & 1 \\ 1 & -0.25 & -0.5 \\ 1 & -0.75 & -0.5 \end{vmatrix}$$

$$= \begin{vmatrix} Z+2M & 0 & 0 \\ 0 & Z-M & 0 \\ 0 & 0 & Z-M \end{vmatrix}$$

This gives the same impedance transformation and decoupling as in Equation 4.37. *Then, what is so special about the symmetrical component transformation vectors?* The vectors arbitrarily chosen are further examined.

Let us denote the voltages in a balanced three-phase system as

$$V_a = V < 0°$$

$$V_b = V < -120° \qquad (4.51)$$

$$V_c = V < 120°$$

Then, apply symmetrical component transformation,

$$\begin{vmatrix} V_0 \\ V_1 \\ V_2 \end{vmatrix} = \dfrac{1}{3} \begin{vmatrix} 1 & 1 & 1 \\ 1 & -0.5+j0.866 & -0.5-j0.866 \\ 1 & -0.5-j0.866 & -0.5+j0.866 \end{vmatrix} \begin{vmatrix} V \\ V(-0.5-j0.866) \\ V(-0.5+j0.866) \end{vmatrix} = \begin{vmatrix} 0 \\ V \\ 0 \end{vmatrix} \quad (4.52)$$

That is, zero and negative sequence components are zero. In the symmetrical components, the operator "*a*" rotates a unity vector by 120 electrical degrees in the counterclockwise direction. By international convention, this is the accepted rotation of the vectors in a three-phase system. Also in a balanced system, each current and voltage vector is displaced by 120 electrical degrees. Thus, for a balanced system, these vectors can be called as positive sequence vectors (components), and the negative sequence and zero sequence components will be zero, as illustrated.

Now apply the transformation that we have arbitrarily chosen in this example:

$$\begin{vmatrix} V_0 \\ V_1 \\ V_2 \end{vmatrix} = \begin{vmatrix} \dfrac{1}{3} & \dfrac{1}{3} & \dfrac{1}{3} \\ 0 & 2 & -2 \\ \dfrac{2}{3} & -2\dfrac{1}{3} & 1\dfrac{2}{3} \end{vmatrix} \begin{vmatrix} V \\ V(-0.5-j0.866) \\ V(-0.5+j0.866) \end{vmatrix} = \begin{vmatrix} 0 \\ -j3.5V \\ V(1+j3.5) \end{vmatrix} \quad (4.53)$$

Meaningful interpretation cannot be applied to the end result of above Equation.

In symmetrical component transformation, the eigenvectors are selected to provide meaningful physical interpretation. The scientific and engineering logic of symmetrical components lies in selection of appropriate eigenvectors.

150   Short-Circuits in AC and DC Systems

4.6 Characteristics of Symmetrical Components

Matrix Equations 4.32 and 4.33 are written in the expanded form as follows:

$$V_a = V_0 + V_1 + V_2$$

$$V_b = V_0 + a^2 V_1 + a V_2 \tag{4.54}$$

$$V_c = V_0 + a V_1 + a^2 V_2$$

and

$$V_0 = \frac{1}{3}(V_a + V_b + V_c)$$

$$V_1 = \frac{1}{3}(V_a + a V_b + a^2 V_c) \tag{4.55}$$

$$V_2 = \frac{1}{3}(V_a + a^2 V_b + a V_c)$$

These relations are graphically represented in Figure 4.5, which clearly shows that phase voltages V_a, V_b, and V_c can be resolved into three voltages: V_0, V_1, and V_2 defined as follows:

- V_0 is the zero sequence voltage. It is of equal magnitude in all the three phases and is cophasal.
- V_1 is the system of balanced positive sequence voltages, *of the same phase sequence* as the original unbalanced system of voltages. It is of equal magnitude in each phase, but displaced by 120°, the component of phase b lagging the component of phase a by 120°, and the component of phase c leading the component of phase a by 120°.
- V_2 is the system of balanced negative sequence voltages. It is of equal magnitude in each phase, and there is a 120° phase displacement between the voltages, the component of phase c lagging the component of phase a, and the component of phase b leading the component of phase a.

Therefore, the positive and negative sequence voltages (or currents) can be defined as "the order in which the three phases attain a maximum value." For the positive sequence, the order is *abca*, whereas for the negative sequence, it is *acba*. We can also define positive and negative sequence by the order in which the phasors pass a *fixed point* on the vector plot. *Note that the rotation is counterclockwise for all three sets of sequence components, as was assumed for the original unbalanced vectors* [Figure 4.5d]. Sometimes, this is confused and negative sequence rotation is said to be the *reverse of* positive sequence. *The negative sequence vectors do not rotate in a direction opposite to the positive sequence vectors*, though the negative phase sequence is opposite to the positive phase sequence.

Example 4.2

An unbalanced three-phase system has the following voltages:

$$V_a = 0.9 < 0° \text{ per unit}$$

$$V_b = 1.25 < 280° \text{ per unit}$$

$$V_c = 0.6 < 110° \text{ per unit}$$

Short-Circuit Currents and Symmetrical Components   151

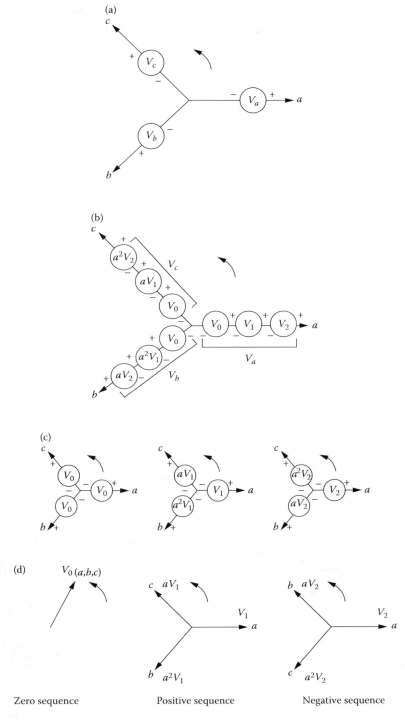

FIGURE 4.5
(a–d) Progressive resolution of voltage vectors into sequence components.

The phase rotation is *abc*, counterclockwise. The unbalanced system is shown in Figure 4.6a. Resolve into symmetrical components and sketch the sequence voltages.

Using the symmetrical component transformation, the resolution is shown in Figure 4.6b. One can verify this as an exercise and then convert back from the calculated sequence vectors into original *abc* voltages, graphically and analytically.

In a symmetrical system of three phases, the resolution of voltages or currents into a system of zero, positive, and negative components is equivalent to three separate systems. Sequence voltages act in isolation and produce zero, positive, and negative sequence currents, and the theorem of superposition applies. The following generalizations of symmetrical components can be made:

1. In a three-phase unfaulted system, in which all loads are balanced and generators produce positive sequence voltages, only positive sequence currents flow, resulting in balanced voltage drops of the same sequence. There are no negative sequence or zero sequence voltage drops.
2. In symmetrical systems, the currents and voltages of different sequences do not affect each other, i.e., positive sequence currents produce only positive sequence voltage drops. By the same analogy, the negative sequence currents produce only negative sequence drops, and zero sequence currents produce only zero sequence drops.

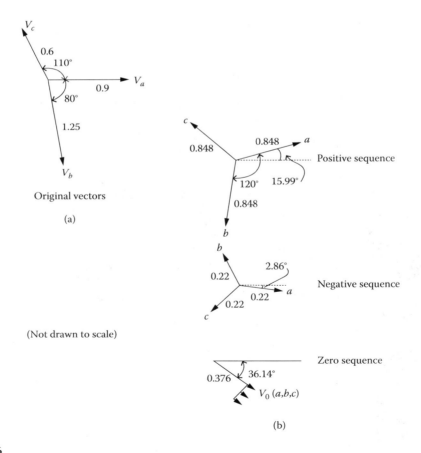

FIGURE 4.6
(a) Unbalanced voltage vectors and (b) resolution into symmetrical components.

Short-Circuit Currents and Symmetrical Components   153

3. Negative and zero sequence currents are set up in circuits of unbalanced impedances only, i.e., a set of unbalanced impedances in a symmetrical system may be regarded as a source of negative and zero sequence current. Positive sequence currents flowing in an unbalanced system produce positive, negative, and possibly zero sequence voltage drops. The negative sequence currents flowing in an unbalanced system produce voltage drops of all three sequences. The same is true about zero sequence currents.

4. In a three-phase three-wire system, no zero sequence currents appear in the line conductors. This is so because $I_0 = (1/3)(I_a + I_b + I_c)$ and, therefore, there is no path for the zero sequence current to flow. In a three-phase four-wire system with neutral return, the neutral must carry out-of-balance current, i.e., $I_n = (I_a + I_b + I_c)$. Therefore, it follows that $I_n = 3I_0$. At the grounded neutral of a three-phase wye system, positive and negative sequence voltages are zero. The neutral voltage is equal to the zero sequence voltage or product of zero sequence current and three times the neutral impedance, Z_n.

5. From the above statement, phase conductors emanating from ungrounded wye- or delta-connected transformer windings cannot have zero sequence current. In a delta winding, zero sequence currents, if present, set up circulating currents in the delta winding itself. This is because the delta winding forms a closed path of low impedance for the zero sequence currents; each phase zero sequence voltage is absorbed by its own phase voltage drop and there are no zero sequence components at the terminals.

4.7 Sequence Impedance of Network Components

The impedance encountered by the symmetrical components depends on the type of power system equipment, i.e., a generator, a transformer, or a transmission line. The sequence impedances are required for component modeling and analysis. We derived the sequence impedances of a symmetrical coupled transmission line in Equation 4.37. Zero sequence impedance of overhead lines depends on the presence of ground wires, tower footing resistance, and grounding. It may vary between two and six times the positive sequence impedance. The line capacitance of overhead lines is ignored in short-circuit calculations. Three-phase matrix models of transmission lines, bundle conductors, and cables, and their transformation into symmetrical components are covered in Volume 2 of this series. While estimating sequence impedances of power system components is one problem, constructing the zero, positive, and negative sequence impedance networks is the first step for unsymmetrical fault current calculations.

4.7.1 Construction of Sequence Networks

A sequence network shows how the sequence currents, if present, will flow in a system. Connections between sequence component networks are necessary to achieve this objective. The sequence networks are constructed as viewed from the *fault point*, which can be defined as the point at which the unbalance occurs in a system, i.e., a fault or load unbalance.

The voltages for the sequence networks are taken as line-to-neutral voltages. The only active network containing the voltage source is the positive sequence network. Phase *a* voltage is taken as the reference voltage, and the voltages of the other two phases are expressed with reference to phase *a* voltage, as shown in Figure 4.7.

The sequence networks for positive, negative and zero sequence will have per phase impedance values, which may differ. Normally, the sequence impedance networks are constructed on the basis of per unit values on a common MVA base, and a base MVA of 100 is in common use. For nonrotating equipment like transformers, the impedance to negative sequence currents will be the same as that to the positive sequence currents. The impedance to negative sequence currents of rotating equipment will be different from the positive sequence impedance. And, in general, the impedance to zero sequence currents will be different from the positive or negative sequence impedances for all the apparatuses. For a study involving sequence components, the sequence impedance data can be (1) calculated by using subroutine computer programs, (2) obtained from manufacturers' data, (3) calculated by long-hand calculations, or (4) estimated from tables in published references.

The positive directions of current flow in each sequence network are outward at the faulted or unbalanced point. This means that the sequence currents flow in the same direction in all three sequence networks.

Sequence networks are shown schematically in boxes in which the fault points from which the sequence currents flow outward are marked as F_1, F_2, and F_0, and the neutral buses are designated as N_1, N_2, and N_0, respectively, for the positive, negative, and zero sequence impedance networks. Each network forms a two-port network with Thévenin sequence voltages across sequence impedances. Figure 4.7 illustrates this basic formation. Note the direction of currents. The voltage across the sequence impedance rises from N to F. As stated before, only the positive sequence network has a voltage source, which is the Thévenin equivalent. With this convention, appropriate signs must be allocated to the sequence voltages:

$$V_1 = V_a - I_1 Z_1$$
$$V_2 = -I_2 Z_2 \quad (4.56)$$
$$V_0 = -I_0 Z_0$$

The sequence voltages can be represented in matrix form as follows:

$$\begin{vmatrix} V_0 \\ V_1 \\ V_2 \end{vmatrix} = \begin{vmatrix} 0 \\ V_a \\ 0 \end{vmatrix} - \begin{vmatrix} Z_0 & 0 & 0 \\ 0 & Z_1 & 0 \\ 0 & 0 & Z_2 \end{vmatrix} \begin{vmatrix} I_0 \\ I_1 \\ I_2 \end{vmatrix} \quad (4.57)$$

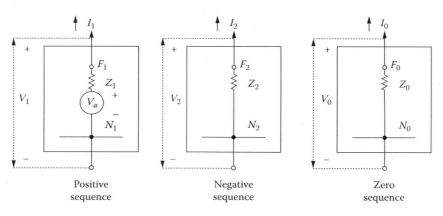

FIGURE 4.7
Positive, negative, and zero sequence network representation.

Short-Circuit Currents and Symmetrical Components 155

Based on the discussions so far, we can graphically represent the sequence impedances of various system components.

4.7.2 Transformers

The positive and negative sequence impedances of a transformer can be taken to be equal to its leakage impedance. As the transformer is a static device, the positive or negative sequence impedances do not change with phase sequence of the applied balanced voltages. The zero sequence impedance can, however, vary from an open circuit to a low value depending on the transformer winding connection, method of neutral grounding, and transformer construction, i.e., core or shell type.

We will briefly discuss the shell and core form of construction, as it has a major impact on the zero sequence flux and impedance. Referring to Figure 4.8a, in a three-phase core-type transformer, the sum of the fluxes in each phase in a given direction along the cores is zero; however, the flux going up one leg of the core must return through the other two, i.e., the magnetic circuit of a phase is completed through the other two phases in parallel. The magnetizing current per phase is that required for the core and part of the yoke. This means that in a three-phase core-type transformer, the magnetizing current will be different in each phase. Generally, the cores are long compared to yokes and the yokes are of greater cross section. The yoke reluctance is only a small fraction of the core and the variation of magnetizing current per phase is not appreciable. However, now consider the zero sequence flux, which will be directed in one direction, in each of the core-legs. The return path lies, not through the core legs, but through insulating medium and tank.

In three separate single-phase transformers connected in three-phase configuration or in shell-type three-phase transformers, the magnetic circuits of each phase are complete in themselves and do not interact as shown in Figure 4.8b. Due to advantages in short circuit and transient voltage performance, the shell form is used for larger transformers. The variations in shell form have five- or seven-legged cores. Briefly, we can say that, in a core type, the windings surround the core, and in the shell type, the core surrounds the windings.

4.7.2.1 Delta–Wye or Wye–Delta Transformer

In a delta–wye transformer with the wye winding grounded, zero sequence impedance will be approximately equal to the positive or negative sequence impedance, viewed from the wye connection side. Impedance to the flow of zero sequence currents in the core-type transformers is lower as compared to the positive sequence impedance. This is so, because there is no return path for zero sequence exciting flux in core-type units except through insulating medium and tank, a path of high reluctance. In groups of three single-phase transformers or in three-phase shell-type transformers, the zero sequence impedance is higher.

The zero sequence network for a wye–delta transformer is constructed as shown in Figure 4.9a. The grounding of the wye neutral allows the zero sequence currents to return through the neutral and circulate in the windings to the source of unbalance. Thus, the circuit on the wye side is shown connected to the L sideline. On the delta side, the circuit is open, as no zero sequence currents appear in the lines, though these currents circulate in the delta windings to balance the ampère-turns in the wye windings. The circuit is open on the H sideline, and the zero sequence impedance of the

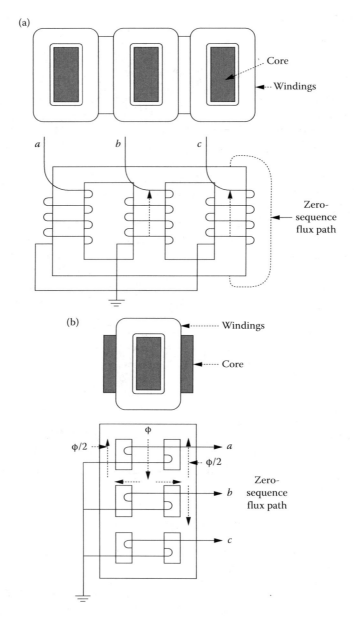

FIGURE 4.8
Core form of three-phase transformer, flux paths for phase and zero sequence components and (b) shell form of three-phase transformer.

transformer seen from the high side is an open circuit. If the wye winding neutral is left isolated (Figure 4.9b), the circuit will be open on both sides, indicating an infinite impedance.

Three-phase current flow diagrams can be constructed based on the convention that current always flows to the unbalance and that the ampère-turns in primary windings must be balanced by the ampère-turns in the secondary windings.

Short-Circuit Currents and Symmetrical Components 157

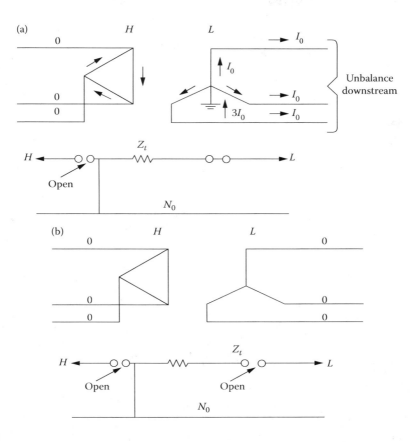

FIGURE 4.9
(a) Derivation of equivalent zero sequence circuit for a delta–wye transformer, wye neutral solidly grounded and (b) zero sequence circuit of a delta–wye transformer, wye neutral isolated.

4.7.2.2 Wye–Wye Transformer

In a wye–wye-connected transformer, with both neutrals isolated, no zero sequence currents can flow. The zero sequence equivalent circuit is open on *both* sides and indicates an infinite impedance to the flow of zero sequence currents. When one of the neutrals is grounded, still no zero sequence currents can be transferred from the grounded side to the ungrounded side. With one neutral grounded, there are no balancing ampère-turns in the ungrounded wye windings to enable current to flow in the grounded neutral windings. Thus, neither of the windings can carry a zero sequence current. Both neutrals must be grounded for the transfer of zero sequence currents.

A wye–wye-connected transformer with isolated neutrals is not used, due to the phenomenon of the oscillating neutral. This is discussed in Volume 2. Due to saturation in transformers and the flat-topped flux wave, a peak EMF is generated, which does not balance the applied sinusoidal voltage and generates a resultant third (and other) harmonics. These distort the transformer voltages as the neutral oscillates at thrice the supply frequency, a phenomenon called the "oscillating neutral." A tertiary delta is added to circulate the third harmonic currents and stabilize the neutral. It may also be designed as a load winding, which may have a rated voltage distinct from high- and low-voltage windings. This is further discussed in Section 4.7.2.5. When provided for zero sequence current circulation and harmonic suppression, the terminals of the tertiary-connected delta winding may not be brought out of the transformer tank. Sometimes core-type transformers are provided with five-legged cores to circulate the harmonic currents.

4.7.2.3 Delta–Delta Transformer

In a delta–delta connection, no zero currents will pass from one winding to another. On the transformer side, the windings are shown connected to the reference bus, allowing the circulation of currents within the windings.

4.7.2.4 Zigzag Transformer

A zigzag transformer is often used to derive a neutral for grounding of a delta–delta-connected system. This is shown in Figure 4.10. Windings a_1 and a_2 are on the same core

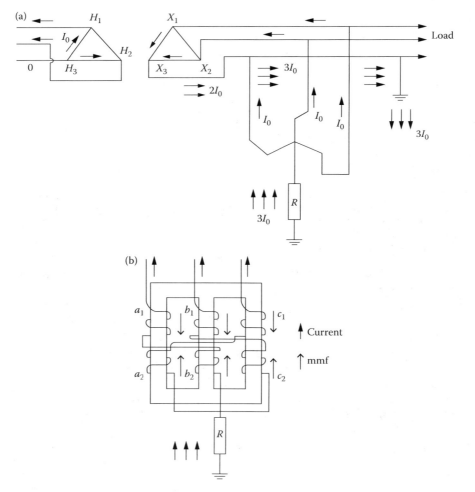

FIGURE 4.10
(a) Current distribution in a delta–delta system with zigzag grounding transformer for a single line-to-ground fault and (b) zigzag transformer winding connections.

Short-Circuit Currents and Symmetrical Components 159

leg and have the same number of turns but are wound in the opposite direction. The zero sequence currents in the two windings on the same core leg have canceling ampère-turns. Referring to Figure 4.10b, the currents in the winding sections a_1 and c_2 must be equal as they are in series. By the same analogy, all currents must be equal, balancing the magnetomotive forces in each leg:

$$i_{a1} = i_{a2} = i_{b1} = i_{b2} = i_{c1} = i_{c2}$$

The impedance to the zero sequence currents is that due to leakage flux of the windings. For positive or negative sequence currents, neglecting magnetizing current, the connection has infinite impedance. Figure 4.10a shows the distribution of zero sequence current and its return path for a single line to ground fault on one of the phases. The ground current divides equally through the zigzag transformer; one-third of the current returns directly to the fault point and the remaining two-thirds pass through two phases of the delta-connected windings to return to the fault point. Two phases and windings on the primary delta must carry current to balance the ampère-turns of the secondary winding currents as shown in Figure 4.10b. Impedance can be added between the artificially derived neutral and ground to limit the ground fault current.

Table 4.1 shows the sequence equivalent circuits of three-phase two-winding transformers. When the transformer neutral is grounded through an impedance Z_n, a term $3Z_n$ appears in the equivalent circuit. We have already proved that $I_n = 3I_0$. The zero sequence impedance of the high- and low-voltage windings is shown as Z_H and Z_L, respectively. The transformer impedance $Z_T = Z_H + Z_L$ on a per unit basis. This impedance is specified by the manufacturer as a percentage impedance on transformer MVA base, based on OA (natural liquid cooled for liquid-immersed transformers) or AA (natural air cooled, without forced ventilation for dry-type transformers) rating of the transformer. For example, a 138–13.8 kV transformer may be rated as follows:

40 MVA, OA rating at 55°C rise

44.8 MVA, OA rating at 65°C rise

60 MVA, FA (forced air, i.e., fan cooled) rating at first stage of fan cooling, 65°C rise

75 MVA, FA second-stage fan cooling, 65°C rise

These ratings are normally applicable for an ambient temperature of 40°C, with an average of 30°C over a period of 24 h. The percentage impedance will be normally specified on a 40 MVA or possibly a 44.8 MVA base.

The difference between the zero sequence impedance circuits of wye–wye-connected shell- and core-form transformers in Table 4.1 is noteworthy. Connections 8 and 9 are for a core-type transformer and connections 7 and 10 are for a shell-type transformer. The impedance Z_M accounts for magnetic coupling between the phases of a core-type transformer.

4.7.2.5 Three-Winding Transformers

The theory of linear networks can be extended to apply to multiwinding transformers. A linear network having n terminals requires $\frac{1}{2}n(n+1)$ quantities to specify it completely for a given frequency and EMF. Figure 4.11 shows the wye equivalent circuit of a

TABLE 4.1

Equivalent Positive, Negative, and Zero Sequence Circuits of Two-Winding Transformers

No	Winding Connections	Zero Sequence Circuit	Positive or Negative Sequence Circuit
1			
2			
3			
4			
5			
6			
7			
8			
9			
10			

Short-Circuit Currents and Symmetrical Components

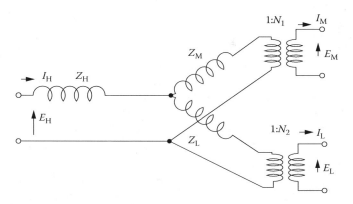

FIGURE 4.11
Wye equivalent circuit of a three-winding transformer.

three-winding transformer. One method to obtain the necessary data is to designate the pairs of terminals as 1, 2,..., n. All the terminals are then short-circuited except terminal one and a suitable EMF is applied across it. The current flowing in each pair of terminals is measured. This is repeated for all the terminals. For a three-winding transformer,

$$Z_H = \frac{1}{2}(Z_{HM} + Z_{HL} - Z_{ML})$$
$$Z_M = \frac{1}{2}(Z_{ML} + Z_{HM} - Z_{HL}) \quad (4.58)$$
$$Z_L = \frac{1}{2}(Z_{HL} + Z_{ML} - Z_{HM})$$

where Z_{HM} = leakage impedance between the H and X windings, as measured on the H winding with M winding short-circuited and L winding open-circuited,

Z_{HL} = leakage impedance between the H and L windings, as measured on the H winding with L winding short-circuited and M winding open-circuited, and

Z_{ML} = leakage impedance between the M and L windings, as measured on the M winding with L winding short-circuited and H winding open-circuited.

Equation 4.58 can be written as

$$\begin{vmatrix} Z_H \\ Z_M \\ Z_L \end{vmatrix} = 1/2 \begin{vmatrix} 1 & 1 & -1 \\ 1 & -1 & 1 \\ -1 & 1 & 1 \end{vmatrix} \begin{vmatrix} Z_{HM} \\ Z_{HL} \\ Z_{ML} \end{vmatrix}$$

We also see that

$$Z_{HL} = Z_H + Z_L$$
$$Z_{HM} = Z_H + Z_M \quad (4.59)$$
$$Z_{ML} = Z_M + Z_L$$

Table 4.2 shows the equivalent sequence circuits of a three-winding transformer.

TABLE 4.2

Equivalent Positive, Negative, and Zero Sequence Circuits of Three-Winding Transformers

No	Winding Connections	Zero Sequence Circuit	Positive or Negative Sequence Circuit
1			
2			
3			
4			
5			
6			

Short-Circuit Currents and Symmetrical Components   163

Example 4.3

A three-phase three-winding transformer nameplate reads as follows:

High voltage winding: 138 kV, wye connected rated at 60 MVA; medium-voltage winding: 69 kV, wye connected rated at 50 MVA; 13.8 kV winding (tertiary winding), delta connected and rated 10 MVA. $Z_{HM} = 9\%$, $Z_{HL} = 6\%$, $Z_{ML} = 3\%$. Calculate Z_H, Z_M, and Z_L for representation in the circuit of Figure 4.11.

Before Equation 4.59 is applied, the percentage impedances given on respective MVA base of the windings are converted to a common base, say on the primary winding rating of 60 MVA. This gives $Z_{HM} = 9\%$, $Z_{HL} = 7.2\%$, $Z_{ML} = 18\%$. Then, $Z_H = -0.9\%$, $Z_M = 9.9\%$, $Z_L = 8.1\%$. Note that Z_H is negative. In a three-winding transformer, a negative impedance means that the voltage drop on load flow can be reduced or even made to rise.

4.7.3 Static Load

Consider a static three-phase load connected in a wye configuration with the neutral grounded through an impedance Z_n. Each phase impedance is Z. The sequence transformation is

$$\begin{vmatrix} V_a \\ V_b \\ V_c \end{vmatrix} = \begin{vmatrix} Z & 0 & 0 \\ 0 & Z & 0 \\ 0 & 0 & Z \end{vmatrix} \begin{vmatrix} I_a \\ I_b \\ I_c \end{vmatrix} + \begin{vmatrix} I_n Z_n \\ I_n Z_n \\ I_n Z_n \end{vmatrix}$$

$$T_s \begin{vmatrix} V_0 \\ V_1 \\ V_2 \end{vmatrix} = \begin{vmatrix} Z & 0 & 0 \\ 0 & Z & 0 \\ 0 & 0 & Z \end{vmatrix} T_s \begin{vmatrix} I_0 \\ I_1 \\ I_2 \end{vmatrix} + \begin{vmatrix} 3I_0 Z_n \\ 3I_0 Z_n \\ 3I_0 Z_n \end{vmatrix}$$

$$\tag{4.60}$$

$$\begin{vmatrix} V_0 \\ V_1 \\ V_2 \end{vmatrix} = T_s^{-1} \begin{vmatrix} Z & 0 & 0 \\ 0 & Z & 0 \\ 0 & 0 & Z \end{vmatrix} T_s \begin{vmatrix} I_0 \\ I_1 \\ I_2 \end{vmatrix} + T_s^{-1} \begin{vmatrix} 3I_0 Z_n \\ 3I_0 Z_n \\ 3I_0 Z_n \end{vmatrix}$$

$$\tag{4.61}$$

$$= \begin{vmatrix} Z & 0 & 0 \\ 0 & Z & 0 \\ 0 & 0 & Z \end{vmatrix} \begin{vmatrix} I_0 \\ I_1 \\ I_2 \end{vmatrix} + \begin{vmatrix} 3I_0 Z_n \\ 0 \\ 0 \end{vmatrix} = \begin{vmatrix} Z + 3Z_n & 0 & 0 \\ 0 & Z & 0 \\ 0 & 0 & Z \end{vmatrix} \begin{vmatrix} I_0 \\ I_1 \\ I_2 \end{vmatrix}$$

$$\tag{4.62}$$

This shows that the load can be resolved into sequence impedance circuits. This result can also be derived by merely observing the symmetrical nature of the circuit.

4.7.4 Synchronous Machines

Negative and zero sequence impedances are specified for synchronous machines by the manufacturers on the basis of the test results. The negative sequence impedance is measured with the machine driven at rated speed and the field windings short-circuited.

A balanced negative sequence voltage is applied and the measurements taken. The zero sequence impedance is measured by driving the machine at rated speed, field windings short-circuited, all three phases in series, and a single-phase voltage applied to circulate a single-phase current. The zero sequence impedance of generators is low, while the negative sequence impedance is approximately given by

$$\frac{X_d'' + X_q''}{2} \tag{4.63}$$

where X_d'', and X_q'' are the direct axis and quadrature axis subtransient reactances. An explanation of this averaging is that the negative sequence in the stator results in a double-frequency negative component in the field. The negative sequence flux component in the air gap may be considered to alternate between poles and interpolar gap, respectively.

The following expressions can be written for the terminal voltages of a wye-connected synchronous generator, neutral grounded through an impedance Z_n:

$$V_a = \frac{d}{dt}[L_{af}\cos\theta I_f - L_{aa}I_a - L_{ab}I_b - L_{ac}I_c] - I_aR_a + V_n$$

$$V_b = \frac{d}{dt}[L_{bf}\cos(\theta - 120°)I_f - L_{ba}I_a - L_{bb}I_b - L_{bc}I_c] - I_aR_b + V_n \tag{4.64}$$

$$V_c = \frac{d}{dt}[L_{cf}\cos(\theta - 240°)I_f - L_{ca}I_a - L_{cb}I_b - L_{cc}I] - I_aR_c + V_n$$

The first term is the generator internal voltage, due to field linkages, and L_{af} denotes the field inductance with respect to phase A of stator windings and I_f is the field current. These internal voltages are displaced by $120°$ and may be termed E_a, E_b, and E_c. The voltages due to armature reaction, given by the self-inductance of a phase, i.e., L_{aa}, and its mutual inductance with respect to other phases, i.e., L_{ab} and L_{ac}, and the IR_a drop is subtracted from the generator internal voltage and the neutral voltage is added to obtain the line terminal voltage V_a.

For a symmetrical machine,

$$\begin{aligned} L_{af} &= L_{bf} = L_{cf} = L_f \\ R_a &= R_b = R_c = R \\ L_{aa} &= L_{bb} = L_{cc} = L \\ L_{ab} &= L_{bc} = L_{ca} = L' \end{aligned} \tag{4.65}$$

Thus,

$$\begin{vmatrix} V_a \\ V_b \\ V_c \end{vmatrix} = \begin{vmatrix} E_a \\ E_b \\ E_c \end{vmatrix} - j\omega \begin{vmatrix} L & L' & L' \\ L' & L & L' \\ L' & L' & L \end{vmatrix} \begin{vmatrix} I_a \\ I_b \\ I_c \end{vmatrix} - \begin{vmatrix} R & 0 & 0 \\ 0 & R & 0 \\ 0 & 0 & R \end{vmatrix} \begin{vmatrix} I_a \\ I_b \\ I_c \end{vmatrix} - Z_n \begin{vmatrix} I_n \\ I_n \\ I_n \end{vmatrix} \tag{4.66}$$

Short-Circuit Currents and Symmetrical Components

Transform using symmetrical components,

$$T_s \begin{vmatrix} V_0 \\ V_1 \\ V_2 \end{vmatrix} = T_s \begin{vmatrix} E_0 \\ E_1 \\ E_2 \end{vmatrix} - j\omega \begin{vmatrix} L & L' & L' \\ L' & L & L' \\ L' & L' & L \end{vmatrix} T_s \begin{vmatrix} I_0 \\ I_1 \\ I_2 \end{vmatrix} - \begin{vmatrix} R & 0 & 0 \\ 0 & R & 0 \\ 0 & 0 & R \end{vmatrix} T_s \begin{vmatrix} I_0 \\ I_1 \\ I_2 \end{vmatrix} - 3Z_n \begin{vmatrix} I_0 \\ I_0 \\ I_0 \end{vmatrix}$$

$$\begin{vmatrix} V_0 \\ V_1 \\ V_2 \end{vmatrix} = \begin{vmatrix} E_0 \\ E_1 \\ E_2 \end{vmatrix} - j\omega \begin{vmatrix} L_0 & 0 & 0 \\ 0 & L_1 & 0 \\ 0 & 0 & L_2 \end{vmatrix} \begin{vmatrix} I_0 \\ I_1 \\ I_2 \end{vmatrix} - \begin{vmatrix} R & 0 & 0 \\ 0 & R & 0 \\ 0 & 0 & R \end{vmatrix} \begin{vmatrix} I_0 \\ I_1 \\ I_2 \end{vmatrix} - \begin{vmatrix} 3I_0 Z_n \\ 0 \\ 0 \end{vmatrix} \quad (4.67)$$

where

$$L_0 = L + 2L' \quad (4.68)$$
$$L_1 = L_2 + L - L'$$

Therefore, the equation can be written as

$$\begin{vmatrix} V_0 \\ V_1 \\ V_2 \end{vmatrix} = \begin{vmatrix} 0 \\ E_1 \\ 0 \end{vmatrix} - \begin{vmatrix} Z_0 + 3Z_n & 0 & 0 \\ 0 & Z_1 & 0 \\ 0 & 0 & Z_2 \end{vmatrix} \begin{vmatrix} I_0 \\ I_1 \\ I_2 \end{vmatrix} \quad (4.69)$$

The equivalent circuit is shown in Figure 4.12. This can be compared with the static three-phase load equivalents. Even for a cylindrical rotor machine, the assumption $Z_1 = Z_2$ is not strictly valid. The resulting generator impedance matrix is nonsymmetrical.

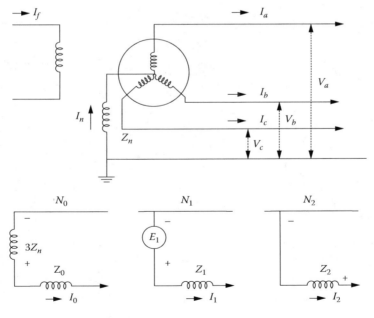

FIGURE 4.12
Sequence components of a synchronous generator impedance.

Example 4.4

Figure 4.13a shows a single line diagram, with three generators, three transmission lines, six transformers, and three buses. It is required to construct positive, negative, and zero sequence networks looking from the fault point marked F. Ignore the load currents.

The positive sequence network is shown in Figure 4.13b. There are three generators in the system, and their positive sequence impedances are clearly marked in Figure 4.13b. The generator impedances are returned to a common bus. The Thévenin voltage at the fault point is shown to be equal to the generator voltages, which are all equal. This has to be so as all load currents are neglected, i.e., all the shunt elements representing loads

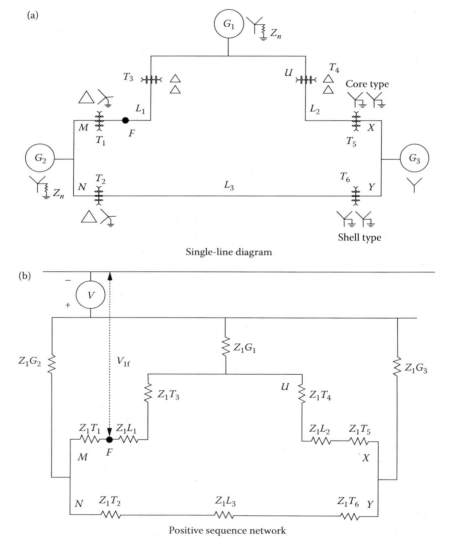

FIGURE 4.13
(a) A single line diagram of distribution and (b–d) positive, negative, and zero sequence networks of the distribution system in (a).

(Continued)

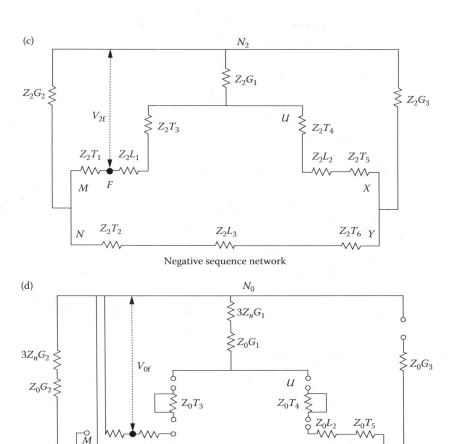

FIGURE 4.13 (CONTINUED)
(a) A single line diagram of distribution and (b–d) positive, negative, and zero sequence networks of the distribution system in (a).

are open-circuited. Therefore, the voltage magnitudes and phase angles of all three generators must be equal. When load flow is considered, generation voltages will differ in magnitude and phase, and the voltage vector at the chosen fault point, prior to the fault, can be calculated based on load flow. We have discussed that the load currents are normally ignored in short-circuit calculations. Fault duties of switching devices are calculated based on the rated system voltage rather than the actual voltage, which varies with load flow. This is generally true, unless the prefault voltage at the fault point remains continuously above or below the rated voltage.

Figure 4.13c shows the negative sequence network. Note the similarity with the positive sequence network with respect to interconnection of various system components.

Figure 4.13d shows zero sequence impedance network. This is based on the transformer zero sequence networks shown in Table 1.1. The neutral impedance is multiplied by a factor of three.

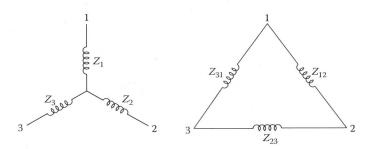

FIGURE 4.14
Wye–delta and delta–wye transformation of impedances.

Each of these networks can be reduced to a single impedance using elementary network transformations. Referring to Figure 4.14, wye-to-delta and delta-to-wye impedance transformations are given by

$$Z_1 = \frac{Z_{12}Z_{31}}{Z_{12}+Z_{23}+Z_{31}}$$
$$Z_2 = \frac{Z_{12}Z_{23}}{Z_{12}+Z_{23}+Z_{31}} \qquad (4.70)$$
$$Z_3 = \frac{Z_{23}Z_{31}}{Z_{12}+Z_{23}+Z_{31}}$$

and from wye-to-delta,

$$Z_{12} = \frac{Z_1Z_2+Z_2Z_3+Z_3Z_1}{Z_3}$$
$$Z_{23} = \frac{Z_1Z_2+Z_2Z_3+Z_3Z_1}{Z_1} \qquad (4.71)$$
$$Z_{31} = \frac{Z_1Z_2+Z_2Z_3+Z_3Z_1}{Z_2}$$

4.8 Computer Models of Sequence Networks

Referring to the zero sequence impedance network of Figure 4.13d, a number of discontinuities occur in the network, depending on transformer winding connections and system grounding. These disconnections are at nodes marked T, U, M, and N. These make a node disappear in the zero sequence network, while it exists in the models of positive and negative sequence networks. The integrity of the nodes should be maintained in all the sequence networks for computer modeling. Figure 4.15 shows how this discontinuity can be resolved.

Short-Circuit Currents and Symmetrical Components

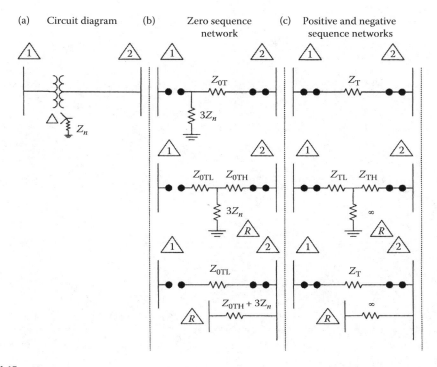

FIGURE 4.15
(a) Representation of a delta–wye transformer, (b) and (c) zero and positive sequence network representation maintaining integrity of nodes.

Figure 4.15a shows a delta–wye transformer, wye side neutral grounded through an impedance Z_n, connected between buses numbered 1 and 2. Its zero sequence network, when viewed from the bus 1 is an open circuit.

Two possible solutions in computer modeling are shown in Figure 4.15b and c. In Figure 4.15b, a fictitious bus R is created. The positive sequence impedance circuit is modified by dividing the transformer positive sequence impedance into two parts: Z_{TL} for the low voltage winding and Z_{TH} for the high voltage winding. Infinite impedance between the junction point of these impedances to the fictitious bus R is connected. In computer calculations, this infinite impedance will be simulated by a large value, i.e., $999 + j9999$, on a per unit basis.

The zero sequence network is treated in a similar manner, i.e., the zero sequence impedance is split between the windings and the equivalent grounding resistor $3R_N$ is connected between the junction point and the fictitious bus R.

Figure 4.15c shows another approach for the creation of a fictitious bus R to preserve the integrity of nodes in the sequence networks. For the positive sequence network, a large impedance is connected between bus 2 and bus R, while for the zero sequence network an impedance equal to $Z_{0TH} + 3R_N$ is connected between bus 2 and bus R.

This chapter provides the basic concepts related to short-circuit calculations. The discussions about symmetrical components, construction of sequence networks, and fault current calculations have been covered in Chapter 5.

Problems

4.1. A short transmission line of inductance 0.05 H and resistance 1 ohm is suddenly short-circuited at the receiving end, while the source voltage is $480(\sqrt{2})\sin(2\pi ft + 30°)$. At what instant of the short-circuit will the DC offset be zero? At what instant will the DC offset be maximum?

4.2. Figure 4.1 shows a nondecaying AC component of the fault current. Explain why this is not correct for a fault close to a generator.

4.3. Explain similarity transformation. How is it related to the diagonalization of a matrix?

4.4. Find the eigenvalues of the matrix:

$$\begin{bmatrix} 6 & -2 & 2 \\ -2 & 3 & -1 \\ 2 & -2 & 3 \end{bmatrix}$$

4.5 A power system is shown in Figure P4.1. Assume that loads do not contribute to the short-circuit currents. Convert to a common 100 MVA base and form sequence impedance networks. Redraw zero sequence network to eliminate discontinuities.

4.6 Three unequal load resistances of 10, 20, and 20 ohms are connected in delta 10 ohms between lines a and b, 20 ohms between lines b and c, and 20 ohms between lines c and a. The power supply is a balanced three-phase system of 480 V rms between the lines. Find symmetrical components of line currents and delta currents.

4.7 In Figure 4.10, the zigzag transformer is replaced with a wye–delta-connected transformer. Show the distribution of the fault current for a phase-to-ground fault on one of the phases.

4.8 Resistances of 6, 6, and 5 ohms are connected in a wye configuration across a balanced three-phase supply system of line-to-line voltage of 480 V rms (Figure P4.2). The wye point of the load (neutral) is not grounded. Calculate the neutral voltage with respect to ground.

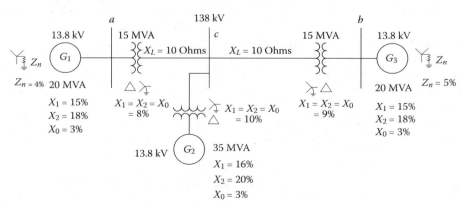

FIGURE P4.1
Power system with impedance data for Problem 4.5.

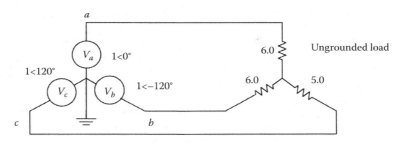

FIGURE P4.2
Network for Problem 4.8.

4.9 Write equations for a symmetrical three-phase fault in a three-phase wye-connected system, with balanced impedances in each line.

4.10 The load currents are generally neglected in short-circuit calculations. Do these have any effect on the DC component asymmetry? (1) Increase it, (2) decrease it, (3) have no effect. Explain.

4.11 Figure 4.9a shows the zero sequence current flow for a delta–wye transformer, with the wye neutral grounded. Construct a similar diagram for a three-winding transformer, wye–wye connected, with tertiary delta and both wye neutrals solidly grounded.

4.12 Convert the sequence impedance networks of Example 4.8 to single impedances as seen from the fault point. Use the following numerical values on a per unit basis (all on a common MVA base). Neglect resistances.

Generators G_1, G_2, and G_3: $Z_1 = 0.15$, $Z_2 = 0.18$, $Z_0 = 0.08$, Z_n (neutral grounding impedance) = 0.20;

Transmission lines L_1, L_2, and L_3: $Z_1 = 0.2$, $Z_2 = 0.2$;

Transformers T_1, T_2, T_3, T_4, T_5, and T_6: $Z_1 = Z_2 = 0.10$, transformer T_1: $Z_0 = 0.10$

4.13 Prove Equation 4.2 with methods of differential equations and Laplace transform.

Bibliography

1. CF Wagner, RD Evans. *Symmetrical Components*. McGraw-Hill, New York, 1933.
2. E Clarke. *Circuit Analysis of Alternating Current Power Systems*, vol. 1. Wiley, New York, 1943.
3. JO Bird. *Electrical Circuit Theory and Technology*. Butterworth Heinemann, Oxford, 1997.
4. GW Stagg, A Abiad. *Computer Methods in Power Systems Analysis*. McGraw-Hill, New York, 1968.
5. WE Lewis, DG Pryce. *The Application of Matrix Theory to Electrical Engineering*. E&FN Spon, London, 1965, Chapter 6.
6. LJ Myatt. *Symmetrical Components*. Pergamon Press, Oxford, 1968.
7. CA Worth, Ed. *J&P Transformer Book*, 11th ed., Butterworth, London, 1983.
8. *Westinghouse Electric Transmission and Distribution Handbook*, 4th ed., Westinghouse Electric Corp., East Pittsburgh, PA, 1964.
9. AE Fitzgerald, C Kingsley, A Kusko. *Electric Machinery*, 3rd ed., McGraw-Hill, New York, 1971.
10. MS Chen, WE Dillon. Power system modeling. *Proc IEEE*, 62, 901–915, 1974.

11. CL Fortescu. Method of symmetrical coordinates applied to the solution of polyphase networks. *AIEE Trans*, 37, 1027–1140, 1918.
12. IEEE Std. 399, IEEE Recommended Practice for Power Systems Analysis, 1997.
13. HW Dommel, WS Meyer. Computation of electromagnetic transients. *Proc IEEE*, 62, 983–993, 1974.
14. ATP Rule Book, ATP User Group, Portland, OR, 1992.
15. A Greenwood. *Electrical Transients in Power Systems*, 2nd ed., John Wiley, New York, 1991.
16. JC Das. *Transients in Electrical Systems*, McGraw-Hill, New York, 2010.

5

Unsymmetrical Fault Calculations

Chapter 4 discussed the nature of sequence networks and how three distinct sequence networks can be constructed as seen from the fault point. Each of these networks can be reduced to a single Thévenin positive, negative, or zero sequence impedance. Only the positive sequence network is active and has a voltage source, which is the prefault voltage. For unsymmetrical fault current calculations, the three separate networks can be connected in a certain manner, depending on the type of fault.

Unsymmetrical fault types involving one or two phases and ground are as follows:

- A single line-to-ground fault
- A double line-to-ground fault
- A line-to-line fault

These are called shunt faults. A three-phase fault may also involve ground. The unsymmetrical series type faults are the following:

- One conductor opens
- Two conductors open

The broken conductors may be grounded on one side or on both sides of the break. An open conductor fault can occur due to operation of a fuse in one of the phases.

Unsymmetrical faults are more common. The most common type is a line-to-ground fault. Approximately 70% of the faults in power systems are single line-to-ground faults.

While applying symmetrical component method to fault analysis, we will ignore the load currents. This makes the positive sequence voltages of all the generators in the system identical and equal to the pre-fault voltage.

In the analysis to follow, Z_1, Z_2, and Z_0 are the positive, negative, and zero sequence impedances as seen from the fault point; V_a, V_b, and V_c are the phase to ground voltages at the fault point, prior to fault, i.e., *if the fault does not exist*; and V_1, V_2, and V_0 are corresponding sequence component voltages. Similarly, I_a, I_b, and I_c are the line currents and I_1, I_2, and I_0 their sequence components. A fault impedance of Z_f is assumed in every case. For a bolted fault, $Z_f = 0$.

5.1 Line-to-Ground Fault

Figure 5.1a shows that phase *a* of a three-phase system goes to ground through an impedance Z_f. The flow of ground fault current depends on the method of system grounding. A solidly grounded system with zero ground resistance is assumed. The Z_f ground resistance

173

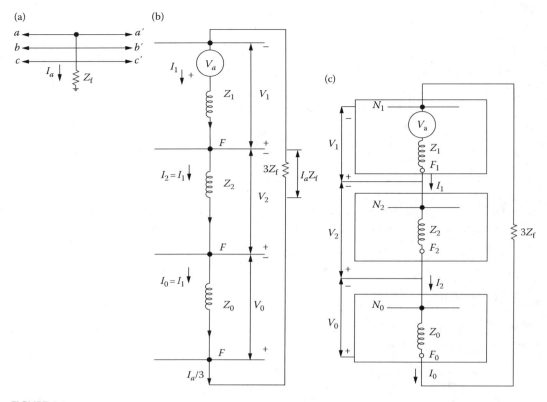

FIGURE 5.1
(a) Line-to-ground fault in a three-phase system, (b) line-to-ground fault equivalent circuit, and (c) sequence network interconnections.

can be added in series with the fault impedance Z_f. The ground fault current must have a return path through the grounded neutrals of generators or transformers. If there is no return path for the ground current, $Z_0 = \infty$ and ground fault current is zero. This is an obvious conclusion.

The phase a is faulted in Figure 5.1a. As the load current is neglected, currents in phases b and c are zero, and the voltage at the fault point, $V_a = I_a Z_f$. The sequence components of the currents are given by

$$\begin{vmatrix} I_0 \\ I_1 \\ I_2 \end{vmatrix} = \frac{1}{3} \begin{vmatrix} 1 & 1 & 1 \\ 1 & a & a^2 \\ 1 & a^2 & a \end{vmatrix} \begin{vmatrix} I_a \\ 0 \\ 0 \end{vmatrix} = \frac{1}{3} \begin{vmatrix} I_a \\ I_a \\ I_a \end{vmatrix} \qquad (5.1)$$

Also,

$$I_0 = I_1 = I_2 = \frac{1}{3} I_a \qquad (5.2)$$

$$3I_0 Z_f = V_0 + V_1 + V_2 = -I_0 Z_0 + (V_a - I_1 Z_1) - I_2 Z_2 \qquad (5.3)$$

Unsymmetrical Fault Calculations 175

that gives

$$I_0 = \frac{V_a}{Z_0 + Z_1 + Z_2 + 3Z_f} \tag{5.4}$$

The fault current I_a is

$$I_a = 3I_0 = \frac{3V_a}{(Z_1 + Z_2 + Z_0) + 3Z_f} \tag{5.5}$$

This shows that the equivalent fault circuit using sequence impedances can be constructed as shown in Figure 5.1b. In terms of sequence impedances' network blocks the connections are shown in Figure 5.1c.

The following result could also have been arrived from Figure 5.1b:

$$(V_a - I_1 Z_1) + (-I_2 Z_2) + (-I_0 Z_0) - 3Z_f I_0 = 0$$

that gives the same Equations 5.4 and 5.5. The voltage of phase b to ground under fault conditions is

$$\begin{aligned} V_b \quad &= a^2 V_1 + a V_2 + V_0 \\ &= V_a \frac{3a^2 Z_f + Z_2(a^2 - a) + Z_0(a^2 - 1)}{(Z_1 + Z_2 + Z_0) + 3Z_f} \end{aligned} \tag{5.6}$$

Similarly, the voltage of phase c can be calculated.

An expression for the ground fault current used in grounding grid designs and system grounding is as follows:

$$I_a = \frac{3V_a}{(R_0 + R_1 + R_2 + 3R_f + 3R_G) + j(X_0 + X_1 + X_2)} \tag{5.7}$$

where
R_f is the fault resistance
R_G is the resistance of the grounding grid
R_0, R_1, and R_2 are the sequence resistances
X_0, X_1 and X_2 are the sequence reactances.

5.2 Line-to-Line Fault

Figure 5.2a shows a line-to-line fault. A short circuit occurs between phases b and c, through a fault impedance Z_f. The fault current circulates between phases b and c, flowing back to source through phase b and returning through phase c; $I_a = 0$, $I_b = -I_c$. The sequence components of the currents are

$$\begin{vmatrix} I_0 \\ I_1 \\ I_2 \end{vmatrix} = \frac{1}{3} \begin{vmatrix} 1 & 1 & 1 \\ 1 & a & a^2 \\ 1 & a^2 & a \end{vmatrix} \begin{vmatrix} 0 \\ -I_c \\ I_c \end{vmatrix} = \frac{1}{3} \begin{vmatrix} 0 \\ -a+a^2 \\ -a^2+a \end{vmatrix} \quad (5.8)$$

From Equation 5.8, $I_0 = 0$ and $I_1 = -I_2$.

$$V_b - V_c = \begin{vmatrix} 0 & 1 & -1 \end{vmatrix} \begin{vmatrix} V_a \\ V_b \\ V_c \end{vmatrix} = \begin{vmatrix} 0 & 1 & -1 \end{vmatrix} \begin{vmatrix} 1 & 1 & 1 \\ 1 & a^2 & a \\ 1 & a & a^2 \end{vmatrix} \begin{vmatrix} V_0 \\ V_1 \\ V_2 \end{vmatrix}$$

$$= \begin{vmatrix} 0 & a^2-a & a-a^2 \end{vmatrix} \begin{vmatrix} V_0 \\ V_1 \\ V_2 \end{vmatrix} \quad (5.9)$$

Therefore,

$$\begin{aligned} V_b - V_c &= (a^2 - a)(V_1 - V_2) \\ &= (a^2 I_1 + a I_2) Z_f \\ &= (a^2 - a) I_1 Z_f \end{aligned} \quad (5.10)$$

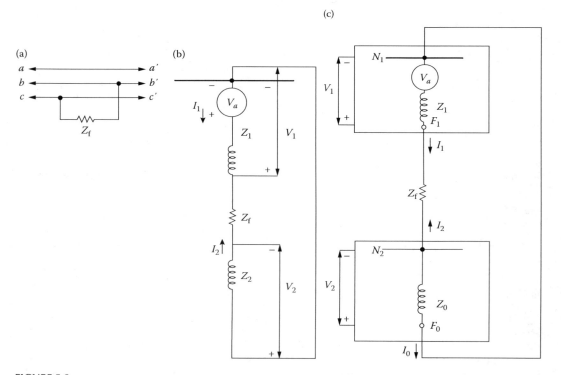

FIGURE 5.2
(a) Line-to-line fault in a three-phase system, (b) line-to-line fault equivalent circuit, and (c) sequence network interconnections.

Unsymmetrical Fault Calculations 177

This gives

$$(V_1 - V_2) = I_1 Z_f \tag{5.11}$$

The equivalent circuit is shown in Figure 5.2b and c.
  Also,

$$I_b = (a^2 - a)I_1 = -j\sqrt{3}I_1 \tag{5.12}$$

and

$$I_1 = \frac{V_a}{Z_1 + Z_2 + Z_f} \tag{5.13}$$

The fault current is

$$I_b = -I_c = \frac{-j\sqrt{3}V_a}{Z_1 + Z_2 + Z_f} \tag{5.14}$$

5.3 Double Line-to-Ground Fault

A double line-to-ground fault is shown in Figure 5.3a. The phases b and c go to ground through a fault impedance Z_f. The current in the ungrounded phase is zero, i.e., $I_a = 0$. Therefore, $I_1 + I_2 + I_0 = 0$.

$$V_b = V_c = (I_b + I_c)Z_f \tag{5.15}$$

Thus,

$$\begin{vmatrix} V_0 \\ V_1 \\ V_2 \end{vmatrix} = \frac{1}{3} \begin{vmatrix} 1 & 1 & 1 \\ 1 & a & a^2 \\ 1 & a^2 & a \end{vmatrix} \begin{vmatrix} V_a \\ V_b \\ V_b \end{vmatrix} = \frac{1}{3} \begin{vmatrix} V_a + 2V_b \\ V_a + (a+a^2)V_b \\ V_a + (a+a^2)V_b \end{vmatrix} \tag{5.16}$$

that gives $V_1 = V_2$ and

$$\begin{aligned} V_0 &= \frac{1}{3}(V_a + 2V_b) \\ &= \frac{1}{3}[(V_o + V_1 + V_2) + 2(I_b + I_c)Z_f] \tag{5.17} \\ &= \frac{1}{3}[(V_0 + 2V_1) + 2(3I_0)Z_f] \\ &= V_1 + 3Z_f I_0 \end{aligned}$$

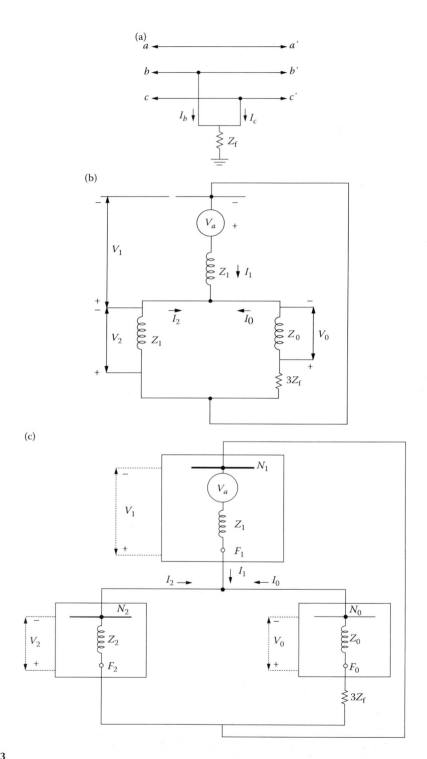

FIGURE 5.3
(a) Double line-to-ground fault in a three-phase system, (b) double line-to-ground fault equivalent circuit, and (c) sequence network interconnections.

Unsymmetrical Fault Calculations

This gives the equivalent circuit of Figure 5.3b and c. The fault current is

$$I_1 = \frac{V_a}{Z_1 + [Z_2 \| (Z_0 + 3Z_f)]}$$
$$= \frac{V_a}{Z_1 + \dfrac{Z_2(Z_0 + 3Z_f)}{Z_2 + Z_0 + 3Z_f}} \qquad (5.18)$$

5.4 Three-Phase Fault

The three phases are short-circuited through equal fault impedances Z_f (Figure 5.4a). Since a symmetrical fault is considered and there is no path to ground, the vectorial sum of fault currents is zero.

$$I_a = 0, \; I_a + I_b + I_c = 0 \qquad (5.19)$$

As the fault is symmetrical,

$$\begin{vmatrix} V_a \\ V_b \\ V_c \end{vmatrix} = \begin{vmatrix} Z_f & 0 & 0 \\ 0 & Z_f & 0 \\ 0 & 0 & Z_f \end{vmatrix} \begin{vmatrix} I_a \\ I_b \\ I_c \end{vmatrix} \qquad (5.20)$$

The sequence voltages are given by

$$\begin{vmatrix} V_0 \\ V_1 \\ V_2 \end{vmatrix} = [T_s]^{-1} \begin{vmatrix} Z_f & 0 & 0 \\ 0 & Z_f & 0 \\ 0 & 0 & Z_f \end{vmatrix} [T_s] \begin{vmatrix} I_0 \\ I_1 \\ I_2 \end{vmatrix} = \begin{vmatrix} Z_f & 0 & 0 \\ 0 & Z_f & 0 \\ 0 & 0 & Z_f \end{vmatrix} \begin{vmatrix} I_0 \\ I_1 \\ I_2 \end{vmatrix} \qquad (5.21)$$

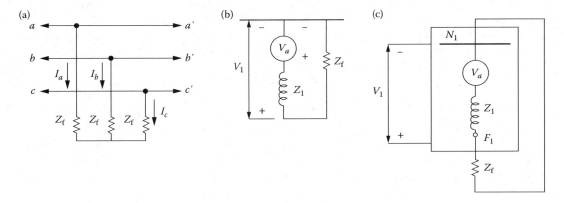

FIGURE 5.4
(a) Three-phase symmetrical fault, (b) equivalent circuit, and (c) sequence network.

180 *Short-Circuits in AC and DC Systems*

This gives the equivalent circuit of Figure 5.4b and c.

$$I_a = I_1 = \frac{V_a}{Z_1 + Z_f}$$
$$I_b = a^2 I_1 \tag{5.22}$$
$$I_c = a I_1$$

5.5 Phase Shift in Three-Phase Transformers

5.5.1 Transformer Connections

Transformer windings can be connected in wye, delta, zigzag, or open delta. The transformers three-phase units or three-phase banks can be formed from single-phase units. Autotransformer connections should also be considered. The variety of winding connections is, therefore, large [1]. It is not the intention to describe these connections completely. The characteristics of a connection can be estimated from the vector diagrams of the primary and secondary electromotive forces (EMFs). There is a phase shift in the secondary voltages with respect to the primary voltages, depending on the connection. This is important while paralleling transformers. A vector diagram of the transformer connections can be constructed based on the following:

1. The voltages of primary and secondary windings on the same leg of the transformer are in opposition, while the induced EMFs are in the same direction. (Refer to Appendix C for further explanation.)
2. The induced EMFs in three phases are equal, balanced, and displaced mutually by a one-third period in time. These have a definite phase sequence.

Delta–wye connections are discussed, as these are most commonly used. Figure 5.5 shows polarity markings and connections of delta–wye transformers. For all liquid-immersed transformers, the polarity is subtractive according to American National Standard Institute (ANSI) standard [2]. (Refer to Appendix C for an explanation.) Two-winding transformers have their windings designated as high voltage (H) and low voltage (X). Transformers with more than two windings have their windings designated as H, X, Y, and Z. External terminals are distinguished from each other by marking with a capital letter, followed by a subscript number, i.e., H_1, H_2, and H_3.

5.5.2 Phase Shifts in Winding Connections

The angular displacement of a polyphase transformer is the time angle expressed in degrees between the line-to-neutral voltage of the reference identified terminal and the line-to-neutral voltage of the corresponding identified low-voltage terminal. In Figure 5.5a, wye-connected side voltage vectors lead the delta-connected side voltage vectors by 30°, for counterclockwise rotation of phasors. In Figure 5.5b, the delta-connected side leads the wye-connected side by 30°. For transformers manufactured according to the ANSI/IEEE (Institute of Electrical and Electronics Engineers, Inc., USA), standard [3], the *low-voltage side,*

Unsymmetrical Fault Calculations

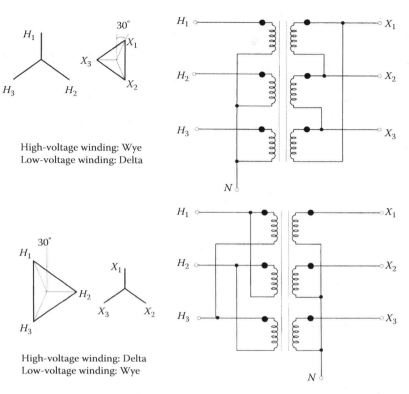

FIGURE 5.5
Winding connections and phase displacement for voltage vectors for delta–wye-connected transformers.

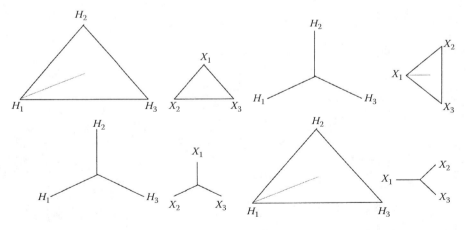

FIGURE 5.6
Phase designations of terminal markings in three-phase transformers according to ANSI/IEEE standard.

whether in wye or delta connection, has a phase shift of 30° lagging with respect to the high-voltage side phase-to-neutral voltage vectors. Figure 5.6 shows ANSI/IEEE [3] transformer connections and a phasor diagram of the delta side and wye side voltages. These relations and phase displacements are applicable to positive sequence voltages.

The International Electrotechnical Commission (IEC) allocates vector groups, giving the type of phase connection and the *angle of advance* turned though in passing from the vector representing the high-voltage side EMF to that representing the low-voltage side EMF at the corresponding terminals. The angle is indicated much like the hands of a clock, the high-voltage vector being at 12 o'clock (zero) and the corresponding low-voltage vector being represented by the hour hand. The total rotation corresponding to hour hand of the clock is 360°. Thus, Dy11 and Yd11 symbols specify 30° lead (11 being the hour hand of the clock) and Dy1 and Yd1 signify 30° lag. Table 5.1 shows some IEC vector groups of transformers and their winding connections.

TABLE 5.1

Transformer Vector Groups, Winding Connections, and Vector Diagrams

Vector Group and Phase Shift	Winding Connections	Vector diagram
Yy0 0°	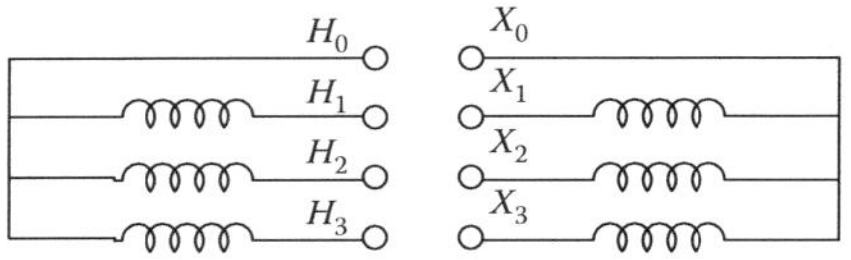	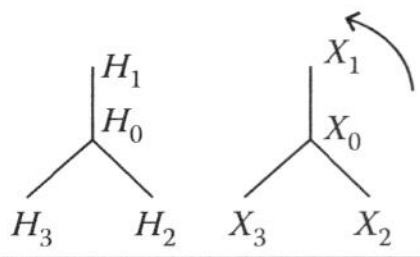
Yy6 180°	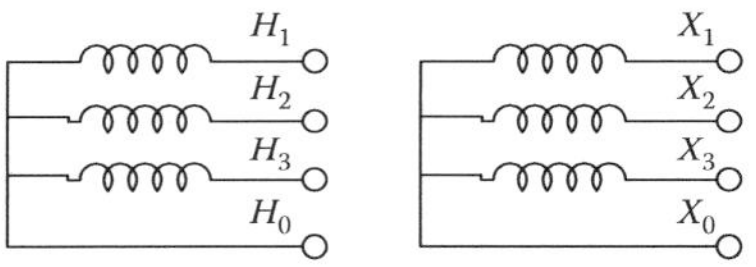	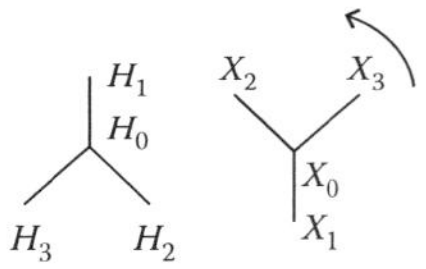
Dd0 0°	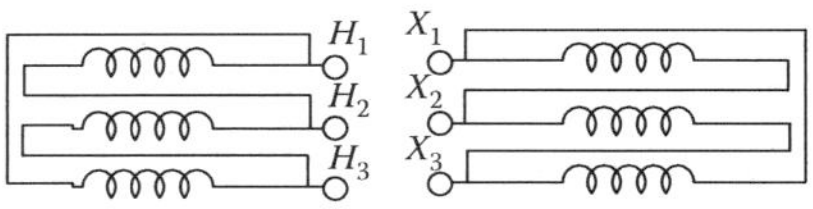	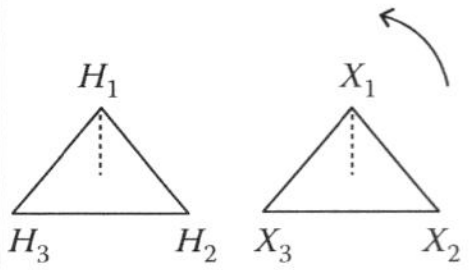
Dz0 0°	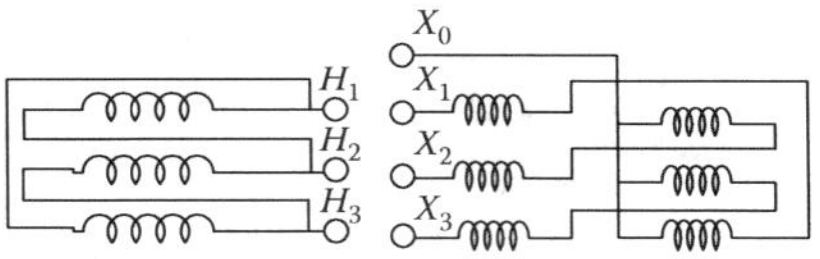	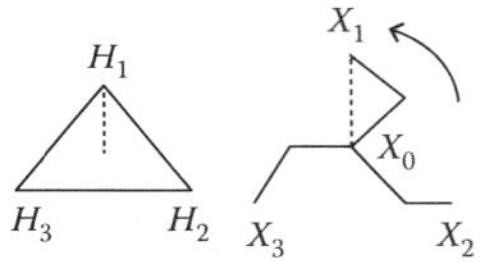
Dz6 180°	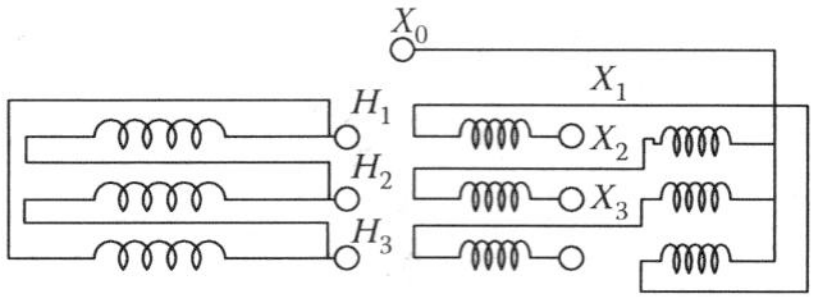	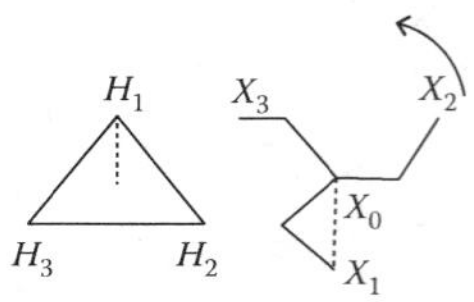
Dd6 180°	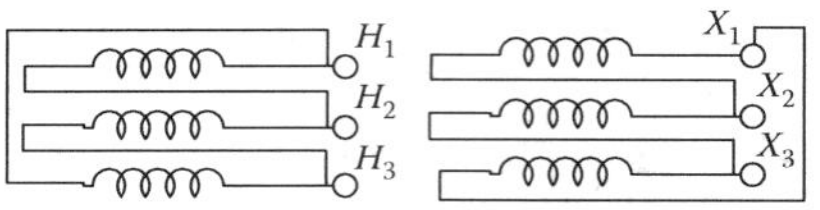	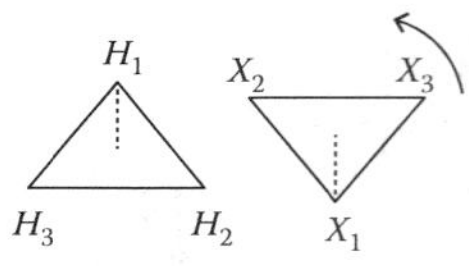

(Continued)

Unsymmetrical Fault Calculations 183

TABLE 5.1 (*Continued*)

Transformer Vector Groups, Winding Connections, and Vector Diagrams

Vector Group and Phase Ghift	Winding Connections	Vector diagram
Dy1 −30°		
Yd1 −30°		
DY11 +30°		
Yd11 +30°		
Yz1 −30°		
Yz11 +30°		

5.5.3 Phase Shift for Negative Sequence Components

The phase shifts described above are applicable to positive sequence voltages or currents. If a voltage of negative phase sequence is applied to a delta–wye-connected transformer, the phase angle displacement will be equal to the positive sequence phasors but in the opposite direction. Therefore, when the positive sequence currents and voltages on one side lead the positive sequence current and voltages on the other side by 30°, the corresponding negative sequence currents and voltages will lag by 30°. In general, if the positive sequence voltages and currents on one side lag the positive sequence voltages and currents by 30°, then on the other side the negative sequence voltages and currents will lead by 30°.

Example 5.1

Consider a balanced three-phase delta load connected across an unbalanced three-phase supply system, as shown in Figure 5.7. The currents in lines a and b are given.

The currents in the delta-connected load and also the symmetrical components of line and delta currents are required to be calculated. From these calculations, the phase shifts of positive and negative sequence components in delta windings and line currents can be established.

The line current in c is given by

$$I_c = -(I_a + I_b)$$
$$= -30 + j6.0 \text{ A}$$

The currents in delta windings are

$$I_{AB} = \frac{1}{3}(I_a - I_b) = -3.33 + j4.67 = 5.735 < 144.51° \text{ A}$$

$$I_{BC} = \frac{1}{3}(I_b - I_c) = 16.67 - j5.33 = 17.50 < -17.7° \text{ A}$$

$$I_{CA} = \frac{1}{3}(I_c - I_a) = -13.33 + j0.67 = 13.34 < 177.12° \text{ A}$$

Calculate the sequence component of the currents I_{AB}. This calculation gives

$$I_{AB1} = 9.43 < 89.57° \text{ A}$$
$$I_{AB2} = 7.181 < 241.76° \text{ A}$$
$$I_{AB0} = 0 \text{ A}$$

Calculate the sequence component of current I_a. This calculation gives

$$I_{a1} = 16.33 < 59.57° \text{ A}$$
$$I_{a2} = 12.437 < 271.76° \text{ A}$$
$$I_{a0} = 0 \text{ A}$$

This shows that the positive sequence current in the delta winding is $1/\sqrt{3}$ times the line positive sequence current, and the phase displacement is +30°, i.e.,

$$I_{AB1} = 9.43 < 89.57° = \frac{I_{a1}}{\sqrt{3}} < 30° = \frac{16.33}{\sqrt{3}} < (59.57° + 30°) \text{ A}$$

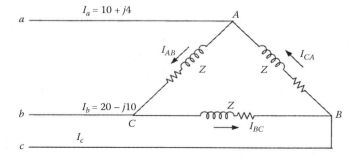

FIGURE 5.7
Balanced delta-connected load on an unbalanced three-phase power supply.

Unsymmetrical Fault Calculations

The negative sequence current in the delta winding is $1/\sqrt{3}$ times the line negative sequence current, and the phase displacement is $-30°$, i.e.,

$$I_{AB2} = 7.181 < 241.76° = \frac{I_{a2}}{\sqrt{3}} < -30° = \frac{12.437}{\sqrt{3}} < (271.76° - 30°) \text{ A}$$

This example illustrates that the negative sequence currents and voltages undergo a phase shift, which is the reverse of the positive sequence currents and voltages.

The relative magnitudes of fault currents in two-winding transformers for secondary faults are shown in Figure 5.8 on a per unit basis. The reader can verify the fault current flows as shown in the following figure.

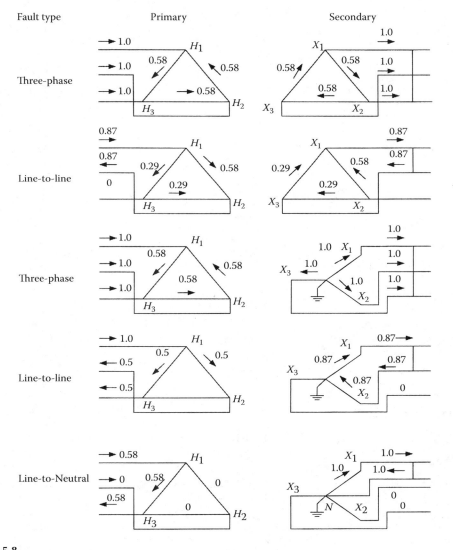

FIGURE 5.8
Three-phase transformer connections and fault current distribution for secondary faults.

(Continued)

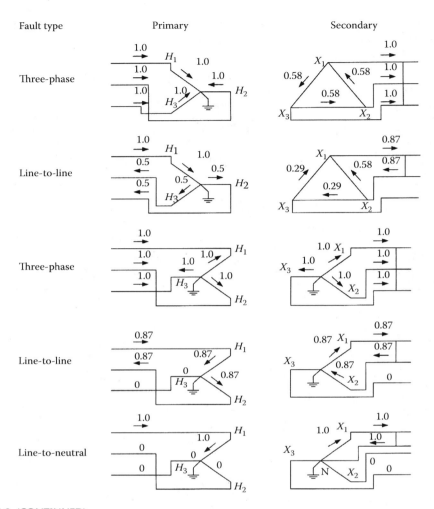

FIGURE 5.8 (CONTINUED)
Three-phase transformer connections and fault current distribution for secondary faults.

5.6 Unsymmetrical Fault Calculations

Example 5.2

The calculations using symmetrical components can best be illustrated with an example. Consider a subtransmission system as shown in Figure 5.9. A 13.8-kV generator G_1 voltage is stepped up to 138 kV. At the consumer end, the voltage is stepped down to 13.8 kV, and generator G_2 operates in synchronism with the supply system. Bus B has a 10,000-hp motor load. A line-to-ground fault occurs at bus B. It is required to calculate the fault current distribution throughout the system and also the fault voltages. The resistance of the system components is ignored in the calculations.

Unsymmetrical Fault Calculations

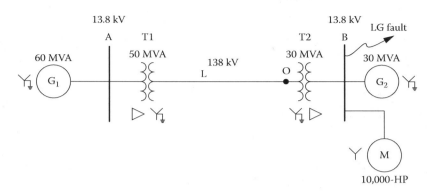

FIGURE 5.9
A single line diagram of power system for Example 5.2.

Impedance Data

The impedance data for the system components are shown in Table 5.2. Generators G_1 and G_2 are shown solidly grounded, which will not be the case in a practical installation. A high-impedance grounding system is used by utilities for grounding generators in step-up transformer configurations. Generators in industrial facilities, directly connected to the load buses are low resistance grounded, and the ground fault currents are limited to 200–400 A. The simplifying assumptions in the example are not applicable to a practical installation but clearly illustrate the procedure of calculations.

The first step is to examine the given impedance data. Generator-saturated subtransient reactance is used in the short-circuit calculations and this is termed as positive sequence reactance; 138-kV transmission line reactance is calculated from the given data for conductor size and equivalent conductor spacing. The zero sequence impedance of the transmission line cannot be completely calculated from the given data and is estimated on the basis of certain assumptions, i.e., a soil resistivity of $100\,\Omega\cdot m$.

Compiling the impedance data for the system under study from the given parameters, from manufacturers' data, or by calculation and estimation can be time-consuming. Most computer-based analysis programs have extensive data libraries and companion programs for calculation of system impedance data and line constants, which has partially removed the onus of generating the data from step-by-step analytical calculations. Appendix B provides models of line constants for coupled transmission lines, bundle conductors, and line capacitances. References [3, 4] provide analytical and tabulated data.

Next, the impedance data are converted to a common mega volt amp (MVA) base. An assumption of familiarity with per unit system is considered. The voltage transformation ratio of transformer T_2 is 138–13.2 kV, while a bus voltage of 13.8 kV is specified, which should be considered in transforming impedance data on a common MVA base. Table 5.2 shows raw impedance data and their conversion into sequence impedances.

For a single line-to-ground fault at bus B, the sequence impedance network connections are shown in Figure 5.10, with the impedance data for components clearly marked. This figure is based on the fault equivalent circuit shown in Figure 5.1b, with fault impedance $Z_f = 0$. The calculation is carried out per unit, and the units are not stated in every step of the calculation.

TABLE 5.2
Impedance Data for Example 5.2

Equipment	Description	Impedance Data	Per Unit impedance 100 MVA Base
G_1	13.8 kV, 60 MVA, 0.85 power factor generator	Subtransient reactance = 15%	$X_1 = 0.25$
		Transient reactance = 20%	$X_2 = 0.28$
		Zero sequence reactance = 8%	$X_0 = 0.133$
		Negative sequence reactance = 16.8%	
T_1	13.8–138 kV step-up transformer, 50/84 MVA, delta-wye connected. Neutral Neutral solidly grounded	$Z = 9\%$ on 50 MVA base	$X_1 = X_2 = X_0$ $= 0.18$
L	Transmission line, 5 miles long, 266.8 KCMIL, ACSR	Conductors at 15 ft (4.57 m) equivalent spacing	$X_1 = X_2 = X_0$ $= 0.15$
T_2	138–13.2 kV, 30 MVA step-down transformer, wye–delta connected	$Z = 8\%$	$X_1 = X_2 = X_0$ $= 0.24$
G_2	13.8 kV, 30 MVA, 0.85 power factor generator	Subtransient reactance = 11 %	$X_1 = 0.37$
		Transient reactance = 15%	$X_2 = 0.55$
		Zero sequence reactance = 6%	$X_0 = 0.20$
		Negative sequence reactance = 16.5%	
M	10,000 hp induction motor load	Locked rotor reactance = 16.7% on motor base kVA (consider hp ≈ 1 kVA)	$X_1 = 1.67$ $X_2 = 1.80$ $X_0 = \infty$

Note: Resistances are neglected in the calculations.
KCMIL: kilo-circular mils, the same as MCM.
ACSR: aluminum conductor steel reinforced.

The positive sequence impedance to the fault point is

$$Z_1 = \frac{j(0.25 + 0.18 + 0.04 + 0.24) \times \dfrac{j0.37 \times j1.67}{j(0.37 + 1.67)}}{j(0.25 + 0.18 + 0.04 + 0.24) + \dfrac{j0.37 \times j1.67}{j(0.37 + 1.67)}}$$

This gives $Z_1 = j0.212$.

$$Z_2 = \frac{j(0.28 + 0.18 + 0.04 + 0.24) \times \dfrac{j0.55 \times j1.8}{j(0.55 + 1.8)}}{j(0.28 + 0.18 + 0.04 + 0.24) + \dfrac{j0.55 \times j1.8}{j(0.55 + 1.8)}}$$

This gives $Z_2 = j0.266$.
$Z_0 = j0.2$. Therefore,

$$I_1 = \frac{E}{Z_1 + Z_2 + Z_0} = \frac{1}{j0.212 + j0.266 + j0.2} = -j1.475$$

$$I_2 = I_0 = -j1.475$$

$$I_a = I_0 + I_1 + I_2 = 3(-j1.475) = -j4.425 \text{ pu}$$

Unsymmetrical Fault Calculations

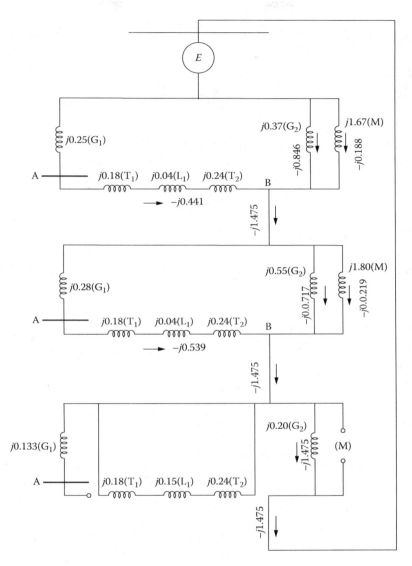

FIGURE 5.10
Sequence network connections for single line-to-ground fault (Example 5.2).

The fault currents in phases b and c are zero:

$$I_b = I_c = 0$$

The sequence voltages at a fault point can now be calculated:

$$V_0 = -I_0 Z_0 = j1.475 \times j0.2 = -0.295$$
$$V_2 = -I_2 Z_2 = j1.475 \times j0.266 = -0.392$$
$$V_1 = E - I_1 Z_1 = I_1(Z_0 + Z_2) = 1 - (-j1.475 \times j0.212) = 0.687$$

A check of the calculation can be made at this stage; the voltage of the faulted phase at fault point B is zero:

$$V_a = V_0 + V_1 + V_2 = -0.295 - 0.392 + 0.687 = 0$$

The voltages of phases b and c at the fault points are

$$
\begin{aligned}
V_b &= V_0 + a V_1 + a^2 V_2 \\
&= V_0 - 0.5(V_1 + V_2) - j0.866(V_1 - V_2) \\
&= -0.295 - 0.5(0.687 - 0.392) - j0.866(0.687 + 0.392) \\
&= -0.4425 - j0.9344 \\
|V_b| &= 1.034 \text{ pu}
\end{aligned}
$$

Similarly,

$$
\begin{aligned}
V_c &= V_0 - 0.5(V_1 + V_2) + j0.866(V_1 - V_2) \\
&= -0.4425 + j0.9344 \\
|V_c| &= 1.034 \text{ pu}
\end{aligned}
$$

The distribution of the sequence currents in the network is calculated from the known sequence impedances. The positive sequence current contributed from the right side of the fault, i.e., by G_2 and motor M is

$$-j1.475 \frac{j(0.25 + 0.18 + 0.04 + 0.24)}{j(0.25 + 0.18 + 0.04 + 0.24) + \dfrac{j0.37 \times j1.67}{j(0.37 + 1.67)}}$$

This gives $-j1.0338$. This current is composed of two components, one from the generator G_2 and the other from the motor M. The generator component is

$$(-j1.0338) \frac{j1.67}{j(0.37 + 1.67)} = -j0.8463$$

The motor component is similarly calculated and is equal to $-j0.1875$.
The positive sequence current from the left side of bus B is

$$-j1.475 \frac{\dfrac{j0.37 \times j1.67}{j(0.37 + 1.67)}}{j(0.25 + 0.18 + 0.04 + 0.24) + \dfrac{j0.37 \times j1.67}{j(0.37 + 1.67)}}$$

This gives $-j0.441$. The currents from the right side and the left side should sum to $-j1.475$. This checks the calculation accuracy.
The negative sequence currents are calculated likewise and are as follows:

In generator $G_2 = -j0.7172$
In motor $M = -j0.2191$
From left side, bus $B = -j0.5387$
From right side $= -j0.9363$

Unsymmetrical Fault Calculations 191

The results are shown in Figure 5.10. Again, verify that the vectorial summation at the junctions confirms the accuracy of calculations.

Currents in generator G_2

$$
\begin{aligned}
I_a(G_2) &= I_1(G_2) + I_2(G_2) + I_0(G_2) \\
&= -j0.8463 - j0.7172 - j1.475 \\
&= -j3.0385
\end{aligned}
$$

$$|I_a(G_2)| = 3.0385 \text{ pu}$$

$$
\begin{aligned}
I_b(G_2) &= I_0 - 0.5(I_1 + I_2) - j0.866(I_1 - I_2) \\
&= -j1.475 - 0.5(-j0.8463 - j0.7172) - j0.866(-j0.8463 + j0.7172) \\
&= -0.1118 - j0.6933
\end{aligned}
$$

$$|I_b(G_2)| = 0.7023 \text{ pu}$$

$$
\begin{aligned}
I_c(G_2) &= I_0 - 0.5(I_1 + I_2) + j0.866(I_1 - I_2) \\
&= 0.1118 - j0.6933
\end{aligned}
$$

$$|I_c(G_2)| = 0.7023 \text{ pu}$$

This large unbalance is noteworthy. It gives rise to increased thermal effects due to negative sequence currents and results in overheating of the generator rotor. A generator will be tripped quickly on negative sequence currents.

Currents in motor M: The zero sequence current in the motor is zero, as the motor wye-connected windings are not grounded as per industrial practice in the USA. Thus,

$$
\begin{aligned}
I_a(M) &= I_1(M) + I_2(M) \\
&= -j0.1875 - j0.2191 \\
&= -j0.4066
\end{aligned}
$$

$$|I_a(M)| = 0.4066 \text{ pu}$$

$$I_b(M) = -0.5(-j0.4066) - j0.866(0.0316) = 0.0274 + j0.2033$$
$$I_c(M) = -0.0274 + j0.2033$$
$$|I_b(M)| = |I_c(M)| = 0.2051 \text{ pu}$$

The summation of the line currents in the motor M and generator G_2 are

$$I_a(G_2) + I_a(M) = -j3.0385 - j0.4066 = -j3.4451$$
$$I_b(G_2) + I_b(M) = -0.118 - j0.6993 + 0.0274 + j0.2033 = -0.084 - j0.490$$
$$I_c(G_2) + I_c(M) = 0.1118 - j0.6933 - 0.0274 + j0.2033 = 0.084 - j0.490$$

Currents from the left side of the bus B are

$$
\begin{aligned}
I_a &= -j0.441 - j0.5387 \\
&= -j0.98
\end{aligned}
$$

$$
\begin{aligned}
I_b &= -0.5(-0.441 - j0.5387) - j0.866(-0.441 + j0.5387) \\
&= 0.084 + j0.490
\end{aligned}
$$

$$I_c = -0.084 + j0.490$$

192   *Short-Circuits in AC and DC Systems*

These results are consistent as the sum of currents in phases b and c at the fault point from the right and left side is zero and the summation of phase a currents gives the total ground fault current at $b = -j4.425$. The distribution of currents is shown in a three-line diagram (Figure 5.11).

Continuing with the example, the currents and voltages in the transformer T_2 windings are calculated. We should correctly apply the phase shifts for positive and negative sequence components when passing from delta secondary to wye primary of the transformer. The positive and negative sequence currents on the wye side of transformer T_2 are

$$I_{1(p)} = I_1 < 30° = -j0.441 < 30° = 0.2205 - j0.382$$
$$I_{2(p)} = I_2 < -30° = -j0.5387 < -30° = -0.2695 - j0.4668$$

Also, the zero sequence current is zero. The primary currents are

$$I_{a(p)} = I_{1(p)} + I_{2(p)}$$
$$= 0.441 < 30° + 0.5387 < -30° = -0.049 - j0.8487$$
$$I_{b(p)} = a^2 I_{1(p)} + a I_{2(p)} = -0.0979$$
$$I_{c(p)} = a I_{1(p)} + a^2 I_{2(p)} = -0.049 - j0.8487$$

Currents in the lines on the delta side of the transformer T_1 are similarly calculated. The positive sequence component, which underwent a 30° positive shift from delta to wye in transformer T_2, undergoes a −30° phase shift; as for an ANSI connected transformer, it is the low-voltage vectors, which lag the high-voltage side vectors. Similarly, the negative sequence component undergoes a positive phase shift. The currents on the delta side of transformers T_1 and T_2 are identical in amplitude and phase. Note that 138 kV line is considered lossless. Figure 5.11 shows the distribution of currents throughout the distribution system.

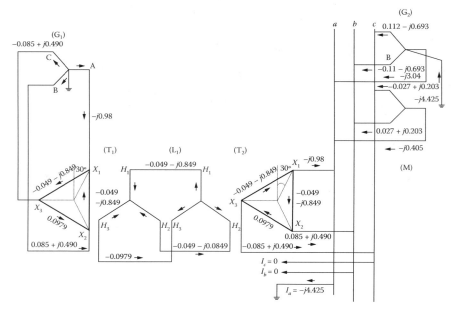

FIGURE 5.11
Three-line diagram of fault current distribution (Example 5.2).

Unsymmetrical Fault Calculations 193

The voltage on the primary side of transformer T_2 can be calculated. The voltages undergo the same phase shifts as the currents. Positive sequence voltage is the base fault positive sequence voltage, phase shifted by 30° (positive) minus the voltage drop in transformer reactance due to the positive sequence current:

$$V_1(p) = 1.0 < 30° - jI_{1(p)}X_2$$
$$= 1.0 < 30° - (j0.441 < 30°)(j0.24)$$
$$= 0.9577 + j0.553$$
$$V_2(p) = 0 - I_{2(p)}X_2$$
$$= -(0.539 < -30°)(j0.24)$$
$$= 0.112 - j0.0647$$

Thus,

$$V_{a(p)} = 0.9577 + j0.553 + 0.112 - j0.0647 = 1.0697 + j0.4883 = 1.17 < 24.5°$$
$$V_{b(p)} = -0.5(V_{1(p)} + V_{2(p)}) - j0.866(V_{1(p)} - V_{2(p)})$$
$$= -j0.9763$$
$$V_{c(p)} = 0.5(V_{1(p)} + V_{2(p)}) - j0.866(V_{2(p)} - V_{1(p)})$$
$$= 1.0697 + j0.4883 = 1.17 < 155.5°$$

Note the voltage unbalance caused by the fault.

5.7 System Grounding

System grounding refers to the electrical connection between the phase conductors and ground, and dictates the manner in which the neutral points of wye-connected transformers and generators or artificially derived neutral systems through delta–wye or zig-zag transformers are grounded. The equipment grounding refers to the grounding of the exposed metallic parts of the electrical equipment, which can become energized and create a potential to ground, say due to breakdown of insulation or fault, and can be a potential safety hazard. The safety of the personnel and human life is of importance. The safety grounding is to establish an equipotential surface in the work area to mitigate shock hazard. The utility systems at high-voltage transmission level, subtransmission level, and distribution level are solidly grounded. The utility generators connected through step-up transformers are invariably high resistance grounded (HRG) through a distribution transformer with secondary loading resistor. The industrial systems at medium-voltage level are low resistance grounded. The implications of system grounding are as follows:

- Enough ground fault current should be available to selectively trip the faulty section with minimum disturbance to the system. Ground faults are cleared even faster than the phase faults to limit equipment damage.
- Line-to-ground fault is the most important cause of system temporary overvoltage that dictates the selection of surge arresters.

- Grounding should prevent high overvoltages of ferroresonance, overvoltages of arcing type ground fault, and capacitive–inductive resonant couplings. If a generator neutral is left ungrounded, there is a possibility of generating high voltages through inductive–capacitive couplings. Ferroresonance can also occur due to the presence of generator PT's (Potential Transformers). In ungrounded systems, a possibility of resonance with high voltage generation, approaching five times or more of the system voltage exists for values of X_0/X_1 between 0 and −40. For the first phase-to-ground fault, the continuity of operations can be sustained, though unfaulted phases have $\sqrt{3}$ times the normal line-to-ground voltage. All unremoved faults, thus, put greater than normal voltage on system insulation and increased level of conductor and motor insulation may be required. The grounding practices in the industry have withdrawn from this method of grounding.
- The relative magnitude of the fault currents depends upon sequence impedances.
- In industrial systems, the continuity of processes is important. The current industry practice is to have HRG systems for low-voltage distributions.

Figure 5.12 shows the system grounding methods. A brief discussion follows.

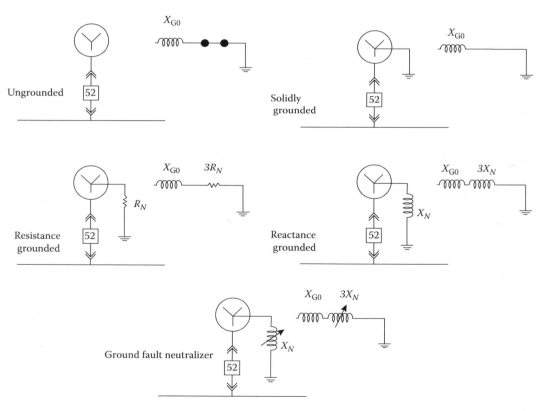

FIGURE 5.12
Methods of system grounding.

Unsymmetrical Fault Calculations 195

5.7.1 Solidly Grounded Systems

In a solidly grounded system, no intentional impedance is introduced between the system neutral and ground. These systems meet the requirement of "effectively grounded" systems in which the ratio X_0/X_1 is positive and less than 3.0, and the ratio R_0/X_0 is positive and less than 1, where X_1, X_0, and R_0 are the positive sequence reactance, zero sequence reactance, and zero sequence resistance, respectively. The coefficient of grounding (COG) is defined as a ratio of E_{LG}/E_{LL} in percentage, where E_{LG} is the highest root mean square (rms) voltage on a sound phase, at a selected location, during a fault affecting one or more phases to ground, and E_{LL} is the rms phase-to-phase power frequency voltage that is obtained at the same location with the fault removed. Calculations in Example 5.2 show the fault voltage rises on unfaulted phases. Solidly grounded systems are characterized by a COG of 80%. By contrast, *for ungrounded systems*, definite values cannot be assigned to ratios X_0/X_1 and R_0/X_0. The ratio X_0/X_1 is negative and may vary from low to high values. The COG approaches 120%. For values of X_0/X_1 between 0 and −40, a possibility of resonance with consequent generation of high voltages exists. The overvoltages based on relative values of sequence impedances are plotted in Reference [4].

The COG affects the selection of rated voltage of the surge arresters and stresses on the insulation systems. Solidly grounded systems are generally characterized by COG of 80%. Approximately, a surge arrester with its rated voltage calculated on the basis of the system voltage multiplied by 0.8 can be applied.

The low-voltage systems in industrial power distribution systems used to be solidly grounded. However, this trend is changing and high resistance grounding is being adopted.

The solidly grounded systems have an advantage of providing effective control of overvoltages, which become impressed on or are self-generated in the power system by insulation breakdowns and restriking faults. Yet, these give the highest arc fault current and consequent damage and require immediate isolation of the faulty section. Single-line-to-ground fault currents can be higher than the three-phase fault currents. These high magnitudes of fault currents have two-fold effects that are as follows:

- Higher burning or equipment damage
- Interruption of the processes, as the faulty section must be selectively isolated without escalation of the fault to unfaulted sections

The arcing faults are caused by insulation failure, loose connections, or accidents. The arc behavior is difficult to predict and model because of spasmodic nature of the arc fault. This is due to elongation and blowout effects and arc reignition. Arc travels from point to point and physical flexing of cables and structures can occur. Arcing faults can exhibit low levels because of the impedance of the arc fault circuit itself. Arcing faults can be discontinuous requiring a certain minimum voltage for reignition. The limits of the acceptable damage to material for arc fault currents of 3,000–26,000 A in 480 V systems have been established by testing [5, 6] and are given by the following equation:

$$\text{Fault damage} \propto (I)^{1.5} t \tag{5.23}$$

where I is the arc fault current and t is the duration in seconds.

$$V_D = K_s(I)^{1.5}t(\text{in})^3 \tag{5.24}$$

where K_s is the burning rate of material in $\text{in}^3/\text{As}^{1.5}$, V_D is acceptable damage to material in in^3, I is the arc fault current, t is the duration of flow of fault current, and K_s depends upon type of material and is given by

$$
\begin{aligned}
K_s &= 0.72 \times 10^{-6} \text{ for copper} \\
&= 1.52 \times 10^{-6} \text{ for aluminium} \\
&= 0.66 \times 10^{-6} \text{ for steel}
\end{aligned} \tag{5.25}
$$

NEMA [7] assumes a practical limit for the ground fault protective devices, so that

$$(I)^{1.5}t \triangleleft 250I_r \tag{5.26}$$

I_r is the rated current of the conductor, bus, disconnect or circuit breaker to be protected.
Combining these equations, we can write:

$$V_D = 250K_sI_r \tag{5.27}$$

As an example, consider a circuit of 4000 A. Then the NEMA practical limit is 1.0×10^6 (A)$^{1.5}$·s and the permissible damage to copper from Equation (5.24) is 0.72 in^3. To limit the arc fault damage to this value, the maximum fault clearing time can be calculated. Consider that the arc fault current is 20 kA. Then, the maximum fault clearing time including the relay operating time and breaker interrupting time is 0.35 s. It is obvious that vaporizing 0.72 in^3 of copper on a ground fault, which is cleared according to established standards, is still the determining factor to the operation of the equipment. A shutdown and repairs will be needed after the fault incidence.

The arc fault current is not of the same magnitude as the three-phase fault current, due to voltage drop in the arc. In the low-voltage 480 V systems, it may be 50%–60% of the bolted three-phase current, while for medium-voltage systems, it will approach three-phase bolted fault current but somewhat lower.

Due to high arc fault damage and interruption of processes, the solidly grounded systems are not in much use in the industrial distribution systems. However, AC circuits of less than 50 V and circuits of 50–1000 V for supplying premises wiring systems and single-phase 120/240 V control circuits must be solidly grounded according to the National Electrical Code (NEC) [8].

5.7.2 Resistance Grounding

An impedance grounded system has a resistance or reactance connected in the neutral circuit to ground, as shown in Figure 5.12b. In a low-resistance grounded system, the resistance in the neutral circuit is selected so that the ground fault is limited to approximately full load current or even lower, typically 200–400 A. The arc fault damage is

Unsymmetrical Fault Calculations 197

reduced and these systems provide an effective control of the overvoltages generated in the system by resonant capacitive–inductive couplings and restriking ground faults. Though the ground fault current is much reduced, it cannot be allowed to be sustained and selective tripping must be provided to isolate the faulty section. For a ground fault current limited to 400 A, the pickup sensitivity of modern ground fault devices can be even lower than 5 A. Considering the available fault current of 400 A and the relay pickup of 5 A, approximately 98.75% of the transformer or generator windings from the line terminal to neutral are protected. This assumes a linear distribution of voltage across the winding. (Practically the pickup will be higher than the low set point of 5 A.) The incidence of ground fault occurrence toward the neutral decreases as square of the winding turns. Medium-voltage distribution systems in industrial distributions are commonly low resistance grounded.

The low-resistance grounded systems are adopted at medium voltages, 13.8, 4.16, and 2.4 kV for industrial distribution systems. Also, industrial bus-connected generators were commonly low resistance grounded. A recent trend in industrial bus-connected medium-voltage generator grounding is hybrid grounding systems [9].

5.7.2.1 High-Resistance Grounded Systems

HRG systems limit the ground fault current to a low value, so that an immediate disconnection on occurrence of a ground fault is not required. It is well documented that to control over voltages in the HRG systems, the grounding resistor should be selected so that

$$R_n = \frac{V_{ln}}{3I_c} \tag{5.28}$$

where V_{ln} is the line to neutral voltage and I_c is the stray capacitance current of each phase conductor. Figure 5.13 shows transient voltage in percent of normal line-to-ground crest voltage versus the resistor kW/charging capacitive kVA. The transients are a minimum when this ratio is unity [4]. This leads to the requirement of accurately calculating the stray capacitance currents in the system [9]. Cables, motors, transformers, surge arresters, generators—all contribute to the stray capacitance current. Surge capacitors connected line-to-ground must be considered in the calculations. Once the system stray capacitance is determined, then the charging current per phase, I_c, is given by

$$I_c = \frac{V_{ln}}{X_{co}} \tag{5.29}$$

where X_{co} is the capacitive reactance of each phase, stray capacitance considered lumped together.

This can be illustrated with an example. A high-resistance grounding system for a wye-connected neutral of a 13.8 kV–0.48 transformer is shown in Figure 5.14a. This shows that the stray capacitance current per phase of all the distribution system connected to the secondary of the transformer is 0.21 A per phase, assumed to be balanced in each phase. Generally for low-voltage distribution systems, a stray capacitance current of 0.1 A per MVA of transformer load can be taken, even though this rule of thumb is no substitute for

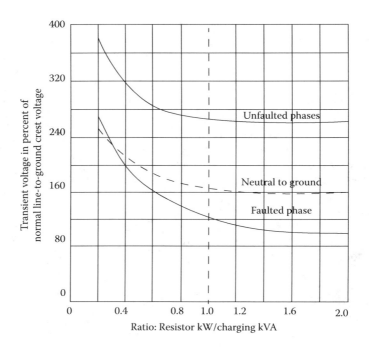

FIGURE 5.13
Overvoltages versus ratio of resistor kW/charging kVA.

accurate calculations of stray capacitance currents. Figure 5.14a shows that under no fault condition, the vector sum of three capacitance currents is zero, as these are 90° displaced with respect to each voltage vector and therefore 120° displaced with respect to each other. Thus, the grounded neutral does not carry any current and the neutral of the system is held at the ground potential (Figure 5.14b). As

$$I_{c1} + I_{c2} + I_{c3} = 0 \tag{5.30}$$

no capacitance current flows into the ground. On occurrence of a ground fault, say in phase a, the situation is depicted in Figure 5.14c and d. The capacitance current of faulted a phase is short-circuited to ground. The faulted phase, assuming zero fault resistance is at the ground potential (Figure 5.14d) and the other two phases have line-to-line voltages with respect to ground. Therefore, the capacitance current of the unfaulted phases b and c increases proportional to voltage increase, i.e., $\sqrt{3} \times 0.21 = 0.365$ A. Moreover this current in phase b and c reverses, flows through the transformer windings, and sums up in the transformer winding of phase a. Figure 5.14e shows that this vector sum = 0.63 A.

Now consider that the ground current through the grounding resistor is limited to 1 A only. This is acceptable according to Equation 5.28 as the stray capacitance current is 0.63 A. This resistor ground current also flows through the transformer phase winding a to the fault and the total ground fault current is $I_g = \sqrt{1^2 + 0.63^2} = 1.182$ A (Figure 5.14e).

Unsymmetrical Fault Calculations

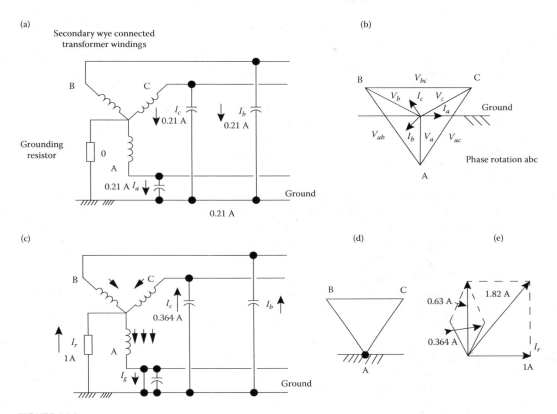

FIGURE 5.14
(a and b) Stray currents under no fault conditions, (c) flow of phase capacitance and ground currents for a phase-to-ground fault phase *a*, (d) voltages to ground, and (e) summation of capacitance and resistor currents.

The above analysis assumes a full neutral shift, ignores the fault impedance itself, and assumes that the ground grid resistance and the system zero sequence impedances are zero. Practically, the neutral shift will vary.

Some obvious advantages are the following:

- The resistance limits the ground fault current and, therefore, reduces burning and arcing effects in switchgear, transformers, cables, and rotating equipment.
- It reduces mechanical stresses in circuits and apparatus carrying fault current.
- It reduces arc blast or flash hazard to personnel who happen to be in close proximity of ground fault.
- It reduces line-to-line voltage dips due to ground fault and three phase loads can be served.
- Control of transient overvoltages is secured by proper selection of the resistor.

The ground fault protection in a *n* interconnected large system with cogenerators and utility ties is illustrated in Reference [10].

5.7.2.1.1 Limitations of HRG Systems

The limitation of the system is that the capacitance current should not exceed approximately 10 A to prevent immediate shutdowns. As the system voltage increases, so does the capacitance currents. This limits the applications to systems of rated voltages of 4.16 kV and below.

Though immediate shutdown is prevented, the fault situation should not be prolonged. The fault should be localized and removed. There are three reasons for this that are as follows:

1. Figure 5.14d shows that the unfaulted phases have voltage rise by a factor of $\sqrt{3}$ to ground. This increases the normal insulation stresses between phase to ground. This may be of special concern for low-voltage cables. If the time required to de-energize the system is indefinite, 173% insulation level for the cables must be selected [11]. However, NEC does not specify 173% insulation level and for 600 V cables insulation levels correspond to 100% and 133%. Also, Reference [7] specifies that the actual operating voltage on cables should not exceed 5% during continuous operation and 10% during emergencies. This is of importance when 600 V nominal three-phase systems are used for power distributions. The DC loads served through 6-pulse converter systems will have a DC voltage of 648 and 810 V, respectively, for 480 and 600 V rms AC systems.

2. Low levels of fault currents if sustained for a long time may cause irreparable damage. Though the burning rate is slow, the heat energy released over the course of time can damage cores and windings of rotating machines even for ground currents as low as 3-4 A. This has been demonstrated in test conditions [12].

3. A first ground fault remaining in the system increases the probability of a second ground fault on other phase. If this happens, then it amounts to a two-phase to ground fault with some interconnecting impedance depending upon the fault location. The potentiality of equipment damage and burnout increases.

Example 5.3

Figure 5.15a shows a 5 MVA 34.5–2.4 kV delta–delta transformer serving industrial motor loads. It is required to derive an artificial neutral through a wye–delta-connected transformer and high resistance ground the 2.4 kV secondary system through a grounding resistor. The capacitance charging current of the system is 8 A. Calculate the value of the resistor to limit the ground fault current through the resistor to 10 A. Neglect transformer resistance and source resistance.

The connection of sequence networks is shown in Figure 5.15b and the given impedance data reduced to a common 100 MVA base are shown in Table 5.3. Motor wye-connected neutrals are left ungrounded in the industrial systems, and therefore motor zero sequence impedance is infinite. (This contrasts with grounding practices in some European countries, where the motor neutrals are grounded.) The source zero sequence impedance can be calculated based on the assumption of equal positive and negative sequence reactances. The motor voltage is 2.3 kV and, therefore it's per unit reactance on 100 MVA base is given by

$$\left(\frac{16.7}{1.64}\right)\left(\frac{2.3}{2.4}\right)^2 = 9.35$$

Similarly, the grounding transformer per unit calculations should be adjusted for correct voltages:

$$X_0 = \left(\frac{1.5}{0.06}\right)\left(\frac{2.4}{2.4/\sqrt{3}}\right)^2 = 75$$

Unsymmetrical Fault Calculations

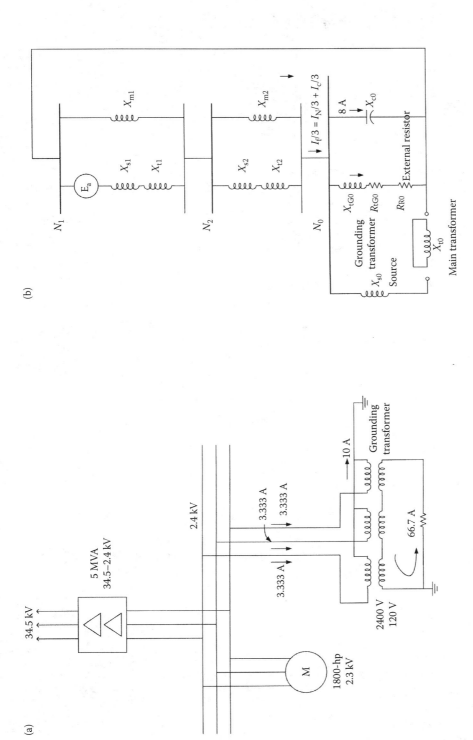

FIGURE 5.15
(a) Artificially derived neutral grounding in a delta–delta system through a wye–delta grounding transformer (Example 5.3) and (b) connection of sequence impedances for high-resistance fault calculations (Example 5.3).

TABLE 5.3

Impedance Data for Example 5.3—High Resistance Grounding

Equipment	Given Data	Per Unit Impedance on 100 MVA Base
34.5 kV source	Three-phase fault = 1500 MVA Line-to-ground fault = 20 kA sym	$X_{s1} = X_{s2} = 0.067$ $X_{s0} = 0.116$
34.5–2.4 kV, 5 MVA transformer, delta–delta connected	$X_1 = X_2 = X_0 = 8\%$	$X_{t1} = X_{t2} = X_{t0} = 1.60$
2.3 kV 1800 hp (1640 kVA) induction motor load	Locked rotor reactance = 16.7% (on motor base kVA)	$X_{m1} = X_{m2} = 9.35$
Grounding transformer, 60 kVA, wye–delta connected 2400:120 V	$X_0 = 1.5\%$ $R_0 = 1.0\%$	$X_0 = 75$ $R_0 = 50$

The equivalent positive and negative sequence reactances are 1.41 per unit each. The zero sequence impedance of the grounding transformer is $50 + j75$ per unit. The total fault current should be limited to $10 - j8 = 12.80$ A. Thus, the required impedance is

$$Z_t = \left(\frac{2400 / \sqrt{3}}{12.8 / 3} \right) = 324.8 \text{ ohms}$$

The base ohms (100 MVA base) = 0.0576. The required $Z_t = 324.8/$base ohms = 5638.9 per unit. This shows that the system positive and negative sequence impedances are low compared to the desired total impedance in the neutral circuit. The system positive and negative sequence impedances can, therefore, be neglected.

$I_{R0} = 10/3 = 3.33$ A. Therefore, $Z_{R0} = (2400/\sqrt{3})/3.33 = 416.09$ ohms = 416.09/base ohms = 7223.9 per unit. The additional resistor to be inserted:

$$R_{R0} = \sqrt{Z_{R0} - X_{tG0}} - R_{tG0}$$
$$= \sqrt{7223.9^2 - 75^2} - 50$$
$$= 7173.5 \text{ pu}$$

where R_{R0} is added resistor, Z_{R0} is the total impedance in the ground circuit, and X_{tG0} and R_{tG0} are reactance and resistance of the grounding transformer, Figure 5.15b. Multiplying by base ohms, the required resistance = 413.2 ohms.

These values are in symmetrical component equivalents. In actual values, referred to 120 V secondary, the resistance value is

$$R_R = \left(\frac{120}{2400} \right)^2 413.2 \times 3 = 3.1 \text{ ohms}$$

If we had ignored all the sequence impedances, including that of the grounding transformer, the calculated value is 3.12 ohms. This is often done in the calculations for grounding resistance for HRG systems, and all impedances including that of the grounding transformer can be ignored without appreciable error in the final results. The grounding transformer should be rated to permit continuous operation, with a ground fault on the system. The per phase grounding transformer kVA requirement is 2.4 (kV) × 3.33 A = 8 kVA, *i.e., a total of* $8 \times 3 = 24\,kVA$. The grounding transformer of the example is, therefore, adequately rated.

Unsymmetrical Fault Calculations

5.7.2.2 Coefficient of Grounding

Simplified equations can be applied for calculation of COG. Sometimes we define EFF (IEC standards, earth fault factor). It is simply

$$EFF = \sqrt{3}COG \tag{5.31}$$

COG can be calculated by the equations described below and more rigorously by the sequence component matrix methods as illustrated above.

Single line-to-ground fault:

$$COG\ (phase\ b) = -\frac{1}{2}\left(\frac{\sqrt{3}k}{2+k} + j1\right)$$
$$COG\ (phase\ c) = -\frac{1}{2}\left(\frac{\sqrt{3}k}{2+k} - j1\right) \tag{5.32}$$

Double line-to-ground fault:

$$COG\ (phase\ a) = \frac{\sqrt{3}k}{1+2k} \tag{5.33}$$

where k is given by

$$k = \frac{Z_0}{Z_1} \tag{5.34}$$

To take into account of fault resistance, k is modified as follows:

Single line-to-ground fault:

$$k = \left(R_0 + R_f + jX_0\right)/\left(R_1 + R_f + jX_1\right) \tag{5.35}$$

For double line-to-ground fault:

$$k = \left(R_0 + 2R_f + jX_0\right)/\left(R_1 + 2R_f + jX_1\right) \tag{5.36}$$

If R_0 and R_1 are zero, then the above equations reduce to the following:

For single line-to-ground fault:

$$COG = \frac{\sqrt{k^2 + k + 1}}{k + 2} \tag{5.37}$$

For double line-to-ground fault:

$$COG = \frac{\sqrt{3}k}{2k + 1} \tag{5.38}$$

$$\text{where } k \text{ is now} = X_0/X_1 \tag{5.39}$$

In general, fault resistance will reduce COG, except in low resistance systems. The numbers on the curves indicate COG for *any type of fault* in percent of unfaulted line-to-line voltage for the area bounded by the curve and the axes. All impedances must be on the same MVA base.

Example 5.4

Calculate the COG at the faulted bus B in Example 5.2. If generator G_2 is grounded through a 400 A resistor, what is the COG?

In Example 5.2, all resistances are ignored. A voltage of 1.034 per unit was calculated on the unfaulted phases, which gives a COG of 0.597.

If the generator is grounded through 400 A resistor, then $R_0 = 19.19$ ohms, the positive sequence reactance is 0.4 ohms, and the zero sequence reactance is 0.38 ohms, which is much smaller than R_0. In fact, in a resistance grounded or HRG system, the sequence components are relatively small and the ground fault current can be calculated based upon the grounding resistor alone. The total ground fault current at bus 4 will reduce to approximately 400 A. This gives a COG of approximately 100%. This means that phase-to-ground voltage on unfaulted phases will be equal to line-to-line voltage.

5.8 Open Conductor Faults

Symmetrical components can also be applied to the study of open conductor faults. These faults are in series with the line and are called series faults. One or two conductors may be opened, due to mechanical damage or by operation of fuses on unsymmetrical faults.

5.8.1 Two-Conductor Open Fault

Consider that conductors of phases b and c are open-circuited. The currents in these conductors then go to zero.

$$I_b = I_c = 0 \tag{5.40}$$

The voltage across the unbroken phase conductor is zero, at the point of break, Figure 5.16a.

$$V_{a0} = V_{a01} + V_{a02} + V_{a0} = 0$$
$$I_{a1} = I_{a2} = I_{a0} = \frac{1}{3} I_a \tag{5.41}$$

This suggests that sequence networks can be connected in series as shown in Figure 5.16b.

5.8.2 One-Conductor Open

Now consider that phase a conductor is broken (Figure 5.17a):

$$I_a = 0 \quad V_{b0} - V_{c0} = 0 \tag{5.42}$$

Unsymmetrical Fault Calculations

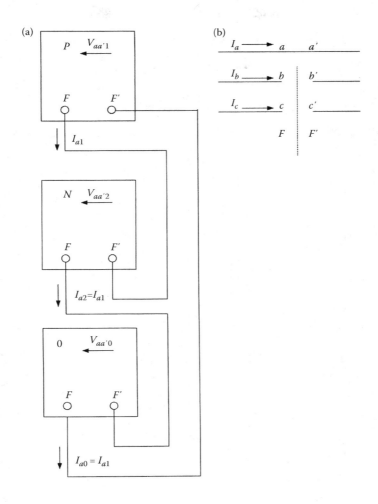

FIGURE 5.16
(a) Two-conductor open series fault and (b) connection of sequence networks.

Thus,

$$V_{a01} = V_{a02} = V_{a00} = \frac{1}{3}V_{a0} \qquad (5.43)$$

$$I_{a1} + I_{a2} + I_{a0} = 0$$

This suggests that sequence networks are connected in parallel (Figure 5.17b).

Example 5.5

Consider that one conductor is broken on the high-voltage side at the point marked O in Figure 5.9. The equivalent circuit is shown in Figure 5.18.

An induction motor load of 10,000 hp was considered in the calculations for a single line-to-ground fault in Example 5.2. All other *static loads*, i.e., lighting and resistance heating loads, were ignored, as these do not contribute to short-circuit currents. Also,

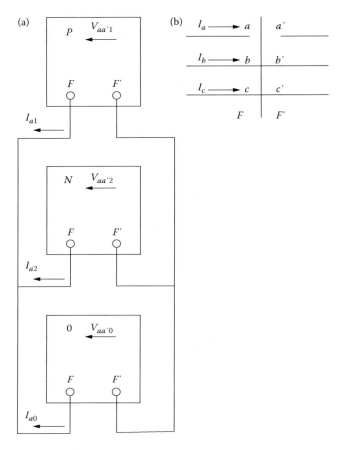

FIGURE 5.17
(a) One-conductor open series fault and (b) connection of sequence networks.

all drive system loads, connected through converters are ignored, unless the drives are in a regenerative mode. If there are no loads and a broken conductor fault occurs, no load currents flow.

Therefore, for broken conductor faults, all loads, irrespective of their types, should be modeled. For simplicity of calculations, again consider that a 10,000 hp induction motor is the only load. Its positive and negative sequence impedances for load modeling will be entirely different from the impedances used in short-circuit calculations.

Chapter 9 and also Volume 2 discuss induction motor equivalent circuits for positive and negative sequence currents. The range of induction motor impedances in per unit (based upon motor kVA base) are

$$X_{1r} = 0.14 - 0.2 < 83° - 75°$$
$$X_{load}^+ = 0.9 - 0.95 < 20° - 26° \tag{5.44}$$
$$X_{load}^- \approx X_{1r}$$

where X_{1r} = induction motor locked rotor reactance at its rated voltage, X_{load}^+ = positive sequence load reactance, and X_{load}^- = negative sequence load reactance.

Unsymmetrical Fault Calculations

The load impedances for the motor are as shown in Figure 5.18. For an open conductor fault as shown in this figure, the load is not interrupted. Under normal operating conditions, the motor load is served by generator G_2, and in the system of Figure 5.18 no current flows in the transmission line L. If an open conductor fault occurs, generator G_2, operating in synchronism, will trip on operation of negative sequence current relays. To proceed with the calculation, assume that G_2 is out of service, when the open conductor fault occurs.

The equivalent impedance across an open conductor is

$$[j0.25 + j0.18 + j0.04 + j0.24 + 9.9 + j4.79]_{pos}$$
$$+\{[j0.28 + j0.18 + j0.04 + j0.24 + 0.20 + j1.65]_{neg}$$
$$\text{in parallel with } [j0.18 + j0.15 + j0.24]_{zero}\}$$
$$= 9.918 + j6.771 = 12.0 < 34.32°$$

The motor load current is

$$0.089 < -25.84° \text{ pu (at 0.9 power factor [PF])}$$

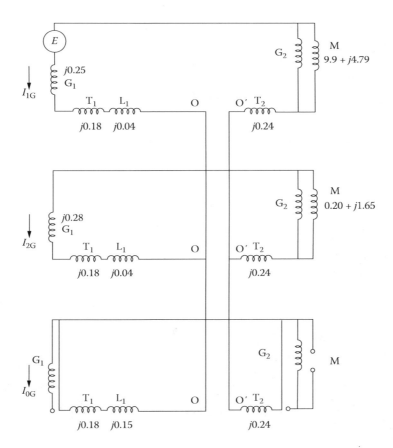

FIGURE 5.18
Equivalent circuit of an open conductor fault (Example 5.5).

The load voltage is assumed as the reference voltage. Thus, the generated voltage is

$$V_g = 1 < 0° + (0.089 < -25.84°)(j0.25 + j0.18 + j0.04 + j0.24)$$
$$= 1.0275 + j0.0569 = 1.0291 < 3.17°$$

The positive sequence current is

$$I_{1g} = V_G / Z_t = \left[\frac{1.0291 < 3.17°}{12.00 < 34.32°} \right]$$
$$= 0.0857 < 31.15° = 0.0733 + j0.0443$$

The negative sequence and zero sequence currents are

$$I_{2g} = -I_{1g} \frac{Z_0}{Z_2 + Z_0}$$
$$= (0.0857 < 31.15°) \left[\frac{0.53 < 90°}{2.897 < 86.04°} \right]$$
$$0.0157 < 215.44°$$

$$I_{0g} = -I_{1g} \frac{Z_2}{Z_2 + Z_0}$$
$$= 0.071 < 210.33°$$

Calculate line currents:

$$I_{ag} = I_{1g} + I_{2g} + I_{0g} = 0$$
$$I_{bg} = a^2 I_{1g} + a I_{2g} + I_{0g} = 0.1357 < 250.61°$$
$$I_{cg} = 0.1391 < -8.78°$$

The line currents in the two phases are increased by 52%, indicating serious overheating. A fully loaded motor will stall. The effect of negative sequence currents in the rotor is simulated by the following equation:

$$I^2 = I_1^2 + k I_2^2 \tag{5.45}$$

where k can be as high as 6. The motors are disconnected from service by anti-single phasing devices and protective relays.

The "long way" of calculation using symmetrical components, illustrated by the examples, shows that, even for simple systems, the calculations are tedious and lengthy. For large networks consisting of thousands of branches and nodes, these are impractical. There is an advantage in the hand calculations, in the sense that verification is possible at each step and the results can be correlated with the expected final results. For large systems, matrix methods and digital simulation of the systems are invariable. This gives rise to an entirely new challenge for analyzing the cryptical volumes of analytical data, which can easily mask errors in system modeling and output results.

Problems

5.1 A double line-to-ground fault occurs on the high-voltage terminals of transformer T_2 in Figure 5.9. Calculate the fault current distribution and the fault voltages throughout the system, similar to Example 5.2.

5.2 Repeat Problem 1 for a line-to-line fault and then for a line-to-ground fault.

5.3 Calculate the percentage reactance of a 60/100 MVA, 13.8–138 kV transformer in Figure P5.1 to limit the three-phase fault current at bus A to 28 kA symmetrical, for a three-phase symmetrical fault at the bus. Assume only nondecaying AC component of the short-circuit current and neglect resistances.

5.4 In Problem 5.3, another similar generator is to be added to bus A. What is the new short-circuit current? What can be done to limit the three-phase short-circuit level at this bus to 36 kA?

5.5 Calculate the three-phase and single line-to-ground fault current for a fault at bus C, for the system shown in Figure P4.1, Chapter 4. As all the generators are connected through delta–wye transformers, and delta windings block the zero sequence currents. Does the presence of generators (1) increase, (2) decrease, or (3) have no effect on the single line-to-ground fault at bus C?

5.6 In Problem 5.5, list the fault types in the order of severity, i.e., the magnitude of the fault current.

5.7 Calculate the three-phase symmetrical fault current at bus 4 in the system configuration of Figure P5.2. Neglect resistances.

5.8 Figure P5.3 shows an industrial system motor load being served from a 115 kV utilities system through a step-down transformer. A single line-to-ground fault occurs at the secondary of the transformer terminals. Specify the sequence network considering the motor load. Consider a load operating power factor of 0.85 and an overall efficiency of 94%. Calculate the fault current.

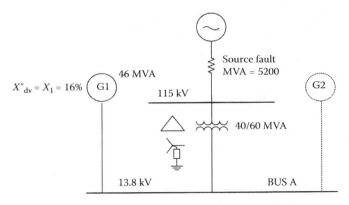

FIGURE P5.1
Distribution system for Problem 5.3.

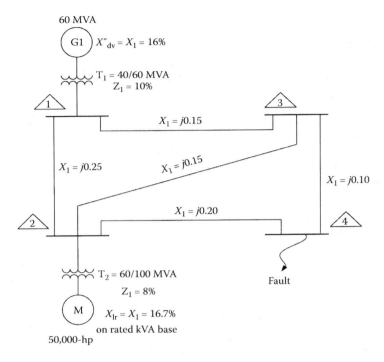

FIGURE P5.2
System configuration for Problem 5.7.

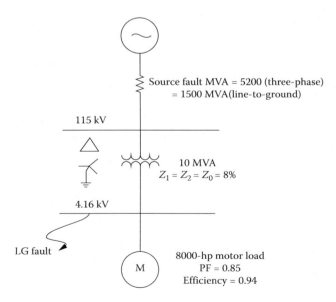

FIGURE P5.3
System configuration for Problem 5.8.

Unsymmetrical Fault Calculations 211

5.9 A wye–wye-connected transformer, with neutral isolated and a tertiary close-circuited delta winding serves a single phase load between two phases of the secondary windings. A 15 ohm resistance is connected between two lines. For a three-phase balanced supply voltage of 480 V between the primary windings, calculate the distribution of currents in all the windings. Assume a unity transformation ratio.

5.10 Calculate the COG factor in Example 5.2 for all points where the fault voltages have been calculated.

5.11 Why is it permissible to ignore all the sequence impedances of the system components and base the fault current calculations only on the system voltage and resistance to be inserted in the neutral circuit when designing a high-resistance grounded system?

Bibliography

JL Blackburn. *Symmetrical Components for Power Systems Engineering*. Marcel Dekker, New York, 1993.

GO Calabrase. *Symmetrical Components Applied to Electric Power Networks*. Ronald Press Group, New York, 1959.

JC Das, Grounding of AC and DC low-voltage and medium-voltage drive systems. *IEEE Ind Appl*, 34(1), 295–216, 1998.

CA Gross. *Power System Analysis*, John Wiley, New York, 1979.

IEEE Std. 142, IEEE Recommended Practice for Grounding of Industrial and Commercial Power Systems, 1991.

LJ Myatt. *Symmetrical Components*, Pargaon Press, Oxford, London, 1968.

WF Robertson, JC Das. Grounding medium voltage mobile or portable equipment. *Ind Appl Mag*, 6(3), 33–42, 2000.

DR Smith. Digital simulation of simultaneous unbalances involving open and faulted conductors. *IEEE Trans Power App Syst*, 89(8), 1826–1835, 1970.

WD Stevenson. *Elements of Power System Analysis*, 4th ed., McGraw-Hill, New York, 1982.

References

1. Transformer Connections (Including Auto-transformer Connections). General Electric, Publication no. GET-2H, Pittsfield, MA, 1967.
2. ANSI/IEEE. General Requirements of Liquid Immersed Distribution, Power and Regulating Transformers. Standard C57.12.00-2006.
3. ANSI. Terminal Markings and Connections for Distribution and Power Transformers. Standard C57.12.70-1978.
4. *Electrical Transmission and Distribution Reference Book*, 4th ed., Westinghouse Electric Corp., East Pittsburgh, PA, 1964.
5. HL Stanback. Predicting damage from 277 volt single-phase-to-ground arcing faults. *IEEE Ind Appl*, 13(4), 307–314, 1977.
6. RH Kaufman, JC Page. Arcing fault protection for low voltage power distribution systems—Nature of the problem. *IEEE Ind Appl*, 79, 160–167, 1960.

7. NEMA PB1-2. Application Guide for Ground Fault Protective Devices for Equipment, 1977.
8. ANSI/NFPA 70, National Electric Code, 2014.
9. DS. Baker. Charging current data for guess work-free design of high resistance grounded systems. *IEEE Trans Ind Appl*, IA-15(2), 136–140, 1979.
10. JC Das. Ground fault protection of bus-connected generators in an interconnected 13.8 kV system. *IEEE Ind Appl*, 43(2), 453–461, 2007.
11. ICEA Pub. S-61-40, NEMA WCS, Thermoplastic Insulated Wire for Transmission and Distribution of Electrical Energy, 1979.
12. JR Dunki-Jacobs. The reality of high resistance grounding. *IEEE Ind Appl*, IA-13, 469–475, 1977.

6

Matrix Methods for Network Solutions

Calculations for the simplest of power systems, i.e., Example 5.2, are involved and become impractical. For speed and accuracy, modeling on digital computers is a must. The size of the network is important. Even the most powerful computer may not be able to model all the generations, transmissions, and consumer connections of a national grid, and the network of interest is "islanded" with boundary conditions represented by current injection or equivalent circuits. Thus, for performing power system studies on a digital computer, the first step is to construct a suitable mathematical model of the power system network and define the boundary conditions. As an example, for short-circuit calculations in industrial systems, the utility's connection can be modeled by sequence impedances, which remain invariant. This generalization may not, however, be valid in every case. For a large industrial plant, with cogeneration facilities, and the utility's generators located close to the industrial plant, it will be necessary to extend the modeling into the utility's system. The type of study also has an effect on the modeling of the boundary conditions. For the steady-state analysis, this model describes the characteristics of the individual elements of the power system and also the interconnections.

A transmission or distribution system network is an assemblage of a linear, passive, bilateral network of impedances connected in a certain manner. The points of connections of these elements are described as buses or nodes. The term bus is more prevalent and a bus may be defined as a point where shunt elements are connected between line potential and ground, though it is not a necessary requirement. In a series circuit, a bus may be defined as the point at which a system parameter, i.e., current or voltage, needs to be calculated. The generators and loads are also connected to buses or nodes.

Balanced three-phase networks can be described by equivalent positive sequence elements with respect to a neutral or ground point. An infinite conducting plane of zero impedance represents this ground plane, and all voltages and currents are measured with reference to this plane. If the ground is not taken as the reference plane, a bus known as a slack or swing bus is taken as the reference bus, and all the variables are measured with reference to this bus.

6.1 Network Models

Mathematically, the network equations can be formed in the bus (or nodal) frame of reference, in the loop (or mesh) frame of reference, or in the branch frame of reference. The bus frame of reference is important. The equations may be represented using either impedance or admittance parameters.

213

In the bus frame of reference, the performance is described by $n-1$ linear independent equations for n number of nodes. As stated earlier, the reference node, which is at ground potential, is always neglected. In the admittance form, the performance equation can be written as

$$\bar{I}_B = \bar{Y}_B \bar{V}_B \tag{6.1}$$

where $\bar{I}_B$ is the vector of injection bus currents. The usual convention for the flow of current is that it is positive when flowing toward the bus and negative when flowing away from the bus. $\bar{V}_B$ is the vector of bus or nodal voltages measured from the reference node, and $\bar{Y}_B$ is the bus admittance matrix. Expanding Equation 6.1, we get

$$\begin{vmatrix} I_1 \\ I_2 \\ \cdot \\ I_{(n-1)} \end{vmatrix} = \begin{vmatrix} Y_{11} & Y_{12} & \cdot & Y_{1,n-1} \\ Y_{21} & Y_{22} & \cdot & Y_{2,n-1} \\ \cdot & \cdot & \cdot & \cdot \\ Y_{(n-1),1} & Y_{(n-1),2} & \cdot & Y_{(n-1),(n-1)} \end{vmatrix} \begin{vmatrix} V_1 \\ V_2 \\ \cdot \\ V_{n-1} \end{vmatrix} \tag{6.2}$$

$\bar{Y}_B$ is a nonsingular square matrix of order $(n-1)(n-1)$. It has an inverse:

$$\bar{Y}_B^{-1} = \bar{Z}_B \tag{6.3}$$

where $\bar{Z}_B$ is the bus impedance matrix. Equation 6.3 shows that this matrix can be formed by inversion of the bus admittance matrix; $\bar{Z}_B$ is also of the order $(n-1)(n-1)$. It also follows that

$$\bar{V}_B = \bar{Z}_B \bar{I}_B \tag{6.4}$$

6.2 Bus Admittance Matrix

We noticed the similarity between the bus impedance and admittance matrices; however, there are differences in their formation and application as we examine further. In the impedance matrix, the voltage equations are written in terms of known constant voltage sources, known impedances, and unknown loop currents. In the admittance matrix, current equations are written in terms of known admittances and unknown node voltages. The voltage source of Thévenin branch equivalent, acting through a series impedance Z, is replaced with a current source equal to EY, in parallel with an admittance $Y = 1/Z$, according to Norton's current equivalent. The two circuits are essentially equivalent, and the terminal conditions remain unaltered. These networks deliver at their terminals a specified current or voltage irrespective of the state of the rest of the system. This may not be true in every case. A generator is neither a true current nor a true voltage source.

The formation of a bus admittance matrix for a given network configuration is straightforward. Figure 6.1a shows a five-node impedance network with three voltage sources,

Matrix Methods for Network Solutions

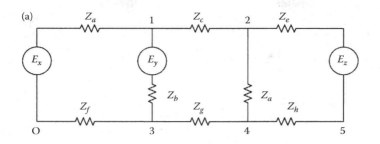

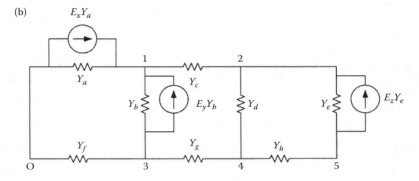

FIGURE 6.1
(a) Network with voltage sources and (b) identical network with Norton equivalent current sources.

E_x, E_y, and E_z; Figure 6.1b shows the admittance equivalent network derived from the impedance network. Following current equations can be written for each of the nodes 1–5. Note that node O is the reference node. Five independent node-pair voltages are possible, measured from node O to the other nodes. As node 0 is taken as the reference node, there is one node–voltage pair less than the number of the nodes. The current equation at node 1 is

$$E_x Y_a + E_y Y_b = V_{01} Y_a + (V_{01} - V_{03}) Y_b + (V_{01} - V_{02}) Y_c$$
$$E_x Y_a = V_{01}(Y_a + Y_b + Y_c) - V_{02} Y_c - V_{03} Y_b \tag{6.5}$$

Similarly, for node 2

$$E_z Y_e = V_{02}(Y_c + Y_d + Y_e) - V_{01} Y_c - V_{04} Y_d - V_{05} Y_e \tag{6.6}$$

and equations for nodes 3–5 are

$$\text{Node } 3: -E_y Y_b = V_{03}(Y_b + Y_f + Y_g) - V_{01} Y_b - V_{04} Y_g \tag{6.7}$$

$$\text{Node } 4: 0 = V_{04}(Y_d + Y_g + Y_h) - V_{02} Y_d - V_{03} V_g - V_{05} Y_h \tag{6.8}$$

$$\text{Node } 5: -E_z Y_e = V_{05}(Y_h + Y_e) - V_{02} Y_e - V_{04} Y_h \tag{6.9}$$

While writing these equations, the direction of current flow must be properly accounted for by change of sign. If a source current arrow is directed away from the node, a minus sign is associated with the term. The above equations can be written in the matrix form as follows:

$$
\begin{vmatrix}
E_x Y_a + E_y Y_b \\
E_Z Y_e \\
-E_y Y_b \\
0 \\
-E_z Y_e
\end{vmatrix}
=
\begin{vmatrix}
(Y_a + Y_b + Y_c) & -Y_c & -Y_b & 0 & 0 \\
-Y_c & (Y_e + Y_d + Y_c) & 0 & -Y_d & -Y_e \\
-Y_b & 0 & (Y_b + Y_f + Y_g) & -Y_g & 0 \\
0 & -Y_d & -Y_g & (Y_d + Y_g + Y_h) & -Y_h \\
0 & -Y_e & 0 & -Y_h & (Y_h + Y_e)
\end{vmatrix}
\times
\begin{vmatrix}
V_{01} \\
V_{02} \\
V_{03} \\
V_{04} \\
V_{05}
\end{vmatrix}
$$

(6.10)

For a general network with $n + 1$ nodes

$$
\bar{Y} =
\begin{vmatrix}
Y_{11} & Y_{12} & \cdot & Y_{1n} \\
Y_{21} & Y_{22} & \cdot & Y_{2n} \\
\cdot & \cdot & \cdot & \cdot \\
Y_{n1} & Y_{n2} & \cdot & Y_{nn}
\end{vmatrix}
$$

(6.11)

where each admittance Y_{ii} ($i = 1, 2, 3, 4, \ldots$) is the self-admittance or driving point admittance of node i, given by the diagonal elements, and it is equal to an algebraic sum of all admittances terminating at that node. Y_{ik} ($i,k = 1, 2, 3, 4, \ldots$) is the mutual admittance between nodes i and k or transfer admittance between nodes i and k and is equal to the negative of the sum of all admittances directly connected between those nodes. The current entering a node is given by

$$
I_k = \sum_{n=1}^{n} Y_{kn} V_n
$$

(6.12)

To find an element, say, Y_{22}, the following equation can be written:

$$
I_2 = Y_{21} V_1 + Y_{22} V_2 + Y_{23} V_3 + \cdots + Y_{2n} V_n
$$

(6.13)

The self-admittance of a node is measured by shorting all other nodes and finding the ratio of the current injected at that node to the resulting voltage (Figure 6.2):

$$
Y_{22} = \frac{I_2}{V_2} (V_1 = V_3 = \cdots V_n = 0)
$$

(6.14)

Similarly, the transfer admittance is

$$
Y_{21} = \frac{I_2}{V_1} (V_2 = V_3 = \cdots V_n = 0)
$$

(6.15)

Matrix Methods for Network Solutions

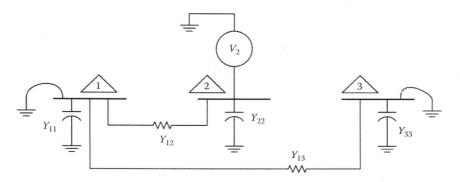

FIGURE 6.2
Calculations of self-admittance in a network, with unit voltage applied at a bus and other buses short-circuited to ground or reference node.

Example 6.1

Figure 6.3a shows a simple network of three buses with series and shunt elements. Numerical values of circuit elements are impedances. The shunt resistors may represent unity power factor loads. Write the bus admittance matrix by examination and by use of Equations 6.14 and 6.15.

The bus admittance matrix is formed by inspection. At node 1, the self-admittance Y_{11} is $1 + 1/j0.2 = 1 - j5$, and the transfer admittance between node 1 and 2 is $Y_{12} = -(1/j0.2) = j5$. Similarly, the other admittance elements are easily calculated as follows:

$$\bar{Y}_B = \begin{vmatrix} Y_{11} & Y_{12} & Y_{13} \\ Y_{21} & Y_{22} & Y_{23} \\ Y_{31} & Y_{32} & Y_{33} \end{vmatrix}$$

$$= \begin{vmatrix} 1-j5 & j5 & 0 \\ j5 & 0.5-j8.33 & j3.33 \\ 0 & j3.33 & 0.33-j3.33 \end{vmatrix} \quad (6.16)$$

Alternatively, the bus admittance matrix can be constructed by use of Equations 6.14 and 6.15. Apply unit voltages, one at a time, to each bus while short-circuiting the other bus voltage sources to ground. Figure 6.3b shows unit voltage applied to bus 1, while buses 2 and 3 are short-circuited to ground. The input current I to bus 1 gives the driving point admittance. This current is given by

$$\frac{V}{1.0} + \frac{V}{j0.2} = 1 - j5 = Y_{11}$$

Bus 2 is short-circuited to ground; the current flowing to bus 2 is

$$\frac{-V}{j0.2} = Y_{12} = Y_{21}$$

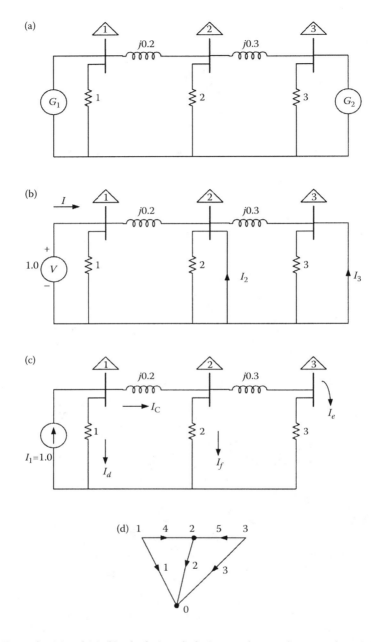

FIGURE 6.3
(a) Network for Examples 6.1 and 6.2, (b) calculation of admittance elements by unit voltage injection, (c) calculation of impedance elements by unit current injection, and (d) diagram of network.

Only buses directly connected to bus 1 have current contributions. The current flowing to bus 3 is zero:

$$Y_{13} = Y_{31} = 0$$

The other elements of the matrix can be similarly found and the result obtained is the same as in Equation 6.16.

6.3 Bus Impedance Matrix

The bus impedance matrix for $(n + 1)$ nodes can be written as

$$\begin{vmatrix} V_1 \\ V_2 \\ . \\ V_m \end{vmatrix} = \begin{vmatrix} Z_{11} & Z_{12} & . & Z_{1m} \\ Z_{21} & Z_{22} & . & Z_{2m} \\ . & . & . & . \\ Z_{m1} & Z_{m2} & . & Z_{mm} \end{vmatrix} \begin{vmatrix} I_1 \\ I_2 \\ . \\ I_m \end{vmatrix} \quad (6.17)$$

Unlike the bus admittance matrix, the bus impedance matrix cannot be formed by simple examination of the network circuit. The bus impedance matrix can be formed by the following methods:

- Inversion of the admittance matrix
- Open-circuit testing
- Step-by-step formation
- Graph theory

Direct inversion of the Y matrix is rarely implemented in computer applications. Certain assumptions in forming the bus impedance matrix are as follows:

1. The passive network can be shown within a closed perimeter Figure 6.4. It includes the impedances of all the circuit components, transmission lines, loads, transformers, cables, and generators. The nodes of interest are brought out of the bounded network, and it is excited by a unit-generated voltage.

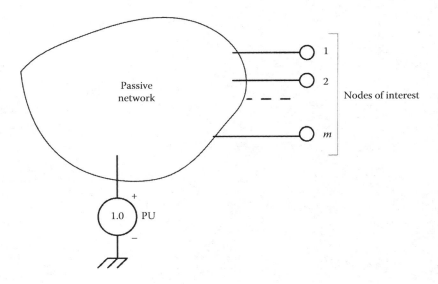

FIGURE 6.4
Representation of a network as passive elements with loads and faults excluded. The nodes of interest are pulled out of the network and unit voltage is applied at the common node.

2. The network is passive, i.e., no circulating currents flow in the network. Also, the load currents are negligible with respect to the fault currents. For any currents to flow, an external path (a fault or load) must exist.
3. All terminals marked 0 are at the same potential. All generators have the same voltage magnitude and phase angle and are replaced by one equivalent generator connected between 0 and a node. For fault current calculations, a unit voltage is assumed.

6.3.1 Bus Impedance Matrix from Open-Circuit Testing

Consider a passive network with m nodes as shown in Figure 6.5 and let the voltage at node 1 be measured when unit current is injected at bus 1. Similarly, let the voltage at bus 2 be measured when unit current is injected at bus 2. All other currents are zero and the injected current is 1 per unit. The bus impedance matrix Equation 6.17 becomes

$$\begin{vmatrix} V_1 \\ V_2 \\ \cdot \\ V_m \end{vmatrix} = \begin{vmatrix} Z_{11} \\ Z_{21} \\ \cdot \\ Z_{m1} \end{vmatrix} \tag{6.18}$$

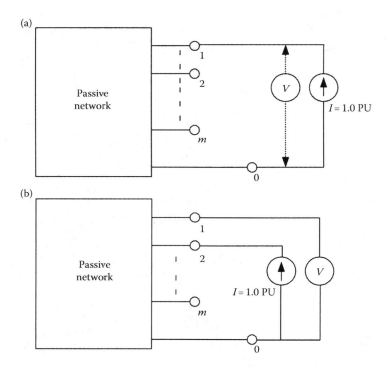

FIGURE 6.5
(a, b) Equivalent networks for calculation of Z_{11} and Z_{12} for formation of bus impedance matrix.

Matrix Methods for Network Solutions 221

where Z_{11} can be defined as the voltage at bus 1 when 1 per unit current is injected at bus 1. This is the *open-circuit driving point impedance*. Z_{12} is defined as voltage at bus 1 when 1 per unit current is injected at bus 2. This is the *open-circuit transfer impedance* between buses 1 and 2. Z_{21} is defined as voltage at bus 2 when 1 per unit current is injected at bus 1. This is the open-circuit transfer impedance between buses 2 and 1. Generally,

$$Z_{12} = Z_{21} \tag{6.19}$$

To summarize, the open-circuit driving point impedance of bus i is determined by injecting a unit current between bus i and the reference node and keeping all other buses open-circuited. This gives the diagonal elements of the bus impedance matrix. The open-circuit transfer impedance between buses i and j is found by applying a unit current between bus i and the reference node and measuring the voltage at bus j, while keeping all other buses open-circuited. This gives the off-diagonal elements.

Example 6.2

Find the bus impedance matrix of Example 6.1 by inversion and also by open-circuit testing.

The inversion of the admittance matrix in Equation 6.16, calculated in Example 6.1, gives

$$\bar{Z}_B = \begin{vmatrix} 0.533 + j0.05 & 0.543 - j0.039 & 0.533 - j0.092 \\ 0.543 - j0.039 & 0.55 + j0.069 & 0.552 + j0.014 \\ 0.533 - j0.092 & 0.522 + j0.014 & 0.577 + j0.258 \end{vmatrix} \tag{6.20}$$

From Equation 6.20, we observe that the zero elements of the bus admittance matrix get populated in the bus impedance matrix. As we will discover, the admittance matrix for power system networks is sparse and this sparsity is lost in the impedance matrix.

The same bus impedance matrix can be constructed from the open-circuit test results. Unit currents are injected, one at a time, at each bus and the other current sources are open-circuited. Figure 6.3c shows unit current injected at bus 1 with bus 3 current source open-circuited. Z_{11} is given by voltage at node 1 divided by current I_1 (which is equal to 1.0 per unit):

$$1 \left\| \left[(2 + j0.02) \| (3 + j0.03) \right] \right] = Z_{11} = 0.533 + j0.05$$

The injected current divides as shown in Figure 6.3c. Transfer impedance $Z_{12} = Z_{21}$ at bus 2 is the potential at bus 2. Similarly, the potential at bus 3 gives $Z_{13} = Z_{31}$. The example shows that this method of formation of bus impedance matrix is tedious.

6.4 Loop Admittance and Impedance Matrices

In the loop frame of reference

$$\bar{V}_L = \bar{Z}_L \bar{I}_L \tag{6.21}$$

where $\bar{V}_L$ is the vector of loop voltages, $\bar{I}_L$ is the vector of unknown loop currents, and $\bar{Z}_L$ is the loop impedance matrix of order $l \times l$; $\bar{Z}_L$ is a nonsingular square matrix and it has an inverse:

$$\bar{Z}_L^{-1} = \bar{Y}_L \tag{6.22}$$

where $\bar{Y}_B$ is the loop admittance matrix. We can write the following:

$$\bar{I}_L = \bar{Y}_L \bar{V}_L \tag{6.23}$$

It is important to postulate the following:

- The loop impedance matrix can be constructed by examination of the network. The diagonal elements are the self-loop impedances and are equal to the sum of the impedances in the loop. The off-diagonal elements are the mutual impedances and are equal to the impedance of the elements common to a loop.
- The loop admittance matrix can only be constructed by inversion of the loop impedance matrix. It has no direct relation with the actual network components.

Compare the formation of bus and loop impedance and admittance matrices.

The loop impedance matrix is derived from basic loop impedance equations. It is based on Kirchoff's voltage law, which states that voltage around a closed loop sums to zero. A potential rise is considered positive and a potential drop, negative. Consider the simple network of Figure 6.6. Three independent loops can be formed as shown in this figure, and the following equations can be written:

$$\begin{aligned} E_1 &= I_1(Z_1 + Z_2) - I_2 Z_2 \\ 0 &= -I_1 Z_2 + I_2(Z_2 + Z_3) - I_3 Z_4 \\ -E_2 &= 0 - I_2 Z_4 + I_3(Z_4 + Z_5) \end{aligned} \tag{6.24}$$

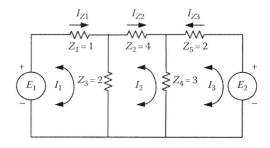

FIGURE 6.6
Network with correct choice of loop currents.

Matrix Methods for Network Solutions

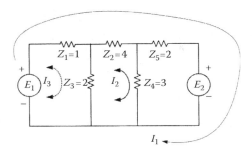

FIGURE 6.7
Incorrect choices of loop currents in the network when only two loops I_1 and I_2 are selected. I_3 must be selected.

In the matrix form, these equations can be written as

$$\begin{vmatrix} E_1 \\ 0 \\ -E_2 \end{vmatrix} = \begin{vmatrix} Z_1+Z_2 & -Z_2 & 0 \\ -Z_2 & Z_2+Z_3 & -Z_4 \\ 0 & -Z_4 & Z_4+Z_5 \end{vmatrix} \begin{vmatrix} I_1 \\ I_2 \\ I_3 \end{vmatrix} \qquad (6.25)$$

The impedance matrix in the above example can be written without writing the loop equations and by examining the network. As stated before, the diagonal elements of the matrix are the self-impedances around each loop, while off-diagonal elements are the impedances common to coupled loops. An off-diagonal element is negative when it carries a current in a loop in the opposite direction to the current in the coupled loop.

6.4.1 Selection of Loop Equations

The selection of loop equations is arbitrary, yet there are certain limitations in forming these equations. The selection should result in a sufficient number of independent voltage equations. As an example, in Figure 6.7, the selection of loop currents, I_1 and I_2, is not adequate. An additional loop current marked I_3 must be selected.

6.5 Graph Theory

Linear network graphs help in the assembly of a network. The problem for large networks is that a minimum number of linearly independent equations of zero redundancy must be selected to provide sufficient information for the solution of the network.

A topographical graph or map of the network is provided by shorting branch electromagnetic forces, opening branch current sources, and considering all branch impedances as zero. The network is, thus, replaced by simple lines joining the nodes. A linear graph depicts the geometric interconnections of the elements of a network. A graph is said to be *connected* only if there is a path between every pair of nodes. If each element is assigned a direction, it is called an *oriented* graph. The direction assignment follows the direction assumed for the current in the element.

Consider the network of Figure 6.8a. It is required to construct a graph. First, the network can be redrawn as shown in Figure 6.8b. Each source and the shunt admittance across it are represented by a single element. Its graph is shown in Figure 6.8c. It has a total of nine branches, which are marked from 1 to 9.

The *node* or *vertex* of a graph is the end point of a branch. In Figure 6.8c, the nodes are marked as 0–4. Node 0 is at ground potential. A route traced out through a linear graph, which goes through a node no more than one time is called a *path*.

The *tree-link* concept is useful when large systems are involved. A tree of the network is formed when it includes all the nodes of a graph, but no closed paths. Thus, a tree in a subgraph of a given connected graph has the following characteristics:

- It is connected.
- It contains all the nodes of the original graph.
- It does not contain any closed paths.

Figure 6.8d shows a tree of the original graph. The elements of a tree are called *tree branches*. The number of branches B is equal to the number of nodes minus 1, which is the reference node:

$$B = n - 1 \tag{6.26}$$

The number of tree branches in Figure 6.8d is four, which is one less than the number of nodes. The elements of the graph that are not included in the tree are called links or link branches and they form a *subgraph*, which may or may not be connected. This graph is called a *cotree*. Figure 6.8e shows the cotree of the network. It has five *links*. Each *link* in the tree will close a *new* loop and a corresponding loop equation is written. Whenever a link closes a loop, a *tie-set* is formed, which contains one link and one or more branches of the established tree. A tree and cotree of a graph are not unique. A set of branches of a connected graph is called a *cut-set*, if the following properties hold:

- If the branches of the cut-set are removed, the graph is split into two parts, which are not connected.
- No subset of the above set has this property. This ensures that no more than the minimum number of branches are included in the cut-set.

Basic cut-sets are those which contain only one branch and, therefore, the number of basic cut-sets is equal to the number of branches. An admittance matrix can be formed from the cut-sets. The network is divided into pieces of cut-sets and Kirchoff's current law must be satisfied for each cut-set line. No reference node need to be given. The various cut-sets for the tree in Figure 6.8d are as shown in Figure 6.8f.

A loop that is formed by closing only one link is called a *basic* loop. The number of basic loops is equal to the number of links. In a metallically coupled network, loops L are given by

$$L = e - B \tag{6.27}$$

where B is the number of tree branches and e is the number of nodes. The graph shown in Figure 6.8c has nine elements and there are four tree branches. Therefore, the number of loops is equal to five.

Matrix Methods for Network Solutions

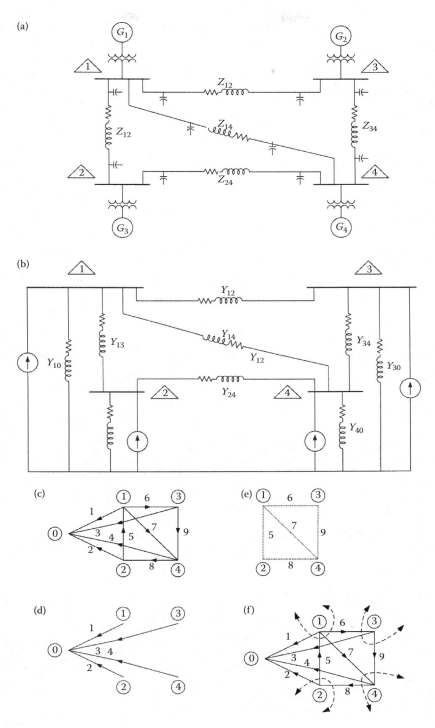

FIGURE 6.8
(a) Equivalent circuit of a network, (b) network redrawn with lumped elements and current injections at the nodes, (c) oriented connected graph of the network, (d) tree of the oriented network, (e) cotree of the oriented network, and (f) cut-sets of the network.

If any two loops, say loops 1 and 2, have a mutual coupling, the diagonal term of loop 1 will have all the self-impedances, whereas off-diagonal terms will have mutual impedance between 1 and 2.

Elimination of loop currents can be done by matrix partitioning. In a large network, the currents in certain loops may not be required and these can be eliminated. Consider a partitioned matrix:

$$\left| \begin{array}{c} \bar{E}_x \\ \bar{E}_y \end{array} \right| = \left| \begin{array}{cc} \bar{Z}_1 & \bar{Z}_2 \\ \bar{Z}_3 & \bar{Z}_4 \end{array} \right| \left| \begin{array}{c} \bar{I}_x \\ \bar{I}_y \end{array} \right| \tag{6.28}$$

The current given by array $\bar{I}_x$ is only of interest. This is given by

$$\bar{I}_x = \left[\bar{Z}_1 - \bar{Z}_2 \bar{Z}_4^{-1} \bar{Z}_3 \right]^{-1} \bar{E}_x \tag{6.29}$$

6.6 Bus Admittance and Impedance Matrices by Graph Approach

The bus admittance matrix can be found by graph approach:

$$\bar{Y}_B = \bar{A} \bar{Y}_p \bar{A}' \tag{6.30}$$

where $\bar{A}$ is the bus incidence matrix, $\bar{Y}_P$ is the primitive admittance matrix, and $\bar{A}'$ is the transpose of the bus incident matrix. Similarly, the loop impedance matrix can be formed by

$$\bar{Z}_L = \bar{B} \bar{Z}_p \bar{B}' \tag{6.31}$$

where $\bar{B}$ is the basic loop incidence matrix, $\bar{B}'$ its transpose, and $\bar{Z}_P$ the primitive bus impedance matrix.

6.6.1 Primitive Network

A network element may contain active and passive components. Figure 6.9 shows impedance and admittance forms of a network element, and equivalence can be established between these two. Consider the impedance form shown in Figure 6.9a. The nodes P and Q have voltages V_p and V_q and let $V_p > V_q$. Self-impedance, Z_{pq}, has a voltage source, e_{pq}, in series. Then

$$V_p + e_{pq} - Z_{pq} i_{pq} = V_q$$

or

$$V_{pq} + e_{pq} = Z_{pq} V_{pq} \tag{6.32}$$

Matrix Methods for Network Solutions

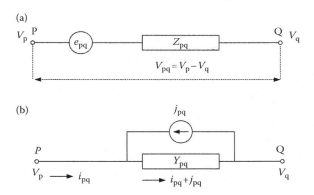

FIGURE 6.9
(a, b) Primitive network, impedance, and admittance forms.

where $V_{pq} = V_p - V_q$ = voltage across P–Q and i_{pq} is current through P–Q. In the admittance form and referring to Figure 6.9b:

$$i_{pq} + j_{pq} = Y_{pq} V_{pq} \tag{6.33}$$

Also

$$j_{pq} = -Y_{pq} e_{pq}$$

where j_{pq} is current source in parallel with P–Q.

The performance equations of the primitive network can be derived from Equations 6.32 and 6.33. For the entire network, the variables become column vector and parameters become matrices. The performance equations in impedance and admittance forms are

$$\bar{V} + \bar{e} = \bar{Z}_p \bar{i}$$
$$\bar{i} + \bar{j} = \bar{Y}_p \bar{V} \tag{6.34}$$

Also

$$\bar{Y}_p = \bar{Z}_p^{-1}$$

The diagonal elements are the impedances or admittances of the element p–q, and the off-diagonal elements are mutual impedances/admittances. When there is no coupling between the elements, the matrices are diagonal.

Example 6.3

Write the primitive admittance matrix of the network shown in Figure 6.8b.

The matrix is written by simply examining the figure. There are no mutual couplings between the elements. The required matrix is

	1	2	3	4	5	6	7	8	9
	0–1	0–2	0–3	0–4	1–2	1–3	1–4	2–4	3–4
0–1	$y10$								
0–2		$y20$							
0–3			$y30$						
0–4				$y40$					
1–2					$y12$				
1–3						$y13$			
1–4							$y14$		
2–4								$y24$	
3–4									$y34$

The top and left-side identifications of the elements between the nodes are help-ful in formation of the matrix. It is a diagonal matrix. If there are mutual couplings between elements of the network, the appropriate off-diagonal elements in the matrix are populated.

6.6.2 Incidence Matrix from Graph Concepts

Consider a graph with n nodes and e elements. The matrix $\bar{A}$ of Equation 6.30 has n rows, which correspond to the n nodes and e columns, which correspond to the e elements. This matrix is known as an *incidence matrix*. The matrix elements can be formed as follows:

$a_{ij} = 1$, if jth element is incident to and directed away from the node i.

$a_{ij} = -1$, if jth element is incident to but directed *toward* the node i.

$a_{if} = 0$, if jth element is not incident to the node.

When a node is taken as the reference node, then the matrix $\bar{A}$ is called a *reduced incidence matrix* or *bus incidence matrix*.

Example 6.4

Form an incidence and reduced incidence matrix from the graph of the network in Figure 6.8a.

The bus incidence matrix is formed from the graph of the network in Figure 6.8c:

e/n	1	2	3	4	5	6	7	8	9
1	1				−1	1	1		
2		1			1			−1	
3			1		−1				1
4				1			−1	1	−1
0	−1	−1	−1	−1					

This matrix is singular. The matrix without the last row pertaining to the reference node, which is highlighted in gray, is a reduced incidence matrix. It can be partitioned as follows:

$e/(n-1)$	Branches	Links
Buses	Ab	AL

Matrix Methods for Network Solutions 229

We can, therefore, write the following:

$$A_{(n-1),e} = (A_b)_{(n-1),(n-1)} \times (A_L)_{(n-1),l} \tag{6.35}$$

Figure 6.8b ignores the shunt capacitance of the lines. Referring to the graph shown in Figure 6.8c, we can write the following:

$$
\begin{vmatrix}
1 & 0 & 0 & 0 & -1 & 1 & 1 & 0 & 0 \\
0 & 1 & 0 & 0 & 1 & 0 & 0 & -1 & 0 \\
0 & 0 & 1 & 0 & 0 & -1 & 0 & 0 & 1 \\
0 & 0 & 0 & 1 & 0 & 0 & -1 & 1 & -1
\end{vmatrix}
\begin{vmatrix}
I_1 \\ I_2 \\ I_3 \\ I_4 \\ I_5 \\ I_6 \\ I_7 \\ I_8 \\ I_9
\end{vmatrix}
=
\begin{vmatrix}
I_{1j} \\ I_{2j} \\ I_{3j} \\ I_{4j}
\end{vmatrix}
\tag{6.36}
$$

where I_{1j}, I_{2j}—are the injected currents at nodes 1, 2, …
In abbreviated form:

$$\bar{A}\bar{I}_{pr} = \bar{I} \tag{6.37}$$

where $\bar{I}_{pr}$ is the column vector of the branch currents and $\bar{I}$ is column vector of the injected currents.
With respect to voltages,

$$
\begin{vmatrix}
V_1 \\ V_2 \\ V_3 \\ V_4 \\ V_5 \\ V_6 \\ V_7 \\ V_8 \\ V_9
\end{vmatrix}
=
\begin{vmatrix}
1 & 0 & 0 & 0 \\
0 & 1 & 0 & 0 \\
0 & 0 & 1 & 0 \\
0 & 0 & 0 & 1 \\
-1 & 1 & 0 & 0 \\
1 & 0 & -1 & 0 \\
1 & 0 & 0 & -1 \\
0 & -1 & 0 & 1 \\
0 & 0 & 1 & 1
\end{vmatrix}
\begin{vmatrix}
V_1 \\ V_2 \\ V_3 \\ V_4
\end{vmatrix}
\tag{6.38}
$$

Referring to Figure 6.8b

$$
\begin{aligned}
V_5 &= V_2 - V_1 \\
V_6 &= V_1 - V_3 \\
V_7 &= V_1 - V_4 \\
V_8 &= V_4 - V_2 \\
V_9 &= V_3 - V_4
\end{aligned}
\tag{6.39}
$$

Equation 6.38 can be written as

$$\bar{V}_{pr} = \bar{A}^t V \tag{6.40}$$

where $\bar{V}_{pr}$ is the column vector of the primitive branch voltage drops and $\bar{V}$ is the vector of the bus voltages measured to the common node.

From these relations, a reader can prove Equation 6.30. Using this equation, the bus admittance matrix can be given as

$$\bar{A}\bar{Y}_P\bar{A}^t = \bar{Y}_B = \begin{vmatrix} Y_{10}+Y_{12}+Y_{13}+Y_{14} & -Y_{12} & -Y_{13} & -Y_{14} \\ -Y_{12} & Y_{20}+Y_{12}+Y_{24} & 0 & -Y_{24} \\ -Y_{13} & 0 & Y_{30}+Y_{13}+Y_{43} & -Y_{43} \\ -Y_{14} & -Y_{24} & -Y_{43} & Y_{40}+Y_{14}+Y_{24}+Y_{43} \end{vmatrix} \tag{6.41}$$

This matrix could be formed by simple examination of the network.

Example 6.5

Form the primitive impedance and admittance matrix of network shown in Figure 6.3a and then the bus incidence matrix. From this, calculate the bus admittance matrix.

The graph of the network is shown in Figure 6.3d. The primitive impedance matrix is

$$\bar{Z}_P = \begin{vmatrix} 1 & 0 & 0 & 0 & 0 \\ 0 & 2 & 0 & 0 & 0 \\ 0 & 0 & 3 & 0 & 0 \\ 0 & 0 & 0 & j0.2 & 0 \\ 0 & 0 & 0 & 0 & j0.3 \end{vmatrix}$$

Then the primitive admittance matrix is

$$\bar{Y}_P = \begin{vmatrix} 1 & 0 & 0 & 0 & 0 \\ 0 & 0.5 & 0 & 0 & 0 \\ 0 & 0 & 0.333 & 0 & 0 \\ 0 & 0 & 0 & -j5 & 0 \\ 0 & 0 & 0 & 0 & -j3.33 \end{vmatrix}$$

None of the diagonal elements are populated.

From the graph of Figure 6.3d, the reduced bus incidence matrix (ignoring node 0) is

$$\bar{A} = \begin{vmatrix} 1 & 0 & 0 & 1 & 0 \\ 0 & 1 & 0 & -1 & -1 \\ 0 & 0 & 1 & 0 & 1 \end{vmatrix}$$

Matrix Methods for Network Solutions 231

The bus admittance matrix is

$$\bar{Y}_B = \bar{A}\bar{Y}_P\bar{A}' = \begin{vmatrix} 1-j5 & j5 & 0 \\ j5 & 0.5-j8.33 & j3.33 \\ 0 & j3.33 & 0.33-j3.33 \end{vmatrix}$$

This is the same matrix as calculated before.

The loop impedance matrix can be similarly formed from the graph concepts. There are five basic loops in the network of Figure 6.8b. The basic loop matrix $\bar{B}_L$ of Equation 6.31 is constructed with its elements as defined below:

$b_{ij} = 1$, if jth element is incident to jth basic loop and is oriented in the same direction.
$b_{ij} = -1$, if jth element is incident to the jth basic loop and is oriented in the opposite direction.
$b_{ij} = 0$, if jth basic loop does not include the jth element.

Branch Frame of Reference

The equations can be expressed as follows:

$$\begin{aligned} \bar{V}_{BR} &= \bar{Z}_{BR}\bar{I}_{BR} \\ I_{BR} &= Y_{BR}\bar{V}_{BR} \end{aligned}$$

(6.42)

The branch impedance matrix is the matrix of the branches of the tree of the connected power system network and has dimensions, $b \times b$; $\bar{V}_{BR}$ is the vector of branch voltages and $\bar{I}_{BR}$ is the vector of currents through the branches.

Example 6.6

In Example 6.1, unit currents $(1 < 0)$ are injected at nodes 1 and 3. Find the bus voltages.
From Example 6.1

$$\begin{vmatrix} 1-j5 & j5 & 0 \\ j5 & 0.5-j8.33 & j3.33 \\ 0 & j3.33 & 0.33-j3.33 \end{vmatrix} \begin{Vmatrix} 1 \\ 0 \\ 1 \end{Vmatrix} = \begin{vmatrix} V_1 \\ V_2 \\ V_3 \end{vmatrix}$$

Taking inverse of the Y matrix, we get

$$\begin{vmatrix} V_1 \\ V_2 \\ V_3 \end{vmatrix} = \begin{vmatrix} 0.552+j0.05 & 0.542-j0.039 & 0.532-j0.092 \\ 0.542-j0.039 & 0.55+j0.069 & 0.551+j0.014 \\ 0.532-j0.092 & 0.551+j0.014 & 0.577+j0.257 \end{vmatrix} \begin{Vmatrix} 1 \\ 0 \\ 1 \end{Vmatrix} = \begin{vmatrix} 1.084-j0.042 \\ 1.093-j0.025 \\ 1.109+j0.164 \end{vmatrix}$$

Example 6.7

Consider that there is a mutual coupling of $Z_M = j0.4$ between the branch elements (4 and 5) in Figure 6.3d. Recalculate the bus admittance matrix.

The primitive impedance matrix is

$$\bar{Z}_P = \begin{vmatrix} 1 & 0 & 0 & 0 & 0 \\ 0 & 2 & 0 & 0 & 0 \\ 0 & 0 & 3 & 0 & 0 \\ 0 & 0 & 0 & j0.2 & j0.4 \\ 0 & 0 & 0 & j0.4 & j0.3 \end{vmatrix}$$

For the coupled branch:

$$\bar{Y}_{jj} = \frac{1}{\bar{Z}_{jj}}$$

Therefore,

$$\begin{vmatrix} j0.2 & j0.4 \\ j0.4 & j0.3 \end{vmatrix}^{-1} = \begin{vmatrix} -j3 & j4 \\ j4 & -j2 \end{vmatrix}$$

Then the primitive Y matrix is

$$\bar{Y}_P = \begin{vmatrix} 1 & 0 & 0 & 0 & 0 \\ 0 & 0.5 & 0 & 0 & 0 \\ 0 & 0 & 0.333 & 0 & 0 \\ 0 & 0 & 0 & j3 & -j4 \\ 0 & 0 & 0 & -j4 & j2 \end{vmatrix}$$

Then the Y bus matrix is

$$\bar{Y}_B = \begin{vmatrix} 1+j3 & j & -4j \\ j & 0.5-j3 & j2 \\ -j4 & 2j & 0.333+j2 \end{vmatrix}$$

If similar currents as in Example 6.1 are injected at buses 1 and 3, then

$$\begin{vmatrix} V_1 \\ V_2 \\ V_3 \end{vmatrix} = \begin{vmatrix} 0.549-j0.015 & 0.54-j0.033 & 0.543-j0.093 \\ 0.54-j0.033 & 0.563+j0.147 & 0.537-j0.123 \\ 0.543-j0.093 & 0.537-j0.123 & 0.565-j0.096 \end{vmatrix} \begin{vmatrix} 1 \\ 0 \\ 1 \end{vmatrix} = \begin{vmatrix} 1.092-j0.079 \\ 1.077-j0.155 \\ 1.108+j0.003 \end{vmatrix}$$

6.6.3 Node Elimination in Y-Matrix

It is possible to eliminate a group of nodes from the nodal matrix equations. Consider a five-node system, and assume that nodes 4 and 5 are to be eliminated. Then the Y-matrix can be partitioned as follows:

Matrix Methods for Network Solutions 233

$$
\begin{vmatrix} I_1 \\ I_2 \\ I_3 \\ - \\ I_4 \\ I_5 \end{vmatrix} = \begin{vmatrix} Y_{11} & Y_{12} & Y_{13} & | & Y_{14} & Y_{15} \\ Y_{21} & Y_{22} & Y_{23} & | & Y_{24} & Y_{25} \\ Y_{31} & Y_{32} & Y_{33} & | & Y_{34} & Y_{35} \\ -- & -- & -- & -- & -- & -- \\ Y_{41} & Y_{42} & Y_{43} & | & Y_{44} & Y_{45} \\ Y_{51} & Y_{52} & Y_{53} & | & Y_{54} & Y_{55} \end{vmatrix} \begin{vmatrix} V_1 \\ V_2 \\ V_3 \\ -- \\ V_4 \\ V_5 \end{vmatrix}
\tag{6.43}
$$

This can be written as

$$
\begin{vmatrix} \overline{I}_x \\ \overline{I}_y \end{vmatrix} = \begin{vmatrix} \overline{Y}_1 & \overline{Y}_2 \\ \overline{Y}_3 & \overline{Y}_4 \end{vmatrix} \begin{vmatrix} \overline{V}_x \\ \overline{V}_y \end{vmatrix}
\tag{6.44}
$$

The current $\overline{I}_y = 0$ for the eliminated nodes. Therefore,

$$
\overline{I}_y = 0 = \overline{Y}_3 \overline{V}_x + \overline{Y}_4 \overline{V}_y
\tag{6.45}
$$

Or

$$
\overline{V}_y = -\overline{Y}_4^{-1} \overline{Y}_3 \overline{V}_x
\tag{6.46}
$$

Also

$$
\overline{I}_x = \overline{Y}_1 \overline{V}_x + \overline{Y}_2 \overline{V}_y
\tag{6.47}
$$

Eliminating $\overline{V}_y$ from Equation 6.47, we get

$$
\overline{I}_x = \left[\overline{Y}_1 - \overline{Y}_2 \overline{Y}_4^{-1} \overline{Y}_3 \right] \overline{V}_x
\tag{6.48}
$$

And the reduced matrix is

$$
\overline{Y}_{\text{reduced}} = \overline{Y}_1 - \overline{Y}_2 \overline{Y}_4^{-1} \overline{Y}_3
\tag{6.49}
$$

6.7 Algorithms for Construction of Bus Impedance Matrix

The $\overline{Z}$ matrix can be formed step-by-step from basic building concepts. This method is suitable for large power systems and computer analysis.

234 *Short-Circuits in AC and DC Systems*

Consider a passive network with m independent nodes. The bus impedance matrix of this system is given as

$$
\begin{vmatrix}
Z_{11} & Z_{12} & . & Z_{1m} \\
Z_{21} & Z_{22} & . & Z_{2m} \\
. & . & . & . \\
Z_{m1} & Z_{m2} & . & Z_{mm}
\end{vmatrix}
\tag{6.50}
$$

The buildup of the impedance matrix can start with an arbitrary element between a bus and a common node and then adding branches and links, one by one. The following section describes the building blocks of the matrix.

6.7.1 Adding a Tree Branch to an Existing Node

Figure 6.10a shows that a branch pk is added at node p. This increases the dimensions of the primitive bus impedance matrix by 1:

$$
\bar{Z} =
\begin{vmatrix}
Z_{11} & Z_{12} & . & Z_{1m} & Z_{1k} \\
. & . & . & . & . \\
Z_{m1} & Z_{m2} & . & Z_{mm} & Z_{mk} \\
Z_{k1} & Z_{k2} & . & Z_{km} & Z_{kk}
\end{vmatrix}
\tag{6.51}
$$

This can be partitioned as shown:

$$
\begin{vmatrix}
\bar{Z}_{xy,xy} & \bar{Z}_{xy,pk} \\
\bar{Z}_{pk,xy} & \bar{Z}_{pk,pk}
\end{vmatrix}
\begin{vmatrix}
\bar{I}_{xy} \\
\bar{I}_{pk}
\end{vmatrix}
=
\begin{vmatrix}
\bar{V}_{xy} \\
\bar{V}_{pk}
\end{vmatrix}
\tag{6.52}
$$

where $\bar{Z}_{xy,\,pk}$ = primitive bus impedance matrix . This remains unchanged as addition of a new branch does not change the voltages at the nodes in the primitive matrix.

$\bar{Z}_{xy,\,pk}$ = mutual impedance matrix between the original primitive matrix and element pq.

$\bar{Z}_{pk,\,xy} = \bar{Z}_{xy,\,pk}$ and $\bar{Z}_{pk,\,pk}$ = impedance of new element.

From Equation 6.52, we get

$$
\begin{vmatrix}
\bar{I}_{xy} \\
\bar{I}_{pk}
\end{vmatrix}
=
\begin{vmatrix}
\bar{Y}_{xy,xy} & \bar{Y}_{xy,pk} \\
\bar{Y}_{pk,xy} & \bar{Y}_{pk,pk}
\end{vmatrix}
\begin{vmatrix}
\bar{V}_{xy} \\
\bar{V}_{pk}
\end{vmatrix}
\tag{6.53}
$$

where

$$
\begin{vmatrix}
\bar{Y}_{xy,xy} & \bar{Y}_{xy,pk} \\
\bar{Y}_{pk,xy} & \bar{Y}_{pk,pk}
\end{vmatrix}
=
\begin{vmatrix}
\bar{Z}_{xy,xy} & \bar{Z}_{xy,pk} \\
\bar{Z}_{pk,xy} & \bar{Z}_{pk,pk}
\end{vmatrix}^{-1}
\tag{6.54}
$$

Matrix Methods for Network Solutions

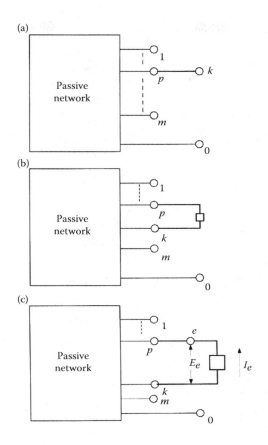

FIGURE 6.10
(a) Adding a tree branch, (b, c) adding a link in the step-by-step formation of bus impedance matrix.

The matrix $\bar{Z}$ has, therefore, to be inverted.
From Equation 6.53, we get

$$\bar{I}_{pk} = \bar{Y}_{pk,xy}\bar{V}_{xy} + \bar{Y}_{pk,pk}\bar{V}_{pk} \tag{6.55}$$

1. If 1 per unit current is injected at any bus other than k and all other currents are zero, then $I_{pk} = 0$. This gives

$$\bar{V}_{pk} = -\frac{\bar{Y}_{pk,xy}\bar{V}_{xy}}{\bar{Y}_{pk,pk}}$$

$$\bar{V}_p - \bar{V}_k = -\frac{\bar{Y}_{pk,xy}(\bar{V}_x - \bar{V}_y)}{\bar{Y}_{pk,pk}} \tag{6.56}$$

$$\bar{Z}_{kj} - \bar{Z}_{pj} = \frac{\bar{Y}_{pk,xy}(\bar{Z}_{xj} - \bar{Z}_{yj})}{\bar{Y}_{pk,pk}}$$

Thus, Z_{kj} (where $j = 1, 2, \ldots, m, j \neq k$) can be found from the following equation:

$$\bar{Z}_{kj} = \bar{Z}_{pj} + \frac{\bar{Y}_{pk,xy}\left(\bar{Z}_{xj} - \bar{Z}_{yj}\right)}{\bar{Y}_{pk,pk}} \tag{6.57}$$

If 1 per unit current is injected at bus k and all other currents are zero, then Equation 6.55 becomes

$$-1 = \bar{Y}_{pk,xy}\bar{V}_{xy} + \bar{Y}_{pk,pk}\bar{V}_{pk} \tag{6.58}$$

This gives Z_{kk}:

$$\bar{Z}_{kk} = \bar{Z}_{pk} + \frac{1 + \bar{Y}_{pk,xy}\left(\bar{Z}_{xk} - \bar{Z}_{yk}\right)}{\bar{Y}_{pk,pk}} \tag{6.59}$$

2. If there is no coupling between pk and any existing branch xy, then

$$\bar{Z}_{kj} = \bar{Z}_{pj} \quad j = 1, 2, \ldots, m, \quad j \neq k \tag{6.60}$$

$$\bar{Z}_{kk} = \bar{Z}_{pk} + \bar{Z}_{pk,pk} \tag{6.61}$$

3. If the new branch is added between p and the reference node 0, then

$$\bar{Z}_{pj} = 0 \tag{6.62}$$

$$\bar{Z}_{kj} = 0 \tag{6.63}$$

$$\bar{Z}_{kk} = \bar{Z}_{pk,pk} \tag{6.64}$$

6.7.2 Adding a Link

A link can be added as shown in Figure 6.10b. As k is not a new node of the system, the dimensions of the bus impedance matrix do not change; however, the elements of the bus impedance matrix change. To retain the elements of the primitive impedance matrix, let a new node e be created by breaking the link pk as shown in Figure 6.10c. If E_e is the voltage of node e with respect to node k, the following equation can be written

$$\begin{vmatrix} V_1 \\ V_2 \\ . \\ V_p \\ . \\ V_m \\ E_e \end{vmatrix} = \begin{vmatrix} Z_{11} & Z_{12} & . & Z_{1m} & Z_{1e} \\ Z_{21} & Z_{22} & . & Z_{2m} & Z_{2e} \\ . & . & . & . & . \\ Z_{p1} & Z_{p2} & . & Z_{pm} & Z_{pe} \\ . & . & . & . & . \\ Z_{m1} & Z_{m2} & . & Z_{mm} & Z_{me} \\ Z_{e1} & Z_{e2} & . & Z_{em} & Z_{ee} \end{vmatrix} \begin{vmatrix} I_1 \\ I_2 \\ . \\ I_p \\ . \\ I_m \\ I_e \end{vmatrix} \tag{6.65}$$

Matrix Methods for Network Solutions 237

In this case, the primitive impedance matrix does not change, as the new branch *pe* can be treated like the addition of a branch from an existing node to a new node, as discussed above. The impedances bearing a subscript *e* have the following definitions:

Z_{1e} = voltage at bus 1 with respect to the reference node when unit current is injected at *k*.

Z_{ei} = voltage at bus *e with reference to k* when unit current is injected at bus 1 from reference bus.

Z_{ee} = voltage at bus *e with respect to k* when unit current is injected to bus *e* from *k*.

Equation 6.65 is partitioned as shown:

$$\left| \begin{array}{c} \bar{I}_{xy} \\ \bar{I}_{pe} \end{array} \right| = \left| \begin{array}{cc} \bar{Y}_{xy,xy} & \bar{Y}_{xy,pe} \\ \bar{Y}_{pe,xy} & \bar{Y}_{pe,pe} \end{array} \right| \left| \begin{array}{c} \bar{V}_{xy} \\ \bar{V}_{pe} \end{array} \right| \tag{6.66}$$

Thus,

$$\bar{I}_{pe} = \bar{Y}_{pe,xy} \bar{V}_{xy} + \bar{Y}_{pe,pe} \bar{V}_{pe} \tag{6.67}$$

If unit current is injected at any node, except node *e*, and all other currents are zero, then

$$0 = \bar{Y}_{pe,xy} \bar{V}_{xy} + \bar{Y}_{pe,pe} \bar{V}_{pe} \tag{6.68}$$

This gives

$$\bar{Z}_{ej} = \bar{Z}_{pj} - \bar{Z}_{kj} + \frac{\bar{Y}_{pe,xy} \left(\bar{Z}_{xj} - \bar{Z}_{yj} \right)}{\bar{Y}_{pe,pe}} \quad j = 1, 2, \ldots, m \quad j \neq e \tag{6.69}$$

If I_e is 1 per unit and all other currents are zero, then

$$-1 = \bar{Y}_{pe,xy} \bar{V}_{xy} + \bar{Y}_{pe,pe} \bar{V}_{pe} \tag{6.70}$$

This gives

$$\bar{Z}_{ee} = \bar{Z}_{pe} - \bar{Z}_{ke} + \frac{1 + \bar{Y}_{pe,xy} \left(\bar{Z}_{xe} - \bar{Z}_{ye} \right)}{\bar{Y}_{pe,pe}} \tag{6.71}$$

Thus, this treatment is similar to that of adding a link. If there is no mutual coupling between *pk* and other branches and *p* is the reference node, then

$$\bar{Z}_{pj} = 0 \tag{6.72}$$

$$\bar{Z}_{ej} = \bar{Z}_{pj} - \bar{Z}_{kj} \tag{6.73}$$

$$\bar{Z}_{ee} = \bar{Z}_{pk,pk} - \bar{Z}_{ke} \tag{6.74}$$

The artificial node can be eliminated by letting voltage at node $e = 0$:

$$\left|\begin{array}{c} \bar{V}_{bus} \\ 0 \end{array}\right| = \left|\begin{array}{cc} \bar{Z}_{bus} & \bar{Z}_{ej} \\ \bar{Z}_{ej} & \bar{Z}_{ee} \end{array}\right| \left|\begin{array}{c} \bar{I}_{bus} \\ I_e \end{array}\right| \qquad (6.75)$$

$$\bar{Z}_{bus,modified} = \bar{Z}_{bus,primitive} - \frac{\bar{Z}_{je}\bar{Z}'_{je}}{\bar{Z}_{ee}} \qquad (6.76)$$

6.7.3 Removal of an Uncoupled Branch

An uncoupled branch can be removed by adding a branch in parallel with the branch to be removed, with an impedance equal to the negative of the impedance of the uncoupled branch.

6.7.4 Changing Impedance of an Uncoupled Branch

The bus impedance matrix can be modified by adding a branch in parallel with the branch to be changed with its impedance given as

$$Z_n = Z \left\| \left(\frac{Z \cdot Z_n}{Z + Z_n} \right) \right. \qquad (6.77)$$

where Z_n is the required new impedance and Z is the original impedance of the branch.

6.7.5 Removal of a Coupled Branch

A branch with mutual coupling M can be modeled as shown in Figure 6.11. We calculated the elements of bus impedance matrix by injecting a current at a bus and measuring the voltage at the other buses. The voltage at buses can be maintained if four currents as shown in Figure 6.11 are injected at either side of the coupled branch:

$$\left|\begin{array}{c} I_{gh} \\ I_{pk} \end{array}\right| = \left|\begin{array}{cc} Y_{gh,gh} & Y_{gh,pk} \\ Y_{pk,gh} & Y_{pk,pk} \end{array}\right| \left|\begin{array}{c} V_g - V_h \\ V_p - V_k \end{array}\right| \qquad (6.78)$$

This gives

$$\left|\begin{array}{c} I \\ -I \\ I' \\ -I' \end{array}\right| = \left|\begin{array}{cccc} Y_{gh,gh} & -Y_{gh,gh} & Y_{gh,pk} & -Y_{gh,pk} \\ -Y_{gh,gh} & Y_{gh,gh} & -Y_{gh,pk} & Y_{gh,pk} \\ Y_{pk,gh} & -Y_{pk,gh} & Y_{pk,pk} - 1/Z_{pk,pk} & Y_{pk,pk} + 1/Z_{pk,pk} \\ -Y_{pk,gh} & Y_{pk,gh} & -Y_{pk,pk} + 1/Z_{pk,pk} & Y_{pk,pk} - 1/Z_{pk,pk} \end{array}\right| \left|\begin{array}{c} V_g \\ V_h \\ V_p \\ V_k \end{array}\right| \qquad (6.79)$$

Matrix Methods for Network Solutions

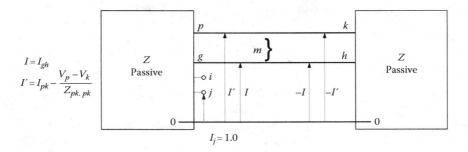

FIGURE 6.11
Adding a coupled branch in the step-by-step formation of impedance matrix.

The above equation is written in abbreviated form as

$$\bar{I}_w = \bar{K}\bar{V}_w \tag{6.80}$$

Partition the original matrix and separate out the coupled portion as follows:

$$\begin{vmatrix} V_1 \\ \cdot \\ V_j \\ \cdot \\ V_{g-1} \\ V_{h+1} \\ \cdot \\ V_{p-1} \\ V_{k+1} \\ \cdot \\ V_m \\ -- \\ V_g \\ V_h \\ V_p \\ V_k \end{vmatrix} = \begin{Vmatrix} Z_{11} & | & \cdot & Z_{11} & Z_{11} & Z_{11} \\ \cdot & | & \cdot & \cdot & \cdot & \cdot \\ Z_{j1} & | & \cdot & Z_{j1} & Z_{j1} & Z_{j1} \\ \cdot & | & \cdot & \cdot & \cdot & \cdot \\ \cdot & | & \cdot & \cdot & \cdot & \cdot \\ \cdot & | & \cdot & \cdot & \cdot & \cdot \\ \cdot & | & \cdot & \cdot & \cdot & \cdot \\ \cdot & | & \cdot & \cdot & \cdot & \cdot \\ \cdot & | & \cdot & \cdot & \cdot & \cdot \\ \cdot & | & \cdot & \cdot & \cdot & \cdot \\ \cdot & | & \cdot & \cdot & \cdot & \cdot \\ - & - & - & - & - & - \\ Z_{g1} & | & Z_{gg} & Z_{gh} & Z_{gp} & Z_{kk} \\ Z_{h1} & | & \cdot & & \cdot & \cdot \\ Z_{p1} & | & \cdot & & \cdot & \cdot \\ Z_{k1} & | & Z_{kg} & Z_{kh} & Z_{kp} & Z_{kk} \end{Vmatrix} \begin{vmatrix} 0 \\ \cdot \\ 1 \\ \cdot \\ \cdot \\ \cdot \\ \cdot \\ \cdot \\ \cdot \\ \cdot \\ \cdot \\ -- \\ I \\ -I \\ I' \\ -I' \end{vmatrix} \tag{6.81}$$

This is written as

$$\begin{vmatrix} \bar{V}_u \\ \bar{V}_w \end{vmatrix} = \begin{Vmatrix} \bar{A} & \bar{B} \\ \bar{C} & \bar{D} \end{Vmatrix} \begin{vmatrix} \bar{I}_u \\ \bar{I}_w \end{vmatrix} \tag{6.82}$$

From Equations 6.80 and 6.82, we get

$$\text{If } \bar{I}_u = 0$$

$$\bar{V}_w = (I - \bar{D}\bar{K})^{-1} \bar{D}\bar{I}_j \qquad (6.83)$$

$$\bar{V}_u = \bar{B}\bar{K}\bar{V}_w + \bar{B}\bar{I}_j \qquad (6.84)$$

$$\text{If } \bar{I}_u \neq 0$$

$$\bar{V}_w = (I - \bar{D}\bar{K})^{-1} \bar{C}\bar{I}_u \qquad (6.85)$$

$$\bar{V}_u = \bar{A}\bar{I}_u + \bar{B}\bar{K}\bar{V}_w \qquad (6.86)$$

If the transformed matrix is defined as

$$\begin{vmatrix} \bar{A}' & \bar{B}' \\ \bar{C}' & \bar{D}' \end{vmatrix} \qquad (6.87)$$

then Equations 6.83 through 6.86 give $\bar{C}'$, $\bar{D}'$, $\bar{A}'$, and $\bar{B}'$, respectively.

Example 6.8

Consider a distribution system with four buses, whose positive and negative sequence networks are shown in Figure 6.12a and zero sequence network in Figure 6.12b. The positive and negative sequence networks are identical and rather than $r + jx$ values, numerical values are shown for ease of hand calculations. There is a mutual coupling between parallel lines in the zero sequence network. It is required to construct bus impedance matrices for positive and zero sequence networks.

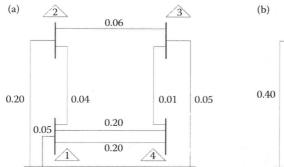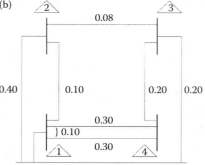

FIGURE 6.12
(a) Positive and negative sequence network for Example 6.6, (b) zero sequence network for Example 6.6.

Matrix Methods for Network Solutions

The primitive impedance or admittance matrices can be written by examination of the network. First, consider the positive or negative sequence network of Figure 6.12a. The following steps illustrate the procedure.

1. The buildup is started with branches 01, 02, and 03 that are connected to the reference node as shown in Figure 6.13a. The primitive impedance matrix can be simply written as

$$\begin{vmatrix} 0.05 & 0 & 0 \\ 0 & 0.2 & 0 \\ 0 & 0 & 0.05 \end{vmatrix}$$

2. Next, add link 1–2 as shown in Figure 6.13b. As this link has no coupling with other branches of the system:

$$Z_{ej} = Z_{pj} - Z_{kj}$$
$$Z_{ee} = Z_{pk,pk} + Z_{pe} - Z_{ke}$$

Substituting $p=1, k=2.$, we get

$$Z_{e1} = Z_{11} - Z_{21} = 0.05$$
$$Z_{e2} = Z_{12} - Z_{22} = 0 - 0.2 = -0.2$$
$$Z_{e3} = Z_{13} - Z_{23} = 0$$
$$Z_{ee} = Z_{12,12} + Z_{1e} - Z_{2e} = 0.04 + 0.05 + 0.2 = 0.29$$

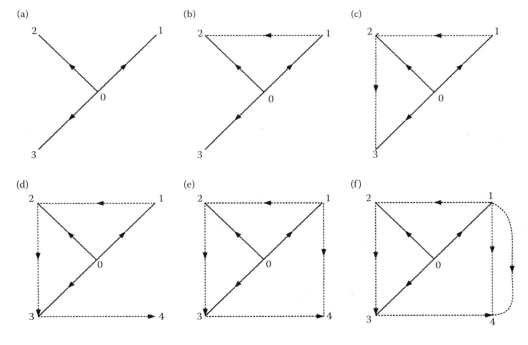

FIGURE 6.13
(a through f) Step-by-step formation of impedance matrix.

The augmented matrix is

$$
\begin{vmatrix}
0.05 & 0 & 0 & 0.05 \\
0 & 0.2 & 0 & -0.2 \\
0 & 0 & 0.05 & 0 \\
0.05 & -0.2 & 0 & 0.29
\end{vmatrix}
$$

3. Eliminate the last row and last column by using Equation 6.76. This gives

$$
\begin{vmatrix}
0.0414 & 0.0345 & 0 \\
0.0345 & 0.0621 & 0 \\
0 & 0 & 0.05
\end{vmatrix}
$$

4. Add link 2–3 as shown in Figure 6.13c:

Substituting $p = 2, k = 3$ and there is no mutual coupling with other branches. This gives

$$
\begin{aligned}
Z_{e1} &= Z_{21} - Z_{31} = 0.345 - 0 = 0.345 \\
Z_{e2} &= Z_{22} - Z_{32} = 0.0621 - 0 = 0.0621 \\
Z_{e3} &= Z_{23} - Z_{33} = 0 - 0.05 = -0.05 \\
Z_{ee} &= Z_{23,23} + Z_{2e} - Z_{3e} = 0.06 + 0.0621 - (-0.05) = 0.1721
\end{aligned}
$$

The augmented matrix is

$$
\begin{vmatrix}
0.0414 & 0.0345 & 0 & 0.0345 \\
0.0345 & 0.0621 & 0 & 0.0621 \\
0 & 0 & 0.05 & -0.05 \\
0.0345 & 0.0621 & -0.05 & 0.1721
\end{vmatrix}
$$

5. Eliminate last row and column. The modified matrix is

$$
\begin{vmatrix}
0.0345 & 0.0221 & 0.010 \\
0.0221 & 0.0397 & 0.018 \\
0.010 & 0.0180 & 0.355
\end{vmatrix}
$$

6. Add branch 3–4 and substitute $p = 3, k = 4$:

$$
\begin{aligned}
Z_{41} &= Z_{31} = 0.01 \\
Z_{42} &= Z_{32} = 0.018 \\
Z_{43} &= Z_{33} = 0.0355 \\
Z_{44} &= Z_{34} + Z_{34,34} = 0.0355 + 0.1 = 0.1355
\end{aligned}
$$

The augmented matrix is

Matrix Methods for Network Solutions 243

$$\begin{vmatrix} 0.0345 & 0.0221 & 0.01 & 0.01 \\ 0.0221 & 0.0397 & 0.018 & 0.018 \\ 0.01 & 0.018 & 0.0355 & 0.0355 \\ 0.01 & 0.018 & 0.0355 & 0.1355 \end{vmatrix}$$

7. Add first parallel link 1–4 as shown in Figure 6.13d:
Substituting $p = 1, k = 4$, we get

$$Z_{e1} = Z_{1e} = 0.0345 - 0.01 = 0.0245$$
$$Z_{e2} = Z_{2e} = 0.0221 - 0.018 = 0.0041$$
$$Z_{e3} = Z_{3e} = 0.01 - 0.0355 = -0.0255$$
$$Z_{e4} = Z_{4e} = 0.01 - 0.1355 = 0.1255$$
$$Z_{ee} = 0.2 + 0.245 - (-0.1255) = 0.350$$

This gives the following matrix:

$$\begin{vmatrix} 0.0345 & 0.0221 & 0.01 & 0.01 & 0.0245 \\ 0.0221 & 0.0397 & 0.018 & 0.018 & 0.0041 \\ 0.01 & 0.018 & 0.0355 & 0.0355 & -0.0255 \\ 0.01 & 0.018 & 0.0355 & 0.1355 & -0.1255 \\ 0.0245 & 0.0041 & -0.0255 & -0.1255 & 0.349 \end{vmatrix}$$

8. Eliminate last row and last column using Equation 6.76:

$$\begin{vmatrix} 0.0328 & 0.0218 & 0.0118 & 0.0188 \\ 0.0218 & 0.0397 & 0.0183 & 0.0195 \\ 0.0118 & 0.0183 & 0.0336 & 0.0264 \\ 0.0188 & 0.0195 & 0.0264 & 0.0905 \end{vmatrix}$$

9. Finally, add second parallel link 1–4 as shown in Figure 6.13e:

$$Z_{e1} = Z_{1e} = 0.0328 - 0.0188 = 0.014$$
$$Z_{e2} = Z_{2e} = 0.0218 - 0.0195 = 0.0023$$
$$Z_{e3} = Z_{3e} = 0.0118 - 0.0264 = -0.0146$$
$$Z_{e4} = Z_{4e} = 0.0188 - 0.0905 = -0.0717$$
$$Z_{ee} = 0.2 + 0.014 - (-0.0717) = 0.2857$$

This gives

$$\begin{vmatrix} 0.0328 & 0.0218 & 0.0118 & 0.0188 & 0.014 \\ 0.0218 & 0.0397 & 0.0183 & 0.0195 & 0.0023 \\ 0.0118 & 0.0183 & 0.0336 & 0.0264 & -0.0146 \\ 0.0188 & 0.0195 & 0.0264 & 0.0905 & -0.0717 \\ 0.014 & 0.0023 & -0.0146 & -0.0717 & 0.2857 \end{vmatrix}$$

10. Eliminate last row and column:

$$\begin{vmatrix} 0.0328 & 0.0218 & 0.0118 & 0.0188 \\ 0.0218 & 0.0397 & 0.0183 & 0.0195 \\ 0.0118 & 0.0183 & 0.0336 & 0.0264 \\ 0.0188 & 0.0195 & 0.0264 & 0.0905 \end{vmatrix}$$

$$- \frac{\begin{vmatrix} 0.014 \\ 0.0023 \\ -0.0416 \\ -0.0717 \end{vmatrix} \begin{vmatrix} 0.0014 & 0.0023 & -0.0146 & -0.0717 \end{vmatrix}}{0.2857}$$

This gives the final positive or negative sequence matrix as

$$\bar{Z}^+, \bar{Z}^- = \begin{vmatrix} 0.0321 & 0.0217 & 0.0125 & 0.0223 \\ 0.217 & 0.0397 & 0.0184 & 0.0201 \\ 0.0125 & 0.0184 & 0.0329 & 0.0227 \\ 0.0223 & 0.0201 & 0.0227 & 0.0725 \end{vmatrix}$$

Zero Sequence Impedance Matrix

The zero sequence impedance matrix is similarly formed, until the last parallel coupled line between buses 1 and 4 is added. The zero sequence impedance matrix, until the coupled branch is required to be added, is formed by a step-by-step procedure as outlined for the positive sequence matrix:

$$\begin{vmatrix} 0.0184 & 0.0123 & 0.0098 & 0.0132 \\ 0.0123 & 0.0670 & 0.0442 & 0.314 \\ 0.0098 & 0.0442 & 0.0806 & 0.0523 \\ 0.0132 & 0.0314 & 0.0523 & 0.1567 \end{vmatrix}$$

Add parallel coupled lines between buses 1 and 4 and substitute $p = 1, k = 4$. The coupled primitive impedance matrix is

$$Z_{\text{pr}} = \begin{vmatrix} 0.3 & 0.1 \\ 0.1 & 0.3 \end{vmatrix}$$

Its inverse is given as

$$Z_{\text{pr}}^{-1} = \begin{vmatrix} 0.3 & 0.1 \\ 0.1 & 0.3 \end{vmatrix}^{-1} = \begin{vmatrix} 3.750 & -1.25 \\ -1.25 & 3.750 \end{vmatrix}$$

$$Y_{pe,pe} = 3.75$$

$$Y_{pe,xy} = -1.25$$

Substituting $p = 1, k = 4$ coupled with 2–3. Thus,

Matrix Methods for Network Solutions 245

$$Z_{x1} = 0.0184, Z_{x2} = 0.0123, Z_{x3} = 0,0098, Z_{x4} = 0.0132$$
$$Z_{y1} = 0.0132, Z_{y2} = 0.0314, Z_{y3} = 0,0523, Z_{y4} = 0.1567$$

This gives

$$Z_{e1} = Z_{11} - Z_{41} + \frac{Y_{pe,pe}(Z_{x1} - Z_{y1})}{Y_{pe,ep}}$$

$$= 0.0184 - 0.0132 + \frac{(-1.25)(0.0184) - (0.00132)}{3.75} = -0.0035$$

Similarly,

$$Z_{e2} = 0.0123 - 0.314 + \left(\frac{-1.25}{3.75}\right)(0.0123 - 0.314) = -0.0127$$

$$Z_{e3} = 0.0098 - 0.523 + \left(\frac{-1.25}{3.75}\right)(0.0098 - 0.0523) = -0.0283$$

$$Z_{e4} = 0.0132 - 0.1567 + \left(\frac{-1.25}{3.75}\right)(0.0132 - 0.1567) = -0.0957$$

Z_{ee} is given as

$$Z_{ee} = Z_{1e} - Z_{4e} + \frac{1 + (-1.25)(Z_{e1} - Z_{4e})}{3.75}$$

$$= 0.0035 - (-0.0957) + \frac{1 + (-1.25)(0.0035 - (-0.0957))}{3.75} = 0.3328$$

The modified impedance matrix will become

$$\begin{vmatrix} 0.0184 & 0.0123 & 0.0098 & 0.0132 & 0.0035 \\ 0.0123 & 0.0670 & 0.0442 & 0.0314 & -0.0127 \\ 0.0098 & 0.0442 & 0.0806 & 0.0523 & -0.0283 \\ 0.0132 & 0.0324 & 0.0523 & 0.1567 & -0.0957 \\ 0.0035 & -0.0127 & -0.0283 & -0.0957 & 0.3328 \end{vmatrix}$$

Finally, eliminate the last row and column:

$$\begin{vmatrix} 0.0184 & 0.0123 & 0.0098 & 0.132 \\ 0.0123 & 0.0670 & 0.0442 & 0.0314 \\ 0.0098 & 0.0442 & 0.0806 & 0.0523 \\ 0.0132 & 0.0314 & 0.0523 & 0.1567 \end{vmatrix}$$

$$- \frac{\begin{vmatrix} 0.0035 \\ -0.0127 \\ -0.0283 \\ -0.0957 \end{vmatrix} \begin{vmatrix} 0.0035 & -0.0127 & -0.0283 & -0.0957 \end{vmatrix}}{0.3328}$$

This gives the final zero sequence bus impedance matrix:

$$\begin{vmatrix} 0.0184 & 0.0124 & 0.0101 & 0.0142 \\ 0.0124 & 0.0665 & 0.0431 & 0.0277 \\ 0.0101 & 0.0431 & 0.0782 & 0.0442 \\ 0.0142 & 0.0277 & 0.0442 & 0.1292 \end{vmatrix}$$

Note that at each last row and column elimination, $Z_{21} = Z_{12}$, $Z_{23} = Z_{32}$, etc. Also, in the final matrix, there are no negative elements. If a duplex reactor is modeled (Appendix C), some elements may be negative; the same is true, in some cases, for modeling of three-winding transformers. Other than that, there is no check on correct formation.

6.8 Short-Circuit Calculations with Bus Impedance Matrix

Short-circuit calculations using bus impedance matrices follow the same logic as developed in Chapter 5. Consider that the positive, negative, and zero sequence bus impedance matrices Z_{ss}^1, Z_{ss}^2, and Z_{ss}^0 are known and a single line-to-ground fault occurs at the rth bus. The positive sequence is then injected only at the rth bus and all other currents in the positive sequence current vector are zero. The positive sequence voltage at bus r is given by

$$V_r^1 = -Z_{rr}^1 I_r^1 \tag{6.88}$$

Similarly, the negative and zero sequence voltages are

$$V_r^2 = -Z_{rr}^2 I_r^2$$
$$V_r^0 = -Z_{rr}^0 I_r^0 \tag{6.89}$$

From the sequence network connections for a line-to-ground fault, we get

$$I_r^1 = I_r^2 = I_r^0 = \frac{1.0}{Z_{rr}^1 + Z_{rr}^2 + Z_{rr}^0 + 3Z_f} \tag{6.90}$$

This shows that the following equations can be written for a shorted bus s.

6.8.1 Line-to-Ground Fault

$$I_s^0 = I_s^1 = I_s^2 = \frac{1}{Z_{ss}^1 + Z_{ss}^2 + Z_{ss}^0 + 3Z_f} \tag{6.91}$$

6.8.2 Line-to-Line Fault

$$I_s^1 = -I_s^2 = \frac{1}{Z_{ss}^1 + Z_{ss}^2 + 3Z_f} \tag{6.92}$$

6.8.3 Double Line-to-Ground Fault

$$I_s^1 = \cfrac{1}{Z_{ss}^1 + \cfrac{Z_{ss}^2\left(Z_{ss}^0 + 3Z_f\right)}{Z_{ss}^2 + \left(Z_{ss}^0 + 3Z_f\right)}} \tag{6.93}$$

$$I_s^0 = \frac{-Z_{ss}^2}{Z_{ss}^2 + (Z_{ss}^0 + 3Z_f)} I_s^1 \tag{6.94}$$

$$I_s^0 = \frac{-(Z_{ss}^0 + 3Z_f)}{Z_{ss}^2 + (Z_{ss}^0 + 3Z_f)} I_s^1 \tag{6.95}$$

The phase currents are calculated by

$$I_s^{abc} = T_s I_s^{012} \tag{6.96}$$

The voltage at bus j of the system is

$$\begin{vmatrix} V_j^0 \\ V_j^1 \\ V_j^2 \end{vmatrix} = \begin{vmatrix} 0 \\ 1 \\ 0 \end{vmatrix} - \begin{vmatrix} Z_{js}^0 & 0 & 0 \\ 0 & Z_{js}^1 & 0 \\ 0 & 0 & Z_{js}^2 \end{vmatrix} \begin{vmatrix} I_s^0 \\ I_s^1 \\ I_s^2 \end{vmatrix} \tag{6.97}$$

where $j = 1, 2, \ldots, s, \ldots, m$.

The fault current from bus x to bus y is given by

$$\begin{vmatrix} I_{xy}^0 \\ I_{xy}^1 \\ I_{xy}^2 \end{vmatrix} = \begin{vmatrix} Y_{xy}^0 & 0 & 0 \\ 0 & Y_{xy}^1 & 0 \\ 0 & 0 & Y_{xy}^2 \end{vmatrix} \begin{vmatrix} V_x^0 - V_y^0 \\ V_x^1 - V_y^1 \\ V_x^2 - V_y^2 \end{vmatrix} \tag{6.98}$$

where

$$I_{xy}^0 = \begin{vmatrix} I_{12}^0 \\ I_{13}^0 \\ . \\ I_{mm}^0 \end{vmatrix} \tag{6.99}$$

and

$$\bar{Y}_{xy}^0 = \begin{vmatrix} Y_{12,12}^0 & Y_{12,13}^0 & . & Y_{12,mn}^0 \\ Y_{13,12}^0 & Y_{13,13}^0 & . & Y_{13,mn}^0 \\ . & . & . & . \\ Y_{mn,12}^0 & Y_{mn,13}^0 & . & Y_{mn,mn}^0 \end{vmatrix} \tag{6.100}$$

where Y_{xy}^0 is the inverse of the primitive matrix of the system. Similar expressions apply to positive sequence and negative sequence quantities.

Example 6.9

The positive, negative, and zero sequence matrices of the system shown in Figure 6.12 are calculated in Example 6.8. A double line-to-ground fault occurs at bus 4. Using the matrices already calculated, it is required to calculate

- Fault current at bus 4
- Voltage at bus 4
- Voltage at buses 1, 2, and 3
- Fault current flows from buses 3 to 4, 1 to 4, 2 to 3
- Current flow in node 0 to bus 3

The fault current at bus 4 is first calculated as follows:

$$I_4^1 = \cfrac{1}{Z_{4s}^1 + \cfrac{Z_{4s}^2 \times Z_{4s}^0}{Z_{4s}^2 + Z_{4s}^0}}$$

$$= \cfrac{1}{0.0725 + \cfrac{0.0725 \times 0.1292}{0.0725 + 0.1292}}$$

$$= 8.408$$

$$I_4^0 = \cfrac{-Z_{4s}^2}{Z_{4s}^2 + Z_{4s}^0} I_4^1$$

$$= \cfrac{-0.0725 \times 8.408}{0.0725 + 0.1292}$$

$$= -3.022$$

$$I_4^2 = \cfrac{-Z_{4s}^0}{Z_{4s}^2 + Z_{4s}^0} I_4^1$$

$$= \cfrac{-0.1292 \times 8.408}{0.0727 + 0.1292}$$

$$= -5.386$$

The line currents are given by

$$I_4^{abc} = T_s I_4^{012}$$

Matrix Methods for Network Solutions

i.e.,

$$
\begin{vmatrix} I_4^a \\ I_4^b \\ I_4^c \end{vmatrix} = \begin{vmatrix} 1 & 1 & 1 \\ 1 & a^2 & a \\ 1 & a & a^2 \end{vmatrix} \begin{vmatrix} I_4^0 \\ I_4^1 \\ I_4^2 \end{vmatrix}
$$

$$
= \begin{vmatrix} 1 & 1 & 1 \\ 1 & a^2 & a \\ 1 & a & a^2 \end{vmatrix} \begin{vmatrix} -3.022 \\ 8.408 \\ -5.386 \end{vmatrix}
$$

$$
= \begin{vmatrix} 0 \\ -4.533 - j11.946 \\ -4.533 + j11.946 \end{vmatrix} = \begin{vmatrix} 0 \\ 12.777 < 249.2° \\ 12.777 < 110.78° \end{vmatrix}
$$

Sequence voltages at bus 4 are given by

$$
\begin{vmatrix} V_4^0 \\ V_4^1 \\ V_4^2 \end{vmatrix} = \begin{vmatrix} 0 \\ 1 \\ 0 \end{vmatrix} - \begin{vmatrix} Z_{4s}^0 & & \\ & Z_{4s}^1 & \\ & & Z_{4s}^2 \end{vmatrix} \begin{vmatrix} I_4^0 \\ I_4^1 \\ I_4^2 \end{vmatrix}
$$

$$
= \begin{vmatrix} 0 \\ 1 \\ 0 \end{vmatrix} - \begin{vmatrix} 0.1292 & 0 & 0 \\ 0 & 0.0725 & 0 \\ 0 & 0 & 0.0725 \end{vmatrix} \begin{vmatrix} -3.022 \\ 8.408 \\ -5.386 \end{vmatrix}
$$

$$
= \begin{vmatrix} 0.3904 \\ 0.3904 \\ 0.3904 \end{vmatrix}
$$

Line voltages are, therefore,

$$
\begin{vmatrix} V_4^a \\ V_4^b \\ V_4^c \end{vmatrix} = \begin{vmatrix} 1 & 1 & 1 \\ 1 & a^2 & a \\ 1 & a & a^2 \end{vmatrix} \begin{vmatrix} V_4^0 \\ V_4^1 \\ V_4^2 \end{vmatrix}
$$

$$
= \begin{vmatrix} 1 & 1 & 1 \\ 1 & a^2 & a \\ 1 & a & a^2 \end{vmatrix} \begin{vmatrix} 0.3904 \\ 0.3904 \\ 0.3904 \end{vmatrix} = \begin{vmatrix} 1.182 \\ 0 \\ 0 \end{vmatrix}
$$

Similarly,

$$\bar{V}_1^{0,1,2} = \begin{vmatrix} 0 \\ 1 \\ 0 \end{vmatrix} - \begin{vmatrix} 0.0142 & 0 & 0 \\ 0 & 0.0223 & 0 \\ 0 & 0 & 0.0223 \end{vmatrix} \begin{vmatrix} -3.022 \\ 8.408 \\ -5.386 \end{vmatrix}$$

$$= \begin{vmatrix} 0.0429 \\ 0.8125 \\ 0.1201 \end{vmatrix}$$

Similarly, sequence voltages at buses 2 and 3 are calculated as

$$\bar{V}_2^{0,1,2} = \begin{vmatrix} 0.0837 \\ 0.8310 \\ 0.1083 \end{vmatrix}$$

$$\bar{V}_3^{0,1,2} = \begin{vmatrix} 0.1336 \\ 0.8091 \\ 0.1223 \end{vmatrix}$$

The sequence voltages are converted into line voltages:

$$\bar{V}_1^{abc} = \begin{vmatrix} 0.976 < 0^0 \\ 0.734 < 125.2^0 \\ 0.734 < 234.8^0 \end{vmatrix} \quad \bar{V}_2^{abc} = \begin{vmatrix} 1.023 < 0^0 \\ 0.735 < 121.5^0 \\ 0.735 < 238.3^0 \end{vmatrix} \quad \bar{V}_3^{abc} = \begin{vmatrix} 1.065 < 0^0 \\ 0.681 < 119.2^0 \\ 0.681 < 240.8^0 \end{vmatrix}$$

The sequence currents flowing between buses 3 and 4 are given by

$$\begin{vmatrix} I_{34}^0 \\ I_{34}^1 \\ I_{34}^2 \end{vmatrix} = \begin{vmatrix} 1/0.2 & 0 & 0 \\ 0 & 1/0.01 & 0 \\ 0 & 0 & 1/0.01 \end{vmatrix} \begin{vmatrix} 0.1336 - 0.3904 \\ 0.8091 - 0.3904 \\ 0.1223 - 0.3904 \end{vmatrix}$$

$$= \begin{vmatrix} -1.284 \\ 4.1870 \\ -2.681 \end{vmatrix}$$

Similarly, the sequence currents between buses 3 and 2 are

$$\begin{vmatrix} I_{32}^0 \\ I_{32}^1 \\ I_{32}^2 \end{vmatrix} = \begin{vmatrix} 1/0.08 & 0 & 0 \\ 0 & 1/0.06 & 0 \\ 0 & 0 & 1/0.06 \end{vmatrix} \begin{vmatrix} 0.1336 - 0.0837 \\ 0.8091 - 0.8310 \\ 0.1223 - 0.1083 \end{vmatrix} = \begin{vmatrix} 0.624 \\ -0.365 \\ 0.233 \end{vmatrix}$$

This can be transformed into line currents.

Matrix Methods for Network Solutions

$$\overline{I}_{32}^{abc} = \begin{vmatrix} 0.492 \\ 0.69 + j0.518 \\ 0.69 + j0.518 \end{vmatrix} \qquad \overline{I}_{34}^{abc} = \begin{vmatrix} 0.222 \\ -2.037 + j5.948 \\ -2.037 + j5.948 \end{vmatrix}$$

The lines between buses 1 and 4 are coupled in the zero sequence network. The $\overline{Y}$ matrix between zero sequence coupled lines is

$$\overline{Y}_{14}^0 = \begin{vmatrix} 3.75 & -1.250 \\ -1.25 & 3.75 \end{vmatrix}$$

Therefore, the sequence currents are given by

$$\begin{vmatrix} I_{14a}^0 \\ I_{14b}^0 \\ I_{14a}^1 \\ I_{14b}^1 \\ I_{14a}^2 \\ I_{14b}^2 \end{vmatrix} = \begin{vmatrix} 3.75 & -1.25 & 0 & 0 & 0 & 0 \\ -1.25 & 3.75 & 0 & 0 & 0 & 0 \\ 0 & 0 & 5 & 0 & 0 & 0 \\ 0 & 0 & 0 & 5 & 0 & 0 \\ 0 & 0 & 0 & 0 & 5 & 0 \\ 0 & 0 & 0 & 0 & 0 & 5 \end{vmatrix} \begin{vmatrix} 0.0429 - 0.3904 \\ 0.0429 - 0.3904 \\ 0.8125 - 0.3904 \\ 0.8125 - 0.3904 \\ 0.1201 - 0.3904 \\ 0.1201 - 0.3904 \end{vmatrix} = \begin{vmatrix} -0.8688 \\ -0.8688 \\ 2.1105 \\ 2.1105 \\ -1.3515 \\ -1.3515 \end{vmatrix}$$

Each of the lines carries sequence currents:

$$\overline{I}_{14a}^{012} = \overline{I}_{14b}^{012} = \begin{vmatrix} -0.8688 \\ 2.1105 \\ -1.3515 \end{vmatrix}$$

Converting into line currents:

$$\overline{I}_{14a}^{012} = \overline{I}_{14b}^{012} = \begin{vmatrix} -0.11 \\ -1.248 - j2.998 \\ -1.248 + j2.998 \end{vmatrix}$$

Also, the line currents between buses 3 and 4 are

$$\overline{I}_{34}^{abc} = \begin{vmatrix} -0.222 \\ -2.037 - j5.948 \\ -2.037 + j5.948 \end{vmatrix}$$

Within the accuracy of calculation, the summation of currents (sequence components as well as line currents) at bus 4 is zero. This is a verification of the calculation. Similarly, the vectorial sum of currents at bus 3 should be zero. As the currents between 3 and 4 and 3 and 2 are already known, the currents from node 0 to bus 3 can be calculated.

Example 6.10

Reform the positive and zero sequence impedance matrices of Example 6.8 after removing one of the parallel lines between buses 1 and 4.

The positive sequence matrix is

$$
\begin{vmatrix}
0.0321 & 0.0217 & 0.0125 & 0.0223 \\
0.0217 & 0.0397 & 0.0184 & 0.0201 \\
0.0125 & 0.0184 & 0.0329 & 0.0277 \\
0.0223 & 0.0201 & 0.0227 & 0.0725
\end{vmatrix}
$$

Removing one of the parallel lines is similar to adding an impedance of -0.2 between 1 and 4:

$$
\begin{aligned}
Z_{1e} &= Z_{e1} = Z_{11} - Z_{41} = 0.0321 - 0.0223 = 0.0098 \\
Z_{2e} &= Z_{e2} = Z_{12} - Z_{42} = 0.0217 - 0.0201 = 0.0016 \\
Z_{3e} &= Z_{e3} = Z_{13} - Z_{43} = 0.0125 - 0.0227 = -0.0102 \\
Z_{4e} &= Z_{e4} = Z_{14} - Z_{44} = 0.0223 - 0.0725 = 0.0502 \\
Z_{ee} &= -0.2 + 0.0098 - (-0.0502) = -0.1400
\end{aligned}
$$

The augmented impedance matrix is, therefore,

$$
\begin{vmatrix}
0.0321 & 0.0217 & 0.0125 & 0.0223 & 0.0098 \\
0.0217 & 0.0397 & 0.0184 & 0.0201 & 0.0016 \\
0.0125 & 0.0184 & 0.0329 & 0.0227 & -0.0102 \\
0.0223 & 0.0201 & 0.0227 & 0.0725 & -0.0502 \\
0.0098 & 0.0016 & -00102 & -0.0502 & -0.1400
\end{vmatrix}
$$

Eliminate the last row and column:

$$
\begin{vmatrix}
0.0321 & 0.0217 & 0.0125 & 0.0223 \\
0.0217 & 0.0397 & 0.0184 & 0.0201 \\
0.0125 & 0.0184 & 0.0329 & 0.0227 \\
0.0223 & 0.0201 & 0.0227 & 0.0725
\end{vmatrix}
$$

$$
- \frac{\begin{vmatrix} 0.0098 \\ 0.0016 \\ -0.0102 \\ -0.0502 \end{vmatrix} \begin{vmatrix} 0.0998 & 0.0016 & -0.0102 & -0.0502 \end{vmatrix}}{-0.14}
$$

This is equal to

$$
\begin{vmatrix}
0.0328 & 0.0218 & 0.0118 & 0.0188 \\
0.218 & 0.0397 & 0.0183 & 0.0195 \\
0.0118 & 0.0183 & 0.0336 & 0.0264 \\
0.0188 & 0.0195 & 0.0264 & 0.0905
\end{vmatrix}
$$

Matrix Methods for Network Solutions 253

The results can be checked with those of Example 6.8. This is the same matrix that was obtained before the last link between buses 1 and 4 was added.

Zero Sequence Matrix

Here, removal of a coupled link is involved. The zero sequence matrix from Example 6.8 is

	1	2	3	4
1	0.0184	0.0124	0.0101	0.0142
2	0.0124	0.0665	0.0431	0.0277
3	0.0101	0.0431	0.0782	0.0442
4	0.0142	0.0277	0.0442	0.1292

Rewrite this matrix as

	2	3	1	4
2	0.0665	0.0431	0.0124	0.0277
3	0.0431	0.0782	0.0101	0.0442
1	0.0124	0.0101	0.0184	0.0142
4	0.0277	0.0442	0.0142	0.1292

Therefore,

$$
\begin{vmatrix} V_2 \\ V_3 \\ V_1 \\ V_4 \end{vmatrix} = \begin{vmatrix} 0.0665 & 0.0431 & 0.0124 & 0.0277 \\ 0.0431 & 0.0782 & 0.0101 & 0.0442 \\ 0.0124 & 0.0101 & 0.0184 & 0.0142 \\ 0.0277 & 0.0442 & 0.0142 & 0.1292 \end{vmatrix} \begin{vmatrix} 0 \\ 0 \\ I+I' \\ -I-I' \end{vmatrix}
$$

or

$$
\begin{vmatrix} \bar{V}_u \\ \bar{V}_w \end{vmatrix} = \begin{vmatrix} \bar{A} & \bar{B} \\ \bar{C} & \bar{D} \end{vmatrix} \begin{vmatrix} \bar{I}_u \\ \bar{I}_w \end{vmatrix}
$$

From adding the coupled link in Example 6.8, we have

$$
\begin{vmatrix} I_{14} \\ I'_{14} \end{vmatrix} = \begin{vmatrix} 3.75 & -1.25 \\ -1.25 & 3.75 \end{vmatrix} \begin{vmatrix} V_1 - V_4 \\ V_1 - V_4 \end{vmatrix}
$$

The matrix $\bar{K}$ is

$$
\bar{K} = \begin{vmatrix} Y_{14,14} & -Y_{14,14} & -Y_{14,14a} & -Y_{14,14a} \\ -Y_{14,14} & Y_{14,14} & -Y_{14,14a} & Y_{14,14a} \\ Y_{14a,14} & -Y_{14a,14} & Y_{14a,14a}-1/Z_{14a,14a} & Y_{14a,14a}+1/Z_{14a,14a} \\ -Y_{14a,14} & -Y_{14a,14a} & -Y_{14a,14a}+1/Z_{14a,14a} & Y_{14a,14a}-1/Z_{14a,14a} \end{vmatrix}
$$

Substituting the following values:

$$
\bar{K} = \begin{vmatrix} 3.75 & 3.75 & -1.25 & 1.25 \\ -3.75 & 3.75 & 1.25 & -1.25 \\ -1.25 & 1.25 & 0.4167 & -0.4167 \\ 1.25 & -1.25 & -0.4167 & 0.4167 \end{vmatrix}
$$

From Equation 6.80, we get

$$\bar{I}_w = \begin{vmatrix} I \\ -I \\ I' \\ -I' \end{vmatrix} = \bar{K} \begin{vmatrix} V_1 \\ V_4 \\ V_1 \\ V_4 \end{vmatrix}$$

This gives

$$\begin{vmatrix} I + I' \\ -I - I' \end{vmatrix} = \begin{vmatrix} 1.6667 & -1.6667 \\ -1.6667 & 1.6667 \end{vmatrix} \begin{vmatrix} V_1 \\ V_4 \end{vmatrix}$$

Therefore, from equivalence of currents, we apply Equations 6.83 through 6.86. New $\bar{C}', \bar{I}_u \neq 0$

$$\bar{V}_w = (1 - \bar{D}\bar{K})^{-1} \overline{CI}_u$$

$$\begin{vmatrix} V_1 \\ V_4 \end{vmatrix}_2 = \left[\begin{vmatrix} 1 & 0 \\ 0 & 1 \end{vmatrix} - \begin{vmatrix} 0.0184 & 0.0142 \\ 0.0142 & 0.1292 \end{vmatrix} \begin{vmatrix} 1.6667 & -1.6667 \\ -1.6667 & 1.6667 \end{vmatrix} \right]^{-1} \times \begin{vmatrix} 0.0124 & 0.0101 \\ 0.0277 & 0.0442 \end{vmatrix} \begin{vmatrix} 1 \\ 0 \end{vmatrix}$$

This is equal to

$$\begin{vmatrix} 1.0087 & -0.0087 \\ -0.2392 & 1.2392 \end{vmatrix} \begin{vmatrix} 0.0124 \\ 0.0277 \end{vmatrix} = \begin{vmatrix} 0.0123 \\ 0.0314 \end{vmatrix}$$

Similarly,

$$\begin{vmatrix} V_1 \\ V_4 \end{vmatrix}_3 = \left[\begin{vmatrix} 1 & 0 \\ 0 & 1 \end{vmatrix} - \begin{vmatrix} 0.0184 & 0.0142 \\ 0.0142 & 0.1292 \end{vmatrix} \begin{vmatrix} 1.6667 & -1.6667 \\ -1.6667 & 1.6667 \end{vmatrix} \right]^{-1}$$

$$\times \begin{vmatrix} 0.0124 & 0.0101 & 0.0277 & 0.0442 \end{vmatrix} \begin{vmatrix} 0 \\ 1 \end{vmatrix}$$

$$= \begin{vmatrix} 1.0087 & -0.0087 \\ -0.2392 & 1.2392 \end{vmatrix} \begin{vmatrix} 0.0101 \\ 0.0442 \end{vmatrix} = \begin{vmatrix} 0.0098 \\ 0.0523 \end{vmatrix}$$

Therefore, the new $\bar{C}^t$ is

$$\bar{C}' = \begin{vmatrix} 0.0123 & 0.0098 \\ 0.0314 & 0.0523 \end{vmatrix}$$

New $\bar{A}', \bar{I}_u \neq 0$

$$\bar{V}_u = \bar{A}\bar{I}_u + \bar{B}\bar{K}\bar{V}_w$$

Matrix Methods for Network Solutions

Thus,

$$\begin{vmatrix} V_2 \\ V_3 \end{vmatrix}_2 = \begin{vmatrix} 0.0665 & 0.0431 \\ 0.0431 & 0.0782 \end{vmatrix} \begin{vmatrix} 1 \\ 0 \end{vmatrix} + \begin{vmatrix} 0.0124 & 0.0277 \\ 0.0101 & 0.0442 \end{vmatrix} \begin{vmatrix} 1.6667 & -1.6667 \\ -1.6667 & 1.6667 \end{vmatrix} \times \begin{vmatrix} 0.0123 \\ 0.0314 \end{vmatrix} = \begin{vmatrix} 0.067 \\ 0.0442 \end{vmatrix}$$

Similarly,

$$\begin{vmatrix} V_2 \\ V_3 \end{vmatrix}_3 = \begin{vmatrix} 0.0442 \\ 0.0806 \end{vmatrix}$$

Thus, the new $\bar{A}'$ is

$$\bar{A}' = \begin{vmatrix} 0.0670 & 0.0442 \\ 0.0442 & 0.0806 \end{vmatrix}$$

New $\bar{D}', \bar{I}_u = 0$

$$\bar{V}_u = (1 - \bar{D}\bar{K})^{-1} \bar{D}\bar{I}_j$$

$$\begin{vmatrix} V_1 \\ V_4 \end{vmatrix}_1 = \left[\begin{vmatrix} 1 & 0 \\ 0 & 1 \end{vmatrix} - \begin{vmatrix} 0.0184 & 0.0142 \\ 0.0142 & 0.1292 \end{vmatrix} \begin{vmatrix} 1.6667 & -1.6667 \\ -1.6667 & 1.6667 \end{vmatrix} \right]^{-1} \times \begin{vmatrix} 0.0184 & 0.0142 \\ 0.0142 & 0.1292 \end{vmatrix} \begin{vmatrix} 1 \\ 0 \end{vmatrix} = \begin{vmatrix} 0.0184 \\ 0.0132 \end{vmatrix}$$

Similarly,

$$\begin{vmatrix} V_1 \\ V_4 \end{vmatrix}_4 = \begin{vmatrix} 0.0132 \\ 0.1567 \end{vmatrix}$$

Therefore, $\bar{D}'$ is

$$\bar{D}' = \begin{vmatrix} 0.0184 & 0.0132 \\ 0.0132 & 0.1567 \end{vmatrix}$$

New $\bar{B}', \bar{I}_u = 0$

$$\bar{V}_u = \bar{B}\bar{K}\bar{V}_w + \bar{B}\bar{I}_j$$

$$\begin{vmatrix} V_2 \\ V_3 \end{vmatrix}_1 = \begin{vmatrix} 0.01214 & 0.0277 \\ 0.0101 & 0.0442 \end{vmatrix} \begin{vmatrix} 1.6667 & -1.6667 \\ -1.6667 & 1.6667 \end{vmatrix} \begin{vmatrix} 0.0184 \\ 0.0132 \end{vmatrix} + \begin{vmatrix} 0.0124 \\ 0.0101 \end{vmatrix}$$

$$= \begin{vmatrix} 0.0123 \\ 0.0098 \end{vmatrix}$$

Similarly,

$$\begin{vmatrix} V_2 \\ V_3 \end{vmatrix}_4 = \begin{vmatrix} 0.0314 \\ 0.0523 \end{vmatrix}$$

The new $\bar{B}'$ is

$$\bar{B}' = \begin{vmatrix} 0.0123 & 0.0314 \\ 0.0098 & 0.0523 \end{vmatrix}$$

Substituting these values, the impedance matrix after removal of the coupled line is

	2	3	1	4
2	0.0670	0.0442	0.0123	0.0314
3	0.0442	0.0806	0.0098	0.0523
1	0.0123	0.0098	0.0184	0.0132
4	0.0314	0.0523	0.0132	0.1567

Rearranging in the original form:

	2	3	1	4
2	0.0670	0.0442	0.0123	0.0314
3	0.0442	0.0806	0.0098	0.0523
1	0.0123	0.0098	0.0184	0.0132
4	0.0314	0.0523	0.0132	0.1567

Referring to Example 6.8, this is the same matrix before the coupled link between 1 and 4 was added. This verifies the calculation.

6.9 Solution of Large Network Equations

The bus impedance method demonstrates the ease of calculations throughout the distribution system, with simple manipulations. Yet it is a full matrix and requires storage of each element. The Y-matrix of a large network is very sparse and has a large number of zero elements. In a large system, the sparsity may reach 90% because each bus is connected to only a few other buses. The sparsity techniques are important in matrix manipulation and are covered in Appendices A and B. Some of these matrix techniques are

- Triangulation and factorization: Crout's method, Bifactorization, and Product Form
- Solution by forward–backward substitution
- Sparsity and optimal ordering

A matrix can be factored into lower, diagonal, and upper form called LDU form. This is of special interest. This formation always requires less computer storage. The sparse techniques exhibit a distinct advantage in computer time required for the solution of a network and can be adapted to system changes, without rebuilding these at every step. Also see Appendix B

Problems

6.1 For the network in Figure P6.1, draw its graph and specify the total number of nodes, branches, buses, basic loops, and cut-sets. Form a tree and a cotree. Write the bus admittance matrix by direct inspection. Also, form a reduced bus incidence admittance matrix, and form a bus admittance matrix using Equation 6.30. Write the basic loop incidence impedance matrix and form loop impedance matrix, using Equation 6.31.

6.2 For the network shown in Figure P6.2, draw its graph and calculate all the parameters, as specified in Problem 6.1. The self-impedances are as shown in Figure P6.2. The mutual impedances are as follows:

Buses 1–3: 0.2 Ω; buses 2–5 = 0.3 Ω

6.3 Figure P6.3 shows the positive and negative sequence network of a power system. Form the bus impedance matrix by a step-by-step buildup process, as illustrated in Example 6.6.

6.4 A double line-to-ground fault occurs at bus 2 in the network of Figure P6.3. Find the fault current. Assume for the simplicity of hand calculations that the zero sequence impedance network is the same as the positive and negative sequence impedance.

6.5 In Problem 6.4, calculate all the bus voltages.

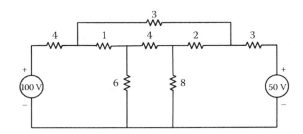

FIGURE P6.1
Network for Problem 6.1.

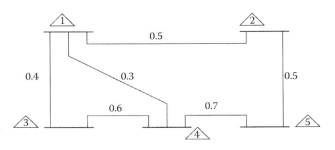

FIGURE P6.2
Network for Problem 6.2.

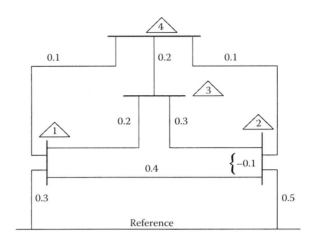

FIGURE P6.3
Network for Problem 6.8.

6.6 In Problem 6.4, calculate the fault currents flowing between nodes 1–2, 4–2, 3–1, and 3–4, also the current flowing between node 0 and bus 1.

6.7 Remove the coupled element of 0.4 between buses 3 and 4 and reform the bus impedance matrix.

6.8 A single line-to-ground fault occurs at bus 4 in Problem 6.7, after the impedance matrix is reformed. Calculate the fault current and fault voltages at buses 1, 2, 3, and 4, and the currents flowing between all buses and the current in the ground circuit from buses 1 and 2.

Bibliography

PM Anderson. *Analysis of Faulted Power Systems*. Iowa State Press, Ames, IA, 1973.

J Arrillaga, CP Arnold, BJ Harker. *Computer Modeling of Electrical Power Systems*. John Wiley & Sons, New York, 1983.

R Bergen, V Vittal. *Power System Analysis*, 2nd ed. Prentice Hall, Upper Saddle River, NJ, 1999.

HE Brown, CE Parson. Short-circuit studies of large systems by the impedance matrix method. *Proc PICA*, 1967, pp. 335–346.

HE Brown. *Solution of Large Networks by Matrix Methods*. John Wiley & Sons, New York, 1975.

Z Dong, P Zhang, J Ma, J Zhao. *Engineering Techniques in Power Systems Analysis*. Springer, Berlin, 2010.

GT Heydt. *Computer Analysis Methods for Power Systems*. Macmillan, New York, 1986.

WJ Maron. *Numerical Analysis*. Macmillan, New York, 1987.

GW Stagg, AH El-Abiad. *Computer Methods in Power Systems Analysis*. McGraw-Hill, New York, 1968.

7

Current Interruptions in AC Networks

Current interruption in high-voltage AC networks has been intensively researched since the introduction of high-voltage transmission lines. These advances can be viewed as an increase in the breaker interrupting capacity per chamber or a decrease in the weight with respect to interrupting capacity. Fundamental electrical phenomena occurring in the electrical network and the physical aspects of arc interruption processes need to be considered simultaneously. The phenomena occurring in an electrical system and the resulting demands on the switchgear can be well appreciated and explained theoretically, yet no well-founded and generally applicable theory of the processes in a circuit breaker itself exists. Certain characteristics have a different effect under different conditions, and care must be applied in generalizations.

The interruption of short circuits is not the most frequent duty a circuit breaker has to perform, but this duty subjects it to the greatest stresses for which it is designed and rated. As a short circuit represents a serious disturbance in the electrical system, a fault must be eliminated rapidly by isolating the faulty part of the system. Stresses imposed on the circuit breaker also depend on the system configuration, and it is imperative that the fault isolation takes place successfully, and that the circuit breaker itself is not damaged during fault interruption and remains serviceable.

This chapter explores the various fault types, their relative severity, effects on the electrical system, and the circuit breaker itself. The basic phenomenon of interruption of short-circuit currents is derived from the electrical system configurations and the modifying effect of the circuit breaker itself. This chapter shows that short-circuit calculations according to empirical methods in IEC or ANSI/IEEE standards do not and cannot address all the possible applications of circuit breakers. It provides a background in terms of circuit interruption concepts and paves a way for better understanding of the calculation methods to follow in the following chapters.

The two basic principles of interruption are (1) high-resistance interruption or an ideal rheostatic circuit breaker and (2) low resistance or zero-point arc extinction. Direct current circuit breakers also employ high-resistance arc interruption; however, emphasis in this chapter is on AC current interruption.

7.1 Rheostatic Breaker

An ideal rheostatic circuit breaker inserts a constantly increasing resistance in the circuit until the current to be interrupted drops to zero. The arc is extinguished when the system voltage can no longer maintain the arc, because of high-voltage drops. The arc length is increased and the arc resistance acquires a high value. The energy stored in the

259

system is gradually dissipated in the arc. The volt-ampere characteristic of a *steady* arc is given by

$$V_{arc} = \text{anode voltage} + \text{cathode voltage} + \text{voltage across length of arc}$$

$$A + \frac{C}{I_{arc}} + \left(B + \frac{D}{I_{arc}}\right)d \tag{7.1}$$

where I_{arc} is the arc current; V_{arc} is the voltage across the arc; d is the length of the arc; and A, B, C, and D are constants. For small arc lengths, the voltage across the arc length can be neglected:

$$V_{arc} = A + \frac{C}{I_{arc}} \tag{7.2}$$

The voltage across the arc reduces as the current increases. The energy dissipated in the arc is

$$E_{arc} = \int_0^t iv\,dt \tag{7.3}$$

where $i = i_m \sin\omega t$ is current and $v = ir$ voltage in the arc.

Equation 7.3 can be written as

$$E_{arc} = \int_0^t i_m^2 r \sin^2 \omega t\,dt \tag{7.4}$$

The approximate variation of arc resistance, r, with time, t, is obtained for different parameters of the arc by experimentation and theoretical analysis.

In a rheostatic breaker, if the arc current is assumed to be constant, the arc resistance can be increased by increasing the arc voltage. Therefore, the arc voltage and the arc resistance can be increased by increasing the arc length. The arc voltage increases until it is greater than the voltage across the contacts. At this point, the arc is extinguished. If the arc voltage remains lower, the arc will continue to burn until the contacts are destroyed.

Figure 7.1 shows a practical design of the arc lengthening principle. The arc originates at the bottom of the arc chutes and is blown upward by the magnetic force. It is split by arc

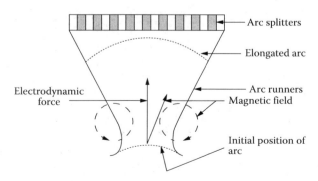

FIGURE 7.1
Principle of a rheostatic breaker and arc elongation.

Current Interruptions in AC Networks

splitters, which may consist of resin-bonded plates of high-temperature fiberglass, placed perpendicular to the arc path. Blow-out coils, in some breaker designs, subject the arc to a strong magnetic field, forcing it upward in the arc chutes.

This principle has been successfully employed in some commercial designs of medium-voltage breakers, and is stated here for reference only. All high-voltage breakers are current-zero breakers and further discussions are confined to this method of arc interruption.

7.2 AC Arc Interruption

The ionization of gaseous mediums and the contact space of circuit breakers depends upon:

- Thermal ionization
- Ionization by collision
- Thermal emission from contact surfaces
- Secondary emission from contact surface
- Photoemission and field emission
- The arc models in circuit breakers are further influenced by the contact geometry, turbulence, gas pressure, and arc extinguishing devices [1]

Deionization can take place by recombination, diffusion, and conduction heat. The arc behaves like nonlinear resistor with arc voltage in phase with the arc current. At low temperature, the arc has falling volt-ampere characteristics. At higher currents, voltage gradient levels out, practically becoming independent of current. At higher currents, arc temperatures are of the order of 20,000 K and tend to be limited by radiant heat transfer. The heat transfer and electrical conductivity have nearly constant values within the arc column.

It is rather perplexing that when the arc is cooled the temperature increases. This happens due to the reduction in the diameter of the core, which results in higher current density (of the order of several thousand amperes/cm^2).

High- and low-pressure arcs can be distinguished. High-pressure arcs exist at pressures above atmosphere and appear as a bright column, characterized by a small burning core at high temperatures, of the order of 20,000 K. In low pressure (vacuum arcs), the arc voltages are low of the order of 40 V and the positive column of the arc is solely influenced by the electrode material, while that of high-pressure arcs it is made up of ionized gasses from arc's surrounding medium.

7.2.1 Arc Interruption Theories

There are a number of arc interruption theories, and the Cassie-Mayr theory briefly discussed below seems to be practical and employed in circuit breaker arc interruption models.

7.2.1.1 Cassie's Theory

A differential equation describing the behavior of arc was presented by A.M. Cassie in 1939 [2]. It assumes a constant temperature across the arc diameter. As the current varies, so does the arc cross section, but not the temperature inside the arc column. Under given assumptions, the conductance G of the model is proportional to current, so that the steady-state voltage gradient E_0 is fixed. Define the time constant

$$\theta = \frac{Q}{N} \tag{7.5}$$

where Q = energy storage capability, and N is the finite rate of energy loss. The following is simplified Cassie's equation:

$$\frac{d}{dt}(G^2) = \frac{2}{\theta}\left(\frac{I}{E_0}\right)^2 \tag{7.6}$$

For high-current region, there is good agreement with the model, but for current-zero region, agreement is good only for high rates of current decay.

7.2.1.2 Mayr's Theory

O. Mayr considered an arc column where arc diameter is constant and where arc temperature varies with time and radial dimensions. The decay of temperature of arc is assumed due to thermal conduction and electrical conductivity of arc is dependent upon temperature:

$$\frac{1}{G}\frac{dG}{dt} = \frac{1}{\theta}\left(\frac{EI}{N_0} - 1\right) \tag{7.7}$$

where,

$$\theta = \frac{Q_0}{N_0} \tag{7.8}$$

The validity of the theory during current-zero periods is acknowledged.

7.2.1.3 Cassie-Mayr Theory

Mayr assumed arc temperature of 6000K, but it is recognized to be in excess of 20,000K. At these high temperatures, there is a linear increase in gas conductivity. Assuming that before current zero, the current is defined by driving circuit and that after current zero, the voltage across the gap is determined by arc circuit. We write the following two equations:

1. Cassie's period prior to current zero:

$$\frac{d}{dt}\left(\frac{1}{R^2}\right) + \frac{2}{\theta}\left(\frac{1}{R^2}\right) = \frac{2}{\theta}\left(\frac{1}{E_0}\right)^2 \tag{7.9}$$

Current Interruptions in AC Networks

2. Mayr's period around current zero:

$$\frac{dR}{dt} - \frac{R}{\theta} = -\frac{e^2}{\theta N_0} \qquad (7.10)$$

where $1/R$ is arc conductance

7.3 Current-Zero Breaker

The various arc quenching mediums in circuit breakers for arc interruption have different densities, thermal conductivities, dielectric strengths, arc time constants, etc. In a current-zero circuit breaker, the interruption takes place during the passage of current through zero. At the same time, the electrical strength of the break-gap increases so that it withstands the recovery voltage stress. All high-voltage breakers, and high-interrupting capacity breakers, whatever may be the arc quenching medium (oil, air, or gas), use current-zero interruption. In an ideal circuit breaker, with no voltage drop before interruption, the arc energy is zero. Modern circuit breakers approach this ideal on account of short arc duration and low arc voltage.

The circuit breakers are called upon to interrupt currents of the order of tens of kilo amperes or even more. Generator circuit breakers having an interrupting capability of 250 kA rms symmetrical are available. At the first current zero, the electrical conductivity of the arc is fairly high and since the current-zero period is very short, the *reignition* cannot be prevented. The time lag between current and temperature is commonly called as arc hysteresis.

Figure 7.2b shows a typical short-circuit current waveform in an inductive circuit of Figure 7.2a, where the short-circuit is applied at $t = 0$. The short-circuit current is limited by the impedance $R + j\omega L$ and is interrupted by breaker B. The waveform shows asymmetry as discussed in Chapter 4. At $t = t_2$, the contacts start parting. The time $t_2 - t_0$ is termed the contact parting time, since it takes some finite time for the breaker operating mechanism to set in motion and the protective relaying to signal a fault condition. The contact parting time is the sum of tripping delay, which is taken in the standards as ½ cycle $(t_1 - t_0)$ and opening time $(t_2 - t_1)$.

As the contacts start parting, an arc is drawn, which is intensely cooled by the quenching medium (air, SF_6, or oil). The arc current varies sinusoidally for a short duration. As the contacts start parting, the voltage across these increases. This voltage is the voltage drop across the arc during the arcing period and is shown exaggerated in Figure 7.2c for clarity. The arc is mostly resistive and the voltage in the arc is in phase with the current. The peculiar shape of the arc voltage shown in Figure 7.2c is the result of the volt-ampere characteristic of the arc, and at a particular current zero, the dielectric strength of the arc space is sufficiently high for the arcing to continue. The contacts part at high speed, to prevent continuation of the arc. When the dielectric strength builds and a current zero occurs, the current is interrupted. In Figure 7.2c, it occurs at $t = t_3$, and the interval $t_3 - t_2$ is the arcing time. The interval $t_3 - t_1$ is the interrupting time. With modern high-voltage breakers, it is as low as 2 cycles or even lower, based on the system power frequency. The ANSI standard [3] defines the rated interrupting time of a circuit breaker as the time between trip circuit energization and power arc interruption on an opening operation, and it is used to classify breakers of different speeds. The rated interrupting time may be exceeded at low values of current and for close–open operations; also, the time for interruption of the resistor current for interrupters equipped with resistors may exceed the rated interrupting time. The increase in interrupting time on close–open operation may be important from the standpoint of line damage or possible instability.

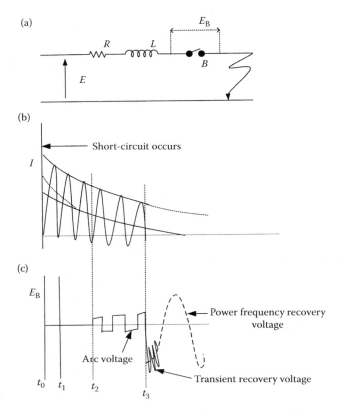

FIGURE 7.2
Current interruption in a current-zero breaker: (a) circuit diagram showing a mainly inductive circuit and short circuit at terminals, (b) current waveform, and (c) voltage during interruption and after fault clearance.

The significance of interrupting time on breaker interrupting duty can be appreciated from Figure 7.2b. As the short-circuit current consists of decaying AC and DC components, the *faster the breaker, the greater the asymmetry and the interrupted current.*

The interrupting current at final arc extinction is asymmetrical in nature, consisting of an AC component and a DC component. The rms value of this asymmetrical current is given by

$$I_i = \sqrt{(\text{rms of ac component})^2 + (\text{dc component})^2} \qquad (7.11)$$

IEC specifications term this as the asymmetrical breaking current.

7.4 Transient Recovery Voltage

The essential problem of current interruption consists in rapidly establishing an adequate electrical strength across the break after current zero, so that restrikes are eliminated. Whatever may be the breaker design, it can be said that it is achieved in most interrupting

Current Interruptions in AC Networks

mediums, i.e., oil, air blast, or SF_6 by an intense blast of gas. The flow velocities are always governed by aerodynamic laws. However, there are other factors that determine the rate of recovery of the dielectric medium: nature of the quenching gases, mode of interaction of pressure and velocity of the arc, arc control devices, contact shape, and number of breaks per phase. Interruption in vacuum circuit breakers is entirely different and discussed in Section 7.16.3.

At the final arc interruption, a high-frequency oscillation superimposed on the power frequency appears across the breaker contacts. A short-circuit current loop is mainly inductive, and the power frequency voltage has its peak at current zero; however, a sudden voltage rise across the contacts is prevented by the inherent capacitance of the system, and in the simplest cases, a transient of the order of some hundreds to 10,000 c/s occurs. It is termed the natural frequency of the circuit. Figure 7.3 shows the recovery voltage profile after final current extinction. The two components of the recovery voltage, (1) a high-frequency damped oscillation, and (2) the power frequency recovery voltages, are shown. The high-frequency component is called the transient recovery voltage (TRV) and sometimes the restriking voltage. Its frequency is given by

$$f_n = \frac{1}{2\pi\sqrt{LC}} \tag{7.12}$$

where f_n is the natural frequency, and L and C are the equivalent inductance and capacitance of the circuit.

If the contact space breaks down within a period of 1/4 cycle of initial arc extinction, the phenomena are called reignitions and if the breakdown occurs after 1/4 cycle, the phenomena are called *restrikes*.

The transient oscillatory component subsides in a few microseconds and the power frequency component continues.

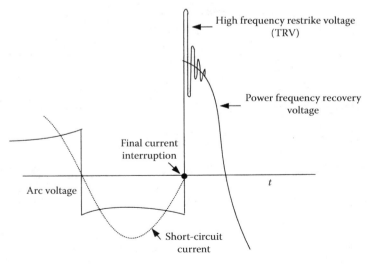

FIGURE 7.3
Final current interruption and the resulting recovery voltages.

7.4.1 First Pole to Clear Factor

TRV refers to the voltage across the first pole to clear, because it is generally higher than the voltage across the other two poles of the breaker, which clear later. Consider a three-phase ungrounded fault; the voltage across the breaker phase, first to clear, is 1.5 times the phase voltage (Figure 7.4a). The arc interruption in three phases is not simultaneous, as the three phases are mutually 120° apart. Thus, theoretically, the power frequency voltage of the first pole to clear is 1.5 times the phase voltage. It may vary from 1.5 to 2, rarely exceeding 3, and can be calculated using symmetrical components. The first pole to clear factor is defined as the ratio of rms voltage between the faulted phase and unfaulted phase and phase-to-neutral voltage with the fault removed. Figure 7.4 shows the first pole to clear factors for three-phase terminal faults. The first pole to clear factor for a three-phase fault with ground contact is calculated as

$$1.5 \frac{2X_0/X_1}{1+2X_0/X_1} \tag{7.13}$$

where X_1 and X_2 are the positive and zero sequence reactances of the source side. Also, in Figure 7.4, Y_1 and Y_2 are the sequence reactances of the load side. Figure 7.5 illustrates the slopes of tangents to three TRV waveforms of different frequencies. As the natural frequency rises, the rate of rise of recovery voltage (RRRV) increases. Therefore, it can be concluded that

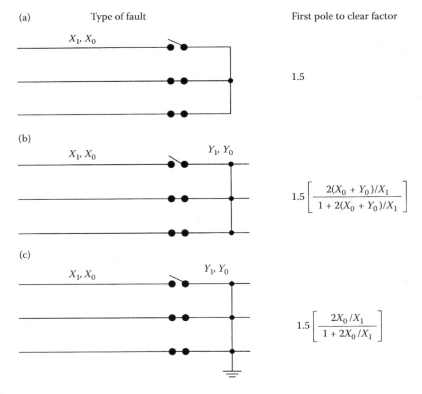

FIGURE 7.4
First pole to clear factor for three-phase faults: (a) a three-phase terminal fault with no connection to ground, (b) a three-phase fault with no contact to ground and an extension of the load side circuit, and (c) a three-phase fault with ground contact.

Current Interruptions in AC Networks 267

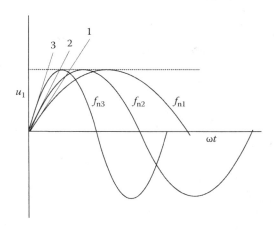

FIGURE 7.5
Effect of frequency of transient recovery voltage (TRV) on the rate of rise of recovery voltage (RRRV).

1. Voltage across breaker contacts rises slowly, as RRRV decreases.
2. There is a swing beyond the recovery voltage value, the amplitude of which is determined by the circuit breaker and breaker damping.
3. The higher is the natural frequency of the circuit, and the lower is the breaker interrupting rating.
4. Interrupting capacity/frequency characteristics of the breaker should not fall below that of the system.

The TRV is affected by many factors, among which the power factor of the current being interrupted is important. At zero power factor, maximum voltage is impressed across the gap at the instant of current zero, which tends to reignite the arc in the hot arc medium.

TRV can be defined by specifying the crest and the time to reach the crest, and alternatively, by defining the segments of lines which envelop the TRV waveform.

The steepest rates of rise in a system are due to short circuits beyond transformers and reactors which are connected to a system of high short-circuit power. In these cases, the capacitance that retards the voltage rise is small; however, the breaking capacity of the breaker for such faults need only be small compared to the short-circuit power of the system, as the current is greatly reduced by the reactance of the transformers and reactors. It means that, in most systems, high short-circuit levels of current and high natural frequencies may not occur simultaneously.

The interrupting capacity of every circuit breaker decreases with an increase in natural frequency. It can, however, be safely said that the interrupting (or breaking) capacity for the circuit breakers decreases less rapidly with increasing natural frequency than the short-circuit power of the system. The simplest way to make the breaking capacity independent of the natural frequency is to influence the RRRV across breaker contacts by resistors, which is discussed further. Yet, there may be special situations where the interrupting rating of a breaker may have to be reduced or a breaker of higher interrupting capacity may be required.

A circuit of a single-frequency transient occurs for a terminal fault in a power system composed of distributed capacitance and inductances. A terminal fault is defined as a fault close to the circuit breaker, and the reactance between the fault and the circuit breaker is negligible. TRV frequency can vary from low to high values, in the range of 20–10,000 Hz.

A circuit with inductance and capacitance on both sides of the circuit breaker gives rise to a double-frequency transient. After fault interruption, both circuits oscillate at their own frequencies, and a composite double-frequency transient appears across the circuit-breaker contacts. This can occur for a short-line fault. Recovery may feature traveling waves, depending on the faulted components in the network.

In the analyses to follow, we will describe IEC methods of estimating the TRV wave shape [4]. The ANSI/IEEE methods [5] are described in the rating structure of ANSI-rated breakers in Chapter 8. The IEC methods are simpler for understanding the basic principles and these are well defined in the IEC standard.

Figure 7.6 shows the basic parameters of the TRV for a terminal fault in a simplified network. Figure 7.6a shows power system constants, i.e., resistance, inductance, and capacitances.

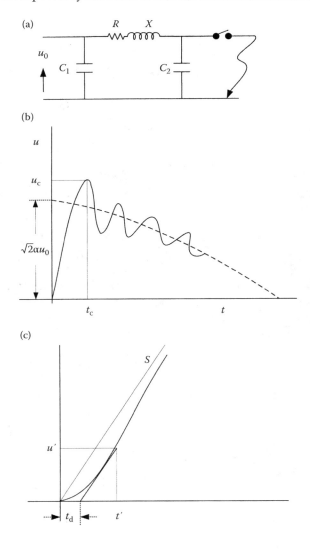

FIGURE 7.6
Basic parameters of the recovery voltage profile in a simplified terminal fault: (a) system configuration—R, X, C_1, and C_2 are system resistance, reactance, and shunt capacitances, respectively, (b) recovery voltage profile, and (c) initial TRV curve, delay line, and RRRV, shown as S.

Current Interruptions in AC Networks 269

The circuit shown may well represent the π model of a transmission line (see Volume 2). Figure 7.6b shows the behavior of the recovery voltage, transient component, and power frequency component. The amplitude of the power frequency component is given by

$$\alpha\sqrt{2}u_0 \tag{7.14}$$

where α depends on the type of fault and the network, and u_0 is the rated system rms voltage. The RRRV (RRRV = S) is the tangent to the TRV starting from the zero point of the unaffected or inherent TRV (ITRV). This requires some explanation. The TRV can change by the circuit breaker design and operation. The TRV measured across terminals of two circuit breakers can be different. The power system characteristics are calculated, ignoring the effect of the breakers. This means that an ideal circuit breaker has zero terminal impedance when carrying its rated current, and when interrupting short-circuit current its terminal impedance changes from zero to infinity instantaneously at current interruption. The TRV is then called ITRV.

Figure 7.6c shows an enlarged view of the slope. Under a few microseconds, the voltage behavior may be described by the time delay, t_d, which is dependent on the ground capacitance of the circuit. The time delay, t_d, in Figure 7.6c is approximated by

$$t_d = CZ_0 \tag{7.15}$$

where C is the ground capacitance and Z_0 is the surge impedance. Measurements show that a superimposed high-frequency oscillation often appears. IEC specifications recommend a linear voltage rise with a surge impedance of $450\,\Omega$ and no time delay, when the faulted phase is the last to be cleared in a single line-to-ground fault. This gives the maximum TRV. It is practical to distinguish between terminal faults and short-line faults for these phenomena.

7.5 The Terminal Fault

This is a fault in the immediate vicinity of a circuit breaker, which may or may not involve ground. This type of fault gives the maximum short-circuit current. There are differences in the system configuration, and the TRV profile differs widely. Two- and four-parameter methods are used in the IEC standard.

7.5.1 Four-Parameter Method

Figure 7.7 illustrates a representation of the TRV wave by the four-parameter method. In systems above 72.5 kV, clearing terminal faults higher than 30% of the rating will result in TRV characteristics that has four-parameter envelope. The wave has an initial period of high rise, followed by a low rate of rise. Such waveforms can be represented by the four-parameter method:

u_1 = first reference voltage (kV)

t_1 = time to reach u_1 (μs)

u_c = second reference voltage, peak value of TRV

t_2 = time to reach u_c (μs)

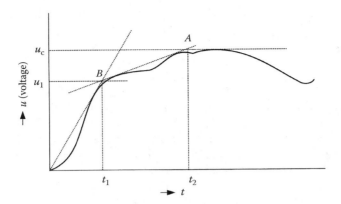

FIGURE 7.7
IEC four-parameter representation of TRV.

IEC specifies values of u_1, u_c, t_1, and t_2 for the circuit breakers. The interrupting current tests are carried out on circuit breakers with specified TRVs. The segments can be plotted, as shown in Figure 7.7, based on the breaker data. A table in IEEE Std. C37.011 provides factors for calculating rated TRV profiles for terminal faults, based upon IEC standards:

$$u_c = k_{af} \times k_{pp} \times \sqrt{2/3} u_r \tag{7.16}$$

$$u_c(T10) = u_c(T100) k u_c \tag{7.17}$$

where u_r is the breaker rated voltage. The amplitude factor k is defined as the ratio of the peak recovery voltage and the power frequency voltage:

$$k = \frac{u_c}{u_1} \tag{7.18}$$

The natural frequency is given by

$$f_n = \frac{10^3}{2 t_2} \text{ kHz} \tag{7.19}$$

For application & explanation of symbols in Eq. (7.16) & (7.17) see chapter 8.

7.5.2 Two-Parameter Representation

Figure 7.8 depicts the representation of TRV wave by two parameter method. TRV can be approximately represented by a single frequency transient[6] for terminal faults ranging between 10% and 30%, for systems above 72.5 kV, and for systems below 72.5 kV:

u_c = peak of TRV wave (kV)
t_3 = time to reach peak (μs)

Current Interruptions in AC Networks

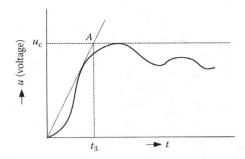

FIGURE 7.8
IEC two-parameter representation of TRV.

The initial rate of rise of TRV is contained within segments drawn according to the two- or four-parameter method by specifying the delay line, as shown in Figure 7.6c. The rate of rise of TRV can be estimated from

$$S = 2\pi f \sqrt{2} I_k Z_r \qquad (7.20)$$

where I_k is the short-circuit current, and Z_r is the resultant surge impedance; Z_0 can be found from sequence impedances. For a three-phase ungrounded fault, and with n equal outgoing lines

$$Z_r = 1.5(Z_1/n)\frac{2Z_0/Z_1}{1+2Z_0/Z_1} \qquad (7.21)$$

where Z_1 and Z_0 are the surge impedances in positive and zero sequences systems of individual lines, and n is the number of lines emanating from the substation. Factor 1.5 in Equation 7.21 can be 1.3. For single-frequency transients, with concentrated components, IEC tangent (rate of rise) can be estimated from

$$S = \frac{2\sqrt{2}kf_n u_0}{0.85} \qquad (7.22)$$

where f_n is the natural frequency and k is the amplitude factor.

The peak value u_c in the four-parameter method cannot be found easily, due to many variations in the system configurations. The traveling waves are partially or totally reflected at the points of discontinuity of the surge impedance. A superimposition of all forward and reflected traveling waves gives the overall waveform, see IEEE standard Reference [5].

7.6 The Short-Line Fault

Faults occurring between a few and some hundreds of kilometers from the breaker are termed short-line faults. A small length of the line lies between the breaker and the fault location, Figure 7.9a. After the short-circuit current is interrupted, the breaker terminal at the line end assumes a saw-tooth oscillation shape, as shown in Figure 7.9c. The rate of rise of voltage is directly proportional to the effective surge impedance (which can vary between 35 and 450 Ω, the lower value being applicable to cables) and to the rate of rise of current at current zero. The component on the supply side exhibits the same waveform as for a terminal fault, Figure 7.9b. The circuit breaker is stressed by the difference between these two voltages, Figure 7.9d. Because of the high-frequency oscillation of the line- side terminal, the TRV has a very steep initial rate of rise. In many breaker designs, the short-line fault may become a limiting factor of the current-interrupting capability of the breaker.

It is possible to reduce the voltage stresses of TRV by incorporating resistances or capacitances in parallel with the contact break. SF_6 circuit breakers of single break design up to 345 kV and 50 kA interrupting rating that have been developed.

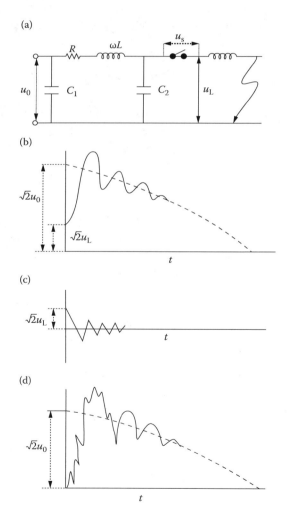

FIGURE 7.9
Behavior of TRV in a short-line fault: (a) system equivalent circuit, (b) recovery voltage on the source side, (c) recovery voltage on the load side, and (d) voltage across the breaker contacts.

Current Interruptions in AC Networks

273

7.7 Interruption of Low Inductive Currents

A circuit breaker is required to interrupt low inductive currents of transformers at no-load, high-voltage reactors, or locked rotor currents of motors. On account of arc instability in a low-current region, *current chopping* can occur, irrespective of the breaker interrupting medium. In a low-current region, the characteristics of the arc decrease, corresponding to a negative resistance that lowers the damping of the circuit. This sets up a high-frequency oscillation, depending on the LC of the circuit.

Figure 7.10a illustrates the circuit diagram for interruption of low inductive currents. The inductance L_2 and capacitance C_2 on the load side can represent transformers and motors. As the arc becomes unstable at low currents, the capacitances C_1 and C_2 partake in an oscillatory process of frequency:

$$f_3 = \frac{1}{2\sqrt{L\dfrac{C_1 C_2}{C_1 + C_2}}} \tag{7.23}$$

Practically, no current flows through the load inductance L_2. This forces a current zero, before the natural current zero, and the current is interrupted, Figure 7.10b. This interrupted current, i_a, is the chopped current at voltage u_a, Figure 7.10c. Thus, the chopped current is not only affected by the circuit breaker, but also by the properties of the circuit. The energy stored in the load at this moment is

$$i_a^2 \frac{L_2}{2} + u_a^2 \frac{C_2}{2} \tag{7.24}$$

which oscillates at the natural frequency of the disconnected circuit:

$$f_2 = \frac{1}{2\pi\sqrt{L_2 C_2}} \tag{7.25}$$

This frequency may vary between 200 and 400 Hz for a transformer. The maximum voltage on the load side occurs when all the inductive energy is converted into capacitive energy:

$$u_{2\max}^2 \frac{C_2}{2} = u_2^2 \frac{C_2}{2} + i_a^2 \frac{L_2}{2} \tag{7.26}$$

The source side voltage builds up with the frequency:

$$f_1 = \frac{1}{2\pi\sqrt{L_1 C_1}} \tag{7.27}$$

The frequency f lies between 1 and 5 kHz. This is shown in Figure 7.10c.

The linear expression of magnetic energy at the time of current chopping is not strictly valid for transformers and should be replaced with

$$\text{Volume of transformer core} \times \int_0^{B_m} H \, dB \tag{7.28}$$

where B is the magnetic flux density and H is the magnetic field intensity (B–H hysteresis curve).

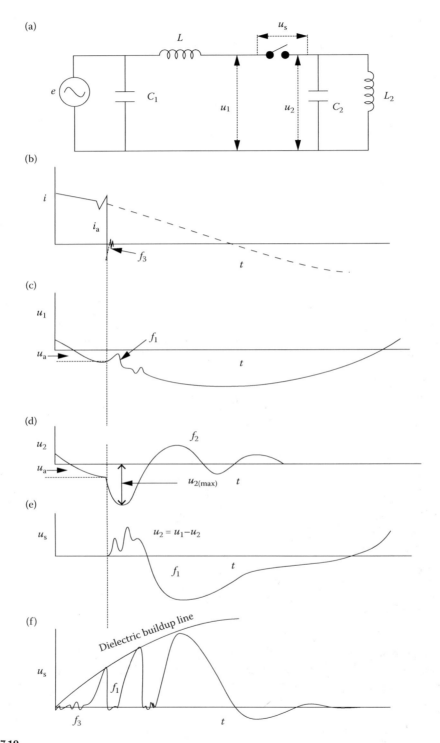

FIGURE 7.10
Interruption of low inductive currents: (a) the equivalent circuit diagram, (b) chopped current i_a at u_a, (c) source side voltage, (d) load side voltage, (e) voltage across breaker contacts, and (f) phenomena of repeated restrikes.

Current Interruptions in AC Networks

The load side voltage decays to zero, on account of system losses. The maximum load side overvoltage is of concern; relating from Figure 7.10d and from the simplified relationship (Equation 7.26), it is given by

$$u_{2\max} = \sqrt{u_a^2 + \eta_m i_a^2 \frac{L_2}{C_2}} \quad (7.29)$$

where η_m is the magnetic energy efficiency

A similar expression applies for the supply side voltage.

Thus, the overvoltage is dependent on the chopped current. If the current is chopped at its peak value, the voltage is zero. The chopped currents in circuit breakers have been reduced with better designs and arc control. The voltage across the supply side of the break, neglecting the arc voltage drop, is u_s and it oscillates at the frequency given by L and C_1. The voltage across the breaker contacts is $u_s = u_2 - u_1$. The supply side frequency is generally between 1 and 5 kHz.

If the circuit breaker voltage intersects the dielectric recovery characteristics of the breaker, reignition occurs and the process is repeated anew, Figure 7.10f. With every reignition, the energy stored is reduced, until the dielectric strength is large enough and further reignitions are prevented. Overvoltages of the order of two to four times may be produced on disconnection of inductive loads.

7.7.1 Virtual Current Chopping

This term is frequently used and needs some explanation. Reignition, restrike, or prestrike in one phase can cause high-frequency currents to flow, which can get coupled to the other two phases and this phenomenon is called virtual current chopping [6–8]. This is clearly illustrated in Figure 7.11. At the instant of reignition in phase a, the high-frequency current flows to ground through the load capacitance of phase a and divides into phases b and c equally, assuming a balanced three-phase system. The power frequency current in phases b and c is approximately 0.87 times the peak value of the 60 Hz currsent. If the magnitude of the high-frequency currents in phases b and c ($I_t/2$)

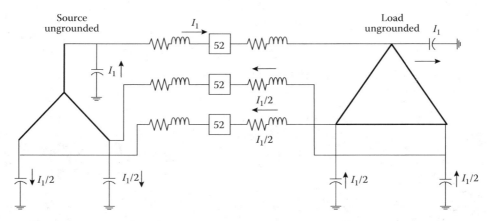

FIGURE 7.11
Circuit illustrating the virtual current chopping due to reignition, restrike, or prestrike.

is greater than the power frequency current; the high-frequency current forces the 60 Hz current to zero. This forced current phenomenon is virtual current chopping and gives rise to overvoltages.

7.8 Interruption of Capacitance Currents

A breaker may be used for line dropping and interrupt charging currents of cables open at the far end or shunt capacitor currents. These duties impose voltage stresses on the breaker. Consider the single-phase circuit of Figure 7.12a. The distributed line capacitance

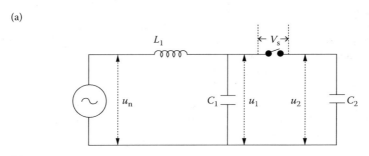

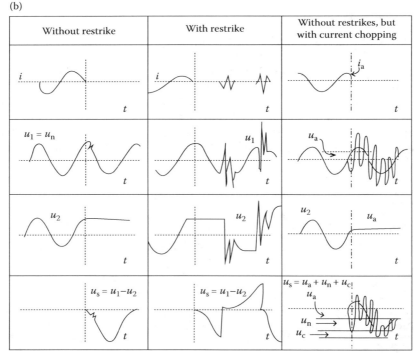

FIGURE 7.12
Interruption of capacitance current; (a) the equivalent circuit diagram and (b) current and voltage waveforms without restrike, with restrike, and with current chopping. i_a is the chopping current and u_a is voltage on the disconnected side on current chopping. λ = damping factor < 1.

Current Interruptions in AC Networks 277

is represented by a lumped capacitance C_2, or C_2 may be a power capacitor. The current and voltage waveforms of capacitance current interruption in a single pole of a circuit breaker under the following three conditions are shown in Figure 7.12b.

- Without restrike
- With restrike
- With restrike and current chopping

After interruption of the capacitive current, the voltage across the capacitance C_2 remains at the peak value of the power frequency voltage:

$$u_2 = \frac{\sqrt{2}u_n}{\sqrt{3}} \tag{7.30}$$

The voltage at the supply side oscillates at a frequency given by supply side C_1 and L_1, about the driving voltage u_n. The difference between these two voltages appears at the breaker pole. This can be more than double the rated voltage, with no prior charge on the capacitors.

If the gap across poles of a circuit breaker has not recovered enough dielectric strength, restrike may occur. As the arc bridges the parting contacts, the capacitor being disconnected is again reconnected to the supply system. This results in a frequency higher than that of the natural frequency of the source side system being superimposed on the 60 Hz system voltage. The current may be interrupted at a zero crossing in the reignition process. Thus, the high-frequency voltage at its crest is trapped on the capacitors. Therefore, after half a cycle following the restrike, the voltage across the breaker poles is the difference between the supply side and the disconnected side, which is at the peak voltage of the equalizing process, and a second restrike may occur. Multiple restrikes can occur, pumping the capacitor voltage to 3, 5, 7,... times the system voltage at each restrike. The multiple restrikes can terminate in two ways: (1) these may cease as the breaker parting contacts increase the dielectric strength and (2) these may continue for a number of cycles, until these are damped out.

A distinction should be made between reignitions in less than 5 ms of current zero and reignitions at 60 Hz power frequency. Reignitions in less than 5 ms have a low voltage across the circuit breaker gap and do not lead to overvoltages.

Disconnecting a three-phase capacitor circuit is more complex. The instant of current interruption and trapped charge level depends on the circuit configuration. In an ungrounded three-phase wye-connected bank, commonly applied at medium- and high-voltage levels, let phase a current be first interrupted. This will occur when the voltage of phase a is at its peak. Figure 7.13a shows that phase a is interrupted first. The charge trapped in phase a is 1 per unit and that trapped in phases b and c is 0.5 per unit.

The interruption of phase a changes the circuit configuration and connects the capacitors in phases b and c in series. These capacitors are charged with equal and opposite polarities. The current in phases b and c will interrupt simultaneously as soon as the phase-to-phase current becomes zero. This will occur at 90° after the current interruption in phase a, at the crest of the phase-to-phase voltage so that an additional charge of $\sqrt{3/2}$ is stored in

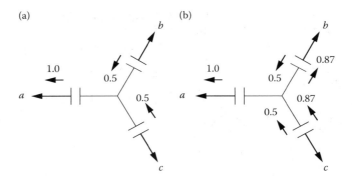

FIGURE 7.13
Sequence of creating trapped charges in a three-phase ungrounded capacitor bank: (a) first phase a clears and (b) phases b and c clear in series.

the capacitors, as shown in Figure 7.13b. These charges will add to those already trapped on the capacitors in Figure 7.13a and thus voltages across capacitor terminals are

$$E_{ab} = 0.634$$
$$E_{bc} = 1.73 \text{ per unit} \qquad (7.31)$$
$$E_{ac} = 2.37 \text{ per unit}$$

Further escalation of voltages occurs if phases b and c are not interrupted after 90° of current interruption in phase a. It is hardly possible to take into account all forms of three-phase interruptions with restrikes, see Reference [8].

7.9 TRV in Capacitive and Inductive Circuits

The TRV on interruption of capacitive and inductive circuits is illustrated in Figure 7.14. In a capacitive circuit when the current passes through zero, Figure 7.14a, the system voltage is trapped on the capacitors. The recovery voltage, the difference between the source and load sides of the breaker reaches a maximum of 2.0 per unit after ½ cycle of current interruption, Figure 7.14b and c. The TRV oscillations are practically absent, as large capacitance suppresses the oscillatory frequency (Equation 7.11) and the rate of rise of the TRV is low. This may prompt a circuit breaker contacts to interrupt, when there is not enough separation between them and precipitate restrikes.

This is not the case when disconnecting an inductive load, Figure 7.14d through f. The capacitance on the disconnected side is low and the frequency of the isolated circuit is high. The TRV is therefore oscillatory. The rate of rise of TRV after disconnection is fairly high.

Reverting to TRV on disconnection of a capacitance, IEEE standard[8] specifies that voltage across capacitor neutral, not ground, during an opening operation should not exceed 3.0 per unit for a general purpose breaker. For the definite purpose breaker rated 72.5 kV and lower, this limit is 2.5 per unit. For breakers rated 121 kV and above capacitor banks are normally grounded, and the voltage is limited to 2.0 per unit. It was observed that

Current Interruptions in AC Networks

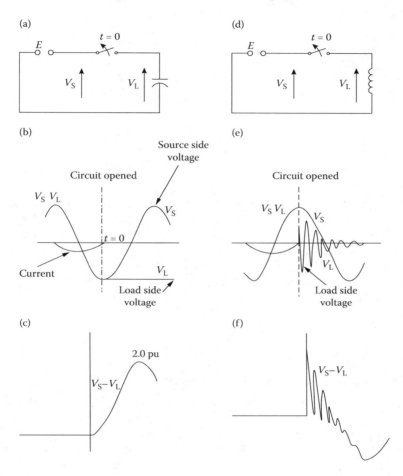

FIGURE 7.14
TRV, interruption of capacitive and inductive currents: (a)–(c) interruption of capacitive current, and (d)–(f) interruption of inductive current.

when interrupting capacitance current 2.0 per unit voltage occurs at small contact separation and this can lead to restrikes. With no damping the voltage will reach 3.0 per unit. This suggests that one restrike is acceptable for general purpose breakers, but not for definite purpose breakers.

Some voltage-controlled methods are resistance switching and surge arresters, see Section 7.13 for resistance switching.

7.10 Prestrikes in Circuit Breakers

A prestrike may occur on closing a circuit breaker, establishing the current flow, before the contacts physically close. A prestrike occurs in a current flow at a frequency given by the inductances and capacitances of the supply circuit and the circuit being closed. In Figure 7.15, this high-frequency current is interrupted at $t = t_1$. Assuming no trapped

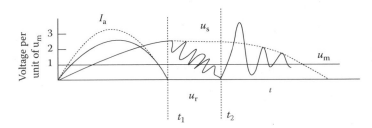

FIGURE 7.15
Voltages due to prestrikes at $t = t_1$ and $t = t_2$. The inrush current is interrupted at $t = t_1$.

charge on the capacitors, the voltage rises to approximately 2 per unit. A high-frequency voltage given by source reactance and stray capacitance is superimposed on the recovering bus voltage. If a second prestrike occurs at $t = t_2$, a further escalation of the bus voltage occurs. Thus, the transient conditions are similar as for restrikes; however, the voltage tends to decrease as the contacts come closer in a closing operation. In Figure 7.15, u_m is the maximum system voltage, u_r is the recovery voltage, and u_s is the voltage across the breaker contacts.

7.11 Overvoltages on Energizing HV Lines

The highest overvoltages occur when unloaded high-voltage transmission lines are energized and reenergized and this imposes voltage stresses on circuit breakers [9].

Figure 7.16a shows the closing of a line of surge impedance Z_0 and of length l, open at the far end. Before the breaker is closed, the voltage on the supply side of the breaker terminal is equal to the power system voltage, while the line voltage is zero. At the moment of closing, the voltage at the sending end must rise from zero to the power frequency voltage. This takes place in the form of a traveling wave on the line with its peak at u_m interacting with the supply system parameters. As an unloaded line has capacitive impedance, the steady-state voltage at the supply end is higher than the system voltage and, due to the Ferranti effect; the receiving end voltage is higher than the sending end, Volume 2.

Overvoltage factor can be defined as follows:

$$\text{Total overvoltage factor} = OV_{\text{tot}} = \frac{u_m}{u_n} \quad (7.32)$$

where u_m is the highest voltage peak at a given point and u_n is the power frequency voltage at supply side of breaker before switching.

$$\text{Power frequency overvoltage factor} = OV_{\text{pf}} = \frac{u_{\text{pf}}}{u_n} \quad (7.33)$$

This is the ratio of the power frequency voltage u_{pf} after closure at a point and power frequency voltage u_n on supply side before closing.

Current Interruptions in AC Networks

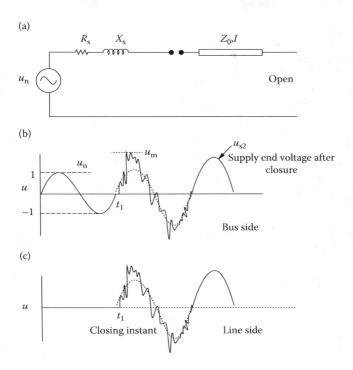

FIGURE 7.16
Overvoltages to ground on closing a transmission line: (a) basic circuit, (b) and (c) voltages on the source and line side, and superimposition of traveling waves occurs at $t = t_1$.

$$\text{Transient overvoltage factor} = OV_{tr} = \frac{u_m}{u_{pf}} \quad (7.34)$$

The power frequency overvoltage factor can be calculated by known line parameters. This is given by

$$\frac{1}{\cos\alpha 1 - X_s/Z_0 \sin\alpha l} \quad (7.35)$$

where the surge impedance and α are given by

$$Z_0 = \sqrt{\frac{L_1}{C_1}} \quad (7.36)$$

$$\alpha = 2\pi f \sqrt{L_1 C_1} \quad (7.37)$$

The relationship between sending and receiving end voltages is $1/\cos\alpha l$.

This shows that the increase in power frequency voltage depends considerably on the line length. The transient voltage is not so simple to determine and depends on the phase angle at the closing instant. At the instant $t = t_1$, the maximum superposition of the transient and power frequency voltages occurs, see Figure 7.16b and c.

Trapped charges occur on the transmission lines in three-pole auto-recloser operations. Contact making of three poles of a circuit breaker is nonsimultaneous. Consider breakers at the sending and receiving ends of a line and a transient ground fault, which needs to be cleared by an auto-recloser operation. The opening of the two breakers is nonsimultaneous and the one which opens later must clear two line phases at no-load. These two phases can, therefore, remain charged at the peak of the power frequency voltage, which is still present when the closure takes place. After the dead time, one breaker has to close with two phases still charged. If the closing instant happens to be such that the trapped charge and the power frequency voltage are of opposite polarity, maximum transient overvoltage will occur.

Switching overvoltages do not cause flashovers to the same extent as caused by lightning. The frequency, point of occurrence, amplitude and characteristics govern the selection of equipment, insulation level, and economical design. The switching surges gain more importance as the system voltage rises and in extra high voltage (EHV) networks, it is the switching surges which are of primary importance in insulation coordination. The switching surges have, generally, a power frequency and a transient component, which bear certain relation to each other and may be of equal amplitude in the absence of damping. The nonsimultaneous closing of breaker poles can increase the transient component. The EHV installations are primarily concerned with the stresses imposed on the insulation by switching surges and the coordination of the insulation is based upon these values. To lower the costs, it is desirable to reduce the insulation levels at high voltages. On the transmission line towers, the flashover voltage cannot be allowed to increase along with service voltage, as beyond a certain point the electrical strength of air gaps can no longer be economically increased by increasing the clearances [9]. Thus, while the external lightning voltages are limited, the switching over voltages become of primary concern.

All equipment for operating voltages above 300 kV is tested for switching impulses. No switching surge values are assigned to equipment below 60 kV.

A reader may refer to Chapter 8 that shows rated line closing switching surge factors for circuit breakers specifically designed to control line switching closing surge. Note that the parameters are applicable for testing with standard reference transmission lines, as shown in Table 8.7.

The lightning surges are responsible for approximately 10% of all short circuits in substations and almost 50% of short circuits on lines and systems above 300 kV. Less than 1% short circuits are caused by switching surges [10].

7.11.1 Overvoltage Control

The power frequency component of the overvoltage is controlled by connecting high-voltage reactors from line to ground at the sending and receiving ends of the transmission lines. The effect of the trapped charge on the line can be eliminated if the closing takes place during that half cycle of the power frequency voltage, which has the same polarity as the trapped charge. The high-voltage circuit breakers may be fitted with devices for polarity-dependent closing. Controlling overvoltages with switching resistors is yet another method.

Lines with trapped charge and no compensation and no switching resistors in breakers may have overvoltages greater than three times the rated voltage. Without trapped charge, this overvoltage will be reduced to 2.0–2.8 times the rated voltage.

Current Interruptions in AC Networks

283

With single-stage closing resistors and compensated line, overvoltages are reduced to less than twice the rated voltage. With two-stage closing resistors or compensated lines with optimized closing resistors, the overvoltage factor is 1.5.

7.11.2 Synchronous Operation

A breaker can be designed to open or close with reference to the system voltage sensing and zero crossing. An electronic control monitors the zero crossing of the voltage wave and controls the shunt release of the breaker. The contacts can be made to touch at voltage zero or voltage crest. In the opening operation, the current zero occurs at a definite contact gap. As the zeroes in three-phase voltages will be displaced, the three poles of the breaker must have independent operating mechanisms. Independent pole operation is a standard feature for breakers of 550 kV. Though it can be fitted to breakers of even lower voltages, say 138 kV. A three-pole device is commercially available for staggering the pole operating sequence.

Switching of unloaded transformers, transmission lines, shunt reactors, and capacitor banks can benefit from the synchronous operation. The operating characteristics of the breaker must be matched with that of the electrical system. Variations can occur due to aging ambient temperature, level of energy stored in the operating mechanism, and control voltage levels. Deviations in the closing time of 1–2 cycles and in the opening time of 2.9 ms can occur. Resistance switching can achieve the same objectives and is less prone to the mechanical variations—a point often put in the forefront by the advocates of resistance switching. Currently, IEC circuit breaker standard IEC 62271-100 [4] excludes circuit breakers with intentional nonsimultaneous pole operation, but this operation may be soon added to the specifications. The resistance switching and controlled switching are not used in combination.

Rate of rise of dielectric strength (RRDS) is defined as the circuit breaker characteristic, which describes the rate of voltage withstand at opening of a circuit breaker. The value defines the maximum arcing time needed for reignition free interruption of inductive loads. The RRDS determination is done at no-load opening operations by determining the flashover limits at different contact distances, which translate into different times after contact separation by applying a rapidly increasing voltage. The manufacturers specify their specific breaker designs with staggered pole closing controllers for controlled switching. Upon closing operation, the RRDS will rapidly decrease and is zero when the contacts touch.

7.11.3 Synchronous Capacitor Switching

Though the restrikes in vacuum circuit breakers are greatly reduced (but not eliminated), synchronous switching can eliminate higher frequency component of the voltage swings. The contacts are closed at a nominal voltage zero.

Energizing a capacitor bank means that there should be zero voltage across the breaker contacts when the contacts touch. A number of studies show that the overvoltages can be controlled if the contacts touch 1 ms before or after the voltage zero, Figure 7.17. The closing can be set slightly after voltage zero in order to minimize the influence of statistical variations.

The closing operation of grounded and ungrounded capacitor banks must be distinguished. When the capacitors are grounded, then each pole must close independently at the voltage zero of that phase. When the capacitor neutrals are ungrounded, two poles

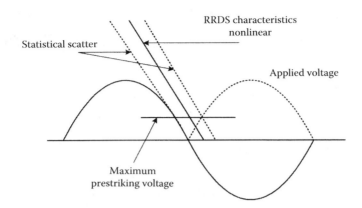

FIGURE 7.17
Synchronous closing operation with respect to voltage zero and RRDS.

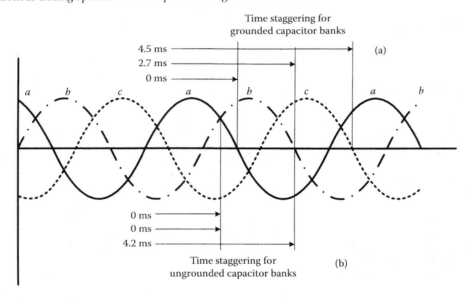

FIGURE 7.18
Synchronous closing of shunt capacitor banks: (a) grounded banks and (b) ungrounded banks.

can close at phase-to-phase voltage of zero, and the last pole will close at 4.2 ms. This is depicted in Figure 7.18a and b.

7.11.4 Shunt Reactors

The shunt reactors may be directly grounded, for systems voltages of 60 kV and above and for systems above 60 kV, these are generally ungrounded or grounded through a reactor. The current to be interrupted is generally less than 300 A; however, shunt reactor switching imposes a severe duty on the circuit breaker and connected system. For grounded shunt reactors, two types of overvoltages are generated [11]:

- Chopping overvoltages with frequencies up to 5 kHz
- Reignition overvoltages with frequencies up to several hundred kilohertz

Current Interruptions in AC Networks

For a single interrupter circuit breaker, the chopping current level is given by

$$i_{ch} = \lambda \sqrt{C_t}$$

where i_{ch} is the chopping current (A), C_t is the total capacitance in parallel with circuit breaker (F), and λ is the chopping number for a single interrupter (AF$^{-0.5}$). The circuit breaker chopping numbers are shown in Table 7.1 and are derived by laboratory tests [11].

The characteristics of shunt reactors depend on their design:

- Three legged gapped iron core
- Five legged gapped iron core
- Shell type gapped iron core
- Air core

Table 7.2 gives the electrical characteristics of shunt reactors for voltages of 60 kV and above.

The general case of shunt reactor switching is shown in Figure 7.19 and general schematic of current chopping and reignition overvoltages is shown in Figure 7.20, taken from CIGRE Technical Brochure 305 [12].

TABLE 7.1

Circuit Breaker Chopping Numbers

Circuit Breaker Type	Chopping Number λ
	AF$^{-0.5}$
Minimum oil	$5.8 \times 10^4 - 10 \times 10^4$
Air blast	$15 \times 10^4 - 20 \times 10^4$
SF$_6$ puffer	$4 \times 10^4 - 19 \times 10^4$
SF$_6$ self-blast	$3 \times 10^4 - 10 \times 10^4$
SF$_6$ rotating arc	$0.39 \times 10^4 - 0.77 \times 10^4$

TABLE 7.2

Shunt Reactor Electrical Characteristics for Voltages of 60 kV and Above

Maximum System Voltage (kV)	Rated Voltage (kV)	Rating (Mvar)	Rated Power Frequency (Hz)	Rated Current (A)	Inductance (H)	Capacitance (nF)	Natural Frequency (kHz)
800	765	150–300	60	113–226	5.17–10.35	1.7–4.0	1.1–1.7
800	735	330	60	259	4.34	4.1	1.2
550	525	135	60	148	5.43	1.8–4.0	1.1–1.6
420	400	120–200	50	173–289	2.55–4.25	1.9–3.2	1.4–2.3
245	236	125	60	306	1.18	2.1	3.2
145	132	55	50	240	1.0	1.3	4.4
121	115	25	60	126	1.4	2.9	2.5
72.5	60	20	60	190	0.48	2.0	5.1

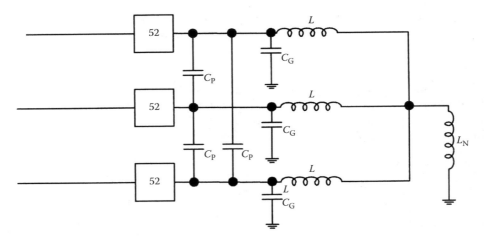

FIGURE 7.19
A general case for shunt reactor switching.

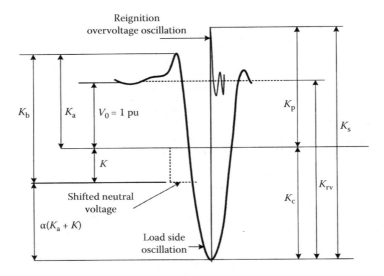

FIGURE 7.20
A general schematic of chopping currents and reignition voltages, reactor switching.

Figure 7.20 shows that prior to interruption of current in any one phase, the neutral of the reactor is at ground potential; when the first pole clears, the neutral potential shifts through a transient oscillation toward a bias voltage K per unit, *depending on the grounding arrangement*. The mean value of shifted neutral voltage decays to zero depending upon grounding. The load side oscillation of the first pole to clear will oscillate around shifted neutral voltage, the degree of chopping determining the magnitude of the overvoltage. The suppression peak overvoltage K_a in per unit of the phase-to-ground peak voltage V_0 is given by

$$K_a = (1+K)\sqrt{1+\frac{1}{1+K}\left(\frac{i_{ch}}{V_0}\right)^2\left(\frac{L}{C_L}\right)} - K \qquad (7.38)$$

Current Interruptions in AC Networks

And

$$K = \frac{1}{2+(L/L_N)} \tag{7.39}$$

where C_L is the effective load side capacitance. For directly grounded reactors, $K=0$ and Equation 7.38 can be applied to any pole. For ungrounded or reactor grounded reactors, Equation 7.38 applies to first pole to clear.

The magnitude of the suppression peak overvoltage relative to shifted neutral is given by

$$K_b = K_a + K \tag{7.40}$$

The magnitude of the overvoltage to ground at recovery peak K_c is

$$K_c = K + \alpha(K_a = K) \tag{7.41}$$

where α is the damping factor associated with chopping overvoltage oscillations.

The peak recovery voltage across the circuit breaker K_{rv} is

$$K_{rv} = 1 + K + \alpha(K_a + K) \tag{7.42}$$

The maximum reignition overvoltage to ground K_p is

$$K_p = 1 + \beta[1 + K + \alpha(K_a + 1)] \tag{7.43}$$

where β is the damping factor associated with the reignition overvoltage oscillation and can be assumed $= 0.5$.

The maximum reignition overvoltage excursion to peak K_s is given by

$$K_s = 1 + \beta[1 + K + \alpha(K_a + K)] \tag{7.44}$$

For specific applications to grounded reactors, the above equations can be modified by putting $K=0$ for solidly grounded reactor, and $K=0.5$ for ungrounded reactors, while for reactors grounded through a reactor, K is retained in these equations.

This can reach 4.0 per unit. Metal-oxide surge arresters can reduce it to 2.4 per unit, and controlled opening applied to the circuit breaker will reduce it to zero.

7.11.4.1 Oscillation Modes

The equations for oscillation modes are as follows:
Load side oscillations

$$f_L = \frac{1}{2\pi\sqrt{LC_L}} \tag{7.45}$$

This is of the order of 1–5 kHz.

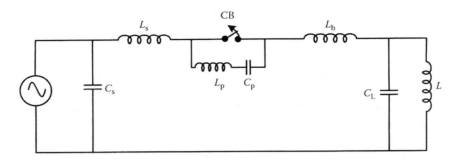

FIGURE 7.21
Schematic equivalent circuit of reignition overvoltage oscillation.

In reignition oscillations, three different oscillation circuits are involved. A first parallel oscillation occurs when C_p (L_p and C_p are the inductance and capacitance in parallel with the circuit breaker) discharges through the circuit breaker:

$$f_{p1} = \frac{1}{2\pi\sqrt{L_p C_p}} \tag{7.46}$$

This is of the order of 1–10 MHz.

The circuit breaker will not interrupt the current associated with first parallel oscillation. A second parallel oscillation (reignition overvoltage oscillation) will follow as a result of which the voltages across C_s and C_L are equalized, see Figure 7.21:

$$f_{p2} = \frac{1}{2\pi}\sqrt{\frac{C_L + C_s}{L_b C_L C_s}} \tag{7.47}$$

This oscillation is in the range of 50–1000 kHz.

The circuit breaker may interrupt the current associated with second parallel oscillation, if it does not, then a main circuit oscillation develops. Neglecting L_b, as it is comparatively small

$$f_m = \frac{1}{2\pi}\sqrt{\frac{L_s + L}{L_s L (C_s + C_L)}} \tag{7.48}$$

This is in the range of 5–20 kHz.

7.12 Out-of-Phase Closing

Figure 7.22 shows two interconnected systems which are totally out of phase. In Figure 7.22a, a voltage equal to three times the system peak voltage appears across the breaker pole, while in Figure 7.22b, a ground fault exists on two different phases at

Current Interruptions in AC Networks

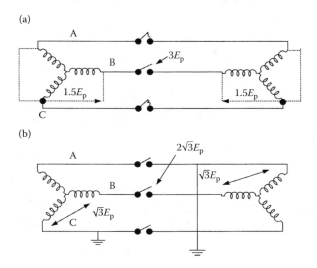

FIGURE 7.22
Overvoltages due to two interconnected systems totally out of phase: (a) unfaulted condition, the maximum voltage equal to three times the peak system voltage and (b) ground faults on different phases on the source and load sides, the maximum voltage equal to $2\sqrt{3}$ times the peak system voltage.

the sending and receiving ends (rather an unusual condition). The maximum voltage across a breaker pole is $2\times\sqrt{3}$ times the normal system peak voltage. The present-day high-speed relaying has reduced the tripping time and, thus, the divergence of generator rotors on fast closing is reduced. Simultaneously, the high-speed auto-reclosing to restore service and remove faults increases the possibility of out-of-phase closing, especially under fault conditions. The effect of the increased recovery voltage when the two systems have drifted apart can be stated in terms of the short-circuit power that needs to be interrupted. If the interrupting capacity of a circuit breaker remains unimpaired up to double the rated voltage, it will perform all events satisfactorily as a tie-line breaker when the two sections of the system are completely out of synchronism. The short-circuit power to be interrupted under out-of-step conditions is approximately equal to the total short-circuit power of the entire system, but reaches this level only if the two systems which are out of phase have the same capacity (Figure 7.23). In this figure, P_1 is the interrupting capacity under completely out-of-phase conditions of two interconnected systems, and P_2 is the total short-circuit capacity; X_a and X_b are the short-circuit reactances of the two systems.

7.13 Resistance Switching

Circuit breaker resistors can be designed and arranged for the following functions:

- To reduce switching surges and overvoltages
- For potential control across multibreaks per phase
- To reduce natural frequency effects

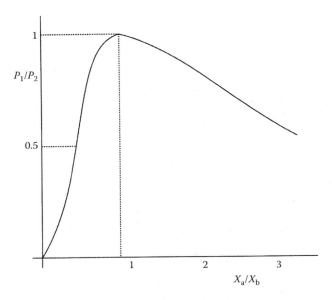

FIGURE 7.23
Interrupting capacity P_1 in the case of out-of-step operation as a function of ratio of short-circuit reactances X_a/X_b and P_2, the total short-circuit capacity.

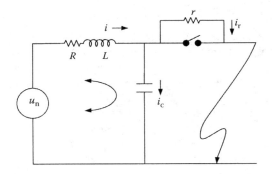

FIGURE 7.24
Resistance "r" connected in parallel with the circuit breaker contacts (resistance switching) on a short-circuit interruption.

Figure 7.24 shows a basic circuit of resistance switching. A resistor "r" is provided in parallel with the breaker pole, and R, L, and C are the system parameters on the source side of the break. Consider the current loops in this figure. The following equations can be written:

$$u_n = iR + L\frac{di}{dt} + \frac{1}{C}\int i_c\, dt \qquad (7.49)$$

$$\frac{1}{C}\int i_c dt = i_r r \qquad (7.50)$$

$$i = i_r + i_c \tag{7.51}$$

This gives

$$\frac{d^2 i_r}{dt^2} + \left(\frac{R}{L} + \frac{1}{rC}\right)\frac{di_r}{dt} + \left(\frac{1}{LC} + \frac{R}{rLC}\right)i_r = 0 \tag{7.52}$$

The frequency of the transient is given by

$$f_n = \frac{1}{2}\sqrt{\frac{1}{LC} - \frac{1}{4}\left(\frac{R}{L} - \frac{1}{rC}\right)^2} \tag{7.53}$$

In power systems, R is $\ll L$. If a parallel resistor across the contacts of value $r < \frac{1}{2}\sqrt{L/C}$ is provided, the frequency reduces to zero. The value of r at which frequency reduces to zero is called the critical damping resistor. The critical resistance can be evaluated in terms of the system short-circuit current, I_{sc}:

$$r = \frac{1}{2}\sqrt{\frac{u_n}{I_{sc}\omega C}} \tag{7.54}$$

Figure 7.25 shows the effect of the resistors on the recovery voltage. Opening resistors are also called switching resistors and are in parallel with the main break and in series with an auxiliary resistance break switch. On an opening operation, the resistor switch remains closed and opens with a certain delay after the main contacts have opened. The resistance switch may be formed by the moving parts in the interrupter or striking of an arc dependent on the circuit breaker design.

Figure 7.26 illustrates the sequence of opening and closing in a circuit breaker provided with both opening and closing resistors. The closing resistors control the switching overvoltage on energization of, say, long lines. An interrupting and closing operation is shown. The main break is shown as SB, the breaking resistor as RB. On an opening operation, as the main contact start arcing, the current is diverted through the resistor RB, which is interrupted by contacts SC. In Figure 7.26d, the breaker is in open position. Figure 7.26e and f shows the closing operation. Finally, the closed breaker is shown in Figure 7.26a.

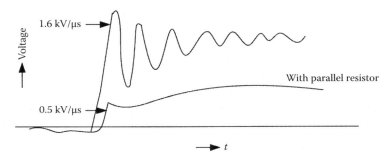

FIGURE 7.25
Reduction of RRRV with parallel resistance switching.

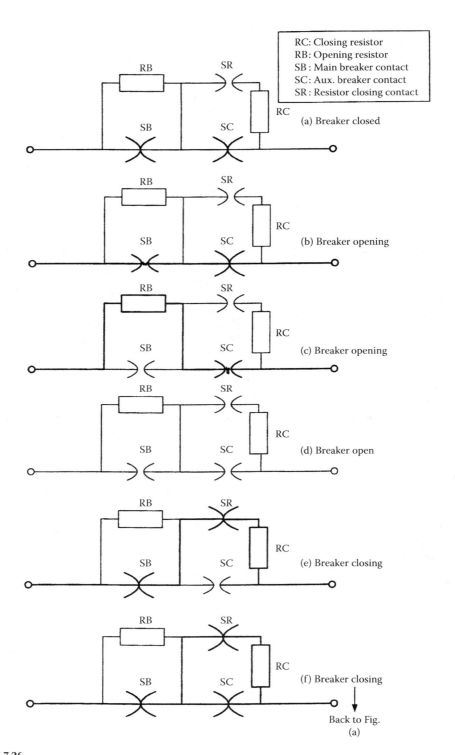

FIGURE 7.26
Sequence of closing and opening operation of a high-voltage circuit breaker provided with opening and closing resistors.

7.14 Failure Modes of Circuit Breakers

In AC circuit breakers, the phenomena of arc interruption are complex. Arc plasma temperatures of the order of 25,000–5000 K are involved, with conductivity changing a billion times as fast as temperature in the critical range associated with thermal ionization. Supersonic turbulent flow occurs in changing flow and contact geometry at speeds from 100 to 1000 m/s in the arc. The contact system should accelerate from a stationary condition to high speeds in a very short time.

With parabolic pressure distribution in the contact zone of a double-nozzle configuration, a cylindrical arc with temperatures nearing 25,000 K exists. Due to the low density of gas at high temperatures, the plasma is strongly accelerated by an axial pressure gradient. A so-called thermal arc boundary at a temperature of 300–2000 K exists. The arc downstream expands into nozzles and in this region, the boundary layer between arc and gas is turbulent with formation of vortices.

Two types of failures can occur: (1) dielectric failure that is usually coupled with a terminal fault and (2) thermal failure that is associated with a short-line fault. If after a current zero, the RRRV is greater than a critical value, the decaying arc channel is reestablished by ohmic heating. This period, which is controlled by the energy balance in the arc is called the thermal interruption mode. Figure 7.27a shows a successful thermal interruption and Figure 7.27b shows a thermal failure. Within 2 μs after interruption, the voltage deviates from TRV. It decreases and approaches the arc voltage.

Following the thermal mode, a hot channel exists at temperatures from 300 to 5000 K, and a gas zone adjacent to the arc diminishes at a slow rate. The recovering system voltage distorts and sets the dielectric limits. After successful thermal interruption, if the TRV can reach such a high peak value that the circuit breaker gap fails; it is called a dielectric failure mode. This is shown in Figure 7.28. Figure 7.28a shows successful interruption, and Figure 7.28b shows dielectric failure at the peak of the recovery voltage, and rapid voltage decay.

The limit curves for circuit breakers can be plotted on a log u and log I basis, as shown in Figure 7.29. In this figure, u is the system voltage and I the short-circuit current. The portion in thick lines shows dielectric limits, while the vertical portion in thin lines shows thermal limits. In thermal mode

1. Metal vapor production from contact surfaces.
2. di/dt, i.e., the rate of decrease of the current at current zero.

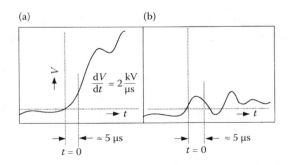

FIGURE 7.27
Thermal failure mode of a circuit breaker, while opening: (a) successful interruption and (b) failure in the thermal mode.

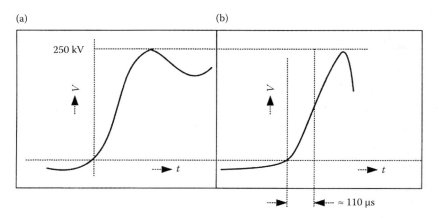

FIGURE 7.28
Dielectric failure mode of a high-voltage circuit breaker: (a) successful interruption and (b) failure at the peak of TRV.

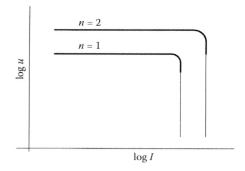

FIGURE 7.29
General form of limiting curve in a circuit breaker plotted as log u versus log I, where V is the rated voltage of the breaker and I is the short-circuit current; n indicates the number of interrupting chambers in series. Thick lines show dielectric mode and thin lines show thermal mode of possible failure.

3. Arc constrictions in the nozzle due to finite velocity.
4. Nozzle configurations.
5. Presence of parallel capacitors and resistors.
6. Type of quenching medium and pressures are of importance. In the dielectric mode, the generation of electrons in an electric field is governed by Townsend's equation:

$$\frac{\partial n_e}{\partial t} = -\frac{\partial n_e V_e}{\partial d} + (\alpha - \eta) n_s V_e \qquad (7.55)$$

where n_e is the number of electrons, α is the Townsend coefficient (number of charged particles formed by negatively charged ions), d is the spacing of electrodes, η is the attachment coefficient, and V_e is the electron drift velocity.

Current Interruptions in AC Networks

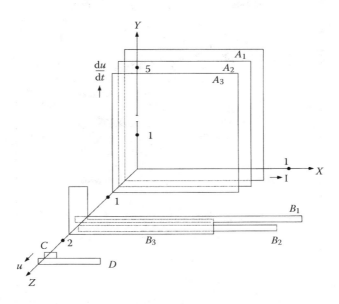

FIGURE 7.30
Stresses in a high-voltage circuit breaker in terms of short-circuit current, RRRV, and maximum overvoltage. A_1, A_2, and A_3 short-line faults; B_1, B_2, and B_3 terminal faults; C capacitance currents; D out-of-phase or asynchronous conditions.

7.15 Stresses in Circuit Breakers

The stresses in a circuit breaker under various operating conditions are summarized in Figure 7.30. These stresses are shown in terms of three parameters, current, voltage, and du/dt, in a three-dimensional plane. Let the current stress be represented along the x-axis, the du/dt, stress along the y-axis, and the voltage stress along the z-axis. We see that a short-line fault (A_1, A_2, A_3) gives the maximum RRRV stress, though the voltage stress is low. A terminal fault (B_1, B_2, B_3) results in the maximum interrupting current, while capacitor switching (C) and out-of-phase switching (D) give the maximum voltage stresses. All the stresses do not occur simultaneously in an interrupting process.

7.16 Classification of Circuit Breakers according to Interrupting Medium

The interrupting mediums in circuit breakers are as follows:

- SF_6, gas circuit breakers
- No medium, i.e., vacuum
- Air blast
- Magnetic air break
- Bulk oil

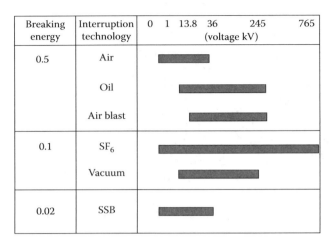

FIGURE 7.31
Depiction of various interrupting technologies.

Figure 7.31 shows the current interruption technologies. In air break designs, the energy expended is equal to about half the electromagnetic energy of the circuit to be broken. In low-voltage application, the technique holds a monopoly—though for medium voltages also it is possible to obtain successful breaking with an apparatus having high time constant, as long as cooling capacity is sufficient to prevent thermal racing after the current zero. Then, the circuit breaker can terminate the interruption gently. A great advantage is that switching overvoltages are avoided. The breaker is designed so that the arc length is short so long as current is high and then elongated in ceramic plates toward current zero. The circuit breaker does not try to interrupt the current at once when the zero is reached, but goes from a low to high arc resistance and increasing this resistance. This operation is kind to the recovery voltage after the current interruption. At high voltages, the limitations are the size determined by dimensions of the arcing chambers and insulation clearances in air.

Bulk oil circuit breakers were popular before 1950. For voltages above 72.5 kV, three tanks were required and for voltages to 36 kV, single tank construction was popular with phase barriers. Arcing in oil gives rise to about 70% of hydrogen which is an excellent arc quenching medium having low ionization time constant. Acetylene and some other gases are also produced. Care had to be taken to avoid *bubbles* coalescing or reaching the walls of the tank to prevent ignition, the so-called "bubble theory" was devised and used to determine the size of the tank. With increasing voltages and currents, this equipment became quite large and for 200 kV, 20,000 Liters of oil per phase. In the minimum oil content breakers, the bubble is confined, so as to reduce its volume, in an insulated arc chamber which can withstand high pressures. Very less oil is used. Both bulk oil and minimum oil content circuit breakers are no longer used.

7.16.1 SF$_6$ Circuit Breakers

The superior interrupting properties of SF$_6$ have led to application in high-power high-voltage circuit breakers, high-power medium-voltage circuit breakers, EHV breakers, large interrupting capability in-line generator circuit breakers with 200 kA interrupting and 50 kA continuous current carrying capabilities, load break switches, gas-insulated substations, bus bars, transformers and the technology has recently trickled down to even low-voltage circuit breakers.

SF_6 is an electronegative gas: if the dielectric strength of helium tested in uniform electric filed under 13.6 atm is taken as unity, the dielectric strength of nitrogen is 7.2, carbon dioxide is 8, and SF_6 is 16.

Townsend suggested that phenomena leading to electrical rupture in an insulating gas can be expressed by

$$\gamma e^{\alpha d - 1} = 1 \tag{7.56}$$

where γ is the number of charged particles formed as a result of positive ion collision, also designated as second Townsend coefficient, and d is the spacing between electrodes.

He studied that at low values of x/p (where x is the electrical gradient and p is the pressure), the current increases as a linear function of gap distance:

$$i = i_0 e^{\alpha x} \tag{7.57}$$

where i_0 is the current at zero plate separation. $1/\alpha$ also represents the average distance traveled by an electron to produce a new ion pair. The theory has been further modified by Loeb [13].

7.16.1.1 Electronegativity of SF_6

Molecule of SF_6 is octahedral with six fluorine atoms arranged symmetrically around sulfur atom. The molecular shapes can be explained in terms of mutual repulsion of covalent bonds. When two atoms are bonded by covalent bonds, both of them share a pair of electrons. The attraction that one of the atoms exerts on this shared pair of electrons is called electronegativity. Atoms with nearly filled shells of electrons (halogens) tend to have higher electronegativity.

The electronegativity of SF_6 is a major characteristic in electrical behavior and breakdown. Its ability to seize an electron and *form a negative ion is reverse of ionization process*. Thus, if attachment rate is equal to that of ionization, no breakdown is possible:

$$SF_6 \rightarrow SF_6^+ + F^- \tag{7.58}$$

In the attachment process, the free electrons get attached to molecules forming negative ions. These negative ions are too massive to produce collisional ionization. These ions can also reduce positive space charge around an electrode requiring a higher voltage to produce an arc across the gap.

Townsend's criterion for breakdown in an electronegative gas is written as follows:

$$\frac{\gamma \alpha}{\alpha - \eta} \left[e^{(\alpha - \mu)d - 1} \right] = 1 \tag{7.59}$$

The effective ionization coefficient $(\alpha - \eta)$ plays an important role in the breakdown voltage.

Figure 7.32a shows the superior breakdown properties of SF$_6$ with respect to air, Figure 7.32b shows the corona onset voltage related to pressure for SF$_6$ and air, Figure 7.32c depicts the breakdown strength of transformer oil, air, and SF$_6$, and Figure 7.32d shows the heat transfer coefficient of SF$_6$, air, and transformer oil under natural convection.

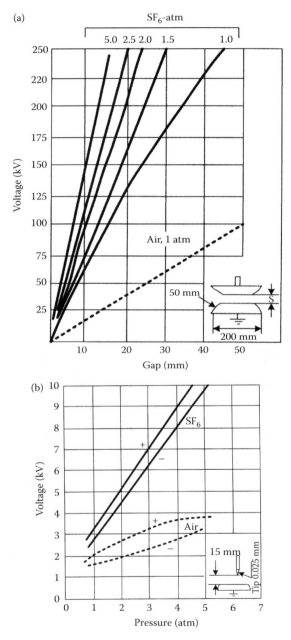

FIGURE 7.32
(a) Superior breakdown properties SF$_6$ with respect to air, (b) the corona onset voltage related to pressure for SF$_6$ and air, (c) breakdown strength of transformer oil, air, and SF$_6$, and (d) heat transfer coefficient of SF$_6$, air, and transformer oil under natural convection.

Current Interruptions in AC Networks

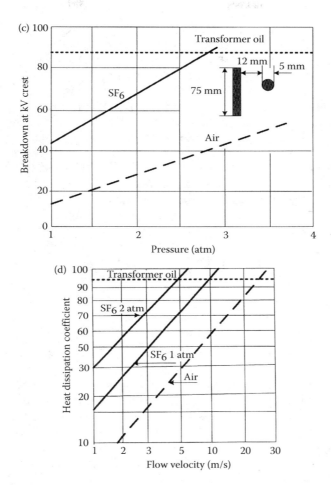

FIGURE 7.32 (CONTINUED)
(a) Superior breakdown properties SF$_6$ with respect to air, (b) the corona onset voltage related to pressure for SF$_6$ and air, (c) breakdown strength of transformer oil, air, and SF$_6$, and (d) heat transfer coefficient of SF$_6$, air, and transformer oil under natural convection.

During arcing, decomposition products are formed; these are reaction products formed inside the arc and products formed after arcing which are affected by electrodes and container, etc. Metal fluorides such as CuF$_2$ and WF$_6$, sometimes containing sulfur may be formed. Sulfur fluorides such as SF$_4$ and S$_2$F$_2$ may be formed. These decomposition products are good dielectrics, and deposits on insulating surfaces do not impair their operation. However, the presence of moisture with aforesaid substances produces hydrogen fluoride (HF) and water. HF is highly corrosive. It severely attacks materials containing silicon, glass, and porcelain. Thus, penetration of water and leakage is carefully controlled in SF$_6$ circuit breakers.

7.16.2 Operating Mechanisms

The operating energy mechanisms of circuit breakers have evolved into energy efficient and lighter designs, made possible by computer simulations and laboratory models.

With respect to SF_6 breakers, the earlier designs of double-pressure breakers are obsolete, and "puffer type" designs were introduced in 1967. The gas is sealed in the interrupting chamber at a low pressure and the contact movement itself generates a high pressure—part of the moving contact forms a cylinder and it moves against a stationary piston, thus the terminology puffer type—meaning blowing out the arc with a puff. The last 20 years have seen the development of "self-blast technology" applicable to 800 kV. A valve was introduced between the expansion and compression volume. When the interrupting current is low, the valve opens under effect of overpressure and the interruption phenomena is similar to puffer design. At high-current interruption, the arc energy produces a high overpressure in the expansion volume and the valve remains closed. The overpressure for breaking is obtained by the optimal use of thermal effect and nozzle clogging effect to heat the gas and raise its pressure. The cross section of the arc significantly reduces the exhaust of gas in the nozzle. The self-blast technology is further optimized by double-motion principle; this consists of displacing the two arcing contacts in opposite directions. This leads to further reduction in the operating energy. In "thermal blast chamber with arc-assisted opening," arc energy is used to generate the blast by thermal expansion and also to accelerate the moving part of the circuit breaker when interrupting high currents [14, 15].

7.16.3 Vacuum Interruption

Vacuum circuit breakers are extensively applied in the voltage range of 5–38 kV, though vacuum breakers have been built to higher voltages. World's first 145 kV, 3500 MVA vacuum circuit breaker was installed by AEI (CEGB) in England in 1967. Vacuum contactors are used for motor starting and on-load tap changing—they must be suitable for repeatedly switching the motor load currents, though the interrupting currents are low in 5–10 kA. The vacuum breakers are currently available to 63 kA interrupting at 13.8 kV and 40 kA at 38 kV. At these voltages, the vacuum technology has replaced magnetic air circuit breakers, which are no longer in use.

Pressures below 10–3 mm of mercury are considered high vacuum. The charged particles moving from one electrode to another electrode at such a low pressure are unlikely to cause collision with a residual gas molecule, though electrons can be emitted from cold metal surfaces due to high electrical field intensity, called field emission. This electron current may vary according to

$$I = Ce^{-B/E} \tag{7.60}$$

where B and C are constants. As the contact separate, at a lower gap, the electrical field intensity is high. As the voltage is increased, the current suddenly increases and gap breaks down, the phenomena called vacuum spark. This depends upon the surface conditions and material of the electrodes. Secondary emission takes place by bombardment of high energy on electrode surfaces. The current leaves the electrodes from a few bright spots, and the current densities are high. The core of the arc has high temperatures of the order of 6000–15,000 K. At such temperatures, thermal emission takes place. The dielectric recovery after sparkover occurs at a rate of 5–10 kV/μs.

An arc cannot persist in ideal vacuum, but separation of contacts causes vapor to be released from the contact surface, giving rise to plasma. At low currents, there are several parallel paths each originating and sinking in a hot spot, a diffused arc. At high currents above 15 kA, the arc becomes concentrated on a small region and becomes self-sustaining.

Current Interruptions in AC Networks 301

The transition from diffused arc to concentrated arc depends upon contact material and geometry. As the current wave approaches zero, rate of release of vapor is reduced, the arc becomes diffused and the medium tends to regain the dielectric strength provided vapor density around the contacts is substantially reduced. Thus, extinction process is related to the material and shape of the contacts and techniques used to condense the metal vapor. Contact geometry is designed so that the root of the arc keeps moving. The axial magnetic field decreases the arc voltage.

The vacuum interrupters interrupt the small current before natural current zero, causing current chopping. Again this is a function of contact material and the chopping current of modern circuit breakers has been reduced to 3 A, 50% cumulative probability. The interrupting capability depends upon the contact material, size, and type of magnetic field produced.

A number of restrikes and insulation failures were reported during initial development of vacuum interrupter technology. With innovations in contact materials and configurations, these are rare in the modern designs. However, we must distinguish between the vacuum circuit breakers and contactors, the latter are more prone to restrikes. An area of possible multiple ignitions occurs with respect to locked rotor current of the motor (for application of motors up to 6.6 kV). The frequency is given by [16]

$$f_n = \frac{1}{2\pi\sqrt{LC}} \tag{7.61}$$

where L = ungrounded motor inductance at locked rotor condition, l = length of the cable connection from the contactor to the motor, and C_v = capacitance of the cable per unit length ($C = lC_v$). Essentially, Equation 7.61 is a resonance formula of cable capacitance with motor locked rotor reactance. Therefore, a certain length of cable, if exceeded can lead to reignition. This shows that: (1) motor insulation is more vulnerable to stresses on account of lower basic insulation level (BIL) and (2) possibility of restrikes in the switching devices. As an example, for a 500 hp motor overvoltage protection device is needed, if #4 shielded cable length, used to connect the motor to the switching vacuum contactor, exceeds approximately 470 ft. This is based upon the cable data.

Further descriptions of constructional features, interruptions in other mediums are not discussed. It is sufficient to highlight that the interruption of currents in AC circuits is not an independent phenomena depending upon the circuit conditions alone, it is seriously modified by the physical and electrical properties of the interrupting mediums in circuit breakers and the system to which these are connected.

The failure rate of circuit breakers all over the world is decreasing on account of better designs and applications.

7.16.3.1 Current Chopping and Multiple Ignitions

The switching surge phenomena associated with vacuum interruption is

- Current chopping
- Multiple reignitions
- Voltage escalations
- Virtual current chopping—see Section 7.7.1

As discussed earlier, current chopping is the premature suppression of natural current zero, due to arc instability. The diffused arc in vacuum is unstable; each spot on electrodes requires a minimum intensity in order to emit the electron stream. The current in vacuum interrupter can chop to zero instantaneously, but not so in load inductance. When current

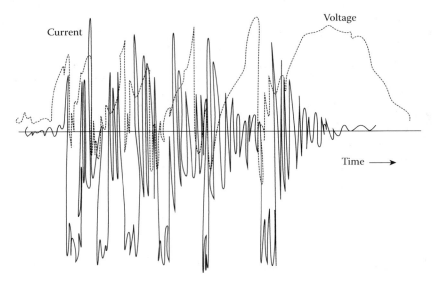

FIGURE 7.33
Five reignitions with high dv/dt and escalation of voltage, vacuum interruption.

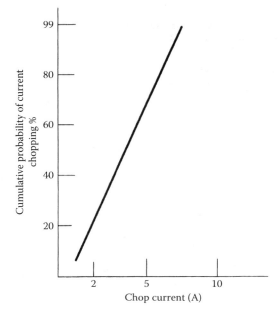

FIGURE 7.34
Cumulative probability of current chopping in vacuum circuit breakers, see text.

chop occurs, the energy stored in effective load inductance is transferred to available load side capacitance to produce chop voltage given by

$$I_c \sqrt{\frac{(1-\gamma)L_b}{C_s}} \qquad (7.62)$$

where $\sqrt{L_b/C_s}$ may be called load side surge impedance, and γ represents circuit losses, especially, iron loss.

Thus, long cable lengths, or a surge capacitor will reduce the surge impedance.

We have discussed the multiple reignitions in Section 7.7. This sequence of successive reignitions and high-frequency current interruption can give rise to voltage escalation. Figure 7.33 shows five reignitions and that each time the system is subjected to high dv/dt. Much experimentation and spastics show that medium normal frequency breakdown field is 33 kV/mm. Ninety-nine percent of all reignitions occur with a field less than 66 kV/mm and one percent of all reignitions occur with a field less than about 15 kV/mm. Cumulative probability of current chopping in a vacuum circuit breaker is shown in Figure 7.34 [17].

7.16.3.2 Switching of Unloaded Dry-Type Transformers

An unloaded transformer is highly inductive and the no-load current may be 1%–3% of the transformer full load current. Chopping of magnetizing current traps energy proportional to magnetizing inductance and the square of the peak current.

Figure 7.35 shows the maximum voltages to ground, from Reference [18], transformer rated voltage of 13.8 kV, the voltages vary according to the transformer rating and the cable length. Tables 7.3 and 7.4 illustrate the winding impulse tests on liquid filled and dry-type transformers according to IEEE standards.

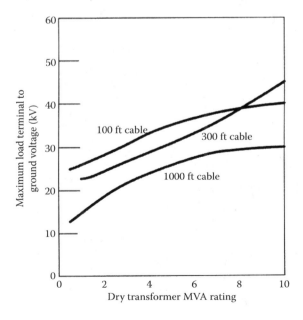

FIGURE 7.35
Maximum load terminal to ground voltage on current chopping, switching of dry-type transformers to 10 MVA, connected through varying cable lengths, vacuum circuit breakers.

TABLE 7.3

Winding Impulse Test Levels for Liquid Immersed Distribution Transformers

Nominal System Voltage	Applied Voltage Test	Chopped Wave Minimum Time to Flashover		BIL Full Wave 1.2/50 µs
kV rms	kV rms	kV Crest	µs	kV Crest
5	19	66	1.5	60
8.7	26	83	1.5	75
15	34	105	1.8	95
25	40	138	2.0	125

TABLE 7.4

Winding Impulse Test Levels for Dry-Type Distribution Transformers

Insulation Class	High Pot Tests	Chopped Wave Minimum Time to Flashover		BIL Full Wave 1.2/50 µs
kV rms	kV rms	kV Crest	µs	kV Crest
5	12	30	1.0	30
8.7	19	45	1.25	45
15	34	60	1.5	60
25	50	110	1.8	110

Even if reduced BIL level of dry-type transformer is related with Figure 7.35, no transformer damage should take place. However, transformer failures due to vacuum switching have been reported from all parts of the world and these failures are not limited to dry-type transformers. A user can specify higher BIL levels for dry-type transformers, same as for liquid filled transformers.

7.17 Part Winding Resonance in Transformers

Part winding resonance in transformers occurs mostly due to switching surges, current chopping in circuit breakers (low inductive currents) and very fast transients (VFTs) in gas-insulated substations. Failure of four autotransformers of 500 and 765 kV systems of American Electric Power in 1968, 1971 led to the investigations of winding resonance phenomena [19–21]. When an exciting oscillating overvoltage arise due to line or cable switching or faults and coincides with one of the fundamental frequencies of part of a winding in the transformer, high overvoltages and dielectric stresses can occur. *Low-amplitude and high-oscillatory switching transients cannot be suppressed by surge arrestors*, but can couple to the transformer windings.

Terminal resonance can be defined to occur at maximum current and minimum impedance, it is also called a series resonance. Here, the reactive component of the terminal impedance is zero. The terminal antiresonance is defined as the minimum current and maximum impedance. The internal resonance escalates the voltage and leads to possible insulation failure. *The terminal resonance and internal resonance may not necessarily bear a direct relation.* A part winding resonance may significantly influence transient oscillations of a major part of the transformer winding, but its effect may not show up in the terminal impedance plot. A *terminal* reactance and resistance plot is shown in Figure 7.36.

Generally, the first natural resonant frequency of the transformer is above 5 kHz, the resonant frequencies in core-type transformer lie between 5 kHz and few hundred kilohertz. The resonant frequencies may not vary much between transformers of different manufacturers and cannot be altered much, though efforts have been directed in this direction too. Generally, efforts are made to avoid network conditions that can produce oscillating voltages. In the design stage, a lumped equivalent circuit of winding inductances and capacitance can be analyzed and for the transformers in operation the frequency response can be measured.

The resonant overvoltages can be determined by the winding design and damping due to frequency-dependent winding resistance. Oscillations are significantly impacted by internal damping, winding and core losses, and external damping, i.e., the line resistance.

The winding resonance response can be determined by actual measurements using conductively coupled probes. The disadvantage is that the winding response cannot be predicted at the design stage, and after the transformer is manufactured, the insulation is pierced to install inductive probes. Capacitance coupled nondestructive probes can be used after careful scrutiny of accuracy [22]. Another method is to build an electromagnetic scale model of the transformer [23], in which a scaled model is used to

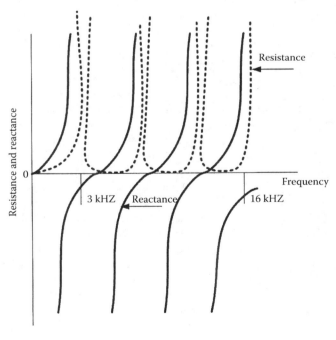

FIGURE 7.36
A terminal resonance plot of a transformer-resistance and reactance.

determine the natural frequencies and voltage response. This is a high cost and time-consuming alternative.

In this simulation of the transformer response, two important factors are (1) accurate model of the transformer and (2) accurate model of the circuit breaker. As stated earlier, a circuit breaker is not an ideal switch. Recent attempts have been made to develop a model of vacuum switching in electromagnetic transient program (EMTP). Also see Appendix C for extended models of the transformers. It may be necessary to model each layer of the windings and the mutual inductance between all sections of the windings must be taken into account to determine correctly the natural frequencies.

7.17.1 Snubber Circuits

The simulation and the determination of the resonant frequencies are complex, and it has been suggested that the vacuum circuit breakers can be provided with snubbers to prevent damage due to switching. The snubber consists of a resistance and capacitance combination connected in series. It slows down the high dv/dt rise and the peak of the transient. Metal-oxide surge arresters can be installed in addition, to limit overvoltage to BIL rating, but these cannot modify the rate of rise of current, di/dt. Also surge arresters are connected line-to-ground, while the switching transients are more related to line-to-line phenomena. Thus, metal-oxide surge arrester protection alone is not considered adequate. Surge capacitors are recommended for vacuum interrupters when using lower BIL, say for dry-type transformers, and rotating machines. This lowers the rate of rise di/dt and claimed to be of much importance for rotating machines, especially when long cable lengths are involved, greater than about 50 m.

Note that reignitions in SF_6 circuit breakers are practically nonexistent. Although virtual current chopping is a rare phenomenon, it is even less likely to occur in SF_6 interrupters.

7.18 Solid-State Circuit Breakers

Solid-state circuit breakers (SSBs) offer considerable advantage over mechanical breakers. The short-circuit fault current is reduced. The voltage dips due to three-phase fault clearance (lasting for about 100 ms or more) can be reduced to 100 μs. The impediments in the development have been material costs and on-state losses.

At 15 kV level, the SSBs can be built using GTOs or SCRs. The SCRs have better blocking voltages, lesser losses, and higher current ratings. GTOs can interrupt current with negligible delay, and can turn off the fault current during first half cycle when the overload condition is detected. This interruption must take place before exceeding the current interrupting rating of GTOs. Figure 7.37a shows a hybrid solution applied to a 15 kV breaker. It consists of parallel branches, composed of GTOs and SCRs. The GTO section conducts load current in the steady state. It is rated for maximum line currents but not for fault currents. It opens rapidly when the preset level of the current is exceeded, say 3000 A. A number of GTO switches are connected in series with snubber circuits and metal-oxide arresters. The SCR section that incorporates pairs of antiparallel connected thyristor devices in series is normally kept open and has no continuous current rating. Figure 7.37b shows a waveform with practically bump less transfer. It conducts fault currents (say 15 kA) for a period of

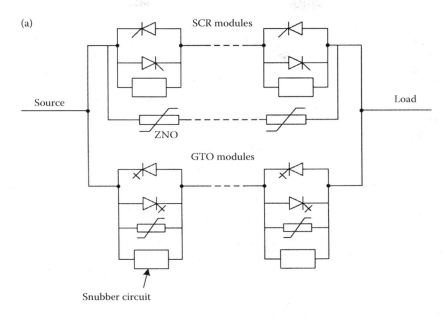

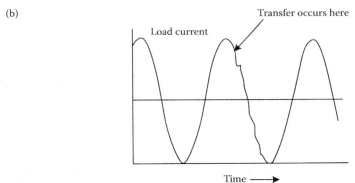

FIGURE 7.37
(a) A hybrid solution applied to a 15 kV breaker, consisting of parallel branches, composed of GTO's and SCR's and (b) waveform of transfer, showing practically a bump less transfer to load.

10–15 cycles, so that the downstream devices on the load side may coordinate. Further topologies of SSBs are described in Reference [24].

SSBs can be used as a bus section switch or static transfer switches. Figure 7.37b shows the load transfer characteristics of two SSBs connected in a system configuration of Figure 7.37a, normal breakers replaced with SSBs. The load current is basically unaffected due to transfer.

The rated capabilities of an SSB developed under Electric Power Research Institute (EPRI) projects are shown in Table 7.5. SSBs operate fast and therefore downstream coordination is a consideration. These cannot be used as incoming breakers. The breaker described in Table 7.5 operates to limit the short-circuit current and then allow the downstream protection to coordinate and operate.

TABLE 7.5

13.8 kV SSB Specifications

Characteristics	Data
Objective	Instantaneously interrupt the current in a 13.8 kV three-phase, grounded neutral distribution line. After the interruption, limit the fault current to a lower value to allow downstream coordination
Nominal voltage class	13.8 kV
Nominal three-phase MVA	1000
Rated maximum voltage	15 kV
Rated continuous current	670 A
Maximum peak current interruption level	2500 A peak. The SSB interrupts instantaneously thereby preventing the fault current exceeding the peak current instantaneous level
Fault let-through current rating	8000 A rms for 15 cycles, followed by
	Open for 15 cycles, followed by
	8000 A rms for 15 cycles, followed by
	Open for 15 cycles, followed by
	8000 A rms for 15 cycles, followed by
	Open
Rated interrupting time	30 μs (approximately 1/500 of a cycle)
Maximum number of fault interruptions	Unlimited
Rated closing time	1 cycle
Rated closing capability (momentary)	8000 A rms for 30 cycles, when operating in current limiting mode
	2500 A peak for three successive current interruptions, when operating in the interruption mode
Insulation level	95 kV crest, BIL
	36 kV rms, low-frequency withstand test
Losses	Maximum 40 kW at 670 A continuous current
Design features	10% redundancy in individual power electronic devices
	Forced air cooling devices, with a minimum of one redundant fan or blower
	Replacement of individual power electronic devices requires no more than 2 h
Assembly	Outdoor metal enclosure
	Bushings for connections to main power leads
	Dimensions, excluding bushings: 118″ H × 70″ W × 70″ L (300 × 178 × 178 mm)
	Weight 2270 kg
Control power requirement	5 kVA, 120 V, 60 Hz, single phase

Problems

7.1 Distinguish between reignitions, restrikes, and current chopping in high-voltage circuit breakers.

7.2 What is a delay line in TRV representation by two- and four-parameter representation? Describe the parameters on which it depends and its calculation.

Current Interruptions in AC Networks

7.3 Find the recovery voltage across the breaker contacts while interrupting the 4 A (peak) magnetizing current of a 138 kV, 20 MVA transformer. Assume a capacitance of 4 nF to ground and an inductance of 4 μH.

7.4 What is the value of a switching resistor to eliminate the restriking transient in Problem 7.3?

7.5 On the source side of a generator breaker, $L = 1.5$ mH and $C = 0.005$ μF. The breaker interrupts a current of 20 kA. Find (1) RRRV, (2) time to reach peak recovery voltage, and (3) frequency of oscillation.

7.6 Comment on the correctness of these statements: (1) interrupting an asymmetrical current gives rise to higher TRV than interrupting a symmetrical current; (2) as the current to be interrupted reduces, so does the initial rate of rise of the recovery voltage; (3) the thermal mode of a failure of breaker is excited when interrupting a capacitor current, due to higher TRV; (4) an oscillatory TRV occurs for a fault on a transformer connected to a transmission line; and (5) selecting a breaker of higher interrupting rating is an assurance that, in general, its TRV capability is better.

7.7 Describe a simple circuit element to control the RRRV, when interrupting a highly magnetizing current.

7.8 Why there is a high probability of flashover across breaker contacts when interrupting capacitive currents? How this can be avoided.

7.9 What are the common measures to control the effect of restrikes in vacuum contactors applied for switching long cables to motors?

Bibliography

A Braun, A Eidinger, E Rouss. Interruption of short-circuit currents in high-voltage AC networks. *Brown Boveri Rev*, 66. Baden, 1979.

TE Browne Jr (Ed.). *Circuit Interruption—Theory and Techniques*. Marcel Dekker, New York, 1984.

CIGRE Working Group 13-02. Switching overvoltages in EHV and UHV systems with special reference to closing and reclosing transmission lines. *Electra*, 30, 70–122, 1973.

CIGRE Working Group A3.12. Failure survey on circuit breaker controls systems. *Electra*, 251, 17–31, 2007.

JC Das. SF_6 in high-voltage outdoor switchgear. *Proceedings IEEE*, 61, pt. EL2, 1–7, 1980.

RD Garzon. *HV Circuit Breakers—Design and Applications*, 2nd ed., Marcel Dekker, New York, 2002.

A Greenwood. *Electrical Transients in Power Systems*. Wiley Interscience, New York, 1991.

W Hermann, K Ragaller. Theoretical description of current interruption in gas blast circuit breakers. *IEEE Trans Power App Syst*, 96, 1546–1555, 1977.

T Itoh, T Muri, T Ohkura, T Yakami. Voltage escalation in switching of the motor control circuit by the vacuum contactors. *IEEE Trans Power App Syst*, 91, 1897–1903, 1972.

V Koschik, SR Lambert, RG Rocamora, CE Wood, G Worner. Long line single phase switching transients and their effect on station equipment. *IEEE Trans Power App Syst*, 97, 857–964, 1978.

LB Loeb. *Fundamental Processes of Electrical Breakdown in Gases*, John Wiley, New York, 1975.

E Slamecka. Interruption of small interrupting currents. CIGRE Working Group 13.02, *Electra*, 72, 73–103, 1980, and *Electra*, 5–30, March 1981.

H Toda, Y Ozaki, I Miwa. Development of 800-kV gas insulated switchgear. *IEEE Trans Power Delivery*, 7, 316–322, 1992.

T Ushio, I Shimura, S Tominaga. Practical problems of SF_6 gas circuit breakers. *IEEE Trans Power App Syst*, 90(5), 2166–2174, 1971.

L van der Sluis, ALJ Jansen. Clearing faults near shunt capacitor banks. *IEEE Trans Power Delivery*, 5(3), 1346–1354, 1990.

References

1. K Ragaller. *Current Interruption in High Voltage Networks*. Plenum Press, New York, 1978.
2. AM Cassie. Arc Rupture and Circuit Theory. CIGRE Report n. 102, 1939.
3. ANSI/IEEE Std. C37.010. IEEE Application Guide for High-Voltage Circuit Breakers Rated on Symmetrical Current Basis, 1999.
4. IEC Std. 62271-100, High Voltage Alternating Current Circuit Breakers, 2001.
5. IEEE Std. C37.011, IEEE Application Guide for Transient Recovery Voltage for AC High Voltage Circuit Breakers, 2005.
6. J Panek, KG Fehrie. Overvoltage phenomena associated with virtual current chopping in three-phase circuits. *IEEE Trans Power App Syst*, 94, 1317–1325, 1975.
7. JF Perkins. Evaluation of switching surge overvoltages on medium voltage power systems. *IEEE Trans Power App Syst*, 101(6), 1727–1734, 1982.
8. IEEE Std. C37.012. Application Guide for Capacitor Current Switching for AC High Voltage Circuit Breakers, 2005.
9. H Glavitsch. Problems associated with switching surges in EHV networks. *BBC Rev*, 53, 267–277, 1966.
10. CIGRE Working Group 13-02, Switching Surge Phenomena in EHV systems. Switching overvoltages in transmission lines with special reference to closing and reclosing transmission lines. *Electra*, 30, 70–122, 1973.
11. IEEE Std. C37.015. IEEE Guide for Application of Shunt Reactor Switching, 2009.
12. CIGRE Technical Brochure 305: Guide for Application of IEC 62271-100 and IEC 62271-1—Part 2: Making and Breaking Tests, 2006.
13. LB Loeb. *Fundamental Process of Electric Breakdown in Gases*, John Wiley, New York, 1975.
14. D Dufournet, F Sciullo, J Ozil, A Ludwig. New interrupting and drive techniques to increase high voltage circuit breakers performance and reliability. CIGRE Session 1998, paper 13–104.
15. HE Spindle, TF Garrity, CL Wagner. Development of 1200 kV circuit breaker for GIS system—Requirements and circuit breaker parameters. CIGRE, Rep. 13-07, Paris, 1980.
16. SF Frag, RG Bartheld. Guidelines for application of vacuum contactors. *IEEE Trans Ind Appl*, 22(1), 102–108, 1986.
17. T Itoh, T Muri, T Ohkura, T Yakami. Voltage escalation in switching of motor control circuit by vacuum contactors. *IEEE Trans Power App Syst*, 91, 1897–1903, 1972.
18. JF Perkins, D Bhasavanich. Vacuum switchgear application study with reference to switching surge protection. *IEEE Trans Ind Appl*, 19(5), 879–888, 1983.
19. JC Das. Analysis and control of large-shunt-capacitor-bank switching transients. *IEEE Trans Ind Appl*, 41(6), 1444–1451, 2005.
20. HB Margolis, JD Phelps, AA Carlomagno, AJ McElroy. Experience with part-winding resonance in EHV autotransformers; diagnosis and corrective measures. *IEEE Trans Power App Syst*, 94(4), 1294–1300, 1975.

21. AJ McElroy. On significance of recent transformer failures involving winding resonance. *IEEE Trans Power App Syst*, 94(4), 1301–1309, 1975.
22. PA Abetti, FJ Maginniss. Natural frequencies of coils and windings determined by equivalent circuit. *AIEE Trans*, 72(3), 495–503, 1953.
23. BI Gururaj. Natural frequencies of 3-phase transformer windings. *AIEE Trans*, 318–329, 1963.
24. RK Smith, PG Slade, M Sarkozi, EJ Stacy, JJ Bonk, H Mehta. Solid state distribution current limiter and circuit breaker application requirements and control strategies. *IEEE PES Summer Meeting*, 1992, Paper no. 92SM 572-8.24.

8

Application and Ratings of Circuit Breakers and Fuses according to ANSI Standards

In Chapter 7, we discussed current interruption in AC circuits and the stresses that can be imposed on the circuit breakers, depending on the nature of fault or the switching operation. We observed that the system modulates the arc interruption process and the performance of a circuit breaker. In this chapter, we review the ratings of circuit breakers and fuses according to ANSI, mainly from the short-circuit and switching point of view, and examine their applications. While a *general-purpose* circuit breaker may be adequate in most applications, yet higher duties may be imposed in certain systems that should be carefully considered. This chapter also forms a background for the short-circuit calculation procedures in Chapter 10.

There are attempts that have been made to harmonize ANSI/IEEE standards with International Electrotechnical Commission (IEC). Institute of Electrical and Electronics Engineering (IEEE) unapproved draft standard IEEE PC37.06/D11 for AC high-voltage circuit breakers rated on symmetrical current basis – preferred ratings and required related capabilities for voltage above 1000 V [1]. A joint task force group was established to harmonize the requirements of Transient Recovery Voltage (TRV) with that in IEC62271-100 [2], including amendments 1 and 2. This draft standard also publishes new rating tables and completely revises the ratings and the nomenclature of circuit breakers, which has been in use in the USA for many years:

- Classes S1 and S2 are used to denote traditional terms of "indoor" and "outdoor", respectively. The class S1 circuit breaker is for cable systems; indoor circuit breakers are predominantly used with cable distribution systems. Class S2 is for overhead line systems.

- The term "peak" is used—the term "crest" has been dropped from the usage. All tables show "prospective" or "inherent" characteristics of the current and voltages. The word "prospective" is used in conformance to international standard.

- Two and four parameters of TRV representations are adopted in-line with IEC standards.

- For class C_0, general-purpose circuit breakers, no ratings are assigned for back-to-back capacitor switching. For this class, exposed to transient currents for nearby capacitor banks during fault conditions, the capacitance transient current on closing shall not exceed lesser of either 1.41 times rated short-circuit current or 50 kA peak. The product of transient inrush current peak and frequency shall not exceed 20 kA kHz. *Definite-purpose* circuit breakers are now identified as Classes C_1 and C_2. Here, the manufacturer shall specify the inrush current and frequency at which Class C_1 or C_2 performance is met.

313

8.1 Total and Symmetrical Current Basis

Prior to 1964, high-voltage circuit breakers were rated on a total current basis [3, 4]. At present, these are rated on a symmetrical current basis [5–8]. Systems of nominal voltage less than 1000 V are termed low voltage, from 1000 to 100,000 V as medium voltage, and from 100 to 230 kV as high voltage. Nominal system voltages from 230 to 765 kV are extra-high voltage (EHV), and those higher than 765 kV are termed ultra-high voltage (UHV). ANSI covers circuit breakers rated above 1–800 kV. The difference in total and symmetrical ratings depends on how the asymmetry in short-circuit current is addressed. The symmetrical rating takes asymmetry into account in the rating structure of the breaker itself. The asymmetrical profile of the short-circuit current shown in Figure 8.1 is the same as that shown in Figure 4.1, except that a decaying AC component is shown. The root mean square (rms) value of a symmetrical sinusoidal wave, at any instant, is equal to the peak-to-peak value divided by 2.828. The rms value of an asymmetrical wave shape at any instant is given by Equation 7.11. For circuit breakers rated on a total current rating basis, short-circuit interrupting capabilities are expressed in the total rms current. For circuit breakers rated on a symmetrical current basis, the interrupting capabilities are expressed as the rms symmetrical current at contact parting time. The symmetrical capacity for polyphase is the highest value of the symmetrical component of the short-circuit current in rms ampères at the instant of primary arcing contact separation, which the circuit breaker will be required to interrupt at a specified operating voltage on the standard operating duty and *with a direct current component of less than 20% of the current value of the symmetrical component* [6, 1999 revision]. Note that the 1978 edition of this standard stated "irrespective of the total DC component of the total short-circuit current.

Figure 8.2 shows an asymmetrical current waveform, where t is the instant of contact parting. The peak-to-peak value of the asymmetrical AC current is given by ordinate A. This is the sum of two ordinates A' and B', called the major and minor ordinates, respectively, as measured from the zero line. The DC component D is given as follows:

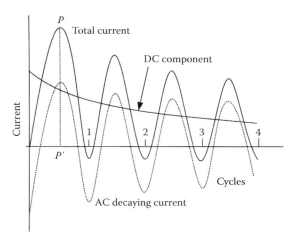

FIGURE 8.1
Asymmetrical current wave with decaying AC component.

Application and Ratings of Circuit Breakers and Fuses according to ANSI Standards 315

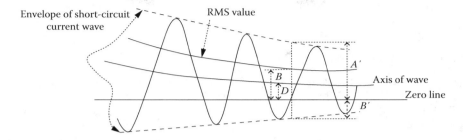

FIGURE 8.2
Evaluation of rms value of an offset wave.

$$D = \frac{A' - B'}{2} \quad (8.1)$$

The rms value of AC component B is given as follows:

$$B = \frac{A' + B'}{2.828} \quad (8.2)$$

Thus, the total interrupting current in rms is

$$\sqrt{B^2 + D^2} \quad (8.3)$$

This is an asymmetrical current. Equation 8.3 forms the total current rating basis, and Equation 8.2 forms the symmetrical current basis. Figure 8.3 shows that the total rms asymmetrical current is higher than the rms of the AC component alone. This does not mean that the effect of the DC component at the contact parting time is ignored in the symmetrical rating of the breakers. It is considered in the testing and rating of the breaker. Breakers rated on a symmetrical current basis are discussed in the rest of this chapter.

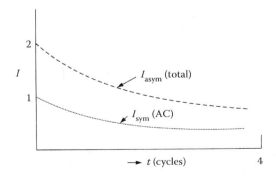

FIGURE 8.3
Symmetrical AC and total current profiles, rms values.

8.2 Asymmetrical Ratings

8.2.1 Contact Parting Time

The asymmetry in polyphase and phase-to-phase interrupting is accounted for by testing the breaker at higher total current (asymmetrical current) based on a minimum contact parting time. This includes the *tripping delay*. The contact parting is the sum of the tripping delay and breaker opening time. The ANSI standard [6, 1999 revision] specifies that the primary contact parting time will be considered equal to the sum of one-half cycle (tripping delay) and the *lesser of* (1) the actual opening time of the particular breaker, or (2) 1.0, 1.5, 2.5, or 3.5 cycles for breakers having a rated interrupting time of 2, 3, 5, or 8 cycles, respectively. This means that 2, 3, 5, and 8 cycle breakers have contact parting times of 1.5, 2, 3, and 4 cycles, respectively, unless the test results show a *lower* opening time plus a half-cycle tripping delay.

The asymmetrical rating in the 1979 standard was expressed as a ratio S, which is the required asymmetrical interrupting capability per unit of the symmetrical interrupting capability. Ratio S is found by multiplying the symmetrical interrupting capability of the breaker determined for the operating voltage with an appropriate factor. The value of S was specified as 1.4, 1.3, 1.2, 1.1, and 1.0 for breakers having a contact parting time of 1, 1.5, 2, 3, and 4 or more cycles, respectively.

The factor S curve is now replaced by the curve shown in Figure 8.4 [5, revision 2005]. The percentage DC is given by the following expression:

$$\%DC = 100e^{-t/45} \tag{8.4}$$

where t is the contact parting time in ms.

The required asymmetrical capability for the three-phase faults is the value of the total rms current at the instant of arcing contact separation that the circuit breaker will be required to interrupt at a specified operating voltage, on the standard operating duty cycle:

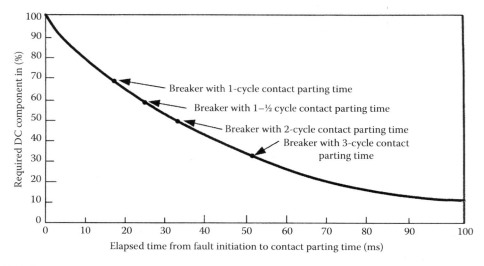

FIGURE 8.4
DC component at contact parting time.

Application and Ratings of Circuit Breakers and Fuses according to ANSI Standards   317

$$I_t = I_{sym}\sqrt{1 + 2\left(\frac{\%DC}{100}\right)^2} \tag{8.5}$$

Consider a breaker, with an interrupting time of 5 cycles and, therefore, a contact parting time of 3 cycles. From Figure 8.4, the percentage DC is 32.91%. Then,

$$\frac{I_t}{I_{rms}} = \sqrt{1 + 2(0.3291)^2} \approx 1.1$$

As specified in the earlier (1979) edition of Reference [5], S is equal to 1.1 for 5-cycle breakers. Thus, the new revised standards have converted the S curve to percentage DC curve. The following observations are pertinent:

- If the actual relay time is less than 0.5 cycles, the fact should be considered in calculations of asymmetrical ratings. The curve in Figure 8.4 is based upon ½ cycle tripping delay.
- With X/R greater than 17 at 60 Hz, the decrement rate of DC will be slower than that incorporated in rating structures in standards [6, 7]. E/X method of AC and DC adjustments can be used, see Chapter 10.
- The longer DC time constants can cause problems with some types of SF_6 puffer breakers, see Chapter 10. The E/X adjustment can be used provided that the X/R does not exceed 45 at 60 Hz (DC time constant is 120 ms).
- For time constants beyond 120 ms, consultation with manufacturer is recommended.
- In some application of large generators, current zeros may not be obtained due to high asymmetry, see Chapter 10.

8.3 Voltage Range Factor K

The 2000 revision of ANSI/IEEE standard [8] has changed the K factor of indoor oil-less circuit breakers to 1 (Table 8.1). Prior to that, K factor was higher than 1, see Table 8.2. As the K factor-rated breakers have been extensively used in the industry over the past years, and may continue to be in service, manufacturers can still supply K factor-rated breakers, though retrofitting with newly rated $K = 1$ breakers is a possibility, if replacements and upgrades are required.

Note that both Tables 8.1 and 8.2 show a rated interrupting time of 5 cycles; however, 3-cycle breakers are available and can be used as advantage for faster fault-clearing time, for example, improving the transient stability limits and reducing arc flash hazard. IEEE draft standard [1] shows that 5- and 3-cycle breakers have interrupting times of 83 and 50 ms, respectively.

The maximum symmetrical interrupting capability of a circuit breaker is K times the rated short-circuit current. Between the rated maximum voltage and $1/K$ times the rated maximum voltage, the symmetrical interrupting capacity is defined as follows:

$$\text{rated short circuit current} \times \frac{\text{rated maximum voltage}}{\text{operating voltage}} \tag{8.6}$$

TABLE 8.1

Preferred Ratings of Indoor Oil-Less Circuit Breakers, $K = 1$

Rated Maximum Voltage (kV, rms)	Rated Voltage Range Factor (K)	Rated Continuous Current at 60 Hz (A, rms)	Rated Short-Circuit Current at Rated Maximum kV (kA, rms)	Rated Transient Recovery Voltage		Rated Interrupting Time (m)s	Closing and Latching Current (kA, Peak)
				Rated Peak Voltage E2 kV, Peak	Rated Time to Peak T2 (µs)		
4.76	1.0	1200, 2000	31.5	8.9	50	83	82
4.76	1.0	1200, 2000	40	8.9	50	83	104
4.76	1.0	1200, 2000, 3000	50	8.9	50	83	130
8.25	1.0	1200, 2000, 3000	40	15.5	60	83	104
15	1.0	1200, 2000	20	28	75	83	52
15	1.0	1200, 2000	25	28	75	83	65
15	1.0	1200, 2000	31.5	28	75	83	82
15	1.0	1200, 2000, 3000	40	28	75	83	104
15	1.0	1200, 2000, 3000	50	28	75	83	130
15	1.0	1200, 2000, 3000	63	28	75	83	164
27	1.0	1200	16	51	105	83	42
27	1.0	1200, 2000	25	51	105	83	65
38	1.0	1200	16	71	125	83	42
38	1.0	1200, 2000	25	71	125	83	65
38	1.0	1200, 2000, 3000	31.5	71	125	83	82
38	1.0	1200, 2000, 3000	40	71	125	83	104

Source: ANSI Std. C37.06. AC High-Voltage Circuit Breakers Rated on a Symmetrical Current Basis—Preferred Ratings and Related Capabilities. 2000 (revision of 1987).

TABLE 8.2

Preferred Ratings of Indoor Oil-Less Circuit Breakers, $K > 1$

Rated Maximum Voltage (kV, rms)	Rated Voltage Range Factor (K)	Rated Continuous Current at 60 Hz (A rms)	Rated Short-Circuit Current at Rated Maximum kV (kA, rms)	Rated Interrupting Time Cycles	Rated Maximum Voltage Divided by K (kV, rms)	Maximum Symmetrical Interrupting Capability and Rated Short-Time Current (kA, rms)	Closing and Latching Capability 2.7K Times Rated Short-Circuit Current (kA, Peak)
4.76	1.36	1200	8.8	5	3.5	12	32
4.76	1.24	1200, 2000	29	5	3.85	36	97
4.76	1.19	1200, 2000, 3000	41	5	4.0	49	132
8.25	1.25	1200, 2000	33	5	6.6	41	111
15.0	1.3	1200, 2000	18	5	11.5	23	62
15.0	1.3	1200, 2000	28	5	11.5	36	97
15.0	1.3	1200, 2000, 3000	37	5	11.5	48	130
38.0	1.65	1200, 2000, 3000	21	5	23.0	35	95
38.0	1.0	1200, 3000	41	5	38.0	40	108

Source: ANSI Std. C37.06. AC High-Voltage Circuit Breakers Rated on a Symmetrical Current Basis—Preferred Ratings and Related Capabilities. 2000 (revision of 1987).

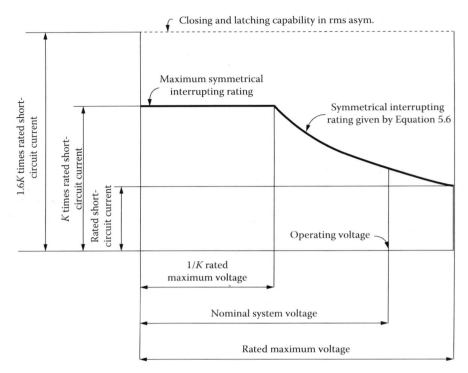

FIGURE 8.5
Relationship of interrupting capacity, reclosing and latching capability, and current carrying capability to rated short-circuit current for earlier K-rated indoor circuit breakers.

This is illustrated in Figure 8.5. The interrupting rating at lower voltage of application cannot exceed the rated short-circuit current multiplied by K. With new K = 1 rated breakers, a 15 kV breaker rated at 40 kA interrupting duty can be applied at 2.4 kV, and the interrupting rating will be still 40 kA.

Example 8.1

A 15 kV circuit breaker (maximum rated voltage) has a K factor of 1.30 and a rated short-circuit current of 37 kA rms. What is its interrupting rating at 13.8 kV? What is the breaker interrupting rating if applied at 2.4 kV?

The rated interrupting current at 13.8 kV is given by Equation 8.6, and therefore, it will be $37 \times (15/13.8) = 40.2$ kA. The maximum symmetrical interrupting capability of this breaker is $37 K = 48$ kA rms.

Thus, the interrupting rating if applied at 2.4 kV will be 48 kA.

Compare this to a 15 kV 40 kA rated breaker K = 1. The interrupting rating if applied at 13.8 kV or 2.4 kV is 40 kA.

8.4 Circuit Breaker Timing Diagram

Figure 8.6 shows the operating time diagram of a circuit breaker. We have already defined ½ cycle relay time, contact opening time, contact parting time, arcing time, and interrupting

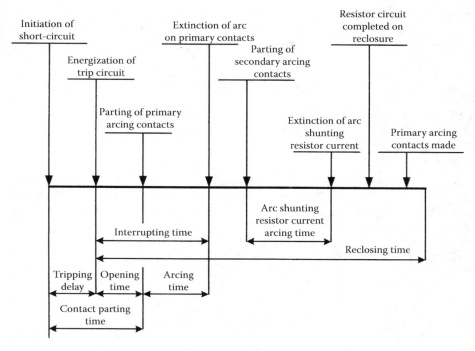

FIGURE 8.6
Operating time diagram of circuit breakers.

time. The interrupting time is not a constant parameter, and variations occur depending upon the fault. For a line-to-ground fault, it may increase by 0.1 cycle, and for asymmetrical faults, it may exceed rated interrupting time by 0.2 cycles.

High-speed reclosing is applied to radial and tie lines to (1) clear a ground fault of the transient nature, (2) improve system stability through fast reclosing; the phase angles and voltages across breaker contacts may not reach beyond the capability to maintain synchronism, and (3) during single-pole reclosing, some synchronizing power is transferred and stability improved. Before a circuit is re-energized, there has to be some dead time in the circuit breaker for arc path to become de-ionized. A dead time of 135 ms is normally required at 115–138 kV for breakers without resistors across the interrupters. A manufacturer will supply this data for the specific breaker type. The rated interrupting capability has to be de-rated for reclosing [5].

For utility services to industrial consumers having induction and synchronous motor loads, the reclosing is not used. A sudden uncontrolled return of power can damage the motors. Synchronous motors are not suitable for reclosing; and for induction motor transfer schemes, special electronic and control devices are available to affect (1) fast transfer, (2) in-phase transfer, and (3) residual voltage transfer, discussed in Volume 4.

8.5 Maximum Peak Current

The asymmetrical peak *does not* occur at ½ cycle of the 60 Hz wave, considering that the fault occurs at zero point on the voltage wave, see Chapter 10. The peak will occur at ½ cycle for purely inductive circuits only. IEEE standard [9] tabulates per unit peak currents using exact calculations, IEC equations, ½ cycle values, and also the equations recommended in this standard. The peak given at ½ cycles by the following equation is *non-conservative*:

$$I_{\text{peak,hakfcycle}} = I_{\text{ac peak}} + \left(1 + e^{\frac{2\pi t}{(X/R)}}\right) t = 0.5 \text{ s} \tag{8.7}$$

See Chapter 10 for further discussions.

Example 8.2

Consider a circuit breaker of 15 kV, $K = 1.3$, Table 8.1, a rated short-circuit current of 37 kA applied at 13.8 kV. Its interrupting duty at voltage of application is 40.2 kA, and the peak close and latch capability is 130 kA. The breakers thus rated had a peak close and latch 2.7K times the short-circuit rating = $2.7 \times 1.3 \times 37 = 128.87$ kA ≈ 130 kA, as read from Table 8.1.

Now consider a 15 kV, 50 kA interrupting rating breaker, $K = 1$, see Table 8.2. Its close and latch is also 130 kA. This is so because in the new standards the close and latch has been reduced to 2.6 times the rated short-circuit current = $2.6 \times 50 = 130$ kA.

This factor should be considered when applying circuit breakers according to the revised standards. A circuit breaker of same interrupting rating has a corresponding lower close and latch capability. Close to the generating stations, due to high X/R ratios, the close and latch capability may be the limiting factor in selecting the ratings of the circuit breakers. In fact, manufacturer's marketed breakers of close and latch capability higher than the interrupting rating. For example, referring to Table 8.2, a 15.5 kV breaker of a rated short-circuit rating of 28 kA, but a close and latch capability of 130 kA peak, was available. Also see Chapter 10.

8.6 Permissible Tripping Delay

Values of the maximum tripping delay Y are specified in an ANSI standard [8], i.e., for circuit breakers of 72.5 kV and below, the rated permissible delay is 2 s, and for circuit breakers of 121 kV and above, it is 1 s. The tripping delay at lower values of current can be increased and within a period of 30 min should not increase the value given by

$$\int_0^t i^2 \mathrm{d}t = Y[\text{rated short-circuit current}]^2 \tag{8.8}$$

It will then be capable of interrupting any short-circuit current, which at the instant of primary arcing contact separation has a symmetrical value not exceeding the required asymmetrical capability.

8.7 Service Capability Duty Requirements and Reclosing Capability

ANSI-rated circuit breakers are specified to have the following interrupting performance:

1. Between 85% and 100% of asymmetrical interrupting capability at operating voltage, there are three standard duty cycles.

Application and Ratings of Circuit Breakers and Fuses according to ANSI Standards   323

O–15 s–CO–3 min–CO

CO–3 s–CO for circuit breakers for rapid reclosing.

For generator breakers: rated short-circuit duty cycle shall be two operations with a 30 min interval between operations CO–30min–CO

2. Between rated continuous current and 85% of required asymmetrical capability, there are a number of operations in which the sum of interrupted currents is a minimum of 800% of the required asymmetrical interrupting capability of the breaker at rated voltage.

CO: close–open

Whenever a circuit breaker is applied having more operations or a shorter time interval between operations, other than the standard duty cycle, the rated short-circuit current and related required capabilities are reduced by a reclosing capability factor R, determined as follows:

$$R = (100 - D)\%$$ (8.9)

$$D = d_1(n-2) + d_1 \frac{(15 - t_1)}{15} + d_1 \frac{(15 - t_2)}{15} + \cdots$$ (8.10)

where D is the total reduction factor in percent, d_1 is specified in an ANSI standard [5], n is the total number of openings, t_1 is the first time interval (<15 s), and t_2 is the second time interval (<15 s).

Interrupting duties thus calculated are subject to further qualifications. These should be adjusted for X/R ratios. All breakers are not rated for reclosing duties. Breakers rated more than 1200 A and below 100 kV are not intended for reclosing operations. Breakers rated 100 kV and above have reclosing capabilities irrespective of the current ratings. Table 8.3 shows preferred ratings for outdoor circuit breakers 121 kV and above.

8.7.1 Transient Stability on Fast Reclosing

High-speed reclosing is used to improve transient stability and voltage conditions in a grid system. On a step variation of the shaft power (increase), the torque angle of the synchronous machine (Chapter 9) will overshoot, which may pass the peak of the stability curve. It will settle down to *new torque angle demanded by the changed conditions* after a series of oscillations. If these oscillations damp out, we say that the stability will be achieved. If these oscillations diverge, then the stability will be lost.

The basic concept of equal area criteria of stability is illustrated with reference to Figure 8.7. Note that the acceleration area due to variation of the kinetic energy of the rotating masses is

$$A_{\text{accelerating}} = ABC = \int_{\sigma_1}^{\sigma_2} (T_{\text{shaft}} - T_e) d\sigma$$ (8.11)

The area CDE is the de-acceleration area:

$$B_{\text{deaccelerating}} = CDE = \int_{\sigma_2}^{\sigma_3} (T_{\text{shaft}} - T_e) d\sigma$$ (8.12)

TABLE 8.3

Preferred Ratings for Outdoor Circuit Breakers 121 kV and above including Circuit Breakers Applied in Gas-Insulated Substations

Rated Maximum Voltage (kV)	Rated Voltage Range Factor (K)	Rated Continuous Current at 60 Hz (A, rms)	Rated Short-Circuit Current at Rated Maximum Voltage (kA, rms)	Rated Time to Point P ($T2$, µs)	Rated Rate R (kV/µs)	Rated Delay Time (T_1, µs)	Rated Interrupting Time Cycles	Maximum Permissible Tripping Delay	Closing and Latching Capability, 2.6 K Times Rated Short-Circuit Current (kA, peak)
123	1.0	1200, 2000	31.5	260	2.0	2	3 (50 ms)	1	82
123	1.0	1600, 2000, 3000	40	260	2.0	2	3	1	104
123	1.0	2000, 3000	63	260	2.0	2	3	1	164
145	1.0	1200, 2000	31.5	330	2.0	2	3	1	82
145	1.0	2000, 3000	63	310	2.0	2	3	1	164
145	1.0	2000, 3000	80	310	2.0	2	3	1	208
170	1.0	1600, 2000, 3000	31.5	360	2.0	2	3	1	82
170	1.0	2000	40	360	2.0	2	3	1	104
170	1.0	2000	63	360	2.0	2	3	1	164
245	1.0	1600, 2000, 3000	31.5	520	2.0	2	3	1	82
245	1.0	2000, 3000	63	520	2.0	2	3	1	164
362	1.0	2000, 3000	40	775	2.0	2	2 (33 ms)	1	104
362	1.0	2000, 3000	63	775	2.0	2	2	1	164
550	1.0	2000, 3000	50	1325	2.0	2	2	1	130
550	1.0	2000, 3000	63	1325	2.0	2	2	1	164
800	1.0	3000, 4000	40	1530	2.0	2	2	1	104
800	1.0	2000, 3000	40	1530	2.0	2	2	1	104
800	1.0	3000, 4000	63	1530	2.0	2	2	1	164

Source: ANSI Std. C37.06. AC High-Voltage Circuit Breakers Rated on a Symmetrical Current Basis—Preferred Ratings and Related Capabilities. 2000 (revision of 1987).

Application and Ratings of Circuit Breakers and Fuses according to ANSI Standards

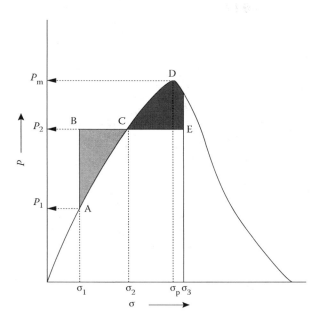

FIGURE 8.7
To illustrate the concept of equal area criteria of stability.

In accelerating and decelerating, the machine passes the equilibrium point given by δ_2 giving rise to oscillations, which will either increase or decrease. If the initial impact is large enough, the machine will be unstable in the very first swing.

$$A_{\text{accelerating}} = B_{\text{deaccelerating}} \tag{8.13}$$

If the above equation holds true, then there are chances of the generator remaining in synchronism [10]. The asynchronous torque produced by the dampers has been neglected in this analysis, also the synchronizing power is assumed to remain constant during the disturbance. It is clear that at point C, the accelerating power is zero, and assuming a generator connected to an infinite bus, the speed continues to increase. It is more than the speed of the infinite bus, and at point E, the relative speed is zero, and the torque angle ceases to increase, but the output is more than the input and the torque angle starts decreasing; rotor decelerates. But for damping, these oscillations can continue. This is the concept of "equal area criterion of stability."

Figure 8.8a and b illustrates the effect of high-speed, single-phase reclosing on transient stability using equal area criteria of stability. A transient single line-to-ground fault occurs on the tie line; the tie line breaker opens and then closes within a short time delay, called the dead time of the breaker. Some synchronizing power flows through two unfaulted phases, during a single line-to-ground fault, and no power flows during the dead time. The dead time of the circuit breaker is σ_1–σ_2; synchronous motors and power capacitors tend to prolong the arcing time. Applying equal area criteria of stability, if shaded area 1 + C is equal to shaded area 2; stability is possible. Also see Volume 4.

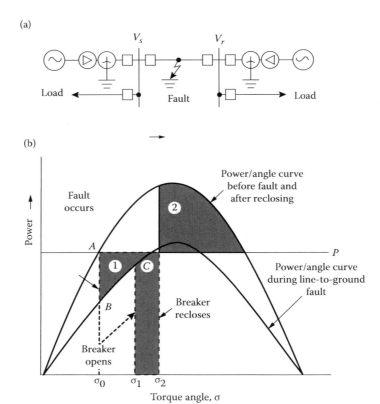

FIGURE 8.8
Transient stability of a tie line circuit with fast reclosing on a single line-to-ground fault: (a) equivalent system representation and (b) equal area criteria of stability. Fault occurs at torque angle σ_0; breaker opens at σ_1 and recloses at σ_2 to remove the transient fault.

8.8 Shunt Capacitance Switching

The capacitance switching is encountered when switching overhead lines and cables at no load and when switching shunt capacitor banks and filters. With respect to the application of circuit breakers for capacitance switching, consider (1) the type of application that is overhead line, cable or capacitor bank or shunt capacitor filter switching; (2) power frequency and system grounding; and (3) the presence of single or two phases to ground faults.

ANSI ratings [8] distinguish between general- and definite-purpose circuit breakers, and there is a vast difference between their capabilities for capacitance current switching. Definite-purpose circuit breakers may have different constructional features, i.e., a heavy-duty closing and tripping mechanism and testing requirements. Table 8.4 shows the capacitance current switching ratings. General-purpose circuit breakers do not have any back-to-back capacitance current switching capabilities. A 121 kV general-purpose circuit breaker of rated current 2 and 63 kA symmetrical short circuit has overhead line charging current or isolated capacitor switching current capability of 50 A rms and no back-to-back switching capability.

TABLE 8.4

Preferred Capacitance Current Switching Rating for Outdoor Circuit Breakers 121 kV and above, including Circuit Breakers Applied in Gas-Insulated Substations

Rated Maximum Voltage (kV)	Rated Short-Circuit Current at Rated Maximum Voltage (kA, rms)	Rated Continuous Current at 60 Hz (A, rms)	General-Purpose Circuit Breakers Rated Overhead Line Current (A, rms)	General-Purpose Circuit Breakers, Rated Isolated Current (A, rms)	Definite-Purpose Breakers Rated Capacitance Switching Current Shunt Capacitor Bank or Cable				
					Overhead Line Current (A, rms)	Rated Isolated Current (A, rms)	Back-to-Back Switching		
							Current (A, rms)	Inrush Current Peak Current (kA)	Frequency (Hz)
123	31.5	1200, 2000	50	50	160	315	315	16	4250
123	40	1600, 2000, 3000	50	50	160	315	315	16	4250
123	63	2000, 3000	50	50	160	315	315	16	4250
145	31.5	1200, 2000	80	80	160	315	315	16	4250
145	40	1600, 2000, 3000	80	80	160	315	315	16	4250
145	63	2000, 3000	80	80	160	315	315	16	4250
145	80	2000, 3000	80	80	160	315	315	16	4250
170	31.5	1600, 2000	100	100	160	400	400	20	4250
170	40	2000, 3000	100	100	160	400	400	20	4250
170	50	2000, 3000	100	100	160	400	400	20	4250
170	63	2000, 3000	100	100	160	400	400	20	4250
245	31.5	1600, 2000, 3000	160	160	200	400	400	20	4250
245	40	2000, 3000	160	160	200	400	400	20	4250
245	50	2000, 3000	160	160	200	400	400	20	4250
245	63	2000, 3000	160	160	200	400	400	20	4250
362	40	2000, 3000	250	250	315	500	500	25	4250
362	63	2000, 3000	250	250	315	500	500	25	4250
550	40	2000, 3000	400	400	500	500	500	25	4250
550	63	3000, 4000	400	400	500	500	500	25	4250
800	40	2000, 3000	500	500	500	500	500	—	—
800	63	3000, 4000	500	500	500	500	500	—	—

Source: ANSI Std. C37.06. AC High-Voltage Circuit Breakers Rated on a Symmetrical Current Basis—Preferred Ratings and Related Capabilities. 2000 (revision of 1987).

Rated transient inrush current is the highest magnitude, which the circuit breaker will be required to close at any voltage up to the rated maximum voltage and will be measured by the system and unmodified by the breaker. Rated transient inrush frequency is the highest natural frequency, which the circuit breaker is required to close at 100% of its rated back-to-back shunt capacitor or cable switching current.

For systems below 72.5 kV, shunt capacitors may be grounded or ungrounded, and for systems above 121 kV, both the shunt capacitors and the systems will be solidly grounded. If the neutral of the system, the capacitor bank, or both, are ungrounded, the manufacturer should be consulted for circuit breaker application. The first phase to interrupt affects the recovery voltage.

The following definitions are applicable:

1. Rated open wire line charging current is the highest line charging current that the circuit breaker is required to switch at any voltage up to the rated voltage.

2. Rated isolated cable charging and isolated shunt capacitor bank switching current is the highest isolated cable or shunt capacitor current that the breaker is required to switch at any voltage up to the rated voltage.

3. The cable circuits and switched capacitor bank are considered isolated if the rate of change of transient inrush current, di/dt, does not exceed the maximum rate of change of symmetrical interrupting capability of the circuit breaker at the applied voltage [11]:

$$\left(\frac{di}{dt}\right)_{max} = \sqrt{2}\omega[\text{rated maximum voltage/operating voltage}]I \tag{8.14}$$

where I is the rated short-circuit current in ampères.

4. Cable circuits and shunt capacitor banks are considered switched back-to-back if the highest rate of change of inrush current on closing exceeds that for which the cable or shunt capacitor can be considered isolated.

The oscillatory current on back-to-back switching is limited only by the impedance of the capacitor bank and the circuit between the energized bank and the switched bank.

The inrush current and frequency on capacitor current switching can be calculated by solution of the following differential equation:

$$iR + L\frac{di}{dt} + \int \frac{idt}{C} = E_m\sin \omega t \tag{8.15}$$

The solution to this differential equation is discussed in many texts and is of the following form:

$$i = A \sin(\omega t + \alpha) + Be^{-Rt/2L} \sin(\omega_0 t - \beta) \tag{8.16}$$

where

$$\omega_0 = \sqrt{\frac{1}{LC} - \frac{R^2}{4L^2}} \tag{8.17}$$

Application and Ratings of Circuit Breakers and Fuses according to ANSI Standards 329

The first term is a forced oscillation, which in fact is the steady-state current, and the second term represents a free oscillation, and has a damping component $e^{-rt/2L}$. Its frequency is given by $\omega_0/2\pi$. Resistance can be neglected; hence, the solution is simplified. The maximum inrush current is given at an instant of switching when $t = \sqrt{(LC)}$. For the purpose of evaluation of switching duties of circuit breakers, the maximum inrush current on switching an isolated bank is

$$I_{\text{peak}} = \frac{\sqrt{2}}{\sqrt{3}} E_{\text{rms}} \sqrt{\frac{C}{L}} \tag{8.18}$$

where E is the line-to-line voltage, and C and L are in H and F, respectively. The inrush frequency is given as follows:

$$f_{\text{inrush}} = \frac{1}{2\pi\sqrt{LC}} \tag{8.19}$$

For back-to-back switching, i.e., energizing a bank on the same bus when another energized bank is present, the inrush current is entirely composed of interchange of currents between the two banks. The component supplied by the source is of low frequency and can be neglected. This will not be true if the source impedance is comparable to the impedance between the banks being switched back-to-back. The back-to-back switching current is given as follows:

$$I_{\text{inrush}} = \frac{\sqrt{2}}{\sqrt{3}} E_{\text{rms}} \sqrt{\frac{C_1 C_2}{(C_1 + C_2)(L_{\text{eq}})}} \tag{8.20}$$

where L_{eq} is the equivalent reactance between the banks being switched. The inrush frequency is given as follows:

$$f_{\text{inrush}} = \frac{1}{2\pi\sqrt{\dfrac{L_{\text{eq}} C_1 C_2}{(C_1 + C_2)}}} \tag{8.21}$$

The 2005 revisions of IEEE standards [11, 12] and also IEEE standard [1] classify the breakers for capacitance switching as classes C_0, C_1, and C_2, as well as mechanical endurance class M_1 or M_2 are assigned. These classes have specific type of testing duties, coordinated as per standards [13, 14]. Table 8.5 from IEEE standard [8] shows the capacitance current switching ratings of breakers from 121 to 800 kV, while Table 8.7 shows similar ratings for Classes C_0, C_1, and C_2 from IEEE standard [1]. These two tables can be compared.

The presence of single- and two-phase faults is an important factor that determines the recovery voltage across the breaker contacts. See Chapter 7. Another phenomena that has been observed primarily in vacuum circuit breakers is called Non-Sustained Disruptive Discharge (NSDD), which is defined as the disruptive discharge associated with current interruption that does not result in current at natural frequency of the circuit. For other types of breakers, this can be ignored.

Class C_1 circuit breaker is acceptable for medium voltage circuit breakers and for circuit breakers applied for infrequent switching of transmission lines and cables. Class C_2 is

TABLE 8.5

Preferred Capacitance Current Switching Ratings for Circuit Breakers Rated 100 kV and above including Circuit Breakers in Gas-Insulated Substations

Rated Maximum Voltage	Class C₀ Circuit Breakers			Rated Isolated Capacitor Bank or Cable Current (A, rms)	Rated Overhead Line Current (A, rms)	Rated Current (A, rms)	Class C₁ or C₂ Circuit Breakers							
							Back to Back Capacitor Bank Switching							
							Rated Inrush Current							
	Rated Continuous Current (Applicable to All Continuous Current Ratings) (A, rms)	Rated Overhead Line Current (A, rms)	Rated Isolated Capacitor Bank or Cable Current (A, rms)				Preferred Ratings		Alternate 1 Rating		Alternate 2 Rating		Alternate 3 rating	
							Peak Value (kA)	Frequency (kHz)	Peak Value (kA)	Frequency (kHz)	Peak Value (kA)	Frequency (kHz)	Peak Value (kA)	Frequency (kHz)
123	As in Table 8.4	50	50	1200	160	700	16	4.3	6	2	25	13	60	8.5
145		80	80	1200	160	700	16	4.3	6	2	25	13	60	8.5
170		100	100	1200	160	700	20	4.3	6	2	25	13	60	8.5
245		160	160	1200	200	700	20	4.3	6	2	25	13	60	8.5
362		250	250	1200	315	800	25	4.3	6	2	20	21	65	8.5
550		400	400	1000	500	800	25	4.3	6	2	20	21	65	8.5
800		900	500	1000	900	800	25	4.3	6	2	20	21	65	8.5

Source: IEEE PC37.06/D11. Draft Standard AC High-Voltage Circuit Breakers Rated on Symmetrical Current Basis-Preferred Ratings and Related Required Capabilities for Voltages above 1000 volts. 2009.

Application and Ratings of Circuit Breakers and Fuses according to ANSI Standards 331

recommended for frequent switching of transmission lines and cables [11]. An important consideration is the transient overvoltages that may be generated by restrikes during opening operation, see Figure 7.12. The effect of these transients will be local as well as remote.

8.8.1 Switching of Cables

The cable charging currents depend upon the length of cable, cable construction, system voltage, and insulation dielectric constants. Much akin to switching devices for capacitor banks, a cable is considered isolated if the maximum rate of change with respect to time of transient inrush current on energizing an uncharged cable does not exceed the rate of change associated with maximum symmetrical interrupting current of the switching device.

The cables may be switched back-to-back, much akin to capacitor banks. Transient currents of high magnitude and initial high rate of change flow between cables when the switching circuit breaker is closed or restrikes on opening.

For isolated cable switching, we can use the following expression:

$$i = \frac{u_m - u_t}{Z}\left[1 - \exp\left(-\frac{Z}{L}t\right)\right] \tag{8.22}$$

where u_m is the supply system voltage, u_t is the trapped voltage on the cable being switched, Z is the cable surge impedance, and L is the source inductance. From Equation 8.22, the maximum inrush current is given as follows:

$$i_p = \frac{u_m - u_t}{Z} \tag{8.23}$$

Equation 8.23 can be modified for back-to-back switching as follows:

$$i = \frac{u_m - u_t}{Z_1 + Z_2}\left[1 - \exp\left(-\frac{Z_1 + Z_2}{L}t\right)\right] \tag{8.24}$$

where Z_1 and Z_2 are the surge impedances of cables 1 and 2, respectively.

Again, akin to back-to-back switching of capacitors, the source reactance can be ignored. The peak current is given as follows:

$$i_p = \frac{u_m - u_t}{Z_1 + Z_2} \tag{8.25}$$

And its frequency is given by

$$f_{eq} = f\left[\frac{u_m - u_t}{\omega(L_1 + L_2)I_{ir}}\right] \tag{8.26}$$

where f_{eq} is the inrush frequency; I_{ir} is the charging current of one cable; L_1 and L_2 are the inductances of cables 1 and 2, respectively.

The switching of single cable forms a series circuit, and the transient will be oscillatory, damped, or critically damped. The following equations can be written for the inrush current:

$$Z = \sqrt{4L/C} \qquad i_p = 0.368\frac{u_m - u_t}{Z}$$

$$Z < \sqrt{4L/C} \qquad i_p = \frac{u_m - u_t}{\sqrt{L/C - Z^2/4}}\left[\exp\left(-\frac{Z\pi}{4L}\frac{1}{\sqrt{L/C - Z^2/4}}\right)\right] \qquad (8.27)$$

$$Z > \sqrt{4L/C} \qquad i_p = \frac{u_m - u_t}{\sqrt{L/C - Z^2/4}}\left[\exp(-\alpha t_m) - \exp(-\beta t_m)\right]$$

where

$$t_m = \frac{\ln \alpha/\beta}{\alpha - \beta}$$

$$\alpha = \frac{Z}{L} - \sqrt{\frac{Z^2}{L^2} - \frac{4}{LC}} \qquad (8.28)$$

$$\beta = \frac{Z}{L} + \sqrt{\frac{Z^2}{L^2} - \frac{4}{LC}}$$

The magnetic fields due to high inrush currents during back-to-back switching can induce voltage in control cables by capacitive and magnetic couplings. This is minimized by shielding the control cables.

Example 8.3

Consider the system shown in Figure 8.9a. Two capacitor banks C_1 and C_2 are connected on the same 13.8 kV bus. The inductances in the switching circuit are calculated in Table 8.6. Let C_1 be the first switched. The inrush current is mostly limited by the source inductance, which predominates. The inrush current magnitude and frequency, calculated using the expressions in Equations 8.18 and 8.19, are 5560 A peak and 552.8 Hz, respectively. The maximum rate of change of current is

$$2\pi(552.81)(5560)\,10^{-6} = 19.31\,\text{A/}\mu\text{s}$$

Consider that a definite-purpose indoor 15 kV breaker of 2 kA continuous rating and 40.2 kA interrupting at 13.8 kA. From Equation 8.14, the breaker di/dt is

$$2\pi(60)(40200)10^{-6} = 21.43\,\text{A/}\mu\text{s}$$

This is more than 19.283 A/μs as calculated above. Thus, the capacitor bank can be considered isolated.

Now calculate the inrush current and frequency on back-to-back switching, i.e., capacitor C_2 is switched when C_1 is already connected to the bus. The equivalent inductance

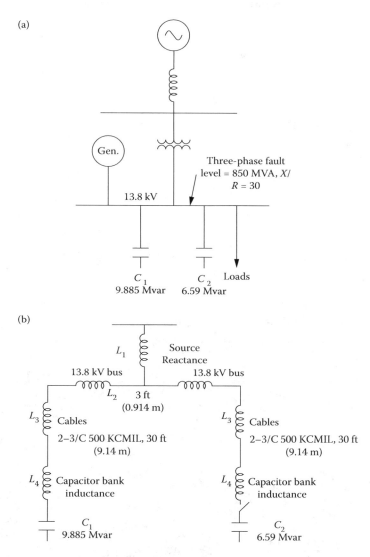

FIGURE 8.9
(a) Connection diagram for capacitor bank switching and (b) equivalent reactance diagram, Example 8.3.

on back-to-back switching consists of a small length of bus between the banks and their cable connections. The source inductance is ignored, as practically no current is contributed from the source. The inductance in the back-to-back switching circuit is 14.46 µH, as shown in Figure 8.9b. From Equations 8.20 and 8.21, the inrush current is 22.7 kA, and the inrush frequency is 5653 Hz. As per the data from Reference [8], the intended 15.5 kV definite-purpose breaker to be used has a maximum peak inrush current of 18 kA and an inrush frequency of 2.4 kHz. In this example, even a definite-purpose circuit breaker will be applied beyond its rating. In order to reduce the inrush current and frequency, an additional reactance should be introduced into the circuit. An inductance of 70 µH will reduce the inrush current to 9.4 kA and the frequency to 2332 Hz. Inrush current-limiting reactors are generally required when the capacitor banks are switched back-to-back on the same bus. In the case where power capacitors are applied as shunt-tuned

TABLE 8.6

Capacitor Switching (Example 5.3): Calculation of Inductances and Capacitances

No.	System Data	Calculated Inductance or Capacitance
1	Three-phase short-circuit level at 13.8-kV bus, 850 MVA, $X/R = 30$	L_1 source = 593.97 µH
2	3′ (0.914 m) of 13.8-kV bus	L_2 bus = 0.63 µH
3	30′ (9.14 m) of 2–3/C 500 KCMIL cables	L_3 cable = 1.26 µH
4	Inductance of the bank itself	L_4 bank = 5 µH
5	Total inductance, when capacitor C_1 is switched = $L_1 + L_2 + L_3 + L_4$	600.86 µH
6	Capacitance of bank C_1, consisting of 9 units in parallel, one series group, wye connected, rated voltage 8.32 kV, 400 kvar each, total three-phase kvar at 13.8 kV = 9.885 Mvar	$C_1 = 0.138 \times 10^{-3}$ F
7	Total inductance when C_2 is switched and C_1 is already energized = inductance of 9 ft of 13.8 kV bus, 60 ft of cables and inductances of banks themselves	14.42 µH
8	Capacitance of bank C_2, consisting of 6 units in parallel, one series group, wye connected, rated voltage 8.32 kV, 400 kvar each, total three-phase kvar at 13.8 kV = 6.59 Mvar	$C_2 = 0.092 \times 10^{-3}$ F

filters, the filter reactors will reduce the inrush current and its frequency, so that the breaker duties are at acceptable levels.

An Electromagnetic Transients Program (EMTP) simulation of the back-to-back switching of inrush current in Example 8.3 is shown in Figure 8.10.

The switching of capacitor currents is associated with restrikes, see Chapter 7. The overvoltage control is also discussed in Chapter 7. Special applications may exist where the circuit breaker duties need to be carefully evaluated [8, 11]. These applications may be as follows:

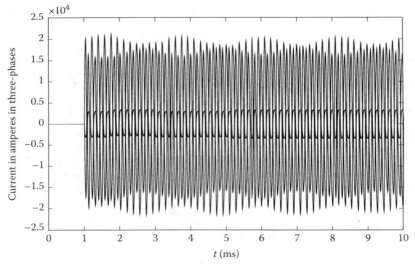

FIGURE 8.10
Results of EMTP simulation for back-to-back switching of Example 8.3.

Application and Ratings of Circuit Breakers and Fuses according to ANSI Standards 335

1. Switching through a transformer of turns ratio greater than one will have the effect of increasing the switching current. De-energizing no-load EHV and UHV lines through low-voltage circuit breakers can increase the effective line charging current in the 750–1000 A range. The capacitive switching rating of circuit breaker should be carefully examined.
2. The effect of the capacitive discharge currents on voltage induced in the secondary of the bushing-type current transformer should be considered. In certain system configurations, i.e., when a number of capacitors are connected to a bus, for a fault on a feeder circuit, all bus connected capacitors will discharge into the fault. The bushing current transformer (BCT) secondary voltage may reach high values. This secondary voltage can be estimated from

$$\left(\frac{1}{\text{BCT ratio}}\right)(\text{crest transient current})$$

$$\left((\text{relay reactance})\left(\frac{\text{transient frequency}}{\text{system frequency}}\right)\right) \tag{8.29}$$

Higher than the normal inrush currents are possible on fast reclosing of power capacitors. Reclosing is, generally, not attempted on power capacitor banks. Capacitors over 600 V are provided with internal discharge devices (resistors) to reduce the residual charge to 50 V or less within 5 min.

3. When parallel banks of capacitors are located on a bus section, caution must be taken in fault switching sequence, so that the last circuit breaker to clear the fault is not subjected to a capacitive switching duty beyond its capability.
4. Switching capacitor banks under faulted conditions give rise to high recovery voltages, depending on the grounding and fault type. A phase-to-ground fault produces the most severe conditions when the source is ungrounded and the bank neutral is grounded. If an unfaulted phase is first to clear, the current may reach 1.73 times the rated current and the recovery voltage 3.46 E_{max}, phase-to-ground. When the faulted phase is first to interrupt, the current is 3.06 times the rated current and the recovery voltage 3.0 E_{max}.
5. Switching of transformers and capacitors can bring harmonic resonance, increase the inrush current and its time duration.

8.9 Line Closing Switching Surge Factor

The rated line closing switching surge factors are specified in ANSI for breakers of 362 kV and above, specifically designed to control the switching overvoltages, and are shown in Table 8.7 [8]. The rating designates that the breaker is capable of controlling the switching surge voltages so that the probability of not exceeding the rated overvoltage factor is 98% or higher when switching the standard *reference transmission line* from a *standard reference source* [13, 14].

Switching surge overvoltages are discussed in Section 7.11. ANSI takes a statistical approach. Random closing of circuit breaker will produce line closing switching surge maximum voltages that vary in magnitude according to the instantaneous value of the source voltage, the parameters of the connected system, and the time difference between

TABLE 8.7

Rated Line Closing Switching Surge Factors for Circuit Breakers Specifically Designed to Control Line Closing Switching Surge Maximum Voltage, and Parameters of Standard Reference Transmission Lines

Rated Maximum Voltage (kV, rms)	Rated Line Closing Switching Surge Factor	Line Length (miles/km)	Percentage Shunt Capacitance Divided Equally at Line Ends	L_1	L_0/L_1	R_1	R_0	C_1	C_1/C_0
362	2.4	150 (241)	0	1.6	3	0.05	0.5	0.02	1.5
500	2.2	200 (322)	0	1.6	3	0.05	0.5	0.02	1.5
800	2.0	200 (322)	60	1.4	3	0.05	0.5	0.02	1.5

L_1 = positive and negative sequence inductance in mH per mile (1.609 km)
L_0 = zero sequence inductance in mH per mile.
R_1 = positive and negative sequence resistance in ohms per mile.
R_0 = zero sequence resistance in ohms per mile.
C_1 = positive and negative sequence capacitance in microfarads per mile.
C_0 = zero sequence capacitance in microfarads per mile.
Source: ANS1/IEEE Std. C37.04, Rating structure for AC high voltage circuit breakers, 1999 (revision of 1979).

completions of a circuit path by switching traveling waves in each phase. These variations will be governed by the laws of probability, and the highest and lowest overvoltages will occur infrequently.

The assumptions are that the circuit breaker connects the overhead line directly to a power source, open at the receiving end and not connected to terminal apparatus such as a power transformer, though it may be connected to an open switch or circuit breaker. The system does not include surge arresters, shunt reactors, potential transformers, or series or shunt capacitors.

The reference power source is a three-phase, wye-connected voltage source with the neutral grounded and with each of the three-phase voltages in series with an inductive reactance, which represents the short-circuit capability of the source. The maximum source voltage, line-to-line, is the rated voltage of the circuit breaker. The series reactance is the one that produces the rated short-circuit current of the circuit breaker, both three phase and single phase at rated maximum voltage with the short circuit applied at the circuit breaker terminals.

The standard transmission line is a perfectly transposed three-phase transmission line with parameters as listed in the ANSI/IEEE standard [13]. Any power system that deviates significantly from the standard reference power system may require that a simulated study be made.

8.9.1 Switching of Transformers

Certain switching operations can excite high-range frequencies of transformer windings. This can cause excessive stress on the inter-turn windings of the transformer. Traditional surge arresters applied at the circuit breakers or transformer terminals are ineffective to counteract this phenomenon. Resistance and capacitance "snubber" [5] networks connected across the transformer windings, phase-to-ground are shown to be effective. Also see Chapter 7.

Application and Ratings of Circuit Breakers and Fuses according to ANSI Standards   337

The term "sympathetic inrush" is applied when a transformer is switched on the same bus to which another transformer is connected. The transformer that is on line will experience inrush currents when the parallel transformer on the same bus is switched.

8.10 Out-of-Phase Switching Current Rating

The assigned out-of-phase switching rating is the maximum out-of-phase current that can be switched at an out-of-phase recovery voltage specified in ANSI and under prescribed conditions. If a circuit breaker has an out-of-phase switching current rating, it will be 25% of the maximum short-circuit current in kiloampères, unless otherwise specified. The duty cycles are specified in ANSI/IEEE standard [13]. The conditions for out-of-phase switching currents are as follows:

1. Opening and closing operations in conformity with manufacturers' instructions, closing angle limited to a maximum out-of-phase angle of 90° whenever possible.
2. Grounding conditions of the neutral corresponding to that for which the circuit breaker is tested.
3. Frequency within ±20% of the rated frequency of the breaker.
4. Absence of fault on either sides of the circuit breaker.

In instances where frequent out-of-phase operations are anticipated, the actual system recovery voltages should be evaluated, see Section 7.12. A special circuit breaker, or one rated at a higher voltage, may sometimes be required. As an alternative solution, the severity of out-of-phase switching can be reduced in several systems by using relays with coordinated impedance sensitive elements to control the tripping instant, so that interruption will occur substantially after or substantially before the instant the phase angle is 180°. Polarity sensing and synchronous breakers are discussed in Chapter 7.

8.11 Transient Recovery Voltage

As discussed in Chapter 7, the interrupting capability of the circuit breaker is related to TRV. If the specified TRV withstand boundary is exceeded in any application, a different circuit breaker should be used or the system should be modified. The addition of capacitors to a bus or line is one method of improving the recovery voltage characteristics. For proper application,

$$TRV_{breaker} > TRV_{system} \qquad (8.30)$$

To calculate system TRV, dynamic simulation is required though simplified equations can be used for hand calculations. IEEE std. C37.011 [15] is the application guide for TRV for AC high-voltage circuit breakers. The TRV ratings are given for interrupting three-phase-to-ground faults at the rated symmetrical short-circuit current and maximum rated voltage of the circuit breaker. For values of fault currents other than rated and for line faults, related TRV capabilities are given. The TRV when interrupting asymmetrical current values are generally less severe than the ones that occur when interrupting the rated symmetrical current, because the instantaneous value of the supply voltage is less than the peak value. Rated and related TRV capabilities are defined by parameters given in ANSI C37.06.

The following calculation procedure is based on C37.04 [6]

8.11.1 Circuit Breakers Rated Below 100 kV

For circuit breakers rated below 100 kV, the rated transient voltage is defined as the envelope formed by a 1 − cosine curve using the values of E_2 and T_2 defined in the ANSI standard [8]; E_2 is the peak of TRV, and its value is 1.88 times the maximum rated voltage. This value of 1.88 considers a first-pole-to-clear factor of 1.5, as the systems below 100 kV may be ungrounded. The time T_2, specified in microseconds, to reach the peak is variable, depending on short-circuit type, circuit breaker, and voltage rating. For indoor oil-less circuit breakers up to 38 kV, T_2 varies from 50 to 125 µs.

The plot of this response curve for first half cycle of the oscillatory component of TRV is shown in Figure 8.11. The supply voltage is considered at its peak during this interval and is represented by a straight line in Figure 8.11, i.e., the power frequency component of TRV is constant. This definition of TRV by two parameters, E_2 and T_2, is akin to that of Figure 7.8, i.e., IEC representation by two parameters. The curve of Figure 8.11 is called *one-minus-cosine* curve.

8.11.2 Circuit Breakers Rated 100 kV and Above

For breakers rated 100 kV and above [6] or 72.5 kV [15], the rated TRV is defined by higher of an exponential waveform and 1 − cosine waveform, Figure 8.12. Mostly, for systems above 100 kV, the systems will be grounded and a first-pole-to-clear factor of 1.3 is considered. $E_1 = 1.06\,V$ and $E_2 = 1.49\,V$. Envelope formed by the exponential cosine curve obtained by using the rated values of E_1, R, T_1, E_2, and T_2 from the standards, and applying these values at the rated short-circuit current of the breaker. R is defined as the rated TRV rate, ignoring the effect of the bus side lumped capacitance, at which the recovery voltage rises across the terminals of a first-pole-to-interrupt for a three-phase, ungrounded load-side terminal fault under the specified rated conditions. The rate is a close approximation of the

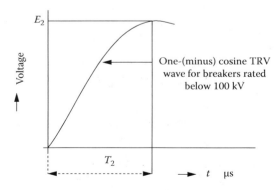

FIGURE 8.11
One-minus-cosine TRV wave for breakers rated below 100 kV.

Application and Ratings of Circuit Breakers and Fuses according to ANSI Standards

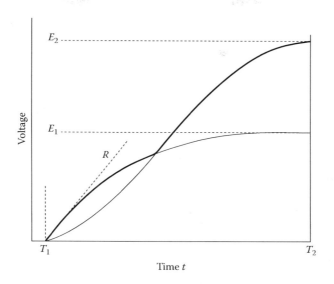

FIGURE 8.12
TRV profile for breakers for system voltages rated above 100 kV.

maximum de/dt in the rated envelope, but is slightly higher because the bus side capacitance is ignored.

The exponential cosine envelope is defined by whichever of e_1 and e_2 is larger:

$$e_1 = E_1(1 - e^{-t/\tau}) \text{ with a time delay } T_1 \, \mu s \qquad (8.31)$$

$$\tau = E_1/R \qquad (8.32)$$

$$e_2 = \frac{E_2}{2}(1 - \cos(\pi t/T_2)) \qquad (8.33)$$

Table 8.8 showing the rated TRV parameters is from IEEE C7.04 [6].

Example 8.4

Consider a 550 kV breaker. The ratings are as follows:

K factor = 1
Current rating = 2 kA

TABLE 8.8

Rated TRV Parameters

Breaker Rating	Envelope	E_2	T_2	R	E_1	T_1
Below 100 kV	1-Cos Figure 8.8	1.88 V	ANSI C37.06-1997	NA	NA	NA
100 kV and above	Expo-cos	1.49 V	ANSI C37.06-1997	ANSI C37.06-1997	1.06 V	ANSI C37.06-1997 Figure 8.8

Rated short-circuit current = 40 kA
Rated time to point P, T_2 μs = 1325
R, rate of rise of recovery voltage = 2 kV/μs
Rated time delay T_1 = 2 μs
E_2 = 1.49 × rated maximum voltage, from Table 8.8
E_1 = 1.06 × rated maximum voltage

Then,

E_1 = 583 kV
E_2 = 819.5 kV
$\tau = E_1/R = 583/2 = 291.5$ μs

Substituting in Equations 8.31 and 8.33, we get

$e_1 = 583(1 - e^{-t/291.5})$
$e_2 = 409.80(1 - \cos 0.1358 t^0)$

The calculated TRV, for a rated fault current, is shown in Figure 8.13.
Fault currents other than the rated fault current

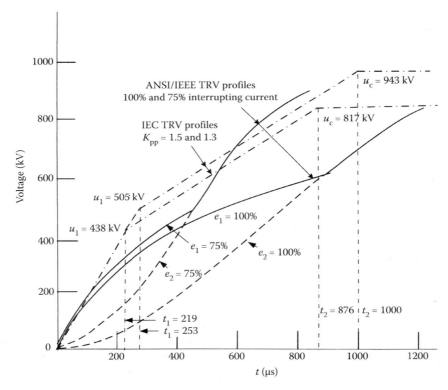

FIGURE 8.13
Calculated TRV wave shapes (Example 8.4).

Application and Ratings of Circuit Breakers and Fuses according to ANSI Standards

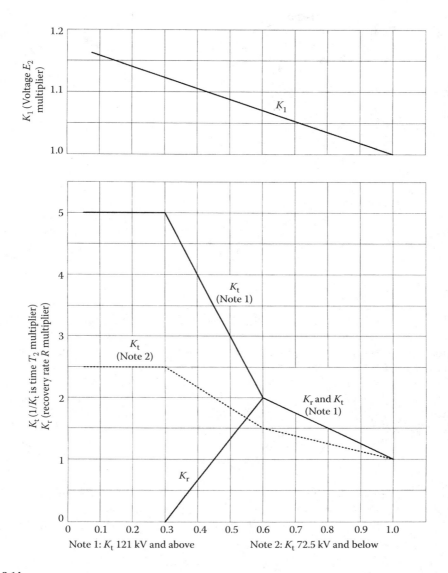

FIGURE 8.14
TRV rates and voltage multipliers for fractions f rated interrupting current. (From ANSI/IEEE Std. C37.04. Rating Structure for AC High Voltage Circuit Breakers. 1999 (revision of 1979).)

Circuit breakers are required to interrupt short-circuit currents that are less than the rated short-circuit currents. This increases E_2, and T_2 time is shorter. The adjustment factors are given in curves as shown in Figure 8.14 from the 1979 edition of Reference [6]. The revision to this standard does not reproduce these curves but refers to the 1979 edition.

Example 8.5

Now, consider that the TRV is required to be calculated for 75% of the rated fault current. This requires calculation of adjustment factors from Figure 8.14.

$K_r = 1.625$ (rate of rise multiplying factor)
$K_1 = 1.044$ (E_2 multiplying factor)
$K_t = 1.625$ (T_2 dividing factor)

The adjusted parameters are as follows:

$E_1 = 583\,\text{kV}$
$E_2 = (819.5)(K_1) = 855.5\,\text{kV}$
$R = (2)(K_r) = 3.25\ \text{kV/μs}$
$T_2 = (1325)/(K_t) = 815.4\ \text{μs}$
$T_1 = 2\ \text{μs}$
$\tau = E_1/R = (583)/3.25 = 179.4\ \text{μs}$

The TRV for 75% interrupting fault duty is superimposed in Figure 8.13 for comparison, and it is higher than the TRV for 100% interrupting current. TRV at lower short-circuit current is calculated to 10% of the rated short-circuit current.

8.11.3 Short-Line Faults

We discussed sawtooth TRV waveform for a short-line fault in Section 7.6. Initial TRV can be defined as an initial ramp and plateau of voltage added to the initial front of an exponential cosine wave shape. This TRV is due to relatively close inductance and capacitance associated with substation work. For breakers installed in gas-insulated substations, the initial TRV can be neglected because of low bus surge impedance and small distance to the first major discontinuity. However, for other systems at low levels of fault current, the initial rate of TRV may exceed the envelope defined by the standards. In such cases, the short-line initial TRV capability can be superimposed on the calculated TRV curve, and the results can be examined. The sawtooth voltage at the input line terminals can be calculated by ladder diagrams.

The circuit breakers rated 15.5 kV and above will be capable of interrupting single-phase line faults at any distance from the circuit breaker on a system in which

- The TRV on a terminal fault is within the rated or related transient voltage envelope.
- The voltage on the first ramp of the sawtooth wave is equal to or less than that in an ideal system in which surge impedance and amplitude constant are 450 ohms and 1.6, respectively.
- There is a time delay of 0.5 μs for circuit breakers rated 245 kV and above and 0.2 μs for the circuit breakers rated below 245 kV.

The amplitude constant d is the peak of the ratio of the sawtooth component, which will appear across the circuit breaker terminal at the instant of interruption. The SLF TRV capability up to the first peak of TRV is defined as follows:

$$e = e_L + e_S$$

where e_L is the line-side contribution to TRV, e_S is the source side contribution to TRV, and e is the first peak of TRV.

$$e_L = d(1-M)\sqrt{\frac{2}{3}}E_{max} \tag{8.34}$$

Application and Ratings of Circuit Breakers and Fuses according to ANSI Standards

$$e_S = 2M(t_L - t_d) \tag{8.35}$$

$$R_L = \sqrt{2}\omega MIZ \times 10^{-6}\,\text{kV}/\mu\text{s} \tag{8.36}$$

$$t_L = \frac{e}{R_L}\,\mu\text{s} \tag{8.37}$$

where R_L is the rate of rise, t_L is the time to peak, M is the ratio of fault current to rated short-circuit current, I is the rated short-circuit current in kA, and Z is the surge impedance.

It is not necessary to calculate SLF TRV, as long as terminal fault TRV are within rating and transmission line parameters are within those specified in Table 8.9.

Example 8.6

For a 550 kV breaker, whose TRV wave shapes are plotted in Figure 8.13, plot the short-line capability for a 75% short-circuit current and a surge impedance of 450 ohms.
$M = 0.75$, $I = 40\,\text{kA}$, $V = 550$, $d = 1.6$, and $Z = 450$ ohms. This gives:

$$e_L = 1.6(0.75)(550) = 179.6\,\text{kV}$$

Also,

$$R_L = \sqrt{2} \times 377 \times 0.75 \times 40 \times 450 \times 10^{-6} = 7.2\,\text{kV}/\mu\text{s}$$

and $t_L = 25\,\mu\text{s}$ (given)

This is shown in Figure 8.15, the contribution of source side TRV, until t_L can be calculated, assuming $t_d = 2\,\mu\text{s}$ and RRRV = 2 kV/μs, when interrupting 100% of rated

TABLE 8.9

Related Required TRV Capabilities of Circuit Breakers at Various Interrupting Levels for Terminal Faults

	Multipliers for Rated Parameters							
Percent Interrupting Capability	**72.5 kV and Below**		**72.5 kV and Below**					
	Indoor/Cable System		Outdoor/Line System				100 kV and Above	
Note 1	Ku_c	Kt_3	Ku_c	Kt_3	Ku_1	Kt_1	Ku_c	Kt_2 or Kt_3
T100	1	1	1	1	1	1	1	1/—
T60	1.07	0.44	1.07	0.67	1	0.67	1.07	0.5/—
T30	1.14	0.22	1.13	0.4	—	—	1.13	—/0.211
T10	1.21	0.22	1.17	0.4	—	—	1.17/1.26	—/0.156 or 0.168
							Note 2	Note 3

Note 1: For other interrupting capabilities interpolate from Figure 8.17.
Note 2: Multiplier Ku_c is 1.17 for applications with $K_{pp} = 1.5$ and 1.26 for applications with $K_{pp} = 1.3$.
Note 3: Multiplier Kt_3 is 0.156 for applications with $K_{pp} = 1.5$ and 0.168 for applications with $K_{pp} = 1.3$.

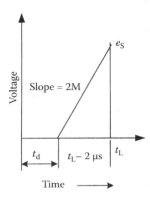

FIGURE 8.15
TRV profile first peak only, source side, Example 8.6.

short-circuit current. When interrupting a SLF with reduced fault current, the RRRV on supply side is reduced to $2M = 1.5$ kV/μs.

Then, the source side contribution to TRV is

$$e_s = 1.5(25-2) = 34.5 \text{ kV}$$

The initial TRV peak is

$$e_T = 180 + 34.5 = 214.5 \text{ kV} \quad \text{at } t_L = 25 \text{ μs}$$

The SLF in some situations can exceed the test values. Generally, this higher TRV peak is associated with lower RRRV.

8.11.4 Oscillatory TRV

Figure 8.16 shows an example of an underdamped TRV, where the system TRV exceeds the breaker TRV capability curve. Such a waveform can occur when a circuit breaker clears a low-level, three-phase ungrounded fault, limited by a transformer on the source side or a reactor, as shown in Figure 8.16a and b. Figure 8.16c shows that the circuit breaker TRV capability is exceeded. When this happens, the following choices exist:

1. Use a breaker with higher interrupting rating.
2. Add capacitance to the circuit breaker terminals to reduce the rate of rise of TRV.
3. Consult the manufacturer concerning the application.

A computer simulation using EMTP of the TRV will be required.

8.11.4.1 Exponential (Overdamped) TRV

Assuming that a network can be reduced to a simple parallel RLC circuit, the TRV is exponential if

$$R \le \frac{1}{2}\sqrt{\frac{L}{C}} \tag{8.38}$$

Application and Ratings of Circuit Breakers and Fuses according to ANSI Standards

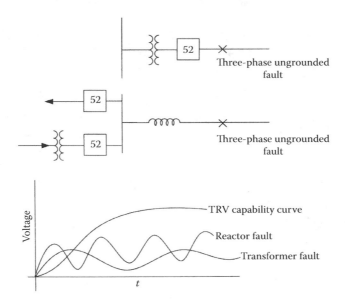

FIGURE 8.16
(a and b) Power system configuration where TRV may exceed the breaker capabilities for a fault limited by the transformer or reactor and (b) oscillatory TRV.

8.11.5 Initial TRV

Circuit breakers rated 100 kV and a short-circuit rating of 31.5 kV and above will have an initial TRV capability for phase-to-ground fault. This rises linearly from origin to first peak voltage E_i and time T_i [6]. The E_i is given by the following expression:

$$E_i = \omega \times \sqrt{2} \times I \times Z_b \times T_i \times 10^{-6} \text{ kV} \qquad (8.39)$$

where Z_b is the bus surge impedance = 450 ohms for outdoor substations, phase-to-ground faults, and I is in the fault current in kA. The term initial TRV refers to the conditions during the first microsecond or so, following a current interruption, when the recovery voltage on the source side of the circuit breaker is influenced by proximity of buses, capacitors, isolators, etc. Akin to short-line fault voltage oscillation is produced, but this oscillation has a lower voltage peak magnitude. The traveling wave will move down the bus where the first discontinuity occurs. The initial slope depends upon the surge impedance and di/dt and the peak of ITRV appears at a time equal to twice the traveling wave time. There can be as many variations of ITRV, as the station layouts. T_i in μs, time to peak is given in IEEE standard [6] with respect to maximum system voltage: For 121, 145, 169, 362, 350, and 500, it is 0.3, 0.4, 0.5, 0.6, 0.8, 1.0, and 1.1, respectively.

8.11.6 Adopting IEC TRV Profiles in IEEE Standards

In the draft revision of IEEE standard [1], the TRV capability is defined by two-parameter and four-parameter envelopes, much akin to IEC standards (Chapter 7). These revisions are in the draft form and not yet approved, at the time of writing this edition of the book. The TRV profiles from this draft standard are shown in Tables 8.10 and 8.11. Note that TRV profiles are distinct for the type of operation. TF stands for Terminal Fault, OS stands for Out-of-Step switching, and SLF stands for short-line fault.

TABLE 8.10

Preferred Ratings of Prospective TRV for Class S1 Circuit Breakers Rated below 100 kV, for Cable Systems Noneffectively Grounded (T100, T60, T10 Test Duties) TRV Representation by Two-Parameter Method

Rated Maximum Voltage (kV, rms)	Test Duty	First Pole-to-Clear Factor K_{pp} (per unit)	Amplitude Factor K_{af} (per unit)	TRV Peak Value u_c	Time t_3 (µs)	Time Delay t_d (µs)	Reference Voltage u' (kV)	Time t' (µs)	RRRV u_c/t_3 (kV/µs)
4.76	TF	1.5	1.4	8.2	44	7	2.7	21	0.19
	OS	2.5	1.25	12.1	88	13	4.0	43	0.14
8.25	TF	1.5	1.4	14.1	52	8	4.7	25	0.27
	OS	2.5	1.25	21.1	104	16	7.0	50	0.20
15	TF	1.5	1.4	25.7	66	10	8.6	32	0.39
	OS	2.5	1.25	38.3	32	20	12.8	4	0.29
27	TF	1.5	1.4	46.3	92	14	15.4	45	0.50
	OS	2.5	1.25	68.9	184	28	23.0	90	0.37
38	TF	1.5	1.4	65.2	109	16	21.7	53	0.60
	OS	2.5	1.25	97.0	218	33	32.3	105	0.45
72.5	TF	1.5	1.4	124	165	25	41.4	80	0.75
	OS	2.5	1.25	185	330	50	61.7	160	0.56

Source: IEEE PC37.06/D11. Draft Standard AC High-Voltage Circuit Breakers Rated on Symmetrical Current Basis-Preferred Ratings and Related Required Capabilities for Voltages above 1000 volts, 2009.

TABLE 8.11

Preferred Ratings of Prospective TRV for Class S1 Circuit Breakers Rated 100 kV and above, including Circuit Breakers Applied in Gas-Insulated Substations, for Effectively Grounded Systems and Grounded Faults with a First Pole-to-Clear Factor of 1.3, T100

Rated Maximum Voltage (kV, rms)	Test Duty	First Pole-to-Clear Factor K_{PP} (per unit)	Amplitude Factor k_{af} (per unit)	First Reference Voltage u_1 (kV)	Time t_1 (μs)	TRV Peak Value (kV)	Time T_2 (μs)	Time Delay T_d (μs)	Voltage u' (kV)	Time t' (μs)	RRRV u_c/t_3 (kV/μs)
123	TF	1.3	1.4	98	49	183	196	2	49	27	2
	SLF	1.0	1.4	75	38	141	152	2	38	21	2
	OS	2.0	1.25	151	98	251	392	2	75	51	1.54
145	TF	1.3	1.4	115	58	215	232	2	58	31	2
	SLF	1.0	1.4	89	44	166	176	2	44	24	2
	OS	2.0	1.25	178	116	296	464	2	89	60	1.54
170	TF	1.3	1.4	135	68	253	272	2	68	36	2
	SLF	1.0	1.4	104	52	194	208	2	52	28	2
	OS	2.0	1.25	208	136	347	544	2	104	70	1.54
245	TF	1.3	1.4	195	98	364	392	2	98	51	2
	SLF	1.0	1.4	150	75	280	300	2	75	40	2
	OS	2.0	1.25	300	196	500	784	2	150	99	1.54
362	TF	1.3	1.4	288	144	538	576	2	144	74	2
	SLF	1.0	1.4	222	111	414	444	2	111	57	2
	OS	2.0	1.25	443	288	739	1152	2	222	146	1.54
550	TF	1.3	1.4	438	219	817	876	2	219	112	2
	SLF	1.0	1.4	337	168	629	672	2	168	86	2
	OS	2.0	1.25	674	438	1120	1752	2	337	221	1.54
800	TF	1.3	1.4	637	318	1190	1272	2	319	161	2
	SLF	1.0	1.4	490	245	914	980	2	245	124	2
	OS	2.0	1.25	980	636	1630	2544	2	490	326	1.54

Source: IEEE PC37.06/D11. Draft Standard AC High-Voltage Circuit Breakers Rated on Symmetrical Current Basis–Preferred Ratings and Related Required Capabilities for Voltages above 1000 volts, 2009.

Interestingly, the TRV capability envelope of a 550 kV breaker at 100% of its short-circuit current rating from standard Reference [15] is superimposed upon the calculated curves in Figure 8.13. The values for 100% interrupting rating can be straightaway read from Table 8.11, and no calculation is required. These values are as follows:

$u_1 = 438\,\text{kV}$

$t_1 = 219\,\mu\text{s}$

$u_c = 817\,\text{kV}$

$t_2 = 876\,\mu\text{s}$

At 75% of the interrupting rating, multiplying factors are applicable as shown in Table 8.9 and Figure 8.17.

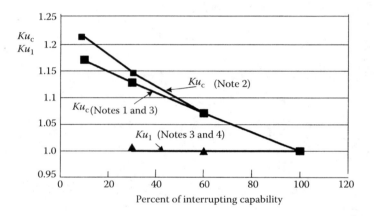

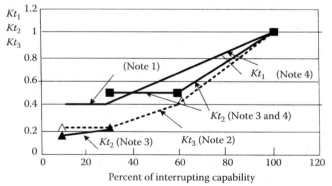

Note 1: For outdoor circuit breakers and/or line systems 72.5 kV and below
Note 2: For indoor circuit breakers and/or cable systems 72.5 kV and below
Note 3: Rated voltages above 72.5kV, values shown for $K_{pp} = 1.5$
Note 4: Kt_1, Ku_1, Kt_2 are applicable for currents higher than 30% of interrupting capability
TRV parameters u_c (or u_1) and t_3 (or t_1 or t_2) obtained by multiplying the values given in ANSI C37.06 by corresponding K_u and K_t multipliers

FIGURE 8.17
TRV parameter multipliers for fraction of rated breaking current. (From IEEE Std. C37.011 Application Guide for Transient Recovery Voltage for AC High-Voltage Circuit Breakers. 2005.)

The following multipliers are obtained:

$Ku_1 = 1$
$Kt_1 = 0.794$
$Ku_c = 1.044$
$Kt_2 = 0.687$

Therefore, the TRV profile is given as follows:

$u_1 = 438\,kV$
$t_1 = 174\,\mu s$
$u_c = 853\,kV$
$t_2 = 602\,\mu s$

This is also shown in Figure 8.13. The profile for first-pole-to-clear factor of 1.3 is also shown.

Considering SLF, the TRV at 75% of the interrupting capacity of the breaker is more clearly shown in Figure 8.18.

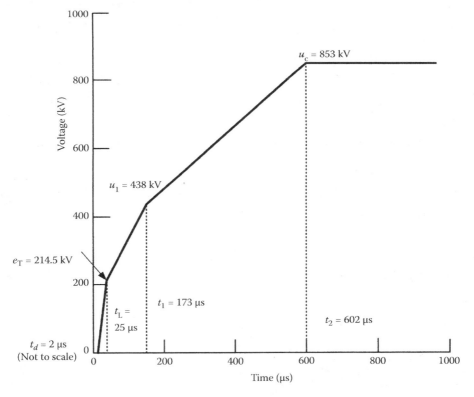

FIGURE 8.18
Calculated TRV profile according to IEEE standard [15], including SLF profile.

Note that these calculations show the TRV capability of the breaker. TRV must be calculated when the breaker is used in a system and the two values of TRV must be compared for safe application.

The standard Reference [15] contains an example of TRV calculations in a practical electrical system, which an interested reader may refer to. The calculations involve the following:

1. Start with a circuit configuration.
2. Put a fault on the system to be studied, for example, a three-phase-to-ground fault.
3. Draw the circuit on each side (line and source side) of the breaker poles.
4. Reduce this circuit to an equivalent circuit. The transmission lines and cables can be modeled with their surge impedances. The capacitances on either side of the breaker are important.
5. An equivalent circuit with positive zero and negative sequence impedances can be constructed.
6. Calculate the fault current at the circuit breaker.
7. Consider first-pole-to-clear factor.

The TRV profiles can be calculated based upon the simplified equations in Reference [15]. The standard also provides typical values of surge impedances and first-pole-to-clear factor for various fault types. The TRV at reduced short-circuit currents can be calculated using adjustment factors provided in this standard. A rigorous calculation is through EMTP and similar transient simulation programs [16].

8.11.7 Definite-Purpose TRV Breakers

ANSI standard C37.06.1 [17] is for "Definite purpose circuit breakers for fast transient recovery voltages." This is somewhat akin to specifications of definite-purpose breakers for capacitor switching. The standard qualifies that

1. No fast T_2 values or tests are proposed for fault currents >30% of the rated short-circuit current.
2. The proposed T_2 values are chosen to meet 90% of the known TRV circuits, but even these fast values do not meet the requirements of all fast TRV applications.
3. A circuit breaker that meets the requirements of definite purpose for fast TRV may or may not meet the requirements of definite-purpose circuit breakers for capacitor switching.

8.11.8 TRV Calculation Techniques

In case of transmission voltages, three-phase-to-ground faults are the basis of ratings because it is recognized that three-phase ungrounded faults have a very low probability of occurrence. During interruption of three-phase faults, the circuits shown in Figure 8.19a through c are valid till the reflection returns from the remote buses.

The equivalent inductance is given as follows:

Application and Ratings of Circuit Breakers and Fuses according to ANSI Standards 351

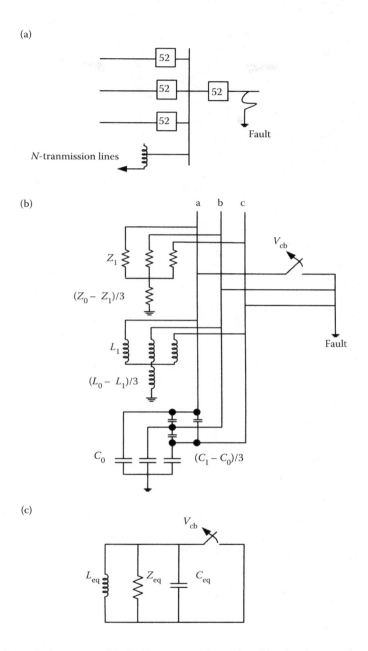

FIGURE 8.19
Interruption of a three-phase-to-ground fault: (a) system configuration, (b) equivalent sequence network, and (c) equivalent circuit. (Circuits valid till the reflected wave returns.)

$$L_{eq} = \frac{3L_0 L_1}{L_1 + 2L_0} \tag{8.40}$$

For three-phase grounded faults in effectively grounded systems, $L_0 = 3L_1$, $L_{eq} = 9L_1/7 = 1.3L_1$, and for three-phase-to-ground faults in ungrounded systems, L_0 is infinite and $L_{eq} = 1.5L_1$

Equivalent surge impedance is given as follows:

$$Z_{eq} = \frac{3}{n} \frac{Z_0 Z_1}{Z_1 + 2Z_0} \qquad (8.41)$$

where $Z_0 = 1.6Z_1$ and $Z_{eq} = 1.14(Z_1/n)$.

Equivalent capacitance is given as follows:

$$C_{eq} = C_0 + \frac{2(C_1 - C_0)}{3} = \frac{C_0 + 2C_1}{3} \qquad (8.42)$$

If

$$C_0 = C_1 \quad \text{then}$$
$$C_{eq} = C_0 = C_1 \qquad (8.43)$$

where Z_1 and Z_0 are positive and zero sequence surge impedances of n transmission lines terminating in the substation, L_1 and L_0 are the positive sequence and zero sequence reactance, representing all other parallel sources terminating at the substation; and C_1 and C_0 are the positive and zero sequence capacitance.

For special case of three-phase ungrounded fault in three-phase effectively grounded systems:

$$L_{eq} = 1.5L_1, \quad Z_{eq} = (1.5Z_1)/n, \quad C_{eq} = C_1/1.5 \qquad (8.44)$$

Exponential overdamped TRV is given as follows:

$$V_{cb} = E_1 \left(1 - e^{-\alpha t} \left(\cosh \beta t + \frac{\alpha}{\beta} \sinh \beta t \right) \right) kV \qquad (8.45)$$

where

$$E_1 = \sqrt{2} I \omega L_{eq}$$

I = short-circuit current

$$\alpha = \frac{1}{2 Z_{eq} C_{eq}}$$
$$\beta = \sqrt{\alpha^2 - 1/(L_{eq} C_{eq})} \qquad (8.46)$$

Derivative of equation gives RRRV as follows:

$$R = \sqrt{2}I\omega Z_{eq} \times 10^{-6} \, \text{kV}/\mu\text{s} \qquad (8.47)$$

Oscillatory (underdamped) TRV is given as follows:

$$V_{cb} = E_1\left[1 - \cos\left(\frac{t}{\sqrt{L_{eq}C_{eq}}}\right)\right] \text{kV} \qquad (8.48)$$

To be oscillatory, the surge impedance of source side line has to be such that

$$Z_{eq} \geq 0.5\sqrt{\frac{L_{eq}}{C_{eq}}} \qquad (8.49)$$

8.12 Generator Circuit Breakers

There is no other national or international standard on generator circuit breakers, other than IEEE standard C37.013 [18, 19]; the 1997 revision of this standard is adapted to international practice. The specified ratings of the generator breakers are as follows:

Required symmetrical interrupting capability for three-phase faults: For a three-phase fault, the generator breaker shall be capable of interrupting the rated three-phase symmetrical short-circuit current for the rated duty cycle, irrespective of the direct current component of the total short-circuit current, at the instant of primary arcing contact separation for operating voltages equal to rated maximum voltage.

Required asymmetrical interrupting capability for three-phase faults: The required asymmetrical *system source* interrupting rating, at maximum operating voltage and rated duty cycle, is composed of the rms symmetrical current and the percent DC component. This value of the DC component in percent of peak value of the symmetrical short-circuit current is given by Figure 8.20 for primary arcing contact parting time in ms. This figure is applicable for system short-circuit currents. The curve is based upon time constant of decay of DC component of 133 ms. The primary contact parting time is considered equal to ½ cycle plus minimum opening time of the particular breaker. In Section 8.2.1, for a five-cycle breaker with contact parting time of three cycle, DC component of 32.91% was calculated. From Figure 8.20, for a generator breaker, the DC component is 54.7%. For time constants different from 133 ms, the following expressions are used:

$$\alpha = \frac{I_{dc}}{I_{ac\,peak}} \qquad (8.50)$$

where α is the degree of asymmetry calculated at the contact part time, t_{cp} (in ms) of the breaker, and the DC component is

$$I_{dc} = I_{ac\,peak}e^{-t_{cp}/133} \qquad (8.51)$$

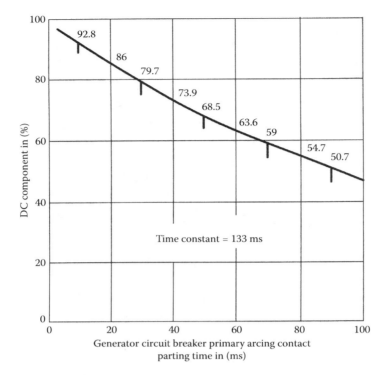

FIGURE 8.20
Asymmetrical interrupting capability of generator circuit breakers: DC component in percentage of peak value of the system symmetrical three-phase short-circuit currents.

Generator source symmetrical interrupting capability for three-phase faults: No specific value is assigned to generator source symmetrical interrupting short-circuit current for three-phase faults, because its maximum value is usually less than the short-circuit current from the power system. If a rating is assigned, the generator breaker shall be tested for the three-phase, short-circuit interrupting current solely contributed by the generator source.

Generator source asymmetrical interrupting capability for three-phase faults: The required generator source asymmetrical interrupting capability for three-phase faults at rated maximum voltage and duty cycle is composed of rms generator source symmetrical current and a DC component. A DC component value of 110% of the peak value of the symmetrical generator source short-circuit current is specified.

Generator source asymmetrical interrupting capability for maximum required degree of asymmetry: The maximum required degree of asymmetry of the current required for the maximum required degree of asymmetry is 130% of the peak value of symmetrical current for this condition. The symmetrical component of the current for this condition of maximum asymmetry is only 74% of the value of required generator source symmetrical interrupting capability. This is further discussed in Chapter 10, with an example of calculation.

Required interrupting capability for single-phase-to-ground faults: The generator circuit breakers are designed for use on high-resistance grounded systems where single phase-to-ground short-circuit current will not exceed 50 A.

The first-pole-to-clear factor shall be 1.5, and the amplitude factor shall be 1.5.

Closing and latching: (1) The generator circuit breaker is capable of closing and latching any power frequency current whose crest does not exceed 2.74 times rated symmetrical

Application and Ratings of Circuit Breakers and Fuses according to ANSI Standards   355

short-circuit current or the maximum peak (peak making current) and the generator source short-circuit current, whichever is higher. No numerical value is given for generator source peak current as it depends upon the generator characteristic data. (2) The circuit breaker shall carry the short-circuit current for 0.25 s.

Short-time current carrying capability: The short-time carrying is for a period of 1 s, any short-circuit current determined from the envelope of current wave at the time of maximum peak, whose value does not exceed 2.74 times the rated short-circuit current is given as follows:

$$I = \sqrt{\int_0^1 i^2 \, dt} \tag{8.52}$$

TRV: The standard specifies TRV values for system source faults, generator source faults, load current switching and out-of-phase current switching. Both power frequency recovery voltage and inherent transient recovery voltage (unmodified by the presence of generator breaker) should be considered. The power frequency recovery voltage across generator breaker contacts consists of a sum of voltage variations at each side of the generator breaker. Maximum voltage is

$$1.5 \frac{V}{\sqrt{3}} \times (X_d'' + X_t) \tag{8.53}$$

where V is the rated maximum voltage in per unit; X_d'' and X_t are the generator and transformer per unit reactance, respectively. Their sum does not exceed 0.5 per unit even for large machines, and therefore, recovery voltage that appears across generator breaker contacts after a short-circuit current interruption is standardized at 0.43 V, from Equation 8.53.

The TRV for load current switching is normally a dual frequency curve, field tests for accurate estimate rather than theoretical calculations.

Short-circuit currents with delayed current zero: A generator breaker will be required to interrupt generator source currents with delayed current zeros. The magnitude of these currents is considerably lower than the rated short-circuit currents. The standard recommends that capability to interrupt delayed current zeros can be ascertained by computations that consider effect of arc voltage on prospective short-circuit current.

Out of phase current switching capability: When out-of-phase switching capability is assigned, it is based upon an out-of-phase angle of 90° at rated maximum voltage. The maximum out-of-phase current will be 50% of the symmetrical system short-circuit current. The out-of-phase switching current can be calculated from the following expression:

$$I_{oph} = \frac{\delta I_n}{X_d'' + X_t + X_s} \tag{8.54}$$

where X_s is the system reactance. All reactances are based on generator rated mega volts ampere (MVA) in per unit. I_n is the rated generator current and $\delta = 1.4$ for a 90° out-of-phase angle and 2 for 180° out-of-phase angle.

A generator breaker may be specified for out-of-phase current rating and also capacitance switching capability, which should be demonstrated by tests.

In the utility systems, the synchronous generators are connected directly through a step-up transformer to the transmission systems, as shown in Figure 8.21. In this figure, the ratings of generators and transformers are not shown for generality. Note that generators 1 and 4 do not have a generator breaker and the generator and transformer are protected as a unit with overlapping zones of differential protection, (discussed in vol. 4). Generators 2 and 3 have generator breakers, which allow generator step-up (GSU) transformers to be used as step-up down transformers during cold startup. The power to the generation auxiliary loads is duplicated with auto-switching of bus section breaker and also there is a third standby power source. Utilizing the GSU transformer as generator step-down transformer may result in cost savings; the addition of generator breaker may be more economical than a dedicated high-voltage step-down transformer. Inline generator breakers of 100 kA. An advantage is that generator and transformer can be protected by separate differential zones of protection compared to protecting the two as a unit. A disadvantage that is noted is as follows:

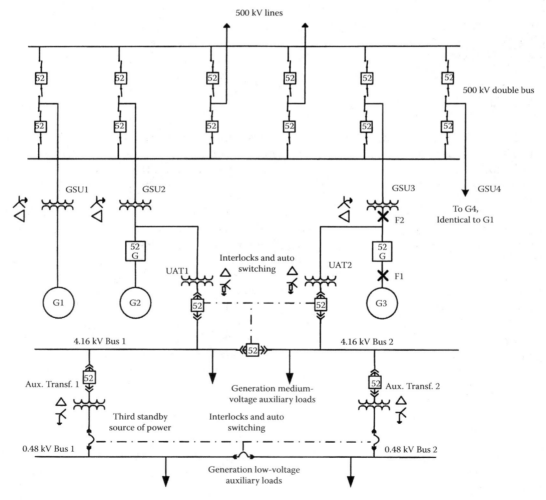

FIGURE 8.21
Interconnections of large generators in utility systems, with and without a generator breaker.

1. During cold start and when the transformer is used as a step-down transformer, with generator breaker open, the adjustment of load voltage profiles requires negative tap adjustment on the transformer primary windings to counteract the voltage drops on load flow in the auxiliary distribution.
2. After the startup with generator is synchronized, positive transformer taps are required to utilize the generator reactive power capability.

These are two conflicting requirements. Sometimes, the generator reactive power may remain trapped due to this limitation. GSU transformers are sometimes provided additional positive tap adjustment range. See Volume 2, reactive power flow and control.

Figure 8.22 shows a generator of 81.82 MVA, 12.47 kV, and 0.85 power factor directly connected to a 12.47 kV bus, which is also powered by a 30/40/50 MVA, 115–12.47 kV utility transformer. The two sources are run in synchronism, and the plant running load is 45 MVA; the excess generated power is supplied into the utility system. The size of a generator that can be bus connected in an industrial distribution, primary distribution voltage of 13.8 kV, is approximately limited to 100 MVA, as an acceptable level of short circuit should be maintained at the medium voltage switchgear and the downstream distributions.

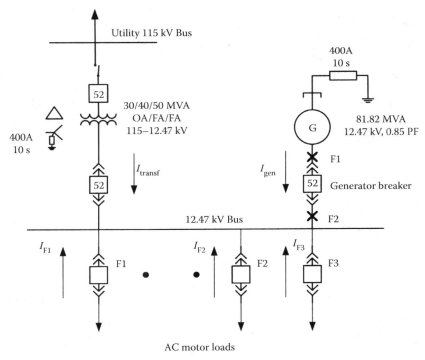

FIGURE 8.22
Interconnection of a large bus connected generator in an industrial system.

8.13 Specifications of High-Voltage Circuit Breakers

The discussions in the above sections are limited to current interrupting performance of the high voltage circuit breakers. There are a number of other relevant specifications, which depend upon a specific installation. As an example, ambient temperature, short-time overload capability, location in environmentally adverse conditions, outdoor installations in high seismic zones, and dielectric strength of external and internal insulation systems should form a part of the comprehensive specifications. Standards specify power frequency tests (dry and wet), impulse tests (full wave and chopped wave), switching impulse tests (voltages 362 kV and higher), minimum creepage distances to ground, etc. Current transformers of relaying and metering accuracy can be located in the circuit breakers (for outdoor installations in the outdoor bushings, for indoor installation bus mounted or window type on draw-out contacts spouts of metal-clad switchgear). A breaker may be fitted with more than one trip coil. Specifications of power operating mechanism and host of safety interlocks are required. For SF_6 breakers, a gas density monitor is normally provided. Breakers requiring single pole operation at higher voltages must be specified for this application and the manufacturer should supply the expected pole scatter.

A circuit breaker manufactured according to ANSI/IEEE must comply with the rating structure and other specifications laid down in the standard, yet specific applications will require additional data to be made available to the manufacturers. The short-circuit ratings are an important parameter and must be calculated accurately, see Chapters 10 and 11. Electrical power systems invariably expand and a conservative margin in short-circuit ratings can save much higher costs of replacements when the short-circuit ratings of the circuit breakers become inadequate due to system growth. The capacitance current switching performance, TRV performance will require definite-purpose breakers, specified based upon the study results. Out-of-phase switching duty, single pole closing capability, limiting switching overvoltages, and synchronous and resistance switching will require specific calculations. These switching transients can be calculated based upon EMTP type programs [17], and are not discussed in this book.

8.14 Low-Voltage Circuit Breakers

The three classifications of low-voltage circuit breakers are (1) molded case circuit breakers (MCCB), (2) insulated case circuit breakers (ICCB), and (3) low-voltage power circuit breakers (LVPCB) [20–25].

8.14.1 Molded Case Circuit Breakers

In MCCBs, the current carrying parts, mechanism, and trip devices are completely contained in a molded-case insulating material and these breakers are not maintainable. Available frame sizes range from 15 to 6000 A, interrupting ratings from 10 to 100 kA symmetrical without current-limiting fuses and to 200 kA symmetrical with current-limiting fuses. These can be provided with electronic trip units, and have limited short-time delay

Application and Ratings of Circuit Breakers and Fuses according to ANSI Standards   359

and ground fault sensing capability. When provided with thermal magnetic trips, the trips may be adjustable or nonadjustable, and are instantaneous in nature. Motor circuit protectors (MCPs) may be classified as a special category of MCCBs and are provided with instantaneous trips only. MCPs do not have an interrupting rating by themselves and are tested in conjunction with motor starters. All MCCBs are fast enough to limit the amount of prospective current let-through and some are fast enough to be designated as current-limiting circuit breakers. For breakers claimed to be current limiting, peak current and I^2t are tabulated for the threshold of current-limiting action.

8.14.2 Insulated Case Circuit Breakers (ICCBs)

The ICCBs utilize characteristics of design from both the power and MCCBs, are not fast enough to qualify as current-limiting type, and are partially field maintainable. These can be provided with electronic trip units and have short-time ratings and ground fault sensing capabilities. These utilize stored energy mechanisms similar to LVPCB.

MCCBs and ICCBs are rated and tested according to UL 489 standard [24]. Both MCCBs and ICCBs are tested in the open air without enclosure and are designed to carry 100% of their current rating in open air. When housed in an enclosure there is 20% de-rating, though some models and frame sizes may be listed for application at 100% of their continuous current rating in an enclosure. MCCBs are fixed mounted in switchboards and bolted to bus bars. ICCBs can be fixed mounted or provided in draw-out design.

8.14.3 Low-Voltage Power Circuit Breakers (LVPCBs)

LVPCBs are rated and tested according to ANSI C37.13 [21, 22] and are used primarily in draw-out switchgear. These are the largest in physical size and are field maintainable. Electronic trip units are almost standard with these circuit breakers and these are available in frame sizes from 800 to 6000 A, interrupting ratings of 40–100 kA symmetrical without current-limiting fuses.

All three types of circuit breakers have different ratings, short-circuit test requirements, and applications. The short-circuit ratings and fault current calculation considerations are of interest here.

The symmetrical interrupting rating of the circuit breaker takes into account the initial current offset due to circuit X/R ratio. The value of the standard X/R ratio is used in the test circuit. For LVPCBs, this standard is $X/R = 6.6$, corresponding to a 15% power factor. Table 8.12 shows the multiplying factor (MF) for other X/R ratios. The recommended MFs for unfused circuit breakers are based on the highest peak current and can be calculated from

$$\mathrm{MF} = \frac{\sqrt{2}\left[1 + e^{-\pi/(X/R)}\right]}{2.29} \tag{8.55}$$

The MF for the fused breaker is based on the total rms current (asymmetrical) and is calculated from

$$\mathrm{MF} = \frac{\sqrt{1 + 2e^{-2\pi/(X/R)}}}{1.25} \tag{8.56}$$

TABLE 8.12

Multiplying Factors for Low-Voltage LVPCBs

System Short-Circuit Power Factor (%)	System X/R Ratio	Multiplying Factors for the Calculated Current	
		Unfused Circuit Breakers	Fused Circuit Breakers
20	4.9	1.00	1.00
15	6.6	1.00	1.07
12	8.27	1.04	1.12
10	9.95	1.07	1.15
8.5	11.72	1.09	1.18
7	14.25	1.11	1.21
5	20.0	1.14	1.26

Source: ANSI/IEEE Std. C37.13. Standard for Low-Voltage AC Power Circuit Breakers used in Enclosures. 2008.

In general, when X/R differs from the test power factor, the MF can be approximated by

$$\mathrm{MF} = \frac{1 + e^{-\pi(X/R)}}{1 + e^{-\pi/\tan\phi}} \tag{8.57}$$

where ϕ is the test power factor.

MCCBs and ICCBs are tested in the prospective fault test circuit according to UL 489 [24]. Power factor values for the test circuit are different from LVPCBs and are given in Table 8.13. If a circuit has an X/R ratio that is equal to or lower than the test circuit, no corrections to interrupting rating are required. If the X/R ratio is higher than the test circuit X/R ratio, the interrupting duty requirement for that application is increased by a MF from Table 8.14. The MF can be interpreted as a ratio of the offset peak of the calculated system peak (based on X/R ratio) to the test circuit offset peak.

While testing the breakers, the actual trip unit type installed during testing should be the one represented by referenced specifications and time-current curves. The short-circuit ratings may vary with different trip units, i.e., a short-time trip only (no instantaneous) may result in reduced short-circuit interrupting rating compared to testing with instantaneous trips. The trip units may be rms sensing or peak sensing, electronic or electromagnetic, and may include ground fault trips.

IEC standards do not directly correspond to the practices and standards in use in North America for single-pole duty, thermal response, and grounding. A direct comparison is not possible.

TABLE 8.13

Test Power Factors of MCCBs

Interrupting Rating (kA, rms Symmetrical)	Test Power Factor Range	X/R
10 or less	0.45–0.50	1.98–1.73
10–20	0.25–0.30	3.87–3.18
Over 20	0.15–0.20	6.6–4.9

Application and Ratings of Circuit Breakers and Fuses according to ANSI Standards 361

TABLE 8.14

Short-Circuit Multiplying Factors for MCCBs and ICCBs

Power Factor (%)	X/R Ratio	Interrupting Rating Multiplying Factor		
		10 kA or less	10–20 kA	>20 kA
5	19.97	1.59	1.35	1.22
6	16.64	1.57	1.33	1.20
7	14.25	1.55	1.31	1.18
8	12.46	1.53	1.29	1.16
9	11.07	1.51	1.28	1.15
10	9.95	1.49	1.26	1.13
13	7.63	1.43	1.21	1.09
15	6.59	1.39	1.18	1.06
17	5.80	1.36	1.15	1.04
20	4.90	1.31	1.11	1.00
25	3.87	1.24	1.05	1.00
30	3.18	1.18	1.00	1.00
35	2.68	1.13	1.00	1.00
40	2.29	1.08	1.00	1.00
50	1.98	1.04	1.00	1.00

8.14.3.1 Single-Pole Interrupting Capability

A single-pole interruption connects two breaker poles in series, and the maximum fault current interrupted is 87% of the full three-phase fault current. The interrupting duty is less severe as compared to a three-phase interruption test, where the first-pole-to-clear factor can be 1.5. Therefore, the three-phase tests indirectly prove the single-pole interrupting capability of three-pole circuit breakers. For the rated X/R, every three-pole circuit breaker intended for operation on a three-phase circuit can interrupt a bolted single-phase fault. LVPCBs are single-pole tested with maximum line-to-line voltage impressed across the single-pole and at the theoretical maximum single-phase fault current level of 87% of maximum three-phase bolted fault current. Generally, single-pole interrupting is not a consideration. Nevertheless, all MCCBs and ICCBs do not receive the same 87% test at full line-to-line voltage. In a corner grounded delta system (not much used in the industry), a single line-to-ground fault on the load side of the circuit breaker will result in single-phase fault current flowing through only one pole of the circuit breaker, but full line-to-line voltage impressed across that pole. A rare fault situation in ungrounded or high-resistance grounded systems can occur with two simultaneous bolted faults on the line side and load side of a circuit breaker and may require additional considerations. Some manufacturers market circuit breakers rated for a corner grounded system.

Thus, normally, the three-phase faults calculated at the point of application gives the maximum short-circuit currents on which the circuit breaker rating can be based, adjusted for fault point X/R. But in certain cases, a line-to-ground fault in solidly grounded system can exceed the three-phase symmetrical fault. Care needs to be exercised in such applications.

8.14.3.2 Short-Time Ratings

MCCBs, generally, do not have short-time ratings. These are designed to trip and interrupt high-level faults without intentional delays. When provided with electronic trip units, capabilities of these breakers are utilized for short-delay tripping. ICCBs do have some short-time capability, typically 15 cycles. Yet these are provided with high set instantaneous trips. LVPCBs are designed to have short-time capabilities, typically 30 cycles, and can withstand short-time duty cycle tests.

Short-time rating becomes of concern when two devices are to be coordinated in series and these see the same magnitude of fault current. If an upstream device has a short-time withstand capability, a slight delay in the settings can ensure coordination. This is an important concept from time-current coordination point of view, see Volume 4.

For an unfused LVPCB, the rated short-time current is the designated limit of prospective current at which it will be required to perform its short-time duty cycle of two periods of 0.5 s current flow separated by 15 s intervals of zero current at rated maximum voltage under prescribed test conditions. This current is expressed in rms symmetrical ampères. The unfused breakers will be capable of performing the short-time current duty cycle with all degrees of asymmetry produced by three-phase or single-phase circuits having a short-circuit power factor of 15% or greater. Fused circuit breakers do not have a short-time current rating, though the unfused circuit breaker elements have a short-time rating as described above.

8.14.3.3 Series Connected Ratings

Series connection of MCCBs or MCCBs and fuses permits a downstream circuit breaker to have an interrupting rating less than the calculated fault duty, and the current limiting characteristics of the upstream device "protects" the downstream lower-rated devices. Series combination is recognized for application by testing only. The upstream device is fully rated for the available short-circuit current and protects a downstream device, which is not fully rated for the available short-circuit current by virtue of its current-limiting characteristics. The series rating of the two circuit breakers makes it possible to apply the combination as a single device, the interrupting rating of the combination being that of the higher-rated device. As an example, a single upstream incoming breaker of 65 kA interrupting may protect a number of downstream feeder breakers of 25 kA interrupting and the complete assembly will be rated for 65 kA interrupting. The series rating should not be confused with cascading arrangement. IEC also uses this term for their series rated breakers [25]. A method of cascading that is erroneous and has been in use in the past is shown in Figure 8.23.

Consider a series combination of an upstream current-limiting fuse of 1200 A and a downstream MCCB. The available short-circuit current is 50 kA symmetrical, while the MCCB is rated for 25 kA. Figure 8.23 shows the let-through characteristics of the fuse. The required interrupting capability of the system, i.e., 50 kA is entered at the point A, and moving upwards the vertical line is terminated at the 1200 A fuse let-through characteristics. Moving horizontally, the point C is intercepted and then moving vertically down the point D is located. The symmetrical current given by D is read off, which in Figure 8.23 is 19 kA. As this current is less than the interrupting rating of the downstream device to be protected, the combination is considered safe. This method can lead to erroneous results, as the combination may not be able to withstand the peak let-through current given by point E in Figure 8.23 on the y axis. Calculations of series ratings is not permissible and these can only be established by testing.

A disadvantage of series combination is lack of selective co-ordination. On a high fault current magnitude, both the line-side and load-side circuit breakers will trip. A series

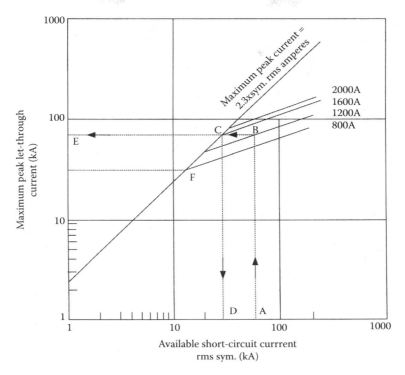

FIGURE 8.23
Let-through curves of current-limiting fuses.

combination should not be applied if motors or other loads that contribute to short-circuit current are connected between the line-side and load-side MCCBs. NEC (240.86 (c)) [26] specifies that series rating will not be used where

- Motors are connected on load side of higher-rated overcurrent device and on the line side of the lower-rated overcurrent device
- The sum of motor full load currents exceeds 1% of the interrupting rating of the lower-rated circuit breaker

8.15 Fuses

Fuses are fault sensing and interrupting devices, while circuit breakers must have protective relays as sensing devices before these can operate to clear short-circuit faults. Fuses are direct acting, single-phase devices, which respond to magnitude and duration of current. Electronically actuated fuses are a recent addition and these incorporate a control module that provides current sensing, electronically derived time-current characteristics, energy to initiate tripping, and an interrupting module that interrupts the current.

8.15.1 Current-Limiting Fuses

A current-limiting fuse is designed to reduce equipment damage by interrupting the rising fault current before it reaches its peak value. Within its current-limiting range, the fuse operates within ¼ to ½ cycle. The total clearing time consists of melting time, sometimes called the pre-arcing time and the arcing time. This is shown in Figure 8.24. The let-through current can be much lower than the prospective fault current peak and the rms symmetrical available current can be lower than the let-through current peak. The prospective fault current can be defined as the current that will be obtained if the fuse was replaced with a bolted link of zero impedance. By limiting the rising fault current, the I^2t let-through to the fault is reduced because of two counts: (1) high speed of fault clearance in ¼ cycle typically in the current-limiting range, and (2) fault current limitation. This reduces the fault damage.

Current-limiting fuses have a fusible element of nonhomogeneous cross section. It may be perforated or notched and while operating it first melts at the notches, because of reduced cross-sectional area. Each melted notch forms an arc that lengthens and disperses the element material into the surrounding medium. When it is melted by current in the specified current-limiting range, it abruptly introduces a high resistance to reduce the current magnitude and duration. It generates an internal arc voltage, much greater than the system voltage, to force the current to zero before the natural current zero crossing. Figure 8.25 shows the current interruption in a current-limiting fuse. Controlling the arcs in series controls the rate of rise of arc voltage and its magnitude. The arc voltages must be controlled to levels specified in the standards [28–30], i.e., for 15.5 kV fuses of 0.5–12 A, the maximum arc voltage is 70 kV peak, and for fuses greater than 12 A, the arc voltage is 49 kV peak.

The current-limiting action of a fuse becomes effective only at a certain magnitude of the fault current, called the critical current or threshold current. It can be defined as the first peak of a fully asymmetrical current wave at which the current-limiting fuse will melt. This can be determined by the fuse let-through characteristics and is given by the

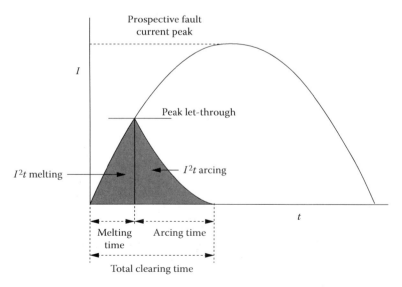

FIGURE 8.24
Current interruption by a current-limiting fuse.

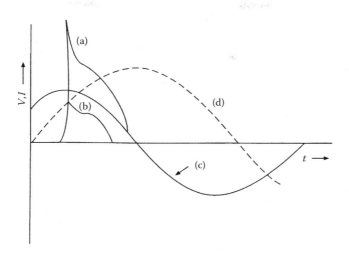

FIGURE 8.25
Arc voltage generated by a current-limiting fuse during interruption: (a) arc voltage, (b) interrupted current, (c) system voltage, and (d) perspective fault current.

inflection point on the curve where the peak let-through current begins to increase less steeply with increasing short-circuit current, i.e., point F in Figure 8.23 for a 800 A fuse. The higher the rated current of the fuse, the greater the value of the threshold current at which the current-limiting action starts.

8.15.2 Low-Voltage Fuses

Low-voltage fuses can be divided into two distinct classes: current-limiting type, and noncurrent-limiting type. The current-limiting fuses are types CC, T, K, G, J, L, and R. Noncurrent-limiting fuses, i.e., class H fuses, have a low interrupting rating of 10 kA, are not in much use in industrial power systems, and are being replaced with current-limiting fuses. Current-limiting fuses have interrupting capabilities up to 200 kA rms symmetrical. The various classes of current-limiting fuses are designed for specific applications, have different sizes and mounting dimensions, and are not interchangeable. As an example, classes J, RK1, and RK5 may be used for motor controllers, control transformers, and back-up protection. Class L (available in current ratings up to 6 kA) is commonly used as a current-limiting device in series rated circuits. Class T is a fast-acting fuse that may be applied to load-center, panel-board, and circuit-breaker backup protection. Also see Volume 4.

8.15.3 High-Voltage Fuses

High-voltage fuses can be divided into two distinct categories: distribution fuse cut-outs, and power fuses. Distribution cut-outs are meant for outdoor pole or cross arm mounting (except distribution oil cut-outs), have basic insulation levels (BILs) at distribution levels, and are primarily meant for distribution feeders and circuits. These are available in voltage ratings up to 34.5 kV. The interrupting ratings are relatively low, 5.00 kA rms symmetrical at 34.5 kV. The power fuses are adapted to station and substation mounting, have BILs at power levels and are meant primarily for applications in stations and substations. These are of two types: expulsion-type fuses and current-limiting fuses. Expulsion-type fuses can again be of two types: (1) fiber-lined fuses having voltage ratings up to 169 kV and (2) solid

TABLE 8.15

Short-Circuit Interrupting Ratings of High-Voltage Fuses

Fuse Type	Current Ratings	Nominal Voltage Rating in kV—Maximum Short-Circuit Interrupting Rating (kA rms Symmetrical)
Distributions fuse cut-outs	Up to 200 A	4.8–12.5, 7.2–15, 14.4–13.2, 25–8, 34.5–5
Solid-material boric acid fuses	Up to 300 A	17.0–14.0, 38–33.5, 48.3–31.5, 72.5–25, 121–10.5, 145–8.75
Current-limiting fuses	Up to 1350 A for 5.5 kV, up to 300 A for 15.5 kV, and 100 A for 25.8 and 38 kV	5.5–50, 15.5–50 (85 sometimes), 25.8–35, 38.0–35

boric acid fuses having voltage ratings up to 145 kV. The solid boric acid fuse can operate without objectionable noise or emission of flame and gases. High-voltage current-limiting fuses are available up to 38 kV, and these have comparatively much higher interrupting ratings. Table 8.15 shows comparative interrupting ratings of distribution cut-outs, solid boric acid, and current-limiting fuses. While the operating time of the current-limiting fuses is typically one-quarter of a cycle in the current-limiting range, the expulsion-type fuses will allow the maximum peak current to pass through and interrupt in more than one cycle. This can be a major consideration in some applications where a choice exists between the current-limiting and expulsion-type fuses.

Class E fuses are suitable for protection of voltage transformers, power transformers, and capacitor banks, while class R fuses are applied for medium-voltage motor starters. All class E fuses are not current limiting; E rating merely signifies that class E-rated power fuses in ratings of 100E or less will open in 300 s at currents between 200% and 240% of their E rating. Fuses rated above 100E open in 600 s at currents between 220% and 264% of their E ratings. Also see Volume 4.

8.15.4 Interrupting Ratings

The interrupting ratings relate to the maximum rms asymmetrical current available in the first half cycle after fault, which the fuse must interrupt under the specified conditions. The interrupting rating itself has no direct bearing on the current-limiting effect of the fuse. Currently, the rating is expressed in maximum rms symmetrical current and thus the fault current calculation based on an E/Z basis can be directly used to compare the calculated fault duties with the short-circuit ratings. Many power fuses and distribution cut-outs were earlier rated on the basis of maximum rms asymmetrical currents; rms asymmetrical rating represents the maximum current that the fuse has to interrupt because of its fast-acting characteristics. For power fuses, the rated asymmetrical capability is 1.6 times the symmetrical current rating. The asymmetrical rms factor can exceed 1.6 for high X/R ratios or low-power factors short-circuit currents. Figure 8.26 from Reference [28] relates rms multiplying factors and peak multiplying factors [29–30].

Note that the test X/R ratio is 25 only for expulsion-type and current-limiting type fuses [31]. For distribution class fuse, cut-outs interrupting tests (except current-limiting and open-link cut-outs) the minimum X/R ratio varies from 1.5 to 15 [28]. It is important to calculate the interrupting duty based upon the actual system X/R and apply proper adjustment factors.

A basic understanding of the ratings and problems of application of circuit breakers and fuses for short-circuit and switching duties can be gained from this chapter. The treatment is not exhaustive and an interested reader would like to explore further.

Application and Ratings of Circuit Breakers and Fuses according to ANSI Standards

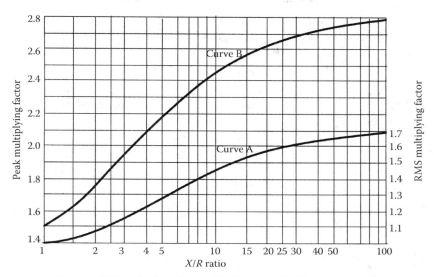

Curve A: RMS multiplying factor = RMS max sym/RMS sym
Curve B: Peak multiplying factor = Peak max sym/RMS sym

FIGURE 8.26
Relation of X/R to rms and peak multiplying factors. (From IEEE Std. C37.41 IEEE Standard Design Tests for High-Voltage (>1000 V) Fuses, Fuse and Disconnecting Cut-outs, Distribution Enclosed Single-Pole Air Switches, Fuse Disconnecting Switches, and Fuse Links and Accessories Used with these Devices, 2016.)

Problems

8.1. A 4.76 kV-rated breaker has a rated short-circuit current of 41 kA symmetrical and a K factor of 1.19. Without referring to tables calculate its (1) maximum symmetrical interrupting capability, (2) short-time current rating for 3 s, and (3) close and latch capability in asymmetrical rms and peak. If the breaker is applied at 4.16 kV, what is its interrupting capability and close and latch capability? How will these values change if the breaker is applied at 2.4 kV? Calculate similar ratings for 40 kA, $K = 1$ rated circuit breaker according to ANSI year 2000 revision.

8.2. The breaker of Problem 1 has a rated interrupting time of five cycles. What is its symmetrical and asymmetrical rating for phase faults, when applied at 4.16 and 2.4 kV, respectively?

8.3. A 15 kV circuit breaker applied at 13.8 kV has a rated short-circuit current of 28 kA rms, K factor = 1.3, and permissible tripping delay $Y = 2$ s. What is its permissible delay for a short-circuit current of 22 kA?

8.4. In Example 8.3, reduce all reactances by 10% and increase all capacitances by 10%. Calculate the inrush current and frequency on (1) isolated capacitor bank switching and (2) back-to-back switching. Find the value of reactor to be added to limit the inrush current magnitude and frequency to acceptable levels for a definite-purpose breaker.

8.5. Provide two examples of power system configurations, where TRV is likely to increase the standard values. Why can it be reduced by adding capacitors?

8.6. Plot the TRV characteristics of a 121 kV breaker, from the data in Table 8.4, at (1) a rated interrupting current of 40 kA and (2) a 50% short-circuit current. Also, plot the initial profile of TRV for a short-line fault.

8.7. Why is the initial TRV not of concern for gas-insulated substations?

8.8. A LVPCB, ICCB, and MCCB are similarly rated at 65 kA symmetrical interrupting. What other short-circuit rating is important for their application and protection coordination?

8.9. Each type of breaker in Problem 8.8 is subjected to a fault current of 50 kA, $X/R = 7.0$. Calculate the interrupting duty multiplying factors from the tables in this chapter.

8.10. What are the advantages and disadvantages of current-limiting fuses as compared to relayed circuit breakers for short-circuit interruption? How do these compare with expulsion-type fuses?

8.11. Explain the series interrupting ratings of two devices. What are the relative advantages and disadvantages of this configuration? Why should the series rating of two devices not be calculated?

8.12. Construct a table and show the major ratings of a 13.8 kV general purpose breaker, definite-purpose breaker for capacitor switching, definite-purpose breaker for TRV and a generator circuit breaker.

References

1. IEEE PC37.06/D11. Draft Standard AC High-Voltage Circuit Breakers Rated on Symmetrical Current Basis-Preferred Ratings and Related Required Capabilities for Voltages above 1000 volts, 2009.
2. IEC Std. 62271-100. Standard Test Procedures for AC High-Voltage Circuit Breakers Rated on Symmetrical Current Basis-Amendment 2 To Change Description of Transient Recovery Voltage for Harmonizing with IEC-2911.
3. ANSI Std. C37.010e. IEEE Application Guide for High-Voltage Circuit Breakers. Rated on Symmetrical Current Basis (Supplement to ANSI/IEEE C37.010-1979), 1985.
4. ANSI/IEEE Std. C37.5 Guide for Calculation of Fault Currents for Application of AC High-Voltage Circuit Breakers Rated on a Total Current Basis, 1979.
5. ANSI/IEEE Std. C37.010. Application Guide for AC High-Voltage Circuit Breakers Rated on a Symmetrical Current Basis, 1999 (R-2005).
6. ANS1/IEEE Std. C37.04. Rating Structure for AC High Voltage Circuit Breakers, 1999 (revision of 1979).
7. IEEE Std. PC37.04-1999/Cor 1/Draft B3 Draft IEEE Standard Rating Structure for AC High Voltage Circuit Breakers-Corrigendum 1, May 2009.
8. ANSI Std. C37.06. AC High-Voltage Circuit Breakers Rated on a Symmetrical Current Basis—Preferred Ratings and Related Capabilities, 2000 (revision of 1987).
9. IEEE Std. 551 (Violet Book), IEEE Recommended Practice for Calculating Short-Circuit Currents in Industrial and Commercial Power Systems, 2006.
10. P Kundur. *Power System Stability and Control*, McGraw-Hill, New York, 1993.
11. ANSI/IEEE Std. C37.012. Application Guide for Capacitance Current Switching for AC High-Voltage Circuit Breakers Rated on a Symmetrical Current Basis, 2005 (revision of 1979).

Application and Ratings of Circuit Breakers and Fuses according to ANSI Standards

12. IEEE Std. C37.04a. IEEE Standard Rating Structure for AC High-Voltage Circuit Breakers Rated on Symmetrical Current Basis, Amendment 1: Capacitance Current Switching, 2003.
13. IEEE Std. C37.09. Test Procedure for AC High-Voltage Circuit Breakers Rated on a Symmetrical Current Basis, 1999 (R2007).
14. IEEE Std. C37.09 Test Procedure for AC High-Voltage Circuit Breakers Rated on a Symmetrical Current Basis. Corrigendum 1, 2007.
15. IEEE Std. C37.011 Application Guide for Transient Recovery Voltage for AC High-Voltage Circuit Breakers, 2005.
16. JC Das. *Transients in Electrical Systems—Analysis, Recognition and Mitigation*, McGraw-Hill, New York, 2010.
17. IEEE Std. C37.06.1 Guide for High Voltage Circuit Breakers Rated on Symmetrical Current Basis Designated. Definite Purpose for Fast Transient Voltage Recovery Times, 2000.
18. IEEE Std. C37.013 IEEE Standard for AC High-Voltage Generator Circuit Breakers Rated on a Symmetrical Current Basis, 2008.
19. IEEE Std. C37.013a IEEE Standard for AC High-Voltage Generator Circuit Breakers Rated on a Symmetrical Current Basis Amendment 1: Supplement for Use with Generators Rated 10–100 MVA, 2007.
20. IEEE Std. 1015 Applying Low-Voltage Circuit Breakers Used in Industrial and Commercial Power Systems, 1997.
21. ANSI/IEEE Std. C37.13. Standard for Low-Voltage AC Power Circuit Breakers used in Enclosures, 2008.
22. IEEE Std. C37.13.1 IEEE Standard for Definite-Purpose Switching Devices for use in Metal-Enclosed Low-Voltage Power Circuit Breaker Switchgear, 2006.
23. NEMA. Molded Case Circuit Breakers and Molded Case Switches, 1993, Standard AB-1.
24. UL Std. 489. Molded Case Circuit Breakers and Circuit-Breaker Enclosures, 1991.
25. IEC Std. 60947-2. Low-voltage Switchgear and Control Gear—Part 2: Circuit Breakers, 2009.
26. NEC, National Electric Code, NFPA 70, 2014.
27. NEMA Standard SG2 High-Voltage Fuses, 1981.
28. IEEE Std. C37.41 IEEE Standard Design Tests for High-Voltage (>1000 V) Fuses, Fuse and Disconnecting Cut-outs, Distribution Enclosed Single-Pole Air Switches, Fuse Disconnecting Switches, and Fuse Links and Accessories Used with these Devices, 2016.
29. IEEE Std. C37.42 IEEE Standard Specifications for High-Voltage (>1000 V) Expulsion-Type Distribution-Class Fuses, Fuse and Disconnecting Cut-outs, Fuse Disconnecting Switches, and Fuse Links, and Accessories Used with These Devices, 2009.
30. IEEE Std. C37.46 IEEE Standard Specifications for High-Voltage Expulsion and Current Limiting Power Class Fuses and Fuse Disconnecting Switches, 2010.
31. IEEE Std. C37.47. IEEE Standard for High-Voltage Distribution Current-Limiting Fuses and Fuse Disconnecting Switches, 2011.

9

Short Circuit of Synchronous and Induction Machines and Converters

A three-phase short circuit on the terminals of a generator has twofold effects. One, large disruptive forces are brought into play in the machine itself and the machine should be designed to withstand these forces. Two, short circuits should be removed quickly to limit fault damage and improve stability of the interconnected systems. The circuit breakers for generator application sense a fault current of high asymmetry and must be rated to interrupt successfully the short-circuit currents. This is discussed in Chapters 7 and 8.

According to NEMA [1] specifications, a synchronous machine shall be capable of withstanding, without injury, a 30-s, three-phase short circuit at its terminals when operating at rated kVA and power factor, at 5% overvoltage, with fixed excitation. With a voltage regulator in service, the allowable duration t, in seconds, is determined from the following equation, where the regulator is designed to provide a ceiling voltage continuously during a short circuit:

$$t = \left(\frac{\text{Norminal field voltage}}{\text{Exciter ceiling voltage}} \right)^2 \times 30\,\text{s} \tag{9.1}$$

The generator should also be capable of withstanding, without injury, *any other* short circuit at its terminals for 30 s provided that

$$I_2^2 t \leq 40 \text{ for salient-pole machines} \tag{9.2}$$

$$I_2^2 t \leq 30 \text{ for air-cooled cylindrical rotor machines} \tag{9.3}$$

and the maximum current is limited by external means so as not to exceed the three-phase fault; I_2 is the negative sequence current due to unsymmetrical faults.

Synchronous generators are major sources of short-circuit currents in power systems. The fault current depends on the following:

1. The instant at which the short circuit occurs
2. The load and excitation of the machine immediately before the short circuit
3. The type of short circuits, i.e., whether three phases or one or more than one phase and ground are involved
4. Constructional features of the machine, especially leakage and damping
5. The interconnecting impedances between generators

An insight into the physical behavior of the machine during a short circuit can be made by considering the theorem of constant flux linkages. For a closed circuit with resistance

371

r and inductance L, $ri + Ldi/dt$ must be zero. If resistance is neglected, $L\,di/dt = 0$, i.e., the flux linkage Li must remain constant. In a generator, the resistance is small in comparison with the inductance, the field winding is closed on the exciter, and the stator winding is closed due to the short circuit. During the initial couple of cycles following a short circuit, the flux linkages with these two windings must remain constant. On a terminal fault, the generated electromagnetic force (EMF) acts on a closed circuit of stator windings and is analogous to an EMF being suddenly applied to an inductive circuit. Dynamically, the situation is more complex, i.e., the lagging stator current has a demagnetizing effect on the field flux, and there are time constants associated with the penetration of the stator flux and decay of short-circuit current.

9.1 Reactances of a Synchronous Machine

The following definitions are applicable.

9.1.1 Leakage Reactance X_l

The leakage reactance can be defined but cannot be tested. It is the reactance due to flux setup by armature windings, but not crossing the air gap. It can be divided into end-winding leakage and slot leakage. A convenient way of picturing the reactances is to view these in terms of permeances of various magnetic paths in the machine, which are functions of dimensions of iron and copper circuits and independent of the flux density or the current loading. The permeances thus calculated can be multiplied by a factor to consider the flux density and current. For example, the leakage reactance is mainly given by the slot permeance and the end-coil permeance.

9.1.2 Subtransient Reactance X_d''

Subtransient reactance equals the leakage reactance plus the reactance due to the flux setup by stator currents crossing the air gap and penetrating the rotor as far as the damper windings in a laminated pole machine, or as far as the surface damping currents in a solid pole machine. The subtransient conditions last for 1–5 cycles on a 60 Hz basis.

9.1.3 Transient Reactance X_d'

Transient reactance is the reactance after all damping currents in the rotor surface or amortisseur windings have decayed, but before the damping currents in the field winding have decayed. The transient reactance equals the leakage reactance plus the reactance due to flux setup by the armature, which penetrates the rotor to the field windings. Transient conditions last for 5–200 cycles on a 60 Hz basis.

9.1.4 Synchronous Reactance X_d

Synchronous reactance is the steady-state reactance after all damping currents in the field windings have decayed. It is the sum of leakage reactance and a fictitious armature reaction reactance, which is much larger than the leakage reactance. Ignoring resistance, the synchronous reactance per unit is the ratio of voltage per unit on an open circuit divided by per unit armature current on a short circuit for a given field excitation. This gives *saturated* synchronous reactance. The unsaturated value of the synchronous reactance is given by the voltage per unit on air gap open-circuit line divided by per unit armature current on short circuit. If 0.5 per unit field excitation produces full-load armature current on short circuit, the saturated synchronous reactance is 2.0 per unit. The saturated value may be only 60%–80% of the unsaturated value.

9.1.5 Quadrature Axis Reactances X_q'', X_q', and X_q

Quadrature axis reactances are similar to direct axis reactances, except that they involve the rotor permeance encountered when the stator flux enters one pole tip, crosses the pole, and leaves the other pole tip. The direct axis permeance is encountered by the flux crossing the air gap to the center of one pole, then crossing from one pole to the other pole, and entering the stator from that pole. Figure 9.1 shows the armature reaction components. The total armature reaction F_a can be divided into two components, F_{ad} and F_{aq}; F_{ad} is directed across the direct axis and F_{aq} across the quadrature axis. As these magnetomotive forces (MMFs) act on circuits of different permeances, the flux produced varies. If damper windings across pole faces are connected, X_q'' is nearly equal to X_d''.

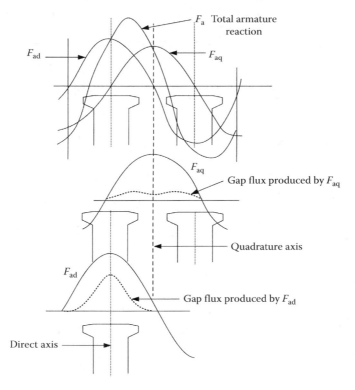

FIGURE 9.1
Armature reaction components in a synchronous machine in the direct and quadrature axes.

9.1.6 Negative Sequence Reactance X_2

The negative sequence reactance is the reactance encountered by a voltage of reverse-phase sequence applied to the stator, with the machine running. Negative sequence flux revolves opposite to the rotor and is at twice the system frequency. Negative sequence reactance is practically equal to the subtransient reactance as the damping currents in the damper windings or solid pole rotor surface prevent the flux from penetrating further. The negative sequence reactance is generally taken as the average of subtransient direct axis and quadrature axis reactances (Equation 4.60).

9.1.7 Zero Sequence Reactance X_0

The zero sequence reactance is the reactance effective when rated frequency currents enter all three terminals of the machine simultaneously and leave at the neutral of the machine. It is approximately equal to the leakage reactance of the machine with full-pitch coils. With two-thirds pitch stator coils, the zero sequence reactance will be a fraction of the leakage reactance.

9.1.8 Potier Reactance X_p

Potier reactance is a reactance with numerical value between transient and subtransient reactances. It is used for the calculation of field current when open circuit and zero power factor curves are available. Triangle ABS in Figure 9.2 is a Potier triangle. As a result of the different slopes of open circuit and zero power factor curves, A' B' in Figure 9.2 is slightly larger than AB and the value of reactance obtained from it is known as the Potier reactance.

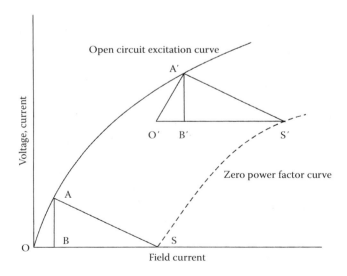

FIGURE 9.2
Open circuit, zero power factor curves, and Potier triangle and reactance.

Short Circuit of Synchronous and Induction Machines and Converters 375

9.2 Saturation of Reactances

Saturation varies with voltage, current, and power factor. For short-circuit calculations, according to ANSI/IEEE methods described in Chapter 7, saturated subtransient reactance must be considered. The saturation factor is usually applied to transient and synchronous reactances, though all other reactances change slightly with saturation. The saturated and unsaturated synchronous reactances are already defined above. In a typical machine, transient reactances may reduce from 5% to 25% on saturation. Saturated reactance is sometimes called the rated voltage reactance and is denoted by subscript "v" added to the "d" and "q" subscript axes, i.e., X''_{dv} and X_{qv} denote saturated subtransient reactances in direct and quadrature axes, respectively.

9.3 Time Constants of Synchronous Machines

9.3.1 Open-Circuit Time Constant T'_{do}

The open-circuit time constant expresses the rate of decay or buildup of field current when the stator is open-circuited and there is zero resistance in the field circuit.

9.3.2 Subtransient Short-Circuit Time Constant T''_d

The subtransient short-circuit time constant expresses the rate of decay of the subtransient component of current under a bolted (zero resistance), three-phase short circuit at the machine terminals.

9.3.3 Transient Short-Circuit Time Constant T'_d

The transient short-circuit time constant expresses the rate of decay of the transient component of the current under a bolted (zero resistance), three-phase short circuit at the machine terminals.

9.3.4 Armature Time Constant T_a

The armature time constant expresses the rate of decay of the DC component of the short-circuit current under the same conditions.

Table 9.1 shows electrical data, reactances, and time constants of a 13.8 kV, 112.1 MVA at 0.85 power factor generator.

9.4 Synchronous Machine Behavior on Short Circuit

The time immediately after a short circuit can be divided into three periods:

- The subtransient period lasting from 1 to 5 cycles.
- The transient period may last up to 100 cycles or more.
- The final or steady-state period. Normally, the generator will be removed from service by protective relaying, much before the steady-state period is reached.

TABLE 9.1

Generator Data

Description	Symbol	Data
Generator		
112.1 MVA, 2-pole, 13.8 kV, 0.85 PF, 95.285 MW, 4690 A, SCR, 235 field V, wye connected	0.56	
Per Unit Reactance Data, Direct Axis		
Saturated synchronous	X_{dv}	1.949
Unsaturated synchronous	X_d	1.949
Saturated transient	X'_{dv}	0.207
Unsaturated transient	X'_d	0.278
Saturated subtransient	X''_{dv}	0.164
Unsaturated subtransient	X''_d	0.193
Saturated negative sequence	X_{2v}	0.137
Unsaturated negative sequence	X_{2I}	0.185
Saturated zero sequence	X_{0v}	0.092
Leakage reactance, overexcited	X_{0I}	0.111
Leakage reactance, under excited	$X_{LM,OXE}$	0.164
	$X_{LM,UEX}$	0.164
Per Unit Reactance Data, Quadrature Axis		
Saturated synchronous	X_{qv}	1.858
Unsaturated synchronous	X_q	1.858
Unsaturated transient	X'_q	0.434
Saturated subtransient	X''_{qv}	0.140
Unsaturated subtransient	X''_q	0.192
Field Time Constant Data, Direct Axis		
Open circuit	T'_{d0}	5.615
Three-phase short-circuit transient	T'_{d3}	0.597
Line-to-line short-circuit transient	T'_{d2}	0.927
Line-to-neutral short-circuit transient	T'_{d1}	1.124
Short-circuit subtransient	T''_d	0.015
Open-circuit subtransient	T''_{d0}	0.022
Field Time Constant Data Quadrature Axis		
Open circuit	T'_{q0}	0.451
Three-phase short-circuit transient	T'_q	0.451
Short-circuit subtransient	T''_q	0.015
Open-circuit subtransient	T''_{q0}	0.046
Armature dc Component Time Constant data		
Three-phase short circuit	T_{a3}	0.330
Line-to-line short circuit	T_{a2}	0.330
Line-to-neutral short circuit	T_{a1}	0.294

Short Circuit of Synchronous and Induction Machines and Converters

In the subtransient period, the conditions can be represented by the flux linking the stator and rotor windings. Any sudden change in the load or power factor of a generator produces changes in the MMFs, both in direct and quadrature axes. A terminal three-phase short circuit is a large disturbance. At the moment of short circuit, the flux linking the stator from the rotor is trapped to the stator, giving a stationary replica of the main-pole flux. The rotor poles may be in a position of maximum or minimum flux linkage, and as these rotate, the flux linkages tend to change. This is counteracted by a current in the stator windings. The short-circuit current is, therefore, dependent on rotor angle. As the energy stored can be considered as a function of armature and field linkages, the torque fluctuates and reverses cyclically. The DC component giving rise to asymmetry is caused by the flux trapped in the stator windings at the instant of short circuit, which sets up a DC transient in the armature circuit. This DC component establishes a component field in the air gap, which is stationary in space, and therefore induces a fundamental frequency voltage and current in the synchronously revolving rotor circuits. Thus, an increase in the stator current is followed by an increase in the field current. The field flux has superimposed on it a new flux pulsating with respect to field windings at normal machine frequency. The single-phase induced current in the field can be resolved into two components, one stationary with respect to the stator which counteracts the DC component of the stator current, and the other component travels at twice the synchronous speed with respect to the stator and induces a second harmonic in it.

The armature and field are linked through the magnetic circuit, and the AC component of lagging current creates a demagnetizing effect. However, some time must elapse before it starts becoming effective in reducing the field current and reaching the steady-state current. The protective relays will normally operate to open the generator breaker and simultaneously the field circuit for suppression of generated voltage.

The above is rather an oversimplification of the transient phenomena in the machine on short circuit. In practice, a generator will be connected in an interconnected system. The machine terminal voltage, rotor angle, and frequency all change depending on the location of the fault in the network, the network impedance, and the machine parameters. The machine output power will be affected by the change in the rotor winding EMF and the rotor position in addition to any changes in the impedance seen at the machine terminals. For a terminal fault, the voltage at the machine terminals is zero, and therefore power supplied by the machine to load reduces to zero, while the prime mover output cannot change suddenly. Thus, the generator accelerates. In a multimachine system, with interconnecting impedances, the speed of all the machines changes, so that they generate their share of synchronizing power in the overall impact, as these strive to reach a mean retardation through oscillations due to stored energy in the rotating masses.

In a dynamic simulation of a short circuit, the following may be considered:

- Network before, during, and after the short circuit
- Induction motors' dynamic modeling, with zero excitation
- Synchronous machine dynamic modeling, considering saturation
- Modeling of excitation systems
- Turbine and governor models

Figure 9.3 shows the transients in an interconnected system on a three-phase short circuit lasting for 5 cycles. Figure 9.3a shows the torque angle swings of two generators, which are stable after the fault; Figure 9.3b shows speed transients; and Figure 9.3c shows the field voltage response of a high-response excitation system. A system having an excitation

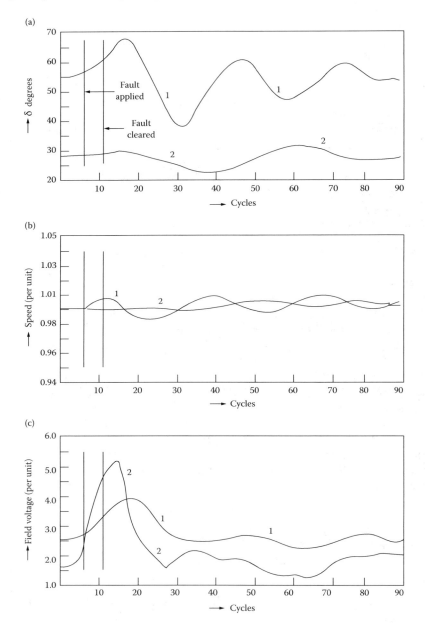

FIGURE 9.3
Transient behavior of two generators in an interconnected system for a three-phase fault cleared in 5 cycles: (a) torque angle swings, (b) speed transients, (c) field voltage, (d) reactive and active power swings, and (e) voltage dip and recovery profile.

(Continued)

Short Circuit of Synchronous and Induction Machines and Converters

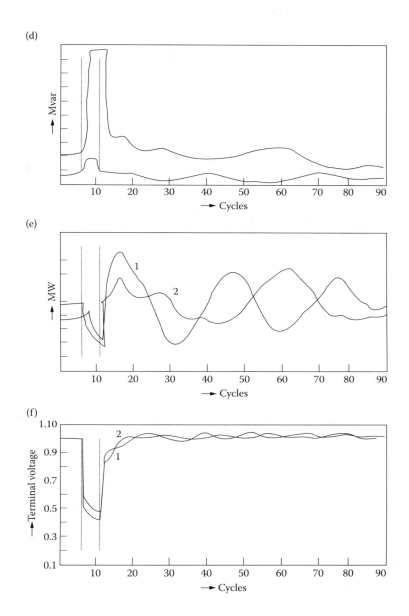

FIGURE 9.3 (CONTINUED)
Transient behavior of two generators in an interconnected system for a three-phase fault cleared in 5 cycles: (a) torque angle swings, (b) speed transients, (c) field voltage, (d) and (e) reactive and active power swings, and (f) voltage dip and recovery profile.

system voltage response of 0.1 s or less is defined as the high-response excitation system [2]. The excitation systems may not affect the first cycle momentary currents but are of consideration for interrupting duty and 30 cycle currents. The reactive and active power transients are shown in Figure 9.3d and e, respectively. The voltage dip and recovery characteristics are shown in Figure 9.3f. A fault voltage dip of more than 60% occurs. Though the generators are stable after the fault removal, the large voltage dip can precipitate shutdown of consumer loads, i.e., the magnetic contactors in motor controllers can drop out during the first-cycle voltage dip. This is of major consideration in continuous process plants. Figure 9.3 is based on system transient stability study. Transient analysis programs such as Electromagnetic Transients Program (EMTP) [3] can be used for dynamic simulation of the short-circuit currents.

For practical calculations, the dynamic simulations of short-circuit currents are rarely carried out. The generator is replaced with an equivalent circuit of voltage and certain impedances intended to represent the worst conditions, after the fault (Chapter 7). The speed change is ignored. The excitation is assumed to be constant, and the generator load is ignored. That this procedure is safe and conservative has been established in the industry by years of applications, testing, and experience.

9.4.1 Equivalent Circuits during Fault

Figure 9.4 shows an envelope of decaying AC component of the short-circuit current wave, neglecting the DC component. The extrapolation of the current envelope to zero time gives the peak current. Note that, immediately after the fault, the current decays rapidly and then more slowly.

Transformer equivalent circuits of a salient pole synchronous machine in the direct and quadrature axis at the instant of short circuit and during subsequent time delays helps to derive the short-circuit current equations and explain the decaying AC component of

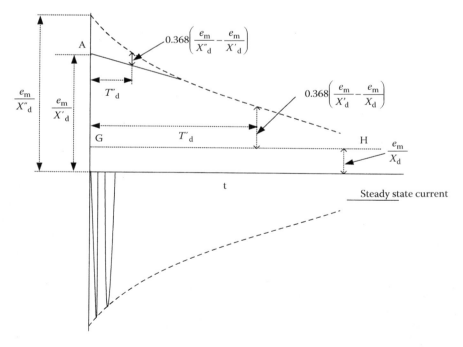

FIGURE 9.4
Decaying AC component of the short-circuit current, and subtransient, transient, and steady-state currents.

Figure 9.4. Based on the above discussions, these circuits are shown in Figure 9.5 in the subtransient, transient, and steady-state periods. As the flux penetrates into the rotor, and the currents die down, it is equivalent to opening a circuit element, i.e., from subtransient to transient state, the damper circuits are opened.

The direct axis subtransient reactance is given by

$$X_d'' = X_l + \frac{1}{1/X_{ad} + 1/X_f + 1/X_{kD}} \tag{9.4}$$

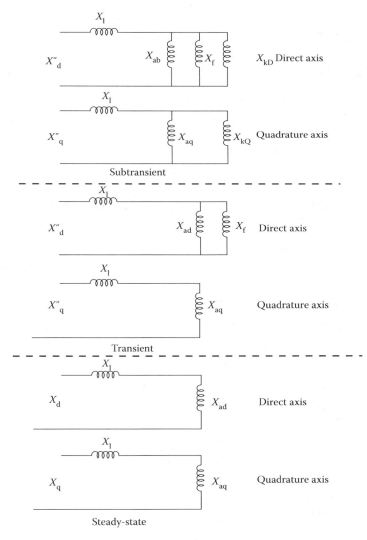

FIGURE 9.5
Equivalent transformer circuits of a synchronous generator during subtransient, transient, and steady-state periods, after a terminal fault.

where X_{ad} is the reactance corresponding to the fundamental space wave of the armature in the direct axis, X_f is the reactance of the field windings, X_{kD} is the reactance of the damper windings in the direct axis, and X_1 is the leakage reactance. X_f and X_{kD} are also akin to leakage reactances. Similarly, the quadrature axis subtransient reactance is given by

$$X_q'' = X_1 + \frac{1}{1/X_{aq} + 1/X_{kQ}} \tag{9.5}$$

where X_{aq} is the reactance corresponding to the fundamental space wave in the quadrature axis, X_{kQ} is the reactance of the damper winding in the quadrature axis, and X_1 is identical in the direct and quadrature axes. The quadrature axis rotor circuit does not carry a field winding, and this circuit is composed of damper bars or rotor iron in the interpolar axis of a cylindrical rotor machine.

The direct axis and quadrature axis short-circuit time constants associated with decay of the subtransient component of the current are

$$T_d'' = \frac{1}{\omega r_D} \left[\frac{X_{ad}X_fX_1}{X_{ad}X_f + X_fX_1 + X_{ad}X_1} + X_{kD} \right] \tag{9.6}$$

$$T_q'' = \frac{1}{\omega r_Q} \left[\frac{X_{aq}X_1}{X_{aq} + X_1} + X_{kQ} \right] \tag{9.7}$$

where r_D and r_Q are resistances of the damper windings in the direct and quadrature axis, respectively. When the flux has penetrated the air gap, the effect of the eddy currents in the pole face ceases after a few cycles given by short-circuit subtransient time constants. The resistance of the damper circuit is much higher than that of the field windings. This amounts to opening of the damper winding circuits, and the direct axis and quadrature axis transient reactances are given by

$$X_d' = X_1 + \frac{1}{1/X_{ad} + 1/X_f} = \left[\frac{X_{ad}X_f}{X_{ad} + X_f} + X_1 \right] \tag{9.8}$$

$$X_q' = X_1 + X_{aq} \tag{9.9}$$

The direct axis transient time constant associated with this decay is

$$T_d' = \frac{1}{\omega r_F} \left[\frac{X_{ad}X_1}{X_{ad} + X_1} + X_f \right] \tag{9.10}$$

where r_F is the resistance of the field windings.

Finally, when the currents in the field winding have also died down, given by the transient short-circuit time constant, the steady-state short-circuit current is given by the synchronous reactance:

$$X_d = X_1 + X_{ad} \tag{9.11}$$

$$X_q = X_1 + X_{aq} \tag{9.12}$$

Short Circuit of Synchronous and Induction Machines and Converters   383

Equations 9.9 and 9.12 show that X'_q is equal to X_q. The relative values of X''_q, X'_q, and X_q depend on machine construction. For cylindrical rotor machines, $X_q \gg X'_q$. Sometimes one or two damper windings are modeled in the q axis.

Reverting to Figure 9.4, the direct axis transient reactance determines the initial value of the symmetrical transient envelope and the direct axis short-circuit time constant T'_d determines the decay of this envelope. The direct axis time constant T'_d is the time required by the transient envelope to decay to a point where the difference between it and the steady-state envelope GH is $1/e$ (which is equal to 0.368) of the initial difference GA. A similar explanation applies to decay of the subtransient component in Figure 9.4.

9.4.2 Fault Decrement Curve

Based on Figure 9.4, the expression for a decaying AC component of the short-circuit current of a generator can be written as

i_{ac} = Decaying subtransient component + decaying transient component + steady-state component

$$= (i''_d - i'_d)e^{-t/T''_d} + (i'_d - i_d)e^{-t/T'_d} + i_d \tag{9.13}$$

The subtransient current is given by

$$i''_d = \frac{E''}{X''_d} \tag{9.14}$$

where E'' is the generator internal voltage behind subtransient reactance:

$$E'' = V_a + X''_d \sin\phi \tag{9.15}$$

where V_a is the generator terminal voltage and ϕ is the load power factor angle, prior to fault.

Similarly, the transient component of the current is given by

$$i'_d = \frac{E'}{X'_d} \tag{9.16}$$

where E' is the generator internal voltage behind transient reactance:

$$E' = V_a + X'_d \sin\phi \tag{9.17}$$

The steady-state component is given by

$$i_d = \frac{V_a}{X_d}\left(\frac{i_F}{i_{Fg}}\right) \tag{9.18}$$

where i_F is the field current at given load conditions (when regulator action is taken into account) and i_{Fg} is the field current at no-load rated voltage.

The DC component is given by

$$i_{dc} = \sqrt{2}\, i_d'' e^{-t/T_a} \tag{9.19}$$

where T_a is the armature short-circuit time constant, given by

$$T_a = \frac{1}{\omega r}\left[\frac{2X_d'' X_q''}{X_d'' + X_q''}\right] \tag{9.20}$$

where r is the stator resistance.

The open-circuit time constant describes the decay of the field transient; the field circuit is closed and the armature circuit is open:

$$T_{do}'' = \frac{1}{\omega r_D}\left[\frac{X_{ad} X_f}{X_{ad} + X_f} + X_{kD}\right] \tag{9.21}$$

and the quadrature axis subtransient open-circuit time constant is

$$T_{qo}'' = \frac{1}{\omega r_Q}\left(X_{aq} + X_{kQ}\right) \tag{9.22}$$

The open-circuit direct axis transient time constant is

$$T_{do}' = \frac{1}{\omega r_F}\left(X_{ad} + X_f\right) \tag{9.23}$$

The short-circuit direct axis transient time constant can be expressed as

$$T_d' = T_{do}\left[\frac{X_d'}{(X_{ad} + X_l)}\right] = T_{do}'\frac{X_d'}{X_d} \tag{9.24}$$

It may be observed that the resistances have been neglected in the above expressions. In fact these can be included, i.e., the subtransient current is

$$i_d'' = \frac{E''}{r_D + X_d''} \tag{9.25}$$

where r_D is defined as the resistance of the armortisseur windings on salient pole machines and analogous body of cylindrical rotor machines. Similarly, the transient current is

$$i_d'' = \frac{E'}{r_f + X_d'} \tag{9.26}$$

Short Circuit of Synchronous and Induction Machines and Converters

Example 9.1

Consider a 13.8 kV, 100 MVA at 0.85 power factor generator. Its rated full-load current is 4184 A. Other data are

Saturated subtransient reactance X_{dv}''	=	0.15 per unit
Saturated transient reactance X_{dv}'	=	0.2 per unit
Synchronous reactance X_d	=	2.0 per unit
Field current at rated load i_f	=	3 per unit
Field current at no-load rated voltage i_{fg}	=	1 per unit
Subtransient short-circuit time constant T_d''	=	0.012 s
Transient short-circuit time constant T_d'	=	0.35 s
Armature short-circuit time constant T_a	=	0.15 s
Effective resistance*	=	0.0012 per unit
Quadrature axis synchronous reactance*	=	1.8 per unit

A three-phase short circuit occurs at the terminals of the generator, when it is operating at its rated load and power factor. It is required to construct a fault decrement curve of the generator for (1) the AC component, (2) DC component, and (3) total current. Data marked with an asterisk are intended for Example 9.5.

From Equation 9.15, the voltage behind subtransient reactance at the generator rated voltage, load, and power factor is

$$E'' = V + X_d'' \sin\phi = 1 + (0.15)(0.527) = 1.079 \text{ per unit}$$

From Equation 9.14, the subtransient component of the current is

$$i_d'' = \frac{E''}{X_{dv}''} = \frac{1.079}{0.15} \text{ per unit} = 30.10 \text{ kA}$$

Similarly, from Equation 9.17, E', the voltage behind transient reactance is 1.1054 per unit, and from Equation 9.16, the transient component of the current is 23.12 kA.

From Equation 9.18, current i_d at constant excitation is 2.09 kA rms. For a ratio of $i_f / i_{Fg} = 3$, current $i_d = 6.28$ kA rms. Therefore, the following equations can be written for the AC component of the current:

With constant excitation:

$$i_{ac} = 6.98 e^{-t/0.012} + 20.03 e^{-t/0.35} + 2.09 \text{ kA}$$

With full load excitation:

$$i_{ac} = 6.98 e^{-t/0.012} + 16.84 e^{-t/0.35} + 6.28 \text{ kA}$$

The AC decaying component of the current can be plotted from these two equations, with the lowest value of $t = 0.01$ to $t = 1000$ s. This is shown in Figure 9.6. The DC component is given by Equation 9.19:

$$i_{dc} = \sqrt{2} i_d'' e^{-t/T_a} = 42.57 e^{-t/0.15} \text{ kA}$$

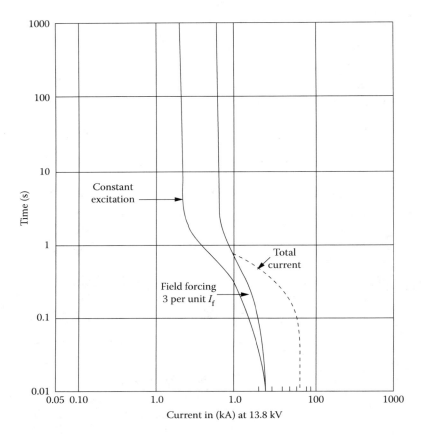

FIGURE 9.6
Calculated fault decrement curves of generator (Example 6.1).

This is also shown in Figure 9.6. At any instant, the total current is

$$\sqrt{i_{ac}^2 + i_{dc}^2}\ \text{kA rms}$$

The fault decrement curves are shown in Figure 9.6. Short-circuit current with constant excitation is 50% of the generator full-load current. This can occur for a stuck voltage regulator condition. Though this current is lower than the generator full-load current, it cannot be allowed to be sustained. Voltage restraint or voltage-controlled overcurrent generator backup relays (ANSI/IEEE device number 51 V) or distance relays (device 21) are set to pick up on this current. The generator fault decrement curve is often required for appropriate setting and coordination of these relays with the system relays.

9.5 Circuit Equations of Unit Machines

The behavior of machines can be analyzed in terms of circuit theory, which makes it useful not only for steady-state performance but also for transients like short circuits. The circuit of machines can be simplified in terms of coils on the stationary (stator) and rotating (rotor) parts and these coils interact with each other according to fundamental electromagnetic laws. The circuit of a unit machine can be derived from consideration of generation of EMF in coupled coils due to (1) transformer EMF, also called pulsation EMF, and (2) the EMF of rotation.

Consider two magnetically coupled, stationary, coaxial coils as shown in Figure 9.7. Let the applied voltages be v_1 and v_2 and the currents i_1 and i_2, respectively. This is, in fact, the circuit of a two-winding transformer, the primary and secondary being represented by single-turn coils. The current in the primary coil (any coil can be called a primary coil) sets up a total flux linkage Φ_{11}. Change of current in this coil induces an EMF given by

$$e_{11} = -\frac{d\Phi_{11}}{d_t} = -\frac{d\Phi_{11}}{di_1}\cdot\frac{di_1}{dt} = -L_{11}\frac{di_1}{dt} = -L_{11}pi_1 \tag{9.27}$$

where $L_{11} = -d\Phi_{11}/di_1$ is the total primary self-inductance and the operator $p = d/dt$. If Φ_1 is the leakage flux and Φ_{12} is the flux linking with the secondary coil, then the variation of current in the primary coil induces in the secondary coil an EMF:

$$e_{12} = -\frac{d\Phi_{12}}{dt} = -\frac{d\Phi_{12}}{di_1}\cdot\frac{di_1}{dt} = -L_{12}\frac{di_1}{dt} = -L_{12}pi_1 \tag{9.28}$$

where $L_{12} = -d\Phi_{12}/di_1$ is the mutual inductance of the primary coil winding with the secondary coil winding. Similar equations apply for the secondary coil winding. All the flux produced by the primary coil winding does not link with the secondary. The leakage inductance associated with the windings can be accounted for as

$$L_{11} = L_{12} + L_1 \tag{9.29}$$

$$L_{22} = L_{21} + L \tag{9.30}$$

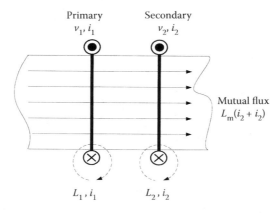

FIGURE 9.7
Representation of magnetically coupled coils of a two-winding transformer.

The mutual inductance between coils can be written as

$$L_{12} = L_{21} = L_m = \sqrt{[(L_{11} - L_1)(L_{22} - L_2)]} = k\sqrt{L_{11}L_{22}} \qquad (9.31)$$

Thus, the equations of a unit transformer are

$$v_1 = r_1 i_1 + (L_m + L_1)pi_1 + L_m pi_2$$
$$v_2 = r_2 i_2 + (L_m + L_2)pi_2 + L_m pi_1$$

Or in the matrix form:

$$\begin{vmatrix} v_1 \\ v_2 \end{vmatrix} = \begin{vmatrix} r_1 + (L_1 + L_m) & L_m p \\ L_m p & r_2 + (L_2 + L_m) \end{vmatrix} \begin{vmatrix} i_1 \\ i_2 \end{vmatrix} \qquad (9.32)$$

If the magnetic axes of the coupled coils are at right angles, no mutually induced pulsation or transformer EMF can be produced by variation of currents in either of the windings. However, if the coils are free to move, the coils with magnetic axes at right angles have an EMF of rotation, e_r, induced when the winding it represents rotates:

$$e_r = \omega_r \Phi \qquad (9.33)$$

where ω_r is the angular speed of rotation and Φ is the flux. This EMF is maximum when the two coils are at right angles to each other and zero when these are cophasal.

To summarize, a pulsation EMF is developed in two coaxial coils and there is no rotational EMF. Conversely, a rotational EMF is developed in two coils at right angles, but no pulsation EMF. If the relative motion is at an angle θ, the EMF of rotation is multiplied by $\sin \theta$:

$$e_r = \omega_r \Phi \sin \theta \qquad (9.34)$$

The equations of a unit machine may be constructed based on the above simple derivation of EMF production in coils. Consider a machine with direct and quadrature axis coils as shown in Figure 9.8. Note that the armature is shown rotating and has two coils D and Q at right angles in the d–q axes. The field winding F and the damper winding KD are shown stationary in the direct axis. All coils are single turn coils. In the direct axis, there are three mutual inductances, i.e., of D with KD, KD with F, and F with D. A simplification is to consider these equal to inductance L_{ad}. Each coil has a leakage inductance of its own. Consequently, the total inductances of the coils are

$$\text{Coil D} : (l_d + L_{ad})$$

$$\text{Coil KD} : (l_{kD} + L_{ad}) \qquad (9.35)$$

$$\text{Coil F} : (l_f + L_{ad})$$

Short Circuit of Synchronous and Induction Machines and Converters

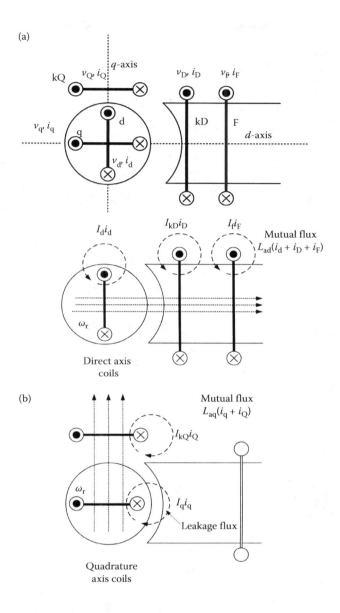

FIGURE 9.8
(a) Development of the circuit of a unit machine and (b) flux linkages in the direct and quadrature axes' coils.

The mutual linkage with armature coil D when all three coils carry currents is

$$\Phi_d = L_{ad}(i_F + i_D + i_d) \tag{9.36}$$

where i_F, i_D, and i_d are the currents in the field, damper, and direct axis coils. Similarly, in the quadrature axis:

$$\Phi_q = L_{aq}(i_q + i_Q) \tag{9.37}$$

The EMF equations in each of the coils can be written on the basis of these observations.

Field coil: no q-axis circuit will affect its flux, nor do any rotational voltages appear. The applied voltage v_f is given as

$$v_f = r_F i_F + (L_{ad} + l_f) p i_F + L_{ad} p i_D + L_{ad} p i_d \tag{9.38}$$

Stator coil KD is located similarly to coil F:

$$v_D = r_D i_D + (L_{ad} + l_{kD}) p i_D + L_{ad} p i_F + L_{ad} p i_d \tag{9.39}$$

Coil KQ has no rotational EMF but will be affected magnetically by any current i_q in coil Q:

$$v_Q = r_Q i_Q + (L_{aq} + l_{kQ}) p i_Q + L_{aq} p i_q \tag{9.40}$$

Armature coils D and Q have the additional property of inducing rotational EMF:

$$v_d = r_d i_d + (L_{ad} + l_d) p i_d + L_{ad} p i_F + L_{ad} p i_D + L_{aq} \omega_r i_Q + (L_{aq} + l_q) \omega_r i_q \tag{9.41}$$

$$v_q = r_q i_q + (L_{aq} + l_q) p i_q + L_{aq} p i_Q - L_{ad} \omega_r i_F - L_{ad} \omega_r i_D - (L_{ad} + l_d) \omega_r i_d \tag{9.42}$$

These equations can be written in a matrix form:

$$\begin{vmatrix} v_f \\ v_D \\ v_Q \\ v_d \\ v_q \end{vmatrix} = \begin{vmatrix} r_F + (L_{ad} + l_f)p & L_{ad} p & \cdot & L_{ad} p & \cdot \\ L_{ad} p & r_D + (L_{ad} + l_{kD})p & \cdot & L_{ad} p & \cdot \\ \cdot & \cdot & r_Q + (L_{aq} + l_{kQ})p & \cdot & L_{aq} p \\ L_{ad} p & L_{ad} p & L_{aq} \omega_r & r_d + (L_{ad} + l_d)p & (L_{aq} + l_q)\omega_r \\ -L_{ad}\omega_r & -L_{ad}\omega_r & L_{aq} p & -(L_{ad} + l_d)\omega_r & r_q + (L_{aq} + l_q)p \end{vmatrix} \begin{vmatrix} i_F \\ i_D \\ i_Q \\ i_d \\ i_q \end{vmatrix}$$

$$\tag{9.43}$$

9.6 Park's Transformation

Park's transformation [4,5] greatly simplifies the mathematical model of synchronous machines. It describes a new set of variables, such as currents, voltages, and flux linkages, obtained by transformation of the actual (stator) variables in three axes: 0, d, and q. The d and q axes are already defined, whereas the 0 axis is the stationary axis.

9.6.1 Reactance Matrix of a Synchronous Machine

Consider the normal construction of a three-phase synchronous machine, with three-phase stationary AC windings on the stator, and the field and damper windings on the rotor (Figure 9.9). The stator inductances vary, depending on the relative position of the stator and rotor. Consider that the field winding is cophasal with the direct axis and also the direct axis carries a damper winding. The q axis also has a damper winding. The phase

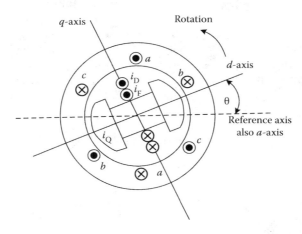

FIGURE 9.9
Representation of d–q axes, reference axes, and windings in a synchronous machine.

windings are distributed but are represented by single turn coils aa, bb, and cc as shown in Figure 9.9. The field flux is directed along the d axis, and therefore, the machine-generated voltage is at right angles to it, along the q axis. For generator action, the generated voltage vector E leads the terminal voltage vector V by an angle δ, and from basic machine theory, we know that δ is the torque angle. At $t = 0$, the voltage vector V is located along the axis of phase a, which is the reference axis as shown in Figure 9.9. The q axis is at an angle δ and the d axis is at an angle $\delta + \pi/2$. For $t > 0$, the reference axis is at an angle $\omega_r t$ with respect to the axis of phase a. The d axis of the rotor is, therefore, at

$$\theta = \omega_r t + \delta + \pi/2 \tag{9.44}$$

For synchronous operation, $\omega_r = \omega_0$ = constant.

Consider phase a inductance, it is a combination of its own self-inductance L_{aa}, and its mutual inductances L_{ab} and L_{bc} with phases b and c. All three inductances vary with the relative position of the rotor with respect to the stator because of saliency of the air gap. When the axis of phase a coincides with the direct axis (Figure 9.9), i.e., $\theta = 0$ or π, the resulting flux of coil aa is maximum in the horizontal direction and its self-inductance is maximum. When at right angles to the d axis, $\theta = \pi/2$ or $3\pi/2$ and its inductance is a minimum. Thus, L_{aa} fluctuates twice per revolution and can be expressed as

$$L_{aa} = L_s + L_m \cos 2\theta \tag{9.45}$$

Similarly, self-inductance of phase b is maximum at $\theta = 2\pi/3$ and that of phase c at $\theta = -2\pi/3$:

$$L_{bb} = L_s + L_m \cos 2\left(\theta - 2\frac{\pi}{3}\right) \tag{9.46}$$

$$L_{cc} = L_s + L_m \cos 2\left(\theta + 2\frac{\pi}{3}\right) \tag{9.47}$$

Phase-to-phase mutual inductances are also a function of θ; L_{ab} is negative and is maximum at $\theta = -\pi/6$. For the direction of currents shown in coils aa and bb, L_{ab} is negative. When the angle is $-\pi/3$, the current in the phase b coil generates the maximum flux, but the linkage with the phase a coil is better when the angle is zero degrees. However, at this angle the flux is reduced. The maximum flux linkage can be considered to take place at an angle, which is an average of these two angles, i.e., $\pi/6$. This can be explained as follows:

$$L_{ab} = -\left[M_s + L_m \cos 2\left(\theta + \frac{\pi}{6} \right) \right] \tag{9.48}$$

$$L_{bc} = -\left[M_s + L_m \cos 2\left(\theta - \frac{\pi}{2} \right) \right] \tag{9.49}$$

$$L_{ca} = -\left[M_s + L_m \cos 2\left(\theta + 5\frac{\pi}{6} \right) \right] \tag{9.50}$$

Stator-to-rotor mutual inductances are the inductances between stator windings and field windings, between stator windings and direct axis damper windings, and between stator windings and quadrature axis damper windings. These reactances are as follows:

From stator phase windings to field windings

$$L_{aF} = M_F \cos \theta \tag{9.51}$$

$$L_{bF} = M_F \cos(\theta - 2\pi/3) \tag{9.52}$$

$$L_{cF} = M_F \cos(\theta + 2\pi/3) \tag{9.53}$$

From stator phase windings to direct axis damper windings

$$L_{aD} = M_D \cos \theta \tag{9.54}$$

$$L_{bD} = M_D \cos(\theta - 2\pi/3) \tag{9.55}$$

$$L_{cD} = M_D \cos(\theta + 2\pi/3) \tag{9.56}$$

From stator phase windings to damper windings in the quadrature axis

$$L_{aQ} = M_Q \sin \theta \tag{9.57}$$

$$L_{bQ} = M_Q \sin(\theta - 2\pi/3) \tag{9.58}$$

$$L_{cQ} = M_Q \sin(\theta + 2\pi/3) \tag{9.59}$$

Short Circuit of Synchronous and Induction Machines and Converters

The rotor self-inductances are L_F, L_D, and L_q. The mutual inductances are

$$L_{DF} = M_R \quad L_{FQ} = 0 \quad L_{DQ} = 0 \tag{9.60}$$

The mutual inductances between field windings and direct axis damper windings are constant and do not vary. Also, the d and q axes are displaced by 90° and the mutual inductances between the field and direct axis damper windings and quadrature axis damper windings are zero.

The inductance matrix can, therefore, be written as

$$\bar{L} = \begin{vmatrix} \bar{L}_{aa} & \bar{L}_{aR} \\ \bar{L}_{Ra} & \bar{L}_{RR} \end{vmatrix} \tag{9.61}$$

where $\bar{L}_{aa}$ is a stator-to-stator inductance matrix:

$$\bar{L}_{aa} = \begin{vmatrix} L_s + L_m \cos 2\theta & -M_s - L_m \cos 2(\theta + \pi/6) & -M_s - L_m \cos 2(\theta + 5\pi/6) \\ -M_s - L_m \cos 2(\theta + \pi/6) & L_s + L_m \cos 2(\theta - 2\pi/3) & -M_s - L_m \cos 2(\theta - \pi/2) \\ -M_s - L_m \cos 2(\theta + 5\pi/6) & -M_s - L_m \cos 2(\theta - \pi/2) & L_s + L_m \cos 2(\theta + 2\pi/3) \end{vmatrix} \tag{9.62}$$

$\bar{L}_{aR} = \bar{L}_{Ra}$ is the stator to-rotor inductance matrix:

$$\bar{L}_{aR} = \bar{L}_{Ra} = \begin{vmatrix} M_F \cos\theta & M_D \cos\theta & M_Q \sin\theta \\ M_F \cos(\theta - 2\pi/3) & M_D \cos(\theta - 2\pi/3) & M_Q \sin(\theta - 2\pi/3) \\ M_F \cos(\theta + 2\pi/3) & M_D \cos(\theta + 2\pi/3) & M_Q \sin(\theta + 2\pi/3) \end{vmatrix} \tag{9.63}$$

$\bar{L}_{RR}$ is the rotor-to-rotor inductance matrix:

$$\bar{L}_{RR} = \begin{vmatrix} L_F & M_R & 0 \\ M_R & L_D & 0 \\ 0 & 0 & L_Q \end{vmatrix} \tag{9.64}$$

The inductance matrix of Equation 9.61 shows that the inductances vary with the angle θ. By referring the stator quantities to rotating rotor dq axes through Park's transformation, this dependence on θ is removed and a constant reactance matrix emerges.

9.6.2 Transformation of Reactance Matrix

Park's transformation describes a new set of variables, such as currents, voltages, and flux linkages in $0dq$ axes. The stator parameters are transferred to the rotor parameters. For the currents, this transformation is

$$\begin{vmatrix} i_0 \\ i_d \\ i_q \end{vmatrix} = \sqrt{\frac{2}{3}} \begin{vmatrix} \dfrac{1}{\sqrt{2}} & \dfrac{1}{\sqrt{2}} & \dfrac{1}{\sqrt{2}} \\ \cos\theta & \cos\left(\theta - 2\dfrac{\pi}{3}\right) & \cos\left(\theta + 2\dfrac{\pi}{3}\right) \\ \sin\theta & \sin\left(\theta - 2\dfrac{\pi}{3}\right) & \sin\left(\theta + 2\dfrac{\pi}{3}\right) \end{vmatrix} \begin{vmatrix} i_a \\ i_b \\ i_c \end{vmatrix} \tag{9.65}$$

Using matrix notation:

$$\bar{i}_{0dq} = \bar{P}\bar{i}_{abc} \tag{9.66}$$

Similarly,

$$\bar{v}_{0dq} = \bar{P}\bar{v}_{abc} \tag{9.67}$$

$$\bar{\lambda}_{0dq} = \bar{P}\bar{\lambda}_{abc} \tag{9.68}$$

where $\bar{\lambda}$ is the flux linkage vector. The a–b–c currents in the stator windings produce a synchronously rotating field, stationary with respect to the rotor. This rotating field can be produced by constant currents in the fictitious rotating coils in the dq axes; P is nonsingular and $\bar{P}^{-1} = \bar{P}'$:

$$\bar{P}^{-1} = \bar{P}' = \sqrt{\frac{2}{3}} \begin{vmatrix} \dfrac{1}{\sqrt{2}} & \cos\theta & \sin\theta \\ \dfrac{1}{\sqrt{2}} & \cos\left(\theta - \dfrac{2\pi}{3}\right) & \sin\left(\theta - \dfrac{2\pi}{3}\right) \\ \dfrac{1}{\sqrt{2}} & \cos\left(\theta + \dfrac{2\pi}{3}\right) & \sin\left(\theta + \dfrac{2\pi}{3}\right) \end{vmatrix} \tag{9.69}$$

To transform the stator-based variables into rotor-based variables, define a matrix as follows:

$$\begin{vmatrix} i_0 \\ i_d \\ i_q \\ i_F \\ i_D \\ i_Q \end{vmatrix} = \begin{vmatrix} \bar{P} & 0 \\ 0 & 1 \end{vmatrix} \begin{vmatrix} i_a \\ i_b \\ i_c \\ i_F \\ i_D \\ i_Q \end{vmatrix} = \bar{B}\bar{i} \tag{9.70}$$

Short Circuit of Synchronous and Induction Machines and Converters   395

where $\bar{1}$ is a 3×3 unity matrix and $\bar{0}$ is a 3×3 zero matrix. The original rotor quantities are left unchanged. The time-varying inductances can be simplified by referring all quantities to the rotor frame of reference:

$$
\begin{vmatrix} \bar{\lambda}_{odq} \\ \bar{\lambda}_{FDQ} \end{vmatrix} = \begin{vmatrix} \bar{P} & \bar{0} \\ \bar{0} & \bar{1} \end{vmatrix} \begin{vmatrix} \bar{\lambda}_{abc} \\ \bar{\lambda}_{FDQ} \end{vmatrix} = \begin{vmatrix} \bar{P} & \bar{0} \\ \bar{0} & \bar{1} \end{vmatrix} \begin{vmatrix} \bar{L}_{aa} & \bar{L}_{aR} \\ \bar{L}_{Ra} & \bar{L}_{RR} \end{vmatrix} \begin{vmatrix} \bar{P}^{-1} & \bar{0} \\ \bar{0} & \bar{1} \end{vmatrix} \begin{vmatrix} \bar{P} & \bar{0} \\ \bar{0} & \bar{1} \end{vmatrix} \begin{vmatrix} \bar{i}_{abc} \\ \bar{i}_{FDQ} \end{vmatrix}
$$

(9.71)

This transformation gives

$$
\begin{vmatrix} \lambda_0 \\ \lambda_d \\ \lambda_q \\ \lambda_F \\ \lambda_D \\ \lambda_Q \end{vmatrix} = \begin{vmatrix} L_0 & 0 & 0 & 0 & 0 & 0 \\ 0 & L_d & 0 & kM_F & kM_D & 0 \\ 0 & 0 & L_q & 0 & 0 & kM_q \\ 0 & kM_F & 0 & L_F & M_R & 0 \\ 0 & kM_D & 0 & M_R & L_D & 0 \\ 0 & 0 & kM_Q & 0 & 0 & L_Q \end{vmatrix} \begin{vmatrix} i_0 \\ i_d \\ i_q \\ i_F \\ i_D \\ i_Q \end{vmatrix}
$$

(9.72)

Define:

$$
L_d = L_s + M_s + \frac{3}{2}L_m
$$

(9.73)

$$
L_q = L_s + M_s - \frac{3}{2}L_m
$$

(9.74)

$$
L_0 = L_s - 2M_s
$$

(9.75)

$$
k = \sqrt{\frac{3}{2}}
$$

(9.76)

The inductance matrix is sparse, symmetric, and constant. It decouples the $0dq$ axes and the same will be illustrated further.

9.7 Park's Voltage Equation

The voltage equation [4,5] in terms of current and flux linkages is

$$
\bar{v} = -\bar{R}\bar{i} - \frac{d\bar{\lambda}}{dt}
$$

(9.77)

or

$$\begin{vmatrix} \upsilon_a \\ \upsilon_b \\ \upsilon_c \\ -\upsilon_F \\ \upsilon_D \\ \upsilon_Q \end{vmatrix} = - \begin{vmatrix} r & 0 & 0 & 0 & 0 & 0 \\ 0 & r & 0 & 0 & 0 & 0 \\ 0 & 0 & r & 0 & 0 & 0 \\ 0 & 0 & 0 & r_F & 0 & 0 \\ 0 & 0 & 0 & 0 & r_D & 0 \\ 0 & 0 & 0 & 0 & 0 & r_Q \end{vmatrix} \begin{vmatrix} i_a \\ i_b \\ i_c \\ i_F \\ i_D \\ i_Q \end{vmatrix} - \frac{\mathrm{d}i}{\mathrm{d}t} \begin{vmatrix} \lambda_a \\ \lambda_b \\ \lambda_c \\ \lambda_F \\ \lambda_D \\ \lambda_Q \end{vmatrix} \tag{9.78}$$

This can be partitioned as

$$\begin{vmatrix} \bar{\upsilon}_{abc} \\ \bar{\upsilon}_{FDQ} \end{vmatrix} = - \begin{vmatrix} \bar{r}_s \\ \bar{r}_{FDQ} \end{vmatrix} \begin{vmatrix} \bar{i}_{abc} \\ \bar{i}_{FDQ} \end{vmatrix} - \frac{\mathrm{d}i}{\mathrm{d}t} \begin{vmatrix} \bar{\lambda}_{abc} \\ \bar{\lambda}_{FDQ} \end{vmatrix} \tag{9.79}$$

The transformation is given by

$$\bar{B}^{-1}\bar{\upsilon}_B = -\bar{R}\bar{B}^{-1}\bar{i}_B - \frac{\mathrm{d}}{\mathrm{d}t}\left(\bar{B}^{-1}\bar{\lambda}_B\right) \tag{9.80}$$

where

$$\begin{vmatrix} \bar{P} & \bar{0} \\ \bar{0} & 1 \end{vmatrix} = \bar{B}, \qquad \bar{B}\begin{vmatrix} \bar{i}_{abc} \\ \bar{i}_{FDQ} \end{vmatrix} = \begin{vmatrix} \bar{i}_{odq} \\ \bar{i}_{FDQ} \end{vmatrix} = \bar{i}_B, \qquad \bar{B}\begin{vmatrix} \bar{\lambda}_{abc} \\ \bar{\lambda}_{FDQ} \end{vmatrix} = \begin{vmatrix} \bar{\lambda}_{odq} \\ \bar{\lambda}_{FDQ} \end{vmatrix} = \bar{\lambda}_B$$

$$\bar{B}\begin{vmatrix} \bar{\upsilon}_{abc} \\ \bar{\upsilon}_{FDQ} \end{vmatrix} = \begin{vmatrix} \bar{\upsilon}_{0dq} \\ \bar{\upsilon}_{FDQ} \end{vmatrix} = \bar{\upsilon}_B \tag{9.81}$$

Equation 9.80 can be written as

$$\upsilon_B = -\bar{B}R\bar{B}^{-1}\bar{i}_B - \bar{B}\frac{\mathrm{d}}{\mathrm{d}t}(\bar{B}^{-1}\bar{\lambda}_B) \tag{9.82}$$

First evaluate:

$$\bar{B}\frac{\mathrm{d}\bar{B}^{-1}}{\mathrm{d}\theta} = \begin{vmatrix} \bar{P} & \bar{0} \\ \bar{0} & 1 \end{vmatrix} \begin{vmatrix} \dfrac{\mathrm{d}\bar{P}^{-1}}{\mathrm{d}\theta} & \bar{0} \\ \bar{0} & \bar{0} \end{vmatrix} = \begin{vmatrix} \bar{P}\dfrac{\mathrm{d}\bar{P}^{-1}}{\mathrm{d}\theta} & \bar{0} \\ \bar{0} & \bar{0} \end{vmatrix} \tag{9.83}$$

where it can be shown that

$$\bar{P}\frac{\mathrm{d}\bar{P}^{-1}}{\mathrm{d}\theta} = \begin{vmatrix} 0 & 0 & 0 \\ 0 & 0 & 1 \\ 0 & -1 & 0 \end{vmatrix} \tag{9.84}$$

As we can write

$$\frac{\mathrm{d}\bar{B}^{-1}}{\mathrm{d}t} = \frac{\mathrm{d}\bar{B}^{-1}}{\mathrm{d}\theta}\frac{\mathrm{d}\theta}{\mathrm{d}t} \tag{9.85}$$

Short Circuit of Synchronous and Induction Machines and Converters 397

$$\bar{B}\frac{d\bar{B}^{-1}}{d\theta} = \begin{vmatrix} 0 & 0 & 0 & 0 & 0 & 0 \\ 0 & 0 & 1 & 0 & 0 & 0 \\ 0 & -1 & 0 & 0 & 0 & 0 \\ 0 & 0 & 0 & 0 & 0 & 0 \\ 0 & 0 & 0 & 0 & 0 & 0 \\ 0 & 0 & 0 & 0 & 0 & 0 \end{vmatrix} \tag{9.86}$$

The voltage equation becomes

$$\bar{v}_B = \bar{R}\bar{i}_B - \frac{d\theta}{dt}\begin{vmatrix} 0 \\ \lambda_q \\ -\lambda_d \\ 0 \\ 0 \\ 0 \end{vmatrix} - \frac{d\bar{\lambda}_B}{dt} \tag{9.87}$$

When the shaft rotation is uniform, $d\theta/dt$ is a constant and Equation 9.87 is linear and time invariant.

9.8 Circuit Model of Synchronous Machines

From the above treatment, the following decoupled voltage equations can be written as

Zero sequence

$$v_0 = ri_0 - \frac{d\lambda_0}{dt} \tag{9.88}$$

Direct axis

$$v_d = -ri_d - \frac{d\theta}{dt}\lambda_q - \frac{d\lambda_d}{dt} \tag{9.89}$$

$$v_F = r_F i_F + \frac{d\lambda_F}{dt} \tag{9.90}$$

$$v_D = r_D i_D + \frac{d\lambda_D}{dt} = 0 \tag{9.91}$$

Quadrature axis

$$v_q = -ri_q + \frac{d\theta}{dt}\lambda_d - \frac{d\lambda_q}{dt} \tag{9.92}$$

$$v_Q = r_Q i_Q + \frac{d\lambda_Q}{dt} = 0 \tag{9.93}$$

The decoupled equations relating to flux linkages and currents are

Zero sequence

$$\lambda_0 = L_0 i_0 \qquad (9.94)$$

Direct axis

$$\begin{vmatrix} \lambda_d \\ \lambda_F \\ \lambda_D \end{vmatrix} = \begin{vmatrix} L_d & kM_F & kM_D \\ kM_F & L_F & M_R \\ kM_D & M_R & L_D \end{vmatrix} \begin{vmatrix} i_d \\ i_F \\ i_D \end{vmatrix} \qquad (9.95)$$

Quadrature axis

$$\begin{vmatrix} \lambda_q \\ \lambda_Q \end{vmatrix} = \begin{vmatrix} L_q & kM_Q \\ kM_Q & L_Q \end{vmatrix} \begin{vmatrix} i_q \\ i_Q \end{vmatrix} \qquad (9.96)$$

This decoupling is shown in equivalent circuits as shown in Figure 9.10. Note that l_d, l_f, l_D are the self-inductances in armature, field, and damper circuits and L_{ad} is defined as

$$L_{ad} = L_D - l_D = L_d - l_d = L_F - l_f = kM_F = kM_D = M_R$$

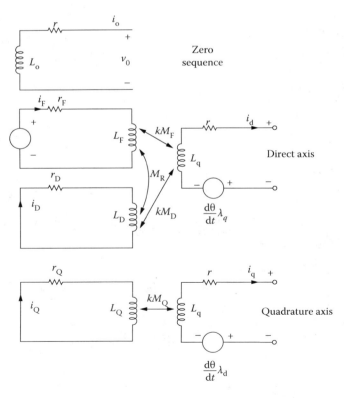

FIGURE 9.10
Synchronous generator decoupled circuits in d–q axes.

Short Circuit of Synchronous and Induction Machines and Converters 399

9.9 Calculation Procedure and Examples

The following three steps are involved:

1. The problem is normally defined in stator parameters, which are of interest. These are transformed into $0dq$ axes variables.
2. The problem is solved in $0dq$ axes, parameters, using Laplace transform or other means.
3. The results are transformed back to a–b–c variables of interest.

These three steps are inherent in any calculation using transformations. For simpler problems, it may be advantageous to solve directly in stator parameters.

Example 9.2

Calculate the time variation of the direct axis, quadrature axis voltages, and field current, when a step function of field voltage is suddenly applied to a generator at no load. Neglect damper circuits.

As the generator is operating without load, $i_{abc} = i_{0dq} = 0$. Therefore, from Equations 9.94 to 9.96:

$$\lambda_0 = 0 \quad \lambda_d = kM_F i_F \quad \lambda_F = L_F i_F \quad \lambda_q = 0$$

From Equations 9.88 to 9.92:

$$v_0 = 0$$

$$v_d = -\frac{d\lambda_d}{dt} = -kM_F \frac{di_F}{dt}$$

$$v_F = r_F i_F + L_F \frac{di_F}{dt}$$

Therefore, as expected, the time variation of field current is

$$i_F = \frac{1}{r_F}\left(1 - e^{(-r_F/L_F)t}\right)$$

The direct axis and quadrature axis voltages are given by

$$v_d = -\frac{kM_F}{L_F}e^{-(r_F/L_F)t}$$

$$v_q = -\frac{\omega kM_F}{r_F}\left(1 - e^{-(r_F/L_F)t}\right)$$

The phase voltages can be calculated using Park's transformation.

Example 9.3

A generator is operating with balanced positive sequence voltage of $\upsilon_a = \sqrt{2}|V|\cos(\omega_0 t + \angle V)$ across its terminals. The generator rotor is described by

$$\theta = \omega_1 t + \frac{\pi}{2} + \delta$$

Find υ_0, υ_d, and υ_q.

This is a simple case of transformation using Equation 9.65:

$$
\begin{vmatrix} \upsilon_0 \\ \upsilon_d \\ \upsilon_q \end{vmatrix} = \frac{2|V|}{\sqrt{3}}
\begin{vmatrix} \dfrac{1}{\sqrt{2}} & \dfrac{1}{\sqrt{2}} & \dfrac{1}{\sqrt{2}} \\ \cos\theta & \cos\left(\theta - \dfrac{2\pi}{3}\right) & \cos\left(\theta + \dfrac{2\pi}{3}\right) \\ \sin\theta & \sin\left(\theta - \dfrac{2\pi}{3}\right) & \sin\left(\theta + \dfrac{2\pi}{3}\right) \end{vmatrix}
\begin{vmatrix} \cos(\omega_0 t + \angle V) \\ \cos\left(\omega_0 t + \angle V - \dfrac{2\pi}{3}\right) \\ \cos\left(\omega_0 t + \angle V - \dfrac{4\pi}{3}\right) \end{vmatrix}
$$

A solution of this equation gives:

$$
\begin{aligned}
\upsilon_d &= \sqrt{3}|V|\sin\left[(\omega_0 - \omega_1)t + <V - \delta\right] \\
\upsilon_q &= \sqrt{3}|V|\cos\left[(\omega_0 - \omega_1)t + <V - \delta\right]
\end{aligned}
\tag{9.97}
$$

These relations apply equally well to derivation of i_d, i_q, λ_d, and λ_q.

For synchronous operation, $\omega_1 = \omega_0$ and the equations are reduced to

$$
\begin{aligned}
\upsilon_q &= \sqrt{3}|V|\cos(\angle V - \delta) \\
\upsilon_d &= \sqrt{3}|V|\sin(\angle V - \delta)
\end{aligned}
\tag{9.98}
$$

Note that υ_q and υ_d are now constant and do not have the slip frequency term $\omega_1 - \omega_0$. We can write:

$$\upsilon_q + j\upsilon_d = \sqrt{3}|V|e^{j(\angle V - \delta)} = \sqrt{3}V_a e^{-j\delta} \tag{9.99}$$

Therefore, V_a can be written as

$$V_a = \left(\frac{\upsilon_q}{\sqrt{3}} + j\frac{\upsilon_d}{\sqrt{3}}\right)e^{j\delta} = (V_q + jV_d)e^{j\delta} \tag{9.100}$$

where

$$V_q = \upsilon_q / \sqrt{3} \text{ and } V_d = \upsilon_d / \sqrt{3} \tag{9.101}$$

This is shown in the phasor diagram of Figure 9.11.

Short Circuit of Synchronous and Induction Machines and Converters

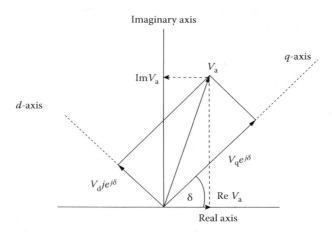

FIGURE 9.11
Vector diagram illustrating relationship of direct axes and quadrature axes' voltages to the terminal voltage.

We can write these equations in the following form:

$$\begin{vmatrix} \text{Re } V_a \\ \text{Im } V_a \end{vmatrix} = \begin{vmatrix} \cos\delta & -\sin\delta \\ \sin\delta & \cos\delta \end{vmatrix} \begin{vmatrix} V_q \\ V_d \end{vmatrix}$$

$$\begin{vmatrix} V_q \\ V_d \end{vmatrix} = \begin{vmatrix} \cos\delta & \sin\delta \\ -\sin\delta & \cos\delta \end{vmatrix} \begin{vmatrix} \text{Re } V_a \\ \text{Im } V_a \end{vmatrix} \quad (9.102)$$

Example 9.4

Steady-State Model of Synchronous Generator

Derive a steady-state model of a synchronous generator and its phasor diagram.

In the steady state, all the currents and flux linkages are constant. Also, $i_0 = 0$ and rotor damper currents are zero. Therefore, Equations 9.89 through 9.94 are reduced to

$$\begin{aligned} v_d &= -r i_d - \omega_0 \lambda_q \\ v_q &= -r i_q + \omega_0 \lambda_d \\ v_F &= r_F i_F \end{aligned} \quad (9.103)$$

where

$$\begin{aligned} \lambda_d &= L_d i_d + k M_F i_F \\ \lambda_F &= k M_F i_d + L_F i_F \\ \lambda_q &+ L_q i_q \end{aligned} \quad (9.104)$$

Substitute values of λ_d and λ_q from Equations 9.104 and 9.103; then from Example 9.3 and then from Equations 9.100 and 9.101, we can write the following equation:

$$V_a = -r(I_q + jI_d)e^{j\delta} + \omega_0 L_d I_d e^{j\delta} - j\omega_0 L_q I_q e^{j\delta} + \frac{1}{\sqrt{2}} \omega_0 M_F i_F e^{j\delta}$$

where $i_d = \sqrt{3}I_d$ and $i_q = \sqrt{3}I_q$.
Define:

$$\sqrt{2}E = \omega_0 M_F i_f e^{j\delta} \tag{9.105}$$

This is the no load voltage or the open-circuit voltage with generator current = 0. We can then write:

$$E = V_a + rI_a + jX_d I_d e^{j\delta} + jX_q I_q e^{j\delta} \tag{9.106}$$

The phasor diagram is shown in Figure 9.12a. The open-circuit voltage on no load is a q-axis quantity and is equal to the terminal voltage.

As the components I_d and I_q are not initially known, the phasor diagram is constructed by first laying out the terminal voltage V_a and line current I_a at the correct phase angle ϕ, then add the resistance drop and reactance drop IX_q. At the end of this vector, the quadrature axis is located. Now the current is resolved into direct axis and quadrature axis components. This allows the construction of vectors $I_q X_q$ and $I_q X_q$. This is shown in Figure 9.12b.

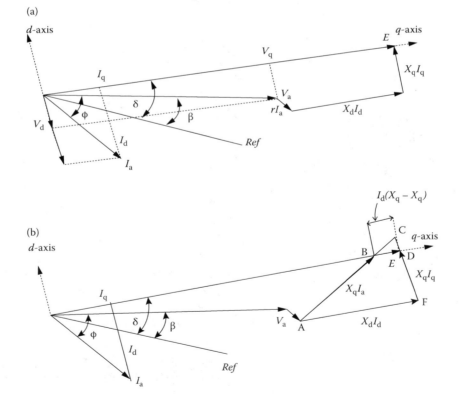

FIGURE 9.12
(a) Phasor diagram of a synchronous generator operating at lagging power factor and (b) to illustrate the construction of phasor diagram from known stator parameters.

Example 9.5

Consider the generator data of Example 9.1. Calculate the direct and quadrature axis components of the currents and voltages and machine voltage E when the generator is delivering its full-load rated current at its rated voltage. Also, calculate all the angles shown in the phasor diagram as shown in Figure 9.12a. If this generator is connected to an infinite bus through an impedance of 0.01 + j0.1 per unit (100 MVA base), what is the voltage of the infinite bus?

The generator operates at a power factor of 0.85 at its rated voltage of 1.0 per unit. Therefore, $\phi = 31.8°$. The generator full-load current is 4183.8 A = 1.0 per unit. The terminal voltage vector can be drawn to scale, the Ir drop (which is equal to 0.0012 per unit) is added, and the vector $IX_q = 1.8$ per unit is drawn to locate the q axis. Current I can be resolved into direct axis and quadrature axis components, and the phasor diagram is completed as shown in Figure 9.12b and the values of V_d, I_d, V_q, I_q, and E are read from it. The analytical solution is as follows:

The load current is resolved into active and reactive components $I_r = 0.85$ per unit and $I_x = 0.527$ per unit, respectively. Then, from the geometric construction shown in Figure 9.13:

$$(\delta - \beta) = \tan^{-1}\left(\frac{X_q I_r + r I_x}{V_a + r I_r - X_q I_x}\right)$$

$$= \tan^{-1}\left(\frac{(1.8)(0.85) + (0.0012)(0.527)}{1 + (0.0012)(0.85) + (1.8)(0.527)}\right) = 38.14°$$

(9.107)

From the above calculation, resistance can even be ignored without an appreciable error. Thus, $\delta - \beta + \phi = 69.93°$; this is the angle of the current vector with the q axis. Therefore,

$$I_q = I_a \cos(\delta - \beta - \phi) = 0.343 \text{ per unit}, \quad i_q = 0.594 \text{ per unit}$$
$$I_d = -I_a \sin(\delta - \beta - \phi) = -0.939 \text{ per unit}, \quad i_d = -1.626 \text{ per unit}$$
$$V_q = V_a \cos(\delta - \beta) = 0.786 \text{ per unit}, \quad \upsilon_q = 1.361 \text{ per unit}$$
$$V_d = -V_a \sin(\delta - \beta) = -0.618 \text{ per unit}, \quad \upsilon_d = 1.070 \text{ per unit}$$

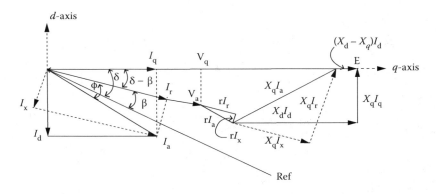

FIGURE 9.13
Phasor diagram of the synchronous generator for Example 9.5.

The machine-generated voltage is

$$E = V_q + rI_q - X_d I_d = 2.66 \text{ per unit}$$

The infinite bus voltage is simply the machine terminal voltage less the IZ drop subtracted vectorially:

$$V_\infty = V_a \angle 0° - I_a \angle 31.8° Z \angle 84.3° = 0.94 \angle -4.8°$$

The infinite bus voltage lags the machine voltage by 4.8°. Practically, the infinite bus voltage will be held constant and the generator voltage changes, depending on the system impedance and generator output.

Example 9.6

Symmetrical Short Circuit of a Generator at No Load

Derive the short-circuit equations of a synchronous generator for a balanced three-phase short circuit at its terminals. Ignore damper circuit and resistances and neglect the change in speed during short circuit. Prior to short circuit, the generator is operating at no load. Ignoring the damper circuit means that the subtransient effects are ignored. As the generator is operating at no load, $i_{abc} = i_{odq} = 0$, prior to the fault.

From Equations 9.89 to 9.94:

Zero sequence

$$v_0 = -L_0 \frac{di_0}{dt} = 0$$

Direct axis

$$v_d = -\omega_0 \lambda_q - \frac{d\lambda_d}{dt} = 0$$

$$v_f = \frac{d\lambda_F}{dt}$$

Quadrature axis

$$v_q = \omega_0 \lambda_d - \frac{d\lambda_q}{dt} = 0$$

The flux linkages can be expressed in terms of currents by using Equations 9.95 and 9.96:

$$\omega_0 L_q i_q + L_d \frac{di_d}{dt} + kM_F \frac{di_F}{dt} = 0$$

$$v_F = kM_F \frac{di_d}{dt} + L_F \frac{di_F}{dt}$$

$$-\omega_0 L_d i_d - \omega_0 kM_F i_F + L_q \frac{di_q}{dt} = 0$$

These equations can be solved using Laplace transformation. The initial conditions must be considered. In Example 9.4, we demonstrated that at no load, prior to fault, the terminal voltage is equal to the generated voltage and this voltage is

Short Circuit of Synchronous and Induction Machines and Converters

$\sqrt{2}E = \omega_0 M_F i_F e^{j\delta}$ and this is a quadrature axis quantity. Also, $\upsilon_d = 0$. The effect of short circuit is, therefore, to reduce υ_q to zero. This is equivalent to applying a step function of $-\upsilon_q 1$ to the q axis. The transient currents can then be superimposed on the currents prior to fault. Except for current in the field coil, all these currents are zero. The solution for i_F will be superimposed on the existing current i_{F0}.

If we write:

$$kM_F = L_{ad} = X_{ad}/\omega$$
$$L_F = (X_f + X_{ad})/\omega$$

The expressions converted into reactances and using Laplace transform, $dx/dt = sX(s) - x(0^-)$ (where $X(s)$ is Laplace transform of $x(t)$) reduces to

$$0 = (1/\omega)(X_{ad} + X_f)si_F + (1/\omega)X_{ad}si_d \tag{9.108}$$

$$0 = (1/\omega)(X_d)si_d + (1/\omega)X_{ad}si_f + (X_q)i_q \tag{9.109}$$

$$-\upsilon_q = (1/\omega)X_q si_q - X_{ad}i_F - X_d i_d \tag{9.110}$$

The field current from Equation 9.108 is

$$i_F = -i_d X_{ad}/(X_{ad} + X_f) \tag{9.111}$$

The field current is eliminated from Equations 9.109 and 9.110 by substitution method. The quadrature axis current is

$$i_q = \frac{1}{X_{aq} + X_l} \left| \frac{X_{aq}X_f}{X_{aq} + X_f} + X_l \right| \frac{s}{\omega} i_d$$

$$= -\left(\frac{X_d'}{X_q}\right)\frac{s}{\omega} i_d$$

and

$$i_d = \frac{\omega^2}{X_d'}\left[\frac{1}{s^2 + \omega^2}\right]\upsilon_q$$

Solving these equations gives

$$i_d = \frac{(\sqrt{3}|E|)}{X_d'}(1 - \cos\omega t)$$

$$i_q = -\frac{(\sqrt{3}|E|)}{X_q}\sin\omega t$$

Note that $k = \sqrt{3/2}$. Apply:
$\overline{i}_{abc} = \overline{P}\overline{i}_{0dq}$ with $\theta = \omega t + \pi/2 + \delta$, the short-circuit current in phase a is

$$i_a = \sqrt{2}|E|\left[\left(\frac{1}{X_d'}\right)\sin(\omega t + \delta) + \frac{X_q + X_d'}{2X_d'X_q}\sin\delta - \frac{X_q - X_d'}{2X_d'X_q}\sin(2\omega t + \delta)\right] \tag{9.112}$$

The first term is normal frequency short-circuit current, the second is constant asymmetric current, and the third is double-frequency short-circuit current. The 120 Hz component imparts a nonsinusoidal characteristic to the short-circuit current waveform. It rapidly decays to zero and is ignored in the calculation of short-circuit currents.

When the damper winding circuit is considered, the short-circuit current can be expressed as

$$i_a = \sqrt{2}E\left[\left(\frac{1}{X_d}\right)\sin(\omega t + \delta) + \left(\frac{1}{X'_d} - \frac{1}{X_d}\right)e^{-t/T'_d}\sin(\omega t + \delta)\right.$$

$$+ \left(\frac{1}{X''_d} - \frac{1}{X'_d}\right)e^{-t/T''_d}\sin(\omega t + \delta) \tag{9.113}$$

$$\left. - \frac{(X''_d + X''_q)}{2X''_d X''_q}e^{-t/T_a}\sin\delta - \frac{(X''_d - X''_q)}{2X''_d X''_q}e^{-t/T_a}\sin(2\omega t + \delta)\right]$$

- The first term is final steady-state short-circuit current.
- The second term is normal frequency decaying transient current.
- The third term is normal frequency decaying subtransient current.
- The fourth term is asymmetric decaying DC current.
- The fifth term is double-frequency decaying current.

Example 9.7

Calculate the component short-circuit currents at the instant of three-phase terminal short-circuit of the generator (particulars as shown in Table 9.1). Assume that phase a is aligned with the field at the instant of short circuit, maximum asymmetry, i.e., $\delta = 0$. The generator is operating at no load prior to short circuit.

The calculations are performed by substituting the required numerical data from Table 9.1 into Equation 9.113:

Steady-state current = 2.41 kA rms
Decaying transient current = 20.24 kA rms
Decaying subtransient current = 5.95 kA rms
Decaying DC component = 43.95 kA
Decaying second harmonic component = 2.35 kA rms

Note that the second harmonic component is zero if the direct axis and quadrature axis subtransient reactances are equal. Also, the DC component in this case is 40.44 kA.

9.9.1 Manufacturer's Data

The relationship between the various inductances and the data commonly supplied by a manufacturer for a synchronous machine is not obvious. The following relations hold:

$$L_{ad} = L_d - l_1 = kM_F = kM_D = M_R \tag{9.114}$$

$$L_{aq} = L_q - l_1 = kM_Q \tag{9.115}$$

Here, l_1 is the leakage reactance corresponding to X_1 and L_{ad} is the mutual inductance between the armature and rotor = mutual inductance between field and rotor = mutual

Short Circuit of Synchronous and Induction Machines and Converters 407

inductance between damper and rotor. Similar relations are applicable in the quadrature axis. Some texts have used different symbols. Field leakage reactance, l_f, is

$$l_f = \frac{L_{ad}(L'_d - l_1)}{(L_d - L'_d)} \tag{9.116}$$

and

$$L_F = l_f + L_{ad} \tag{9.117}$$

The damper leakage reactance in the direct axis is

$$l_{kD} = \frac{L_{ad} l_f (L''_d - l_1)}{L_{ad} l_f - L_F (L''_d - l_1)} \tag{9.118}$$

and

$$L_D = l_{kD} + L_{ad} \tag{9.119}$$

$$l_{kQ} = \frac{L_{aq}(L''_q - l_1)}{(L_q - L''_q)} \tag{9.120}$$

In the quadrature axis, the damper leakage reactance is

$$L_Q = l_{kQ} + L_{aq} \tag{9.121}$$

The field resistance is

$$r_F = \frac{L_F}{T'_{do}} \tag{9.122}$$

The damper resistances in direct axis can be obtained from

$$T''_d = \frac{(L_D L_F - L^2_{ad})}{r_D L_F} \left(\frac{L''_d}{L'_d} \right) \tag{9.123}$$

and in the quadrature axis:

$$T''_q = \frac{L''_q L_Q}{L_a r_Q} \tag{9.124}$$

Example 9.8

Using the manufacturer's data in Table 9.1, calculate the machine parameters in the d–q axes. Applying the equations in Section 9.9.1:

$$L_{ad} = L_d - l_l = 1.949 - 0.164 = 1.785 \text{ per unit} = kM_F = kM_D = M_R$$

$$L_{aq} = L_q - l_l = 1.858 - 0.164 = 1.694 \text{ per unit} = kM_Q$$

$$l_f = (1.758)(0.278 - 0.164) / (1.964 - 0.278) = 0.121 \text{ per unit}$$

$$L_F = 0.121 + 1.785 = 1.906 \text{ per unit}$$

$$l_{kd} = (1.785)(0.121)(0.193 - 0.164) / \{(1.785)(0.164) - (1.096)(0.193 - 0.164)\} =$$
$$0.026 \text{ per unit}$$

$$L_D = 0.026 + 1.785 = 1.811 \text{ per unit}$$

$$l_{kq} = (1.694)(0.192 - 0.164) / (1.858 - 0.192) = 0.028 \text{ per unit}$$

$$L_Q = 0.028 + 1.694 = 1.722 \text{ per unit}$$

$$T'_{do} = 5.615s = 2116.85 \text{ rad}$$

$$r_F = 1.906 / 2116.85 = 1.005 \times 10^{-5} \text{ per unit}$$

$$r_D = \frac{(1.811)(1.906) - 1.785^2}{(0.015)(377)(1.906)} \left(\frac{0.193}{0.278} \right) = 0.0131 \text{ per unit}$$

$$r_Q = \left(\frac{0.192}{1.858} \right) \left(\frac{1.722}{0.015 \times 377} \right) = 0.031 \text{ per unit}$$

Note that in Table 9.1, as the data is in per unit, we consider $X_d = L_d$, and $X_l = l_l$, etc. The per unit system for synchronous machines is not straightforward, and variations in the literature exist. References [6–9] provide further reading.

9.10 Short Circuit of Synchronous Motors and Condensers

The equations derived above for synchronous generators can be applied to synchronous motors and condensers. Synchronous condensers are used for reactive power support in transmission systems and sometimes in industrial systems also, and are not connected to any load. Large synchronous motors are used to drive certain loads in the industry, for example, single units of 30,000–40,000 hp have been used in pulp industry. Figure 9.14 shows V curves of the synchronous machines, generators, and motors, which are obtained with machines operating at different excitations. A phasor diagram of synchronous motor under steady-state operation can be drawn similar to that of a synchronous generator.

With respect to short-circuit calculations according to empirical methods of ANSI/IEEE, the excitation systems and prior loads are ignored—Chapter 7. The subtransient and transient reactances of synchronous motors are much higher compared to that of synchronous generators of the same ratings. Manufacturers may not supply all the reactances and time constants of the synchronous motors, unless specifically requested. Thus, for the same ratings, the synchronous motors contribute smaller magnitude of short-circuit currents. Also, these currents decay faster.

Short Circuit of Synchronous and Induction Machines and Converters

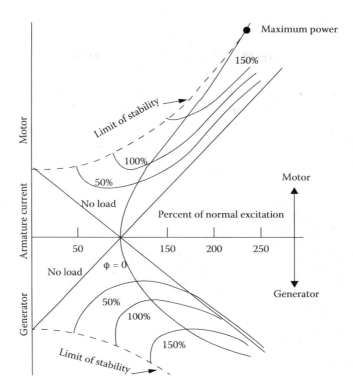

FIGURE 9.14
V-curves of synchronous machines

9.11 Induction Motors

A rotating machine can be studied in the steady state depending upon the specific type of the machine, for example, synchronous, induction, or DC machine. Generalized machine theory [7,8] attempts to unify the piecemeal treatment of rotating machines—after all each machine type consists of some coils on the stationary and rotating part, which interact with each other. The treatment of synchronous machine above amply demonstrates this concept. The ideas were further developed by Kron and others and treatment of electrical machines using tensors or matrices and linear transformation is the basis of the generalized machine theory. Though there are some limitations that saturation, commutation effects, surge phenomena, eddy current losses cannot be accounted for and also the theory is not applicable if there is saliency in both rotor and stator pole construction, yet it is a powerful analytical tool for machine modeling.

The following transformations are used to derive the model of an induction motor:

$$3-\phi|_{abc} \Leftrightarrow 2-\phi|_{\alpha\beta 0} \Leftrightarrow 2-\phi|_{dq0} \tag{9.125}$$

In terms of d–q axes, both the stator and rotor are cylindrical and symmetrical. The d-axis is chosen arbitrarily as the axis of the stator phase a. The three-phase winding is converted into a two-phase winding so that the axis of the second phase becomes the q axis. The two stator phases are then the fixed axis coils, 1D and 1Q, respectively (Figure 9.15).

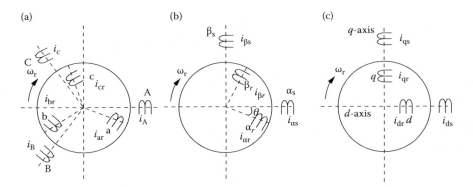

FIGURE 9.15
Transformation of an induction machine.

The rotor circuit, whether of wound type or cage type, is also represented by *d*- and q-axis coils, though a squirrel cage rotor is more complex and space harmonics are neglected. These coils are 2D and 2Q. The impedance matrix can be set up as for a synchronous machine, as follows:

$$\begin{vmatrix} v_{1d} \\ v_{1q} \\ v_{2d} \\ v_{2q} \end{vmatrix} = \begin{vmatrix} r_1+(L_m+L_1)p & & -L_m p & \\ & r_1+(L_m+L_1)p & & L_m p \\ L_m p & -L_m\omega_r & -r_2-(L_m+L_2)p & (L_m+L_2)\omega_r \\ -L_m\omega_r & L_m p & -(L_m+L_2)\omega_r & -r_2-(L_m+L_2)p \end{vmatrix} \begin{vmatrix} i_{1d} \\ i_{1q} \\ i_{2d} \\ i_{2q} \end{vmatrix}$$

(9.126)

where r_1 and r_2 are stator and rotor resistances, respectively, and L_1 and L_2 are stator and rotor reactances, respectively. 1D and 1Q have mutual inductance L_m, which is also the mutual inductance between 1Q and 2Q. Here, $\omega_r = (1-s_1)\omega$, where s_1 is the motor slip. The rotor is short-circuited on itself, therefore, $v_{2d} = v_{2q} = 0$. Equation 9.126 can be written as

$$\begin{vmatrix} v_{1d} \\ v_{1q} \\ 0 \\ 0 \end{vmatrix} = \begin{vmatrix} r_1+X_s & & -jX_m & \\ & r_s+jX_s & & -jX_m \\ jX_m & -(1-s)X_m & -(r_2+jX_r) & (1-s)X_m \\ (1-s)X_m & jX_m & -(1-s)X_r & -(r_2+jX_r) \end{vmatrix} \begin{vmatrix} i_{1d} \\ i_{1q} \\ i_{2d} \\ i_{2q} \end{vmatrix}$$

There is no difference in the *d* and *q* axes except for time, and it is possible to write

$$\begin{aligned} v_{1d} &= v_1 & v_{1q} &= -jv_1 \\ i_{1d} &= i_1 & i_{1q} &= -ji_1 \\ i_{2d} &= i_2 & i_{2q} &= -ji_2 \end{aligned}$$ (9.127)

For a terminal fault, whether an induction machine is operating as an induction generator or motor, the machine feeds into the fault due to trapped flux linkage with the rotor. This fault current will be a decaying transient. A DC decaying component also occurs to maintain the flux linkage constant.

Short Circuit of Synchronous and Induction Machines and Converters 411

By analogy with a synchronous machine, reactances X_a, X_f, and X_{ad} are equivalent to X_1, X_2, and X_m in the induction machine. The transient reactance of the induction machine is

$$X' = X_1 + \frac{X_m X_2}{X_m + X_2} \tag{9.128}$$

This is also the motor-locked rotor reactance. The equivalent circuit of the induction motor is shown in Volume 2. The open-circuit transient time constant is

$$T'_0 = \frac{X_2 + X_m}{\omega r_2} \tag{9.129}$$

The short-circuit transient time constant is

$$T' = T'_0 \frac{X'}{X_1 + X_m} \tag{9.130}$$

This is approximately equal to

$$T' = \frac{X'}{\omega r_2} \tag{9.131}$$

and the time constant for the decay of DC component is

$$T_{dc} = \frac{X'}{\omega r_1} \tag{9.132}$$

AC symmetrical short-circuit current is

$$i_{ac} = \frac{E}{X'} e^{-t/T'} \tag{9.133}$$

and DC current is

$$i_{dc} = \sqrt{2}\, \frac{E}{X'} e^{-t/T_{dc}} \tag{9.134}$$

where E is the prefault voltage behind the transient reactance X'. At no load, E is equal to the terminal voltage.

Example 9.9

Consider a 4 kV, 5000 hp four-pole motor, with full-load kVA rating = 4200. The following parameters in per unit on motor-base kVA are specified:

$$r_1 = 0.0075$$
$$r_2 = 0.0075$$
$$X_1 = 0.0656$$
$$X_2 = 0.0984$$
$$R_m = 100$$
$$X_m = 3.00$$

Calculate the motor short-circuit current equations for a sudden terminal fault. The following parameters are calculated using Equations 9.128 through 9.132:

$$X' = 0.1608$$

$$T' = 0.107 \text{ s}$$

$$T_{dc} = 0.057 \text{ s}$$

$$T_0' = 2.05 \text{ s}$$

The AC component of the short-circuit current is

$$i_{ac} = 6.21 e^{-t/0.107}$$

At $t = 0$, the AC symmetrical short-circuit current is 6.21 times the full-load current. The DC component of the short-circuit current is

$$i_{dc} = 8.79 e^{-t/0.057}$$

The nature of short-circuit currents is identical to that of synchronous machines; however, the currents decay more rapidly. The decay is a function of the motor rating, inertia, load-torque characteristics. Typically, the effect of short-circuit currents from induction machines can be ignored after 6 cycles. Based on above calculations, the short-circuit current decay profile is calculated in Figure 9.16.

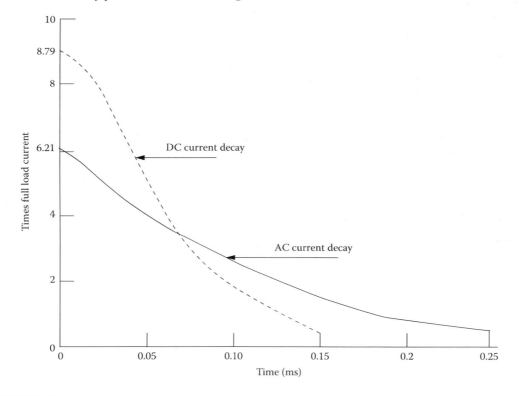

FIGURE 9.16
AC and DC short-circuit decay currents of a 5000 hp induction motor (Example 9.9).

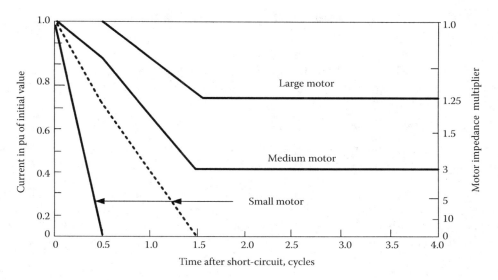

FIGURE 9.17
Short-circuit current decay on induction motors for a terminal fault (see text).

Figure 9.17 shows the symmetrical rms current contributed by an induction motor to a three-phase short circuit at its terminals; solid lines according to Reference [10] (see Chapter 10 for calculations according to ANSI/IEEE standards). Accurately, the short-circuit profile can be calculated based on motor parameters.

9.12 Capacitor Contribution to the Short-Circuit Currents

IEEE standard, Reference [10] provides detailed models of capacitor current discharge into a fault for various types of shunt capacitor installations, namely,

- Shunt capacitors
- Harmonic filters
- Large medium voltage capacitors switched with the motors
- Low-voltage capacitors

This aspect of capacitor current discharge into a nearby short circuit has not been previously discussed or evaluated in other IEEE standards.

As we have seen in Chapter 8, the capacitor currents are at a high frequency. When a fault occurs, system voltage is changed and capacitor discharges at a rapid rate. The current is maximum if the capacitor is charged to maximum at voltage peak. The circuit impedance between the capacitor and fault point impacts the current. Standard capacitor connections show high transient currents that damp quickly before ¼ of a cycle. The capacitor discharge takes place in the initial $\frac{1}{30} - \frac{1}{8}$ cycles depending on the time constant of the system.

With harmonic filters, the transient currents are reduced due to presence of a reactor.

A simulation with one Mvar bank in parallel with a four-pole induction motor shows motor and source current are essentially unchanged, and the amplitude of capacitor current is very small, compared to total fault current.

Low-voltage motor capacitors are small enough and show no significant increase in normal system fault levels.

The standard summarizes that capacitor discharge takes place in $\frac{1}{30} - \frac{1}{8}$ cycles. Since breaker protective device and contacts cannot operate in this time frame, the discharge takes place into closed contacts. The electromagnetically induced forces of the discharge current are instantaneously proportional to the square of the current. Since the close and latch rating of a circuit breaker is the maximum fundamental frequency rms fault current the breaker can withstand, it can also be considered a measure of forces, which may be safely imposed on various physical members of the breaker during a rated frequency fault conditions.

Based on transient simulations in this reference, it recommends that the capacitor discharge currents will have no effect on breaker parting or clearing conditions. This is the current practice that capacitors are not modeled for fault duties on circuit breakers (also see Reference [11] for some transient simulations).

9.13 Static Converters Contribution to the Short-Circuit Currents

The four main types of converters are

- Rectifier
- Inverter
- Cycloconverter
- Chopper

The DC system contributes current to an AC short-circuit current only when the converter operates *in an inverter mode*. Because nonregenerative drives cannot operate in inverter mode, these are not considered for short circuit current. The DC sources of interest are batteries, Uninterruptible Power Supplies (UPS), photovoltaic arrays, Chapter 3, and inverters, say of DC transmission lines. The static power converters can be with a grid control or without a grid control, the one's with grid control being most popular. These are also called controlled or uncontrolled converters. The short-circuit damage in a controlled converter is significantly limited by the grid control systems. The grid control protection schemes enable the blocks the pulses, limiting the short circuit current to *1 cycle*. The converter DC side faults are discussed in Chapter 12.

Arc-back short circuits occur due to failure of a semiconducting device—these are the most common faults in the converters. The calculation of arc-back faults is one important aspect of the theory and application of converter systems.

Example 9.10

Figure 9.18 is a typical circuit of a converter with its equivalent circuit, based on Reference [10]. Consider the parameters of the circuit as shown in Table 9.2, then,

E_m = maximum secondary line-to-neutral voltage = 543 V
X_s referred to 665 V = 0.3809 Ω
L_s = 2.35 mH
$X_t = 5.83 \times 10^{-3}$ ohms, and $R_t = R_\gamma = 4.86 \times 10^{-4}$ Ω
$X_\gamma = X_s + X_t = 6.71 \times 10^{-3}$ ohms
$I_m = E_m / X_\gamma = 80.87$ kA
$X_\gamma / R_\gamma = 13.82$

Calculate MF (multiplying factor) from:

$$I_{ac,peak} = I_{sym,peak}\left[1 + e^{-2\pi\tau(X/R)}\right]$$

where

$$\tau = 0.49 - 0.1e^{-(X/R)/3}$$

At ½ cycle = 0.0083 s (see also Chapter 10)
    Then MF = 1.804

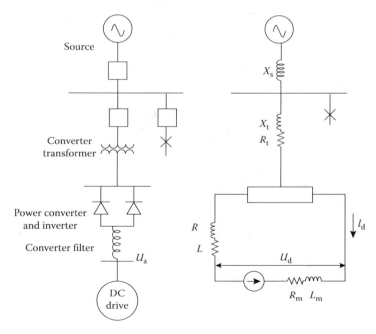

FIGURE 9.18
A converter contribution to AC side short-circuit current (Example 9.10).

TABLE 9.2

Specified Data for Example 9.10

Parameter	Symbol	Value
System voltage	V_1	13.8 kV, 60 Hz
Available short-circuit current	S_{SC}	500 MVA
Transformer rating	S	6.45 MVA
Primary voltage	V_1	13.8 kV
Secondary voltage	V_2	665 V
Transformer impedance	Z_t	8.5%
Transformer X/R ratio	X_t/R_t	12
Converter filter resistance	R	0.0188 Ω
Converter filter reactance	L	1.28×10^{-3} H
DC motor drive		6000 HP
DC motor rated voltage	U_d	700 V
DC motor rated current	I_d	4830 A
DC motor inductance	L_m	0.852×10^{-3} H
DC motor resistance	R_m	0.01248 Ω at T_o
Normal operating temperature	T_o	

$$E_d = U_d + R_m \times I_d = 760\,\text{V}$$

Under normal operating conditions, the magnitude of phase AC rms current is equal to DC current = 4830 A.

For a fault at F1, Figure 9.18, the final maximum DC current is

$$I_\Sigma = E_d/R_\Sigma$$

$$R_\Xi = 1.5 \times R_t + R + R_m = 1.5 \times 4.86 \times 10^{-4} + 1.88 \times 10^{-2} + 1.248 \times 10^{-2} = 3.20 \times 10^{-2}$$

(For a full wave, two commutating intervals $R_\Sigma = 1.5R_t + R$.)

Then

$$I_\Sigma = 23.75\,\text{kA1}$$

If there is no filter, this current will be 57.54 kA.

$$X_\Sigma = X_t + X_m = 0.330\,\Omega/\text{phase}$$

Then

$$\frac{X_\Sigma}{R_\Sigma} = 24.98$$

The magnitude of the DC fault current contribution i_c at a time equal to ½ cycle at system frequency is given by

$$i_c = I_d - (I_d - i(0))e^{-\pi(X_\Sigma/R_\Sigma)}$$

Short Circuit of Synchronous and Induction Machines and Converters 417

Substituting the values and considering that the converter is not provided with a filter, i.e., $I_\Sigma = 23.75$ kA, $i_c = 11.06$ kA.

This is like a single phase current and considering a delta connected winding of the rectifier transformer, the current reflected on 13.8 kV side is

$$1.15 \times i_c \times \frac{V_2}{V_1} = 0.61 \text{ kA}$$

The source short-circuit level is 500 MVA, and therefore symmetrical short-circuit current is 20.92 kA.

For

$$\frac{X_\Sigma}{R_\Sigma} = 24.98$$

$$\tau = 0.4897$$

And the peak current is 55.75 kA.

Adding the DC contribution, it is equal to 56.36 kA, that is, an increase of 1.1%.

We will see in Chapter 11, as a simplification, the IEC recommends adding a current equal to three times the drive rating only to the first cycle or asymmetrical current for the breaker duties. For interrupting calculations, this is ignored. Using this simplification, the short-circuit contribution from DC drive is 0.735 kA versus 0.61 kA calculated above with rigorous calculations.

9.14 Practical Short-Circuit Calculations

For practical short-circuit calculations, dynamic simulation or analytical calculations are rarely carried out. Chapter 10 describes the ANSI empirical calculation procedures and shows that the machine models are simple and represented by a voltage behind impedance, which changes with the type of calculations. The detailed machine models and calculation of time variation of short-circuit currents are sometimes required to validate the empirical results [12]. These form a background to the empirical methods to be discussed in Chapters 10 and 11.

Problems

9.1 Calculate the fault decrement curves of the generator, data as given in Table 9.1. Calculate (1) the AC decaying component, (2) DC component, and (3) total current. Plot the results in a similar manner to those of Figure 9.6.

9.2 Consider the system and data shown in Figure P9.1. Calculate (i) prefault voltages behind reactances X_d, X'_d, and X''_d for faults at G and F, and (ii) the largest possible DC component for faults at G and F.

9.3 Calculate the field current in Problem 9.2 on application of short circuit.

9.4 Calculate the three-phase short-circuit subtransient and transient time constants in Problem 9.2 for a fault at F.

9.5 Write a numerical expression for the decaying AC component of the current for faults at G and F in Problem 9.2. What is the AC component of the fault current at 0.05 and 0.10 s?

9.6 Transform the calculated direct axis and quadrature axis voltages derived in Problem 9.2 into stator voltages using Park's transformation.

9.7 Draw a general steady-state phasor diagram of a synchronous motor operating at (1) leading power factor and (2) lagging power factor.

9.8 Construct a simplified dynamic phasor diagram (ignoring damper circuit and resistances) of a synchronous generator using Park's transformations. How does it differ from the steady-state phasor diagram?

9.9 Show that first column of P' is an eigenvector of L_{11} corresponding to eigenvalue $L_0 = L_s - 2M_s$.

9.10 Form an equivalent circuit similar to Figure 9.10 with numerical values using the generator data from Table 9.1.

9.11 13.8 kV, 10,000 hp four-pole induction motor has a full-load efficiency of 96% and a power factor of 0.93. The locked rotor current is six times the full-load current at a power factor of 0.25. Calculate the time variation of AC and DC components of the current. Assume equal stator and rotor resistances and reactances. The magnetizing resistance and reactance are equal to 130 and 3.0 per unit, respectively, on machine MVA base.

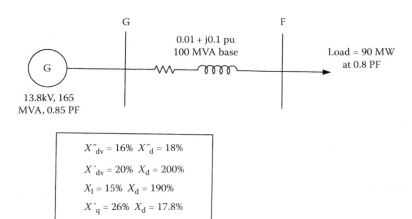

FIGURE P9.1
Circuit diagram and system data for Problem 9.2.

References

1. NEMA. Large Machines—Synchronous Generators. MG-1, Part 22.
2. IEEE Standard 421A. *IEEE Guide for Identification and Evaluation of Dynamic Response of Excitation Control Systems.* 1978.
3. *ATP Rule Book.* Portland, OR: Canadian/American EMTP Group, 1987–1992.
4. RH Park. Two reaction theory of synchronous machines, Part I, *AIEE Trans,* 48, 716–730, 1929.
5. RH Park. Two reaction theory of synchronous machines, Part II, *AIEE Trans,* 52, 352–355, 1933.
6. PM Anderson, A Fouad. *Power System Control and Stability,* IEEE Press, New York, 1991.
7. CV Jones. *The Unified Theory of Electrical Machines,* Pergamon Press, Oxford, UK, 1964.
8. AT Morgan. *General Theory of Electrical Machines,* Heyden & Sons Ltd., London, 1979.
9. IEEE Std. C37.010. *Application Guide for AC High Voltage Circuit Breakers Rated on Symmetrical Current Basis,* 1999 (Reaff 2005).
10. IEEE Std. 551. *IEEE Recommended Practice for Calculating Short-Circuit Currents in Industrial and Commercial Power Systems,* 2006.
11. JC Das. *Transients in Electrical Systems, Analysis Recognition and Mitigation,* McGraw-Hill, New York, 2010.
12. JR Dunki-Jacobs, P Lam, P Stratford. A comparison of ANSI-based and dynamically rigorous short-circuit current calculation procedures, *IEEE Trans Ind Appl,* 24, 1180–1194, 1988.

Bibliography

B Adkins. *The General Theory of Electrical Machines,* Chapman and Hall, London, 1964.

PM Anderson. *Analysis of Faulted Power Systems,* Iowa State University Press, Ames, IA, 1973.

I Boldea. *Synchronous Generators,* CRC Press, Boca Raton, FL, 2005.

C Concordia. *Synchronous Machines,* John Wiley, New York, 1951.

AE Fitzgerald Jr., SD Umans, C Kingsley. *Electrical Machinery,* McGraw-Hill Higher Education, New York, 2002.

NN Hancock. *Matrix Analysis of Electrical Machinery,* Pergamon Press, Oxford, UK, 1964.

IEEE Committee Report. Recommended phasor diagram for synchronous machines, *IEEE Trans Power Apparatus Syst* 88, 1593–1610, 1963.

10

Short-Circuit Calculations according to ANSI Standards

American National Standards Institute (ANSI) methods of short-circuit calculations are used all over North America and are accepted in many other countries. These standards have been around for a much longer time than any other standard in the world. The International Electrotechnical Commission (IEC) [1] standard for a short-circuit calculation was published in 1988, which is now revised [2], and the calculation procedures according to IEC are discussed in Chapter 11. A Verband der Elektrotechnik (VDE) [3] (Deutsche Electrotechnische Kommission) standard has been around since 1971. There has been a thrust for analog methods too. Nevertheless, for all equipment manufactured and applied in the USA industry, ANSI/Institute of Electrical and Electronics Engineers (IEEE) standards prevail. Most foreign equipment for use in the US market has been assigned ANSI ratings.

We will confine our discussions to ANSI/IEEE symmetrical rating basis of the circuit breakers. The USA is the only country that rates the circuit breakers on symmetrical current basis. The interpretations, theory, and concepts governing short-circuit calculations according to the latest ANSI/IEEE standards are discussed with illustrative examples.

10.1 Types of Calculations

In a multivoltage system, four types of short-circuit calculations may be required. These are as follows:

1. First-cycle (momentary) duties for fuses and low-voltage circuit breakers.
2. First-cycle (momentary) duties for medium- or high-voltage circuit breakers.
3. Contact parting (interrupting) duties for high-voltage circuit breakers (circuit breakers rated above 1 kV).
4. Short-circuit currents for time-delayed relaying devices.

The first-cycle, momentary duty, and close and latch duty are all synonymous. However, the close and latch term is applicable to high-voltage circuit breakers only, while the first cycle term can be applied to low-voltage breakers and fuses too. Irrespective of the type of fault current calculation, the power system is reduced to single Thevénin equivalent impedance behind the source voltage.

421

10.1.1 Assomptions

The source voltage or prefault voltage is the system-rated voltage, though a higher or lower voltage can be used in the calculations. The worst short-circuit conditions occur at maximum loads, because the rotating loads contribute to the short-circuit currents. It is unlikely that the operating voltage will be above the rated voltage at the maximum loading. Under light-load conditions, the operating voltage may be higher, but the load contributions to the short-circuit currents will also be reduced. The effect of higher voltage at a reduced load is offset by the reduced contributions from the loads. Therefore, the short-circuit calculations are normally carried out at the rated voltage. Practically, the driving voltage will not remain constant; it will be reduced, and it varies with the machine loading and time subsequent to short circuit. The fault current source is assumed sinusoidal; all harmonics and saturation are neglected. All circuits are linear; the non-linearity associated with rotating machines, transformer, and reactor modeling is neglected. As the elements are linear, the theorem of superimposition is applicable.

Loads prior to the short circuit are neglected, and the short circuit occurs at zero crossing of the voltage wave. At the instant of fault, the direct current (DC) value is equal in magnitude to the fault alternating current (AC) value but opposite in sign.

10.1.2 Maximum Peak Current

The maximum peak current *does not occur* at ½ cycle in the phase that has the maximum initial DC component, unless short circuit occurs on purely inductive circuits with resistance equal to zero. The maximum peak occurs before ½ cycle and before the symmetrical current peak. The IEC equation for peak is given in Chapter 8, Equation 8.3. IEEE standard [4] provides the following equations that give better approximation to peak, as compared to IEC equations:

$$I_{\text{peak}} = I_{\text{AC peak}} + I_{\text{DC}} = \sqrt{2} I_{\text{AC,rms}} \left(1 + e^{-\frac{2\pi\tau}{(X/R)}} \right) = I_{\text{AC peak}} \left(1 + e^{-\frac{2\pi\tau}{(X/R)}} \right) \tag{10.1}$$

where

$$\tau = 0.49 - 0.1e^{-\frac{(X/R)}{3}} \tag{10.2}$$

For the root mean square (RMS) first-cycle RMS current, equation

$$I_{\text{rms}} = \sqrt{I_{\text{AC,rms}}^2 + I_{\text{DC}}^2} \tag{10.3}$$

is only correct if DC is constant; but it is not and decays exponentially. Many times, DC component is calculated at ½ cycle, though this does not correspond to the peak value [5]. The following equations are provided in Reference [4]:

$$\text{IEC}_{\text{rms}} = I_{\text{AC,rms}} \sqrt{1 + 2\left(1.02 + 0.98e^{-\frac{3}{(X/R)}} \right)^2} \tag{10.4}$$

Short-Circuit Calculations according to ANSI Standards 423

$$\text{Half cycle}_{\text{rms}} = I_{\text{AC,rms}} \sqrt{1 + 2\left(e^{-\frac{\pi}{(X/R)}}\right)^2} \tag{10.5}$$

The more exact equation from Reference [4] is

$$I_{\text{rms,total}} = I_{\text{AC rms}} \sqrt{\left(1 + 2e^{-\frac{4\pi\tau}{(X/R)}}\right)} \tag{10.6}$$

Reference [4] tabulated the peak and RMS value calculations using the above equations. Consider an $X/R = 10$, then peak values in per unit are as follows:

Exact calculations: Time to peak = 0.4735 cycles, DC = 0.7368, maximum peak = 1.7368

IEC equations: Peak = 1.6935, % error = 0.53

½ cycle equations: Peak = 1.7304, error = −0.37%

Reference [4] equations: Peak = 1.7367, error = −0.01%

A peak multiplier of 2.6 is often used for simplicity, when calculating the duties of medium- and high-voltage circuit breakers above 1 kV. This 2.6 factor corresponds to $X/R = 17$ for 60 Hz systems, or equivalently a DC component decay governed by L/R time constant of 45 ms for 60 Hz system ($X/R = 14$ for 50 Hz system). When larger X/R ratios are encountered, higher multipliers result. *When selecting a breaker, it is important that the peak is adjusted for the actual fault point X/R ratio.*

For older high-voltage circuit breakers, calculate asymmetrical RMS current based upon ½ cycle peak, given:

$$I_{\text{asym}} = I_{\text{sym}} \sqrt{1 + 2e^{-2\pi/(X/R)}} \tag{10.7}$$

This essentially calculated total asymmetrical current at ½ cycle. An asymmetrical multiplier of 1.6 has been used, which corresponds to $X/R = 25$. Higher factors will be obtained with higher X/R. *When selecting a breaker, it is important that the peak is adjusted for the actual fault point X/R ratio.* Also see Reference [5].

10.2 Accounting for Short-Circuit Current Decay

We have amply discussed that short-circuit currents are decaying transients. In the short-circuit calculations according to standards, a step-by-step account of the decay cannot be considered. Depending on the type of calculation, the ANSI/IEEE standards consider that the dynamic (rotating equipment) reactances are multiplied by factors given in Table 10.1 [6, 7]. The static equipment impedances are assumed to be time invariant, i.e., harmonics and saturation are neglected. Maintaining a constant electromotive force (EMF) and artificially increasing the equivalent impedance to model a machine during short circuit has the same effect as the decay of the flux trapped in the rotor circuit. In Table 10.1,

TABLE 10.1

Impedance Multiplier Factors for Rotating Equipment for Short-Circuit Calculations

	Positive Sequence Reactance for Calculating	
Type of Rotating Machine	**Interrupting Duty** (per unit)	**Closing and Latching Duty** (per unit)
All turbogenerators, all hydrogenerators with amortisseur windings, and all condensers	$1.0X_d''$	$1.0X_d''$
Hydrogenerators without amortisseur windings	$0.75X_d'$	$0.75X_d'$
All synchronous motors	$1.5X_d''$	$1.0X_d''$
Induction Motors		
Above 1000 hp at 1800 r/min or less	$1.5X_d''$	$1.0X_d''$
Above 250 hp at 3600 r/min	$1.5X_d''$	$1.0X_d''$
From 50 to 1000 hp at 1800 r/min or less	$3.0X_d''$	$1.2X_d''$
From 50 to 250 hp at 3600 r/min	$3.0X_d''$	$1.2X_d''$
Neglect all three-phase induction motors below 50 hp and all single-phase motors		
Multivoltage-level calculations including low-voltage systems		
Induction motors of above 50 hp	$3.0X_d''$	$1.2X_d''$
Induction motors below 50 hp	∞	$1.67X_d''$

X_d'' of synchronous rotating machines is the rated voltage (saturated) direct-axis subtransient reactance.

X_d' of synchronous rotating machines is rated-voltage (saturated) direct-axis transient reactance.

X_d'' of induction motors equals 1.00 divided by per unit locked rotor current at rated voltage.

Source: ANSI/IEEE Std. C37.010. Application Guide for AC High-Voltage Circuit Breakers Rated on a Symmetrical Current Basis, 1999 (revision of 1979 standard); IEEE Std. 141. IEEE Recommended Practice for Electrical Power Distribution for Industrial Plants, 1993.

manufacturer's data for the transient and subtransient and locked rotor reactances should be used in the calculations. Some industries may use large motors of the order of thousands of hp rating, i.e., the pulp mill industry may use a single synchronous motor of the order of 40,000 hp for a refiner application. Reactances and time constants need to be accurately modeled for such large machines (Example 10.6). The decay of short-circuit currents of the motors depends upon the size of the motors. For induction machines, a simulation will show high initial current decay followed by fairly rapid decay to zero. For synchronous machines, there is a high initial decay followed by a slower rate of decay to a steady-state value, Chapter 9.

10.2.1 Low-Voltage Motors

For calculation of short-circuit duties on low-voltage systems, a modified subtransient reactance for a group of low-voltage induction and synchronous motors fed from a low-voltage substation can be used. If the total motor horse-power rating is approximately equal to or less than the self-cooled rating in kVA of the substation transformer, a reactance equal to 0.25 per unit, based on the transformer self-cooled rating, may be used as

Short-Circuit Calculations according to ANSI Standards 425

a single impedance to represent the motors. This means that the combined motor loads contribute a short-circuit current equal to four times the rated current. This estimate is based on the low-voltage circuit breaker application guide, IEEE Std. C37.13 [8]. It assumes a typical motor group having 75% induction motors, which contribute short-circuit currents equal to 3.6 times their rated current, and 25% synchronous motors, which contribute 4.8 times the rated current. At present, low-voltage synchronous motors are not in much use in industrial distribution systems; however, higher-rated induction motors in a group may contribute a higher amount of short-circuit current, compensating for the absence of synchronous motors. Overall, four times the full load contribution can be retained.

For calculations of short-circuit duties for comparison with medium- or high-voltage circuit breakers closing and latching capabilities (or momentary ratings according to pre-1964 basis, now no longer in use), motors smaller than 50 hp can be ignored. However, these are required to be considered for low-voltage circuit breaker applications. To simplify multivoltage distribution system short-circuit calculations and obviate the necessity of running two first-cycle calculations, one for the low-voltage circuit breakers and the other for medium- and high-voltage circuit breakers, a single first-cycle network can replace the two networks [7]. This network is constructed by

1. Including all motors <50 hp using a multiplying factor of 1.67 for subtransient reactance or an estimate of first-cycle impedance of 0.28 per unit based on the motor rating.

2. Including all motors ≥50 hp and using a multiplying factor of 1.2 for the subtransient reactance or an estimate of first-cycle impedance of 0.20 per unit based on the motor rating.

This single-combination, first-cycle network adds conservatism to both low- and high-voltage calculations. A typical short-circuit contribution for a terminal fault of an induction motor is six times the full load current. Thus, an estimate of 0.20 per unit for larger low-voltage motors or a multiplying factor of 1.2 is equivalent to a fault current contribution of 4.8 times the rated full-load current. Similarly, for motors ≥50 hp, a multiplying factor of 1.67 or an estimate of 0.28 per unit impedance means a short-circuit contribution of approximately 3.6 times the rated current. These factors are shown in Table 10.1.

Though this simplification can be adopted, where the medium- and high-voltage breakers are applied close to their first-cycle ratings, it is permissible to ignore all low-voltage motors rated <50 hp. Depending on the extent of low-voltage loads, this may permit retaining the existing medium- or high-voltage breakers in service, if these are overdutied from the close and latch capability considerations. Close to a generating station, the close and latch capabilities may be the limiting factor in the application of circuit breakers.

10.3 Rotating Machine Model

The rotating machine model for the short-circuit calculations is shown in Figure 10.1. The machine reactances are modeled with suitable multiplying factors from Table 10.1. The multiplying factors are applicable to resistances as well as reactances, so that the X/R ratio remains the same. The voltage behind the equivalent transient reactance at no load will be

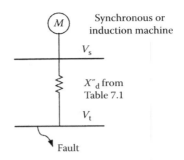

FIGURE 10.1
Equivalent machine model for short-circuit calculations.

equal to the terminal voltage, i.e., $V_s = V_t$, as no prefault currents need be considered in the calculations. A justification of neglecting the prefault currents is indirectly discussed in Chapter 6, i.e., the concept of constant flux linkages. The total current before and after the transition (pre- and post-fault) should not change. The DC component is equal in magnitude to the AC component, but of the opposite polarity. Thus, the AC and DC components of the current summate to zero to maintain a constant flux linkage, i.e., no load conditions. If preloading is assumed, this balance is no longer valid.

10.4 Type and Severity of System Short Circuits

The short-circuit currents in decreasing order of severity are as follows:

- Three-phase ungrounded
- Three-phase grounded
- Phase-to-phase, ungrounded
- Phase-to-phase grounded
- Single phase to ground

This order is also applicable to transient stability studies, a three-phase fault giving the worst situation. If the system is shown to hold together for a three-phase fault, it can be concluded that it will be stable for *other fault types cleared in the same time duration* as the three-phase fault.

A three-phase power system may be subjected to symmetrical and unsymmetrical faults. Generally, three-phase ungrounded faults impose the most severe duty on a circuit breaker, since the first phase to interrupt has a normal frequency recovery voltage of approximately 87% of the system phase-to-phase voltage. A single phase-to-ground fault current can be higher than the three-phase current. This condition exists when the zero sequence impedance at the fault point is less than the positive sequence impedance. Depending on the relative magnitude of sequence impedances, it may be necessary to investigate other types of faults. For calculations of short-circuit duties, the fault resistance is ignored. This gives conservatism to the calculations.

Short-Circuit Calculations according to ANSI Standards 427

10.5 Calculation Methods

Two calculation procedures are as follows:

1. E/X or E/Z simplified method.
2. E/X or E/Z method with adjustments for AC and DC decrements.

10.5.1 Simplified Method $X/R \leq 17$

The results of the E/X calculation can be directly compared with the circuit breaker symmetrical interrupting capability, provided that the circuit X/R ratio for three-phase faults and $(2X_1 + X_0)/(2R_1 + R_0)$ for single line-to-line ground fault is 17 or less. (For single phase-to-ground faults X_1 is assumed equal to X_2.) This is based on the rating structure of the breakers. When the circuit X/R ratio is 17 or less, the asymmetrical short-circuit duty never exceeds the symmetrical short-circuit duty by a proportion greater than that by which the circuit breaker asymmetrical rating exceeds the symmetrical capability. It may only be slightly higher at four-cycle contact parting time.

10.5.2 Simplified Method $X/R > 17$

A further simplification of the calculations is possible when the X/R ratio exceeds 17. For X/R ratios higher than 17, the DC component of the short-circuit current may increase the short-circuit duty beyond the compensation provided in the rating structure of the breakers. A circuit breaker can be immediately applied without calculation of system resistance, X/R ratio, or remote/local considerations, if the E/X calculation does not exceed 80% of the breaker symmetrical interrupting capability. This means that resistance component of the system need not be determined.

10.5.3 *E/X* Method for AC and DC Decrement Adjustments

Where a closer calculation is required, and the current exceeds 80% of the breaker symmetrical rating, AC and DC decrement adjustments should be considered. This method is also recommended when a single line-to-ground fault supplied predominantly by generators, at generator voltage, exceeds 70% of the circuit breaker interrupting capability for single line-to-ground faults. For calculations using this method, the fault point X/R ratio is necessary. Two separate networks are constructed: (1) a resistance network, with complete disregard of the reactance; and (2) a reactance network with complete disregard of the resistance. The fault point X/R ratio is calculated by reducing these networks to an equivalent resistance and an equivalent reactance at the fault point. This gives more accurate results than any other reasonably simple procedure, including the phasor representation at the system frequency.

The resistance values for various system components are required, and for accuracy of calculations, these should be obtained from the manufacturer's data. In the absence of these data, Table 10.2 and Figures 10.2 through 10.4 provide typical resistance data. The variations in X/R ratio between an upper and a lower bound are shown in the ANSI/IEEE standard [6].

Once the E/X calculation is made and the X/R ratio is known, the interrupting duty on the high-voltage circuit breakers can be calculated by multiplying the calculated short-circuit currents with an appropriate multiplying factor. This multiplying factor is based

TABLE 10.2
Resistance of System Components for Short-Circuit Calculations

System Component	Approximate Resistance
Turbine generators and condensers	Effective resistance
Salient pole generators and motors	Effective resistance
Induction motors	1.2 times the DC armature resistance
Power transformers	AC load loss resistance (not including no-load losses or auxiliary losses)
Reactors	AC resistance
Lines and cables	AC resistance

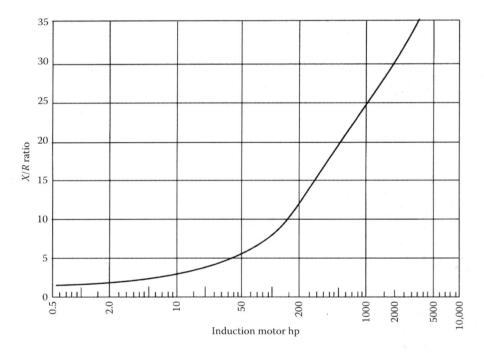

FIGURE 10.2
Typical X/R ratios for induction motors based on induction motor hp rating.

on (1) contact parting time of the circuit breaker, (2) calculated X/R ratio, (3) effects of DC decay (remote sources), and (4) effects of AC and DC decay (local sources).

An important qualification is that the method is applicable for X/R not exceeding 45, and manufacturer should be consulted for X/R .45. *Unfortunately, this has been ignored in the industry in many applications.*

10.5.4 Fault Fed from Remote Sources

If the short-circuit current is fed from generators through (1) two and more transformations, or (2) a per unit reactance external to the generator that is equal to or exceed 1.5 times the generator per unit subtransient reactance on a common megavolt amp (MVA) base, i.e., it supplies less than 40% of its terminal short-circuit current, it is considered a

Short-Circuit Calculations according to ANSI Standards

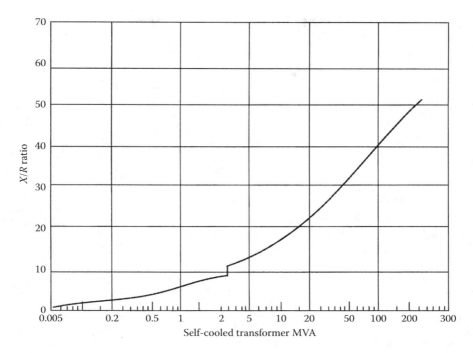

FIGURE 10.3
Typical X/R ratios for transformers based on transformer self-cooled MVA rating.

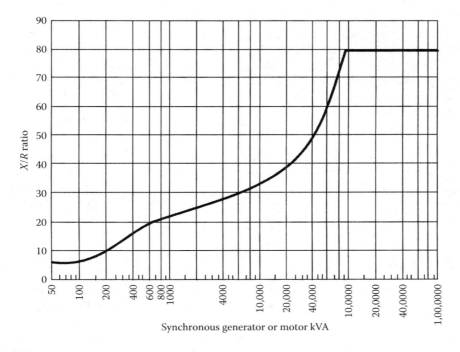

FIGURE 10.4
Typical X/R ratios for synchronous generators and synchronous motors based on their kVA rating.

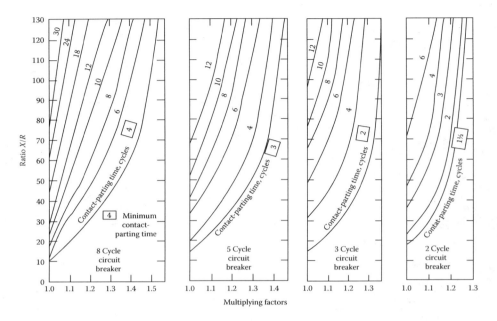

FIGURE 10.5
Three-phase and line-to-ground faults, E/X multiplying factors, DC decrement only (remote sources). (From ANSI/IEEE Std. C37.010. Application Guide for AC High-Voltage Circuit Breakers Rated on a Symmetrical Current Basis, 1999 (revision of 1979 standard).)

remote source. In this case, the effect of AC decay need not be considered, and the curves of multiplying factors include only DC decay. These curves are shown in Figure 10.5. The decrement factor for the standard contact parting time of the breakers is shown within a rectangle, which includes a half-cycle tripping delay. Factors for higher contact parting time, applicable when the tripping delay is increased above a half cycle, are also shown. Interpolation between the curves is possible. The multiplying factor for the remote curves is calculable and is given by

$$I_{asym} / I_{sym} = \sqrt{1 + 2e^{-4\pi C/(X/R)}} \tag{10.8}$$

where C is the contact parting time in cycles at 60 Hz. As an example, the remote multiplying factor for five-cycle breaker, which has a contact parting time of 3 cycles, and for a fault point X/R of 40, from Equation 10.8, is 1.21, which can also be read from the curves in Figure 10.5.

10.5.5 Fault Fed from Local Sources

When the short-circuit current is predominantly fed through no more than one transformation or a per unit reactance external to the generator, which is less than 1.5 times the generator per unit reactance on the same MVA base, i.e., it supplies more than 40% of its maximum terminal fault current, it is termed a local source. The effect of AC and DC decrements should be considered. The multiplying factors are applied from separate curves, reproduced in Figures 10.6 and 10.7.

Short-Circuit Calculations according to ANSI Standards

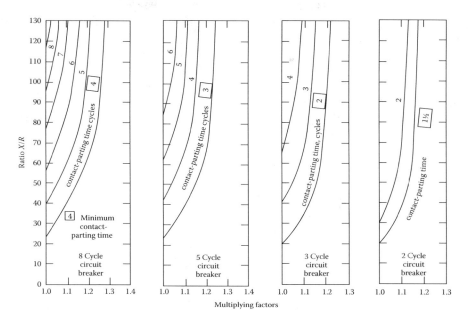

FIGURE 10.6
Three-phase faults, E/X multiplying factors, AC and DC decrements (local sources). (From ANSI/IEEE Std. C37.010. Application Guide for AC High-Voltage Circuit Breakers Rated on a Symmetrical Current Basis, 1999 (revision of 1979 standard).)

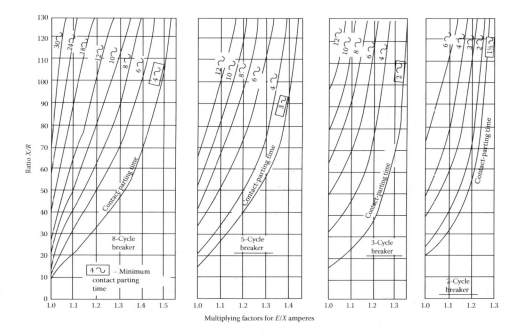

FIGURE 10.7
Line-to-ground faults, E/X multiplying factors, AC and DC decrements (local sources). (From ANSI/IEEE Std. C37.010. Application Guide for AC High-Voltage Circuit Breakers Rated on a Symmetrical Current Basis, 1999 (revision of 1979 standard).)

The asymmetrical multiplying factors for the *local* curves are an unknown equation. A number of sources may contribute to a fault through varying impedances. Each of these contributions has a different AC and DC decay rate. The impedance through which a fault is fed determines whether it is considered a local or remote source. The AC decay in electrically remote sources is slower, as compared to the near sources. The time constant associated with the AC decay is a function of rotor resistance, and added external reactance prolongs it (Chapter 9). An explanation and derivation of the multiplying factors for AC and DC decrements, provided in Reference [6], is as follows.

Figure 10.8 shows the relationship of fault current $[(I_{asym}/I_{sym})_{nacd}]$ (the subscript "nacd" means that there is no AC decrement), as a function of X/R ratio for various contact parting times. The curves of this figure are modified so that the decrement of the symmetrical component of the fault current is taken into consideration. Figure 10.9a shows the

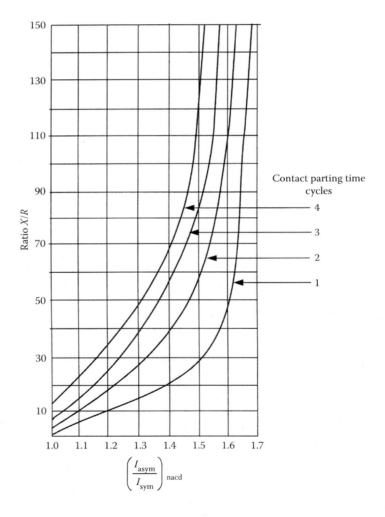

FIGURE 10.8
Ratio $(I_{asym}/I_{sym})_{nacd}$ versus X/R ratio for breaker contact parting times. (From ANSI/IEEE Std. C37.010. Application Guide for AC High-Voltage Circuit Breakers Rated on a Symmetrical Current Basis, 1999 (revision of 1979 standard).)

Short-Circuit Calculations according to ANSI Standards

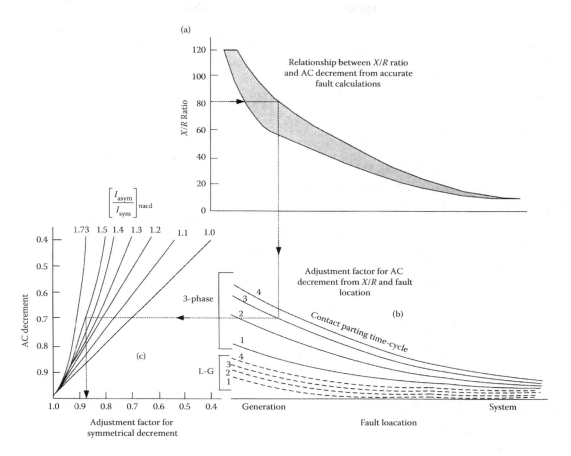

FIGURE 10.9
(a) AC decrement for faults away from the sources, (b) adjustment factors for AC decrement from X/R and fault location for breaker contact parting times (solid lines: three-phase faults; dotted lines: line-to-ground faults), (c) adjustment factors for AC decrement. (From ANSI/IEEE Std. C37.010. Application Guide for AC High-Voltage Circuit Breakers Rated on a Symmetrical Current Basis, 1999 (revision of 1979 standard).)

general relationship of X/R to the AC decrement as the fault location moves away from the generating station. This empirical relationship is shown as a band, based on small to large machines of various manufacturers. Figure 10.9b shows the decay of the symmetrical component (AC component) of the fault current at various times after fault initiation, as a function of the contact parting time and the type of fault. Figure 10.9c establishes reduction factors that can be applied to $(I_{asym}/I_{sym})_{nacd}$ to obtain this effect. The reduction factor is obtained from the following relationship:

$$\text{Reduction factor} = \frac{\sqrt{I_{AC}^2 + I_{DC}^2}/(E/X)}{[I_{asym}/I_{sym}]_{nacd}} \tag{10.9}$$

As an example, consider an X/R ratio of 80 and a contact parting time of 3 cycles; the factor $(I_{sym}/I_{sym})_{nacd}$, as read from Figure 10.8, is 1.5. Enter curve in Figure 10.9a at X/R of 80, follow down to contact parting time curve in Figure 10.9b, and go across to Figure 10.9c, curve labeled $(I_{asym}/I_{asym})_{nacd} = 1.5$. A reduction factor of 0.885 is obtained. The modifier $(I_{sym}/I_{sym})_{nacd}$ ratio for an X/R of 80 is calculated as $0.885 \times 1.5 = 1.33$, and this establishes one point on a three-phase modified decrement curve, as shown in Figure 10.10. This curve is constructed by following the procedure outlined above. Finally, E/X multipliers for the breaker application are obtained through the use of a modified X/R decrement curve and the breaker capability. Continuing with the above calculation, the breaker asymmetric capability factor for a 3-cycle parting time is 1.1. The E/X multiplier required to ensure that sufficient breaker capability is, therefore, $1.33/1.1 = 1.21$. This establishes one point in Figure 10.6.

In a digital computer-based calculation, the matrix equations can be used to calculate voltages at buses other than the faulted bus, and current contributions from individual sources can be calculated (Chapter 6). These currents can then be labeled as remote or local.

In certain breaker applications, a breaker contact parting time in excess of the contact parting time, with a half-cycle tripping delay assumed for the rating structure, may be used. If a breaker with a minimum contact parting time of 2 cycles is relayed such that it actually parts contacts after 4 cycles and after fault initiation, the E/X multiplier for breaker selection can be reduced to account for the fault current decay during the 2-cycle period. This will reduce the interrupting duty because of decaying nature of short-circuit currents.

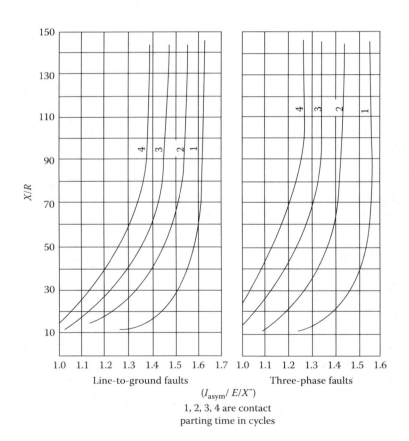

FIGURE 10.10
Relationship of I_{asym}/I_{sym} to X/R for several breaker contact parting times, AC decrement included. (From ANSI/IEEE Std. C37.010. Application Guide for AC High-Voltage Circuit Breakers Rated on a Symmetrical Current Basis, 1999 (revision of 1979 standard).)

Short-Circuit Calculations according to ANSI Standards 435

Sometimes, it can prevent costly replacement of circuit breakers that are applied close to their ratings, and short-circuit duties increase because of system growth [9]. However, the related aspects of delayed opening of the breaker by introducing addition relay time delay have other adverse effects on the system: (1) increase in fault damage and (2) jeopardizing the transient stability in some cases. This does not impact the closing and latch duty.

10.5.6 Weighted Multiplying Factors

For a system with several short-circuit sources, which may include generators that may be classified local or remote, depending on the interconnecting impedances, neither the remote nor the local multiplying factors can be exclusively applied. It is logical to make use of both local and remote multiplying factors in a weighting process. This weighting consists of applying a remote multiplying factor to that part of the E/X symmetrical short-circuit current that is contributed by remote sources. Similarly, the local multiplying factor is applied to the local component of the fault current contribution. The fraction of interrupting current that is contributed by remote sources is identified as the NACD ratio:

$$NACD\,ratio = \frac{\sum NACD\,source\,currents}{E/X\,for\,the\,interrupting\,network} \tag{10.10}$$

This computation requires additional calculations of remote and total current contributed at the fault point from various sources, and is facilitated by digital computers. Figure 10.11 shows interpolated multiplying factors for various NACD ratios [10].

For the short-circuit current contribution from motors, *irrespective of their type and rating and location in the system*, the AC decay is built into the pre-multiplying impedance factors in Table 10.1. Thus, it is assumed that the motors, howsoever remote in the system, will continue contributing to the fault. In an actual system, the postfault recovering voltage may return the motor to normal motoring function. The magnetic contactors controlling the motors may drop in the first cycle of the voltage dip on 30%–70% of their rated voltage, disconnecting the motors from service. This is conservative from the short-circuit calculation point of view, but may give increased arc flash hazard calculations, see Volume 4.

10.6 Network Reduction

Two methods of network reduction are as follows:

1. Short-circuit current can be determined by complex impedance network reduction, and this gives the E/X complex method. It is permissible to replace E/X with E/Z. In fact in all commercially available software, the default calculations are based upon E/Z method, though E/X method can be selected.

2. Short-circuit current can be determined from R and X calculations from separate networks and treating them as a complex impedance at the fault point.

In either case, the X/R ratio is calculated from separate resistance and reactance networks. The X/R ratio thus calculated is used to ascertain multiplying factors and also for

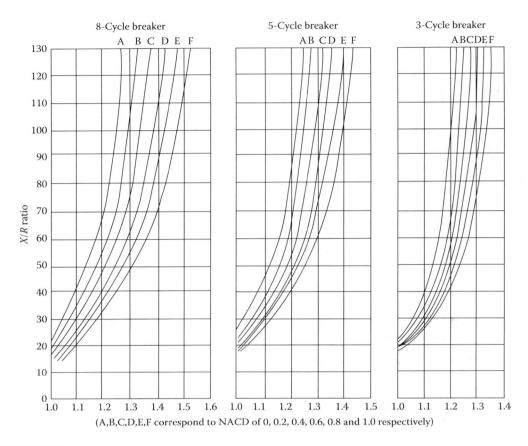

FIGURE 10.11
Multiplying factors for E/X ampères, three-phase faults, for various NACD ratios and breaker interrupting times. (From WC Huening Jr. "Interpretation of New American Standards for power circuit breaker applications," IEEE Transactions on Industry General Applications vol. IGA-5, pp. 121–143, 1969.)

calculation of asymmetry factors for the first-cycle calculation. The X/R ratio for single line-to-ground faults is $(X_1+X_2+X_0)/(R_1+R_2+R_0)$.

The E/Z calculation from separate networks is conservative, and results of sufficient accuracy are obtained. Branch current flows and angles may have greater variation as compared to the complex network solution. Contributions from the adjacent buses may not correlate well. However, this procedure results in much simpler computing algorithms.

There could be a difference of 5%–6% in the calculated results of short-circuit currents between the complex impedance reduction method and the calculations from the separate R and X networks. The separate R and jX calculations give higher values, as compared to the $R+jX$ complex calculation.

10.6.1 *E/X* or *E/Z* Calculation

The E/X calculation will give conservative results, and as the X/R at the fault point is high, there may not be much difference between E/X and E/Z calculations. This may not be always true. For low-voltage systems, it is appropriate to perform E/Z calculations, as

Short-Circuit Calculations according to ANSI Standards 437

the X/R ratios are low and the difference in the results between E/Z and E/X calculations can be significant. Generally, E/Z calculations using the complex method are the standard in industry.

10.7 Breaker Duty Calculations

Once the results of E/X or E/Z calculation for the interrupting network are available and the weighted multiplying factor is ascertained, the adequacy of the circuit breaker for interrupting duty application is given by

$$\text{MF} \times E/X (E/Z)(\text{interrupting network})$$
$$< \text{breaker interrupting rating kA sym} \tag{10.11}$$

For calculations of close and latch capability or the first-cycle calculations, *no considerations of local and remote are required*. Old high-voltage breakers are rated at 1.6K times the rated short-circuit current in RMS asymmetrical. And, new breakers are rated at 2.6 times the rated short-circuit current in peak. The peak current based upon X/R is given by Equation 10.6. The results are compared with the breaker close and latch capability. The adequacy of the breaker for first-cycle or close and latch capability is given by

$$\text{Calculated } I_{\text{peak}} \text{ at the fault point } X/R < \text{breaker close and latch capability} \tag{10.12}$$

Reference [6] cautions that E/X method of calculations with AC and DC adjustments described in Section 10.5.3 can be applied, provided that X/R ratio does not exceed 45 at 60 Hz (DC time constant not greater than 120 ms). For higher X/R ratios, it recommends consulting the manufacturer. The interruption process can be affected, and the interruption window, which is the time difference between the minimum and maximum arcing times of SF_6 puffer breakers, may exceed due to delayed current zero. We will examine in the calculations to follow that the current zeros may be altogether missing for a number of cycles. This qualification has been ignored in the commercially available short-circuit calculations programs, claiming to follow IEEE standards.

10.8 Generator Source Asymmetry

For a generator circuit breaker, the highest value of asymmetry occurs, when prior to fault the generator is operating underexcited with a leading power factor. The DC component may be higher than the symmetrical component of the short-circuit current and may lead to delayed current zeros. An analysis of a large number of generators resulted in a maximum asymmetry of 130% of the actual generator current [11, 12]. The symmetrical component of the short-circuit current is 74% of generator current. Consequently, the ratio of the asymmetrical to symmetrical short-circuit current rating is 1.55.

α is a factor of asymmetry given by

$$\alpha = \frac{I_{DC}}{\sqrt{2}I_{sym}} \tag{10.13}$$

From Equation 10.13,

$$I_{rms\,asym} / I_{rms\,sym} = 1/\sqrt{2\alpha^2 + 1}$$

Thus, for $\alpha=1.3$, $I_{asym}/I_{sym}=2.09$. For $I_{sym}=0.74 I_{gen}$, ratio I_{asym}/I_{gen} can be written as

$$I_{asym} / I_{gen} = (I_{asym}/I_{sym})(I_{sym}/I_{gen}) = 1.55 \tag{10.14}$$

The asymmetry can be calculated by considering the DC component at the contact parting time. Depending on generator subtransient and transient short-circuit time constants in the direct and quadrature axes and armature time constant T_a, the AC component may decay faster than the DC component, leading to delayed current zeros.

Additional resistance in series with the armature resistance forces the DC component to decay faster. The time constant with added resistance is

$$T_a = X_d'' / [2\pi f(r+R_e)] \left[\frac{2X_d''\,X_q''}{X_d''+X_q''}\right] \tag{10.15}$$

where R_e is the external resistance, see Equation 9.20. If there is an arc at the fault point, the arc resistance further reduces the time constant of the DC component. Figure 10.12 shows that at the contact parting time, the DC component changes suddenly due to the influence of the arc voltage of the generator circuit breaker and a current zero is obtained within 1 cycle.

As we have seen generator characteristics, X/R ratio, time constants and the subtransient component of the current influence the degree of asymmetry (Figure 9.4). Also see Reference [13]. It is also dependent upon initial loading conditions. The AC component of the short-circuit current is greater in case of over-excited generator as compared to under-excited generator, but DC component is almost identical [14].

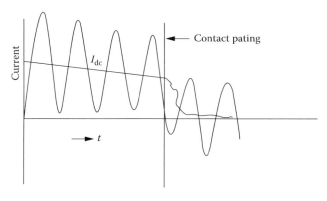

FIGURE 10.12
Fault of high asymmetry: effect of arc resistance to force a current zero at contact separation.

Short-Circuit Calculations according to ANSI Standards 439

The interruption of AC current without a current zero is equivalent to interruption of DC current, and HV breakers have very limited capabilities to interrupt DC currents. Delaying the opening of contacts may ultimately bring a current zero, but note that *both* AC and DC components are decaying, and a number of cycles may elapse before current zero is obtained. Practically, delaying the opening of contacts is not implemented.

The generator breakers are designed and tested to interrupt currents of high asymmetry. The arc interruption medium and arc control devices, i.e., arc rotation, have an effect on the interruption process and introducing an arc resistance to force current zero (Figure 10.12). Large generators and transformers are often protected as a unit and generator circuit breaker eliminated. When required, a generator circuit breaker should be carefully selected and applied. The actual asymmetry can be analytically calculated, see Example 10.2.

Generators circuit breakers (in-line circuit breakers) mounted on phase-isolated buses connecting large generators to step-up transformers are available with short-circuit capabilities of 250 kA RMS symmetrical and a continuous current rating of 100 kA. The manufacturers confirm compliance to IEEE Std. C37-013 [11] and can supply test certificates with 130% asymmetrical current interruption.

10.9 Calculation Procedure

The calculation procedure is described for hand calculations, which is instructive. Reproduction of computer printouts is avoided. Not much data preparation is required for present-day computer-based calculations. Most programs will accept *raw* impedance data, through a graphic user interface; the calculation algorithms tag it with respect to the short-circuit type and apply appropriate impedance multiplying factors, depending on the calculation type. Depending upon the in-built libraries, positive, negative, and zero sequence impedances for each component are generated and converted to a common MVA base. Transformer ratios and tap settings can be accounted for, and the adjusted impedances can be calculated. Matrix equations are solved, and results are presented in a user-friendly format. These include fault-point complex impedance, X/R ratio, and magnitude and phase angles of all the contributions to a bus from the adjacent buses. Remote and local components of the currents, fault voltages, NACD ratio, and interrupting duty and asymmetrical multiplying factors are tabulated for each faulted bus. Based on the input short-circuit ratings of the switching devices, all devices, which are overdutied from short-circuit considerations, can be flagged and the percentage over duty factors plotted. A variety of user-friendly output formats are available including tabulation of all input data, conversions to common MVA base, all generator and motor data, their subtransient reactances, outputs of sequence impedances, NACD ratios, and calculated weighted multiplying factors for breakers of different contact parting times.

10.9.1 Necessity of Gathering Accurate Data

Though the digital computer-based calculations have become user friendly and are necessary for large systems consisting of thousands of buses and machines, the accuracy of the input data must be ensured. As an example, omitting even short lengths of cables in low-voltage systems can seriously impact the results. Gathering accurate impedance data may be time consuming. An example ascertaining correct X/R ratio for the current-limiting reactors is important. These rectors have high X/R ratios, and there can be large differences with respect to the actual manufacturer's data and the built-in computer data base. This can impact the short-circuit asymmetry and the calculated results.

10.9.2 Calculation Procedure

This can be summarized in the following steps:

1. A single-line diagram of the system to be studied is required as the first step. It identifies impedances of all system components as pertinent to the short-circuit calculations. For hand calculation, a separate impedance diagram may be constructed, which follows the pattern of a single-line diagram with impedances and their X/R ratios calculated on a common MVA base. The transformer voltage ratios may be different from the base voltages considered for data reduction. The transformer impedance can be adjusted for transformer voltage adjustment taps and voltage ratios.

2. Appropriate impedance multiplying factors are applied from Table 10.1, depending on the type of calculation. For high-voltage breakers, at least two networks are required to be constructed, one for the first-cycle calculations and the other for the interrupting duty calculations.

3. A fault-point impedance positive sequence network (for three-phase faults) is then constructed, depending on the location of the fault in the system. Both resistances and reactances can be shown in this network, or two separate networks, one for resistance and the other for reactance, can be constructed.

4. For unsymmetrical faults, similarly, zero sequence and negative sequence impedance networks can be constructed and reduced to single impedance at the fault point, see Chapters 5 and 6. For rotating machines, positive and negative sequence impedances are not equal, see Chapters 5 and 12. Computer programs will calculate these networks based upon raw equipment data inputs and the built-in databases, though a user can overwrite the numbers with better/manufacturer's data.

5. For E/Z complex calculation, the fault-point positive sequence network is reduced to single impedance using complex phasors. Alternatively, the resistance and reactance values obtained by reducing separate resistance, and reactance networks to a single-point network to calculate the fault-point X/R ratio can also be used for E/Z calculation. This considerably simplifies hand calculations, compared to complex impedance reduction.

6. If there are many sources in the network, NACD is required to be calculated and that sets a limit to the complexity of networks, which can be solved by hand calculations. The currents from NACD sources have to be traced throughout the system to the faulty node to apply proper weighting factors, and this may not be easy in interconnected networks. The calculation of the first-cycle duty does not require considerations of remote or local.

7. The adjusted currents thus calculated can be used to compare with the short-circuit ratings of the existing switching equipment or selection of new equipment.

Short-Circuit Calculations according to ANSI Standards 441

10.9.3 Analytical Calculation Procedure

10.9.4 Hand Calculations

In simple systems, with limited buses not more than 6–8, the currents from the various sources can be vectorially summed at the fault point. No pre-impedance multiplying factors and post-multiplying duty factors are required. The DC component of the short-circuit current is calculated. The time constants associated with AC and DC decay are required to calculate the currents at the contact parting time of the breaker. This may not be always easy. Once each of the components is calculated, the theorem of superimposition applies and the total currents can be calculated for circuit-breaker duties.

10.9.5 Dynamic Simulation

Alternative methods of calculations of short-circuit currents that give accurate results are recognized. The short-circuit calculations can be conducted on ElectroMagnetic Transients Program (EMTP) or other digital computer programs that could emulate the transient behavior of machines during short-circuit conditions. In this sense, these simulations become more like transient stability analyses carried out in the time domain. The behavior of machines as influenced by their varying electrical and mechanical characteristics can be modeled, and the calculation accuracy is a function of the machine models, static and dynamic elements as well as the assumptions. Synchronous machines can be modeled in much greater details, IEEE standard [15] details out the synchronous machine models for transient analysis studies. This may result in a number of differential equations so that harmonics and DC offset effects can be accounted for. The inclusion of voltage regulators and excitation systems is possible. Dynamic simulation puts heavy demand on the computing resources as well as on accurate data entries, yet this forms an important verification tool for the empirical calculations.

Variations in the results of the short-circuit calculations have been studied by various authors [16–18]. An example is provided in Chapter 11. Table 10.3 shows these variations for the same system configurations using different methods of calculations. Table 10.3 shows that generally ANSI/IEEE calculations are more conservative when compared with dynamic simulation or EMTP. This table shows that in some cases, ANSI calculations are 30% higher than the corresponding results obtained with dynamic simulation. Due to large variations documented in this table, a question arises whether the calculated over-duty by a certain percentage can be ignored? This will not be a prudent approach, unless rigorous alternate techniques are explored and are acceptable to the owners of an establishment.

10.9.6 Circuit Breakers with Sources on Either Side

Circuit breakers are often applied when there are short-circuit sources on either side of the circuit breaker. In such cases, short-circuit duty needs to be calculated for the maximum through fault current. This is illustrated with reference to Figures 8.21 and 8.22.

Figure 8.21 shows two fault locations F_1 and F_2. Fault at location F_1 is fed (1) from the generator, (2) from transformer GSU_3, and also (3) from transformer UAT_2. But the generator breaker sees only components (2) and (3) and not its own contribution to fault at F_1. For a fault at location F_2, again there are three components of the fault current as before, but the generator breaker does not see components (2) and (3) and sees only its own component. Thus, its duty should be based upon the highest of these components in either direction. A similar explanation applies to faults at F_1 and F_2 in Figure 8.22. Faults on either side of a tie breaker should be considered and the first-cycle and interrupting duty based upon the

TABLE 10.3

Comparison of Short-Circuit Calculations, ANSI/IEC/EMTP and Dynamic Simulation Methods

Bus (kV)	ANSI	IEC	EMTP/Dynamic Simulation	Remarks	Reference
13.8	23.39		17.81*	Breaking/interrupting current at 50 ms	[16]
	24.87		18.52*	AC+DC	
	25.49		20.16*		
4.16	46.58		45.70*		
	35.37		36.44*		
13.8	15.45	16.31		E/Z symmetrical current	[17]
	15.36	15.21			
2.4	21.47	19.09			
220	20.99	22.57	20.30**	E/Z symmetrical current	[18]
21	74.35	74.43	60.23**		
10	25.98	23.68	17.75**		

All currents are in kA RMS symmetrical.
The dynamic simulation results shown with a single *.
EMTP simulation results shown with **.

largest of the fault currents. Apart from the absolute magnitude of the currents, the X/R ratio is also important. It is possible that smaller of the two currents, which has a high X/R ratio, can give a higher duty compared to the higher magnitude short-circuit current with a lower X/R ratio.

Figure 8.22 shows a number of feeder circuit breakers connected to 12.47 kV bus, which serve motor loads. Then for a fault on the bus, and applying superimposition theorem, the total short-circuit current is

$$\vec{I}_{t,bus} = \vec{I}_{gen} + \vec{I}_{trans} + \vec{I}_{F1} + \vec{I}_{F2} + \cdots \tag{10.16}$$

A feeder circuit breaker selected on this basis is conservatively applied, and this method is appropriate for the new installations. When evaluating an existing system, where the circuit breakers are applied close to their short-circuit ratings, the duties should be calculated based upon the maximum through fault current. Say for calculation of the interrupting duty of a feeder breaker, the entire component short-circuit currents should be considered less than the short-circuit current contributed by the loads connected to that breaker itself:

$$\overline{I}_{F1max} = \overline{I}_{t,bus} - \overline{I}_{F1} \tag{10.17}$$

A highly loaded feeder circuit breaker will have a lower short-circuit duty as compared to a lightly loaded feeder circuit breaker. Practically, for a switchgear lineup, a uniform rating of switching devices will be selected based upon the maximum short-circuit duty that any one breaker may experience. Generally, 10%–15% additional margin is allowed in the selection of the ratings for future growth. Utility source impedances invariably decreases over the course of time.

However, for the first-cycle duty, entire short-circuit components on the bus should be considered, as this bus runs common through the entire lineup of the circuit breakers. Some available

Short-Circuit Calculations according to ANSI Standards 443

commercial programs will calculate the short-circuit duties on individual breakers on a common bus depending upon the maximum duty imposed by the through fault currents in either direction. Hand calculations can also be made, but these will be laborious.

10.9.7 Switching Devices without Short-Circuit Interruption Ratings

A switching device may not be designed for short-circuit current interruption. For example, high-voltage disconnect switches, which are designed to interrupt the transformer magnetizing currents, but not the short-circuit currents. These are interlocked with the appropriate circuit breakers. Another example in industrial distribution systems is National Electrical Manufacturer's Association (NEMA) type E_2 motor starters for medium-voltage motors. These are fused with current-limiting type R fuses, which have a symmetrical interrupting rating of 50 kA, while the contactor itself has short-circuit interrupting rating of only a few kA. The fuse characteristics are coordinated to protect the contactor. The transformer primary fused load-break switches can interrupt the rated load current, but not the short-circuit current. These are protected by the current-limiting fuses in the same enclosure. Now consider a transformer load break switch without fuses. It should still have the first-cycle or momentary rating, though does not have an interrupting rating. It may be closed on to a fault and should have fault closing rating for certain duration. Calculations of first-cycle duties are still required for such devices.

Similarly, bus ducts, cables, and transformers can be cited as examples of the equipment that should withstand through fault currents. Withstand capability curves of transformers categories I through IV are in ANSI/IEEE standard [19]. The short-circuit protection of transformers is achieved through proper relaying practices, see Volume 4. Reference [20] provides guidelines for transformer primary and secondary protection.

10.9.8 Adjustments for Transformer Taps and Ratios

The transformer voltage ratio may differ from power system nominal values of base voltage chosen. Many system conditions are possible that will affect the manner in which transformer per unit impedance and base voltages are represented. Furthermore, the transformer taps, impact the voltage profile on load flow and may be purposely adjusted, see Volume 2. The standard range of the off-load taps on transformers is +5%, +2.5%, rated, −2.5%, and −5%, generally provided on the primary side because currents are low; however, the taps can be installed on the secondary windings also. Consider a 13.8–4.16 kV transformer, taps on the 13.8 kV primary winding, then the tap adjustments at *no load*, assuming that the system voltage remains at 13.8 kV, equal to the transformer primary voltage will give:

Tap +5% (14.49 kV), secondary voltage 3.952 kV

Tap +2.5% (14.145 kV) secondary voltage 4.056 kV

Rated tap (13.8 kV) secondary voltage 4.16 kV

Tap −2.5% (12.455 kV), secondary voltage 4.264 kV

Tap −5% (13.11 kV), secondary voltage 4.368 kV

See Volume 2 for practical tap setting examples.

The *tap value* is the ratio of tapped winding rated voltage divided by the rated tap voltage.

When the transformer tap voltage ratio does not equal the base voltage ratio, then a fictitious tap value can be used to resolve the difference:

$$\text{Tap}_{\text{fictitious}} = \frac{(\text{tapped winding rated tap voltage})(\text{untapped winding base voltage})}{(\text{untapped winding rated voltage})(\text{tapped winding base voltage})}$$

In Table 10.4, the transformer impedance is taken constant over a tap range, IEEE Std. 551 [4]. This may not be true, and the impedance should be obtained from transformer test reports; vendors include this data in the test results of transformers. Another point to be noted is that though the voltage on a bus may differ from its base value, but we generally ignore it for the short-circuit calculations. For comparison with the circuit-breaker ratings, the calculations are made at the rated voltage. Say a 4.16 kV bus may operate at some plus minus variation, but the short-circuit duties are calculated at 4.16 kV; note that the maximum-rated voltage of a 4.16 kV breaker is 4.76 kV.

10.10 Examples of Calculations

Example 10.1

Example 10.1 is for calculations in a multivoltage-level distribution system (Figure 10.13). Short-circuit duties are required to verify the adequacy of ratings of the switching devices shown in this single-line diagram. These devices are as follows:

1. 13.8 kV circuit breakers at buses 1, 2, and 3
2. 138 kV circuit breakers
3. 4.16 kV circuit breakers at bus 6
4. Primary switches of transformers T_3 and T_4
5. Primary fused switches of transformers T_5 and T_6
6. Type R fuses for medium-voltage motor starters, buses 7 and 8
7. Low-voltage power circuit breakers, bus 10
8. ICCB at bus 10 ($10F_5$)
9. Molded case circuit breakers at low-voltage motor control center, bus 14

Also calculate bus bracings, withstand capability of 13.8 kV #4 ACSR overhead line conductors connected to feeder breaker $2F_3$ with transformer T_7 and #4/0 cable C_1 connected to feeder breaker $2F_4$.

In this example, emphasis is upon evaluation of calculated short-circuit duties with respect to the equipment ratings. Three-phase fault calculations are required to be performed.

10.10.1 Calculation of Short-Circuit Duties

The impedance data reduced to a common 100 MVA base are shown in Table 10.5. Note that no impedance multiplying factors are applied—this table is merely a conversion of the raw equipment impedance data to a common 100 MVA base. Depending upon the type of calculation, the impedance multiplying factors from Table 10.1 are applied to the calculated impedances in Table 10.5. The fault point networks for various faulted buses can be constructed, one at a time, and reduced to a single network. As an example, the fault

Short-Circuit Calculations according to ANSI Standards

TABLE 10.4

Representing Transformers with Non-Base Voltage

Single-Line Diagram	Transformer Rated Voltage Ratio	Transformer Tap	Ratio[a]	Zt (per unit)	Zt + Zs (per unit)	Short-Circuit Current
Base kV = 13.8 kV ⋯ Prefault kV = 4.16 kV base kV = 4.16 kV	13.8–4.16 kV	13.8 kV	1.00	0.6	0.8	17.35 kA at 4.16 kV
Base kV = 13.8 kV ⋯ Prefault kV = 4.16 kV basekV = 4.16 kV	13.2–3.98 kV	13.2 kV	1.00	0.549	0.749	18.53 kA at 4.16 kV
Base kV = 13.8 kV ⋯ Prefault kV = 4.16 kV base kV = 4.16 kV	13.2–4.16 kV	13.8 kV	1.045	0.6	0.8	17.35 kA at 4.16 kV
Base kV = 13.8 kV ⋯ Prefault kV = 4.25 kV base kV = 4.16 kV	13.8–4.16 kV	13.8 kV	1.0	0.6	0.8	17.73 kA at 4.16 kV

(Continued)

TABLE 10.4 (*Continued*)

Representing Transformers with Non-Base Voltage

Single-Line Diagram	Transformer Rated Voltage Ratio	Transformer Tap	Ratio[a]	Zt (per unit)	Zt + Zs (per unit)	Short-Circuit Current
Base kV = 13.8 kV Prefault kV = 4.2667 kV base kV = (4.16×13.8)/13.445 = 4.2667	13.8–4.16 kV	13.455 kV	1.00	0.57	0.77	17.57 kA at 4.2667 kV
Base kV = 13.8 kV Prefault kV = 4.16 kV base kV = (4.16×13.8)/13.445 = 4.2667	13.8-4–16 kV	13.455 kV	1.00	0.57	0.77	17.13 kA at 4.2667 kV
Base kV = 13.8 kV Prefault kV = 4.2667 kV base kV = 4.2667 kV	13.8–4.16 kV	4.2667 kV	1.00	0.6	0.8	16.91 kA at 4.2667 kV
Base kV = 13.8 kV Prefault kV = 4.2667 kV base kV = (4.056×13.8)/13.2 = 4.240	13.8–4.16 kV	4.056 kV	1.0	0.549	0.749	17.84 kA at 4.240 kV

Note: MVA = 500 = 0.2 per unit on 100 MVA base, transformer rating = 10 MVA, and % impedance = 6% in all cases.
Ratio: (Base kV ratio)/(transformer kV ratio)

Short-Circuit Calculations according to ANSI Standards

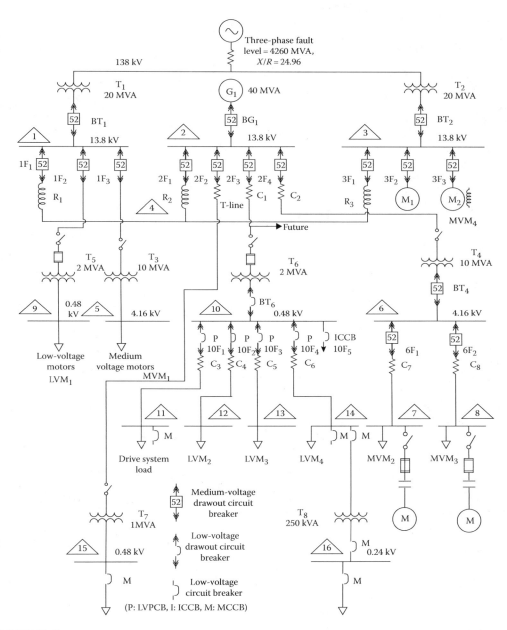

FIGURE 10.13
Single-line diagram of a multivoltage-level distribution system for short-circuit calculations (Example 10.1).

network for the 13.8 kV bus 2 is shown in Figure 10.14. Reducing it to single impedance requires wye-delta impedance transformation. The simplicity, accuracy, and speed of computer methods of solution can be realized from this exercise. The reduced complex impedance for interrupting duty calculations for bus 2 fault is $Z = 0.003553 + j0.155536$, and X/R from separate networks is 47.9; $E/Z = 26.892 < -88.7°$. All the generator contribution of 14.55 kA is a local source as the generator is directly connected to the bus. The remote (utility)

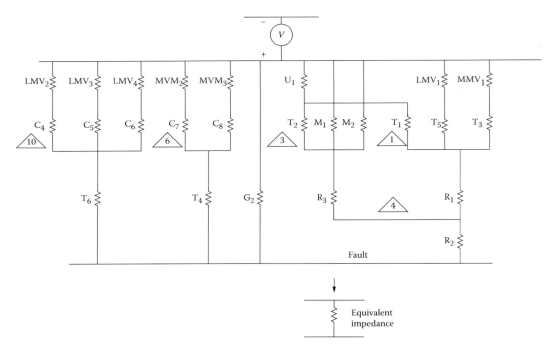

FIGURE 10.14
Positive sequence impedance diagram for a fault at bus 2 (Example 10.1).

source contributes to the fault at bus 2 through transformers T_1 and T_2 and synchronizing bus reactors. The utility's contributions through transformers T_1 and T_2, and synchronizing bus reactors are summed up. This gives 9.01 kA. The remote/total ratio, i.e., the NACD ratio is 0.335. The multiplying factor is 1.163, and the interrupting duty is 31.28 for a 5-cycle symmetrical breaker. If the calculation is based on a separate R–X method, the fault point impedance is $0.003245 + j0.15521$. This gives $E/Z = 26.895$ kA. There is not much difference in the calculations by using the two methods, though a difference up to 5% can occur. The results of calculations are shown in Tables 10.6 through 10.12.

Table 10.6 shows the interrupting and close and latch rating of the breakers according to revised IEEE standards, $K=1$. The peak first-cycle current is calculated using Equation 10.1 for the calculated X/R ratio. *The interrupting duty is based upon bus fault current.* Table 10.7 shows contributions to this bus from all other buses. The vector sum of these currents gives faulted bus 2 total current.

10.10.2 K-Rated 15 kV Breakers

If old ratings of the breakers with $K=1.3$ are considered [21], then for a standard 5-cycle symmetrical breaker rating of 15 kV, rated short-circuit current is 28 kA and close and latch is 97 kA peak or 58 kA RMS asymmetrical. The interrupting duty current at the voltage of application, 13.8 kV will be 30.43 kA RMS symmetrical. The calculated interrupting duty currents shown in Table 10.6, the multiplying factors, the X/R ratios, NACD ratio remains the same for K-rated breakers as for newly rated breakers. This shows that for these K-rated breakers, interrupting duties on bus 2 exceed the breaker ratings by 2.79% (31.28 kA calculated duty, 30.43 kA rating at 13.8 kV). It would be hasty to suggest that the entire bus 2 switchgear be replaced.

Short-Circuit Calculations according to ANSI Standards

TABLE 10.5

Impedance Data (100 MVA Base) Distribution System, Example 10.1

Symbol	Equipment Description	Per Unit Resistance	Per Unit Reactance
U_1	Utility source, 138 kV, 4260 MVA, $X/R = 25$	0.00094	0.02347
G_1	Synchronous generator, 13.8 kV, 40 MVA, 0.85 power factor, saturated subtransient reactance = 11.5%, saturated transient = 15%, $X/R = 56.7$	0.00507	0.28750
R_1, R_2, R_3	Reactors, 13.8 kV, 2 kA, 0.25 ohms, 866 kVA, $X/R = 88.7$	0.00148	0.13127
T_1, T_2	20/33.4 MVA, OA/FA, 138–13.8 kV, delta-wye transformers, $Z = 8.0\%$ on 20 MVA OA rating, $X/R = 21.9$, wye winding low-resistance grounded through 400 A, 10 s resistor	0.01827	0.39958
T_3, T_4	10/14 MVA, OA/FA, 13.8–4.16 kV, delta-wye transformer, $Z = 5.5\%$, $X/R = 15.9$, wye-winding low-resistance grounded through 200 A 10 s resistor	0.03452	0.54892
T_5, T_6	2/2.58 MVA, OA/FA, 13.8–0.48 kV, delta-wye transformer, $Z = 5.75\%$, $X/R = 6.3$, wye-winding high-resistance grounded	0.44754	2.83995
T_7	1/1.29 MVA, AA/FA, 13.8–0.48 kV delta-wye transformer, $Z = 5.75\%$, $X/R = 5.3$, wye-winding high-resistance grounded	1.06494	5.65052
T_8	250 KVA, AA, 0.48–0.24 kV delta-wye transformer, $Z = 4\%$, $X/R = 2.7$, solidly grounded	5.49916	15.02529
C_1	1-3/C #4/0 15 kV grade shielded, MV-90, IAC (interlocked armor), XLPE cable laid in aluminum tray, 200 ft (60.96 m)	0.00645	0.00377
C_2	1-3/C 500 KCMIL, 15 kV grade shielded, MV-90, IAC, XLPE cable laid in aluminum tray, 400 ft (121.92 m)	0.00586	0.00666
C_7, C_8	2-3/C, 350 KCMIL, 5 kV grade shielded, MV-90, XLPE, IAC cable, laid in aluminum tray, 100 ft	0.01116	0.00790
$C_3, C_4,$ C_5, C_6	3-3/C, 500 KCMIL, 0.6 kV grade, THNN, 90°C cables, laid in tray, 60 ft (18.3 m)	0.23888	0.23246
M_1	1 × 12,000 hp, squirrel cage induction motor, 2-pole (10,800 kVA), locked rotor reactance = 16.7%, $X/R = 46$	0.0336	1.54630
M_2	1 × 10,000 hp synchronous motor, 8-pole, 0.8 power factor (10,000 kVA), $X = 20\%$, $X/R = 34.4$	0.05822	2.0000
MVM_1	2 × 1500 hp squirrel cage induction, 2-pole, 5 × 300 hp induction motors, 4-pole, and 1 × 3500 hp, 12-pole, 0.8 power factor synchronous motor	0.21667 0.77958 0.20324	6.1851 (2 × 1500 hp) 11.719 (5 × 300 hp) 5.7142 (3500 hp)
$MVM_2,$ MVM_3	1 × 1800 hp, 12-pole, 0.8 power factor synchronous and 3 × 1100 hp, 6-pole induction motors	0.45904 0.21675	11.111 (1800 hp) 5.6229 (3 × 1100 hp)
LVM_1	Grouped 1350 hp induction motors ≥50 hp and 640 hp induction motors <50 hp	1.4829 5.4362	12.370 (>50 hp) 26.093 (<50 hp)
$LVM_2,$ $LVM_3,$ LVM_4	Grouped 300 hp induction motors ≥50 hp and 172 hp motors, <50 hp	6.0664 20.227	50.606 (>50 hp) 97.093 (<50 hp)
L_1	5200 ft long, GMD = 4 ft, ACSR conductor A-AA class (SWAN), #4 AWG [10]	1.41718	0.41718

Note: 2-3/C: Abbreviated for two three-conductor cables per phase. MV-90, XLPE, THNN: Cable insulation types. See Reference [20]. The impedance of a cable is also a function of its construction and method of installation.

TABLE 10.6

Calculated Duties on 13.8 kV Breakers (Example 10.1)

Bus No.	Breaker Type	Breaker Close and Latch Capability (kA, peak)	Breaker Interrupting Rating (kA Sym. RMS)	Calculate First-Cycle Duties (Close and Latch Capabilities)				Calculate Interrupting Duties			
				Fault Point Z Based on Complex Impedance Reduction	Fault Point X/R Ratio Based upon Separate Networks	Multiplying Factor Based on X/R	Calculated Duty (kA Peak)	Fault Point Z Based on Complex Impedance Reduction	Fault Point X/R Ratio	NACD/ Weighted Multiplying Factor	Calculated Duty (kA RMS Sym.)
1	15 kV, indoor	82	31.5	$0.005767 + j0.175930$	41.61	2.727	64.81	$0.005831 + j0.18426$	41.73	0.6/1.169	26.53
2	Oil-less, 5-cycle	Reference [21], $K = 1$		$0.003578 + j0.149318$	47.84	2.740	76.75	$0.003553 + j0.155536$	47.92	0.335/1.163	31.28
3	Sym. $K = 1$			$0.004903 + j0.159347$	41.05	2.726	71.54	$0.005291 + j0.171520$	41.16	0.571/1.162	28.33

Short-Circuit Calculations according to ANSI Standards

TABLE 10.7

Flow of Component Interrupting Duty Short-Circuit Currents from Connected Buses to 13.8 kV Bus 2, (Example 10.1)

From Bus	Current Amplitude and Angle	X/R
Gen. 2	$24.550 < -88.988°$	56.61
Bus 4	$11.021 < -88.491°$	37.96
Bus 6	$1.244 < -87.338°$	21.51
Bus 10	$0.078 < -82.965°$	8.10
Bus 15	$0 < 0°$	—
Vector sum of currents	$26.9 < -88.805$	47.92

As the duties are calculated for a bus fault, it is necessary to calculate the duties on individual breakers on this bus, by dropping the short-circuit contributions from the loads connected to each of these breakers. Neglecting the load contribution means that the fault point X/R ratio and duty multiplying factors will change. The load current component of the feeder breaker can be vectorially subtracted from the bus fault current. Tables 10.7 and 10.8 are compiled on this basis and show the interrupting duties on the feeder breakers connected to bus 2. It is observed that the feeder breaker $2F_2$ to the transmission line and the feeder breaker $2F_3$ to the 2 MVA transformer are overdutied. As the feeder breaker $2F_2$ has no rotating loads, its interrupting duty is the same as that for a bus fault. It is generally possible to retrofit these two breakers with breakers of higher interrupting rating, depending on the manufacturer and age of the equipment. The newly rated $K=1$ breakers can usually be fitted in the same breaker cubicles with minor modifications. Increasing bus tie reactor impedance by providing another reactor in series with the existing reactor could be another solution, especially if a system expansion is planned. Some remedial measures to the short-circuit problems are as follows:

- Retrofitting overdutied breakers with new breakers
- Replacing with new $K=1$ breakers or entirely new switchgear of higher ratings
- Adding current-limiting series reactors; in generator, feeder, bus, or synchronizing bus tie connections—which is very popular method of reducing short-circuit currents, but any added reactance may impact stability, may result in excessive voltage dips on load flow, see Volume 2

TABLE 10.8

Calculated Duties on Feeder Circuit Breakers on 13.8 kV Bus 2 (Example 10.1)

Breaker ID	Breaker Service	Breaker Rating Interrupting (kA, Sym. RMS)	Breaker Calculated Interrupting Duty (kA Sym. RMS)
BG_2	Generator breaker	30.43	16.94
$2F_1$	Synchronous bus		18.68
$2F_2$	13.8 kV distribution line		31.28 (overexposure = 2.79%)
$2F_3$	2 MVA transformer		31.18 (overexposure = 2.46%)
$2F_4$	10 MVA transformer		30.03

Note: Circuit breakers rated on 1979 basis $K=1.3$, Reference [21].

452   *Short-Circuits in AC and DC Systems*

- Redistribution of loads and reorganization of distribution system
- Short-circuit current limiters, Reference [22], see Section 10.12
- Duplex reactors (see Appendix C)
- Series connected devices for low-voltage systems (see Section 8.14.3.3)
- Current-limiting fuses

A detailed discussion of these topics is not covered.

10.10.3 4.16 kV Circuit Breakers and Motor Starters

Table 10.9 shows 4.16 kV metal-clad circuit-breaker ratings and calculated duties for K factor rated breakers. Table 10.10 shows a similar comparison of R-type fuses in medium-voltage motor starters, NEMA type E_2. These devices are applied much below their short-circuit ratings. Note that for the fuses in the medium-voltage starters, it is not necessary to construct the interrupting duty network.

10.10.4 Transformer Primary Switches and Fused Switches

Short-circuit ratings of transformer primary switches (without fuses) are specified in terms of asymmetrical kA RMS and 10-cycle fault closing. The former rating indicates the maximum asymmetrical withstand current capability, and the latter signifies that the switch can be closed on to a fault for 10 cycles, with the maximum fault limited to specified asymmetrical fault close rating. The upstream protective devices must isolate the fault within 10 cycles. The short-circuit ratings on power fuses are discussed in Chapter 8. First-cycle calculations are required. Table 10.11 shows the comparison and that the equipment is applied within its short-circuit ratings.

10.10.5 Low-Voltage Circuit Breakers

The switching devices in low-voltage distribution should be categorized into low-voltage power circuit breakers (LVPCBs), insulated case circuit breakers (ICCBs), and molded case circuit breakers (MCCBs), as discussed in Chapter 8. These have different test power factors, depending on their type and ratings, and interrupting duty multiplying factors are different. First-cycle calculation is required for ascertaining the duties. Table 10.12 shows these calculations. It is observed that the short-circuit duties on MCCBs at buses 11 and 16 exceed the ratings. A reactor in the incoming service to these buses can be provided. Alternatively, the underrated MCCBs can be replaced or series rated devices (Chapter 8) can be considered, as there are no downstream short-circuit contributions to these buses.

10.10.6 Bus Bracings

Bus bracings are generally specified in peak and symmetrical RMS ampères, and are indicative of mechanical strength under short-circuit conditions. The mechanical stresses are proportional to I^2/d, where d is the phase-to-phase spacing. First-cycle symmetrical current is, therefore, used to compare with the specified bus bracings. In terms of asymmetrical current, the bus bracings are 1.6 times the symmetrical current. Both the symmetrical and

TABLE 10.9

Calculated Duties on 4.16 kV Breakers (Example 10.1)

Bus no.	Breaker Type	Breaker Close and Latch Capability (kA, Asym. RMS)	Breaker Interrupting Rating at Voltage of Application (kA Sym. RMS)	Calculate First-Cycle Duties (Close and Latch Capabilities)				Calculate Interrupting Duties			
				Fault Point Z Based on Complex Impedance Reduction	Fault Point X/R Ratio	Multiplying Factor Based on X/R	Calculated Duty (kA RMS Asym.)	Fault Point Z Based on Complex Impedance Reduction	Fault Point X/R Ratio	NACD/ Weighted Multiplying Factor	Calculated Duty (kA RMS Sym.)
6	4.16 kV, indoor, oil-less, 5-cycle sym.	58 (= 97 kA crest) $K = 1.24$, Reference [21]	30.43	$0.029041 + j0.517160$	18.41	1.649	41.70	$0.032647 + j0.572088$	18.03	0.475/1.00	24.22

TABLE 10.10

Short-Circuit Duties on 4.16 kV MCC (Example 10.1)

Bus No.	Motor Starter Fuse Type	Fuse Interrupting Rating (kA, RMS Sym./ Asym.)	Calculated First-Cycle Duties (Sym. and Asym.)			
			Fault Point Z Based on Complex Impedance Reduction	Fault Point X/R Ratio	Multiplying Factor Based on X/R	Calculated Duty (kA RMS Sym./ Asym.)
7, 8	Current-limiting type R	50/80	$0.037043 + j0.522930$	15.09	1.523	26.47/40.31

TABLE 10.11

Example 10.1: Calculated Duties 13.8 kV Transformer Primary Switches and Fuses

Transformer	Transformer Primary Switch/Fused Switch	Short-Circuit Ratings	Fault-Point Impedance (100 MVA Base)	Fault Point X/R	Symmetrical Current (kA RMS)	Asymmetrical Current (kA RMS)
10 MVA	Switch only	Switch 61 kA RMS asym. Fault closing 10 cycles = 61 kA RMS asym.	$0.008709 + j0.153013$	18.48	26.926	41.92
2 MVA	Fused switch, with current-limiting type class E fuse	Fuse: interrupting = 50 kA RMS, sym. = 80 kA RMS, asym.	$0.009926 + j0.153013$	16.04	27.285	41.84

TABLE 10.12

Short-Circuit Duties on Low-Voltage Circuit Breakers (Example 10.1)

Breaker Identification	Breaker Interrupting Rating (kA Sym.)	Fault Point Z per Unit (100 MVA Base)	Fault Point X/R	Multiplying Factor	E/Z	Calculated Duty (kA Sym.)
Bus 10, LVPCB	50	$0.37625 + j2.491190$	6.65	1.002	$47.73 < -81.36°$	47.82
Bus 10, ICCB	65	$0.37625 + j2.491190$	6.65	1.063	$47.73 < -81.36°$	50.73
Bus 14, MCCB₁	65	$0.588317 + j2.697419$	4.68	1	$43.57 < -77.70°$	43.57
Bus 11, MCCB	35	$0.617505 + j2.723650$	4.44	1	$43.07 < -77.23°$	43.07
Bus 15, MCCB	35	$2.480748 + j6.217019$	2.51	1	$17.97 < -68.25°$	17.97
Bus 16, 240 V MCCB	10	$6.087477 + j17.72271$	2.92	1	$12.84 < 71.04°$	12.84

asymmetrical calculated values should be lower than the ratings. It is sometimes possible to raise the short-circuit capability of the buses in metal-clad switchgear by adding additional bus supports.

The short-circuit ratings (rated short-circuit withstand current and rated short-time current) of non-phase segregated, phase-segregated, and iso-phase metal-enclosed buses are specified in IEEE Std. C37.23 [23]. The rated short-circuit withstand current is specified in

Short-Circuit Calculations according to ANSI Standards 455

kA asym. for 167 ms duration. Rated short-time current for iso-phase buses is specified in kA sym. for 1 s duration.

10.10.7 Power Cables

Power cables should be designed to withstand short-circuit currents so that these are not damaged within the total fault clearing time of the protective devices. During short circuit, approximately, all heat generated is absorbed by the conductor metal, and the heat transfer to insulation and surrounding medium can be ignored. An expression related to the size of copper conductor, magnitude of fault current, and duration of current flow is

$$\left(\frac{I}{CM}\right)^2 tF_{AC} = 0.0297 \log_{10} \frac{T_f + 234}{T_0 + 234} \tag{10.18}$$

where I is the magnitude of fault current in ampères, CM is the conductor size in circular mils, F_{ac} is the skin effect ratio or AC resistance/DC resistance ratio of the conductor, T_f is the final permissible short-circuit conductor temperature, depending on the type of insulation, and T_0 is the initial temperature prior to current change. For aluminum conductors [24], this expression is

$$\left(\frac{I}{CM}\right)^2 tF_{AC} = 0.00125 \log_{10} \frac{T_f + 228}{T_0 + 228} \tag{10.19}$$

where F_{ac} is given in Table 10.13 [25]. The short-circuit withstand capability of 4/0 (211600 CM) copper conductor cable of 13.8 kV breaker $2F_4$ has a short-circuit withstand capability of 0.238 s.

TABLE 10.13

AC/DC Resistance Ratios: Copper and Aluminum Conductors at 60 Hz and 65°C

Conductor Size (KCMIL or AWG)	5–15 kV Nonleaded Shielded Power Cable, 3 Single Concentric Conductors in Same Metallic Conduit	
	Copper	Aluminum
1000	1.36	1.17
900	1.30	1.14
800	1.24	1.11
750	1.22	1.10
700	1.19	1.09
600	1.14	1.07
500	1.10	1.05
400	1.07	1.03
350	1.05	1.03
300	1.04	1.02
250	1.03	1.01
4/0	1.02	1.01
3/0	1.01	<1%
2/0	1.01	<1%

This is based on an initial conductor temperature of 90°C, a final short-circuit temperature for cross-linked polyethylene (XLPE) insulation of 250°C, and a fault current of 31.18 kA sym. (F_{ac} from Table 10.13=1.02). The breaker interrupting time is 5 cycles, which means that the protective relays must operate in less than 9 cycles to clear the fault. Major cable circuits in industrial distribution systems are sized so that these are not damaged even if the first zone of protective relays (instantaneous) fails to operate and the fault has to be cleared in the time-delay zone of the backup device. From these criteria, the cable may be undersized.

10.10.8 Overhead Line Conductors

Calculations of short-circuit withstand ratings for overhead line conductors must also receive similar considerations as cables, i.e., these should be sized not only for load current and voltage drop consideration, but also from short-circuit considerations. For aluminum conductor steel reinforced (ACSR) conductors, a temperature of 100°C (60°C rise over 40°C ambient) is frequently used for normal loading conditions, as the strands retain approximately 90% of rated strength after 10,000 h of operation. Under short circuit, 340°C may be selected as the maximum temperature for all aluminum conductors and 645°C for ACSR, with a sizable steel content. An expression for safe time duration based on this criterion and no heat loss during short circuit for ACSR is [24]

$$t = \left(0.0862 \frac{\text{CM}}{I}\right)^2 \tag{10.20}$$

where t is the duration in seconds, CMs is the area of conductor in circular mils, and I is the current in ampères, RMS.

From Equation 10.20, # 4 (41740 CM) ACSR of the transmission line connected to breaker $2F_2$ has a short-circuit withstand capability of 0.013 s for a symmetrical short-circuit current of 31.28 kA close to bus 2. The conductors, though adequately sized for the load current of a 1 MVA transformer, are grossly undersized from short-circuit considerations.

Example 10.2

This example explores the problems of application of a generator circuit breaker for high X/R ratio fault, when the natural current zero is not obtained because of high asymmetry. The calculation is based upon IEEE standard [11]. Figure 10.15 shows a generating station with auxiliary distribution system. A 112.1 MVA generator is connected through a step-up transformer to supply power to a 138 kV system. The generator data are the same as presented in Table 9.1. Auxiliary transformers of 7.5 MVA, 13.8–4.16 kV, and 1.5 MVA, 4.16–0.48 kV supply medium- and low-voltage motor loads. The generator neutral is high-resistance grounded, through a distribution transformer, the 4.16 kV system is low-resistance grounded, and the 480 V system is high-resistance grounded. Thus, breaker duties are based on three-phase fault currents. The generator auxiliary distribution systems are designed with double-ended substations, for reliability and redundancy. Here, we ignore the alternate source of power.

Three-Phase Fault at F_1 (138 kV), E/Z Method

The fault at F_1 is fed by three sources: utility source, generator, and the motor loads from the auxiliary distribution. The utility contribution predominates, and for selection of 138 kV breaker, the generator contribution and motor contribution can be ignored. However, this breaker will also be required to interrupt the generator and

Short-Circuit Calculations according to ANSI Standards

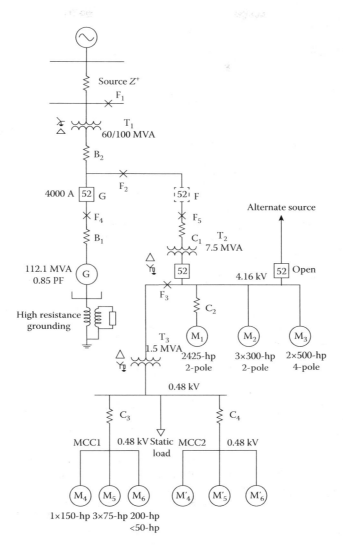

FIGURE 10.15
A generating station single-line diagram for short-circuit calculations (Example 10.2).

motor contribution currents through the step-up transformer and TRV, can be of concern. Table 10.14 gives the impedance data for all the system components broken down in per unit R and X on a common 100 MVA base. The calculation is carried out on per unit basis, and the units are not stated at each step. The effective generator resistance is calculated from the following expression in Table 10.2:

$$\text{Effective resistance} = \frac{X_{2v}}{2\pi f T_{a3}} \tag{10.21}$$

Using the generator data from Table 9.1, $X/R = 130$. Correct input of X/R ratios for generators and large reactors is important, as it may have a pronounced effect on AC and DC decay.

TABLE 10.14

Impedance Data (Example 10.2)

Description of Equipment	Per Unit Resistance on a 100 MVA Base	Per Unit Reactance on a 100 MVA Base
Utility's 138 kV source, three-phase fault level = 4556 MVA, X/R ratio = 13.4	0.00163	0.02195
112.1 MVA generator, saturated sub transient = 16.4% data in Table 9.1	0.001133	0.14630
Transformer T_1, 60/100 MVA, $Z = 7.74\%$, $X/R = 32$	0.00404	0.12894
Transformer T_2, 7.5 MVA, $Z = 6.75\%$, $X/R = 14.1$	0.06349	0.89776
Transformer T_3, 1.5 MVA, $Z = 5.75$, $X/R = 5.9\%$	0.63909	3.77968
13.8 kV cable C_1, 2-1/C per phase, 1000 KCMIL, in steel conduit, 80 ft	0.00038	0.00101
4.16 kV cable C_2, 1-1/C per phase, 500 KCMIL, in steel conduit, 400 ft	0.06800	0.10393
0.48 kV cables C_3 and C_4, 3-1/C per phase, 750 KCMIL, in steel conduit, 150 ft	0.46132	0.85993
M_1, 2425 hp, 2-pole induction motor	0.23485	7.6517
M_2, 300 hp, 2-pole induction motors, 3 each	1.2997	19.532
M_3, 500 hp, 2-pole induction motors, 2 each	0.90995	17.578
M_4 and M'_4, 150 hp, 4-pole induction motor	11.714	117.19
M_5 and M'_5, 75 hp, 4-pole induction motors, 3 each	10.362	74.222
M_6 and M'_6, 200 hp induction motors, lumped, <50 hp	20.355	83.458
B_1, 5 kA bus duct, phase-segregated, 40 ft	0.00005	0.00004
B_2, 5 kA bus duct, phase-segregated, 80 ft	0.00011	0.00008

Appropriate impedance multiplying factors from Table 10.1 are used before constructing the positive sequence fault point network. The impedance multiplying factor for calculation of first-cycle or interrupting duties is 1 for all turbogenerators. The motor impedances after appropriate multiplying factors for the first-cycle and interrupting duty calculation are shown in Table 10.15. A positive sequence impedance network to the fault point under consideration may now be constructed, using modified impedances. Low-voltage motors of <50 hp are ignored for interrupting duty calculations. Figure 10.16 shows the interrupting duty network for a fault at F_1.

The result of reduction of impedance of the network shown in Figure 10.16 with complex phasors gives an interrupting duty impedance of $Z = 0.001426 + j0.020318$ per unit.

The X/R ratio is calculated by separate resistance and reactance networks, i.e., the resistance network is constructed by dropping out the reactances in Figure 10.16, and the reactance network is constructed by dropping out the resistances in Figure 10.16. This gives an X/R ratio of 16.28. The interrupting duty fault current is, therefore, $E/Z = 20.54$ kA symmetrical at 138 kV. To calculate the interrupting duty, NACD is required. This being a radial system, it is easy to calculate the local (generator) contribution through the transformer impedance, which is equal to 1.51 kA. The utility source is considered remote, and it contributes 19.01 kA; NACD = 0.925. A 138 kV breaker will be a 3-cycle breaker, with a contact parting time of 2 cycles. The multiplying factor from Figure 10.11 is, therefore, equal to 1.

The first-cycle network will be similar to that shown in Figure 10.14, except that the motor impedances will change. This gives a complex impedance of $Z = 0.001426 + j0.020309$. The X/R ratio is 16.27, and the first-cycle symmetrical current is 20.55 kA, very close to the interrupting duty current. The peak asymmetrical current is calculated from Equation 10.1 and is equal to 53.1 kA.

The effect of motor loads in this case is small. If the motor loads are dropped and calculations repeated, the interrupting duty current = first-cycle current = 20.525 kA sym.,

Short-Circuit Calculations according to ANSI Standards

TABLE 10.15

Motor Impedances after Multiplying Factors in per Unit at 100 MVA Base

Motor ID	Quantity	Interrupting Duty MF	First-Cycle MF	Interrupting Duty Z per Unit, 100 MVA Base	First-Cycle Z, per Unit 100 MVA Base
M_1	1	1.5	1.0	$0.352 + j11.477$	$0.2348 + j7.6517$
M_2	3	1.5	1.0	$1.949 + j29.298$	$1.2997 + j19.532$
M_3	2	3.0	1.2	$2.7298 + j52.734$	$1.0919 + j21.094$
M_4, M'_4	1	3.0	1.2	$35.142 + j140.628$	$14.057 + j140.628$
M_5, M'_5	3	3.0	1.2	$31.086j + j89.066$	$12.2134 + j89.066$
M_6, M'_6	Group < 50 hp	∞	1.67	∞	$33.99 + j139.380$

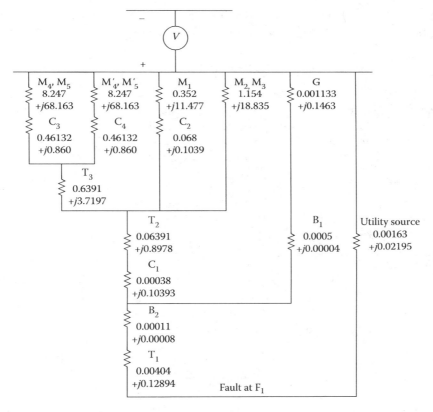

FIGURE 10.16
Positive sequence impedance diagram for a fault at F_1 (Example 10.2).

i.e., a difference of only 0.13%. This is because the low-voltage motors contribute through impedances of transformers T_1, T_2, and T_3 in series, and medium-voltage motors contribute through two-stage transformations. The current contributed by small induction motors and small synchronous motors in utility systems can, usually, be ignored except station service supply systems or at substations supplying industrial distribution systems or locations close to large motors, or both. Motor contributions increase half-cycle current more than the symmetrical interrupting current at the contact parting time.

10.10.9 Generator Source Symmetrical Short-Circuit Current

The generator has a high X/R ratio, and according to qualifications of E/Z method, the calculations with adjustments of AC and DC adjustment factors can be applied provided X/R does not exceed 45, an analytical calculation based upon Example in IEEE standard [11] is carried out. No pre-multiplying impedance factors and post-multiplying duty factors are applicable. The calculation should account for short-circuit current decay at the contact parting time. The symmetrical current at the contact parting time (3 cycles contact parting time = 50.0 ms) can be calculated using Equation 6.13. When an external reactance is added to the generator circuit, the subtransient component of the current is given by

$$i_d'' = \frac{e''}{X_d'' + X_e} \tag{10.22}$$

where X_e is the external reactance. Similar expressions apply for i_d' and i_d''. The time constants are also changed, i.e., the short-circuit transient time constant in Equation 6.24 becomes

$$T_d' = T_{do}' \frac{X_d' + X_e}{X_d + X_e} \tag{10.23}$$

This means that adding an external reactance is equivalent to increasing the armature leakage reactance. The calculations use the generator data in Table 9.1. The short-circuit subtransient time constant is 0.015 s. From Table 10.4, the reactance of the bus duct B_1 is 0.00004 per unit. Its effect on the subtransient time constant form is

$$T_d'' = T_{do}'' \frac{X_d'' + X_e}{X_d' + X_e} = (0.22)\left(\frac{0.193 + 0.00004}{0.278 + 0.00004}\right) = 0.015 \tag{10.24}$$

As the bus B_1 reactance is small, there is not much change in the subtransient time constant. However, this illustrates the procedure. The transient time constant is also practically unchanged at 0.597 s.

The generator voltage behind subtransient or transient reactances is equal to its rated terminal voltage, as the generator is considered at no load and constant excitation. The procedure of calculation is similar to that illustrated in Example 6.1 for calculation of a fault decrement curve. The subtransient current is

$$i_d'' = \frac{E''}{X_d'' + X_e} = \frac{1}{0.14630 + 0.00004} \text{ per unit} = 28.59 \text{ kA} \tag{10.25}$$

We need not drop the resistance. The generator X/R is high, and the bus duct B_1 has a resistance of 0.00005 ohms. The calculated fault point impedance including resistance is $0.001183 + j0.14634$ per unit. Therefore, to be more precise, this gives a current of 28.588 kA at $<-89.543°$ ($X/R = 123.7$)

Similarly, the transient and steady-state components of the currents are 20.17 and 2.15 kA, respectively. The following equation can, therefore, be written for symmetrical current in kA:

$$i_{AC} = 8.42e^{-t/0.015} + 18.02e^{-t/0.597} + 2.15 \tag{10.26}$$

At contact parting time, this gives the symmetrical current component as 18.95 kA.

Short-Circuit Calculations according to ANSI Standards 461

10.10.10 Generator Source Asymmetrical Current

The asymmetrical current at contact parting can also be calculated from Equation 6.113. As the generator subtransient reactances in the direct and quadrature axes are approximately equal, the second frequency term in this equation can be neglected and the DC component at contact parting time needs to be calculated. The decaying DC component is given by

$$i_{DC} = \sqrt{2}\, i_d'' e^{-t/T_a} = 40.43 e^{-t/T_a} \tag{10.27}$$

The effect of external resistance T_a should be considered according to Equation 10.11. Table 9.1 shows an armature time constant of 0.33 s; considering the bus duct B_1 resistance of 0.00005 per unit, the time constant is reduced to approximately 0.32 s. At contact parting time, the DC component has decayed to 34.58 kA. The asymmetry factor α at contact parting time is calculated as follows:

$$\text{Asymmetrical AC current at contact parting time} = \sqrt{2}(18.95) = 26.80\,\text{kA} \tag{10.28}$$
$$\text{DC current at contact parting} = 34.58\,\text{kA}$$

Factor $\alpha = 34.58/26.80 = 1.29$, i.e., the asymmetry at contact parting time is approximately 129% and the current zero is not obtained. The total RMS asymmetrical breaking current at contact parting time is, therefore,

$$\sqrt{18.95^2 + (35.58)^2} = 39.43\,\text{kA}$$

The example in Appendix A of IEEE Std. C37.013 [11] does not calculate the generator source asymmetrical current analytically, though it provides the mathematical equation on which the above calculation is based. The computer simulation shows reduced asymmetry factor by approximately 3%. Further simulation with arc voltage shows a reduction of asymmetry factor to 68%. A free burning arc in air has a voltage of 10 V/cm. Further, demonstrating the capability of a generator breaker to interrupt short-circuit currents with delayed current zeros may be difficult and limited in high power testing stations. In contrast, note that IEEE Std. C37.010 [6] considers bolted fault currents for ascertaining the duties of circuit breakers and any arc fault resistance is neglected. A fault resistance may introduce enough arc resistance to force the current to zero after the contact parting time (Figure 10.12). Simplified empirical/analytical calculations in such cases have limitations, and a dynamic simulation is recommended.

As stated in Chapter 8, symmetrical short-circuit rating of a generator breaker is not specified, but the generator source asymmetrical capability states that the generator breaker should be capable of interrupting up to 110% DC component, based upon the peak value of the symmetrical current. Based upon the above calculation results, the DC component is 34.58 kA. Therefore, the three-phase symmetrical interrupting must be at least 22.2 kA RMS symmetrical.

10.10.11 System Source Symmetrical Short-Circuit Current

We can now conduct analytical calculation for the fault at F_4. Utility source, transformer T_1, and bus duct B_2 impedance in series give $Z = 0.00577 + j0.15097$; $X/R = 26.16$ from the separate X and R networks. This gives a short-circuit current of 27.692 kA at an angle of $<-87.81°$.

462   *Short-Circuits in AC and DC Systems*

The equivalent impedance of low-voltage motors through cables and a 1.5 MVA transformer plus medium-voltage motor loads through a 7.5 MVA transformer, as seen from the fault point F_4, is $0.307 + j4.356$. No impedance multiplying factors to adjust motor impedance from Table 10.1 are used in this calculation. This gives a short-circuit current of 0.958 kA at $<-85.968°$; $X/R = 14.2$. The time constant for the auxiliary distribution is, therefore,

$$\frac{1}{2\pi f}(X/R) = 37.66 \, \text{ms} \tag{10.29}$$

The AC symmetrical component of the auxiliary system decays, and at contact parting time, it can be *assumed* to be 0.7–0.8 times the initial short-circuit current. Therefore, the contribution from the auxiliary system is $= 0.766$ kA. No decay is applicable to the utility source connected through the transformer. The total system source symmetrical current is

$$I_{\text{sym}} = 27.692 + 0.766 = 28.458 \, \text{kA rms}.$$

The total fault point X/R calculated from separate R and X networks is 26.

10.10.12 System Source Asymmetrical Short-Circuit Current

The asymmetrical current is now calculated. The time constant of the utility's contribution for an X/R of 26.16 is 0.0694 s. The time constant of the auxiliary distribution was calculated as 37.66 ms in Equation 10.29. The total DC current at the contact parting time is the sum of components from the utility's source and auxiliary distribution. Thus, the total DC current is given by

$$I_{\text{DC}} = \sqrt{2}[27.692 e^{-50.0/69.4} + 0.958 e^{-50.0/37.66}]$$

This gives $I_{\text{DC}} = 19.14$ kA. Asymmetrical current is given by

$$I_{\text{asym}} = \sqrt{I_{\text{sym}} + I_{\text{DC}}}$$
$$= 34.44 \, \text{kA rms}$$

The asymmetry factor $\alpha = 19.14 / (\sqrt{2} \, 28.458) = 0.476$. There is no problem of not obtaining a current zero. This means that the DC component at the primary arcing contact parting time is 47.6% of the peak value of the symmetrical system-source short-circuit current. Referring to Figure 8.20, a 5-cycle symmetrical rated generator breaker with a contact parting time of 3 cycles has a DC interrupting capability = 68.5%.

10.10.13 Required Closing Latching Capabilities

The ratio of the maximum asymmetrical short-circuit peak current at ½ cycle to the rated short-circuit current of the generator breaker is determined from the following equation [11]:

$$\frac{I_{\text{peak}}}{I_{\text{sym}}} = \sqrt{2} \times \left(1 + e^{\frac{t}{133}}\right) = 2.74 \tag{10.30}$$

Short-Circuit Calculations according to ANSI Standards 463

where time t can be approximately taken as ½ cycle.

We can calculate the close and latch current using Equation 10.1 that requires X/R ratio, based upon separate R and X networks. For the generator source current, $X/R=127.7$, and symmetrical interrupting current is 22.2 kA; for the system source current, $X/R=26$, and symmetrical interrupting current is 28.485 kA. These give close and latch current of

62.04 kA peak, generator source current

76.05 kA peak system source current.

The higher of the two values, that is, 76.05 kA, should be selected.

10.10.14 Selection of the Generator Breaker

Based upon the above calculations, a generator breaker of 30 kA sym. interrupting capability can be selected. This will have $30 \times 2.74=82.2$ kA close and latch capability, 5-cycle symmetrical rating, 3-cycle contact parting time, DC component interrupting capability=68.5%, generator source three-phase fault interrupting capability=25 kA, and generator source DC interrupting capability=110% of the peak of the generator source three-phase fault capability=38.9 kA. Rated continuous current=5000 A, and short-time current for $1\,s = 30$ kA. The TRV profile for a fault at F_1 should be calculated and simulated. Other specifications like out-of-phase switching capabilities are required depending upon the system conditions.

While these specifications may seem adequate, specifying a certain asymmetry is no guarantee that the breaker will be able to interrupt the non-zero currents at the calculated asymmetry. Though the standards state that practically, the asymmetry factor will be reduced on account of arc voltage, it is prudent to add to the specifications the calculated asymmetry. Generator breakers meeting 130% asymmetry requirements are commercially available.

Example 10.3

What are the limitations of using a circuit breaker shown dotted for the primary protection of 7.5 MVA unit auxiliary transformer in Figure 10.15?

For application of a breaker for primary protection of the UAT, through fault currents for faults at F_2 and F_5 can be considered. Fault at F_5 gives higher duties. Only the short-circuit contributions from distribution system rotating loads can be ignored and the short-circuit contributions from utility and generator source should be summed up. E/Z calculation gives a short-circuit current of 56.27 kA at $X/R=75.81$. The calculated interrupting duty becomes 71.69 kA, which is beyond the ratings of any commercially available oil-less metal-clad or cubical-type circuit breaker at 13.8 kV.

Generally, in the generating stations, no breaker is used to selectively isolate the UAT faults. The differential protection is arranged to take care of this fault condition, and both the generator breaker and utility 13.8 kV tie breakers are tripped.

Example 10.4

Construct bus admittance and impedance matrices of the network for Example 10.2 for interrupting duty calculations. Compare the calculated values of self-impedances with the values arrived at in Example 10.2.

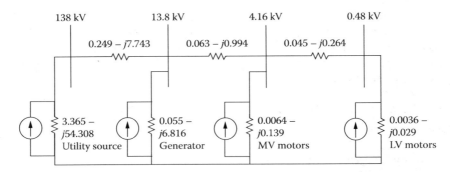

FIGURE 10.17
Equivalent admittance diagram of system of Figure 7.15 (Example 10.2).

The system, in terms of admittances, can be modeled as shown in Figure 10.17. The bus admittance matrix can be written by examination as follows:

$$Y_{bus} = \begin{vmatrix} 3.614-j53.051 & -0.249+j7.743 & 0 & 0 \\ -0.249+j7.743 & 0.367-j15.553 & -0.063+j0.994 & 0 \\ 0 & -0.063+j0.994 & 0.1144-j1.397 & -0.045+j0.264 \\ 0 & 0 & -0.045+j0.264 & 0.0486-j0.293 \end{vmatrix}$$

The bus impedance matrix is

$$Z_{bus} = \begin{vmatrix} 0.0014+j0.0203 & 0.0006+j0.0107 & 0.0005+j0.0092 & 0.00045+j0.0083 \\ 0.0006+j0.0107 & 0.0017+j0.0736 & 0.0014+j0.0632 & 0.0010+j0.0570 \\ 0.0005+j0.0092 & 0.0014+j0.06321 & 0.0551+j0.9135 & 0.0460+j0.8239 \\ 0.00045+j0.0083 & 0.0010+j0.0570 & 0.0460+j0.8239 & 0.5892+j4.0647 \end{vmatrix}$$

The self-impedances Z_{11}, Z_{22}, etc. compare well with the values calculated from positive sequence impedances of the fault point networks, reduced to a single impedance. The X/R ratio should not be calculated from complex impedances, and separate R and X matrices are required. Zero values of elements are not acceptable.

10.11 Deriving an Equivalent Impedance

A section of a network can be reduced to single equivalent impedance. This tool can prove useful when planning a power system. Consider that in Figure 10.13, the 13.8 kV system is of interest for optimizing the bus tie reactors or developing a system configuration. A number of computer runs may be required for this purpose. The distribution connected to each of the 13.8 kV buses can be represented by equivalent impedances, one for interrupting duty calculation and the other for first-cycle calculation. Figure 10.18 shows these equivalent impedances. These can be derived by a computer calculation or from the vectorial summation of the short-circuit currents contributed to the buses.

Short-Circuit Calculations according to ANSI Standards

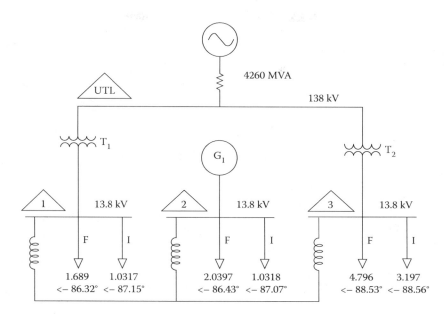

FIGURE 10.18
Equivalent impedances of the distribution system connected to 13.8 kV buses in the distribution system of Figure 10.16. F: first-cycle impedance; I: interrupting-duty impedance.

This concept can be used in subdividing a large network into sections with interfaced impedances at their boundaries, representing the contributions of the connected systems. Attention can then be devoted to the section of interest, with less computer running time and saved efforts in modeling and analyzing the output results. Once the system of interest is finalized, its impact on the interfaced systems can be evaluated by detailed modeling.

Example 10.5

In the majority of cases, three-phase short-circuit currents give the maximum short-circuit duties; however, in some cases, a single line-to-ground fault may give a higher short-circuit duty. Consider the large generating station shown in Figure 10.19. Fault duties are required at point F. The source short-circuit MVA at 230 kV for a three-phase and single line-to-ground fault is the same, 4000 MVA, $X/R=15$. (10.04 kA < $-86.19°$). Each generator is connected through a delta-wye step-up transformer. The high-voltage wye neutral is solidly grounded. This is the usual connection of a step-up transformer in a generating station, as the generator is high-impedance grounded. With the impedance data for the generators and transformers shown in this figure, the positive sequence fault point impedance = $0.00067+j0.0146$ per unit, 100 MVA base. Thus, the three-phase short-circuit current at $F=17.146$ kA sym. Each generator contributes 2.31 kA of short-circuit current. The zero sequence impedance at the fault point = $0.00038+j0.00978$, and the single line-to-ground fault current at $F=19.277$ kA, approximately 12.42% higher than the three-phase short-circuit current.

Example 10.6

Figure 10.20a shows three motors of 10,000 hp, each connected to a 13.8 kV bus, fed from a 138 kV source through a single step-down transformer. For a fault on load side of breaker 52, calculate the motor contributions at the 13.8 kV bus, and the first-cycle and

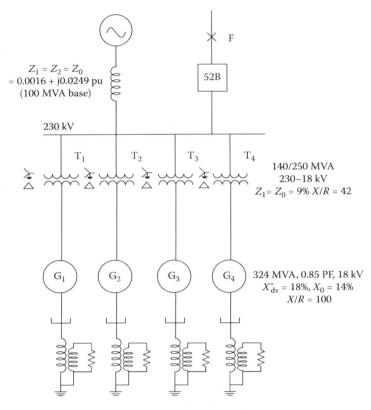

FIGURE 10.19
Single-line diagram of a large generating station for calculation of fault currents (Example 10.4).

interrupting duty currents by simplified ANSI methods, and by analytical calculations for a 3-cycle breaker contact parting time. Repeat the calculations with series impedances added in the motor circuit as shown in Figure 10.20b.

Simplified Calculations

The source impedance in series with the transformer impedance is same for the first-cycle and interrupting duty calculations. This gives Z=0.0139+0.2954 per unit on a 100 MVA base. The source contribution through the transformer is, therefore, 14.147 kA at <−87.306°.

The first-cycle impedance multiplying factor for a 10,000 hp motor is unity. The motor kVA based on a power factor of 0.93 and efficiency of 0.96=8375; X/R=35. Therefore, the per unit locked rotor impedance for three motors in parallel is 0.0189+j0.0667. This gives a first-cycle current of 6.270 kA at <−88.375°. The total first-cycle current is 20.416 kA sym. at <−87.63°; X/R=25.56. This gives an asymmetrical current of 32.69 kA.

For interrupting duty, the impedance multiplying factor is 1.5. The equivalent motor impedance of three motors in parallel is 0.02835+j1.0005 per unit. This gives a motor current contribution of 4.180 kA at <−88.375°. The total current is 18.327 kA < −87.55°; X/R=24.45. The E/X multiplying factor for DC decay at a 3-cycle contact parting time is 1.086. The calculated interrupting duty current is 19.91 kA.

Short-Circuit Calculations according to ANSI Standards

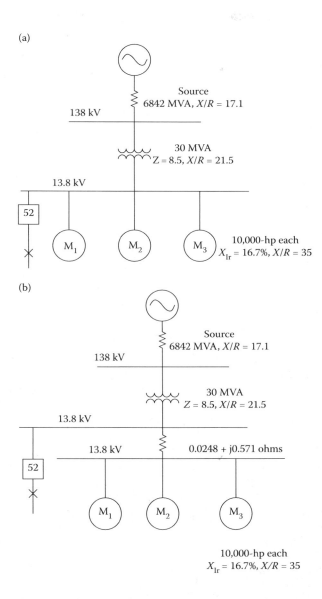

FIGURE 10.20
Example 10.5: calculation of short-circuit current contributions from large motors. (a) Motors directly connected to the 13.8 kV bus and (b) motors connected through common impedance to the 13.8 kV bus.

When a series impedance of 0.015 + 0.30 per unit (100 MVA base) is introduced into the circuit, the following are the results of the calculation:

First-cycle motor contribution = 4.324 kA < −87.992°
Total first-cycle current = 18.47 kA sym. < −87.4°; $X/R = 22.95$
Asymmetry multiplying factor = 1.5888
Asymmetrical current = 29.33 kA RMS
Interrupting duty motor contribution = 3.215 < −88.09°
Total current = 17.326 < −87.45°; $X/R = 22.87$
E/X multiplying factor = 1.070
Calculated interrupting duty current = 18.57 kA

Analytical Calculation

We will recall Equations 9.128 and 9.129 and Example 9.9. The equations for AC and DC components of the short-circuit currents are reproduced below:

$$i_{AC} = \frac{E}{X'}e^{-t/T'}$$

$$i_{DC} = \sqrt{2}\,\frac{E}{X'}e^{-t/T_{DC}}$$

We showed that X' is the locked rotor reactance of the motor. This can be replaced with Z'. $T_{dc}=X'/\omega r_1$ (Equation 9.132), where r_1 is the stator resistance. Therefore, X'/r_1 is in fact the ANSI X/R ratio. When external impedance is added:

$$T_{DC} = \frac{X' + X_e}{\omega(r_1 + R_e)} \tag{10.31}$$

where R_e and X_e are the external resistances and reactances, respectively. Similarly, the time constant T' becomes

$$T' = \frac{X' + X_e}{\omega r_2} \tag{10.32}$$

We will first calculate the motor first-cycle and interrupting current contributions, without series impedance in the motor circuit; $X' = 0.0189 + j0.667$ as before. The time constants of large motors are generally specified by the manufacturers. For the example, $T_{dc}=0.093$ and $T'=0.102$. The AC and DC current equations for the motor are

$$i_{ACm} = 6.27\varepsilon^{-t/0.102} \qquad i_{DCm} = 8.87\varepsilon^{-t/0.093}$$

At half cycle, the AC and DC currents are 5.28 and 8.10 kA, respectively. The source impedance in series with the transformer is $0.0139 + j0.2954$ per unit. Therefore, $X/R=21.25$, and the symmetrical current is 14.147 kA. The equation for the DC current from the source is

$$i_{dcs} = 20e^{-t/0.0564}$$

At half cycle, the DC component from the source is 17.26 kA. Therefore, the total symmetrical current is $5.28+14.147=19.427$ kA, and the DC current is 25.36 kA. This gives an asymmetrical current of 31.94 kA. We calculated 32.69 kA by the simplified method.

The interrupting currents are calculated at 3 cycles. The source DC component is 8.24 kA, the motor DC component is 5.17 kA, the source AC component is 14.147 kA, and the motor AC component is 3.84 kA. This gives a symmetrical current of 17.987 kA and an asymmetrical current of 22.44 kA. The interrupting duty to compare with a 5-cycle breaker is $22.444/1.1=20.39$ kA. We calculated 19.91 kA by the simplified method.

Now consider the effect of series resistance in the motor circuit. The motor impedance plus series impedance is $(0.0189+j0.0667)+(0.015+j0.30)=0.0239+j0.967$ per unit. This gives a current of 4.32 kA.

The time constants are modified to consider external impedance according to Equations 10.31 and 10.32:

$$T_{DC} = \frac{0.667 + 0.30}{\omega(0.0189 + 0.015)} = 0.0757$$

Short-Circuit Calculations according to ANSI Standards 469

Also, T' is

$$T' = \frac{0.667 + 0.30}{\omega(0.0173)} = 0.0148$$

The equations for motor AC and DC components of currents are then

$$i_{acm} = 4.32e^{-t/0.148} \quad i_{dcm} = 6.13e^{-t/0.0757}$$

This gives a total first-cycle current of 29.14 kA RMS (versus 29.33 kA by the simplified method) and a symmetrical interrupting current of 17.23 kA; the DC component at contact parting time is 11.4 kA. The total asymmetrical current is 20.65 kA. The current for comparison with a 5-cycle circuit breaker is 20.65/1.1 = 18.77 kA. We calculated 18.57 kA with the simplified calculations. The results are fairly close in this example, though larger variations can occur.

10.12 Thirty-Cycle Short-Circuit Currents

Thirty-cycle short-circuit currents are required for overcurrent devices co-coordinated on a time-current basis. The 30-cycle short-circuit current is calculated on the following assumptions:

- The contributions from the utility sources remain unchanged.
- The DC component of short-circuit current decays to zero.
- The contribution from the synchronous and induction motors decays to zero.
- The generator subtransient reactance is replaced with transient reactance or a value higher than the subtransient reactance.

The 30-cycle current of the synchronous generators and motors will vary depending upon the excitation systems. If the power for excitation system is taken from the same bus to which the motor is connected, then for a fault close the motor, the voltage reduces practically to zero and the field current will quickly decay depending upon machine time constants. If the excitation power is taken from an independent source, not affected by short circuit, the excitation will be maintained. This can be an important factor for machine stability. For brushless excitation systems of synchronous motors, the excitation power required is small, which can be provided by a battery source or uninterruptible power system (UPS) enhancing the motor stability.

Example 10.7

Calculate 30-cycle currents in the distribution system of Figure 10.13 (Example 10.1).

The generator reactance is changed to transient reactance, and all motor contributions are dropped. Table 10.16 shows the results, which can be compared with interrupting duty currents. The decay varies from 8% to 28%. The buses that serve the motor loads show the highest decay.

TABLE 10.16

Thirty-Cycle Currents, Distribution System (Figure 10.13)

Bus Identification	30-Cycle Currents in kA Sym.
13.8 kV Buses	
1	19.91 < −88.12°
2	21.11 < −88.68°
3	19.91 < −88.12°
4	22.73 < −88.30°
4.16 kV Buses	
5	18.26 < −86.88°
6	18.38 < −86.59°
7	18.17 < −85.79°
8	18.17 < −85.79°
Low-Voltage Buses	
9	39.01 < −81.53°
10	39.10 < −81.43°
11	35.93 < −77.98°
12	35.93 < −77.98°
15	17.85 < −68.39°
16	12.45 < −71.29°

10.13 Fault Current Limiters

Under Example 10.1, we stated the application of current-limiting reactors and other devices for the short-circuit limitations. Even the maximum short-circuit rated breakers may not be adequate in some applications, and system designs, not discussed, should limit the short-circuit duties to acceptable values. This is true for utility systems:

- Electrical demand increases and so the generation.
- Parallel conducting paths are added.
- Interconnections within grid increase.
- Sources of distribution generation are added.

A fault current limiter (FCL) or a triggered current limiter (TCL) was developed under EPRI project RP1142, though a similar device has existed in Europe for a period of more than 40 years [26]. A TCL limits the fault current in time and magnitude, operating in ¼–½ cycles and offering no resistance to the flow of load current.

Figure 10.21 is a schematic diagram of its operation. The main components are the firing and sensing logic, a large copper bar, an explosive charge, and a parallel current-limiting fuse. The firing and sensing logic unit is connected to a current transformer on the copper bar located ahead of parallel fuse. The copper bar carries the load current under normal operation. Explosive material is used to pyrotechnically cut the bar on demand. The copper bar with explosive charge is typically enclosed in a fiberglass housing.

Short-Circuit Calculations according to ANSI Standards

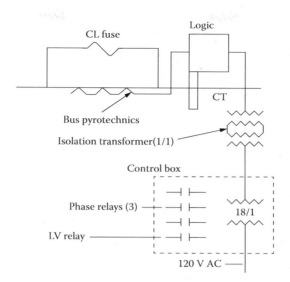

FIGURE 10.21
Schematic of an FCL.

The current transformer and firing logic are mounted on the bar and are at line potential. Alternatively, the electronic measuring and tripping unit may be remotely mounted. The logic unit is connected to the external power supply and controls the unit through the isolating transformer mounted in the copper bar support insulator.

The conductor is cut at high speed, typically 5 mm/µs, while the gas pressure is used to fold the bridging member. The gaps formed in the conductor are not sufficient to interrupt the fault current, and the arcs formed in these gaps generate an arc voltage, which transfers short-circuit current into a parallel small cross-section fusible element. The current-limiting fuse operates in the conventional manner, providing current-limiting action. The total interrupting time is ¼–½ cycles, before the prospective fault reaches its first peak. The proper operation of TCL imposes a number of constraints on fusible element. It must not operate on the fractional current that flows neither through it nor by the inrush and short-circuit currents.

Two types of sensing modes are available, di/dt sensing and threshold sensing; di/dt sensing determines the rate at which the current is increasing, while threshold sensing simply responds to the magnitude of the current. Due to a fast fault clearance, the energy let-though is much reduced. Figure 10.22 illustrates that the post-fault voltage recovers quickly, and shutdowns due to fault voltage dips can even be prevented.

An application in a bypass reactor scheme is shown in Figure 10.23a and b. Consider that a reactor is used to limit the short-circuit current on the 13.8 kV bus so that the existing switchgear can be retained in service. However, the reactor has an adverse impact on the load flow and the voltage regulation. Then, normally, the reactor can be bypassed by the TCL, and the load flow situation improved. We accept much higher short-circuit duties on the circuit breakers in this operation. It is opined that the TCL operates so fast that lower-rated circuit breakers can be retained in service. Now consider a fault downstream, the TCL can operate to introduce the reactor, which will limit the fault current to an acceptable level within the interrupting rating of the switchgear. Due to lack of coordination, this device cannot be used in the main circuits.

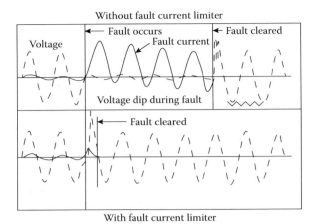

FIGURE 10.22
Arresting the system voltage dip with application of FCL.

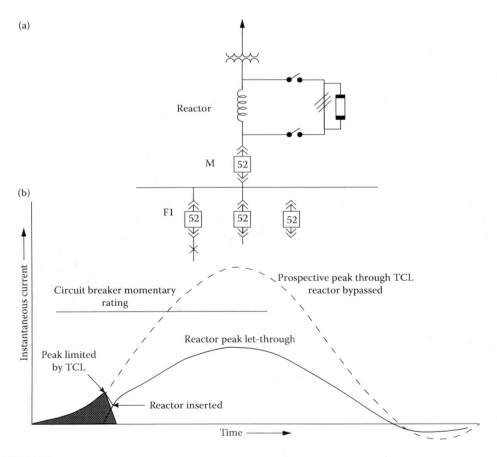

FIGURE 10.23
(a) FCL used in a reactor bypass scheme and (b) time current profile, peak limited by FCL, circuit-breaker momentary rating, peak through reactor also peak prospective current.

A recent discussion is in CIGRE, "Impact of Fault Current Limiters on Existing Protection Schemes" [27].

10.13.1 Superconducting Fault Current Limiters

There are many other technologies of fault current limiters on the horizon, but superconducting fault current limiters utilizing superconducting materials to limit the fault current directly or to supply a DC magnetizing bias current that impacts the level of magnetization of a saturable iron core are in the forefront. These technologies continue to make progress towards commercialization.

The principle of a resistive Superconducting Fault Current Limiter (SFCL) is illustrated in Figure 10.24. In Figure 10.24a, the superconductor is paralleled with an inductance and resistance combination in series. Figure 10.24b shows an abrupt change in the voltage across the superconductor as a function of the current through the device, where I_C is the critical current. When a fault occurs, the current increases and superconductor "quenches" this current rapidly increasing the resistance. This is a function of the superconducting material and temperature. The rapid increase of voltage across superconductor diverts the current through a parallel combination of resistor and inductor, much reducing it. The incipient fault current is limited in less than 1 cycle.

A saturable core SFCL utilizes the magnetic properties of iron to change inductive reactance on AC line. It has two AC windings on two separate cores in series with the source and load side of the line, plus a superconducting constant current DC winding. Under

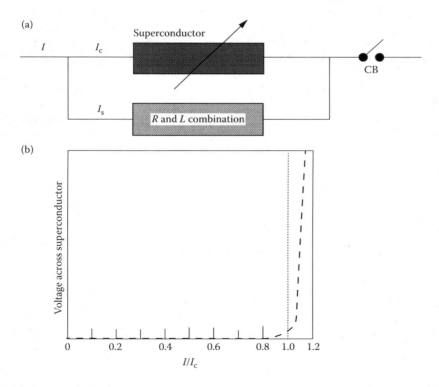

FIGURE 10.24
Schematic diagram of superconducting FEL (SFCL) resistor type.

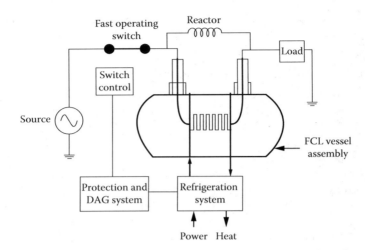

FIGURE 10.25
139 kV SCFCL, see Reference [30].

normal grid currents, the HTS coil fully saturates the iron so that it has a relative permeability of one. Under fault currents, the negative and positive current peaks force the coils out of saturation, resulting in increased line impedance during part of the cycle. This results in considerable reduction in peak fault current.

A super limiter™ design concept is shown in Figure 10.25. It utilizes modular superconducting elements. The fast-acting switch removes the superconductor from service during quench, leaving air core reactor in line that limits the fault current in less than 1 cycle [28]. The 138 kV device being developed has a target capacity of 300 MW (1.2 kA) and limits the fault current from 63 to 40 kA. It is a single-phase device.

In 2000, the IEEE Council for superconductivity (CSC) initiated an internal committee on superconducting standards. Efforts are also being made in IEC to establish standards. The initial effort was in areas of measurements and conductor performance, IEC Technical Committee TC-90. Also see References [29, 30].

Problems

(The problems in this section are constituted so that these can be solved by hand calculations.)

10.1. A circuit breaker is required for 13.8 kV application. The E/X calculation gives 25 kA sym., and the X/R ratio is below 17. Select a suitable breaker from the tables in Chapter 8.

10.2. The X/R ratio in Problem 1 is 25. Without making a calculation, what is the minimum interrupting rating of a circuit breaker for safe application?

10.3. A generating station is shown in Figure P10.1. The auxiliary distribution loads are omitted. Calculate the short-circuit duties on the generator breaker and the 230 kV breaker.

Short-Circuit Calculations according to ANSI Standards

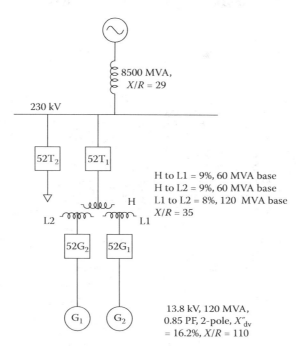

FIGURE P10.1
Power system for Problem 10.3.

10.4. Indicate whether the following sources will be considered remote or local in the ANSI calculation method. What is the NACD ratio? Which E/X multiplying curve shown in this chapter will be used for interrupting duty currents?

- A 50 MVA 0.85 power factor generator; saturated subtransient reactance=16%, $X/R=60$.
- The above generator connected through a 0.4 ohm reactor in series with the generator.
- The generator is connected through a step-up transformer of 40/60 MVA; transformer impedance on a 40 MVA base=10%, $X/R=20$.
- A 10,000 hp synchronous motor, operating at 0.8 power factor leading.
- A 10,000 hp induction motor, operating at 0.93 power factor lagging.

10.5. Figure P10.2 shows the single-line diagram of a multilevel distribution system. Calculate first-cycle and interrupting currents using the ANSI calculation and analytical methods for faults at F_1, F_2, and F_3. Use the impedance data and X/R ratios specified in Table P10.1.

10.6. Calculate the duties on circuit breakers marked 52G and 52T in Figure P10.2. Use separate $R\sim X$ method for E/Z calculation. Each of these breakers is 5-cycle symmetrical rated.

10.7. Calculate the short-circuit withstand rating of a 250 kcmil cable; conductor temperature 90°C, maximum short-circuit temperature 250°C, ambient temperature 40°C, and short-circuit duration 0.5 s.

TABLE P10.1

Impedance Data (Problems 10.5 and 10.6)

Equipment Identification	Impedance Data
138 kV utility's source	3600 MVA, X/R = 20
Generator	13.8 kV, 45 MVA, 0.85 power factor, 2-pole subtransient saturated reactance = 11.7%, X/R = 100, transient saturated = 16.8%, synchronous = 205% on generator MVA base, constant excitation, three-phase subtransient and transient time constants equal to 0.016 and 0.46 s, respectively, armature time constant = 0.18 s
Induction motor M_1	13.2 kV, 5000 hp, 2-pole locked rotor reactance = 16.7%, X/R = 55
Synchronous motor M_2	13.8 kV, 10,000 hp, 4-pole, $X_d'' = 20\%$, X/R = 38
Motor group	4 kV, 200–1000 hp, 4-pole induction motors, locked rotor reactance = 17%, consider an average X/R = 15, total installed kVA = 8000
Transformer T_1	30/50 MVA, 138–13.8 kV, Z = 10%, X/R = 25, data-wye, secondary neutral resistance grounded
Transformers T_2 and T_3	5000 kVA, 13.8 4.16 kV, Z = 5.5%, X/R = 11, transformer taps are set to provide 5% secondary voltage boost
Reactor	0.3 ohms, X/R = 100
Cable C_1	0.002 + j0.018 ohms
Cables C_2 and C_3	0.055 + j0.053 ohms

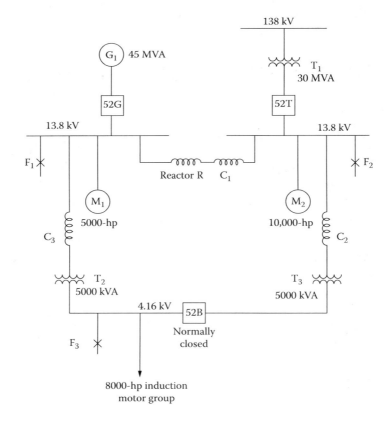

FIGURE P10.2
Power system for Problems 10.5 and 10.6.

Short-Circuit Calculations according to ANSI Standards 477

10.8. The multiplying factors for short-circuit duty calculations for LVPCB and MCCB are 1.095 and 1.19, respectively, though the available short-circuit current is 52.3 kA in each case. What is the fault point X/R?

10.9. In Example 10.1, construct a fault impedance network for the fault at Bus 3. Calculate fault point impedance and the X/R ratio based on separate R–X networks. How does this impedance differ from the values arrived at in Table 10.6 from complex impedance reduction calculation?

10.10. Write five lines on each to describe how the following devices are rated from a short-circuit consideration? LVPCB, ICCB, MCCB, power cables, overhead ACSR conductors, transformer primary fused and unfused switches, bus-bars in switchgear and switchboard enclosures, phase segregated and non-segregated metal-enclosed buses, and transformers and reactors.

10.11. Explain the problems of high X/R faults. How does this affect the duties on the high-voltage generator circuit breakers?

10.12. Which of the following locations of breakers in the same system and on the same bus will have the maximum fault duties: bus tie breaker, incoming breaker, feeder breaker fully loaded, and feeder breaker lightly loaded?

10.13. Represent all the downstream distributions in Figure P10.2 at the 13.8 kV bus with a single interrupting duty and first-cycle impedance.

10.14. Calculate 30-cycle currents in Figure P10.2 at all buses.

10.15. Form a bus admittance and bus impedance matrix of the system in Figure P10.2.

References

1. IEC 60909-0. Short Circuit Calculations in Three-Phase AC Systems, 1988, (now revised).
2. IEC 60909-0. Short Circuit Calculations in Three-Phase AC Systems. Calculation of Currents, 2016.
3. VDE. Calculations of Short-Circuit Currents in Three-Phase Systems. 1971 and 1975, Standard 0102, Parts 1/11.71 and 2/11.75.
4. IEEE Std. 551, Violet Book. IEEE Recommended Practice for Calculating Short-Circuit Currents in Industrial and Commercial Power Systems. 2006.
5. CN Hartman. Understanding asymmetry. *IEEE Trans Ind Appl*, IA-21, 4–85, 1985.
6. ANSI/IEEE Std. C37.010. Application Guide for AC High-Voltage Circuit Breakers Rated on a Symmetrical Current Basis, 1999 (revision of 1979 standard).
7. IEEE Std. 141. IEEE Recommended Practice for Electrical Power Distribution for Industrial Plants, 1993.
8. ANSI/IEEE Std. C37.13. Standard for Low-Voltage AC Power Circuit Breakers in Enclosures, 1997.
9. JC. Das. Reducing interrupting duties of high voltage circuit breakers by increasing contact parting time. *IEEE Trans Ind Appl*, IA-44, 1027–1034, 2008.
10. WC Huening Jr. Interpretation of New American Standards for power circuit breaker applications. *IEEE Trans Ind Gen Appl*, IGA-5, 121–143, 1969.
11. IEEE Std. C37.013. IEEE Standard for AC High-Voltage Generator Breakers Rated on Symmetrical Current Basis, 1999.

12. IEEE Std. C37.013a. IEEE Standard for AC High-Voltage Generator Breakers Rated on Symmetrical Current Basis, Amendment 1: Supplement for Use With Generators Rated 10–100 MVA, 2007.
13. IM Canay, L Warren. Interrupting sudden asymmetric short-circuit currents without zero transition. *Brown Boveri Rev*, 56, 484–493, 1969.
14. IM Canay. Comparison of generator circuit breaker stresses in test laboratory and real service condition. *IEEE Trans Power Delivery*, 16, 415–421, 2001.
15. IEEE Std. 1110. IEEE Guide for Synchronous Generator Modeling Practice and Applications in Power System Stability Analysis, 2002.
16. JR Dunki-Jacobs, BP Lam, RP Stafford. A comparison of ANSI-based and dynamically rigorous short-circuit current calculation procedures. *IEEE Trans Ind Appl*, IA-24, 1180–1194, 1988.
17. A Berizzi, S Massucco, A Silvestri, D Zaninelli. Short-circuit current calculations—A comparison between methods of IEC and ANSI standards using dynamic simulation as reference. *IEEE Transactions on Industry Application*, IA-30, 1099–1106, 1994.
18. AJ Rodolakis. A comparison of North American (ANSI) and European (IEC) fault calculation guidelines. *IEEE Trans Ind Appl*, IA-29, 515–521, 1993.
19. IEEE Std. C37.91. IEEE Guide for Protecting Power Transformers, 2008.
20. NFPA (National Fire Protection Association) National Electric Code, 2014.
21. ANSI Std. C37.06. AC High-Voltage Circuit Breakers Rated on a Symmetrical Current Basis—Preferred Ratings and Related Capabilities, 2000 (revision of 1987).
22. JC Das. Limitations of fault current limiters for expansion of electrical distribution systems. *IEEE Trans Ind Appl*, 33, 1073–1082, 1997.
23. ANSI/IEEE Std. C37.27. Guide for Metal-Enclosed Bus and Calculating Losses in Isolated-Phase Bus, 1987.
24. The Aluminum Association. *Aluminum Electrical Conductor Handbook*, 2nd ed., The Aluminum Association, Washington, DC, 1982.
25. DG Fink, JM Carroll (Eds). *Standard Handbook for Electrical Engineers*, 10th ed., McGraw-Hill, New York, 1968 (see Section 4, Tables 4 through 8).
26. Calor-Emag. Is Limiter. Calor–Emag AG, Mannheim, Germany, Leaflet No. DEACE 204596e, 1996.
27. CIGRE. Impact of fault current limiters on existing protection schemes. CIGRE WG A3.16, Electra 2008.
28. DOE. Development and in-Grid demonstration of a transmission voltage superlimiter™ fault current limiter, 2009, Peer review, Alexandria, VA, 2009.
29. EPRI. Superconductor fault current limiters, Technology watch. EPRI, Palo Alto, CA, Report No. 1017793, 2009.
30. EPRI. Survey of fault current limiter technologies. EPRI, Palo Alto, CA, Report No. 1010760, 2005.

11

Short-Circuit Calculations according to IEC Standards

Since the publication of IEC (International Electrotechnical Commission) 909 for the calculation of short-circuit currents in 1988 and its subsequent revisions [1], it has attracted much attention and the different methodology compared to ANSI methods of calculation has prompted a number of discussions and technical papers. The examples of short-circuit calculations which were included in the 1988 standard have been removed in the revised standard. It is recommended that this standard [1] is read in conjunction with References [2–7]. The IEC standard [1] qualifies that systems at highest voltages of 550 kV and above with long transmission lines need special considerations. Short-circuit currents and impedances can also be determined by system tests, by measurement on a network analyzer or with a digital computer. Calculations of short-circuit currents in installations on board ships and aeroplanes are excluded.

This chapter analyzes and compares the calculation procedures in IEC and ANSI standards. Using exactly the same system configurations and impedance data, comparative results are arrived at, which show considerable differences in the calculated results by the two methods. Some explanation of these variances is provided based on different procedural approaches. Neither standard precludes alternative methods of calculation, which give equally accurate results.

11.1 Conceptual and Analytical Differences

The short-circuit calculations in IEC and ANSI standards are conceptually and analytically different, nor is the rating structure of the circuit breakers identical. There are major differences in the duty cycles, testing, temperature rises, recovery voltages, though there has been attempts to harmonize ANSI/IEEE standards with IEC (Chapters 8 and 10). For the purpose of IEC short-circuit calculations, we will confine our attention to the specifications of interest. Entirely different terminology is used in IEC [1,6] to describe the same phenomena in circuit breakers. The following overview provides a broad picture and correlation with ANSI.

11.1.1 Breaking Capability

The rated breaking capability of a circuit breaker corresponds to the rated voltage and to a reference restriking voltage, equal to the rated value, expressed as (1) rated symmetrical breaking current that each pole of the circuit breaker can break, and (2) rated asymmetrical breaking capability that any pole of the circuit breaker can break. The breaking capacity is expressed in MVA for convenience, which is equal to the product of the rated breaking current in kA and rated voltage multiplied by an appropriate factor, depending on the type

479

of circuit: One (1) for a single-phase circuit, two (2) for a two-phase circuit, and $\sqrt{3}$ for a three-phase circuit.

This is equivalent to the interrupting capability in ANSI standards. In IEC calculations, the asymmetry at the contact parting time must be calculated to ascertain the asymmetrical rating of the breaker. As discussed in Chapter 8, ANSI breakers are rated on a symmetrical current basis and the asymmetry is allowed in the rating structure and postfault correction factors. Unlike ANSI, IEC does not recommend any postmultiplying factors to account for asymmetry in short-circuit currents.

11.1.2 Rated Restriking Voltage

The rated restriking voltage is the reference restriking voltage to which the breaking capacity of the circuit breaker is related. It is recommended that the nameplate of the circuit breaker be marked with the amplitude factor and either the rate of rise of the restriking voltage in volts/μs or natural frequency in kHz/s be stated [6].

11.1.3 Rated Making Capacity

The rated making capacity corresponds to rated voltages and is given by $1.8 \times \sqrt{2} (= 2.55)$ times the rated symmetrical breaking capacity. The making capacity in ampères is inversely proportional to the voltage, when the circuit breaker is dual-voltage rated. For voltages below the lower rated voltage, the making capacity has a constant value corresponding to the lower rated voltage and for voltages higher than the rated voltage no making capacity is guaranteed. This is equivalent to the close and latch capability of ANSI standards.

11.1.4 Rated Opening Time and Break Time

The rated opening time up to separation of contacts is the opening time which corresponds to rated breaking capacity. The rated total breaking time is the total break time which corresponds to the rated breaking capacity. It may be different, depending on whether it refers to symmetrical or asymmetrical breaking capacity.

The minimum time delay t_m is the shortest possible operating time of an instantaneous relay and the shortest opening time of the circuit breaker. It does not take into account adjustable time delays of trapping devices.

11.1.5 Initial Symmetrical Short-Circuit Current

IEC defines I_k'', the initial symmetrical short-circuit current as the alternating current (AC) symmetrical component of a prospective (available) short-circuit current applicable at the instant of short circuit if the impedance remains at zero-time value. This is approximately equal to ANSI first-cycle current in rms symmetrical, obtained in the first cycle at the maximum asymmetry in one of the phases. Note the difference in the specifications. The prospective (available) short-circuit current is defined as the current which will flow if the short-circuit was replaced with an ideal connection of negligible impedance. This is the "bolted" fault current. IEEE adopts the definition of prospective current.

Short-Circuit Calculations according to IEC Standards

11.1.6 Peak Making Current

The peak making current, i_p, is the first major loop of the current in a pole of a circuit breaker during the transient period following the initiation of current during a making operation. This includes the direct current (DC) component. This is the highest value reached in a phase in a polyphase circuit. It is the maximum value of the prospective (available) short-circuit current. The rated peak withstand current is equal to the rated short-circuit making current. This can be reasonably compared with ANSI close and latch capability, though there are differences in the rating structure. Revision of factor 2.7 to 2.6 for 60 Hz circuit breakers and 2.5 for 50 Hz circuit breakers in ANSI standards (see Section 11.1.2) brings these two standards closer, though there are differences. Also, IEC does not have any requirement, similar to that of ANSI, for latching and carrying a current before interrupting.

11.1.7 Breaking Current

The rated short-circuit breaking current, $I_{b,asym}$, is the highest short-circuit current that the circuit breaker shall be capable of breaking (this term is equivalent to ANSI, "asymmetrical interrupting") under the conditions of use and behavior prescribed in IEC, in a circuit asymmetrical having a power frequency recovery voltage corresponding to the rated voltage of the circuit breaker and having a transient recovery voltage equal to the rated value specified in the standards. The breaking current is characterized by (1) the AC component and (2) the DC component. The rms value of the AC component is termed the rated short-circuit current. The symmetrical short-circuit breaking current is defined as the rms value of an integral cycle of symmetrical AC component of the prospective short-circuit current at the instant of contact separation of the fist pole to open of a switching device. The standard values in IEC are 6.3, 8, 10, 12.5, 16, 25, 31.5, 40, 50, 63, 80, and 100 kA. The DC component is calculated at minimum time delay t_m. This is entirely different from ANSI symmetrical ratings and calculations (Chapter 10).

11.1.8 Steady-State Current

The calculations of steady-state fault currents from generators and synchronous motors according to IEC take into consideration the generator excitation, the type of synchronous machine, salient or cylindrical generators, and the excitation settings. The fault current contributed by the generator becomes a function of its rated current using multiplying factors from curves parameterized against saturated synchronous reactance of the generator, excitation settings, and the machine type.

This calculation is more elaborate and departs considerably from ANSI-based procedures for calculation of 30 cycle currents. For the purpose of short-circuit calculations, Table 11.1 shows the equivalence between IEC and ANSI duties, though qualifications apply.

TABLE 11.1

Equivalence between ANSI and IEC Short-Circuit Calculation Types

ANSI Calculation Type	IEC Calculation Type
First-cycle current	Initial short-circuit current, I_k''
Closing-latching duty current, crest	Peak current (making current), i_p
Interrupting duty current	Breaking current $I_{b,sym}$ and $I_{b,asym}$ (symmetrical and asymmetrical)
Time-delayed 30-cycle current	Steady-state current, I_k

The specimen IEC ratings of a typical dual-voltage rated circuit breaker are shown below:

Voltage: 10 kV/11.5 kV
Frequency: 50 Hz
Symmetrical breaking capacity: 20.4 kA/17.8 kA
Asymmetrical breaking capacity: 25.5 kA/22.2 kA
Rated making capacity: 52 kA/45 kA
    Amplitude factor: 1.25
    Rate of rise: 500 V/μs
Rated short-time current (1 s): 20.4 kA
Rated operating duty: O-3m-CO-3m-CO

11.1.9 Highest Short-Circuit Currents

The three-phase, single line-to-ground, double line-to-ground and phase-to-phase fault currents are to be considered. Based on the sequence impedances, Figure 11.1 shows which type of short circuit leads to the highest short-circuit currents. All the sequence impedances have the same impedance angle. This figure is useful for information, but should

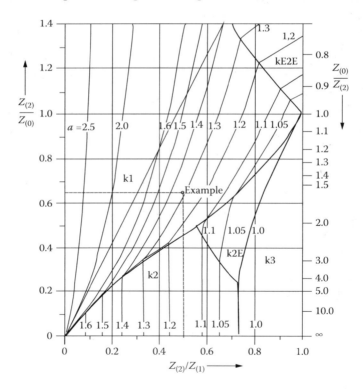

FIGURE 11.1
Diagram to determine the short-circuit type for the highest short-circuit current referred to symmetrical three-phase short-circuit current at the short-circuit location when the impedance angles of positive, negative, and zero sequence impedances are identical.

Short-Circuit Calculations according to IEC Standards 483

not be used instead of calculations. As an example, if $Z_2/Z_1 = 0.5$, and with AC decay $Z_2/Z_0 = 0.65$, the single-line-to-ground fault is the highest.

11.2 Prefault Voltage

IEC defines an equivalent voltage source given in Table 11.2 and states that the operational data on the static loads of consumers, position of tap changers on transformers, excitation of generators, etc. are dispensable; additional calculations about all the different possible load flows at the moment of short circuit are superfluous. The equivalent voltage source is the only active voltage in the system, and all network feeders and synchronous and asynchronous machines are replaced by their internal impedances. This equivalent voltage source is derived by multiplying the nominal system voltage by a factor c given in Table 11.2.

ANSI uses a prefault voltage equal to the system rated voltage, though a higher or lower voltage is permissible, depending on the operating conditions. IEC requires that in *every case* the system voltage be multiplied by factor c from Table 11.2. We will again revert to this c factor.

11.3 Far-From Generator Faults

A "far-from-generator" short-circuit is defined as a short circuit during which the magnitude of the symmetrical AC component of the prospective (available) current remains essentially constant. These systems have no AC component decay. For the duration of a short circuit, there is neither any change in the voltage or voltages that caused the short circuit to develop nor any significant change in the impedance of the circuit, i.e., impedances are considered constant and linear. Far-from-generator is equivalent to ANSI *remote* sources, i.e., no AC decay.

TABLE 11.2

IEC Voltage Factor c

	Voltage Factor c for Calculation of	
Nominal Voltage (U_n)	**Maximum Short-Circuit Current (c_{max})**	**Minimum Short-Circuit Current (c_{min})**
Low Voltage (100–1000 V)		
230 V/400 V	1.00	0.95
Other voltages	1.05	1.00
Medium voltage (>1–35 kV)	1.10	1.00
High voltage (>35–230 kV)	1.10	1.00

Source: IEC 60909-0, Short-circuit calculations in three-phase AC systems, 1st. Ed., 1988, Now revised IEC 60909-0, Short-circuit currents in three-phase AC systems, Calculation of currents, 2001. With permission.

The following equation is supported:

$$I_b = I_k'', \ I_{b2} = I_{k2}'', \ I_{b2E} = I_{k2E}'', \ I_{b1} = I_{k1}'' \tag{11.1}$$

where I_b is the symmetrical breaking current, and I_k'' is the initial symmetrical short-circuit current. The subscripts k1, k2, k2E are line-to-earth short-circuit, line-to-line short-circuit, and line-to-line short-circuit with earth connection. For a single-fed short-circuit current, as shown in Figure 11.2, I_k'' is given by

$$I_k'' = \frac{cU_n}{\sqrt{3}\sqrt{R_k^2 + X_k^2}} = \frac{cU_n}{\sqrt{3}Z_k} \tag{11.2}$$

where U_n is the normal system phase-to-phase voltage in volts and I_k'' is in ampères; R_k and X_k are in ohms and are the sum of the source, transformer, and line impedances, as shown in Figure 11.2.

The peak short-circuit current is given by

$$i_p = \chi\sqrt{2}I_k'' \tag{11.3}$$

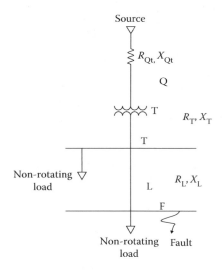

FIGURE 11.2
Calculation of initial short-circuit current, with equivalent voltage source.

Short-Circuit Calculations according to IEC Standards

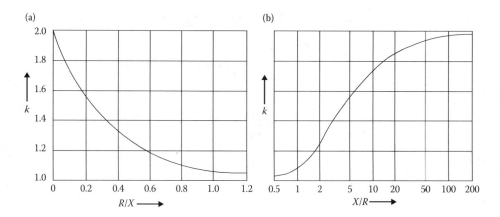

FIGURE 11.3
Factor χ for calculation of peak current. (Reproduced from IEC 60909-0, Short-circuit calculations in three-phase AC systems, 1st Ed., 1988, Now revised IEC 60909-0, Short-circuit currents in three-phase AC systems, Calculation of currents, 2001. With permission.)

where χ can be ascertained from the X/R ratio from the curves in Figure 11.3 or calculated from the expression:

$$\chi = 1.02 + 0.98 e^{-3R/X} \qquad (11.4)$$

The peak short-circuit current, fed from sources which are not meshed with one another, is the sum of the partial short-circuit currents:

$$i_p = \sum i_{pi} \qquad (11.5)$$

11.3.1 Nonmeshed Sources

IEC distinguishes between the types of networks. For nonmeshed sources (Figure 11.4), the initial short-circuit current, the symmetrical breaking current, and the steady-state short-circuit current at fault location F are composed of various separate branch short-circuit

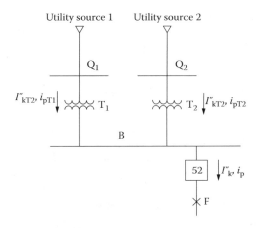

FIGURE 11.4
Short-circuit fed from various sources that are independent of each other.

currents which are *independent of each other*. The branch currents are calculated and summed to obtain the total fault current:

$$I''_k = I''_{kT1} + I''_{kT2} \tag{11.6}$$

That is, the theorem of superimposition applies and the initial short-circuit current is the phasor sum of the individual short-circuit currents.

For calculating the short-circuit currents in Figure 11.5, in case of a power station unit with *on-load tap changer*, the equations for partial initial currents are as follows:

$$I''_{KG} = \frac{cU_{rG}}{\sqrt{3}K_{G,S}Z_G} \tag{11.7}$$

with

$$K_{GS} = \frac{c_{max}}{1 + x''_d \sin \varphi_{rG}} \tag{11.8}$$

and

$$I''_{KT} = \frac{cU_{rG}}{\sqrt{3}\left|Z_{TLV} + \frac{1}{t_T^2}Z_{Qmin}\right|} \tag{11.9}$$

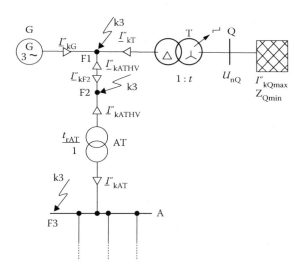

FIGURE 11.5
Short-circuit currents and parallel short-circuit currents for three-phase short circuits between generator and unit transformer with or without load tap changer, or at connection of the auxiliary transformer of the power station unit and at auxiliary bus bar A.

Short-Circuit Calculations according to IEC Standards 487

For the short-circuit current I''_{kF2} feeding into short-circuit location F2 at the high voltage side of the auxiliary transformer AT (Figure 11.5):

$$I''_{kF2} = \frac{cU_{rG}}{\sqrt{3}}\left[\frac{1}{K_{G,S}Z_G} + \frac{1}{K_{T,S}Z_{TLV} + \frac{1}{t_T^2}Z_{Qmin}}\right] = \frac{cU_{rG}}{\sqrt{3}Z_{rs1}}$$

with

$$K_{T,S} = \frac{c_{max}}{1 - x_T \sin\varphi_{rG}} \tag{11.10}$$

where

Z_g = subtransient impedance of the generator = $R_G + jX''_d$
x''_d = subtransient reactance referred to the rated impedance
Z_{TLV} = short-circuit impedance of the transformer referred to low-voltage side
t_T = rated transformation ratio
Z_{Qmin} = minimum value of the impedance of the network feeder

The equations are in actual units, ohms. The calculations can be carried out in per unit system.

For power stations without on-load tap changers, the equations are similar except that the modified factors are

$$K_{G,SO} = \frac{1}{1 + p_G}\frac{c_{max}}{1 + x''_d \sin\varphi_{rG}}$$

and

$$K_{T,SO} = \frac{1}{1 + p_G}\frac{c_{max}}{1 - x_T \sin\varphi_{rG}} \tag{11.11}$$

where p_G is the range of generator voltage regulation.

11.3.2 Meshed Networks

For calculation of i_p in meshed networks (Figure 11.6), three methods (A, B, and C) are described.

11.3.2.1 Method A: Uniform Ratio R/X or X/R Ratio Method

The factor χ in Equation 11.4 is determined from the smallest ratio of R/X of all branches of the network. Only the branches which carry the partial short-circuit currents at the nominal voltage corresponding to the short-circuit location and branches with transformers adjacent to the short-circuit locations are considered. Any branch may be a series combination of several elements.

11.3.2.2 Ratio R/X or X/R at the Short-Circuit Location

The factor $\chi = 1.15\chi_b$, where factor 1.15 is a safety factor to cover inaccuracies caused by using X/R from a meshed network reduction with complex impedances, and χ_b is calculated from curves in Figure 11.3 or mathematically from Equation 11.4. In the low-voltage

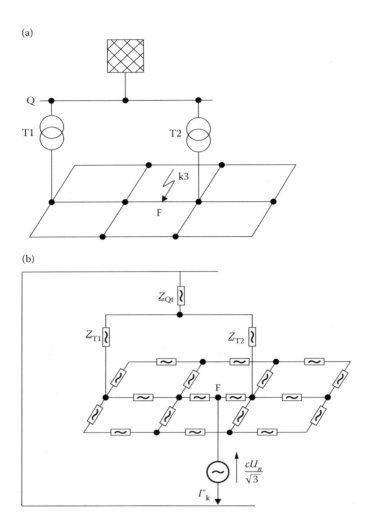

FIGURE 11.6
Calculation of initial short-circuit current in a meshed network. (a) The system diagram; (b) equivalent circuit diagram with equivalent voltage source. (Reproduced from IEC 60909-0, Short-circuit calculations in three-phase AC systems, 1st Ed., 1988, Now revised IEC 60909-0, Short-circuit currents in three-phase AC systems, Calculation of currents, 2001. With permission.)

networks the product of $1.15X_b$ is limited to 1.8 and in the high-voltage networks to 2.0. As long as R/X remains smaller than 0.3 in all branches, it is not necessary to use factor 1.15.

11.3.2.3 Method C: Equivalent Frequency Method

This method provides the equivalent frequency approach. A source of 20 Hz for 50-Hz systems and 24 Hz for 60-Hz systems is considered to excite the network at the fault point. The X/R at the fault point is then given by

$$\frac{X}{R} = \left(\frac{X_c}{R_c}\right)\left(\frac{f}{f_c}\right) \tag{11.12}$$

Short-Circuit Calculations according to IEC Standards 489

where f is the system frequency, f_c is the excitation frequency, and $Z_c = R_c + jX_c$ at the excitation frequency. The factor $\chi = \chi_c$ is used in the calculations for the peak current.

11.4 Near-to-Generator Faults

A "near-to-generator" fault is a short circuit to which at least one synchronous machine contributes a prospective initial symmetrical short-circuit current that is more than twice the generator's rated current, or a short-circuit to which synchronous and asynchronous motors contribute more than 5% of the initial symmetrical short-circuit current I_k'', calculated without motors. These fault types have AC decay. This is equivalent to ANSI *local* faults.

The factor c is applicable to prefault voltages as in the case of far-from-generator faults. The impedances of the generators and power station transformers are modified by additional factors, depending on their connection in the system.

11.4.1 Generators Directly Connected to Systems

When generators are directly connected to the systems, their positive sequence impedance is modified by a factor K_G:

$$Z_{GK} = K_G(R_G + jX_d'') \tag{11.13}$$

K_G is given by

$$K_G = \frac{U_n}{U_{rG}}\left(\frac{C_{max}}{1 + X_d'' \sin\phi_{rG}}\right) \tag{11.14}$$

where U_{rG} is the rated voltage of the generator, U_n is the nominal system voltage, ϕ_{rG} is the phase angle between the generator current I_{rG} and generator voltage U_{rG}, and X_d'' is the subtransient reactance of the generator, at a generator-rated voltage on a generator MVA base. Figure 11.7 shows the applicable phasor diagram.

If the generator voltage is different from U_{rG}, use

$$U_G = U_{rG}(1 + p_G) \tag{11.15}$$

The generator resistance R_G with sufficient accuracy is given by the following expressions:

$$R_{Gf} = 0.05X_d'' \text{ for generators with } U_{rG} > 1\,kV \text{ and } S_{rG} \geq 100\,MVA$$

$$R_{Gf} = 0.07X_d'' \text{ for generators with } U_{rG} > 1\,kV \text{ and } S_{rG} < 100\,MVA \tag{11.16}$$

$$R_{Gf} = 0.15X_d'' \text{ for generators with } U_{rG} \leq 1\,kV$$

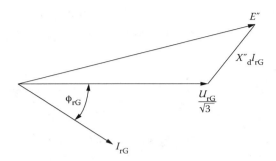

FIGURE 11.7
Phasor diagram of a synchronous generator at rated load and PF.

These values of R_{Gf} are not to be used for the calculations of the peak short-circuit current. These values cannot be used when calculating the aperiodic component I_{DC} of the short-circuit current. The effective stator resistance of synchronous machines is much below the value of R_{Gf} and manufacturer's value of R_G should be used.

This was a limitation in IEC standard of year 1988, that the above values of R_{Gf} gave a depressed value of I_{DC} as compared to other methods of calculations. This anomaly has now been rectified.

11.4.2 Generators and Unit Transformers of Power Station Units

For generators and unit transformers of power stations, the generator and the transformer are considered as a single unit. The following equation is used for the impedance of the whole power station unit for the short circuit on the high side of the unit transformer, *with on-load tap changer* (Figure 11.5):

$$Z_s = K_s \left(t_r^2 Z_G + Z_{THV} \right) \quad (11.17)$$

with

$$K_s = \frac{U_{nQ}^2}{U_{rG}^2} \frac{U_{rTLV}^2}{U_{rTHV}^2} \frac{c_{max}}{1 + \left| x_d'' - x_T \right| \sin \varphi_{rG}} \quad (11.18)$$

Here, U_{nq} is the nominal system voltage at the feeder connection point Q of the power unit.

For calculations *without on-load tap changers*, the following equation can be used for the short circuit on the high side of the transformer unit:

$$Z_s = K_{SO} \left(t_r^2 Z_G + Z_{THV} \right)$$

with $\quad (11.19)$

$$K_{SO} = \frac{U_{nQ}}{U_{rG}(1 + p_G)} \frac{U_{rTLV}}{U_{rTHV}} (1 \pm p_T) \frac{c_{max}}{1 + x_d'' \sin \varphi_{rG}}$$

Short-Circuit Calculations according to IEC Standards 491

$(1 \pm p_T)$ is introduced if the unit transformer has off-load taps and if one of the taps is permanently used. The highest short-circuit current will be given by $1 - p_T$.

11.4.3 Motors

For calculations of I_k'' synchronous motors and synchronous compensators are treated as synchronous generators. The impedance Z_M of asynchronous motors is determined from their locked rotor currents.

The following ratios of resistance to reactance of the motors apply with sufficient accuracy:

$R_M / X_M = 0.10$ with $X_M = 0.995 Z_M$ for high voltage motors with power
$\qquad P_{rm}$ per pairs of poles $\geq 1\,\text{MW}$

$R_M / X_M = 0.15$ with $X_M = 0.989 Z_M$ for high voltage motors with power
$\qquad P_{rm}$ per pairs of poles $< 1\,\text{MW}$

$R_M / X_M = 0.42$ with $X_M = 0.922 Z_M$ for low-voltage motor groups with connection cables.

$$(11.20)$$

11.4.4 Short-Circuit Currents Fed from One Generator

The initial short-circuit current is given by Equation 11.2. The peak short-circuit current is calculated as for far-from-generator faults, considering the type of network. For generator corrected resistance $K_G R_G$ and corrected reactance $K_G X_d''$ are used.

11.4.4.1 Breaking Current

The symmetrical short-circuit breaking current, for single fed or nonmeshed systems is given by

$$I_b = \mu I_k'' \tag{11.21}$$

where factor μ accounts for AC decay. The following values of μ are applicable for medium-voltage turbine generators, salient pole generators, and synchronous compensators excited by rotating exciters or by static exciters, provided that for the static exciters the minimum time delay is less than 0.25 s and the maximum excitation voltage is less than 1.6 times the rated excitation voltage. For all other cases μ is taken to be 1, if the exact value is not known.

When there is a unit transformer between the generator and short-circuit location, the partial short-circuit current at the high side of the transformer is calculated.

$$\mu = 0.84 + 0.26 e^{-0.26 I_{KG}''/I_{rG}} \quad \text{for } t_{min} = 0.02\,\text{s}$$

$$\mu = 0.71 + 0.51 e^{-0.30 I_{KG}''/I_{rG}} \quad \text{for } t_{min} = 0.05\,\text{s}$$

$$\mu = 0.62 + 0.72 e^{-0.32 I_{KG}''/I_{rG}} \quad \text{for } t_{min} = 0.10\,\text{s} \tag{11.22}$$

$$\mu = 0.56 + 0.94 e^{-0.38 I_{KG}''/I_{rG}} \quad \text{for } t_{min} \geq 0.25\,\text{s}$$

If the ratio of the initial short-circuit current and the machine rated current is equal to or less than 2, then the following relation holds:

$$\frac{I''_{kG}}{I_{rG}} \leq 2, \mu = 1 \quad \text{for all values of } t_{min} \tag{11.23}$$

In the case of asynchronous motors, replace

$$\frac{I''_{kG}}{I_{rG}} \text{ by } \frac{I''_{kM}}{I_{rM}} \tag{11.24}$$

The equations can also be used for compound excited low-voltage generators with a minimum time delay not >0.1 s. The calculations of low-voltage breaking currents for a time duration >0.1 s is not included in the IEC standard.

11.4.4.2 Steady-State Current

The maximum and minimum short-circuit currents are calculated as follows:

$$I_{k\max} = \lambda_{\max} I_{rG} \tag{11.25}$$

$$I_{k\min} = \lambda_{\min} I_{rG} \tag{11.26}$$

where $\lambda_{\max}$ and $\lambda_{\min}$ for turbine generators are calculated from the graphs in [1]. Figure 11.8 shows these values for cylindrical rotor generators. In this figure, X_{dsat} is the reciprocal of

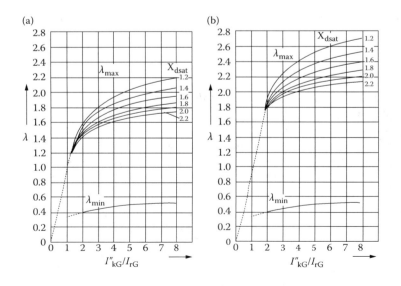

FIGURE 11.8
(a, b) Factors $\lambda_{\max}$ and $\lambda_{\min}$ for cylindrical rotor generators; Series One and Series Two defined in the text. (From IEC 60909-0, Short-circuit calculations in three-phase AC systems, 1st Ed., 1988, Now revised IEC 60909.-0, Short-circuit currents in three-phase AC systems, Calculation of currents, 2001.)

Short-Circuit Calculations according to IEC Standards

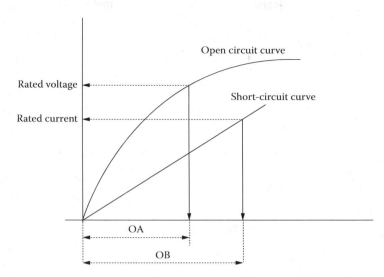

FIGURE 11.9
Open-circuit magnetization and short-circuit curves of a synchronous generator to illustrate short-circuit ratio.

the short-circuit ratio. We have not yet defined the short circuit ratio of a generator. It is given as

$$R_{sc} = \frac{\text{Per unit excitation at normal voltage on open circuit}}{\text{Per unit excitation for rated armature current on short circuit}} \quad (11.27)$$

Referring to the open-circuit and short-circuit curves of the generator, shown in Figure 11.9, the short-circuit ratio is OA/OB. Chapter 9 defined saturated synchronous reactance as the ratio of per unit voltage on open circuit to per unit armature current on short circuit. In Figure 11.9, OA is related to the normal rated voltage and OB is proportional to the rated current of the machine. The short-circuit ratio is, therefore, the reciprocal of synchronous reactance. It is a measure of the stiffness of the machine, and modern generators tend to have lower short-circuit ratios compared to those of their predecessors, of the order of 0.5 or even lower.

For the static excitation systems fed from generator terminals and a short circuit at the terminals, the field voltage collapses with the terminal voltage, and therefore, take $\lambda_{max} = \lambda_{min} = 0$. Maximum λ curves for Series One are based on the highest possible excitation voltage according to either 1.3 times the rated excitation at rated load and power factor (PF) for turbine generators or 1.6 times the rated excitation for a salient-pole machine. The maximum λ curves for Series Two are based on the highest excitation voltage according to either 1.6 times the rated excitation at rated load and PF for turbine generators or 2.0 times the rated excitation for salient-pole machines. The graphs for the salient-pole machines are similar and not shown.

11.4.5 Short-Circuit Currents in Nonmeshed Networks

The procedure is the same as that described for far-from-generator faults. The modified impedances are used. The branch currents are superimposed, as shown in Figure 11.10.

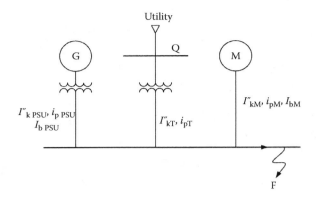

FIGURE 11.10
Calculation of I_k, i_p, I_b, and I_k for a three-phase short-circuit fed from nonmeshed sources.

$$I_k = I_{k,PSU} + I''_{kT} + I''_{kM} + \cdots$$

$$I_p = I_{p,PSU} + I_{pT} + I_{pM} + \cdots$$

$$I_b = I_{b,PSU} + I''_{kT} + I_{bM} + \cdots \quad (11.28)$$

$$I_k = I_{b,PSU} + I''_{kT} + \cdots$$

11.4.6 Short-Circuit Currents in Meshed Networks

Figure 11.11 shows that the initial short-circuit currents in meshed networks can be calculated by using modified impedances and the prefault voltage at the fault point. The peak current i_p is calculated as far-from-generator faults. Methods A, B, and C for meshed networks are applicable. The symmetrical short-circuit breaking current for meshed networks is conservatively given as

$$I_b = I''_k \quad (11.29)$$

A more accurate expression is provided as follows:

$$I_b = I''_k - \sum_i \left(\frac{\Delta U''_{Gi}}{cU_n/\sqrt{3}} \right)(1 - \mu_i)I''_{kGi} \sum_j \left(\frac{\Delta U''_{Mj}}{cU_n/\sqrt{3}} \right)(i - \mu_i q_i)I''_{kMj} \quad (11.30)$$

$$\Delta U''_{Gi} = jX''_{di}I''_{kGi}$$
$$\Delta U''_{Mj} = jX''_{Mj}I''_{kMj} \quad (11.31)$$

where $\Delta U''_{Gi}$ and $\Delta U''_{Mj}$ are the initial voltage differences at the connection points of the synchronous machine i and the asynchronous motor j, and I''_{kGi} and I''_{kMj} are the parts of the initial symmetrical short-circuit currents of the synchronous machine i and the asynchronous motor j; μ is defined in Equation 11.22, and q is defined in Equation 11.36.

Short-Circuit Calculations according to IEC Standards

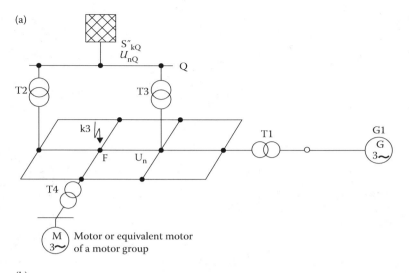

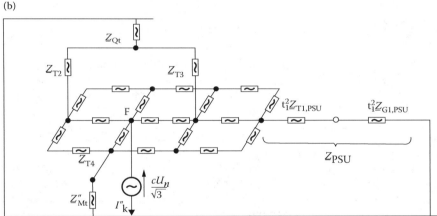

FIGURE 11.11
Calculation of initial short-circuit current in a meshed network fed from several sources. (a) The system diagram; (b) equivalent circuit diagram with equivalent voltage source. (Reproduced from IEC 60909-0, Short-circuit calculations in three-phase AC systems, 1st Ed., 1988, Now revised IEC 60909-0, Short-circuit currents in three-phase AC systems, Calculation of currents, 2001. With permission.)

For the steady-state current, the effect of motors is neglected. It is given as

$$I_k = I''_{kM} \tag{11.32}$$

11.5 Influence of Motors

Synchronous motors and synchronous compensators contribute to I_k, i_p, I_b, and I_k. Asynchronous motors contribute to I_k, i_p, and I_b and for unbalanced faults to I_k. Low-voltage motors in public power supply systems can be neglected. High- and low-voltage motors which are connected through a two-winding transformer can be neglected if

$$\frac{\sum P_{rM}}{\sum S_{rT}} \leq \frac{0.8}{\left| c100 \dfrac{\sum S_{rT}}{S_{kQ}^{''} - 0.3} \right|} \tag{11.33}$$

where $\sum P_{rM}$ is the sum of rated active power of the motors, $\sum S_{rT}$ is the sum of rated apparent power of the transformers, and $S_{kQ}^{''}$ is the symmetrical short-circuit power at the connection point without the effect of motors. This expression is not valid for three-winding transformers.

11.5.1 Low-Voltage Motor Groups

For simplifications of the calculations, groups of low-voltage motors including their connecting cables can be combined into an equivalent motor:

$$I_{rM} = \text{sum of rated currents of the motors in the group}$$
$$\text{Ratio of locked rotor current to full load current} = I_{LR}/I_{rM} = 5$$
$$R_M/X_M = 0.42, \chi_a = 1.3, \text{ and } m = 0.05 \, \text{MW if nothing definite is known}$$
$$\text{Where } m = P_{rm}/P, P = \text{number of pairs of poles} \tag{11.34}$$

The partial short-circuit current of low-voltage motors is neglected if the rated current of the equivalent motor (sum of the ratings of group of motors) is <0.01% of the initial symmetrical short-circuit current at the low-voltage bus to which these motors are directly connected, without the contributions from the motors:

$$I_{rM} \leq 0.01 I_{kM}^{''} \tag{11.35}$$

11.5.2 Calculations of Breaking Currents of Asynchronous Motors

For calculation of breaking short-circuit current from asynchronous motors another factor q (in addition to μ) is introduced; $q = 1$ for synchronous machines. The factor q is given by

$$
\begin{aligned}
q &= 1.03 + 0.12 \ln m \quad \text{for } t_{min} = 0.02 \, \text{s} \\
&= 0.79 + 0.12 \ln m \quad \text{for } t_{min} = 0.05 \, \text{s} \\
&= 0.57 + 0.12 \ln m \quad \text{for } t_{min} = 0.10 \, \text{s} \\
&= 0.26 + 0.10 \ln m \quad \text{for } t_{min} \geq 0.25 \, \text{s}
\end{aligned}
\tag{11.36}
$$

where m is the rated active power of motors per pair of poles.

Short-Circuit Calculations according to IEC Standards 497

Therefore, the breaking current of asynchronous machines is given as

$$I_{b,sym} = \mu q I_k''$$ (11.37)

11.5.3 Static Converter Fed Drives

Static converters for drives as in rolling mills contribute to I_k and i_p only if the rotational masses of the motors and the static equipment provide reverse transfer of energy for deceleration (a transient inverter operation) at the time of short circuit. These do not contribute to I_b. Nonrotating loads and capacitors (parallel or series) do not contribute to the short-circuit currents. Static power converter devices are treated in a similar manner as asynchronous motors with the following parameters:

U_{rM} = rated voltage of the static convertor transformer on the network side

or rated voltage of the static converter if no transformer is present

I_{rM} = rated current of the static convertor transformer on the network side or

rated voltage of the static converter if no transformer is present

$$\frac{I_{LR}}{I_{rM}} = 3$$

$$\frac{R_M}{X_M} = 0.10, \; X_M = 0.995 Z_M$$ (11.38)

11.6 Comparison with ANSI/IEE Calculation Procedures

The following comparison of calculation procedures underlines the basic philosophies in ANSI and IEC calculations [8].

1. IEC requires calculation of initial symmetrical short-circuit current, I_k, in each contributing source. Each component of I_k current form the basis of further calculations. Thus, tracking each contributing source current throughout the network is necessary. Each of these component currents is a function of machine characteristics, R/X ratio, type of network (meshed or radial), type of excitation system for synchronous generators, contact parting time (minimum time delay), and the determination whether contribution is near to or far from the short-circuit location. Multiplying factors on generators, unit transformers, and power station units (PSUs) are applicable before the calculation proceeds. The PSU consisting of a generator and a transformer is considered as a single entity, and separate procedures are applicable for calculation of whether the fault is on the high- or low-voltage side of the transformer. IEC treats each of the above factors differently for each contributing source.

This approach is conceptually different from that of ANSI, which makes no distinction between the type of network, and the network is reduced to a single Thévenin impedance at the fault point, using complex reduction, or from separate R and X networks, though prior impedance multiplying factors are applicable to account for AC decay. ANSI states that there is no completely accurate way of combining two parallel circuits with different values of X/R ratios into a single circuit with one X/R ratio. The current from the several circuits will be the sum of the decaying terms, usually with different exponents, while from a single circuit, it contains just one such term. The standard then advocates separate X and R networks and states that the error for practical purposes is on the conservative side.

In all IEC calculations, therefore, the initial short-circuit current is first required to be calculated as all other currents are based on this current. These initial short-circuit currents from sources must be tracked throughout the distribution system. The peak current and breaking current are then calculated based on factors to be applied to initial short-circuit current. In ANSI calculations, interrupting duty and first-cycle networks can be independently formed with prior impedance multiplying factors. In an application of circuit breakers, one of the two duties, i.e., the first cycle or interrupting may only be the limiting factor.

2. Both standards recognize AC decay, though the treatment is different. ANSI standards model motors with prior multipliers, where the contribution of the large motors is an appreciable portion of the short-circuit current, substitution of tabulated multipliers with more accurate data based on manufacturer's time constants is recommended. But ANSI treats these multipliers on a *global basis* and these do not change with the location of fault points or the contact parting time of the breaker. IEC treats each motor individually and decay must be calculated on the basis of contact parting time, machine type, and its size, speed, and proximity to the fault. Contributions from motors can be ignored in certain cases in IEC, while ANSI considers motor contributions throughout, except as discussed in Section 10.2.1, though effect of motor loads can be insignificant in some cases (Example 10.2).

3. ANSI makes no distinction between remoteness of induction and synchronous motors for the short-circuit calculations. Impedance multiplying factors (Table 10.1) of the motors are considered to account for AC decay, irrespective of their location. IEC considers generators and *motors* as near to or far from the fault location for breaking and steady-state current calculations. Asynchronous machines are considered near if the sum of all motors I_k is greater than 5% of the total I_k without motors, otherwise these are considered remote. Synchronous machines are considered near if their I_k is more than twice the rated current.

4. Both standards recognize the rapid decay of the DC component of the fault current and add a half-cycle of tripping time to arrive at the contact parting time (IEC minimum time delay).

5. IEC calculations require that the DC component be calculated at the contact parting time to calculate the asymmetrical breaking current, i.e.,

$$i_{dc} = \sqrt{2} I_k'' e^{-\omega t / X/R} \tag{11.39}$$

where X/R is computed differently for radial or meshed networks. In radial networks, i_{DC} is the sum of the DC currents calculated with X/R ratios in each of the

Short-Circuit Calculations according to IEC Standards

TABLE 11.3

Equivalent Frequency f_c for Meshed Networks, Method (C)

$f \cdot t$	<1	<2.5	<5	<12.5
f_c/f	0.27	0.15	0.092	0.055

contributing elements. See the qualification with respect to X/R ratio of generators in Section 11.4.1. For meshed networks, the ratio R/X should be determined by method (C). The equivalent f_c should be used as shown in Table 11.3. Methods B and C are applicable for meshed networks for calculation of i_p.

Calculations of the DC component at the contact parting time is not required in ANSI. The DC decay is built into the postfault calculations with E/X or E/Z multipliers from the curves. Also, the rating structure of ANSI takes into account certain asymmetry, depending on the contact parting time as shown in Figure 8.4.

6. ANSI uses a prefault voltage equal to the rated system voltage, unless operating conditions show otherwise. IEC considers a prior voltage multiplying factor c from Table 11.2 *irrespective of the system conditions*. This factor seems to be imported from VDE (Deutsche Electrotechnische Kommission) and is approximately given as

$$cU_n = U_{rG}\left(1 + \frac{I_{rG}X_d'' \sin \phi_{rG}}{U_{rG}}\right) \approx 1.1 \tag{11.40}$$

7. The calculations of steady-state current is materially different and more involved in IEC calculations.

8. IEC specifically details the calculation of i_p and I_k from static converter-fed drive systems and states that the converter-fed motors do not contribute to I_b.

9. IEC calculations are more demanding on the computing resources and require a larger database.

11.7 Examples of Calculations and Comparison with ANSI Methods

Example 11.1

Calculate the fault current contributions of the following synchronous machines, directly connected to a bus, using ANSI and IEC methods. Calculate the first cycle (IEC peak) and interrupting (IEC breaking, symmetrical, and asymmetrical) currents for contact parting times of two cycles and three cycles (IEC minimum time delay = 0.03 and 0.05 s, approximately for a 60-Hz system). Compare the results.

- 110-MVA, 13.8-kV 0.85 PF generator, $X_{dv}'' = 16\%$ on generator MVA base
- 50-MVA, 13.8-kV 0.85 PF generator, $X_{dv}'' = 11\%$ on generator MVA base
- 2000-hp, 10-pole, 2.3-kV 0.8 PF synchronous motor, $X_{lr}'' = 20\%$
- 10,000-hp, 4-pole, 4-kV synchronous motor, 0.8 PF, $X_{lr}'' = 15\%$

Table 11.4 shows the results of ANSI calculations. The first-cycle peak is calculated using Equation 10.1 for new ratings of the breakers. These calculations have already

TABLE 11.4

ANSI Fault Current Calculations from Synchronous Generators and Motors Directly Connected to a Bus

Description	Percentage X_d on Equipment MVA Base	X/R	Impedance Multiplying Factors		First-Cycle Calculations		Interrupting Duty Calculations	
			First Cycle	Interrupting	First-Cycle Current (kA sym.)	First-Cycle Current (kA peak)	3-Cycle Contact Parting Time (kA rms)	2-Cycle Contact Parting Time (kA rms)
110-MVA, 0.85-PF, 13.8-kV generator	16	80.0	1	1	28.76	79.81	34.62 MF = 1.204	34.20 MF = 1.189
50-MVA, 0.85-PF, 13.8-kV generator	11	65.0	1	1	19.02	52.55	22.30 MF = 1.173	22.35 MF = 1.175
2000-hp, 10-pole, 2.3-kV synchronous motor, 0.8 PF (2000 kVA)	20	25.0	1	1.5	2.51	6.67	1.67	1.67
10,000-hp, 4-pole, 4-kV synchronous motor, 0.8 power factor (10,000 kVA)	15	34.4	1	1.5	9.62	25.99	6.41	6.41

Short-Circuit Calculations according to IEC Standards

been discussed in Chapter 10 and the description is not repeated here. Table 11.5 shows all the steps in IEC calculations. We will go through these steps for a sample calculation for a 110-MVA generator.

The percentage subtransient reactances for all the machines are the same in both calculations. The X/R ratio for ANSI calculations is estimated from Figure 10.4, while for IEC calculations it is based on Equation 11.16. For a 110-MVA generator, $R_{Gf} = 0.05\ X_d''$, i.e., $X/R = 20$.

Next, factor K_G is calculated from Equation 11.14. This is based on a rated PF of 0.85 of the generator:

$$K_G = \frac{U_n}{U_{rG}}\left(\frac{c_{max}}{1+X_d''\sin\phi_{rG}}\right)=\frac{1.10}{1+0.16\times0.526}=1.05$$

In the above calculation, X_d'' is in per unit on a machine MVA base at machine rated voltage. From Equation 11.13, the modified generator impedance is

$$Z_{GK} = K_G(R_G + jX_d'')=1.05(0.0073+ j0.1455)\ \text{per unit on 100 MVA base}$$

The initial short-circuit current from Equation 11.2 is

$$\left|I_k''\right| = \frac{c_{max}}{|Z_{GK}|}=7.191\ \text{per unit}=30.09\ \text{kA}$$

TABLE 11.5

IEC Fault Current Calculations from Synchronous Generators and Motors Directly Connected to a Bus

Equipment	110-MVA, 0.85-PF, 13.8-kV Generator	50-MVA, 0.85-PF, 13.8-kV Generator	2000-hp, 10-pole, 0.8-PF, 2.4-kV Synchronous Motor	10,000-hp, 4-pole, 0.8-PF, 4-kV Synchronous Motor
Percentage X_d on equipment kVA base	16	11	20	15
R_G or R_M	$0.05X_d$	$0.07X_d$	$0.07X_d$	$0.07X_d$
X/R for DC component	80	65	—	—
c_{max}	1.1	1.1	1.1	1.1
K_G or K_M	1.015	1.040	0.982	1.010
I_{KG} or I_{kM} kA rms	30.09	20.06	2.68	10.45
K	1.863	1.814	1.814	1.814
i_p kA peak	79.27	51.46	6.88	26.80
μ (0.05 s)	0.78	0.74	0.81	0.73
μ (0.03 s)	0.85	0.82	0.85	0.81
$i_{b,sym}$ (0.05 s)	23.47	14.84	2.17	7.63
$i_{b,sym}$ (0.03 s)	25.57	16.45	2.28	8.46
I_{DC} (0.05 s) kA	34.42	21.22	1.01	3.95
I_{DC} (0.03 s) kA	37.93	23.80	1.72	6.70
$i_{ba,sym}$ (0.05 s) kA	41.74	25.89	2.39	8.50
$i_{ba,sym}$ (0.03 s) kA	45.74	28.90	2.86	10.79

Here, we are interested in magnitude only, as the calculations are from a single source and summations are not involved. For calculation of peak current, the factor χ is calculated from Equation 11.4:

$$\chi = 1.02 + 0.98e^{-3R/X} = 1.02 + 0.98 \times 0.861 = 1.863$$

The peak current from Equation 11.3 is, therefore,

$$i_p = \chi\sqrt{2}I_k'' = 1.863 \times \sqrt{2} \times 30.09 = 79.29 \, \text{kA}$$

The breaking current factor μ is calculated from Equation 11.22. For 0.05 minimum time delay

$$\mu = 0.71 + 0.51e^{-30I_{KG}''/I_{rG}} = 0.71 + 0.51e^{-30 \times 30.09/4.60} = 0.78$$

The calculation for 0.03 minimum time delay is not given directly by Equation 11.22 and interpolation is required. Alternatively, the factor can be estimated from the graphs in the IEC standard.

The symmetrical interrupting current at 0.05 s minimum time delay is

$$i_{b,sym} = \mu I_k'' = 0.78 \times 30.09 = 23.47 \, \text{kA}$$

The DC component is calculated from Equation 11.39. However, to calculate X/R ratio R_G as calculated above is not used, as per qualification stated in section. Using an X/R of 80, same as for the ANSI/IEEE calculation gives

$$i_{DC} = \sqrt{2}I_k''e^{-\omega t/X/R} = \sqrt{2} \times 30.9 \times e^{-377 \times 0.05/80} = 34.52 \, \text{kA}$$

The asymmetrical breaking current at 0.05 s parting time is

$$i_{ba,sym} = \sqrt{i_{b,sym}^2 + i_{DC}^2} = 41.74 \, \text{kA rms}$$

Table 11.5 is compiled similarly for other machines. Synchronous motors are treated as synchronous generators for the calculations. A comparison of the results by two methods of calculation shows some differences. ANSI first-cycle current and IEC peak currents are comparable, with a difference within 3%. ANSI interrupting duty symmetrical currents for generators are higher than IEC breaking currents, i.e., for a 110-MVA generator the ANSI current at 2-cycle contact parting time is 28.76 kA (without multiplying factor), while IEC current is 25.57 kA.

For comparison with IEC asymmetrical breaking current for 5-cycle breaker, from Table 11.5, $i_{ba,sym} = 41.74$ kA and from Table 11.4, ANSI calculation the interrupting sym rating is calculated as 34.62 kA. As per rating structure of the breakers, the asymmetrical interrupting current is $1.1 \times 34.62 = 38.08$ kA. Thus, IEC calculation gives higher asymmetrical breaking current. Similarly for 3-cycle breaker, $i_{ba,sym} = 45.74$ kA, and as per ANSI calculation the corresponding number to compare is $1.2 \times 34.20 = 41.04$. In ANSI calculations, the short-circuit currents from the motors for various contact parting times do not change and no remote or local multiplying factors are applicable to these currents.

Short-Circuit Calculations according to IEC Standards 503

Example 11.2

Calculate the fault current contributions of the following asynchronous machines, directly connected to a bus, using ANSI and IEC methods. Calculate first-cycle (IEC peak) currents and the interrupting (breaking) currents at contact parting times of two and three cycles, 60-Hz basis, and IEC minimum time delays of 0.03 and 0.05 s approximately. Compare the results.

- 320-hp, 2-pole, 2.3-kV induction motor, $X_{lr} = 16.7\%$
- 320-hp, 4-pole, 2.3-kV induction motor, $X_{lr} = 16.7\%$
- 1560-hp, 4-pole 2.3-kV induction motor, $X_{lr} == 16.7\%$

These results of ANSI calculations are shown in Table 11.6, while those of IEC calculations are shown in Table 11.7. Most of the calculation steps for asynchronous motors are in common with those for synchronous motors, as illustrated in Example 11.1. The motor locked rotor reactance of $X_{lr} = 16.7\%$ on a motor kVA base is used in both calculation methods; however, the resistances are based on recommendations in each standard. Factor q must also be calculated for asynchronous motors and it is given by Equation 11.36. This requires m equal to the motor-rated power in megawatts per *pair* of poles to be calculated on the basis of motor PF and efficiency. The symmetrical breaking current is then given by Equation 11.37.

A comparison of results again shows divergence in the calculated currents. In ANSI calculations, the interrupting duty current for a 320-hp two-pole motor is twice that of the four-pole motor, 0.28 kA versus 0.14 kA. This is so because a prior impedance multiplying factor of 1.5 is applicable to a two-pole 320-hp motor and this factor is 3 for a four-pole 320-hp motor. IEC calculation results for a two-pole motor is only slightly higher, 0.235 kA. The IEC calculations consider decay of currents from asynchronous motors, while ANSI/IEEE does not. There is no direct comparison of the asymmetrical motor currents in Table 11.7.

Example 11.3

Example 10.2 of Chapter 10 is repeated with the IEC method of calculation. All the impedance data remain unchanged, except that the resistance components are estimated from IEC equations.

Three-Phase Fault at F1

For a fault at Fl, the generator and transformer are considered as a PSU. The ANSI/IEEE calculations in Example 7.2 are performed at rated system voltages. This means that the adjustments of taps due to tap changing of transformers, variations in the source voltage and generation voltage are neglected. The Equations 11.18 and 11.19 consider variation of the voltages at the fault point due to tap changing, generator voltage regulation or off-load tap settings. As the purpose of the example is to compare results on the same basis, rated voltages are considered. This means that the 60/100 MVA transformer is at rated voltage tap ratio, which is also the nominal system voltage on the high-voltage side and the generator-rated voltage on the low-voltage side For a fault on the high-voltage side, the correction factor is given by Equation 11.18 becomes

$$K_S = \frac{c_{max}}{1 + \left| x_d'' - x_T \right| \sin \phi_{rG}}$$

The ANSI calculations in Example 7.2 are performed with generator at no load. Here consider a PF of 0.85 lagging. Then

$$K_S = \frac{1.1}{1 + (0.164 - 0.0774)(0.527)} = 1.052$$

TABLE 11.6

ANSI Fault Current Calculations from Asynchronous Motors Directly Connected to a Bus

Description	X_{lr} on Equipment MVA Base	X/R	Impedance Multiplying Factors		First-Cycle Calculations		Interrupting Duty Calculations	
			First Cycle	Interrupting	First-Cycle Current (kA sym.)	First-Cycle Current (kA peak)	Three-Cycle Contact Parting Time (kA rms)	Two-Cycle Contact Parting Time (kA rms)
320-hp, 2-pole induction motor, 2.3 kV (kVA = 285)	16.7	15	1	1.5	0.427	1.095	0.28	0.28
320-hp, 4-pole induction motor, 2.3 kV (kVA = 285)	16.7	15	1.2	3	0.356	0.909	0.14	0.14
1560-hp, 4-pole, 2.3-kV induction motor (kVA = 1350)	16.7	28.5	1	1.5	2.028	5.441	1.35	1.35

Short-Circuit Calculations according to IEC Standards

TABLE 11.7

IEC Fault Current Calculations from Asynchronous Motors Directly Connected to a Bus

Description	300-hp, 2-Pole, 2.3-kV Induction Motor (kVA = 285), PF = 0.9, Efficiency = 0.93	300-hp, 4-Pole, 2.3-kV Induction Motor (kVA = 285), Power Factor = 0.9, Efficiency = 0.93	1500-hp, 4-Pole, 2.3-kV Induction Motor (kVA = 1350), Power Factor = 0.92, Efficiency = 0.94
I_{LR}/I_{RM}	6	6	6
I_K''/I_{rM}	6.6	6.6	6.6
m (electric power per pair of poles) (MW)	0.256	0.128	0.621
R_M/X_m	0.15	0.15	0.15
κ_m	1.65	1.65	1.65
μ (0.05 s)	0.79	0.79	0.79
μ (0.03 s)	0.83	0.83	0.83
q (0.05 s)	0.63	0.54	0.73
q (0.03 s)	0.79	0.66	0.86
I_{KM}'' (kA rms)	0.473	0.473	2.229
i_{pM} (kA crest)	1.104	1.104	5.201
$i_{b,sym}$ (0.05 s) kA	0.235	0.202	1.285
$i_{b,sym}$ (0.03 s) kA	0.310	0.259	1.591
I_{DC} (0.05 s) kA	0.029	0.029	0.136
I_{DC} (0.03 s) kA	0.101	0.101	0.479
$i_{ba,sym}$ (0.05 s)	0.237	0.204	1.292
$i_{ba,sym}$ (0.03 s)	0.326	0.278	1.661

The generator and transformer per unit impedances are on their respective MVA base, as shown in Table 10.14. The modified power station impedance is then

$$Z_s = K_S(Z_G + Z_T)$$

The generator resistance calculated is 0.05 times the subtransient reactance. Therefore,

$$Z_s = 1.052(0.007315 + j0.14630 + 0.00404 + j0.12894)$$

$$= 0.01195 + j0.28955$$

in per unit on a 100-MVA base. It is not necessary to consider the asynchronous motor contributions. The initial short-circuit current is then the sum of the source and PSU components, given as

$$I_k'' = I_{k,PSU}'' + I_{kQ}''$$

The utility's contribution, based on the source impedance in per unit from Table 10.14 is given as

$$I_{kQ}'' = \frac{1.1}{Z_{kQ}} = \frac{1.1}{(0.00163 + j0.02195)} = 3.7187 - j49.78 \text{ per unit}$$

$$= |20.90| \text{ kA} \quad \text{at } 13.8 \text{ kV}$$

PSU contribution

$$I_{k,PSU}'' = \frac{1.1}{Z_{PSU}} = \frac{1.1}{0.01195 + j0.28955} = 0.1565 - j3.7925 \text{ per unit}$$

$$= |1.59| \text{ kA} \quad \text{at } 138 \text{ kV}$$

Therefore,

$$I_k'' = 3.7187 - j49.78 + 0.1565 - j3.793 = 3.8875 - j53.573 \text{ per unit} = |22.47| \text{ kA}$$

The peak current i_p is

$$i_p = i_{pPSU} + i_{pQ}$$

Power station $R/X = 0.01195/0.28955 = 0.04128$. Therefore, χ_{PSU} from Equation 11.4 = 1.886. This gives

$$i_{p,PSU} = \chi_{PSU}\sqrt{2}I_{k,PSU}'' = (1.886)\sqrt{2}(1.59) = 4.24 \text{ kA}$$

Utility system $R/X = 0.0746$, $\chi_{PQ} = 1.803$:

$$i_p = 53.29 + 4.24 = 57.53 \text{ kA peak.}$$

The symmetrical breaking current is the summation of two currents—one from the PSU and the other from the source. The equation for the same is given as

$$I_b = I_{bPSU} + I_{bQ} = I_{bPSU} + I_{kQ}''$$

Considering a contact parting time of 0.03 s, for the PSU contribution, μ is calculated from Equation 11.22. This gives $\mu = 0.94$.

$$I_{bPSU} = 0.94 \times 1.59 = 1.495 \text{ kA}$$

Total breaking current = 20.90 + 1.495 = 22.395 kA symmetrical.

The DC components at contact parting time are calculated using Equation 11.39, repeated below:

$$i_{DC} = \sqrt{2}I_k'' e^{-2\pi ftR/X}$$

The DC component of the utility's source, based on $X/R = 14.3$, from Table 10.14, is

$$\sqrt{2}(20.90)e^{-377(0.03)(0.0746)} = 12.71 \text{ kA}$$

Similarly, the DC component from the PSU, based on $X/R = 24.23$ from $Z_{PSU} = 1.41$ kA; the total DC current is 14.12 kA.

Finally, the asymmetrical breaking current is

Short-Circuit Calculations according to IEC Standards 507

$$I_{ba,sym} = \sqrt{I_b^2 + (i_{DC})^2}$$
$$= \sqrt{(22.395)^2 + (14.12)^2}$$
$$= 26.47\,kA$$

Three-Phase Short-Circuit at F2

The adjustment factor $Z_{G\,PSU}$ for the low-voltage side faults for PSUs is given by Equation 11.13:

$$Z_G = R_G + jX_d''$$

$$R_G = 0.05 X_d'', \quad \text{for } U_{rG} > 1\,kV, S_{rG} \geq 100\,MVA$$

$$R_G = 0.05 \times 0.14630 = 0.007315$$

$$Z_G = 0.007315 + j0.14630$$

$$K_{GPSU} = \frac{c_{max}}{1 + X_d'' \sin\phi_{rG}} = \frac{1.1}{1 + 0.164 \times 0.52} = 1.0136$$

Therefore,

$$Z_{GPSU} = K_{GPSU} Z_G = 1.0136(0.007315 + j0.14630)$$
$$= 0.0074 + j0.1483$$

We have ignored small impedances of bus ducts B1 and B2 in the above calculations. We will also ignore contributions from *all* motors. The initial short-circuit currents are the sum of the partial short-circuit currents from the generator and source through transformer T1. The partial short-circuit current from the generator is

$$I_{kG}'' = \frac{1.1}{Z_G} = \frac{1.1}{(0.0074 + j0.1483)}$$

$$= 0.369 - j7.399\,\text{per unit}$$

$$= |30.99|\,kA$$

To calculate the partial short-circuit current of a utility's system through the transformer T1, the transformer impedance is modified. From equation

$$K_{T,S} = \frac{c_{max}}{1 - x_T \sin\phi_{rG}} = \frac{1.1}{(1 - 0.0774 \times 0.5267)} = 1.147$$

The partial short circuit current is given as

$$I_{kt}'' = \frac{1.1}{1.147(0.00404 + j0.12894) + (0.00163 + j0.02195)}$$

$$= \frac{1}{0.00626 + j0.16984} = 0.2167 - j5.8799\,\text{per unit}$$

$$= |24.62|\,kA$$

By summation, the initial short-circuit current is then

$$I_{kF2}{}'' = I_{kG}{}'' + I_{kT}{}''$$

$$= 0.369 - j7.399 + 0.2167 - j5.8799$$

$$= 0.5857 - j13.279 \text{ per unit}$$

$$= |55.56| \text{ kA}$$

As resistance is low, this current could have been calculated using reactances only. The peak current i_p is the sum of the component currents as shown below:

$$i_p = i_{pG} + i_{pT}$$

Factor χ for the generator, based on $R_G/X_d'' = 0.05$, from Equation 11.4 = 1.86. Thus, the peak current contributed by the generator is

$$i_{pG} = \chi_G \sqrt{2} I_{kG}'' = 1.86 \times \sqrt{2} \times 30.99 = 81.52 \text{ kA}$$

Similarly, calculate i_{pT},

$$R/X = 0.00607/0.16378 = 0.0371$$

$$\chi_T = 1.90$$

$$i_{pT1} = 1.90(\sqrt{2})(24.60) = 66.09 \text{ kA}$$

Therefore, the total peak current is 147.61 kA. The symmetrical breaking current I_b is calculated at a minimum time delay of 0.05 s. It is the summation of the currents from the source through T1, which is equal to the initial short-circuit current and the symmetrical breaking current from the generator: $I_b = I_{bG} + I_{bT} = I_{bG} + I_{kT}''$.

$$I_{bG} = \mu I_{kG}''$$

$$\frac{I_{kG}''}{I_{RG}} = \frac{30.99}{4.690} = 6.60$$

$$\mu = 0.78, q = 1$$

$$I_{bG} = (0.78)(30.99) = 24.17 \text{ kA}$$

Total breaking current symmetrical = 48.79 kA.

For asymmetrical breaking current, DC components at a minimum time delay of 0.05 s are calculated. Consider a generator X/R of 125.

Generator component = 37.96 kA
Transformer component = 17.96 kA
Total DC component = 55.92 kA

This gives a total asymmetrical breaking current of 74.21 kA.

These results are compared with the calculations in Example 10.2 and are shown in Table 11.8. IEC currents are higher for a fault at Fl and lower for a fault at F2. Generally, for calculations involving currents contributed mainly by generators, ANSI interrupting

Short-Circuit Calculations according to IEC Standards

TABLE 11.8

Examples 11.3 and 11.4: Comparative Results of Three-Phase Short-Circuit Calculations

Fault Location	Calculation Method	First-Cycle Current kA Asym. Crest (ANSI) or Peak Current i_p (IEC)	Interrupting Duty Current (ANSI) or $I_{ba,sym}$ (IEC)
Fl (138 kV)	ANSI calculation	53.1	$20.54 \times 1.2 = 24.65$
	IEC calculation	57.53	26.47
F2 (13.8 kV)	ANSI calculation	1556.92	$71.69 \times 1.1 = 78.86$
	IEC calculation	150.12	74.21
F3 (4.16 kV)	ANSI calculation	46.23	$16.35 \times 1.1 = 17.99$
	IEC calculation	46.30	19.14

TABLE 11.9

Generator Fault Currents: Example 11.4

Calculation Type	Symmetrical Current (kA rms)	DC Component (kA)	Asymmetrical Current	Asymmetry Factor α
ANSI	18.95	34.58	39.43	1.29
IEC, generator X/R same as in ANSI calculation	24.20	37.96	45.02	1.11

currents are higher than IEC currents. IEC uses an artificially high generator resistance, which is further multiplied by factor K_{gpsu}. As a result, the fault currents comparatively reduces in magnitude and asymmetry. The calculation for faults at F1 is higher in IEC, because of factor c, which increases the source contribution by 10%, while the generator contribution through transformer T1 is comparatively small.

Generator Source Short-Circuit Current

In Example 10.2, we calculated the generator fault current for a fault at F2 and noted high asymmetry. The current zeros are not obtained at the contact parting time and an asymmetry factor of 129% is calculated. The calculation is repeated with IEC methods, the generator X/R ratio for the DC component is the same as for ANSI calculations = 125. The calculations give

$$I_{kG}'' = 30.99 \text{ kA}$$

$$i_{pG} = 81.52 \text{ kA}$$

$$i_{bGsym} = 24.20 \text{ kA}$$

$$i_{GDC} = 37.96 \text{ kA}$$

$$i_{bGasym} = 45.02 \text{ kA}$$

The asymmetry factor is $\alpha = (37.96)/(\sqrt{2} \cdot 24.20) = 111.0\%$.

The results of the calculations are shown in Table 11.9. IEC standard [1], does not discuss the asymmetry at the contact parting time of the breaker. Short-circuit current profiles for "far-from" and "near-to" generators are shown in Figures 11.1 and 11.2 *of this standard*, respectively. There is no discussion of not obtaining a current zero at the contact parting time of the breaker. The IEC standard showing the examples of short-circuit calculations, part-4 is yet to be published. IEC may adopt IEEE standard for the generator breakers.

510

Short-Circuits in AC and DC Systems

Example 11.4

The effect of motors is neglected in Example 11.3. Calculate the partial currents from the motors at 13.8 kV. Do these motor contributions need to be considered in IEC calculations for a fault at F2? Also, calculate the peak current and the asymmetrical breaking current for a fault at F3 on the 4.16-kV bus.

Effect of Motor Contribution at 13.8-kV Bus, Fault F2

For a fault at the 13.8-kV bus F2, an equivalent impedance of the motors through transformers and cables is calculated. The partial currents from medium- and low-voltage motors are calculated in Tables 11.10 and 11.11, respectively. The equivalent impedance of low-voltage motors of two identical groups, from Table 11.11, is $6.45 + j15.235$ per unit.

The per unit impedance of transformer T3 from Table 10.14 is $0.639 + j3.780$. Therefore, the low-voltage motor impedance through transformer T3, seen from the 4.16-kV bus is $7.089 + j19.015$ per unit.

From Table 11.10, the equivalent impedance of medium-voltage motors is $0.58 + j4.55$ per unit. The equivalent impedances of low- and medium-voltage motors in parallel is $(7.089 + j19.015)||(0.58 + j4.55) = 0.637 + j3.707$

TABLE 11.10

Partial Short-Circuit Currents from Asynchronous Medium-Voltage Motors: Example 11.4

Parameter	2425 hp	300 hp	500 hp	Sum (Σ)
Power output, P_{rm} (MW)	1.81	0.224	0.373	
Quantity	1	3	2	
PF ($\cos \phi$)	0.93	0.92	0.92	
Efficiency (η_r)	0.96	0.93	0.94	
Ratio, locked rotor current to full load current (I_{LR}/I_{rM})	6	6	6	
Pair of poles (p)	1	1	2	
Sum of MVA (S_{rM})	2.03	0.78	0.86	
Sum, rated current (I_{rM})	0.28	0.11	0.12	
I_K''/I_{rM}	6.6	6.6	6.6	
Power per pole pair (m)	1.81	0.223	0.186	
R_M/X_M	0.10	0.15	0.15	
κ_m	1.75	1.65	1.65	
μ	0.78	0.78	0.78	
q	0.86	0.61	0.59	
I_{kM}'	1.85	0.73	0.79	$\Sigma = 3.37$
i_{pM}	4.58	1.70	1.84	$\Sigma = 8.12$
i_{bM}	1.24	0.35	0.36	$\Sigma = 1.95$
Z_m	8.23	21.41	19.42	
X_M	$0.995\, Z_M = 8.189$	$0.989\, Z_M = 21.17$	$0.989\, Z_M = 19.21$	
R_M	$0.1\, X_M = 0.82$	$0.15\, X_M = 3.18$	$0.15\, X_M = 2.88$	
Cable C2	$0.068 + j0.104$			
Σ MV motors and cable				$0.58 + j74.55$

Short-Circuit Calculations according to IEC Standards

TABLE 11.11

Low-Voltage Motors, Partial Short-Circuit Current Contributions: Example 11.4

Parameter	Motors M4, M5, and M6 (or Identical Group of Motors M4', M5', and M6,)	Remarks
P_{rm} (MW)	0.43	Calculated active power rating of the motor group
Sum of MVA (S_{rM})	0.52	Active power rating divided by PF
R_M/X_M	0.42	From Equation 11.20 for group of motors connected through cables
χ_m	1.3	From Equation 11.4
Ratio, locked rotor current to full load current (I_{LR}/I_{rM})	6	
Z_M in per unit 100-MVA base	32.12	
X_M in per unit 100-MVA base	$0.922\,Z_M = 29.61$	Equation 11.20
R_M in per unit 100-MVA base	$0.42\,X_M = 12.44$	Equation 11.20
Cables C3 or C4 in per unit 100-MVA base	$0.46 + j0.860$	Table 10.3

To this add the impedance of cable C1 and transformer T2 from Table 10.14, which gives $0.701 + j4.606$ per unit. This is the equivalent impedance as seen from the 13.8-kV bus. Thus, the initial short-circuit current from the motor contribution is $1.1/(0.701 + j4.606)$ per unit = $|0.99|$ kA.

The effect of motors in this example can be ignored and the above calculation of currents from motor contributions is not necessary. From Equation 11.33, ΣP_{rm} = sum of the active powers of all medium- and low-voltage motors = 0.86 MW. Also, ΣS_{rT} = rated apparent power of the transformer = 7.5 MVA. The left-hand side of Equation 11.33 = 0.1147. Symmetrical short-circuit power at the point of connection, without effect of motors, is

$$S_{kQ}'' = \sqrt{3}I_k''U_n = 1350.7\,\text{MVA}$$

The right-hand side of Equation 11.33 gives 2.571 and the identity in Equation 11.33 is satisfied. The effect of motors can be ignored for a fault at 13.8 kV.

If the calculation reveals that motor contributions should be considered, we have to modify i_p at the fault point. This requires calculation of χ, which is not straightforward. High-voltage motors have $\chi = 1.75$ or 1.65 and low-voltage motors have $\chi = 1.65$. For a combination load, $\chi = 1.7$ can be used to calculate i_p approximately.

Fault at F3. For a fault at F3, we will first calculate the motor contributions. The low-voltage motor impedance plus transformer T3 impedance is $7.089 + j19.015$, as calculated above. The initial short-circuit contribution from the low-voltage motor contribution is $1.1/(7.089 + j19.015) = 0.019 - j0.051$ per unit or $|I_k''| = 0.76$ kA. Medium-voltage impedance, from Table 11.10, is $0.58 + j4.55$ per unit. The medium-voltage motor contribution is $0.028 - j0.239$ per unit or $|I_k''| = 3.43$ kA.

To calculate the generator and utility source contributions, the impedances Z_{GPSU} is in parallel with $(Z_{T,PSU} + Z_Q)$, i.e., $0.0074 + j0.1483$ in parallel with $0.006074 + j0.16378$. This gives $0.0034 + j0.0778$ per unit. Add transformer T2 impedance $(0.06349 + j0.89776)$ and

512 *Short-Circuits in AC and DC Systems*

cable C1 impedance (0.00038 + j0.00101) from Table 10.14. This gives an equivalent impedance of 0.0673 + j0.976 per unit. Thus, the initial short-circuit current is 0.077 − j1.122 per unit or $\left|I_K^{''}\right|$ = 15.61 kA. The total initial symmetrical current, considering low- and medium-voltage motor contributions, is 19.80 kA. To calculate i_p, χ must be calculated for the component currents.

For contribution through transformer T2, using Equation 11.4

$$\chi_{AT} = 1.02 + 0.98\varepsilon^{-3(0.06895)}$$
$$= 1.82$$

As this is calculated from a meshed network, a safety factor of 1.15 is applicable, i.e., χ = 1.15 multiplied by 1.82 = 2.093. However, for high-voltage systems χ is not greater than 2.0. This gives $i_{pAT} = (2)(\sqrt{2})(15.65) = 44.27$. For medium-voltage motors, χ can be calculated from Table 11.10:

$$\chi_{MV} = \frac{i_p}{\sqrt{2}I_k^{''}} = \frac{7.39}{\sqrt{2}(3.06)} = 1.71$$

For low-voltage motors through transformer T3

$$\chi_{LVT} = 1.02 + 0.98e^{-3(6.97/19.02)} = 1.346$$
$$i_{pLV} = (1.346)(\sqrt{2})(0.76) = 1.45 \text{ kA}$$

Total peak current by summation = 47.07 kA.

The breaking current is the summation of individual breaking currents:

- Breaking current through transformers at $4.16\text{kV} = \left|I_k^{''}\right| = 15.65\text{kA}$
- Breaking current medium-voltage motors from Table 11.10 = 1.95 kA

For low-voltage motors with $I_{kM}^{''} / I_{RM} = 6.6$ $\mu = 0.78$, q can be conservatively calculated for $m \le 0.3$ and $p = 2$. This gives $q = 0.64$. The component breaking current from low-voltage motors is therefore, $0.78 \times 0.64 \times 0.76 = 0.38\text{kA}$. Total symmetrical breaking current = 17.98 kA.

To calculate the asymmetrical breaking current, the DC components of the currents should be calculated:

- The DC component of the low-voltage motor contribution is practically zero.
- The DC component of the medium-voltage motors at contact parting time of $0.05 \text{ s} = 0.5 \text{ kA}$.
- The DC component of current through transformer T2 = 6.07 kA.

Total DC current at contact parting time = 6.57 kA; this gives asymmetrical breaking current of 19.14 kA.

The results are shown in Table 11.8.

Short-Circuit Calculations according to IEC Standards

Example 11.5: Steady-State Currents

Calculate the steady-state currents on the 13.8-kV bus, fault point F2, in the system of Example 10.2, according to IEC and ANSI methods.

IEC Method

The steady-state current is the summation of the source current through the transformer and the generator steady-state current:

$$I_k = I_{k,PSU} + I_{kG} = I''_{k,PSU} + \lambda_{max} I_{rG}$$

Substitution of λ_{min} gives the minimum steady-state current; λ_{max} and λ_{min} are calculated from Figure 11.8; and $I_{rG} = 4.69$ kA, $I''_{kG} = 30.99$ kA, and ratio $I''_{kG}/I_{rG} = 6.61$. Also from Table 9.1, the generator $X_{d-sat} = 1.949$. From Series One curves in Figure 11.8, $\lambda_{min} = 0.5$ and $\lambda_{max} = 1.78$. Therefore, the maximum generator steady-state current is 8.35 kA, and the minimum generator steady-state current is 2.345 kA. The source steady-state current is equal to the initial short-circuit current (25.53 kA); therefore, the total steady-state short-circuit current is 33.88 kA maximum and 27.88 kA minimum.

ANSI Method

The source current from Example 10.2 is 27.69 kA. From Table 9.1, the generator transient reactance is 0.278 on machine MVA base. The generator contribution is, therefore, 16.87 kA, and the total steady-state current is 44.56 kA.

11.8 Electromagnetic Transients Program Simulation of a Generator Terminal Short Circuit

This example demonstrates the generator source short-circuit calculations using ANSI/IEEE, IEC analytical methods, and rigorous electromagnetic transients programs (EMTPs) simulation. The generator is 234 MVA, 2-pole, 18 kV, 0.85 PF, 198.9 MW, 7505 A stator current 60 Hz, 0.56 SCR, 350 V field volts and wye connected. The manufacturer's data of the generator is shown in Table 11.12. All the data shown in this table is not required for calculations according to Equation 9.113. The generator breaker is a 5-cycle breaker with contact parting time = 50 ms. EMTP routines can accept the manufacturer's data which is converted to dq0 axis using Park's transformation. The results of the calculations are shown in Table 11.13. Figure 11.12 shows the three-phase terminal short-circuit current, and Figure 11.13 shows the current in phase c only for clarity. Zero crossing is not obtained for a number of cycles after the contact parting time of the breaker [9–11]. The results of the comparative calculations are in Table 11.14.

11.8.1 The Effect of PF

The load PF (lagging) in IEC equations does not change the asymmetry at the contact parting time. However, the asymmetry *does* change with the PF and prior load. This is clearly shown in Figure 11.14, with generator absorbing reactive power, i.e., operating at leading PF of 0.29. Generator load prior to short circuit = 28 MW, 92.4 Mvar. It is seen that the

Short-Circuits in AC and DC Systems

TABLE 11.12

Manufacturer's Generator Data, 18 kV, 234 MVA, 0.85 PF, 198.9 MW, 7505 A, 60 Hz, 0.56 SCR

Description	Symbol	Data	Description	Symbol	Data
Saturated synchronous	X_{dv}	2.120	Saturated negative sequence	X_{2v}	0.150
Unsaturated synchronous	X_d	2.120	Unsaturated negative sequence	$_{2I}$	0.195
Saturated transient	X_{dv}	0.230	Saturated zero sequence	X_{0v}	0.125
Unsaturated transient	X_d'	0.260	Unsaturated zero sequence	X_{0I}	0.125
Saturated subtransient	X_{dv}''	0.150	Leakage reactance, overexcited	$X_{LM,OXE}$	0.135
Unsaturated subtransient	X_d''	0.195	Leakage reactance, under excited	$X_{LM,UEX}$	0.150
Saturated synchronous	X_{qv}	1.858	Saturated subtransient	X_{qv}''	0.140
Unsaturated synchronous	X_q	1.858	Unsaturated subtransient	X_q'	0.192
Unsaturated transient	X_q'	0.434	Field time constants Open circuit	T_{d0}'	5.615
Three-phase short circuit transient	T_{d3}'	0.597	Line-to-neutral short-circuit transient	T_{d1}'	1.124
Line-to-line short-circuit transient	T_{d2}'	0.927	Short-circuit subtransient	T_d''	0.015
Open circuit	T_{q0}'	0.451	Open-circuit subtransient	T_{d0}''	0.022
Three-phase short-circuit transient	T_q'	0.451	Short-circuit subtransient	T_q''	0.015
Three-phase short-circuit	T_{a3}	0.330	Open-circuit subtransient	T_{q0}''	0.046
Line-to-line short-circuit	T_{a2}	0.330	Effective X/R	X/R	125

Note: d, direct axis, q, quadrature axis. All reactance's in per unit on machine MVA base. All time constants in seconds.

TABLE 11.13

Generator Parameters in d–q Axis, Park's Transformation

Parameter	Symbol	Value in Ohms	Parameter	Symbol	Value in Ohms
Self-inductance d-axis	L_f	3.463E+02	Mutual inductance q-axis, circuit 2 to armature	L_{akq}	2.822E+01
Mutual inductance field and armature	L_{af}	3.000E+01	Self-inductance of circuit 2 of q-axis	L_{kq}	3.106E+02
Field damper mutual inductance d-axis	L_{fkd}	3.275E+02	Zero-sequence inductance	L_0	1.731E-01
Self-inductance armature d-axis	L_d	2.935E+00	Zero sequence resistance	R_0	3.946E+03
Mutual inductance armature d-axis damper	L_{akd}	3.000E+01	Resistance d-axis field winding	R_f	1.518E-01
Self-inductance of d-axis damper winding	L_{kd}	3.303E+02	Resistance of armature	R_a	1.523E-03
Self-inductance circuit 1 of the q axis	L_g	3.741E+02	Resistance of d-axis damper winding	R_{kd}	1.272E+00
Mutual inductance, q-axis circuit 1 to armature	L_{ag}	2.822E+01	Resistance circuit 1 of the q-axis	R_g	1.679E+00
Mutual inductance circuit 1 to circuit 2	L_{gkq}	3.080E+02	Resistance circuit 2 of q-axis	R_{kq}	1.914E+00
Self-inductance q-axis armature winding	L_q	2.772E+00			

Short-Circuit Calculations according to IEC Standards 515

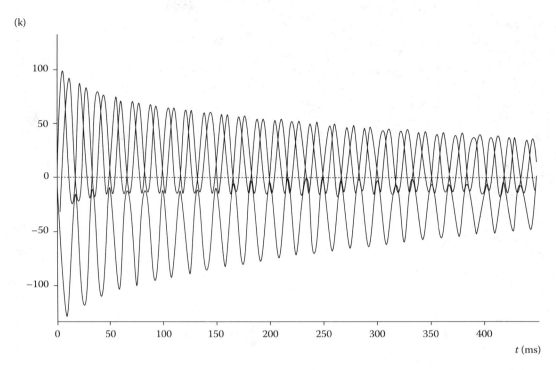

FIGURE 11.12
EMTP simulation results of the terminal three-phase short circuit of 234 MVA generator (Example 8.6).

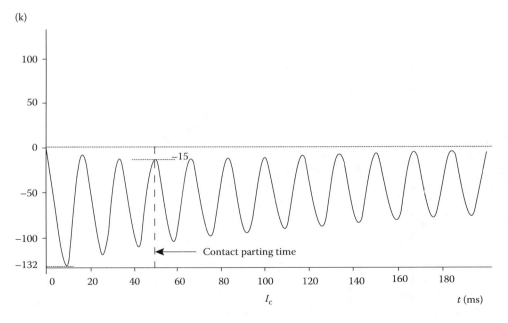

FIGURE 11.13
Short-circuit current profile in phase c with maximum asymmetry, showing delayed current zeros.

TABLE 11.14

Comparison of Calculations Using IEEE/IEC Standards and EMTP Simulations

Calculated Parameter	IEEE	IEC	EMTP
Close and latch, kA peak (IEC peak short-circuit current)	112.2	131.60	132.05
Generator source Interrupting kA sym. RMS (IEC symmetrical breaking current $i_{b,sym}$)	30.90	38.50	33.59
DC component, kA	59.22	60.73	62.50
Total asymmetrical, kA RMS (IEC $i_{ba,sym}$)	66.80	71.90	70.90
Asymmetry factor	135%	112%	131%

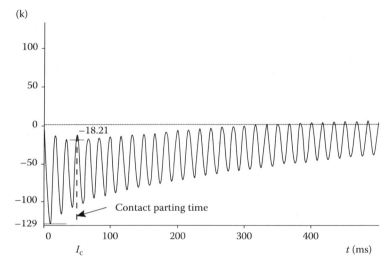

FIGURE 11.14
Effect of generator operating PF (generator loaded to 28 MVA at 0.29 leading PF) prior to short circuit on asymmetry. Current zeros are further delayed.

current zeros are further delayed, as compared to generator operating at no-load; compare with Figure 11.13. The prior leading PF loading further delays the occurrence of current zeros. The example of calculation demonstrates that asymmetry at contact parting time can be even 130% or more. The delayed current zeros can also occur on short-circuits in large industrial systems, with co-generation facilities.

The calculations in this chapter demonstrate that there are differences in results obtained by ANSI and IEC methods and one or the other calculation method can give higher or lower results. The predominant differences are noted in the contributions from motors and generators, which are the major sources of short-circuit currents in power systems. These differences vary with the contact parting times. The factor c in IEC calculations makes the source contributions higher, and this generally results in higher currents. It seems appropriate to follow the calculations and rating structures of breakers in these standards in their entity, i.e., use IEC calculations for IEC-rated breakers and ANSI calculations for ANSI-rated breakers. Example 11.6 demonstrates the comparative results of calculations with EMTP simulation. Other investigators have reached the same conclusions (Table 10.3).

Problems

The problems in this section are constituted so that they can be solved by hand calculations.

11.1 A 13.8-kV, 60-MVA, two-pole 0.8 PF synchronous generator has a subtransient reactance of 11%. Calculate its corrected impedance for a bus fault. If this generator is connected through a step-up transformer of 60 MVA, 13.8–138 kV, transformer $Z = 10\%$, $X/R = 35$, what are the modified impedances for a fault on the 138-kV side and 13.8-kV side?

11.2 In Problem 1, the 138-kV system has a three-phase fault level of 6500 MVA, $X/R = 19$. Calculate initial symmetrical short-circuit current, peak current, symmetrical breaking current, and asymmetrical breaking currents for a three-phase fault on the 138-kV side and 13.8-kV side. The minimum time delay for fault on the 138-kV side is 0.03 s for a fault on the 13.8-kV side is 0.05 s.

11.3 Figure P11.1 shows a double-ended substation with parallel running transformers. A three-phase fault occurs at F. Calculate I_k, i_p'', $I_{b,sym}$, $I_{ba,sym}$, I_{DC}, and k.

11.4 Figure P11.2 shows a generating station with auxiliary loads. Calculate three-phase fault currents for faults at Fl, F2, and F3. Calculate all component currents I_k, i_p'', $i_{b,sym}$, $I_{ba,sym}$, I_{DC}, and I_k in each case. The system data are shown in Table P11.1.

11.5 Repeat Problem 11.4 with typical X/R ratios from Figures 11.2 and 11.4.

11.6 In Example 11.6 using the manufacturer's data of the generator in Table 11.12, mathematically derive the dq0 axis parameters shown in Table 11.13.

11.7 In Example 11.6, verify the results of ANSI/IEEE and IEC short-circuit calculations.

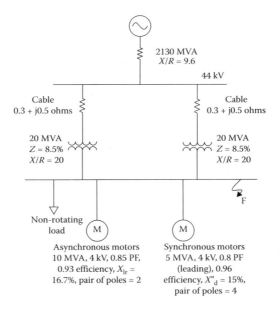

FIGURE P11.1
System configuration for Problem 11.3.

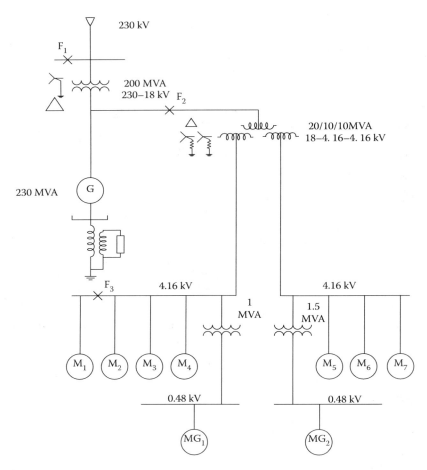

FIGURE P11.2
System configuration for Problems 11.4 and 11.5.

11.8 A 12.47 kV, 81.82 MVA, 0.85 PF generator is directly connected to a 12.47 kV bus. The following generator parameters are applicable:

$X_d'' = 0.162$, $X_d' = 0.223$, $X_d = 2.01$, $X_q'' = 0.159$, $T_d'' = 0.015$, $T_d' = 0.638$, $T_a = 0.476$

Considering a 5-cycle breaker analytically calculate the asymmetry factor using ANSI/IEEE calculations.

Short-Circuit Calculations according to IEC Standards

TABLE P11.1

Impedance Data for Problems 4 and 5

System Component	Description and Impedance Data
230-kV utility source	Three-phase short-circuit = 8690 MVA, X/R = 16.9
Step-up transformer	230–18 kV, 200 MVA, Z = 10%, X/R = 35, HV tap at 23,5750 V
Generator	18 kV, 230 MVA, 0.85 PF, subtransient reactance = 16%
Three-winding auxiliary transformer	18–4.16–4.16 kV, 20/10/10 MVA. impedance 18 kV to each 4.16-kV winding = 8% 20MVA base, impedance 4.16-kV winding to 4.16-kV winding = 15%, 20-MVA base
1-MVA transformer	4.16–0.48 kV, Z = 5.75%, X/R = 5.75
1.5-MVA transformer	4.16–0.48 kV, Z = 5.75%, X/R = 7.5
Medium-voltage motors	Ml = 5000-hp, 6-pole, synchronous, 0.8 PF leading
	M2 = 2500-hp, 8-pole, induction
	M3 = 2500-hp, 2-pole, induction
	M4 = 2500-hp, 12-pole, synchronous, 0.8 PF leading
	M5 = 2000-hp, 4-pole, synchronous, unity PF
	M6 = 1000-hp, 2-pole induction
	M7 = 1000-hp, 6-pole induction
Low-voltage motor groups	MG1 = 2 × 100 hp, and 6 × 40 hp induction 4-pole induction
	MG2 = 4 × 150 hp, and 8 × 75 hp induction 6-pole induction

References

1. IEC 60909-0. Short-circuit calculations in three-phase AC systems, 1st Ed. 1988. Now revised IEC 60909-0. Short-circuit currents in three-phase AC systems, Calculation of currents, 2016.
2. IEC 60909-1. Factors for calculation of short-circuit currents in three-phase AC Systems, 2002.
3. IEC 60909-2. Data for short-circuit current calculations, 2008.
4. IEC 60909-3. Short-circuit current calculations in AC systems—Part 3: Currents during two separate simultaneous single-phase line-to-earth short-circuits and partial short-circuit currents following through earth, 2009.
5. IEC 60909-4. Short-circuit calculations in three-phase systems—Part 4: Examples of calculations of short-circuit currents, 2000.
6. IEC 60056. High voltage alternating current circuit breakers, 1987.
7. IEC 60060-1. High voltage test techniques—Part 1: General definitions and test requirements, 2010.
8. JC Das. Short-circuit calculations—ANSI/IEEE and IEC methods, similarities and differences, *Proceedings of 8th International Symposium on Short-Circuit Currents in Power Systems*, Brussels, Belgium, 1998.
9. JC Das. Study of generator source short-circuit currents with respect to interrupting duty of the generator circuit breakers, EMTP simulation, ANSI/IEEE and IEC methods, *Int J Emerg Electr Power Syst* 9(3), 1–10, 2008.
10. Canay IM, Warren L. Interrupting sudden asymmetrical short-circuit currents without zero transition, *BBC Rev*, 56, 484–493, 1969.
11. Dufournet D, Willieme JM, Montillet GF. Design and implementation of a SF6 interrupting chamber applied to low range generator breakers suitable for interrupting currents having a non-zero passage, *IEEE Trans Power Deliv*, 17, 963–969, 2002.

12

Calculations of Short-Circuit Currents in Direct Current Systems

The calculations of short-circuit currents in direct current (DC) systems is essential for the design and application of distribution and protective apparatuses used in these systems. The DC systems include DC motor drives and controllers, battery power applications, emergency power supply systems for generating stations, data-processing facilities, and computer-based DC power systems and transit systems.

Maximum short-circuit currents should be considered for selecting the rating of the electrical equipment like cables, buses, and their supports. The high-speed DC protective devices may interrupt the current, before the maximum value is reached. It becomes necessary to consider the rise in the rate of current along with interruption time to determine the maximum current that will be actually obtained. Lower speed DC protective devices may permit the maximum value to be reached before current interruption.

Though the simplified procedures for DC short-circuit current calculation are documented in some publications, these are not well established. There is no American National Standards Institute (ANSI)/Institute of Electrical and Electronics Engineers (IEEE) standard for calculation of short-circuit currents in DC systems. A General Electric Company publication [1] and ANSI/IEEE standard C37.14 [2] provide some guidelines. The International Electrotechnical Commission (IEC) standard 61660-1 [3], published in 1997, is the only comprehensive document available on this subject. This standard addresses calculations of short-circuit currents in DC auxiliary installations in power plants and substations and does not include calculations in other large DC power systems, such as electrical railway traction and transit systems.

The IEC standard describes quasi steady-state methods for DC systems. The time variation of the characteristics of major sources of DC short-circuit current from initiation to steady state are discussed and appropriate estimation curves and procedures are outlined.

A dynamic simulation is an option, however, akin to short-circuit current calculations in AC systems; the simplified methods are easy to use and apply, though rigorously these should be verified by an actual simulation.

12.1 DC Short-Circuit Current Sources

Four types of DC sources can be considered:

- Lead acid storage batteries
- DC motors
- Converters in three-phase bridge configuration
- Smoothing capacitors

Figure 12.1 shows the typical short-circuit current–time profiles of these sources, and Figure 12.2 shows the standard approximate function assumed in the IEC standard [3]. The following definitions apply

I_k = quasi steady-state short-circuit current
i_p = peak short-circuit current
T_k = short-circuit duration
t_p = time to peak

τ_1 = rise time constant
τ_2 = decay-time constant

The functions are described by the following equations:

$$i_1(t) = i_p \frac{1-e^{-t/\tau_1}}{1-e^{-t_p/\tau_1}} \tag{12.1}$$

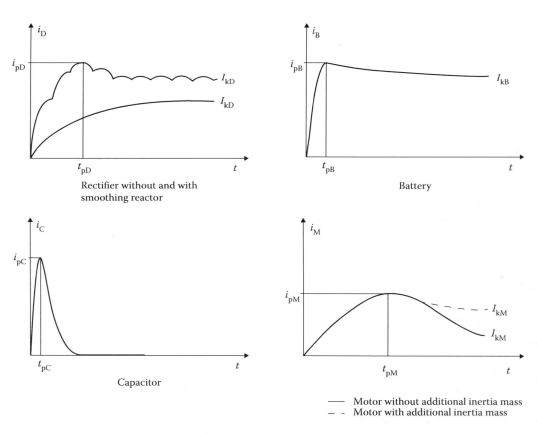

FIGURE 12.1
Short-circuit current–time profile of various DC sources: (a) rectifier without and with smoothing reactor, (b) battery, (c) capacitor, and (d) DC motor with and without additional inertia mass. (From IIEC Standard 61660–1 Short-Circuit Currents in DC Auxiliary Installations in Power Plants and Substations, 1997.

Calculations of Short-Circuit Currents in Direct Current Systems

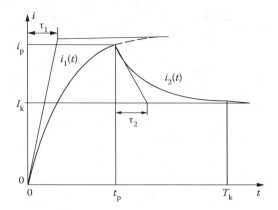

FIGURE 12.2
Standard approximation of short-circuit function. (From IEEE Standard 946. DC Auxiliary Power Systems for Generating Stations, 1996.)

$$i_2(t) = i_p[(1-\alpha)e^{-(t-t_p)\tau_2} + \alpha] \; t \geq t_p \tag{12.2}$$

$$\alpha = I_k/i_p \tag{12.3}$$

The quasi steady-state current, I_k, is conventionally assumed as the value at 1 s after the beginning of short circuit. If no definite maximum is present, as shown in Figure 12.1a for the converter current, then the function is given by Equation 12.1 only.

12.2 Calculation Procedures

12.2.1 IEC Calculation Procedure

Figure 12.3 shows a hypothetical DC distribution system that has all the four sources of short-circuit current, i.e., a storage battery, a charger, a smoothing capacitor, and a DC motor. Two locations of short-circuit are shown: (1) at F_1, without a common branch and (2) at F_2, through resistance and inductance, R_y and L_y of the common branch. The short-circuit current at F_1 is the summation of short-circuit currents of the four sources, as if these were acting alone through the series resistances and inductances. Compare this to the IEC method of AC short-circuit calculations in nonmeshed systems, discussed in Chapter 8.

For calculation of the short-circuit current at F_2, the short-circuit currents are calculated as for F_1 but adding R_y and L_y to the series circuit in each of the sources. Correction factors are introduced and the different time functions are added to the time function of the total current.

Whether it is the maximum or minimum short-circuit current calculation, the loads are ignored (i.e., no shunt branches) and the fault impedance is considered to be zero. For the maximum short-circuit current, the following conditions are applicable:

- The resistance of joints (in bus bars and terminations) is ignored.
- The conductor resistance is referred to 20°C.

524   *Short-Circuits in AC and DC Systems*

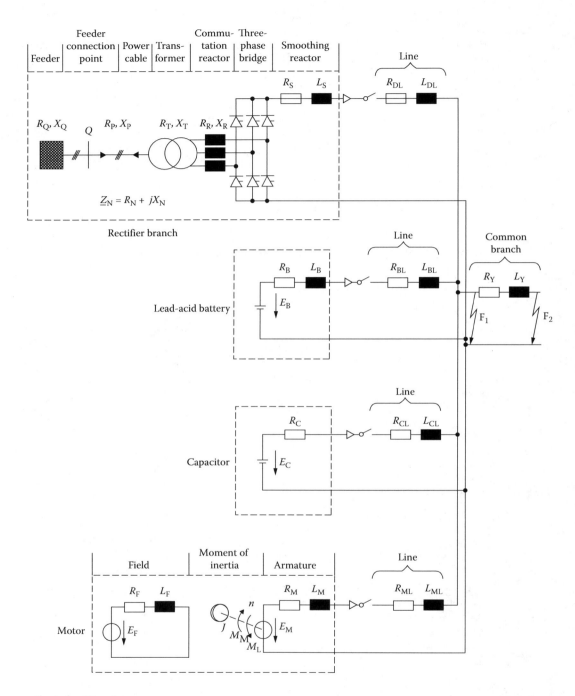

Short-circuit locations:
F_1 Short circuit without common branch
F_2 Short circuit with common branch

FIGURE 12.3
A DC distribution system for calculation of short-circuit currents. (From IEC Standard 61660–1. Short-Circuit Currents in DC Auxiliary Installations in Power Plants and Substations, 1997.)

Calculations of Short-Circuit Currents in Direct Current Systems 525

- The controls for limiting the rectifier current are not effective.
- The diodes for the decoupling part are neglected.
- The battery is fully charged.
- The current limiting effects of fuses or other protective devices are taken into account.

For calculation of the minimum short-circuit current, the following conditions are applicable:

- The conductor resistance is referred to maximum temperature.
- The joint resistance is taken into account.
- The contribution of the rectifier is its rated short-circuit current.
- The battery is at the final voltage as specified by the manufacturer.
- Any diodes in the decoupling parts are taken into account.
- The current-limiting effects of fuses or other protective devices are taken into account.

12.2.2 Matrix Methods

Matrix methods contrast with superimposition techniques. In an example of calculation in Reference [1], three sources of current, i.e., a generator, a rectifier, and a battery, are considered in parallel. The inductances and resistances of the system components are calculated and separate resistance and inductance networks are constructed, much akin to the ANSI/IEEE method for short-circuit current calculations in alternating current (AC) systems. These networks are reduced to a single resistance and inductance, and then the maximum short-circuit current is simply given by the voltage divided by the equivalent resistance and its rate of rise by the equivalent time constant, which is equal to the ratio of equivalent inductance over resistance. This procedure assumes that all sources have the same voltage. When the source voltages differ, then the partial current of each source can be calculated and summed. This is rather a simplification. For calculation of currents from rectifier sources an iterative procedure is required, as the resistance to be used in a Thévenin equivalent circuit at a certain level of terminal voltage during a fault needs to be calculated. This will be illustrated with an example.

12.3 Short-Circuit of a Lead Acid Battery

The battery short-circuit model is shown in Figure 12.4; R_B is the internal resistance of the battery, E_B is the internal voltage, R_C is the resistance of cell connectors, L_{CC} is the inductance of the cell circuit in H, and L_{BC} is the inductance of the battery cells considered as bus bars. The *internal* inductance of the cell itself is zero. The line resistance and inductance are R_L and L_L, respectively. The equivalent circuit is that of a short-circuit of a DC source through equivalent resistance and inductance, i.e.,

$$iR + L\frac{\mathrm{d}i}{\mathrm{d}t} = E_B \tag{12.4}$$

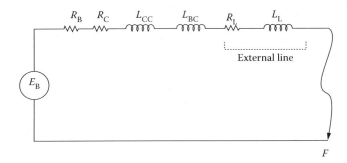

FIGURE 12.4
Equivalent circuit of the short circuit of a battery through external resistance and inductance.

The solution of this equation is given as

$$i = \frac{E_B}{R}\left(1 - e^{-(R/L)t}\right) \qquad (12.5)$$

The maximum short circuit current is then

$$I_{Bsc} = \frac{E_B}{R} \qquad (12.6)$$

and the initial maximum rate of rise of the current is given by di/dt at $t=0$, i.e.,

$$\frac{di_B}{dt} = \frac{E_B}{L} \qquad (12.7)$$

Referring to Figure 12.4, all the resistance and inductance elements in the battery circuit are required to be calculated. The battery internal resistance R_B by the IEEE method [4] is given by

$$R_B = R_{cell} N = \frac{R_p}{N_p} \qquad (12.8)$$

where R_{Cell} is the resistance/per cell, N is the number of cells, R_p is the resistance per positive plate, and N_p is the number of positive plates in a cell; R_p is given by

$$R_p = \frac{V_1 - V_2}{I_2 - I_1} \; \Omega/\text{positive plate} \qquad (12.9)$$

where V_1 is the cell voltage, and I_1 is the corresponding rated discharge current per plate. Similarly, V_2 is the cell voltage, and I_2 is the corresponding rated discharge current per plate at V_2.

The following equation for the internal resistance is from Reference [1]:

$$R_B = \frac{E_B}{100 \times I_{8h}} \; \Omega \qquad (12.10)$$

where I_{8h} is the 8-h ampère rating of the battery to 1.75 V per cell at 25°C. R_B is normally available from manufacturers' data; it is not a constant quantity and depends on the charge state of the battery. A discharged battery will have much higher cell resistance.

Example 12.1

A 60-cell 120 V sealed, valve regulated, lead acid battery has the following electrical and installation details:

Battery rating = 200 Ah (ampere hour), 8-h rate of discharge to 1.75 V per cell. Each cell has the following dimensions: height = 7.95 in. (=200 mm), length = 10.7 in. (=272 mm), and width = 6.8 in. (=173 mm). The battery is rack mounted, 30 cells per row, and the configuration is shown in Figure 12.5. Cell inter-connectors are 250 KCMIL, diameter = 0.575 in. (=1.461 cm).

Calculate the short-circuit current at the battery terminals. If the battery is connected through a cable of approximately 100 ft length to a circuit breaker, cable resistance 5 mΩ and inductance 14 μH, calculate the short-circuit current at breaker terminals.

The battery resistance according to Equation 12.10 and considering a cell voltage of 1.75 V per cell is

$$R_B = \frac{E_B}{100 \times I_{8h}} = \frac{120}{100 \times 200} = 6\,\text{m}\Omega$$

The manufacturer supplies the following equation for calculating the battery resistance:

$$R_B = \frac{31 \times E_B}{I_{8h}}\,\text{m}\Omega \tag{12.11}$$

Substituting the values, this gives a battery resistance of 18.6 mΩ. There is three times the difference in these values, and the manufacturer's data should be used.

From Figure 12.5, battery connectors have a total length of 28 ft (8.53 m), size 250 kcmil. Their resistance from the conductor resistance data is 1.498 mΩ at 25°C. The total resistance in the battery circuit is $R_B + R_C = 20.098$ mΩ. (This excludes the external cable connections.) Therefore, the maximum short-circuit current is $120/(20.098 \times 10^{-3}) = 5970$ A.

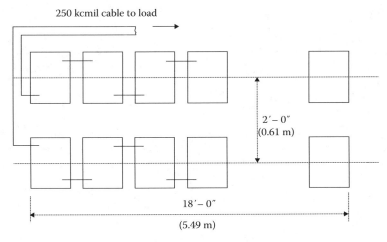

FIGURE 12.5
Battery system layout for calculation of short-circuit current (Examples 12.1 and 12.2).

The inductance L_c of the battery circuit is sum of the inductances of the cell circuit L_{CC} plus the inductance of the battery cells, L_{CB}. The inductance of two round conductors of radius r, spaced at a distance d, is given by

$$L = \frac{\mu_0}{\pi}\left(0.25 + \ln\frac{d}{r}\right) \tag{12.12}$$

where μ_0 is the permeability in vacuum $= 4\pi 10^{-7\text{H/m}}$. From Figure 12.5, the distance $d = 24$ in. and r, the radius of 250 kcmil conductor, is 0.2875 in. Substituting the values in Equation 12.12, the inductance is 1.87 μH/m for the *loop* length. Therefore, for an 18-ft loop length in Figure 12.5, the inductance $L_{CC} = 10.25$ μH.

The inductance of battery cells can be determined by treating each row of cells like a bus bar. Thus, the two rows of cells are equivalent to parallel bus bars at a spacing $d = 24$ in., the height of the bus bar $h =$ height of the cell $= 7.95$ in., and the width of the bus bars $w =$ width of the cell $= 6.8$ in. The expression for inductance of the bus bars in this configuration is given as

$$L = \frac{\mu_0}{\pi}\left(\frac{3}{2} + \ln\frac{d}{h+w}\right) \tag{12.13}$$

This gives inductance in H per meter loop length. Substituting the values, for an 18-ft loop length, inductance $L_{BC} = 4.36$ μH. The total inductance is, therefore, 14.61 μH. The initial rate of rise of the short-circuit current is given as

$$\frac{E_B}{L_C} = \frac{120}{14.61\times 10^{-6}} = 8.21\times 10^6\,\text{A/s}$$

The time constant is given as

$$\frac{L_C}{R_B + R_C} = \frac{14.61\times 10^{-6}}{20.01\times 10^{-3}} = 0.73\,\text{ms}$$

The current reaches 0.63 $\times$ 51970 = 3761 A in 0.73 ms, and in 1.46 ms it will be $0.87 \times 5970 = 5194$ A.

The cable resistance and inductance can be added to the values calculated above, i.e., total resistance $= 25.01$ mΩ and total inductance is 28.61 μH. The maximum short-circuit current is, therefore, 4798 A, and the time constant changes to 1.14 ms. The current profiles can be plotted.

IEC Calculation

To calculate the maximum short-circuit current or the peak current according to IEC, the battery cell resistance R_B is multiplied by a factor 0.9. All other resistances in Figure 12.4 remain unchanged. Also, if the open-circuit voltage of the battery is unknown, then use $E_B = 1.05U_{nB}$, where $U_{nB} = 2.0$ V/cell for lead acid batteries. The peak current is given by

$$i_{pB} = \frac{E_B}{R_{BBr}} \tag{12.14}$$

where i_{pB} is the peak short-circuit current from the battery and R_{BBr} is the total equivalent resistance in the battery circuit, with R_B multiplied by a factor of 0.9. The time to

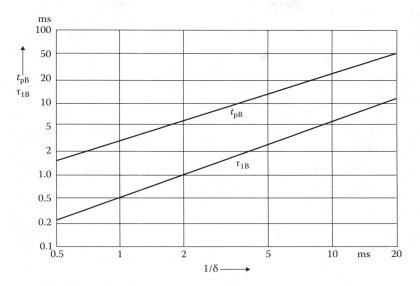

FIGURE 12.6
Time to peak t_{pB} and rise time constant τ_{1B} for short circuit of a battery. (From IEEE Standard 946. DC Auxiliary Power Systems for Generating Stations, 1996.)

peak and the rise time is read from curves in Figure 12.6, based on $1/\delta$, which is defined as follows:

$$\frac{1}{\delta} = \frac{2}{\frac{R_{BBr}}{L_{BBr}} + \frac{1}{T_B}} \tag{12.15}$$

The time constant T_B is specified as equal to 30 ms and L_{BBr} is the total equivalent inductance to the fault point in the battery circuit. The decay time constant $\tau_2 B$ is considered to be 100 ms. The quasi steady-state short-circuit current is given as

$$I_{kB} = \frac{0.95 E_B}{R_{BBr} + 0.1 R_B} \tag{12.16}$$

This expression considers that the battery voltage falls and the internal cell resistance increases after the short circuit. Note that all equations from IEC are in MKS units.

Example 12.2

Calculate the short-circuit current of the battery in Example 12.1, by the IEC method.
The total resistance in the battery circuit, without external cable, is $0.9 \times 18.6 + 1.4128 = 18.238\,\text{m}\Omega$. The battery voltage of 120 V is multiplied by factor 1.05. Therefore, the peak short-circuit current is given as

$$i_{PB} = \frac{E_B}{R_{BBr}} = \frac{1.05 \times 120}{18.238 \times 10^{-3}} = 6908.6\,\text{A}$$

This is 15.7% higher compared to the calculation in Example 12.1; $1/\delta$ is calculated from Equation 12.15.

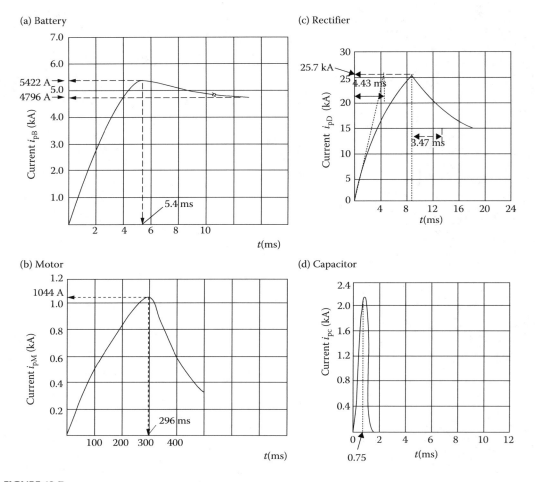

FIGURE 12.7
Calculated short-circuit current–time profiles: (a) battery (Example 12.2), (b) DC motor (Example 12.4), (c) rectifier (Example 12.6), and (d) capacitor (Example 12.7).

$$\frac{1}{\delta} = \frac{2}{\dfrac{18.238 \times 10^{-3}}{14.61 \times 10^{-6}} + \dfrac{1}{30 \times 10^{-3}}} = 1.56\,\text{ms}$$

From Figure 12.6, the time to peak = 4.3 ms and the rise time constant is 0.75 ms. The quasi steady-state short-circuit current is given as

$$I_{kB} = \frac{0.95 \times 120 \times 10^3}{18.238 + 0.1(18.6)} = 5956\,\text{A}$$

The calculations with external cable added are similarly carried out. The cable resistance is 5 mΩ and inductance is 14 μH. Therefore, $R_{BBr} = (0.12)(18.6) + 1.4128 + 5 = 23.24\,\text{m}\Omega$. This gives a peak current of 5422 A, $1/\delta = 2.40$ ms, and the time to peak is 5.4 ms. The rise time constant is 1.3 ms, and the quasi steady-state short-circuit current is 4796 A. The short-circuit current profile is plotted in Figure 12.7a.

Calculations of Short-Circuit Currents in Direct Current Systems 531

12.4 Short-Circuit of DC Motors and Generators

An equation for the short circuit from DC generators and motors [1] is given by

$$i_a = \frac{e_0}{r_d'}(1 - e^{-\sigma_a t}) - \left(\frac{e_0}{r_d'} - \frac{e_0}{r_d}\right)(1 - e^{\sigma_f t}) \qquad (12.17)$$

where

i_a = per unit current

e_0 = internal emf prior to short-circuit in per unit

r_d = steady-state effective resistance of the machine in per unit

r_d' = transient effective resistance of the machine in per unit

σ_a = armature circuit decrement factor

σ_f = field circuit decrement factor

The first part of the equation has an armature time constant, which is relatively short and controls the buildup and peak of the short-circuit current; the second part is determined by the shunt field excitation and it controls the decay of the peak value. The problem of calculation is that the time constants in this equation are not time invariant. Saturation causes the armature circuit decrement factor to increase as the motor saturates. Approximate values suggested for saturated conditions are 1.5–3.0 times the unsaturated value and conservatively a value of 3.0 can be used. The unsaturated value is applicable at the start of the short-circuit current and the saturated value at the maximum current. Between these two extreme values the decrement is changing from one value to another. Figure 12.8 shows the approximate curve of the short-circuit current and its equivalent circuit. For the first two-thirds of the curve, the circuit is represented by machine unsaturated inductance L_a', and for the last one-third L_a' is reduced to one-third with series transient resistance. The peak short-circuit current in per unit is given as

$$i_a' = \frac{e_0}{r_d'} \qquad (12.18)$$

The transient resistance r_d' in per unit requires some explanation. It is the effective internal resistance and is given by

$$r_d' = r_w + r_b' + r_x' \qquad (12.19)$$

where r_w is the total resistance of the windings in the armature circuit, r_x' is the equivalent to flux reduction in per unit, and r_b' is the transient resistance equal to reactance voltage and brush contact resistance in per unit. The flux reduction and distortion are treated as ohmic resistance. The values of transient resistance r_d' in per unit are given graphically

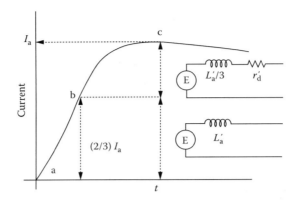

FIGURE 12.8
Short-circuit current–time curve for a DC motor or generator showing two distinct time constants.

in the American Institute for Electrical Engineers (AIEE) Committee Report of 1950 [5], depending on the machine rating, voltage, and speed. The transient resistance is not constant and there is a variation band. The machine load may also affect the transient resistance [6]. There does not seem to be any later publication on the subject. Similarly, the steady-state resistance is defined as

$$r_d = r_w + r_b + r_x \tag{12.20}$$

where r_b is the steady-state resistance equivalent to reactance voltage and brush contact in per unit, and r_x is the steady-state resistance equivalent to flux reduction in per unit.

The maximum rate of rise of the current is dependent on armature unsaturated inductance. The unit inductance is defined as

$$L_{a1} = \frac{V_1}{I_a} \frac{2 \times 60}{2\pi P N_1} \tag{12.21}$$

The per unit inductance is the machine inductance L'_a divided by the unit inductance and is given as

$$C_x = \frac{L'_a}{L_{a1}} = \frac{P N_1 L'_a}{19.1} \frac{I_a}{V_1} \tag{12.22}$$

This can be written as

$$L'_a = \frac{19.1 C_x V_1}{P N_1 I_a} \tag{12.23}$$

where P = number of poles, N_1 = base speed, V_1 = rated voltage, I_a = rated machine current, and C_x varies with the type of machine. Charts of initial inductance plotted against unit

Calculations of Short-Circuit Currents in Direct Current Systems

inductance show a linear relationship for a certain group of machines. For this purpose, the machines are divided into four broad categories as follows:

Motors: $C_x = 0.4$ for motors without pole face windings

Motors: $C_x = 0.1$ for motors with pole face windings

Generators: $C_x = 0.6$ for generators without pole face windings

Generators: $C_x = 0.2$ for generators with pole face windings

The armature circuit decrement factor is given as

$$\sigma_a = \frac{r_d' 2\pi f}{C_x} \tag{12.24}$$

The maximum rate of rise of current in ampéres per second is given as

$$\frac{di_a}{dt} = \frac{V_1 e_0}{L_a'} \tag{12.25}$$

The rate of current rise can also be expressed in terms of per unit rated current:

$$\frac{di_a}{dt} = \frac{PN_1 e_0}{19.1 C_x} \tag{12.26}$$

In Equations 12.25 and 12.26 e_0 can be taken as equal to unity without appreciable error. More accurately, e_0 can be taken as 0.97 per unit for motors and 1.03 for generators. The inductance L_a' is given in tabular and graphical forms in an AIEE publication [7] of 1952. Figure 12.9 shows L_a' values in mH for certain motor sizes, without pole face windings. Again, there does not seem to be a later publication on the subject.

Example 12.3

Calculate the short-circuit current of a 230-V, 150-hp, 1150-rpm motor, for a short circuit at motor terminals. The motor armature current is 541 A.

From graphical data in Reference 1, the transient resistance is 0.068 per unit. Also, the inductance L_a' from the graphical data in Figure 12.9 is 1.0 mH. Then, the peak short-circuit current is given as

$$I_a' = \frac{I_a}{r_d'} = \frac{541}{0.068} = 7956\,A$$

and the initial rate of rise of the current is given as

$$\frac{di_a}{dt} = \frac{V_1}{L_a'} = \frac{230}{1 \times 10^{-3}} = 230\,kA/s$$

As shown in Figure 12.8, the time constant changes at point b.

$$\frac{L_a'}{3 r_d'}$$

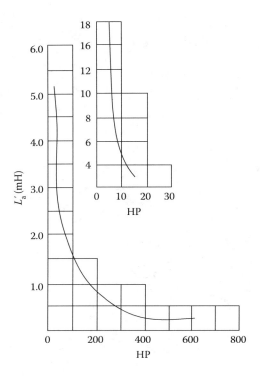

FIGURE 12.9
Inductance of DC motors in mH versus motor hp.

The base ohms are $V_a/I_a = 230/541 = 0.425$. Therefore, r'_d in ohms $= (0.425) \times (0.068) = 0.02812 \, \Omega$. This gives a time constant of 11.52 ms. The short-circuit profile can now be plotted.

IEC Method

The resistance and inductance network of short circuit of a DC machine with a separately excited field is shown in Figure 12.3. The equivalent resistance and inductance are given by

$$R_{MB} = R_M + R_{ML} + R_y$$
$$L_{MB} = L_M + L_{ML} + L_y \tag{12.27}$$

where R_M and L_M are the resistance and inductance of the armature circuit including the brushes, R_{ml} and L_{ml} are the resistance and inductance of the conductor in the motor circuit, and R_y and L_y are the resistance and inductance of the common branch, if present. The time constant of the armature circuit up to the short-circuit location, τM, is given as

$$\tau_M = \frac{L_{MB}}{R_{MB}} \tag{12.28}$$

The quasi steady-state short-circuit current is given by

$$I_{KM} = \frac{L_F}{L_{0F}} \frac{U_{rM} - I_{rM} R_M}{R_{MB}} \quad I_{kM} = 0 \quad \text{when } n \to 0 \tag{12.29}$$

Calculations of Short-Circuit Currents in Direct Current Systems

where

L_F = equivalent saturatad inductance of the field circuit on short-circuit

L_{0F} = equivalent unsaturated inductance of the field circuit at no load

U_{rM} = rated voltage of the motor

I_{rM} = rated current of the motor

n = motor speed

n_n = rated motor speed

The peak short circuit current of the motor is given as

$$i_{pM} = \kappa_M \frac{U_{rM} - I_{rM} R_M}{R_{MB}} \tag{12.30}$$

At normal speed or decreasing speed with $\tau_{mec} \geq 10\tau_F$, the factor $\kappa_M = 1$, where τ_{mec} is the mechanical time constant, given as

$$\tau_{mec} = \frac{2\pi J n_0 R_{MB} I_{rM}}{M_r U_{rM}} \tag{12.31}$$

where J is the moment of inertia, and M_r is the rated torque of the motor. The field circuit time constant τ_F is given as

$$\tau_F = \frac{L_F}{R_F} \tag{12.32}$$

For $\tau_{mec} \geq 10\tau_F$, the time to peak and time constant are given as

$$t_{pM} = \kappa_{1M} \tau_M$$
$$\tau_{1M} = \kappa_{2M} \tau_M \tag{12.33}$$

The factor κ_{1M} and κ_{2M} are taken from Figure 12.10 and are dependent on τ_F/τ_M and L_F/L_{0F}. For decreasing speed with $\tau_{mec} < 10\tau_F$, the factor κ_M is dependent on $1/\delta = 2\tau M$ and ω_0,

$$\omega_0 = \sqrt{\frac{1}{\tau_{mec} \tau_M} \left(1 - \frac{I_{rM} R_M}{U_{rM}} \right)} \tag{12.34}$$

where ω_0 is the undamped natural angular frequency, and δ is the decay coefficient; κ_M is derived from the curves in the IEC standard [3].

For decreasing speed with $\tau_{mec} < 10\tau_F$, the time to peak τ_M is read from a curve in the IEC standard, and the rise time constant is given by

$$\tau_{1M} = \kappa_{3M} \tau_M \tag{12.35}$$

where the factor κ_{3M} is again read from a curve in the IEC standard [3], not reproduced here.

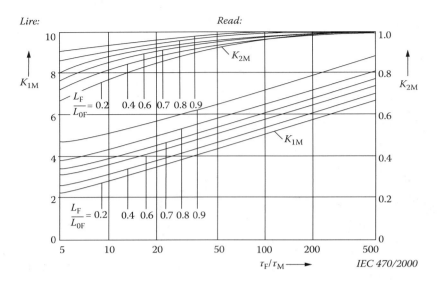

FIGURE 12.10
Factors κ_{1M} and κ_{2M} for determining the time to peak t_{pM} and the rise-time constant $\tau_1 M$ for normal and decreasing speed with $\tau_{mec} \geq 10\tau_F$; short circuit of a DC motor. (From IEEE Standard 946. DC Auxiliary Power Systems for Generating Stations, 1996.)

Decay time constant τ_{2M}
For nominal speed or decreasing speed with $\tau_{mec} \geq 10\tau_F$

$$\tau_{2M} = \tau_F \quad \text{when } n = n_n = \text{constant}$$

$$\tau_{2M} = \frac{L_{0F}}{L_F} \kappa_{4M} \tau_{mec} \quad \text{when } n \to 0 \text{ with } \tau_{mec} \geq 10\tau_F \tag{12.36}$$

For decreasing speed with $\tau_{mec} < 10\tau_F$

$$\tau_{2M} = \kappa_{4M} \tau_{mec} \tag{12.37}$$

where κ_{4M} is again read from a curve in the IEC standard, not reproduced here. Thus, the IEC calculation method requires extensive motor data and use of a number of graphical relations in the standard. The rise and decay time constants are related to $\tau_{mec} < 10\tau_F$ or $\tau_{mec} \geq 10\tau_F$.

Example 12.4

Calculate the short-circuit current for a terminal fault on a 115 V, 1150-rpm, six-pole, 15-hp motor. The armature current = 106 A, the armature and brush circuit resistance = 0.1 Ω, and the inductance in the armature circuit = 8 mH; $\tau_F = 0.8$ s, $\tau_{mec} > 10\tau_F$, $L_F/L_{0F} = 0.5$, and $\tau_{mec} = 20$ s.

There is no external resistance or inductance in the motor circuit. Therefore, $R_{MBr} = R_M = 0.10$ Ω. IEC is not specific about the motor circuit resistance, or how it should be calculated or ascertained.

The time constant is given as

$$\tau_M = \frac{L_M}{R_M} = \frac{8 \times 10^{-3}}{0.10} = 80 \, \text{ms}$$

Calculations of Short-Circuit Currents in Direct Current Systems

The quasi steady-state current from Equation 12.29 is given as

$$0.5\left(\frac{115-(0.10)(106)}{0.10}\right) = 522 \text{ A}$$

From Equation 12.30 the peak current is 1044 A, because for $\tau_{mec} > 10\tau_F$, factor κ_M in Equation 12.30 = 1. The time to peak and time constant are given by Equation 12.33. From Figure 12.10, and for $\tau_F/\tau_M = 10$ and $L_F/L_{0F} = 0.5$, factor $\kappa_1 M = 3.7$ and $\kappa_{2m} = 0.83$. Therefore, the time to peak is 296 ms and the time constant $\kappa_{1M} = 8.3$ ms.

The short-circuit profile is plotted in Figure 12.7b.

12.5 Short-Circuit of a Rectifier

The typical current–time curve for a rectifier short circuit is shown in Figure 12.11. The maximum current is reached at one half-cycle after the fault occurs. The peak at half-cycle is caused by the same phenomenon that creates a DC offset in AC short-circuit calculations. The magnitude of this peak is dependent on X/R ratio, the AC system source reactance, rectifier transformer impedance, and the resistance and reactance through which the current flows in the DC system. The addition of resistance or inductance to the DC system reduces this peak, and depending on the magnitude of these components, the peak may be entirely eliminated, with a smoothing DC reactor, as shown in Figure 12.1a. The region A in Figure 12.11 covers the initial rise of current, the peak current occurs in region B, and region C covers the time after one cycle until the current is interrupted.

The initial rate of rise of the DC short-circuit current for a bolted fault varies with the magnitude of the sustained short-circuit current. The addition of inductance to the DC circuit tends to decrease the rate of rise.

An equivalent circuit of the rectifier short-circuit current is developed with a voltage source and equivalent resistance and inductance. The equivalent resistance varies with

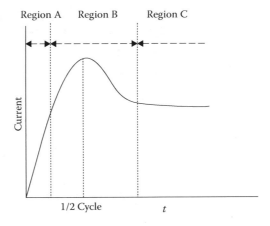

FIGURE 12.11
Short-circuit current profile of a rectifier.

rectifier terminal voltage, which is a function of the short-circuit current. The equivalent resistance is determined from the rectifier regulation curve by an iterative process, which can admirably lend itself to iterative computer solution. The equivalent inductance is determined from a sustained short-circuit current for a bolted fault and rated system voltage. The magnitude of the peak current is determined from AC and DC system impedance characteristics [1].

The following step-by-step procedure can be used:

Calculate total AC system impedance $Z_C = R_C + X_C$ in ohms. Convert to per unit impedance z_C, which may be called the commutating impedance in per unit on a rectifier transformer kVA base. This is dependent on the type of rectifier circuit. For a double-wye, six-phase circuit, the conversion is given by Reference [1]:

$$Z_C = z_C \times 0.6 \times \frac{E_D}{I_D} \, \Omega \tag{12.38}$$

where E_D and I_D are rectifier rated DC voltage and rated DC current.

Assume a value of rectifier terminal voltage e_{da} under faulted condition and obtain factor K_2 from Figure 12.12. The *preliminary* calculated value of the sustained short-circuit current is then given as

$$I_{da} = \frac{K_2}{z_C} I_D \tag{12.39}$$

The equivalent rectifier resistance is then given by

$$R_R = \frac{(E_D - E_{da})}{I_{da}} \, \Omega \tag{12.40}$$

where E_{da} is the assumed rectifier terminal voltage in per unit under fault conditions.

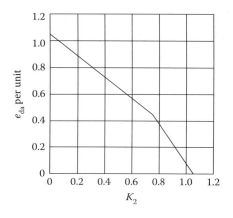

FIGURE 12.12
Sustained fault current factor versus rectifier terminal voltage.

Calculations of Short-Circuit Currents in Direct Current Systems 539

The sustained value of the fault current is given by

$$I_{DC} = \frac{E_D}{R_R + R_D} A \qquad (12.41)$$

where R_D is the resistance external to the rectifier. The rectifier terminal voltage in volts is

$$E_{DC} = E_D - I_{DC} R_R \qquad (12.42)$$

The value of $E_{da} = e_{da} \times E_D$ should be within 10% of the calculated value E_{DC}, which is the rectifier terminal voltage under sustained short-circuit current. The iterative process is repeated until the desired tolerance is achieved.

Example 12.5

Consider a 100-kW source at 125 V DC. The DC resistance of the feeder cable, R_D, is 0.004 Ω. Let the AC source and rectifier transformer impedance $z_C = 0.05$ per unit and $I_D = 800$ A. Calculate the rectifier resistance for a fault at the end of the cable. Assume $e_{da} = 0.5$ per unit, i.e., $E_{da} = 62.5$ V; K_2 from Figure 12.12 = 0.70. Therefore,

$$I_{da} = \frac{K_2}{z_c} I_D = \frac{0.70}{0.05} \times 800 = 11200\,A$$

Then R_R is given as

$$R_R = \frac{E_D - E_{da}}{I_{da}} = \frac{62.5}{11,200} = 0.00558\,\Omega$$

$$I_{DC} = \frac{E_D}{R_R + R_D} = \frac{125}{0.00558 + 0.004} = 13,048\,A$$

$$E_{DC} = 125 - (13,048)(0.00558) = 72.18\ V$$

We can iterate once more for a closer estimate of R_R:

$$E_{DC} = 72.18\ V$$

$$e_{da} = 0.392\ \text{per unit}$$

$$K = 0.77$$

$$I_{da} = 12,320\,A$$

$$R_R = 0.006169\,\Omega$$

$$I_{DC} = 12292.5\ A$$

$$E_{DC} = 49.17\ V$$

This is close enough an estimate, as $E_{DC} = 0.392 \times 125\ V = 49\ V$. To calculate rate of rise of current, the rectifier inductance is required. This is given by

$$L_R = \frac{E_D}{360 \times I_{ds}} H\ (\text{for } 60\,Hz)$$

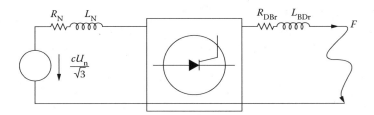

FIGURE 12.13
Equivalent circuit for short-circuit current calculations of a rectifier.

where I_{ds} is the terminal short-circuit current of the rectifier. On a terminal short-circuit, voltage is zero and from Figure 12.12, $K_2 = 1.02$. This gives a short-circuit current of 15,320 A. Thus, inductance is 0.0213 mH. Add to it the inductance of the cable = 0.007126 mH. Then the rate of rise of current is $125/(0.007966 + 0.213) = 4.27$ kA/s. The peak current can be 1.4 to 1.6 times the sustained current depending upon the system parameters [1].

IEC Method

The equivalent short-circuit diagram is shown in Figure 12.13. The maximum DC short-circuit current is given by the minimum impedance Z_{Qmin}, which is obtained from the maximum short-circuit current I''_{kQmax} of the AC system:

$$Z_{Qmin} = \frac{cU_n}{\sqrt{3} I''_{kQmax}} \quad (12.43)$$

The minimum DC current is given as

$$Z_{Qmax} = \frac{cU_n}{\sqrt{3} I''_{kQmin}} \quad (12.44)$$

In Figure 12.13, the resistance and inductances on the AC side are given by

$$R_N = R_Q + R_P + R_T + R_R$$
$$X_N = X_Q + X_P + X_T + X_R \quad (12.45)$$

where R_Q and X_Q are the short-circuit resistance and reactance of the AC source referred to the secondary of the rectifier transformer, R_P and X_P are the short-circuit resistance and reactance of the power supply cable referred to the secondary side of the transformer, R_T and X_T are the short-circuit resistance and reactance of the rectifier transformer referred to the secondary side of the transformer, and R_R and X_R are the short-circuit resistance and reactance of the commutating reactor, if present.

Similarly, on the DC side

$$R_{DBr} = R_S + R_{DL} + R_y$$
$$L_{DBr} = L_S + L_{DL} + L_y \quad (12.46)$$

where R_S, R_{DL}, and R_y, are the resistances of the DC saturated smoothing reactor, the conductor in the rectifier circuit, and the common branch, respectively, and L_S, L_{DL}, and

Calculations of Short-Circuit Currents in Direct Current Systems 541

L_y are the corresponding inductances. The quasi steady-state short-circuit current is given by

$$i_{kD} = \lambda_D \frac{3\sqrt{2}}{\pi} \frac{cU_n}{\sqrt{3}Z_N} \frac{U_{rTLV}}{U_{rTHV}} \tag{12.47}$$

where Z_N is the impedance on the AC side of three-phase network. The factor λ_D as a function of R_N/X_N and R_{DBr}/R_n is estimated from the curves in the IEC standard [3]. Alternatively, it is given by the following equation:

$$\lambda_D = \sqrt{\frac{1+(R_N/X_N)^2}{1+(R_N/X_N)^2(1+0.667(R_{DBr}/R_N))^2}} \tag{12.48}$$

The peak short circuit current is given by

$$i_{pD} = \kappa_D I_{kD} \tag{12.49}$$

where the factor κ_D is dependent on

$$\frac{R_N}{X_N}\left[1+\frac{2R_{DBr}}{3R_N}\right] \text{and} \frac{L_{DBr}}{L_N} \tag{12.50}$$

It is estimated from the curves in the IEC standard [3] (not reproduced) or from the following equation:

$$\kappa_D = \frac{i_{pD}}{I_{kD}} = 1+\frac{2}{\pi}e^{-\left(\frac{\pi}{3}+\phi_D\right)\cot\phi_D} \sin\phi_D\left(\frac{\pi}{2}-\arctan\frac{L_{DBr}}{L_N}\right) \tag{12.51}$$

where

$$\phi_D = \arctan\frac{1}{\dfrac{R_N}{X_N}\left(1+\dfrac{2}{3}\dfrac{R_{DBr}}{R_N}\right)} \tag{12.52}$$

Time to peak t_{pD}, when $\kappa_D \geq 1.05$, is given by

$$\begin{aligned} t_{pD} &= (3\kappa_D+6)\,\text{ms} &&\text{when } \frac{L_{DBr}}{L_N} \leq 1 \\ t_{pD} &= \left[(3\kappa_D+6)+4\left(\frac{L_{DBr}}{L_N}-1\right)\right]\text{ms} &&\text{when } \frac{L_{DBr}}{L_N} > 1 \end{aligned} \tag{12.53}$$

If $\kappa_D < 1.05$, the maximum current, compared with the quasi steady-state short-circuit current, is neglected, and $t_{pD}=T_k$ is used.

The rise time constant for 50 Hz is given as

$$\begin{aligned} \tau_{1D} &= \left[2+(\kappa_D-0.9)\left(2.5+9\frac{L_{DBr}}{L_N}\right)\right]\text{ms} &&\text{when } \kappa_D \geq 1.05 \\ \tau_{1D} &= \left[0.7+\left[7-\frac{R_N}{X_N}\left(1+\frac{2}{3}\frac{L_{DBr}}{L_N}\right)\right]\left(0.1+0.2\frac{L_{DBr}}{L_N}\right)\right]\text{ms} &&\text{when } \kappa_D < 1.05 \end{aligned} \tag{12.54}$$

For simplification

$$\tau_{1D} = \frac{1}{3} t_{pD}$$

(12.55)

The decay time constant τ_{2d} for 50 Hz is given as

$$\tau_{2D} = \frac{2}{\dfrac{R_N}{X_N}\left(0.6 + 0.9\dfrac{R_{DBr}}{R_N}\right)}\, \text{ms}$$

(12.56)

The time constants for a 60-Hz power system are not given in the IEC standard.

Example 12.6

A three-phase rectifier is connected on the AC side to a three-phase, 480–120 V, 100-kVA transformer of percentage $Z_T = 3\%$, $X/R = 4$. The 480-V source short-circuit MVA is 30, and the X/R ratio = 6. The DC side smoothing inductance is 5 µH and the resistance of the cable connections is 0.002 Ω. Calculate and plot the short-circuit current profile at the end of the cable on the DC side.

Based on the AC side data, the source impedance in series with the transformer impedance referred to the secondary side of the rectifier transformer is given as

$$R_Q + jX_Q = 0.00008 + j0.00048 \,\Omega$$
$$R_T + jX_T = 0.001 + j0.00419 \,\Omega$$

Therefore,

$$R_N + jX_N = 0.0011 + j0.004671 \,\Omega$$

On the DC side

$$R_{DBr} = 0.002\,\Omega \quad \text{and} \quad L_{DBr} = 5\,\mu\text{H}$$

This gives

$$\frac{R_N}{X_N} = 0.24 \quad \text{and} \quad \frac{R_{DBr}}{R_N} = 2.0$$

Calculate λ_D from Equation 12.48:

$$\lambda_D = \sqrt{\frac{1 + (0.24)^2}{1 + (0.24)^2(1 + 0.667)(2.0)^2}} = 0.897$$

The quasi steady-state current is, therefore, from Equation 12.47:

$$I_{kD} = (0.897)\left(\frac{3\sqrt{2}}{\pi}\right)\left(\frac{1.05 \times 480}{\sqrt{3} \times 0.0048}\right)\left(\frac{120}{480}\right) = 18.36\,\text{kA}$$

Calculations of Short-Circuit Currents in Direct Current Systems

To calculate the peak current, calculate the ratios as given below:

$$\frac{R_N}{X_N}\left(1+\frac{2}{3}\frac{R_{DBr}}{R_N}\right)=(0.24)(1+0.667\times2)=0.56$$

$$\frac{L_{DBr}}{L_N}=\frac{5\times10^{-6}}{0.0128\times10^{-3}}=0.392$$

Calculate κ_D from Equations 12.51 and 12.52. From Equation 12.52

$$\phi_D=\tan^{-1}\frac{1}{0.24(1+0.667\times2)}=60.75°$$

and from Equation 12.51, $\kappa_D=1.4$. Thus, the peak short-circuit current is

$$i_{pD}=\kappa_D I_{kD}=1.4\times18.36=25.7\,\text{kA}$$

The time to peak is given by Equation 12.53 and is equal to

$$t_{pD}=(3\kappa_D+6)=(3\times1.4+6)=10.17\,\text{ms}$$

The rise time constant is given by Equation 12.54 and is equal to 4.43 ms, and the decay time constant is given by Equation 12.56 and is equal to 3.47 ms.

The current profile is plotted in Figure 12.7c, which shows the calculated values. The intermediate shape of the curve can be correctly plotted using Equations 12.1 and 12.2. Note that, in this example, the IEC equations are for a 50-Hz system. For a 60-Hz system, the peak will occur around 8.3 ms.

12.6 Short-Circuit of a Charged Capacitor

The resistance and inductance in the capacitor circuit from Figure 12.3 are given by

$$R_{CBr}=R_C+R_{CL}+R_y$$
$$L_{CBr}=L_{CL}+L_y \tag{12.57}$$

where R_C is the equivalent DC resistance of the capacitor, and R_{CL} and L_{CL} are the resistance and inductance of a conductor in the capacitor circuit. The steady-state short-circuit current of the capacitor is zero and the peak current is given by

$$i_{pC}=\kappa_C\frac{E_C}{R_{CBr}} \tag{12.58}$$

where E_C is the capacitor voltage before the short circuit, and κ_C is read from curves in the IEC standard [3], based on

$$1/\delta = \frac{2L_{CBr}}{R_{CBr}}$$

$$\omega_0 = \frac{1}{\sqrt{L_{CBr}C}}$$

(12.59)

If $L_{CBr}=0$, then $\kappa_C=1$.

The time to peak t_{pC} is read from curves in the IEC standard [3]. If $L_{CBr}=0$, then $t_{pC}=0$. The rise time constant is given by

$$\tau_{1C} = \kappa_{1C} t_{pC}$$

(12.60)

where κ_{1C} is read from curves in IEC. The decay time constant is given by

$$\tau_{2C} = \kappa_{2C} R_{CBr} C$$

(12.61)

where κ_{2c} is read from curves in IEC standard [3]. The curves for these factors are not reproduced.

Example 12.7

A 120-V, 100-μF capacitor has $R_CB_r=0.05\ \Omega$ and $L_{CBr}=10$ mH. Calculate the terminal short-circuit profile.

From Equation 12.59

$$1/\delta = \frac{2\times 10\times 10^{-3}}{0.05} = 0.4$$

Also,

$$\omega_0 = \frac{1}{\sqrt{10\times 10^{-3} \times 100\times 10^{-6}}} = 1000$$

From curves in the IEC standard, $\kappa_c=0.92$. The peak current from Equation 12.58 is then $(0.92) \times (120/0.05)=2208$ A. The time to peak from curves in IEC$=0.75$ ms, and $\kappa_1C = 0.58$. From Equation 12.60 the rise time constant is $(0.58) \times (0.75)=0.435$ ms. Also, $\kappa_{2C} = 1$, and from Equation 12.61, the decay time constant is $5\,\mu$s. The short-circuit current profile is plotted in Figure 12.7d.

12.7 Total Short-Circuit Current

The total short-circuit current at fault F_1 (Figure 12.3) is the sum of the partial short-circuit currents calculated from the various sources. For calculation of the total short-circuit current at F_2 (Figure 12.3), the partial currents from each source should be calculated by adding the resistance and inductance of the common branch to the equivalent circuit. A correction factor is then applied. The correction factors for every source are obtained from

$$i_{pcorj} = \sigma_j i_{pj}$$

$$I_{kcorj} = \sigma_j I_{kj}$$

(12.62)

where the correction factor σ_j is described in the IEC standard [3].

Calculations of Short-Circuit Currents in Direct Current Systems

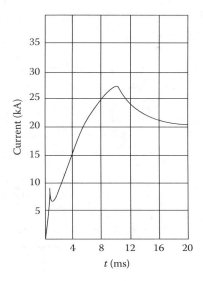

FIGURE 12.14
Total fault current profile of the partial short-circuit currents calculated in Examples 12.1, 12.4, 12.6, and 12.7.

Example 12.8

The sources (rectifier, battery, motor, and capacitor) in Examples 12.2, 12.4, 12.6, and 12.7 are connected together in a system configuration, as shown in Figure 12.3. Plot the total short-circuit current.

The profiles of partial currents shown in Figure 12.7 are summed. As the time to peak, magnitudes, and decay time constants are different in each case, a graphical approach is taken and the total current profile is shown in Figure 12.14. The peak current is approximately 27.3 kA and the peak occurs at approximately 9 ms after the fault.

The short-circuit current from the rectifier predominates. The short-circuit current from the capacitor is a high rise pulse, which rapidly decays to zero. The DC motor short-circuit current rises slowly. Smaller DC motors have higher armature inductance (Figure 12.12) resulting in a slower rate of current rise. The rectifier current peaks approximately in one half-cycle of the power system frequency. The relative magnitudes of the partial currents can vary substantially, depending on the system configuration. This can give varying profiles of total current and time to peak.

12.8 DC Circuit Breakers

The DC breakers may be categorized as follows [2, 8, 9]:

- The general-purpose low-voltage DC power circuit breakers do not limit the peak available current and may not prevent the prospective fault current rise to its peak value. These designs require a peak, short-time, and short-circuit current rating. Circuit breakers having a continuous current rating of 2000 A and below have instantaneous elements set to operate at 15 times the rated continuous current of the breaker. Circuit breakers rated above 2000 A have instantaneous elements set to operate at 12 times the continuous current rating of the breaker.

- A high-speed breaker during interruption limits the peak to a value less than the available (perspective) current and these breakers have a short-circuit rating and short-time rating.
- A semi-high speed circuit breaker does not limit the short-circuit current on circuits with minimal inductance, but becomes current limiting for highly inductive circuits. This design also requires a peak rating.
- Rectifier circuit breakers are a class in themselves and these carry the normal current output of one rectifier, and during fault conditions function to withstand or interrupt abnormal currents as required. This breaker requires short-circuit current rating for n-1 rectifiers and a short-time current rating for its own rectifier. The circuit breakers for rectifier applications are fitted with reverse current trips, set at no more than 50% of the continuous current rating.
- Semi-high speed and high speed circuit breakers are equipped with direct acting instantaneous elements set at no more than 4 times the circuit breaker continuous current rating or at the maximum setting below the available sustained current of the test circuit.

The DC breakers may have thermal magnetic or electronic trip devices, i.e., general-purpose circuit breakers of 2 kA or lower are provided with instantaneous tripping elements set to operate at 15 times the rated continuous current, and breakers rated >2 kA have instantaneous trips set to operate at 12 times the rated current. The rectifier circuit breakers have a reverse-current tripping element set to operate at no more than 50% of the continuous currents rating.

Two or three poles of a breaker may be connected in series for enhanced interrupting rating. The interrupting capacity of a breaker decreases with increasing DC voltage. The maximum inductance for full interrupting rating in micro-henries is specified and the reduced interrupting rating for higher values of inductance can be calculated. When the breakers are rated for AC as well as DC systems, the interrupting rating on DC systems are much lower.

IEEE Standard [9] provides the preferred ratings, related requirements and application recommendations for low-voltage AC (635 V and below) and DC (3200 V and below) power circuit breakers. Table 12.1 is based upon this reference.

Example 12.9

A general-purpose DC circuit breaker for the fault location Fl, the fault current profile shown in Figure 12.14, is selected as follows:

The peak short-circuit current is 28.5 kA, the quasi-steady state current is approximately 22 kA. The continuous load current of all the sources (200 AH DC battery, 15-hp DC motor, and 100 kVA rectifier) is approximately 380 A. Therefore, select a 250 V, circuit breaker frame size 600 A, continuous current=400 A, rated peak current=41 kA, rated maximum short-circuit current or short-time current=25 kA, row 1 of Table 12.1, from Reference [9].

The maximum inductance for full interrupting rating is specified as 160 μH. The actual reactance can be calculated from the time constant in Figure 12.14. The peak short-circuit current is 27.5 kA, therefore the resistance is 4.36 mΩ. The time constant of the current from the rise time is approximately 4 ms, which gives $L = 17.4$ μH.

If the inductance exceeds the value given in Table 12.1, the reduced interrupting rating is obtained from the expression from Reference [9]:

Calculations of Short-Circuit Currents in Direct Current Systems

TABLE 12.1

Preferred Ratings of General Purpose DC Power Circuit Breakers with or without Instantaneous Direct Acting Trip Elements

Circuit Breaker Frame Size (A DC)	System Nominal Voltage (V DC)	Rated Maximum Voltage (V DC)	Rated Peak Current (A) (peak)	Rated Maximum Short-Circuit Current or Rated Short-Circuit Current (A)	Maximum Inductance for Full Interrupting Rating (µH)	Load Circuit Stored Energy Factor (kW-s)	Range of Trip Device Current Ratings (A DC)
600–800	250	300	41,000	25,000	160	50	40–800
1600	250	300	83,000	50,000	80	100	200–1600
2000	250	300	83,000	50,000	80	100	200–2000
3000	250	300	1,24,000	75,000	50	140	2000–3000
4000	250	300	1,65,000	1,00,000	32	160	4000
5000	250	300	1,65,000	1,00,000	32	160	5000
6000	250	300	1,65,000	1,00,000	32	160	6000

The peak current rating is only applicable for circuit breakers in solid-state rectifier applications.
Rated short-circuit current is applicable only to circuit breakers without instantaneous direct acting trip elements (short-time delay elements or remote relay).
Source: IEEE Std. C37.16. IEEE Standard for Preferred Ratings, Related Requirements and Application Recommendations for Low-Voltage AC (635-V and Below) and DC (3200 V and Below) Power Circuit Breakers, 2009.

$$I = 10^4 \sqrt{20W/L}$$

where W is the value of kW-s specified in Table 12.1, L is the actual inductance in μH, and I is in ampéres.

12.9 DC Rated Fuses

DC rated fuses up to 600 V and interrupting ratings of 200 kA are available. The manufacturers may not publish the short-circuit ratings at all the DC voltages, but if a fuse of required interrupting rating and rated for higher voltage is selected, at the lower DC voltage of application, the interrupting capability will be higher.

Example 12.10

A 1000 A class L fuse, rated for 500 V DC, interrupting current 100 kA is applied to a common bus served by the four DC sources, the bolted and arcing current profiles as calculated in Figure 12.14. Calculate the arcing time.

A TCC plot of the fuse is shown in Figure 12.15. This is for the AC current. Manufacturers do not publish characteristics for the DC currents. However, approximately the AC rms current can be considered equivalent to DC current, as far as heating effects are concerned.

When subject to DC currents with certain time constants, the device operating time will be higher and the characteristics curve will shift to the right of the current axis. The 60 cycle AC wave rises to peak in 8.33 ms. Therefore, if the DC short-circuit is peaking in 8–10 ms, no correction to the AC operating time characteristics is required. When subjected too much to slow rising DC currents, the time current characteristics will shift towards right, that is, the operating time will increase. This increased operating time will be a function of the rise time of the DC current, the slower the rise, the more delayed the fuse operation.

The short-circuit profile in Figure 12.14 shows that the peak is reached in approximately 10 ms. Thus, the fuse AC current characteristics can be considered.

A simple procedure will be to calculate the operating time based upon the incremental change in the fault current to which the fuse is subjected. Then take the weighted average to ascertain the average current to which the fuse is subjected.

The peak current is 28.5 kA and occurs at 10 ms, at this current the fuse operates practically instantaneously. Also from any current above 20 kA, the fuse characteristic shows operation in the current limiting zone. Thus, the fuse will operate as the current is rising and from the instant of fault occurrence, the operating time is approximately 6 ms. Thus high speed fault clearance is provided with the selected fuse.

12.10 Protection of the Semi-Conductor Devices

For a short-circuit the semi-conducting devices like diodes, silicon-controlled rectifiers (SCRs), GTOs, cannot be allowed to be damaged. Unlike transformers, motors or cables, these do not have much thermal withstand capability. Invariably these are protected by high speed I^2T limiting current limiting fuses. These operate fast within less than a cycle

Calculations of Short-Circuit Currents in Direct Current Systems 549

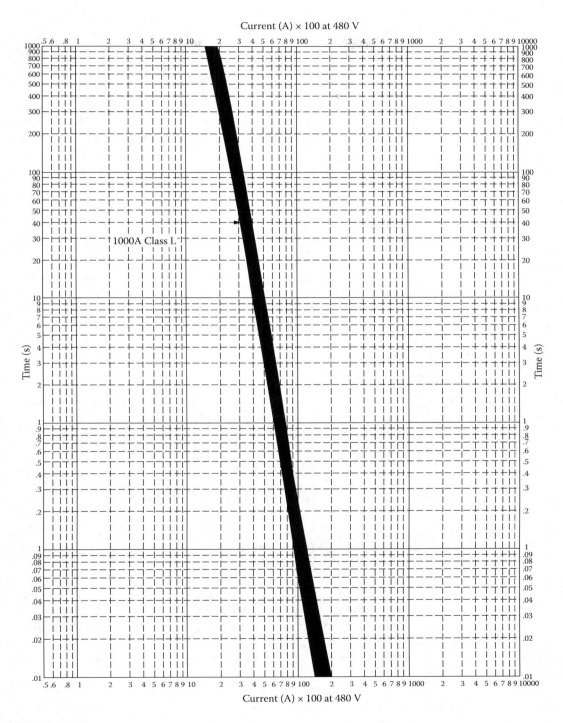

FIGURE 12.15
Time current characteristics of an L-type fuse.

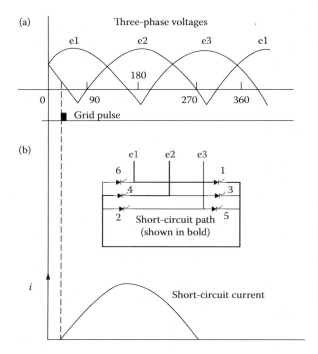

FIGURE 12.16
Short-circuit current profile of a grid controlled converter.

and limit the I^2T let-through. This I^2T of the fuse is coordinated from the device I^2T capability. A manufacturer's data on very fast acting and I^2T limiting fuses gives

- Application voltage 450 V DC, current range 35–1000 A, interrupting rating 79 kA, for a 1000 A fuse clearing I^2T = 500,000 A²s.
- Application voltage 500 V DC, current range 10–1200 A, interrupting rating 100 kA, for a 1200 A fuse clearing I^2T = 100,000 A²s melting, 800,000 A²s clearing.

The short-circuit currents are also limited electronically by gate control. As an example, in UPS systems there may not be enough short-circuit current available to operate instantaneously an overcurrent protective device.

We discussed controlled converters in Chapter 9. The firing angles are adjustable from $0 < \alpha < 180°$ (angles >90° correspond to inverter operation, see Volume 3). The damage for short-circuit currents in a controlled converter system is limited by grid control protection schemes. This enables a grid firing circuit to detect abnormal conditions and block grid pulses. The short-circuit current is limited to 1 cycle. Figure 12.16a and b shows DC short-circuit in a full wave SCR converter, where the grid protection system is operative.

12.11 High-Voltage DC Circuit Breakers

In Chapter 4, we discussed the arc lengthening principle for arc extinguishing. Figure 12.17 shows the DC arc characteristics for different arc lengths. The resistance characteristic is

Calculations of Short-Circuit Currents in Direct Current Systems

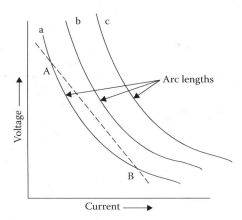

FIGURE 12.17
DC arc characteristics for different lengths.

shown by a straight line. For arc length given by curve a, the arc voltage is lower than the supply system voltage by an amount IR. Only at the points of intersection it is possible to have a stable arc. For the arc of length a, it intersects at point A. The arc voltage is less than the supply voltage and the arc will continue to burn. As the arc length is increased, the arc voltage increases above the supply voltage and the arc is extinguished. During the arcing time the supply source continues to give up energy (energy stored in the inductance of the system), the longer the arc burns the greater is the energy. Thus, the entire interruption process is a question of energy balance.

This principle has been successfully employed in some commercial designs of medium-voltage AC breakers also. In these breaker designs, the current can be easily extinguished at current zero, there is practically no current chopping, and the breaker mechanism is very light, except for a large arc chute with splitters. In case of DC some chopping can be expected as there is no current zero.

Consider a DC current flowing in a parallel LC circuit. If this current is suddenly interrupted (instantaneously), the recovery voltage situation can be simulated by forcing an equal and opposite current through the interrupting breaker. The recovery voltage in this case will be given by the following differential equation written in Laplace transform:

$$V_r(s) = \frac{i}{C}\left(\frac{1}{s^2 + \frac{1}{LC}}\right) \tag{12.63}$$

And its solution is

$$V_r(t) = i\sqrt{\frac{L}{C}} \sin\frac{t}{\sqrt{LC}} \tag{12.64}$$

In case a ramp is used, i.e., the current is gradually brought to zero, the rate of rise of the recovery voltage can be slowed, i.e., slower the chopping rate, slower will be the recovery voltage. A ramp signal in Laplace transform can be written as

$$i = -\left[\frac{a}{s^2} - \frac{a}{s^2}e^{-t_c s}\right] \qquad (12.65)$$

where a = i_0 / t_c = slope of the ramp and the solution with this signal is given as

$$V_r(t) = \frac{-iL}{t_c}\left[\left(t - \cos\frac{t}{\sqrt{LC}}\right) - \left(1 - \cos\frac{t-t_c}{\sqrt{LC}}\right)u(t-t_c)\right] \qquad (12.66)$$

For high-voltage, direct current (HVDC) transmission, high-voltage DC circuit breakers are not necessary and are not used. Thyristor control takes care of the overcurrent, short circuits, and abnormal conditions in HVDC converters. However, metallic return transfer breaker (MRTB) is used in bipolar HVDC transmission and the bipolar mode is changed to monopolar mode during a fault on one of the poles.

In high-voltage DC circuit breakers, the energy stored in the system can be high $0.5LI_d^2 = 10 - 25$ MJ. Artificial current zeros are produced by a reverse flow of current. The main breaker contacts are in parallel with a charged capacitor, reactor, and vacuum gap. This circuit is called a commutating circuit. When the main breaker contacts open, the vacuum gap is triggered. The pre-charged capacitor discharges violently through main circuit causing discharge current to flow in opposition to the main current. The main current drops down and oscillates with current zeros. The current is interrupted by the main circuit breaker at one of the current zeros. Vacuum gap seals in and the commutating circuit is opened.

The energy in the main DC pole is partly absorbed by the commutating circuit and partly by the zinc oxide surge arrester that can absorb up to 20 kJ/kV. The arc in the main circuit is quenched by interrupting medium, air or SF_6. The stresses are much higher because of higher energy associated with DC current interruption and higher transient recovery voltage (TRV) appears across the breaker pole.

The HVDC breaker types identified by the International Council on Large Electric Systems (in French: Conseil International des Grands Réseaux Électriques) (CIGRE) working group [8], are classified on the following basis:

- Switching time
- Current to be interrupted
- TRV
- Switching energy

The classification according to CIGRE is as follows:

- Type A: It is fast breaker, less than 15 ms and does not wait for converter action to reduce DC voltage. The TRV is the highest. Current capability is up to 1.25 I_d
- Type B: It is slow and waits for converter action to reduce DC line current to low value. It is classified into two types B_1 and B_2. Type B_1 has higher TRV compared to type B_2. The operating time of type B_2 is longer 120–120 ms, while that of type B_1 is 60–120 ms. Type A operating time is less than 15 ms.

Calculations of Short-Circuit Currents in Direct Current Systems

Problems

12.1 Calculate and plot the short-circuit current profile for a battery system with details as follows: lead acid battery, 240 V, 120 cells, 400 Ah rating at 8-h rate of 1.75 V per cell at 25°C. Each cell has a length of 15.5 in., width 6.8 in, and height 10 in. The cells are arranged in two-tier configuration in four rows, 30 cells per row. Intercell connectors are of 1 in. × ½ in. cross section and resistance 0.0321 mΩ/ft. Calculate the battery internal resistance by using Equation 12.11. The battery is connected through a cable of resistance 0.002 Ω and inductance 15 μH. The fault occurs at the end of the battery cable. Repeat with the IEC calculation method, as in Example 12.2.

12.2 Calculate and plot the terminal short-circuit current of a DC motor of 50 hp, 230 V, 690 rpm, armature current 178 A, and transient resistance 0.07 Ω, using the method of Example 12.3. Repeat the calculations using the IEC method. Additional motor data: $\tau_f = 1.0$ s, $J = 2$ kg/m^2, $L_{0F}/L_F = 0.3$.

12.3 Calculate and plot the short-circuit current profile for a fault on the DC side of a rectifier system in the following configuration: 480 V, three-phase AC source fault levels 20 kA, $X/R = 5$, 480–230 V, three-phase 300 kVA rectifier transformer, $Z = 3.5\%$, $X/R = 5$, the DC side equivalent resistance and inductance equal to 0.001 Ω and 3 μH, respectively.

12.4 Sum the partial fault currents calculated in Problems 1, 2, and 3 and calculate the maximum short-circuit current and time to peak. What should be the peak short-circuit rating and interrupting rating of a general-purpose DC circuit breaker?

References

1. General Electric Company. GE Industrial Power System Data Book. Schenectady, NY, 1978.
2. ANSI/IEEE Standard C37.14. Standard for Low-Voltage DC Power Circuit Breakers Used in Enclosures, 1972.
3. IEC 61660-1. Short-Circuit Currents in DC Auxiliary Installations in Power Plants and Substations, 1997.
4. IEEE Standard 946. DC Auxiliary Power Systems for Generating Stations, 1996.
5. AIEE Committee Report. Maximum short circuit current of DC motors and generators. Transient characteristics of DC motors and generators. *AIEE Trans*, 69, 146–149, 1950.
6. AT McClinton, EL Brancato, R Panoff. Maximum short circuit current of DC motors and generators. Transient characteristics of DC motors and generators. *AIEE Trans*, 68, 1100–1106, 1949.
7. AG Darling, TM Linville. Rate of rise of short circuit current of DC motors and generators. *AIEE Trans*, 71, 314–325, 1952.
8. CIGRE Joint Working Group 13/14-08. Circuit breakers for meshed multi-terminal HVDC systems, part 1: DC side substation switching under normal and fault conditions. *Electra*, 163, 128–122, 1995.
9. IEEE Std. C37.16. IEEE Standard for Preferred Ratings, Related Requirements and Application Recommendations for Low-Voltage AC (635-V and Below) and DC (3200 V and Below) Power Circuit Breakers, 2009.

Appendix A: Matrix Methods

A.1 Review Summary

A.1.1 Sets

A set of points is denoted by

$$S = (x_1, x_2, x_3) \qquad \text{(A.1)}$$

This shows a set of three points: x_1, x_2, and x_3. Some properties may be assigned to the set:

$$S = \{(x_1, x_2, x_3) \mid x_3 = 0\} \qquad \text{(A.2)}$$

Equation A.2 indicates that the last component of the set $x_3 = 0$. Members of a set are called elements of the set. If a point x, usually denoted by $\bar{x}$, is a member of the set, it is written as

$$\bar{x} \in S \qquad \text{(A.3)}$$

If we write

$$\bar{x} \notin S \qquad \text{(A.4)}$$

then point x is not an element of set S. If all the elements of a set S are also the elements of another set T, then S is said to be a subset of T, or S is contained in T:

$$S \subset T \qquad \text{(A.5)}$$

Alternatively, this is written as

$$T \supset S \qquad \text{(A.6)}$$

The intersection of two sets S_1 and S_2 is the set of all points $\bar{x}$ such that $\bar{x}$ is an element of both S_1 and S_2. If the intersection is denoted by T, we write

$$T = S_1 \cap S_2 \qquad \text{(A.7)}$$

The intersection of n sets is

$$T \equiv S_1 \cap S_2 \cap \ldots \cap S_n \equiv \bigcap_{i=1}^{n} S_i \qquad \text{(A.8)}$$

555

The union of two sets S_1 and S_2 is the set of all points $\bar{x}$ such that $\bar{x}$ is an element of either S_1 or S_2. If the union is denoted by P, we write

$$P = S_1 \cup S_2 \tag{A.9}$$

The union of n sets is written as

$$P \equiv S_1 \cup S_2 \cup \ldots \cup S_n \equiv U_{i=1}^{n} S_i \tag{A.10}$$

A.1.2 Vectors

A vector is an ordered set of numbers, real or complex. A matrix containing only one row or column may be called a vector:

$$\bar{x} = \begin{vmatrix} x_1 \\ x_2 \\ . \\ x_n \end{vmatrix} \tag{A.11}$$

where $x_1, x_2, \ldots, x_n$ are called the constituents of the vector. The transposed form is

$$\bar{x}' = | x_1, x_2, \ldots, x_n | \tag{A.12}$$

In this series the transpose is indicated by a superscript letter t. A null vector $\bar{0}$ has all its components equal to 0 and a sum vector $\bar{1}$ has all its components equal to 1.

The following properties are applicable to vectors:

$$\bar{x} + \bar{y} = \bar{y} + \bar{x}$$

$$\bar{x} + (\bar{y} + \bar{z}) = (\bar{x} + \bar{y}) + \bar{z}$$

$$\alpha_1(\alpha_2 \bar{x}) = (\alpha_1 \alpha_2) \bar{x}$$

$$(\alpha_1 + \alpha_2)\bar{x} = \alpha_1 \bar{x} + \alpha_2 \bar{x} \tag{A.13}$$

$$\bar{0}\bar{x} = \bar{0}$$

Multiplication of two vectors of the same dimensions results in an inner or scalar product:

$$\bar{x}^t \bar{y} = \sum_{i=1}^{n} x_i y_i = \bar{y}^t \bar{x}$$

$$\bar{x}^t \bar{x} = |\bar{x}|^2 \tag{A.14}$$

$$\cos \phi = \frac{\bar{x}^t \bar{y}}{|\bar{x}||\bar{y}|}$$

where φ is the angle between vectors and $|x|$ and $|y|$ are the geometric lengths. Two vectors $\bar{x}_1$ and $\bar{x}_2$ are orthogonal if

$$\bar{x}_1 \bar{x}_2^t = 0 \tag{A.15}$$

Appendix A: Matrix Methods 557

A.1.3 Matrices

1. A matrix is a rectangular array of numbers subject to certain rules of operation and usually denoted by a capital letter within brackets [A], a capital letter in bold, or a capital letter with an overbar. The last convention is followed in this book. The dimensions of a matrix indicate the total number of rows and columns. An element a_{ij} lies at the intersection of row i and column j.

2. A matrix containing only one row or column is called a vector.

3. A matrix in which the number of rows is equal to the number of columns is a square matrix.

4. A square matrix is a diagonal matrix if all off-diagonal elements are zero.

5. A unit or identity matrix $\bar{I}$ is a square matrix with all diagonal elements = 1 and off-diagonal elements = 0.

6. A matrix is symmetric if, for all values of i and j, $a_{ij} = a_{ji}$.

7. A square matrix is a skew symmetric matrix if $a_{ji} = -a_{ij}$ for all values of i and j.

8. A square matrix whose elements below the leading diagonal are zero is called an upper triangular matrix. A square matrix whose elements above the leading diagonal are zero is called a lower triangular matrix.

9. If in a given matrix rows and columns are interchanged, the new matrix obtained is the transpose of the original matrix, denoted by $\bar{A}^t$.

10. A square matrix $\bar{A}$ is an orthogonal matrix if its product with its transpose is an identity matrix:

$$\bar{A}\bar{A}^t = \bar{I} \tag{A.16}$$

11. The conjugate of a matrix is obtained by changing all its complex elements to their conjugates, i.e., if

$$\bar{A} = \begin{vmatrix} 1-i & 3-4i & 5 \\ 7+2i & -i & 4-3i \end{vmatrix} \tag{A.17}$$

then its conjugate is

$$\bar{A}^* = \begin{vmatrix} 1+i & 3+4i & 5 \\ 7-2i & i & 4+3i \end{vmatrix} \tag{A.18}$$

A square matrix is a unit matrix if the product of the transpose of the conjugate matrix and the original matrix is an identity matrix:

$$\bar{A}^{*t}\,\bar{A} = \bar{I} \tag{A.19}$$

12. A square matrix is called a Hermitian matrix if every $i - j$ element is equal to the conjugate complex $j - i$ element, i.e.,

$$\bar{A} = \bar{A}^{*t} \tag{A.20}$$

13. A matrix, such that

$$\bar{A}^2 = \bar{A} \tag{A.21}$$

is called an idempotent matrix.

14. A matrix is periodic if

$$\bar{A}^{k+1} = \bar{A} \tag{A.22}$$

15. A matrix is called nilpotent if

$$\bar{A}^k = 0 \tag{A.23}$$

where k is a positive integer. If k is the least positive integer, then k is called the index of nilpotent matrix.

16. Addition of matrices follows a commutative law:

$$\bar{A} + \bar{B} = \bar{B} + \bar{A} \tag{A.24}$$

17. A scalar multiple is obtained by multiplying each element of the matrix with a scalar. The product of two matrices $\bar{A}$ and $\bar{B}$ is only possible if the number of columns in $\bar{A}$ equals the number of rows in $\bar{B}$.

If $\bar{A}$ is an $m \times n$ matrix and $\bar{B}$ is $n \times p$ matrix, the product $\bar{A}\bar{B}$ is an $m \times p$ matrix, where

$$c_{ij} = a_{i1}b_{1j} + a_{i2}b_{2j} + \cdots + a_{in}b_{nj} \tag{A.25}$$

Consider product of two matrices:

$$\begin{vmatrix} 3 & 1 \\ 0 & 2 \\ 7 & 4 \end{vmatrix} \times \begin{vmatrix} 1 & 3 \\ 0 & 7 \end{vmatrix}$$

A reader can verify that the result is

$$\begin{vmatrix} 3 & 16 \\ 0 & 14 \\ 7 & 49 \end{vmatrix}$$

Multiplication is not commutative

$$\bar{A}\bar{B} \neq \bar{B}\bar{A} \tag{A.26}$$

Multiplication is associative if conformability is assured:

$$\bar{A}(\bar{B}\bar{C}) = (\bar{A}\bar{B})\bar{C} \tag{A.27}$$

Appendix A: Matrix Methods 559

It is distributive with respect to addition:

$$\bar{A}(\bar{B}+\bar{C}) = \bar{A}\bar{B} + \bar{A}\bar{C} \tag{A.28}$$

The multiplicative inverse exists if $|A| \neq 0$. Also,

$$(\bar{A}\bar{B})^t = \bar{B}^t \bar{A}^t \tag{A.29}$$

18. The transpose of the matrix of cofactors of a matrix is called an adjoint matrix. The product of a matrix $\bar{A}$ and its adjoint is equal to the unit matrix multiplied by the determinant of A. If A is a square matrix,

$$\bar{A}_{\text{adj}}\bar{A} = |A|\,\bar{I} \tag{A.30}$$

where the determinant is denoted as $|A|$.

This property can be used to find the inverse of a matrix (see Example A.4). The determinant of a square matrix is a scalar written as $|A|$. It can be calculated as follows:

$$\bar{A} = \begin{vmatrix} a_{11} & a_{12} & a_{13} \\ a_{21} & a_{22} & a_{23} \\ a_{31} & a_{32} & a_{33} \end{vmatrix} = a_{11}a_{22}a_{33} + a_{12}a_{23}a_{31} + a_{21}a_{32}a_{13} - a_{13}a_{22}a_{31} - a_{12}a_{21}a_{33} - a_{11}a_{23}a_{33} \tag{A.31}$$

Exactly one element forms each row and one element from each column appears in each term of the expansion.

The *cofactor* of an element is the signed minor of that element and is given by

$$C_{ij} = (-1)^{(i+j)} M_{ij} \tag{A.32}$$

where M_{ij} is the minor. The minor of matrix $\bar{A}$ is

$$M = \begin{vmatrix} a_{22} & a_{13} \\ a_{32} & a_{33} \end{vmatrix} \tag{A.33}$$

The determinant of a square matrix is evaluated by summing the products of elements and corresponding cofactors:

$$|A| = \sum_{i=1}^{n} A_{ij}C_{ij}$$

or $\tag{A.34}$

$$|A| = \sum_{j=1}^{n} A_{ij}C_{ij}$$

560 *Appendix A: Matrix Methods*

If any two rows or any two columns of a determinant are interchanged, the value of the resulting determinant is the negative of the original.

19. By performing elementary transformations, any nonzero matrix can be reduced to one of the following forms called the normal forms:

$$[I_r], \quad [I_r 0], \quad \begin{vmatrix} I_r \\ 0 \end{vmatrix}, \quad \begin{vmatrix} I_r & 0 \\ 0 & 0 \end{vmatrix} \tag{A.35}$$

The number r is called the rank of matrix $\bar{A}$. The form

$$\begin{vmatrix} I_r & 0 \\ 0 & 0 \end{vmatrix} \tag{A.36}$$

is called the first canonical form of $\bar{A}$. Both row and column transformations can be used here. The rank of a matrix is said to be r if (1) it has at least one nonzero minor of order r and (2) every minor of $\bar{A}$ of order higher than $r = 0$. Rank is a non-zero row (the row that does not have all the elements $= 0$) in the upper triangular matrix.

Example A.1

Find the rank of the matrix:

$$\bar{A} = \begin{vmatrix} 1 & 4 & 5 \\ 2 & 6 & 8 \\ 3 & 7 & 22 \end{vmatrix}$$

This matrix can be reduced to an upper triangular matrix by elementary row operations (see below):

$$\bar{A} = \begin{vmatrix} 1 & 4 & 5 \\ 0 & 1 & 1 \\ 0 & 0 & 12 \end{vmatrix}$$

The rank of the matrix is 3.

A band structured matrix is a square matrix whose nonzero elements appear in diagonal bands parallel to the principal diagonal:

$$\begin{vmatrix} 2 & 3 & 0 & 0 & 0 & 0 \\ 4 & 5 & 6 & 0 & 0 & 0 \\ 0 & 9 & 4 & 8 & 0 & 0 \\ 0 & 0 & 8 & 9 & 2 & 0 \\ 0 & 0 & 0 & 5 & 2 & 1 \\ 0 & 0 & 0 & 0 & 3 & 4 \end{vmatrix} \tag{A.37}$$

Appendix A: Matrix Methods 561

A.2 Characteristic Roots, Eigenvalues, Eigenvectors

For a square matrix $\bar{A}$, the $\bar{A} - \lambda \bar{I}$ matrix is called the characteristic matrix; λ is a scalar and $\bar{I}$ is a unit matrix. The determinant $|A - \lambda I|$ when expanded gives a polynomial, which is called the characteristic polynomial of $\bar{A}$ and the equation $|A - \lambda I| = 0$ is called the characteristic equation of matrix $\bar{A}$. The roots of the characteristic equation are called the characteristic roots or eigenvalues.

Some properties of eigenvalues are as follows:

- Any square matrix $\bar{A}$ and its transpose $\bar{A}^t$ have the same eigenvalues.
- The sum of the eigenvalues of a matrix is equal to the trace of the matrix (the sum of the elements on the principal diagonal is called the trace of the matrix).
- The product of the eigenvalues of the matrix is equal to the determinant of the matrix. If

$$\lambda_1, \lambda_2, \ldots, \lambda_n \tag{A.38}$$

are the eigenvalues of $\bar{A}$, then the eigenvalues of

$$k\bar{A} \text{ are } k\lambda_1, k\lambda_2, \ldots, k\lambda_n$$

$$\bar{A}^m \text{ are } \lambda_1^m, \lambda_2^m, \ldots, \lambda_n^m \tag{A.39}$$

$$\bar{A}^{-1} \text{ are } 1/\lambda_1, 1/\lambda_2, \ldots, 1/\lambda_n$$

- Zero is a characteristic root of a matrix, only if the matrix is singular.
- The characteristic roots of a triangular matrix are diagonal elements of the matrix.
- The characteristic roots of a Hermitian matrix are all real.
- The characteristic roots of a real symmetric matrix are all real, as the real symmetric matrix will be Hermitian.

A.2.1 Cayley–Hamilton Theorem

Every square matrix satisfies its own characteristic equation:

$$\text{If } |\bar{A} - \lambda \bar{I}| = (-1)^n (\lambda^n + a_1 \lambda^{n-1} + a_2 \lambda^{n-2} + \cdots + a_n) \tag{A.40}$$

is the characteristic polynomial of an $n \times n$ matrix, then the matrix equation

$$\bar{X}^n + a_1 \bar{X}^{n-1} + a_2 \bar{X}^{n-2} + \cdots + a_n \bar{I} = 0$$

$$\text{is satisfied by } \bar{X} = \bar{A} \tag{A.41}$$

$$\bar{A}^n + a_1 \bar{A}^{n-1} + a_2 \bar{A}^{n-2} + \cdots + a_n \bar{I} = 0$$

This property can be used to find the inverse of a matrix.

Example A.2

Find the characteristic equation of the matrix:

$$\bar{A} = \begin{vmatrix} 1 & 4 & 2 \\ 3 & 2 & -2 \\ 1 & -1 & 2 \end{vmatrix}$$

and then the inverse of the matrix.

The characteristic equation is given by

$$\begin{vmatrix} 1-\lambda & 4 & 2 \\ 3 & 2-\lambda & -2 \\ 1 & -1 & 2-\lambda \end{vmatrix} = 0$$

Expanding, the characteristic equation is

$$\lambda^3 - 5\lambda^2 - 8\lambda + 40 = 0$$

then, by the Cayley–Hamilton theorem,

$$\bar{A}^2 - 5\bar{A} - 8\bar{I} + 40\bar{A}^{-1} = 0$$
$$40\bar{A}^{-1} = -\bar{A}^2 + 5\bar{A} + 8\bar{I}$$

We can write,

$$40A^{-1} = -\begin{vmatrix} 1 & 4 & 2 \\ 3 & 2 & -2 \\ 1 & -1 & 2 \end{vmatrix}^2 + 5\begin{vmatrix} 1 & 4 & 2 \\ 3 & 2 & -2 \\ 1 & -1 & 2 \end{vmatrix} + 8\begin{vmatrix} 1 & 0 & 0 \\ 0 & 1 & 0 \\ 0 & 0 & 1 \end{vmatrix}$$

The inverse is

$$A^{-1} = \begin{vmatrix} -0.05 & 0.25 & 0.3 \\ 0.2 & 0 & -0.2 \\ 0.125 & -0.125 & 0.25 \end{vmatrix}$$

This is not an effective method of finding the inverse for matrices of large dimensions.

A.2.2 Characteristic Vectors

Each characteristic root λ has a corresponding nonzero vector $\bar{x}$ which satisfies the equation $|\bar{A} - \lambda\bar{I}|\bar{x} = 0$. The nonzero vector $\bar{x}$ is called the characteristic vector or eigenvector. The eigenvector is, therefore, not *unique*.

Appendix A: Matrix Methods 563

A.3 Diagonalization of a Matrix

If a square matrix $\bar{A}$ of $n \times n$ has n linearly independent eigenvectors, then a matrix $\bar{P}$ can be found so that

$$\bar{P}^{-1}\bar{A}\bar{P} \tag{A.42}$$

is a diagonal matrix.

The matrix $\bar{P}$ is found by grouping the eigenvectors of $\bar{A}$ into a square matrix, i.e., $\bar{P}$ has eigenvalues of $\bar{A}$ as its diagonal elements.

A.3.1 Similarity Transformation

The transformation of matrix $\bar{A}$ into $\bar{P}^{-1}\bar{A}\bar{P}$ is called a *similarity transformation*. Diagonalization is a special case of similarity transformation.

Example A.3

$$\text{Let } \bar{A} = \begin{vmatrix} -2 & 2 & -3 \\ 2 & 1 & -6 \\ -1 & -2 & 0 \end{vmatrix}$$

Its characteristic equation is

$$\begin{vmatrix} -2-\lambda & 2 & -3 \\ 2 & 1-\lambda & -6 \\ -1 & -2 & 0-\lambda \end{vmatrix} = 0$$

$$(-2-\lambda)(-\lambda+\lambda^2-12)-2(-2\lambda-6)-3(-4+1-\lambda)=0$$

$$\lambda^3+\lambda^2-21\lambda-45 = 0$$
$$(\lambda-5)(\lambda-3)(\lambda-3)=0$$

$$\lambda = 5, -3, -3$$

The eigenvector is found by substituting the eigenvalues. For $\lambda = 5$,

$$-\begin{vmatrix} -7 & 2 & -3 \\ 2 & -4 & -6 \\ -1 & -2 & -5 \end{vmatrix} \begin{Vmatrix} x \\ y \\ z \end{Vmatrix} = \begin{vmatrix} 0 \\ 0 \\ 0 \end{vmatrix}$$

By manipulation of the rows, this can be reduced to

$$\begin{vmatrix} -1 & -2 & -5 \\ 0 & 16 & 32 \\ 0 & 0 & 0 \end{vmatrix} \begin{Vmatrix} x \\ y \\ z \end{Vmatrix} = \begin{vmatrix} 0 \\ 0 \\ 0 \end{vmatrix}$$

Therefore,

$$-x - 2y - 5z = 0$$

$$16y + 32z = 0$$

As eigenvectors are not unique, by assuming that $z = 1$, and solving, one eigenvector is

$$(-1,\ -2,\ 1)^t$$

Similarly, eigenvectors for $\lambda = -3$ can be found. This gives the following values:

$$\begin{vmatrix} x_1 \\ x_2 \\ (1/3)(x_1 + 2x_2) \end{vmatrix}$$

Choose arbitrarily $x_1 = 2$, $x_2 = -1$ then,

$$(2,\ -1,\ 0)^t$$

is another eigenvector. Similarly,

$$(3,\ 0,\ 1)^t$$

can be the third eigenvector. A matrix formed of these vectors is

$$\bar{P} = \begin{vmatrix} -1 & 2 & 3 \\ -2 & -1 & 0 \\ 1 & 0 & 1 \end{vmatrix}$$

and the diagonalization is obtained:

$$\bar{P}^{-1}\bar{A}\bar{P} = \begin{vmatrix} 5 & 0 & 0 \\ 0 & -3 & 0 \\ 0 & 0 & -3 \end{vmatrix}$$

This contains the eigenvalues as the diagonal elements.

Now choose some other eigenvectors and form a new matrix, say,

$$\bar{P} = \begin{vmatrix} 1 & 1 & 3 \\ 2 & 1 & 0 \\ -1 & 1 & 1 \end{vmatrix}$$

Again with these values $\bar{P}^{-1}\bar{A}\bar{P}$ is

$$\begin{vmatrix} 1 & 1 & 3 \\ 2 & 1 & 0 \\ -1 & 1 & 1 \end{vmatrix}^{-1} \begin{vmatrix} -2 & 2 & -3 \\ 2 & 1 & -6 \\ -1 & -2 & 0 \end{vmatrix} \begin{vmatrix} 1 & 1 & 3 \\ 2 & 1 & 0 \\ -1 & 1 & 1 \end{vmatrix} = \begin{vmatrix} 5 & 0 & 0 \\ 0 & -3 & 0 \\ 0 & 0 & -3 \end{vmatrix}$$

This is the same result as before.

Appendix A: Matrix Methods 565

A.4 Linear Independence and Dependence of Vectors

Vectors $\bar{x}_1, \bar{x}_2, \ldots, \bar{x}_n$ are dependent if all vectors (row or column matrices) are of the same order, and n scalars $\lambda_1, \lambda_2, \ldots, \lambda\eta$ (not all zeros) exist such that

$$\lambda_1 \bar{x}_1 + \lambda_2 \bar{x}_2 + \lambda_3 \bar{x}_3 + \cdots + \lambda_n \bar{x}_n = 0. \tag{A.43}$$

Otherwise they are linearly independent. In other words, if vector $\bar{x}_K + 1$ can be written as a linear combination of vectors $(\bar{x}_1, \bar{x}_2, \ldots, \bar{x}_n)$, then it is linearly dependent, otherwise it is linearly independent. Consider the vectors:

$$\bar{x}_3 = \begin{vmatrix} 4 \\ 2 \\ 5 \end{vmatrix} \quad \bar{x}_1 = \begin{vmatrix} 1 \\ 0.5 \\ 0 \end{vmatrix} \quad \bar{x}_2 = \begin{vmatrix} 0 \\ 0 \\ 1 \end{vmatrix}$$

then

$$\bar{x}_3 = 4\bar{x}_1 + 5\bar{x}_2$$

Therefore, $\bar{x}_3$ is linearly dependent on $\bar{x}_1$ and $\bar{x}_2$

A.4.1 Vector Spaces

If $\bar{x}$ is any vector from all possible collections of vectors of dimension n, then for any scalar α, the vector $\alpha \bar{x}$ is also of dimension n. For any other n vector $\bar{y}$, the vector $\bar{x} + \bar{y}$ is also of dimension n. The set of all n-dimensional vectors are said to form a linear vector space E^n. Transformation of a vector by a matrix is a linear transformation:

$$\bar{A}(\alpha \bar{x} + \beta \bar{y}) = \alpha(\bar{A}\bar{x}) + \beta(\bar{A}\bar{y}) \tag{A.44}$$

One property of interest is

$$\bar{A}\bar{x} = 0 \tag{A.45}$$

i.e., whether any nonzero vector $\bar{x}$ exists, which is transformed by matrix $\bar{A}$ into a zero vector. Equation A.45 can only be satisfied if the columns of $\bar{A}$ are linearly dependent. A square matrix whose columns are linearly dependent is called a singular matrix and a square matrix whose columns are linearly independent is called a nonsingular matrix. In Equation A.45 if $\bar{x} = 0$, then columns of $\bar{A}$ must be linearly independent. The determinant of a singular matrix is zero and its inverse does not exist.

A.5 Quadratic Form Expressed as Product of Matrices

The quadratic form can be expressed as a product of matrices:

$$\text{Quadratic form} = \bar{x}^t A \bar{x} \tag{A.46}$$

where

$$\bar{x} = \begin{vmatrix} x_1 \\ x_2 \\ x_3 \end{vmatrix} \quad A = \begin{vmatrix} a_{11} & a_{12} & a_{13} \\ a_{21} & a_{22} & a_{23} \\ a_{31} & a_{32} & a_{33} \end{vmatrix} \tag{A.47}$$

Therefore,

$$\bar{x}'A\bar{x} = \begin{vmatrix} x_1 & x_2 & x_3 \end{vmatrix} \begin{vmatrix} a_{11} & a_{12} & a_{13} \\ a_{21} & a_{22} & a_{23} \\ a_{31} & a_{32} & a_{33} \end{vmatrix} \begin{vmatrix} x_1 \\ x_2 \\ x_3 \end{vmatrix} \tag{A.48}$$

$$= a_{11}x_1^2 + a_{22}x_2^2 + a_{33}x_3^2 + 2a_{12}x, x_2 + 2a_{23}x_2x_3 + 2a_{13}x_1x_3.$$

A.6 Derivatives of Scalar and Vector Functions

A scalar function is defined as

$$y \cong f(x_1, x_2, \ldots, x_n), \tag{A.49}$$

where $x_1, x_2, \ldots, x_n$ are n variables. It can be written as a scalar function of an n-dimensional vector, i.e., $y = f(\bar{x})$, where $\bar{x}$ is an n-dimensional vector:

$$\bar{x} = \begin{vmatrix} x_1 \\ x_2 \\ . \\ x_n \end{vmatrix} \tag{A.50}$$

In general, a scalar function could be a function of several vector variables, i.e., $y = f(\bar{x}, \bar{u}, \bar{p})$, where $\bar{x}, \bar{u}$, and $\bar{P}$ are vectors of various dimensions. A vector function is a function of several vector variables, i.e., $y = f(\bar{x}, \bar{u}, \bar{p})$.

A derivative of a scalar function with respect to a vector variable is defined as

$$\frac{\partial \bar{f}}{\partial x} = \begin{vmatrix} \dfrac{\partial f}{\partial x_1} \\ \dfrac{\partial f}{\partial x_2} \\ . \\ \dfrac{\partial f}{\partial x_n} \end{vmatrix} \tag{A.51}$$

Appendix A: Matrix Methods 567

The derivative of a scalar function with respect to a vector of n dimensions is a vector of the same dimension. The derivative of a vector function with respect to a vector variable x is defined as

$$\frac{\partial \overline{f}}{\partial x} = \begin{vmatrix} \dfrac{\partial f_1}{\partial x_1} & \dfrac{\partial f_1}{\partial x_2} & \cdot & \dfrac{\partial f_1}{\partial x_n} \\ \dfrac{\partial f_2}{\partial x_1} & \dfrac{\partial f_2}{\partial x_2} & \cdot & \dfrac{\partial f_2}{\partial x_n} \\ \cdot & \cdot & \cdot & \cdot \\ \dfrac{\partial f_m}{\partial x_1} & \dfrac{\partial f_m}{\partial x_2} & \cdot & \dfrac{\partial f_m}{\partial x_n} \end{vmatrix} = \begin{vmatrix} \left[\dfrac{\partial f_1}{\partial x_1}\right]^t \\ \left[\dfrac{\partial f_2}{\partial x_2}\right]^t \\ \cdot \\ \left[\dfrac{\partial f_m}{\partial x_n}\right]^t \end{vmatrix} \tag{A.52}$$

If a scalar function is defined as

$$\overline{s} = \lambda^t f(\overline{x}, \overline{u}, \overline{p})$$

$$= \lambda_1 f_1(\overline{x}, \overline{u}, \overline{p}) + \lambda_2 f_2(\overline{x}, \overline{u}, \overline{p}) + \cdots + \lambda_m f_m(\overline{x}, \overline{u}, \overline{p}) \tag{A.53}$$

Then $\partial s / \partial \lambda$ is

$$\frac{\partial \overline{s}}{\partial \lambda} = \begin{vmatrix} f_1(\overline{x}, \overline{u}, \overline{p}) \\ f_2(\overline{x}, \overline{u}, \overline{p}) \\ \cdot \\ f_m(\overline{x}, \overline{u}, \overline{p}) \end{vmatrix} = f(\overline{x}, \overline{u}, \overline{p}) \tag{A.54}$$

and $\partial s / \partial x$ is

$$\frac{\partial \overline{s}}{\partial x} = \begin{vmatrix} \lambda_1 \dfrac{\partial f_1}{\partial x_1} + \lambda_2 \dfrac{\partial f_2}{\partial x_1} + \cdots + \lambda_m \dfrac{\partial f_m}{\partial x_1} \\ \lambda_1 \dfrac{\partial f_1}{\partial x_2} + \lambda_2 \dfrac{\partial f_2}{\partial x_2} + \cdots + \lambda_m \dfrac{\partial f_m}{\partial x_2} \\ \vdots \\ \lambda_1 \dfrac{\partial f_1}{\partial x_n} + \lambda_2 \dfrac{\partial f_2}{\partial x_n} + \cdots + \lambda_m \dfrac{\partial f_m}{\partial x_n} \end{vmatrix} = \begin{vmatrix} \dfrac{\partial f_1}{\partial x_1} & \dfrac{\partial f_2}{\partial x_1} + \cdots + \dfrac{\partial f_m}{\partial x_1} \\ \dfrac{\partial f_1}{\partial x_2} & \dfrac{\partial f_2}{\partial x_2} + \cdots + \dfrac{\partial f_m}{\partial x_2} \\ \vdots \\ \dfrac{\partial f_1}{\partial x_n} & \dfrac{\partial f_2}{\partial x_n} + \cdots + \dfrac{\partial f_m}{\partial x_n} \end{vmatrix} \begin{vmatrix} \lambda_1 \\ \lambda_2 \\ \vdots \\ \lambda_m \end{vmatrix} \tag{A.55}$$

Therefore,

$$\frac{\partial \overline{s}}{\partial x} = \left| \frac{\partial f}{\partial x} \right|^t \overline{\lambda} \tag{A.56}$$

A.7 Inverse of a Matrix

The inverse of a matrix is often required in the power system calculations, though it is rarely calculated directly. The inverse of a square matrix $\bar{A}$ is defined so that

$$\bar{A}^{-1}\bar{A} = \bar{A}\bar{A}^{-1} = \bar{I} \tag{A.57}$$

The inverse can be evaluated in many ways.

A.7.1 By Calculating the Adjoint and Determinant of the Matrix

$$\bar{A}^{-1} = \frac{\bar{A}_{adj}}{|A|} \tag{A.58}$$

Example A.4

Consider the matrix:

$$\bar{A} = \begin{vmatrix} 1 & 2 & 3 \\ 4 & 5 & 6 \\ 3 & 1 & 2 \end{vmatrix}$$

Its adjoint is

$$\bar{A}_{adj} = \begin{vmatrix} 4 & -1 & -3 \\ 10 & -7 & 6 \\ -11 & 5 & -3 \end{vmatrix}$$

and the determinant of $\bar{A}$ is equal to –9.

Thus, the inverse of $\bar{A}$ is

$$\bar{A}^{-1} = \begin{vmatrix} -\dfrac{4}{9} & \dfrac{1}{9} & \dfrac{1}{3} \\ -\dfrac{10}{9} & \dfrac{7}{9} & -\dfrac{2}{3} \\ \dfrac{11}{9} & -\dfrac{5}{9} & \dfrac{1}{3} \end{vmatrix}$$

A.7.2 By Elementary Row Operations

The inverse can also be calculated by elementary row operations. This operation is as follows:

1. A unit matrix of $n \times n$ is first attached to the right side of matrix $n \times n$ whose inverse is required to be found.
2. Elementary row operations are used to force the augmented matrix so that the matrix whose inverse is required becomes a unit matrix.

Appendix A: Matrix Methods 569

Example A.5

Consider a matrix:

$$\bar{A} = \begin{vmatrix} 2 & 6 \\ 3 & 4 \end{vmatrix}$$

It is required to find its inverse.

Attach a unit matrix of 2×2 and perform the operations as shown:

$$\begin{vmatrix} 2 & 6 \\ 3 & 4 \end{vmatrix} \begin{vmatrix} 1 & 0 \\ 0 & 1 \end{vmatrix} \xrightarrow{\frac{R_1}{2}} \begin{vmatrix} 1 & 3 \\ 3 & 4 \end{vmatrix} \begin{vmatrix} \dfrac{1}{2} & 0 \\ 0 & 1 \end{vmatrix} \to R_2 - 3R_1 \begin{vmatrix} 1 & 3 \\ 0 & -5 \end{vmatrix} \begin{vmatrix} \dfrac{1}{2} & 0 \\ \dfrac{-3}{2} & 1 \end{vmatrix} \to R_1 + \dfrac{5}{3} R_2$$

$$\begin{vmatrix} 1 & 0 \\ 0 & -5 \end{vmatrix} \begin{vmatrix} \dfrac{-2}{5} & \dfrac{3}{5} \\ \dfrac{-3}{2} & 1 \end{vmatrix} \to R_2 - \dfrac{1}{5} \begin{vmatrix} 1 & 0 \\ 0 & 1 \end{vmatrix} \begin{vmatrix} \dfrac{-2}{5} & \dfrac{3}{5} \\ \dfrac{3}{10} & \dfrac{-1}{5} \end{vmatrix}$$

Thus, the inverse is

$$\bar{A}^{-1} = \begin{vmatrix} \dfrac{-2}{5} & \dfrac{3}{5} \\ \dfrac{3}{10} & \dfrac{-1}{5} \end{vmatrix}$$

Some useful properties of inverse matrices are as follows:

The inverse of a matrix product is the product of the matrix inverses taken in reverse order, i.e.,

$$[\bar{A}\,\bar{B}\,\bar{C}]^{-1} = [\bar{C}]^{-1}[\bar{B}]^{-1}[\bar{A}]^{-1} \tag{A.59}$$

The inverse of a diagonal matrix is a diagonal matrix whose elements are the respective inverses of the elements of the original matrix:

$$\begin{vmatrix} A_{11} & & \\ & B_{22} & \\ & & C_{33} \end{vmatrix}^{-1} = \begin{vmatrix} \dfrac{1}{A_{11}} & & \\ & \dfrac{1}{B_{22}} & \\ & & \dfrac{1}{C_{33}} \end{vmatrix} \tag{A.60}$$

A square matrix composed of diagonal blocks can be inverted by taking the inverse of the respective submatrices of the diagonal block:

$$\begin{vmatrix} [\text{block } A] & & \\ & [\text{block } B] & \\ & & [\text{block } C] \end{vmatrix}^{-1} = \begin{vmatrix} [\text{block } A]^{-1} & & \\ & [\text{block } B]^{-1} & \\ & & [\text{block } C]^{-1} \end{vmatrix} \tag{A.61}$$

A.7.3 Inverse by Partitioning

Matrices can be partitioned horizontally and vertically, and the resulting submatrices may contain only one element. Thus, a matrix $\bar{A}$ can be partitioned as shown:

$$
\begin{vmatrix}
a_{11} & a_{12} & a_{13} & \vdots & a_{14} \\
a_{21} & a_{22} & a_{23} & \vdots & a_{24} \\
a_{31} & a_{32} & a_{33} & \vdots & a_{34} \\
\cdots & \cdots & \cdots & \cdots & \cdots \\
a_{41} & a_{42} & a_{43} & \vdots & a_{44}
\end{vmatrix}
=
\begin{vmatrix}
\bar{A}_1 & \bar{A}_2 \\
\bar{A}_3 & \bar{A}_4
\end{vmatrix}
\tag{A.62}
$$

where

$$
\bar{A}_1 =
\begin{vmatrix}
a_{11} & a_{12} & a_{13} \\
a_{21} & a_{22} & a_{23} \\
a_{31} & a_{32} & a_{33}
\end{vmatrix}
\tag{A.63}
$$

$$
\bar{A}_2 =
\begin{vmatrix}
a_{14} \\
a_{24} \\
a_{34}
\end{vmatrix}
\qquad
\bar{A}_3 = \begin{vmatrix} a_{41} & a_{42} & a_{43} \end{vmatrix}
\qquad
\bar{A}_4 = [a_{44}]
\tag{A.64}
$$

Partitioned matrices follow the rules of matrix addition and subtraction. Partitioned matrices $\bar{A}$ and $\bar{B}$ can be multiplied if these are confirmable and columns of $\bar{A}$ and rows of $\bar{B}$ are partitioned exactly in the same manner:

$$
\begin{vmatrix}
\bar{A}_{11_{2\times2}} & \bar{A}_{12_{2\times1}} \\
\bar{A}_{21_{1\times2}} & \bar{A}_{22_{1\times1}}
\end{vmatrix}
\begin{vmatrix}
\bar{B}_{11_{2\times3}} & \bar{B}_{12_{2\times1}} \\
\bar{B}_{21_{1\times3}} & \bar{B}_{22_{1\times1}}
\end{vmatrix}
=
\begin{vmatrix}
\bar{A}_{11}\bar{B}_{11} + \bar{A}_{12}\bar{B}_{21} & \bar{A}_{11}\bar{B}_{12} + \bar{A}_{12}\bar{B}_{22} \\
\bar{A}_{21}\bar{B}_{11} + \bar{A}_{22}\bar{B}_{21} & \bar{A}_{21}\bar{B}_{12} + \bar{A}_{22}\bar{B}_{22}
\end{vmatrix}
\tag{A.65}
$$

Example A.6

Find the product of two matrices $\bar{A}$ and $\bar{B}$ by partitioning

$$
\bar{A} =
\begin{vmatrix}
1 & 2 & 3 \\
2 & 0 & 1 \\
1 & 3 & 6
\end{vmatrix}
\qquad
\bar{B} =
\begin{vmatrix}
1 & 2 & 1 & 0 \\
2 & 3 & 5 & 1 \\
4 & 6 & 1 & 2
\end{vmatrix}
$$

is given by

$$
\bar{A}\bar{B} =
\begin{vmatrix}
\begin{vmatrix} 1 & 2 \\ 2 & 0 \end{vmatrix}
\begin{vmatrix} 1 & 2 & 1 \\ 2 & 3 & 5 \end{vmatrix}
+
\begin{vmatrix} 3 \\ 1 \end{vmatrix}
\begin{vmatrix} 4 & 6 & 1 \end{vmatrix}
&
\begin{vmatrix} 1 & 2 \\ 2 & 0 \end{vmatrix}
\begin{vmatrix} 0 \\ 1 \end{vmatrix}
+
\begin{vmatrix} 3 \\ 1 \end{vmatrix}
|2| \\
\begin{vmatrix} 1 & 3 \end{vmatrix}
\begin{vmatrix} 1 & 2 & 1 \\ 2 & 3 & 5 \end{vmatrix}
+ |6|
\begin{vmatrix} 4 & 6 & 1 \end{vmatrix}
&
\begin{vmatrix} 1 & 3 \end{vmatrix}
\begin{vmatrix} 0 \\ 1 \end{vmatrix}
+ |6||2|
\end{vmatrix}
$$

Appendix A: Matrix Methods 571

$$
= \begin{Vmatrix} \begin{vmatrix} 5 & 8 & 11 \\ 2 & 4 & 2 \\ 7 & 11 & 16 \end{vmatrix} + \begin{vmatrix} 12 & 18 & 3 \\ 4 & 6 & 1 \\ 24 & 36 & 6 \end{vmatrix} & \begin{vmatrix} 2 \\ 0 \\ 3 \end{vmatrix} + \begin{vmatrix} 6 \\ 2 \\ 12 \end{vmatrix} \end{Vmatrix} = \begin{Vmatrix} \begin{vmatrix} 17 & 26 & 14 \\ 6 & 10 & 3 \\ 31 & 47 & 22 \end{vmatrix} & \begin{vmatrix} 8 \\ 2 \\ 15 \end{vmatrix} \end{Vmatrix} = \begin{vmatrix} 17 & 26 & 14 & 8 \\ 6 & 10 & 3 & 2 \\ 31 & 47 & 22 & 15 \end{vmatrix}
$$

A matrix can be inverted by partition. In this case, each of the diagonal submatrices must be square. Consider a square matrix partitioned into four submatrices:

$$
\bar{A} = \begin{vmatrix} \bar{A}_1 & \bar{A}_2 \\ \bar{A}_3 & \bar{A}_4 \end{vmatrix} \tag{A.66}
$$

The diagonal submatrices $\bar{A}_1$ and $\bar{A}_4$ are square, though these can be of different dimensions. Let the inverse of $\bar{A}$ be

$$
\bar{A}^{-1} = \begin{vmatrix} \bar{A}_1'' & \bar{A}_2'' \\ \bar{A}_3'' & \bar{A}_4'' \end{vmatrix} \tag{A.67}
$$

where each of the double-primed submatrices encompasses the inverse of the same dimensions as its corresponding counterpart.
Then,

$$
\bar{A}^{-1}\bar{A} = \begin{vmatrix} \bar{A}_1'' & \bar{A}_2'' \\ \bar{A}_3'' & \bar{A}_4'' \end{vmatrix} \begin{vmatrix} \bar{A}_1 & \bar{A}_2 \\ \bar{A}_3 & \bar{A}_4 \end{vmatrix} = \begin{vmatrix} 1 & 0 \\ 0 & 1 \end{vmatrix} \tag{A.68}
$$

The following relations can be derived from this identity:

$$
\begin{aligned}
\bar{A}_1'' &= [\bar{A}_1 - \bar{A}_2\bar{A}_4^{-1}\bar{A}_3]^{-1} \\
\bar{A}_2'' &= -\bar{A}_1''\bar{A}_2\bar{A}_4^{-1} \\
\bar{A}_4'' &= [-\bar{A}_3\bar{A}_1^{-1}\bar{A}_2 + \bar{A}_4]^{-1} \\
\bar{A}_3'' &= -\bar{A}_4''\bar{A}_3\bar{A}_1^{-1}
\end{aligned} \tag{A.69}
$$

Example A.7

Invert the following matrix by partitioning:

$$
\bar{A} = \begin{vmatrix} 2 & 3 & 0 \\ 1 & 1 & 3 \\ 1 & 2 & 4 \end{vmatrix}
$$

$$
\bar{A}_1 = \begin{vmatrix} 2 & 3 \\ 1 & 1 \end{vmatrix} \quad \bar{A}_2 = \begin{vmatrix} 0 \\ 3 \end{vmatrix} \quad \bar{A}_3 = \begin{vmatrix} 1 & 2 \end{vmatrix} \quad \bar{A}_4 = |4|
$$

Then,

$$A_1'' = (A_1 - A_2 A_4^{-1} A_3)^{-1}$$

$$A_2'' = -A_1'' A_2 A_4^{-1}$$

$$A_3'' = -A_4'' A_3 A_1^{-1}$$

$$A_4'' = (-A_3 A_1^{-1} A_2 + A_4)^{-1}$$

$$\bar{A}_1'' = \left[\begin{vmatrix} 2 & 3 \\ 1 & 1 \end{vmatrix} - \begin{vmatrix} 0 \\ 3 \end{vmatrix} \begin{Vmatrix} 1 \\ 4 \end{Vmatrix} \begin{matrix} 1 & 2 \end{matrix} \right]^{-1} = \begin{vmatrix} \dfrac{2}{7} & \dfrac{12}{7} \\ \dfrac{1}{7} & -\dfrac{8}{7} \end{vmatrix}$$

$$\bar{A}_2'' = -\begin{vmatrix} \dfrac{2}{7} & \dfrac{12}{7} \\ \dfrac{1}{7} & -\dfrac{8}{7} \end{vmatrix} \begin{vmatrix} 0 \\ 3 \end{vmatrix} \begin{Vmatrix} 1 \\ 4 \end{Vmatrix} = \begin{vmatrix} -\dfrac{9}{7} \\ \dfrac{6}{7} \end{vmatrix}$$

$$\bar{A}_3'' = -\begin{bmatrix} \dfrac{1}{7} \end{bmatrix} \begin{matrix} 1 & 2 \end{matrix} \begin{vmatrix} -1 & 3 \\ 1 & -2 \end{vmatrix} = \begin{vmatrix} -\dfrac{1}{7} & \dfrac{1}{7} \end{vmatrix}$$

$$\bar{A}_4'' = \left[-\begin{matrix} 1 & 2 \end{matrix} \begin{vmatrix} -1 & 3 \\ 1 & -2 \end{vmatrix} \begin{vmatrix} 0 \\ 3 \end{vmatrix} + [4] \right]^{-1} = \dfrac{1}{7}$$

$$\bar{A}^{-1} = \begin{vmatrix} \dfrac{2}{7} & \dfrac{12}{7} & -\dfrac{9}{7} \\ \dfrac{1}{7} & -\dfrac{8}{7} & \dfrac{6}{7} \\ -\dfrac{1}{7} & \dfrac{1}{7} & \dfrac{1}{7} \end{vmatrix}$$

Example A.8

A matrix can be evaluated by elementary row manipulations. As an example, evaluate the following matrix by elementary row operation:

$$A = \begin{vmatrix} 2 & 3 & 0 & 2 \\ 1 & 4 & 3 & 1 \\ 5 & 0 & 2 & 7 \\ 2 & 6 & 0 & 3 \end{vmatrix}$$

$$(1/2)R1 = \begin{vmatrix} 1 & 3/2 & 0 & 1 \\ 1 & 4 & 3 & 1 \\ 5 & 0 & 2 & 7 \\ 2 & 6 & 0 & 3 \end{vmatrix}$$

Appendix A: Matrix Methods 573

$$R2 - R1 = \begin{vmatrix} 1 & 3/2 & 0 & 1 \\ 0 & 5/2 & 3 & 0 \\ 5 & 0 & 2 & 7 \\ 2 & 6 & 0 & 3 \end{vmatrix}$$

$$(2/5)R2 = \begin{vmatrix} 1 & 3/2 & 0 & 1 \\ 0 & 1 & 6/5 & 0 \\ 5 & 0 & 2 & 7 \\ 2 & 6 & 0 & 3 \end{vmatrix}$$

$$R3 - 5R1 = \begin{vmatrix} 1 & 3/2 & 0 & 1 \\ 0 & 1 & 6/5 & 0 \\ 0 & 15/2 & 2 & 2 \\ 2 & 6 & 0 & 3 \end{vmatrix}$$

$$(2/15)R3 + R2 = \begin{vmatrix} 1 & 3/2 & 0 & 1 \\ 0 & 1 & 6/5 & 0 \\ 0 & 0 & 22/15 & 4/15 \\ 2 & 6 & 0 & 3 \end{vmatrix}$$

$$(5/22)R3 = \begin{vmatrix} 1 & 3/2 & 0 & 1 \\ 0 & 1 & 6/5 & 0 \\ 0 & 0 & 1 & 2/11 \\ 2 & 6 & 0 & 3 \end{vmatrix}$$

$$R4 - 2R1 = \begin{vmatrix} 1 & 3/2 & 0 & 1 \\ 0 & 1 & 6/5 & 0 \\ 0 & 0 & 1 & 2/11 \\ 0 & 3 & 0 & 1 \end{vmatrix}$$

$$R4 - 3R2 = \begin{vmatrix} 1 & 3/2 & 0 & 1 \\ 0 & 1 & 6/5 & 0 \\ 0 & 0 & 1 & 2/11 \\ 0 & 0 & -18/5 & 1 \end{vmatrix}$$

$$R4 + 18/5R3 = \begin{vmatrix} 1 & 3/2 & 0 & 1 \\ 0 & 1 & 6/5 & 0 \\ 0 & 0 & 1 & 2/11 \\ 0 & 0 & 0 & 91/55 \end{vmatrix}$$

$$(2)\left(\frac{5}{2}\right)\left(\frac{15}{2}\right)\left(\frac{22}{15}\right) = \begin{vmatrix} 1 & 3/2 & 0 & 1 \\ 0 & 1 & 6/5 & 0 \\ 0 & 0 & 1 & 2/11 \\ 0 & 0 & 0 & 91/55 \end{vmatrix}$$

574   *Appendix A: Matrix Methods*

Therefore, $|A| = 2 \times \dfrac{5}{2} \times \dfrac{15}{2} \times \dfrac{22}{5} \times 1 \times 1 \times 1 \times \dfrac{91}{55} = 91$

A.8 Solution of Large Simultaneous Equations

The application of matrices to the solution of large simultaneous equations constitutes one important application in the power systems. Mostly, these are sparse equations with many coefficients equal to zero. A large power system may have more than 3000 simultaneous equations to be solved.

A.8.1 Consistent Equations

A system of equations is consistent if they have one or more solutions.

A.8.2 Inconsistent Equations

A system of equations that has no solution is called inconsistent, i.e., the following two equations are inconsistent:

$$x + 2y = 4$$
$$3x + 6y = 5$$

A.8.3 Test for Consistency and Inconsistency of Equations

Consider a system of n linear equations:

$$a_{11}x_1 + a_{12}x_2 + \cdots + a_{1n}x_{1n} = b_1$$
$$a_{21}x_1 + a_{22}x_2 + \cdots + a_{2n}x_{2n} = b_2$$
$$\vdots$$
$$a_{m1}x_1 + a_{m2}x_2 + \cdots + a_{mn}x_{mn} = b_n$$

        (A.70)

Form an augmented matrix $\bar{C}$,

$$\bar{C} = [\bar{A},\ \bar{B}] = \begin{vmatrix} a_{11} & a_{12} & \cdot & a_1 n & b_1 \\ a_{21} & a_{22} & \cdot & a_2 n & b_2 \\ \cdot & \cdot & \cdot & \cdot & \cdot \\ a_{n1} & a_{n2} & \cdot & a_{nn} & b_n \end{vmatrix}$$

        (A.71)

The following holds for the test of consistency and inconsistency:

- A unique solution of the equations exists if rank of $\bar{A}$ = rank of $\bar{C} = n$, where n is the number of unknowns.
- There are infinite solutions to the set of equations if rank of $\bar{A}$ = rank of $\bar{C} = r$, $r < n$.
- The equations are inconsistent if rank of $\bar{A}$ is not equal to rank of $\bar{C}$.

Appendix A: Matrix Methods 575

Example A.9

Show that the equations

$$\begin{aligned} 2x + 6y &= -11 \\ 6x + 20y - 6z &= -3 \\ 6y - 18z &= -1 \end{aligned}$$

are inconsistent.

The augmented matrix is

$$\bar{C} = \bar{A}\bar{B} = \begin{vmatrix} 2 & 6 & 0 & -11 \\ 6 & 20 & -6 & -3 \\ 0 & 6 & -18 & -1 \end{vmatrix}$$

It can be reduced by elementary row operations to the following matrix:

$$\begin{vmatrix} 2 & 6 & 0 & -11 \\ 0 & 2 & -6 & 30 \\ 0 & 0 & 0 & -91 \end{vmatrix}$$

The rank of A is 2 and that of C is 3. The equations are not consistent.

Equation A.70 can be written as

$$\bar{A}\bar{x} = \bar{b} \tag{A.72}$$

where $\bar{A}$ is a square coefficient matrix, $\bar{b}$ is a vector of constants, and $\bar{x}$ is a vector of unknown terms. If $\bar{A}$ is nonsingular, the unknown vector $\bar{x}$ can be found by

$$\bar{x} = \bar{A}^{-1}\bar{b} \tag{A.73}$$

This requires calculation of the inverse of matrix $\bar{A}$. Large system equations are not solved by direct inversion but by a sparse matrix technique.

Example A.10

This example illustrates the solution by transforming the coefficient matrix to an upper triangular form (backward substitution). The equations

$$\begin{vmatrix} 1 & 4 & 6 \\ 2 & 6 & 3 \\ 5 & 3 & 1 \end{vmatrix} \begin{vmatrix} x_1 \\ x_2 \\ x_3 \end{vmatrix} = \begin{vmatrix} 2 \\ 1 \\ 5 \end{vmatrix}$$

can be solved by row manipulations on the augmented matrix as follows:

$$\begin{vmatrix} 1 & 4 & 6 & 2 \\ 2 & 6 & 3 & 1 \\ 5 & 3 & 1 & 5 \end{vmatrix} \to R_2 - 2R_1 = \begin{vmatrix} 1 & 4 & 6 & 2 \\ 0 & -2 & -9 & -3 \\ 5 & 3 & 1 & 5 \end{vmatrix} \to R_3 - 5R_1$$

$$= \begin{vmatrix} 1 & 4 & 6 \\ 0 & -2 & -9 \\ 0 & -17 & -29 \end{vmatrix} \begin{vmatrix} 2 \\ -3 \\ -5 \end{vmatrix} \rightarrow R_3 - \frac{17}{2} R_2 = \begin{vmatrix} 1 & 4 & 6 \\ 0 & -2 & -9 \\ 0 & 0 & 47.5 \end{vmatrix} \begin{vmatrix} 2 \\ -3 \\ 20.5 \end{vmatrix}$$

Thus,

$$47.5 x_3 = 20.5$$

$$-2 x_2 - 9 x_3 = -3$$

$$x_1 + 4 x_2 + 6 x_3 = 2$$

which gives

$$\bar{x} = \begin{vmatrix} 1.179 \\ -0.442 \\ 0.432 \end{vmatrix}$$

A set of simultaneous equations can also be solved by partitioning:

$$\begin{vmatrix} a_{11}, \ldots, a_{1k} & a_{1m}, \ldots, a_{1n} \\ \vdots & \vdots \\ a_{k1}, \ldots, a_{kk} & a_{km}, \ldots, a_{kn} \\ a_{m1}, \ldots, a_{mk} & a_{mm}, \ldots, a_{mn} \\ \vdots & \vdots \\ a_{n1}, \ldots, a_{nk} & a_{nm}, \ldots, a_{nn} \end{vmatrix} \begin{vmatrix} x_1 \\ \cdot \\ x_k \\ x_m \\ \cdot \\ x_n \end{vmatrix} = \begin{vmatrix} b_1 \\ \cdot \\ b_k \\ b_m \\ \cdot \\ b_n \end{vmatrix} \tag{A.74}$$

Equation A.74 is horizontally partitioned and rewritten as

$$\begin{vmatrix} \bar{A}_1 & \bar{A}_2 \\ \bar{A}_3 & \bar{A}_4 \end{vmatrix} = \begin{vmatrix} \bar{X}_1 \\ \bar{X}_2 \end{vmatrix} \begin{vmatrix} \bar{B}_1 \\ \bar{B}_2 \end{vmatrix} \tag{A.75}$$

Vectors $\bar{X}_1$ and $\bar{X}_2$ are given by

$$\bar{X}_1 = \left[\bar{A}_1 - \bar{A}_2 \bar{A}_4^{-1} \bar{A}_3 \right]^{-1} \left[\bar{B}_1 - \bar{A}_2 \bar{A}_4^{-1} \bar{B}_2 \right] \tag{A.76}$$

$$\bar{X}_2 = \left[\bar{A}_4^{-1} (\bar{B}_2 - \bar{A}_3 \bar{X}_1) \right] \tag{A.77}$$

A.9 Crout's Transformation

A matrix can be resolved into the product of a lower triangular matrix $\bar{L}$ and an upper unit triangular matrix $\bar{U}$, i.e.,

Appendix A: Matrix Methods 577

$$\begin{vmatrix} a_{11} & a_{12} & a_{13} & a_{14} \\ a_{21} & a_{22} & a_{23} & a_{24} \\ a_{31} & a_{32} & a_{33} & a_{34} \\ a_{41} & a_{42} & a_{43} & a_{44} \end{vmatrix} = \begin{vmatrix} l_{11} & 0 & 0 & 0 \\ l_{21} & l_{22} & 0 & 0 \\ l_{31} & l_{32} & l_{33} & 0 \\ l_{41} & l_{42} & l_{43} & l_{44} \end{vmatrix} \begin{vmatrix} 1 & u_{12} & u_{13} & u_{14} \\ 0 & 1 & u_{23} & u_{24} \\ 0 & 0 & 1 & u_{34} \\ 0 & 0 & 0 & 1 \end{vmatrix} \tag{A.78}$$

The elements of $\bar{U}$ and $\bar{L}$ can be found by multiplication:

$$l_{11} = a_{11}$$
$$l_{21} = a_{21}$$
$$l_{22} = a_{22} - l_{21}u_{12}$$
$$l_{31} = a_{31}$$
$$l_{32} = a_{32} - l_{31}u_{12}$$
$$l_{33} = a_{33} - l_{31}u_{13} - l_{32}u_{23}$$
$$l_{41} = a_{41}$$
$$l_{42} = a_{42} - l_{41}u_{12}$$
$$l_{43} = a_{43} - l_{41}u_{13} - l_{42}u_{23}$$
$$l_{44} = a_{44} - a_{41}u_{14} - l_{42}u_{24} - l_{43}u_3 \tag{A.79}$$

and

$$u_{12} = a_{12}/l_{11}$$
$$u_{13} = a_{13}/l_{11}$$
$$u_{14} = a_{14}/l_{11}$$
$$u_{23} = (a_{23} - l_{21}u_{13})/l_{22}$$
$$u_{24} = (a_{24} - l_{21}u_{14})/l_{22}$$
$$u_{34} = (a_{34} - l_{31}u_{14} - l_{32}u_{24})l_{33} \tag{A.80}$$

In general,

$$l_{ij} = a_{ij} - \sum_{k=1}^{k=j-1} l_{ik}u_{kj} \quad i \geq j \tag{A.81}$$

For $j = 1, \ldots, n$

$$u_{ij} = \frac{1}{l_{ii}}\left(a_{ij} - \sum_{k=1}^{k=j-1} l_{ik}u_{kj}\right) \quad i < j \tag{A.82}$$

Example A.11

Transform the following matrix into lower upper (LU) form:

$$
\begin{vmatrix}
1 & 2 & 1 & 0 \\
0 & 3 & 3 & 1 \\
2 & 0 & 2 & 0 \\
1 & 0 & 0 & 2
\end{vmatrix}
$$

From Equations A.81 and A.82,

$$
\begin{vmatrix}
1 & 2 & 1 & 0 \\
0 & 3 & 3 & 1 \\
2 & 0 & 2 & 0 \\
1 & 0 & 0 & 2
\end{vmatrix}
=
\begin{vmatrix}
1 & 0 & 0 & 0 \\
0 & 3 & 0 & 0 \\
2 & -4 & 4 & 0 \\
1 & -2 & 1 & 2.33
\end{vmatrix}
\begin{vmatrix}
1 & 2 & 1 & 0 \\
0 & 1 & 1 & 0.33 \\
0 & 0 & 1 & 0.33 \\
0 & 0 & 0 & 1
\end{vmatrix}
$$

The original matrix has been converted into a product of lower and upper triangular matrices.

A.10 Gaussian Elimination

Gaussian elimination provides a natural means to determine the LU pair:

$$
\begin{vmatrix}
a_{11} & a_{12} & a_{13} \\
a_{21} & a_{22} & a_{23} \\
a_{31} & a_{32} & a_{33}
\end{vmatrix}
\begin{vmatrix}
x_1 \\
x_2 \\
x_3
\end{vmatrix}
=
\begin{vmatrix}
b_1 \\
b_2 \\
b_3
\end{vmatrix}
\tag{A.83}
$$

First, form an augmented matrix:

$$
\begin{vmatrix}
a_{11} & a_{12} & a_{13} & b_1 \\
a_{21} & a_{22} & a_{23} & b_2 \\
a_{31} & a_{32} & a_{33} & b_3
\end{vmatrix}
\tag{A.84}
$$

1. Divide the first row by a_{11}. This is the only operation to be carried out on this row. Thus, the new row is

$$
\begin{array}{cccc}
1 & a'_{12} & a'_{13} & b'_1 \\
\end{array}
$$
$$
a'_{12} = a_{12}/a_{11}, \ a'_{13} = a_{13}/a_{11}, \ b'_1 = b_1/a_{11}
\tag{A.85}
$$

This gives

$$
l_{11} = a_{11}, \ u_{11} = 1, \ u_{12} = a'_{12}, \ u_{13} = a'_{13}
\tag{A.86}
$$

Appendix A: Matrix Methods 579

2. Multiply new row 1 by $-a_{21}$ and add to row 2. Thus, a_{21} becomes zero.

$$0 \quad a'_{22} \quad a'_{23} \quad a'_{33}b'_2$$

$$a'_{22} = a_{22} - a_{21}a'_{12}$$
$$a'_{23} = a_{23} - a_{21}a'_{13}$$
$$b' = b_2 - a_{21}b' \tag{A.87}$$

Divide new row 2 by a'_{22}. Row 2 becomes

$$0 \quad 1 \quad a''_{23} \quad b''_2$$

$$a''_{23} = a'_{23}/a'_{22}$$
$$b''_2 = b'_2/a'_{22} \tag{A.88}$$

This gives

$$l_{21} = a_{21}, l_{22} = a'_{22}, u_{22} = 1, u_{23} = a'_{23} \tag{A.89}$$

3. Multiply new row 1 by $-a_{31}$ and add to row 3. Thus, row 3 becomes

$$0 \quad a'_{32} \quad a'_{33}b'_3$$

$$a'_{32} = a_{32} - a_{32}a'_{12}$$
$$a'_{33} = a_{33} - a_{31}a'_{13} \tag{A.90}$$

Multiply row 2 by $-a_{32}$ and add to row 3. This row now becomes

$$0 \quad 0 \quad a''_{33} \quad b''_3 \tag{A.91}$$

Divide new row 3 by a''_{33}. This gives

$$0 \quad 0 \quad 1 \quad b'''_3$$

$$b'''_3 = b''_3/a''_{33} \tag{A.92}$$

From these relations,

$$l_{33} = a''_{33}, l_{31} = a_{31}, l_{32} = a'_{32}, u_{33} = 1 \tag{A.93}$$

Thus, all the elements of $\bar{L}, \bar{U}$ have been calculated and the process of forward substitution has been implemented on vector $\bar{b}$.

580 *Appendix A: Matrix Methods*

A.11 Forward and Backward Substitution Method

The set of sparse linear equations,

$$\bar{A}\bar{x} = \bar{b} \tag{A.94}$$

can be written as

$$\bar{L}\bar{U}\bar{x} = \bar{b} \tag{A.95}$$

or

$$\bar{L}\bar{y} = \bar{b} \tag{A.96}$$

where

$$\bar{y} = \bar{U}\bar{x} \tag{A.97}$$

$\bar{L}\bar{y} = \bar{b}$ is solved for $\bar{y}$ by forward substitution. Thus, $\bar{y}$ is known. Then $\bar{U}\bar{x} = \bar{y}$ is solved by backward substitution.

Solve $\bar{L}\bar{y} = \bar{b}$ by forward substitution:

$$\begin{vmatrix} l_{11} & 0 & 0 & 0 \\ l_{21} & l_{22} & 0 & 0 \\ l_{31} & l_{32} & l_{33} & 0 \\ l_{41} & l_{42} & l_{43} & l_{44} \end{vmatrix} \begin{vmatrix} y_1 \\ y_2 \\ y_3 \\ y_4 \end{vmatrix} = \begin{vmatrix} b_1 \\ b_2 \\ b_3 \\ b_4 \end{vmatrix} \tag{A.98}$$

Thus,

$$\begin{aligned} y_1 &= b_1/l_{11} \\ y_2 &= (b_2 - l_{21}y_1)/l_{22} \\ y_3 &= (b_3 - l_{31}y_1 - l_{32}y_2)/l_{33} \\ y_4 &= (b_4 - l_{41}y_1 - l_{42}y_2 - l_{43}y_3)/l_{44} \end{aligned} \tag{A.99}$$

Now solve $\bar{U}\bar{x} = \bar{y}$ by backward substitution:

$$\begin{vmatrix} 1 & u_{12} & u_{13} & u_{14} \\ 0 & 1 & u_{23} & u_{24} \\ 0 & 0 & 1 & u_{34} \\ 0 & 0 & 0 & 1 \end{vmatrix} \begin{vmatrix} x_1 \\ x_2 \\ x_3 \\ x_4 \end{vmatrix} = \begin{vmatrix} y_1 \\ y_2 \\ y_3 \\ y_4 \end{vmatrix} \tag{A.100}$$

Appendix A: Matrix Methods 581

Thus,

$$x_4 = y_4$$
$$x_3 = y_3 - u_{34}x_4$$
$$x_2 = y_2 - u_{23}x_3 - u_{24}x_4$$
$$x_1 = y_1 - u_{12}x_2 - u_{13}x_3 - u_{14}x_4 \tag{A.101}$$

The forward–backward solution is generalized by the following equation:

$$\bar{A} = \bar{L}\bar{U} = (\bar{L}_d + \bar{L}_1)(\bar{I} + \bar{U}_u) \tag{A.102}$$

where $\bar{L}_d$ is the diagonal matrix, $\bar{L}_1$ is the lower triangular matrix, $\bar{I}$ is the identity matrix, and $\bar{U}_u$ is the upper triangular matrix.

Forward substitution becomes

$$\bar{L}\bar{y} = \bar{b}$$
$$(\bar{L}_d + \bar{L}_1)\bar{y} = \bar{b}$$
$$\bar{L}_d\bar{y} = \bar{b} - \bar{L}_1\bar{y} \tag{A.103}$$
$$\bar{y} = \bar{L}_d^{-1}(\bar{b} - \bar{L}_1\bar{y})$$

i.e.,

$$\begin{vmatrix} y_1 \\ y_2 \\ y_3 \\ y_4 \end{vmatrix} = \begin{vmatrix} 1/l_{11} & 0 & 0 & 0 \\ 0 & 1/l_{22} & 0 & 0 \\ 0 & 0 & 1/l_{33} & 0 \\ 0 & 0 & 0 & 1/l_{44} \end{vmatrix} \times \left[\begin{vmatrix} b_1 \\ b_2 \\ b_3 \\ b_4 \end{vmatrix} - \begin{vmatrix} 0 & 0 & 0 & 0 \\ l_{21} & 0 & 0 & 0 \\ l_{31} & l_{32} & 0 & 0 \\ l_{41} & l_{42} & l_{43} & l_{44} \end{vmatrix} \begin{vmatrix} y_1 \\ y_2 \\ y_3 \\ y_4 \end{vmatrix} \right] \tag{A.104}$$

Backward substitution becomes

$$(\bar{I} + \bar{U}_u)\bar{x} = \bar{y}$$
$$\bar{x} = \bar{y} - \bar{U}_u\bar{x} \tag{A.105}$$

i.e.,

$$\begin{vmatrix} x_1 \\ x_2 \\ x_3 \\ x_4 \end{vmatrix} = \begin{vmatrix} y_1 \\ y_2 \\ y_3 \\ y_4 \end{vmatrix} - \begin{vmatrix} 0 & u_{12} & u_{13} & u_{14} \\ 0 & 0 & u_{23} & u_{24} \\ 0 & 0 & 0 & u_{34} \\ 0 & 0 & 0 & 0 \end{vmatrix} \begin{vmatrix} x_1 \\ x_2 \\ x_3 \\ x_4 \end{vmatrix} \tag{A.106}$$

A.11.1 Bifactorization

A matrix can also be split into LU form by sequential operation on the columns and rows. The general equations of the bifactorization method are

$$l_{ip} = a_1 p \quad \text{for} \geq p$$

$$u_{pj} = \frac{a_{pj}}{a_{pp}} \quad \text{for } j > p \tag{A.107}$$

$$a_{ij} = a_1 j - l_{ip}u_{pj} \quad i > p, j > p$$

Here, the letter p means the path or the pass. This will be illustrated with an example.

Example A.12

Consider the matrix:

$$\bar{A} = \begin{vmatrix} 1 & 2 & 1 & 0 \\ 0 & 3 & 3 & 1 \\ 2 & 0 & 2 & 0 \\ 1 & 0 & 0 & 2 \end{vmatrix}$$

It is required to convert it into LU form. This is the same matrix of Example A.10.

Add an identity matrix, which will ultimately be converted into a U matrix and the $\bar{A}$ matrix will be converted into an L matrix:

$$\begin{vmatrix} 1 & 2 & 1 & 0 \\ 0 & 3 & 3 & 1 \\ 2 & 0 & 2 & 0 \\ 1 & 0 & 0 & 2 \end{vmatrix} \begin{vmatrix} 1 & 0 & 0 & 0 \\ 0 & 1 & 0 & 0 \\ 0 & 0 & 1 & 0 \\ 0 & 0 & 0 & 1 \end{vmatrix}$$

Step, $p = 1$:

1					1	2	1	0
0	3	3	0		0	1	0	0
2	−4	0	0		0	0	1	0
1	−2	−1	2		0	0	1	0

The shaded columns and rows are converted into L and U matrix column and row and the elements of $\bar{A}$ matrix are modified using Equation A.107, i.e.,

$$a_{32} = a_{32} - l_{31}u_{12}$$
$$= 0 - (2)(2) = -4$$
$$a_{33} = a_{33} - l_{31}u_{31}$$
$$= 2 - (2)(1) = 0$$

Appendix A: Matrix Methods 583

Step 2, pivot column 2, $p = 2$:

1			
0	3	3	0
2	−4	4	1.32
1	−2	1	2.66

1	2	1	0
0	1	1	0.33
0	0	1	0
0	0	0	1

Step 3, pivot column 3, $p = 3$:

1	0	0	0
0	3	0	0
2	−4	4	0
1	−2	1	2.33

1	2	1	0
0	1	1	0.33
0	0	1	0.33
0	0	0	1

This is the same result as derived before in Example A.10.

A.12 LDU Product Form, Cascade or Choleski Form

The individual terms of L, D, and U can be found by direct multiplication. Again, consider a 4×4 matrix:

$$
\begin{vmatrix} a_{11} & a_{12} & a_{13} & a_{14} \\ a_{21} & a_{22} & a_{23} & a_{24} \\ a_{31} & a_{32} & a_{33} & a_{34} \\ a_{41} & a_{42} & a_{43} & a_{44} \end{vmatrix} = \begin{vmatrix} 1 & 0 & 0 & 0 \\ l_{21} & 1 & 0 & 0 \\ l_{31} & l_{32} & 1 & 0 \\ l_{41} & l_{42} & l_{43} & 1 \end{vmatrix} \begin{vmatrix} d_{11} & 0 & 0 & 0 \\ 0 & d_{22} & 0 & 0 \\ 0 & 0 & d_{33} & 0 \\ 0 & 0 & 0 & d_{44} \end{vmatrix} \begin{vmatrix} 1 & u_{12} & u_{13} & u_{14} \\ 0 & 1 & u_{23} & u_{24} \\ 0 & 0 & 1 & u_{34} \\ 0 & 0 & 0 & 1 \end{vmatrix} \quad \text{(A.108)}
$$

The following relations exist:

$$
\begin{aligned}
d_{11} &= a_{11} \\
d_{22} &= a_{22} - l_{21}d_{11}u_{12} \\
d_{33} &= a_{33} - l_{31}d_{11}u_{13} - l_{32}d_{22}u_{23} \\
d_{44} &= a_{44} - l_{41}d_{11}u_{14} - l_{42}d_{22}u_{24} - l_{43}d_{33}u_{34} \\
u_{12} &= a_{12}/d_{11} \\
u_{13} &= a_{13}/d_{11} \\
u_{14} &= a_{14}/d_{11} \\
u_{23} &= (a_{23} - l_{21}d_{11}u_{13})/d_{22} \\
u_{24} &= (a_{24} - l_{21}d_{11}u_{14})/d_{22} \\
u_{34} &= (a_{34} - l_{31}d_{11}u_{14} - l_{32}d_{22}u_{24})/d_{33} \\
l_{21} &= a_{21}/d_{11}
\end{aligned}
$$

$$l_{31} = a_{31}/d_{11}$$
$$l_{32} = (a_{32} - l_{31}d_{11}u_{12})/d_{22}$$
$$l_{41} = a_{41}/d_{11}$$
$$l_{42} = (a_{42} - l_{41}d_{11}u_{12})/d_{22}$$
$$l_{43} = (a_{43} - l_{41}d_{11}u_{13} - l_{42}d_{22}u_{23})/d_{33} \tag{A.109}$$

In general,

$$d_{ii} = a_{11} - \sum_{j=1}^{i=1} l_{ij}d_{jj}u_{ji} \quad \text{for} \quad i = 1,2,\ldots,n$$

$$u_{ik} = \left[a_{ik} - \sum_{j=1}^{i=1} l_{if}d_{jj}u_{jk} \right] / d_{ii} \quad \text{for } k = i+1\ldots,n \quad i = 1,2,\ldots,r \tag{A.110}$$

$$l_{ki} = \left[a_{ki} - \sum_{j=1}^{i=1} l_{kj}d_{jj}u_{ji} \right] / d_{ii} \quad \text{for } k = i+1,\ldots,n \quad i = 1,2,\ldots,r$$

Another scheme is to consider A as a product of sequential lower and upper matrices as follows:

$$A = (L_1 L_2, \ldots, L_n)(U_n, \ldots, U_2 U_1) \tag{A.111}$$

$$\begin{vmatrix} a_{11} & a_{12} & a_{13} & a_{14} \\ a_{21} & a_{22} & a_{23} & a_{24} \\ a_{31} & a_{32} & a_{33} & a_{34} \\ a_{41} & a_{42} & a_{43} & a_{44} \end{vmatrix} = \begin{vmatrix} l_{11} & 0 & 0 & 0 \\ l_{21} & 1 & 0 & 0 \\ l_{31} & 0 & 1 & 0 \\ l_{41} & 0 & 0 & 1 \end{vmatrix} \begin{vmatrix} 1 & 0 & 0 & 0 \\ 0 & a_{22_2} & a_{23_2} & a_{24_2} \\ 0 & a_{32_2} & a_{33_2} & a_{34_2} \\ 0 & a_{42_2} & a_{43_2} & a_{44_2} \end{vmatrix} \begin{vmatrix} 1 & u_{12} & u_{13} & u_{14} \\ 0 & 1 & 0 & 0 \\ 0 & 0 & 1 & 0 \\ 0 & 0 & 0 & 1 \end{vmatrix} \tag{A.112}$$

Here, the second step elements are denoted by subscript 2 to the subscript.

$$l_{21} = a_{21} \quad l_{31} = a_{31} \quad l_{41} = a_{41}$$
$$u_{12} = a_{12}/l_{11} \quad u_{13} = a_{13}/l_{11} \quad u_{14} = a_{14}/l_{11} \tag{A.113}$$
$$a_{ij_2} = a_{1j} - l_{1i}u_{1j} \quad i, j = 2,3,4$$

All elements correspond to Step 1, unless indicated by subscript 2.

In general for the kth step:

$$d_{kk}^k = a_{kk}^k \quad k = 1,2,\ldots,n-1$$
$$l_{ik}^k = a_{ik}^k/a_{kk}^k$$
$$u_{kj} = a_{kj}^k/a_{kk}^k \tag{A.114}$$
$$a_{ij}^{k+1} = (a_{ij}^k - a_{ik}^k a_{kj}^k)/a_{kk}^k$$
$$k = 1,2,\ldots,n-1i, j = k+1,\ldots,n$$

Appendix A: Matrix Methods 585

Example A.13

Convert the matrix of Example A.10 into LDU form:

$$\begin{vmatrix} 1 & 2 & 1 & 0 \\ 0 & 3 & 3 & 1 \\ 2 & 0 & 2 & 0 \\ 1 & 0 & 0 & 2 \end{vmatrix} = l^1 \times l^2 \times l^3 \times D \times u^3 \times u^2 \times u^1$$

The lower matrices are

$$l^1 \times l^2 \times l^3 = \begin{vmatrix} 1 & 0 & 0 & 0 \\ 0 & 1 & 0 & 0 \\ 2 & 0 & 1 & 0 \\ 1 & 0 & 0 & 1 \end{vmatrix} \begin{vmatrix} 1 & 0 & 0 & 0 \\ 0 & 1 & 0 & 0 \\ 0 & -4/3 & 1 & 0 \\ 1 & -2/3 & 0 & 1 \end{vmatrix} \begin{vmatrix} 1 & 0 & 0 & 0 \\ 0 & 1 & 0 & 0 \\ 0 & 0 & 1 & 0 \\ 0 & 0 & 1/4 & 0 \end{vmatrix}$$

The upper matrices are

$$u^3 \times u^2 \times u^1 = \begin{vmatrix} 1 & 0 & 0 & 0 \\ 0 & 1 & 0 & 1/3 \\ 0 & 0 & 1 & 1/3 \\ 0 & 0 & 0 & 1 \end{vmatrix} \begin{vmatrix} 1 & 0 & 0 & 0 \\ 0 & 1 & 1 & 0 \\ 0 & 0 & 1 & 0 \\ 0 & 0 & 0 & 1 \end{vmatrix} \begin{vmatrix} 1 & 2 & 1 & 0 \\ 0 & 1 & 0 & 0 \\ 0 & 0 & 1 & 0 \\ 0 & 0 & 0 & 1 \end{vmatrix}$$

The matrix D is

$$D = \begin{vmatrix} 1 & 0 & 0 & 0 \\ 0 & 3 & 0 & 0 \\ 0 & 0 & 4 & 0 \\ 0 & 0 & 0 & 7/3 \end{vmatrix}$$

Thus, the LDU form of the original matrix is

$$\begin{vmatrix} 1 & 0 & 0 & 0 \\ 0 & 1 & 0 & 0 \\ 2 & -4/3 & 1 & 0 \\ 1 & -2/3 & 1/4 & 1 \end{vmatrix} \begin{vmatrix} 1 & 0 & 0 & 0 \\ 0 & 3 & 0 & 0 \\ 0 & 0 & 4 & 0 \\ 0 & 0 & 0 & 7/3 \end{vmatrix} \begin{vmatrix} 1 & 2 & 1 & 0 \\ 0 & 1 & 1 & 1/3 \\ 0 & 0 & 1 & 1/3 \\ 0 & 0 & 0 & 1 \end{vmatrix}$$

If the coefficient matrix is symmetrical (for a linear bilateral network), then

$$[L] = [U]^t \tag{A.115}$$

Because

$$l_{ip}(\text{new}) = \frac{a_{ip}}{a_{pp}}$$

$$u_{pi} = \frac{a_{pi}}{a_{pp}(a_{ip} = a_{pi})} \tag{A.116}$$

The LU and LDU forms are extensively used in power systems. The bibliography provides further reading.

Bibliography

HE Brown. *Solution of Large Networks by Matrix Methods*. New York: Wiley Interscience, 1975.

PL Corbeiller. *Matrix Analysis of Electrical Networks*. Cambridge, MA: Harvard University Press, 1950.

WE Lewis, DG Pryce. *The Application of Matrix Theory to Electrical Engineering*. London: E. & F.N. Spon, 1965.

RB Shipley. *Introduction to Matrices and Power Systems*. New York: John Wiley & Sons, 1976.

SA Stignant. *Matrix and Tensor Analysis in Electrical Network Theory*. London: Macdonald, 1964.

Appendix B: Sparsity and Optimal Ordering

It is seen that linear simultaneous equations representing a power system are sparse. These give sparse matrices. Consider a 500-node system and assuming that, on average, there are two bus connections to a node; the admittance matrix A will have 500 diagonal elements and $500 \times 2 \times 2 = 2000$ off-diagonal elements. The total number of elements in A are $500 \times 500 = 250{,}000$. The population of nonzero elements is, therefore, 1.0%. The assumptions of two buses per node in a transmission network are high. Typically, it will be 1–1.5, further reducing the percentage of nonzero elements.

When the matrix is factorized in lower upper (LU) or lower, diagonal and upper (LDU) matrices form, nonzero elements are created where none existed before:

$$a_{ij}(\text{new}) = a_{ij}(\text{primitive}) - \frac{a_{ip} a_{pj}}{a_{pp}} \qquad (B.1)$$

If a_{ij} is zero in the original matrix at the beginning of the pth step and both a_{ip} and a_{pj} are nonzero in value, then a_{ij} (new) becomes a new nonzero term.

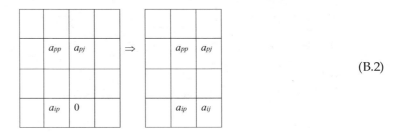

$$(B.2)$$

B.1 Optimal Ordering

Optimal ordering is the rearrangement of the order of sequence of columns and rows in the matrix to minimize nonzero terms. When a programming scheme is used that stores and processes only nonzero terms, saving in operations and computer memory can be achieved by keeping the stored tables of factors as short as possible. A truly optimal ordering scheme should produce the fewest nonzero elements. For large systems, the determination of such an optimal ordering itself may require a long computer time, which may not be justifiable practically, for the benefits in reducing the nonzero elements. An effective algorithm for absolute optimal ordering has not been developed; however, several schemes give near optimal solutions.

B.2 Flow Graphs

Power system matrices are symmetrical and, therefore, their structures can be described by flow graphs. Optimal elimination then translates into a topological problem. The following theorems can be postulated:

1. Any graph that is obtained from another graph by a node-eliminating process is independent of the order in which the nodes are eliminated
2. When any two successive nodes are adjacent in the original graph or in the graph generated by the elimination process, the node with the smaller *valence* occurs first.

The valency of a node connected in a graph is the number of new paths created or added as a result of elimination of the node. Hence, the valency of the node is defined as the number of new links added to the graph, i.e., the new nonzero elements generated in the coefficient matrix because of elimination of the node.

In a connected graph, the nodes communicate between each other. A node can be eliminated if a path exists and the flow in the graph is not interrupted. In the graph of Figure B.1, node 14 can be eliminated. This will not add any new branches nor any new nonzero elements. Similarly, node 13 can be eliminated, as a path exists between node 12 and 5, and, in the coefficient matrix, element $a_{12,5}$ will be nonzero. However, elimination of node 4 will add a new link between 3 and 5, and hence, the element $a_{3,5}$ that was zero earlier will appear as a nonzero element in the matrix. Elimination of node 7 adds four new links, as shown in Figure B.1. The new links are shown dotted.

Consider the network of Figure B.2 and the ordered matrix as shown below with nonzero terms. The ordering here is 1, 2, 3, 4, and 5:

	1	2	3	4	5
1	X	X		X	
2	X	X	X		X
3		X	X		
4	X			X	X
5		X		X	X

(B.3)

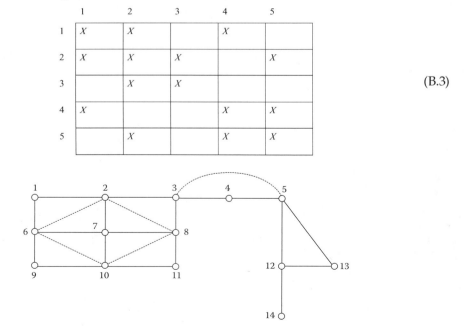

FIGURE B.1
A network for node elimination.

Appendix B: Sparsity and Optimal Ordering

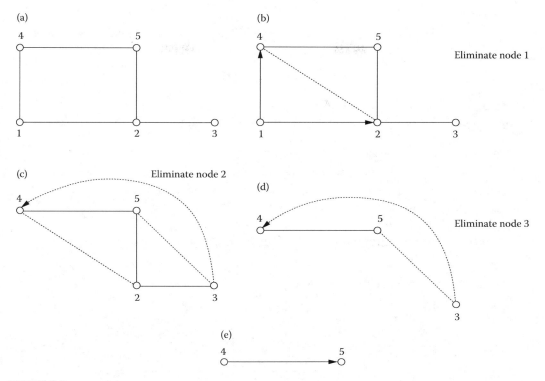

FIGURE B.2
(a) Graph of a five-node network and (b) through (e) successive eliminations of nodes 1, 2, 3, and 4.

Eliminate node 1 (Figure B.2). Two new nonzero elements are introduced:

$$\begin{vmatrix} X & X & & X & \\ X & X & X & X_{new} & X \\ & X & X & & \\ X & X_{new} & & X & X \\ & X & & X & X \end{vmatrix} \quad (B.4)$$

Eliminate node 2 and four new nonzero elements are produced. The matrix, after elimination of node 1 and introduction of new elements due to elimination of node 2, is given as

$$\begin{vmatrix} X & X & X_{new} & X \\ X & X & X_{new} & X_{new} \\ X_{new} & X_{new} & X & X \\ X & X_{new} & X & X \end{vmatrix} \quad (B.5)$$

Eliminate node 3. No new nonzero element is produced. There is already a path between 3 and 5 (dotted). The modified matrix after elimination of node 2 is given as

$$\begin{vmatrix} X & X_{new} & X_{new} \\ X_{new} & X & X \\ X_{new} & X & X \end{vmatrix} \quad (B.6)$$

Node 3 followed by node 4 can now be eliminated.

Now consider the reordered scheme of elimination, as shown in Figure B.3.

1. Node 3 is first eliminated; no new link or nonzero element is created.
2. Node 4 is next eliminated; creates two new nonzero elements.
3. Node 5 is next eliminated; creates no new nonzero element.
4. Node 2 can now be eliminated.

This gives the ordered original matrix as 3, 4, 5, 2, and 1:

	3	4	5	2	1
3	X			X	
4		X	X		X
5		X	X	X	
2	X		X	X	X
1		X		X	X

(B.7)

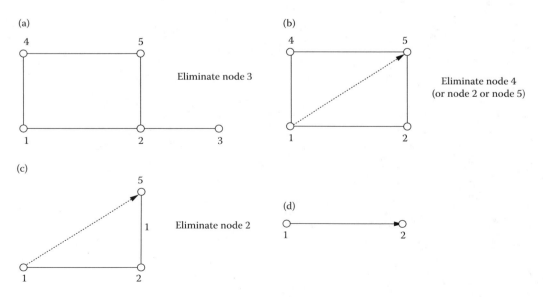

FIGURE B.3
(a) through (d) Graphs with alternate order of node elimination.

Appendix B: Sparsity and Optimal Ordering 591

The number of nonzero elements created at the pth step is given as

$$N_{\text{nonzero}} = \frac{n(n-1)}{2} \tag{B.8}$$

where n is the number of nodes directly connected to the pivot node, but not directly connected to each other.

B.3 Optimal Ordering Schemes

A number of near optimal schemes have been developed:

B.3.1 Scheme 1

The rows of the coefficient matrix are numbered according to the number of nonzero off-diagonal elements, before elimination. The rows with the least number of off-diagonal terms comes first, followed by the next rows in ascending order of nonzero off-diagonal terms. This method is simple and fast to execute, but does not give the minimum number of nonzero terms. In other words, the scheme is to reorder the nodes so that the number of connecting branches to each node is in ascending order in the original network.

B.3.2 Scheme 2

The rows in the coefficient matrix are numbered so that, at each step, the next row to be operated upon is the one with fewest nonzero terms. If more than one row meets this criterion, select any one first. Thus, the nodes are numbered, so that at each step of elimination, the next node to be eliminated is the one having the fewest connected branches, i.e., the minimum degree. This method requires simulation of the elimination process to take into account the changes in the branch connections as each node is eliminated.

B.3.3 Scheme 3

At each step of the elimination process select the node that produces the smallest number of new branches. From the network point of view, at each step, the next node to be eliminated is the one that will produce the fewest row equivalents of every feasible alternative.

B.3.4 Scheme 4

If at any stage of elimination, more than one node has the same degree, then remove the one that creates the minimum off-diagonal zero terms, i.e., new links in the system graph. This exploits the merits of Schemes 2 and 3.

Consider the network of Figure B.4a. Its matrix without prior ordering, in terms of numbered nodes in serial ascending order, is

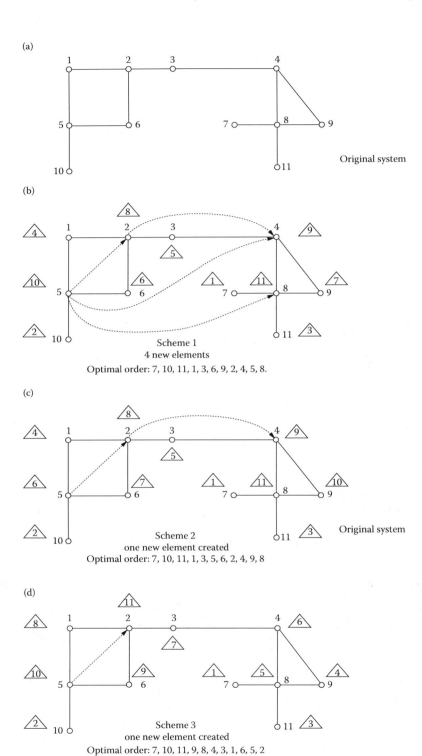

FIGURE B.4
(a) Graph of original 11-node network and (b) through (d) alternate optimal ordering schemes 1, 2, and 3.

Appendix B: Sparsity and Optimal Ordering   593

0	1	2	3	4	5	6	7	8	9	10	11
1	X	X			X						
2	X	X	X								
3		X	X	X							
4			X	X				X	X		
5	X				X	X			X		
6		X			X	X					
7							X	X			
8				X			X	X	X		X
9				X				X	X		
10					X				X		
11								X			X

$$(\text{B.9})$$

An examination of the coefficient matrix shows that:

Nodes 7, 10, and 11 have only one connection with the other nodes and thus one off-diagonal element. According to Scheme 1, the optimal ordering sequence can start with any of these nodes, in any order.

Nodes 1, 3, 6, and 9 come next. These have two connections with the other nodes in the original network and can be eliminated next, in any order.

Nodes 2, 4, and 5 have three connections and should be eliminated next in any order, while node 8 has four connections and should be eliminated last.

Thus, the optimal ordering according to Scheme 1 is 7, 10, 11, 1, 3, 4, 6, 9, 2, 4, 5, and 8. This is shown graphically in Figure B.4b and each step of elimination is marked by the side of the node in a triangle. This scheme generates four new elements.

Next, Scheme 2 is applied and it is shown in Figure B.4c. While optimal ordering could have been written straightaway, without examination of the modified graphs in Scheme 1, or straight from the examination of the coefficient matrix, the optimal ordering according to Scheme 2 is not easily visualized, unless the modified graphs on successive node elimination are examined.

After nodes 7, 10, and 11 are eliminated, nodes 8, 9, 3, 1, 5, and 6 have two branches connected to them. Thus, there is a choice in elimination order. Figure B.4c shows elimination in the order 1, 3, 5, and 6. Elimination of 1 and 3 gives one new element each.

As the network should be examined after each elimination, the node 2 that had three branches has only one branch connected to it after these eliminations. Thus, it comes next in the order of elimination. This leaves nodes 4, 9, and 8 that have two branches connected to each, in a triangle. Thus, any of these can be eliminated first. This gives an optimal order 7, 10, 11, 1, 3, 5, 6, 2, 4, 9, and 8.

This order is not unique. Other orders will satisfy the Scheme 2 criteria.

Figure B.4d shows Scheme 3. Here, only one new element is produced. The optimal order of Scheme 2 could also give the same result. Consider that after elimination of nodes 7, 10, and 11 in Scheme 2, there is a choice to eliminate 8, 9, 5, 6, or 1. Let us eliminate 9. Node 8 then has only one path to node 4 and should be eliminated next. The next node to eliminate will be node 4 (single path) followed by node 3. Thus, Scheme 2 could also give only one new element. As an exercise the reader can construct a graph and optimal order for Scheme 4. The bibliography provides further reading.

Bibliography

AH El-Abiad (Ed.). Solution of network problems by sparsity techniques, Chapter 1. *Proceedings of the Arab School of Science and Technology*, Kuwait, 1983. McGraw-Hill, New York.

HH. Happ. The solution of system problems by tearing, *IEEE Proc* 62(7): 930–940, 1974.

N Sato, WF Tinney. Technique for exploiting the sparsity of the network admittance matrix, *IEEE Trans PAS* 82: 944–950, 1963.

WF Tinney, JW Walker. Direct solutions of sparse network equations by optimally ordered triangular factorization, *IEEE Proc* 55: 1801–1809, 1967.

Appendix C: Transformers and Reactors

A power transformer is an important component of the power system. The transformation of voltages is carried out from generating voltage level to transmission, subtransmission, distribution, and consumer level. The installed capacity of the transformers in a power system may be seven or eight times the generating capacity. The special classes of transformers include furnace, converter, regulating, rectifier, phase shifting, traction, welding, and instrument (current and voltage) transformers. Large converter transformers are installed for high-voltage, direct current (HVDC) transmission.

The transformer models and their characteristics are described in the relevant sections of the book in various chapters. This appendix provides basic concepts and discusses autotransformers, step-voltage regulators, and transformer models not covered elsewhere in the book.

C.1 Model of a Two-Winding Transformer

We represented a transformer model by its series impedance in the load flow and short-circuit studies. We also developed models for tap changing, phase shifting, and reactive power flow control transformers. Concepts of leakage flux, total flux, and mutual and self-reactances in a circuit of two magnetically coupled coils are further discussed in Volume 2. A matrix model can be written as follows:

$$
\begin{vmatrix} v_1 \\ v_2 \\ \cdot \\ v_n \end{vmatrix} = \begin{vmatrix} r_{11} & r_{12} & \cdot & r_{1n} \\ r_{21} & r_{22} & \cdot & r_{2n} \\ \cdot & \cdot & \cdot & \cdot \\ r_{n1} & r_{n2} & \cdot & r_{nn} \end{vmatrix} \begin{vmatrix} i_1 \\ i_2 \\ \cdot \\ i_n \end{vmatrix} + \begin{vmatrix} L_{11} & L_{12} & \cdot & L_{1n} \\ L_{21} & L_{22} & \cdot & L_{2n} \\ \cdot & \cdot & \cdot & \cdot \\ L_{n1} & L_{n2} & \cdot & L_{nn} \end{vmatrix} \frac{\mathrm{d}}{\mathrm{d}t} \begin{vmatrix} i_1 \\ i_2 \\ \cdot \\ i_n \end{vmatrix} \cdot \quad \text{(C.1)}
$$

A two-winding transformer model can be derived from the circuit diagram shown in Figure C.1a and the corresponding phasor diagram (vector diagram) shown in Figure C.2. The transformer supplies a load current I_2 at a terminal voltage V_2 and lagging power factor angle ϕ_2. Exciting the primary winding with voltage V_1 produces changing flux linkages. Though the coils in a transformer are tightly coupled by interleaving the windings and are wound on a magnetic material of high permeability, all the flux produced by primary windings does not link the secondary. The winding leakage flux gives rise to leakage reactances. In Figure C.2, Φ_m is the main or mutual flux, assumed to be constant. The electromotive forces (EMFs) induced in the primary windings is E_1 that lags Φ_m by 90°. In the secondary winding, the ideal transformer produces an EMF E_2 due to mutual flux linkages. There has to be a primary magnetizing current even at no load, in a time phase with its associated flux, to excite the core. The pulsation of flux in the core produces losses. Considering that the no-load current is sinusoidal (which is not true under magnetic saturation), it must have a core loss component due to hysteresis and eddy currents:

595

596   *Appendix C: Transformers and Reactors*

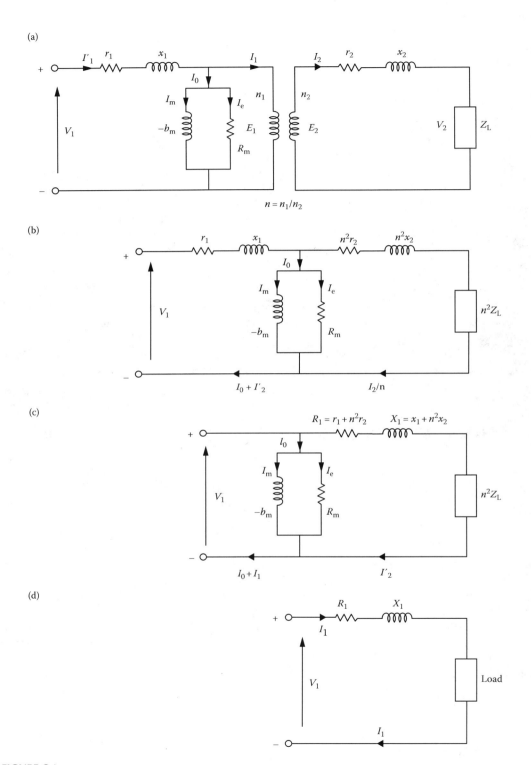

FIGURE C.1
(a) Equivalent circuit of two-winding transformer and (b through d) simplifications to the equivalent circuit.

Appendix C: Transformers and Reactors

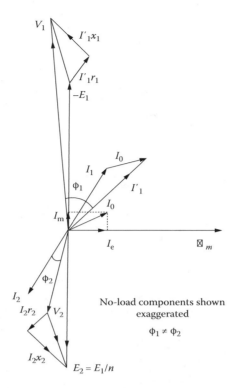

FIGURE C.2
Phasor diagram of two-winding transformer on load.

$$I_0 = \sqrt{I_m^2 + I_e^2} \tag{C.2}$$

where I_m is the magnetizing current, I_e is the core loss component of the current, and I_0 is the no-load current; I_m and I_e are in phase quadrature. The generated EMF because of flux Φ_m is given by

$$E_2 = 4.44 f n_2 \Phi_m \tag{C.3}$$

where E_2 is in volts when Φ_m is in Wb/m², n_2 is the number of secondary turns, and f is the frequency. As primary ampère-turns must be equal to the secondary ampère-turns, i.e., $E_1 I_1 = E_2 I_2$, we can write

$$\begin{aligned} E_1/E_2 &= n_1/n_2 = n \text{ and} \\ I_1/I_2 &\approx n_2/n_1 = 1/n \end{aligned} \tag{C.4}$$

The current relation holds because the no-load current is small. The terminal relations can now be derived. On the primary side, the current is compounded to consider the no-load component of the current, and the primary voltage is equal to $-E_1$ (to neutralize the EMF of induction) and $I_1 r_1$ and $I_1 x_1$ drops in the primary windings. On the secondary side, the terminal voltage is given by the induced EMF, E_2 and I_2r_2 and I_2x_2 drops in the secondary windings. The equivalent circuit is, therefore, as shown in Figure C.1a. The transformer is an ideal lossless transformer of turns ratio, n.

In Figure C.1a, we can refer the secondary resistance and reactance to the primary side or vice versa. The secondary windings of n turns can be replaced with an equivalent winding referred to the primary, where the copper loss in the windings and the voltage drop in reactance is the same as in the actual winding. We can denote the resistance and reactance of the equivalent windings as r_2' and x_2':

$$I_1^2 r_2' = I_2^2 r_2 \qquad r_2' = r_2 \left(\frac{I_2^2}{I_1^2} \right) \approx r_2 \left(\frac{n_1}{n_2} \right)^2 = n^2 r_2$$

$$x_2' = x_2 \left(\frac{I_2 E_1}{I_1 E_2} \right) \approx x_2 \left(\frac{n_1}{n_2} \right)^2 = n^2 x_2.$$

(C.5)

The transformer is a single-phase ideal transformer with no losses and having a turns ratio of unity and no secondary resistance or reactance. By also transferring the load impedance to the primary side, the unity ratio ideal transformer can be eliminated and the magnetizing circuit is pulled out to the primary terminals without appreciable error, Figure C.1b and c. In Figure C.1d the magnetizing and core loss circuit is altogether omitted. The equivalent resistance and reactance are as follows:

$$R_1 = r_1 + n^2 r_2$$

$$X_1 = x_1 + n^2 x_2.$$

(C.6)

Thus, on a simplified basis the transformer positive or negative sequence model is given by its percentage reactance specified by the manufacturer, on the transformer natural cooled Mega Volt Amp (MVA) rating base. This reactance remains fairly constant and is obtained by a short circuit test on the transformer. The magnetizing circuit components are obtained by an open-circuit test.

The expression for hysteresis loss is given by

$$P_h = K_h f B_m^s$$

(C.7)

where K_h is a constant and s is the Steinmetz exponent, which varies from 1.5 to 2.5, depending on the core material, generally, it is $= 1.6$.

The eddy current loss is given by

$$P_e = K_e f^2 B_m^2$$

(C.8)

where K_e is a constant. Eddy current loss occurs in core laminations, conductors, tanks, and clamping plates. The core loss is the sum of the eddy current and hysteresis loss. In Figure C.2, the primary power factor angle ϕ_1 is greater than ϕ_2.

C.1.1 Open-Circuit Test

Figure C.3 shows no-load curves when an open-circuit test at rated frequency and varying voltage is made on the transformer. The test is conducted with the secondary winding open-circuited and rated voltage applied to the primary winding. For high-voltage transformers, the secondary winding may be excited and the primary winding opened. At constant frequency, B_m is directly proportional to applied voltage, and the core loss is approximately proportional to B_m^2. The magnetizing current rises steeply at low flux densities, then more slowly as iron reaches its maximum permeability, and thereafter again steeply, as saturation sets in.

From Figure C.1 the open-circuit admittance is given as

$$Y_{OC} = g_m - jb_m \tag{C.9}$$

This neglects the small voltage drop across r_1 and x_1. Then

$$g_m = \frac{P_0}{V_1^2} \tag{C.10}$$

where P_0 is the measured power and V_1 is the applied voltage. Also,

$$b_m = \frac{Q_0}{V_1^2} = \sqrt{\frac{S_0^2 - P_0^2}{V_1^2}} \tag{C.11}$$

where P_0, Q_0, and S_0 are measured active power, reactive power, and volt–ampères on open circuit. Note that the exciting voltage E_1 is not equal to V_1, due to the drop that no-load current produces through r_1 and x_1. Corrections can be made for this drop.

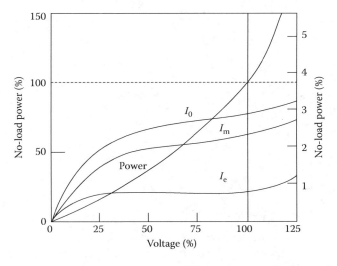

FIGURE C.3
No load test on a transformer.

600 *Appendix C: Transformers and Reactors*

C.1.2 Short-Circuit Test

The short-circuit test is conducted at the rated current of the winding, which is shorted and a reduced voltage is applied to the other winding to circulate a full-rated current:

$$P_{sc} = I_{sc}^2 R_1 = I_{sc}^2 (r_1 + n^2 r_2) \tag{C.12}$$

where P_{sc} is the measured active power on short circuit and I_{sc} is the short-circuit current.

$$Q_{sc} = I_{sc}^2 X_1 = I_{sc}^2 (x_1 + n^2 x_2) \tag{C.13}$$

Example C.1

The copper loss of a 2500 kA, 13.8–4.16 kV delta–wye-connected three-phase transformer is 18 kW on the delta side and 14 kW on the wye side. Find R_1, r_1, r_2, and r_2' for phase values throughout. If the total reactance is 5.5%, find X_1, x_1 x_2, and x_2', assuming that the reactance is divided in the same proportion as resistance.

The copper loss per phase on the 13.8 kV side = 18/3 = 6 kW and the current per phase = 60.4 A. Therefore, $r_1 = 1.645\ \Omega$. Similarly for the 4.16 kV side, the copper loss = 14/3 = 4.67 kW, the current = 346.97 A, and $r_2 = 0.0388\ \Omega$; r_2 referred to the primary side, $r_2' = (0.0388)$ $(13.8 \times \sqrt{3}/4.16)^2 = 1.281\ \Omega$, and $R_1 = 2.926\ \Omega$. A 5.5% reactance on a transformer MVA base of 2.5 = 12.54 Ω on the 13.8 kV side, then $x_1 = (12.54)(1.645)/2.926 = 7.05\ \Omega$ and $x_2' = 5.49$ Ω. Referred to the 4.16 kV side $x_2 = 0.166\ \Omega$. The transformer X/R ratio = 4.28, which is rather low.

Example C.2

The transformer of Example C.1 gave the following results on open-circuit test: open circuit on the 4.16 kV side, rated primary voltage and frequency, input = 10 kW, and no-load current = 2.5 A. Find the magnetizing circuit parameters.

The active component of the current $I_e = 3.33/13.8 = 0.241$ A per phase. Therefore,

$$g_m = \frac{10 \times 10^3}{3 \times (13.8 \times 10^3)^2} = 0.017 \times 10^{-3}\ \text{mhos}$$

The magnetizing current is as follows:

$$I_m = \sqrt{I_0^2 - I_e^2} = \sqrt{1.44^2 - 0.241^2} = 1.42\ \text{A}$$

The power factor angle of the no-load current is 9.63°, and b_m from Equation C.11 is -0.103×10^{-3} mhos per phase.

C.2 Transformer Polarity and Terminal Connections

C.2.1 Additive and Subtractive Polarity

The relative direction of induced voltages, as appearing on the terminals of the windings is dependent on the order in which these terminals are taken out of the transformer tank. As the primary and secondary voltages are produced by the same mutual flux, these must be in the same direction in each turn. The load current in the secondary flows in

Appendix C: Transformers and Reactors

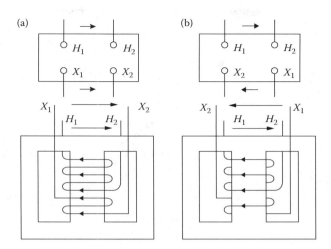

FIGURE C.4
(a) Polarity and polarity markings, subtractive polarity and (b) additive polarity.

a direction so as to neutralize the magnetomotive force (MMF) of the primary. How the induced voltages will appear as viewed from the terminals depends on the relative direction of the windings. The polarity refers to the definite order in which the terminals are taken out of the tank. Polarity may be defined as the voltage vector relations of transformer leads as brought out of the tank. Referring to Figure C.4a the polarity is the relative direction of the induced voltage from H_1 to H_2 as compared with that from X_1 to X_2, both being in the same order. The order is important in the definition of polarity.

When the induced voltages are in the opposite direction, as shown in Figure C.4b, the polarity is said to be additive, and when in the same direction, as in Figure C.4a, it is said to be subtractive. According to the American National Standards Institute (ANSI) standard, all liquid immersed power and distribution transformers have subtractive polarity. Dry-type transformers also have subtractive polarity.

When the terminals of any winding are brought outside the tank and marked so that H_1 and X_1 are adjacent, the polarity is subtractive. The polarity is additive when H_1 is diagonally located with respect to X_1 (Figure C.4b). The lead H_1 is brought out as the right-hand terminal of the high-voltage group as seen when facing the highest voltage side of the case. The polarity is often marked by dots on the windings. If H_1 is dotted, then X_1 is dotted for subtractive polarity. The currents are in phase. Angular displacement and terminal markings for three-phase transformers and autotransformers are discussed in Reference [1].

C.3 Parallel Operation of Transformers

Transformers may be operated in parallel to supply increased loads, and for reliability, redundancy, and continuity of the secondary loads. Ideally, the following conditions must be satisfied:

- The phase sequence must be the same.
- The polarity must be the same.

602 *Appendix C: Transformers and Reactors*

- Voltage ratios must be the same.
- The vector group, i.e., the angle of phase displacement between primary and secondary voltage vectors, should be the same.
- Impedance voltage drops at full load should be the same, i.e., the percentage impedances based on the rated MVA rating must be the same.

It is further desirable that the two transformers have the same ratio of percentage resistance to reactance voltage drops, i.e., the same X/R ratios.

With the above conditions met, the load sharing will be proportional to the transformer MVA ratings. It is basically a parallel circuit with two transformer impedances in parallel and a common terminal voltage as follows:

$$I_1 = \frac{IZ_2}{Z_1 + Z_2}$$
$$I_2 = \frac{IZ_1}{Z_1 + Z_2}$$

(C.14)

where I_1 and I_2 are the current loadings of each transformer and I is the total current. In terms of the total MVA load, S, the equations are as follows:

$$S_1 = \frac{SZ_2}{Z_1 + Z_2}$$
$$S_2 = \frac{SZ_1}{Z_1 + Z_2}$$

(C.15)

While the polarity and vector group are essential conditions, two transformers may be paralleled when they have

- Unequal ratios and equal percentage impedances
- Equal ratios and unequal percentage impedances
- Unequal ratios and unequal percentage impedances

It is not a good practice to operate transformers in parallel when

- Either of the two parallel transformers is overloaded by a significant amount above its rating.
- When the no-load circulating current exceeds 10% of the full-rated load.
- When the arithmetical sum of the circulating current and load current is greater than 110%.

The circulating current means the current circulating in the high- and low-voltage windings, excluding the exciting current.

Appendix C: Transformers and Reactors

Example C.3

A 10-MVA, 13.8–4.16 kV transformer has per unit resistance and reactance of 0.005 and 0.05, respectively. This is paralleled with a 5-MVA transformer of the same voltage ratio, and having per unit resistance and reactance of 0.006 and 0.04, respectively. Calculate how these will share a load of 15 MVA at 0.8 power factor lagging.

Convert Z_1 and Z_2 on any common MVA base and apply Equations C.14 and C.15. The results are

10-MVA transformer $S_1 = 9.255 \angle{-37.93°}$
5-MVA transformer $S_2 = 5.749 \angle{-35.20°}$

The loads do not sum to 15 MVA because of different power factors. The MW and Mvar components of the transformer loads should sum to the total load components:

10-MVA transformer: 7.30 MW and 5.689 Mvar; 5-MVA transformer: 4.698 MW and 3.314 Mvar; total equal to an approximate load MW of 12 MW and 9.0 Mvar.

If the terminal voltages differ, there will be circulating current at no load. With reference to Figure C.5, the load sharing is given by the following equations:

$$I_1 = \frac{E_1 - V}{Z_1} \quad I_2 = \frac{E_2 - V}{Z_2} \tag{C.16}$$

where V is the load voltage. This is given by the following equation:

$$V = (I_1 + I_2)Z_L = \left(\frac{E_1 - V}{Z_1} + \frac{E_2 - V}{Z_2}\right)Z_L \tag{C.17}$$

This can be written as

$$V\left(\frac{1}{Z_1} + \frac{1}{Z_2} + \frac{1}{Z_L}\right) = \left(\frac{E_1}{Z_1} + \frac{E_2}{Z_2}\right) \tag{C.18}$$

For a given load, the calculation is iterative in nature, as shown in Example C.4.

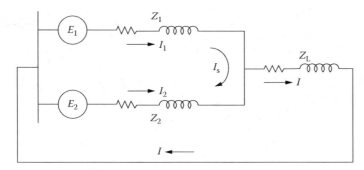

FIGURE C.5
Equivalent circuit of two parallel running transformers with different voltage ratios and percentage impedances.

Example C.4

In Example C.4, the transformers have the same percentage impedances and the same X/R ratios; the secondary voltage of the 10-MVA transformer is 4 kV and that of the 5-MVA transformer is 4.16 kV. Calculate the circulating current at no load.

We will work on per phase basis. The 10-MVA transformer impedance referred to 4 kV secondary is $0.008 + j0.08\,\Omega$, and the 5-MVA transformer impedance at 4.16 kV secondary is $0.0208 + j0.1384\,\Omega$; $Z_1 + Z_2 = 0.0288 + j0.2184$. *Assume that the load voltage is 4 kV*; then, on a per phase basis, the load is 5 MVA at 0.8 power factor and the load impedance is $0.853 + j0.64\,\Omega$.

$$\frac{E_1}{Z_1} + \frac{E_2}{Z_2} = \frac{4000}{\sqrt{3}(0.008 + j0.08)} + \frac{4160}{\sqrt{3}(0.0208 + j0.1384)} = 5409 - j45.550 \text{ kA}$$

Also,

$$\left(\frac{1}{Z_1} + \frac{1}{Z_2} + \frac{1}{Z_L} \right) = 3.048 - j20.003$$

From Equation C.18, the load voltage is $2260 - j74.84$ volts phase-to-neutral. From Equation C.16, the 10-MVA transformer load current is $980.04 - j445.347$ and that of the 5-MVA transformer is $672.17 - j874.42$. The total load current is $1652.2 - j1319.75$ and the single-phase load MVA is 3.645 MW and 3.112 Mvar. This is much different from the desired loading of 4 MW and 3 Mvar. This is due to assumption of the load voltage. The calculation can be repeated with a lower estimate of load voltage and recalculation of load impedance.

C.4 Autotransformers

The circuit of an autotransformer is shown in Figure C.6a. It has windings common to primary and secondary, i.e., the input and output circuits are electrically connected. The primary voltage and current are V_1 and I_1 and the secondary voltage and current are V_2 and I_2. If the number of turns are n_1 and n_2, as shown, then neglecting losses:

$$\frac{V_1}{V_2} = \frac{I_2}{I_1} = \frac{n_1 + n_2}{n_2} = n \tag{C.19}$$

The ampère-turns $I_1 n_1$ oppose ampère-turns $I_2 n_2$, and the common part of the winding carries a current of $I_2 - I_1$. Consequently, a smaller cross section of the conductor is required. Conductor material in the autotransformer as a percentage of conductor material in a two-winding transformer for the same kVA output is

$$\frac{M_{\text{auto}}}{M_{\text{twowinding}}} = \frac{I_1(n_1) + (I_2 - I_1)n_2}{I_1(n_1 + n_2) + I_2 n_2}$$

$$= 1 - \frac{2}{(n) + (I_2/I_2)} = 1 - \frac{V_2}{V_1} \tag{C.20}$$

Appendix C: Transformers and Reactors

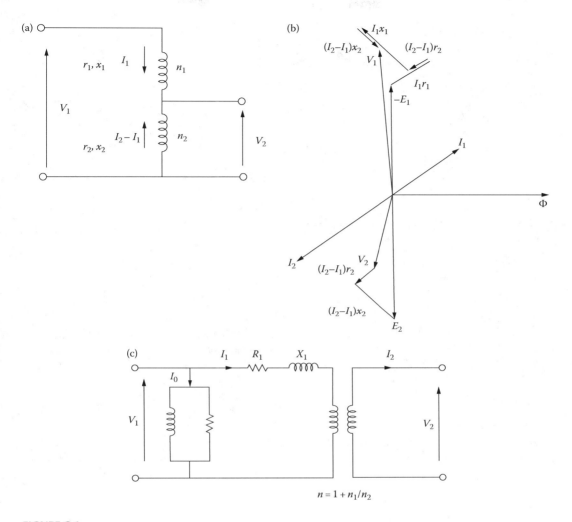

FIGURE C.6
(a) Circuit of an autotransformer, step-down configuration shown. (b) Phasor diagram of the autotransformer on load. (c) Equivalent circuit of an autotransformer.

The savings in material cost are most effective for transformation voltages close to each other. For a voltage ratio of 2, approximately 50% savings in copper could be made.

The vector diagram is shown in Figure C.6b and the equivalent circuit in Figure C.6c. Neglecting the magnetizing current

$$V_1 = -E_1 - I_1(r_1 \cos \phi + x_1 \sin \phi) + (I_2 - I_1)(r_2 \cos \phi + x_2 \sin \phi) \quad \text{(C.21)}$$

where ϕ is the load power factor. Note that the impedance drop in the common winding is added, because the net current is opposed to the direction of I_1. The equation for the secondary voltage is as follows:

$$V_2 = E_2 - (I_2 - I_1)(r_2 \cos \phi + x_2 \sin \phi) \quad \text{(C.22)}$$

Combining these two equations, we write:

$$V_1 = nV_2 + I_1\left[(r_1 + (n-1)^2 r_2)\cos\phi + (x_1 + (n-1)^2 x_2)\sin\phi\right] \tag{C.23}$$

This means that the equivalent resistance and reactance corresponding to a two-winding transformer are

$$R_1 = r_1 + (n-1)^2 r_2$$
$$X_1 = x_1 + (n-1)^2 x_2 \tag{C.24}$$

An autotransformer can be tested for impedances exactly as a two-winding transformer. The resistance and reactance referred to the secondary side are as follows:

$$R_2 = R_1/n^2 = r_2 + r_1/n^2$$
$$X_2 = X_1/n^2 = x_2 + x_1/n^2 \tag{C.25}$$

The kVA rating of the circuit with respect to the kVA rating of the windings is $(n-1)/n$ A 1-MVA, 33–22 kV autotransformer has an equivalent two-winding kVA of $(1.5-1)/1.5 = 0.333 \times 1000 = 333$ kVA. The series impedance is less than that of a two-winding transformer. This is beneficial from the load-flow point of view, as the losses and voltage drop will be reduced, however, a larger contribution to short-circuit current results.

A three-phase autotransformer connection is shown in Figure C.7a. Such banks are usually Y-connected with a grounded neutral, and a tertiary winding is added for third-harmonic circulation and neutral stabilization. This circuit is akin to that of a three-winding transformer, and the positive and zero sequence circuits are as shown. The T-circuit positive sequence parameters Figure C.7b are calculated by shorting one set of terminals and applying positive sequence voltage to the other terminals and keeping the third set of terminals open-circuited,

$$\begin{vmatrix} Z_H \\ Z_X \\ Z_Y \end{vmatrix} = \frac{1}{2}\begin{vmatrix} 1 & 1 & -1 \\ 1 & -1 & 1 \\ 1 & 1 & -1 \end{vmatrix}\begin{vmatrix} Z_{HX} \\ Z_{HY} \\ Z_{XY} \end{vmatrix} \text{pu} \tag{C.26}$$

All impedances are in pu.

The zero sequence impedance, Figure C.7c is given by

$$\begin{vmatrix} Z_{X0} \\ Z_{H0} \\ Z_{n0} \end{vmatrix} = \frac{1}{2}\begin{vmatrix} 1 & -1 & 1 & (n-1)/n \\ 1 & 1 & -1 & -(n-1)/n^2 \\ -1 & 1 & 1 & 1/n \end{vmatrix}\begin{vmatrix} Z_{HX} \\ Z_{HY} \\ Z_{XY} \\ 6Z_n \end{vmatrix} \tag{C.27}$$

where n is defined in Equation C.19. If the neutral of the autotransformer is ungrounded, $Z_n = \infty$ and all impedances in Equation C.27 become infinity. One way to solve this problem

Appendix C: Transformers and Reactors 607

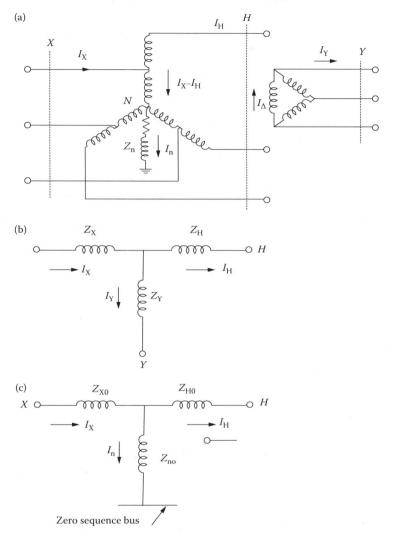

FIGURE C.7
(a) Circuit of a three-phase transformer with tertiary delta and (b, c) positive and zero sequence circuits.

is to convert the Y-equivalent to a delta equivalent and then take the limit as $Z_n = \infty$. The delta equivalent impedances thus calculated are called "resonant delta."

C.4.1 Scott Connection

Two autotransformers with suitable taps can be used in a Scott connection, for three-phase to two-phase conversion and flow of power in either direction. The arrangement is shown in Figure C.8. The line voltage V appears between terminals C and B and also between terminals A and B and A and C. The voltage between A and S is $(V\sqrt{3})/2$; the second autotransformer, called the teaser transformer, has $(\sqrt{3}/2)$ turns. The two secondaries having equal turns produce voltages equal in magnitude and phase quadrature. The

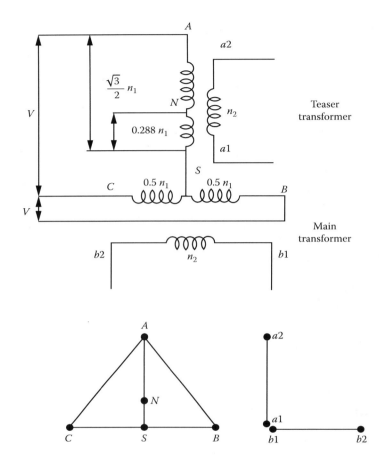

FIGURE C.8
(a) Circuit of a Scott connected transformer and (b) three-phase primary and two-phase secondary voltages.

neutral of the three-phase system can be located on the second or teaser transformer. The neutral must have a voltage of $V/\sqrt{3}$ to terminal A, i.e., the neutral point can be trapped at $V[(\sqrt{3}/2) - 1/\sqrt{3}] = 0.288 n_1$ turns from S. It can be shown that the three-phase side is balanced for a two-phase balanced load, i.e., if the load is balanced on one side it will be balanced on the other.

C.5 Step Voltage Regulators

Step-voltage regulators [2] are essentially autotransformers. The most common voltage regulators manufactured today are of single-phase type with reactive switching, resulting in ±10% voltage regulation in 32 steps, 16 "boosting" and 16 "bucking." The rated voltages are generally up to 19920 kV (line to neutral), 150 kV basic insulation level (BIL), and the current rating ranges from 5 to 2000 A (not at all voltage levels). The general application is in distribution systems and three single-phase voltage regulators can be applied in wye or

Appendix C: Transformers and Reactors

delta connection to a three-phase three-wire or three-phase four-wire system. The winding common to the primary and secondary is designated as a shunt winding, and the winding not common to the primary and secondary is designated as a series winding. The series winding voltage is 10% of the regulator applied voltage. The polarity of this winding is changed with a reversing switch to accomplish buck or boost of the voltage. When the voltage regulation is provided on the load side it is called a type-A connection, Figure C.9a. The core excitation varies as the shunt winding is connected across the source voltage. In a type-B connection, Figure C.9b, the regulation is provided on the load side and the source voltage is applied by way of series taps. Figure C.9d shows the schematic of a tap-changing circuit with current-limiting reactors and equalizer windings.

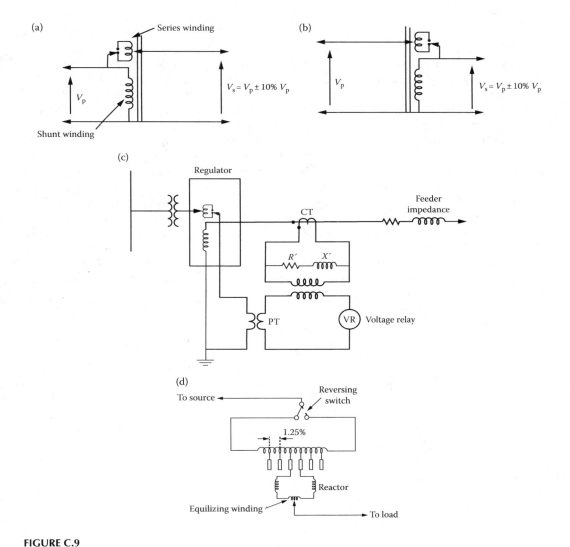

FIGURE C.9
(a, b) Circuits of type A and type B voltage regulators, (c) schematic of line drop compensator, and (d) reactor type tap changer with reversing switch.

C.5.1 Line Drop Compensator

The step regulators are controlled through a line drop compensator; its schematic circuit is shown in Figure C.9c. The voltage drop in the line from the regulator to the load is simulated in a R'–X'' network in the compensator. The settings on these elements are decided on the basis of load flow prior to insertion of the regulator, i.e., the voltage and current at the point of application gives the system impedance to be simulated by R' and X' in the line drop compensator.

C.6 Extended Models of Transformers

A transient transformer model should address saturation, hysteresis, eddy current, and stray losses. Saturation plays an important role in determining the transient behavior of the transformer. Extended transformer models can be very involved and these are not required in every type of study. At the same time, a simple model may be prone to errors. As an example, in distribution system load flow, representing a transformer by series impedance alone and neglecting the shunt elements altogether may not be proper, as losses in the transformers may be considerable. For studies on switching transients, it is necessary to include capacitance of the transformers as high-frequency surges will be transferred more through electrostatic couplings rather than through electromagnetic couplings. For short-circuit calculations, capacitance and core loss effects can be neglected. Thus, the type of selected model depends on the nature of the study. There are many approaches to the models, some of which are briefly discussed.

The equivalent circuit of the shunt branch of a transformer for nonlinearity can be drawn as shown in Figure C.10. The excitation current has half-wave symmetry and contains only odd harmonics. We may consider the excitation current as composed of two components, a fundamental frequency component and a distortion component. The fundamental frequency component is broken into two components, i_e and i_m as discussed before, which give rise to shunt components g_m and $-b_m$ (Figure C.1a). The distortion component may be considered as a number of equivalent harmonic current sources in parallel with the

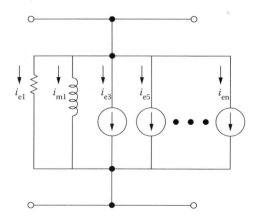

FIGURE C.10
Transformer shunt branch modeled considering nonlinearity.

Appendix C: Transformers and Reactors 611

fundamental frequency components, each of which can be represented in the phasor form, $I_{ei} < \theta_i$. To consider the effect of variation in the supply system voltages, the model parameters at three voltage levels of maximum, minimum, and rated voltage can be approximated by quadratic functions of the supply system voltage:

$$W = a + bV + cV^2$$
$$W = R_m, X_m, I_{e3}, I_{e5}, \ldots, \theta_3, \theta_5, \ldots \quad \text{(C.28)}$$

where $I_{e3}, I_{e5}, \ldots, \theta_3, \theta_5, \ldots$ are the harmonic currents and their angles. The coefficients a, b, and c can be found from

$$\begin{vmatrix} a \\ b \\ c \end{vmatrix} = \begin{vmatrix} 1 & V_{min} & V_{min}^2 \\ 1 & V_{rated} & V_{rated}^2 \\ 1 & V_{max} & V_{max}^2 \end{vmatrix} \begin{vmatrix} W_0 \\ W_1 \\ W_2 \end{vmatrix} \quad \text{(C.29)}$$

where W_0, W_1, W_2 are measured values of W for V_{min}, V_{rated}, and V_{max}, respectively.

C.6.1 Modeling the Hysteresis Loop

A model of the hysteresis loop can be constructed, based on measurements. The locus of the midpoints of the loop is obtained by measurements at four points and its displacement by a *consuming function*, whose maximum value is ob, and ef changes periodically by half-wave symmetry (Figure C.11). The consuming function can be written as $f(x) = -ob \sin \omega t$. The periphery can be then represented by 16 line segments [3]:

$$i = (i_k - m_k \phi_k) + m_k \phi - ob \sin(\omega t)$$
$$\phi_{k-1} < |\phi| \le \phi_k \quad \text{(C.30)}$$
$$k = 1, 2, \ldots, 16$$

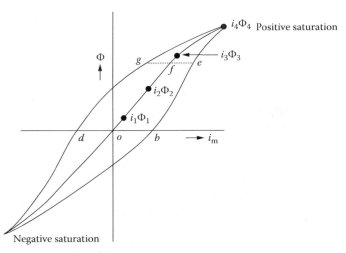

FIGURE C.11 Piecewise hysteresis loop fitting.

C.6.2 EMTP Models

Figure C.12a shows a single-phase model; R remains constant, and it is calculated from excitation losses. The nonlinear inductor is modeled from transformer excitation data and from its nonlinear $V–I$ characteristics. In modern transformers the cores saturate sharply and there is a well-defined knee. Often a two-slope piecewise linear inductor is adequate to model such curves. The saturation curve is not supplied as a flux–current curve, but as a rms voltage–rms current curve. The *satura* routine in ElectroMagnetic Transients Program (EMTP) [4] converts voltage–current input into flux–current data. Figure C.12b and c shows an example of this conversion for a 750-MVA, 420–27 kV five-leg, core type, wye–delta connected transformer. The nonlinear inductance should be connected between the windings closest to the iron core. The input data are presented in per unit values with regard to the winding connections and the base current and voltage. For the delta winding the rms excitation current is $1/\sqrt{3}$ times the excitation current. Also, rms excitation current in delta winding is approximated by

$$I_{\mathrm{m-w}} = \left(I_{\mathrm{ex-w}}^2 - \left(\frac{P_{\mathrm{ex}}}{3V_{\mathrm{ex}}} \right)^2 \right)^{1/2} \tag{C.31}$$

where $I_{\mathrm{m-w}}$ is winding magnetizing current, P_{ex} is the measured excitation power, and V_{ex} is the excitation voltage. Then the data is converted into unit values for the equivalent phase: S base = 250 MVA, V base = 27 kV, and I base = 9259 A. There is linear interpolation between the assumed values and finite-difference approximation to sinusoidal excitation. The hysteresis is ignored.

The EMTP model *hysdat* represents a hysteresis loop in 4–5 points to 20–25 points for a specific core material. The positive and negative saturation points, as shown in Figure C.11a, need only to be specified. The geometric construction leading to positive saturation point A and negative saturation point is specified in Reference [4]. The positive and negative saturation points are shown on the loop. Figure C.13 shows EMTP simulation of the excitation current in a single-phase, 50-kVA transformer. Other EMTP models are BECATRAN and FDBIT, the later model can represent the behavior of a transformer over a wide range of frequencies, though the data for generating this model will not be readily available.

C.6.3 Nonlinearity in Core Losses

Figure C.14 shows a frequency domain approach and considers that winding resistance and leakage reactance remain constant and the nonlinearity is confined to the core characteristics [5]. The core loss is modeled as a superimposition of losses occurring in fictitious harmonic and eddy current resistors. The magnetizing characteristics of the transformer is defined by a polynomial expressing the magnetizing current in terms of flux linkages:

$$i_{\mathrm{M}} = A_0 + A_1\lambda + A_2\lambda^2 + A_3\lambda^3 + \cdots \tag{C.32}$$

Only a specific order of harmonic currents flow to appropriate G_{h} resistors in Figure C.14. From Equations C.7 and C.8 the core loss equation is as follows:

$$P_{\mathrm{fe}} = P_{\mathrm{h}} + P_{\mathrm{e}} = K_{\mathrm{h}}B^s f + K_{\mathrm{e}}B^2 f^2 \tag{C.33}$$

Appendix C: Transformers and Reactors

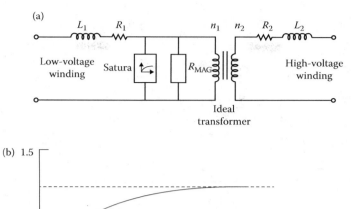

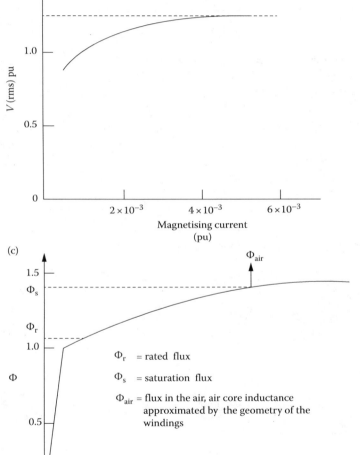

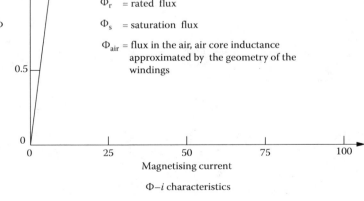

FIGURE C.12
(a) EMTP model *Satura* and (b) Conversion of V–I characteristics into φ – I characteristics.

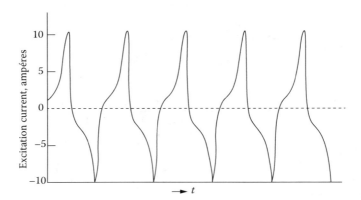

FIGURE C.13
EMTP simulation of inrush current of a 50 kV single-phase transformer.

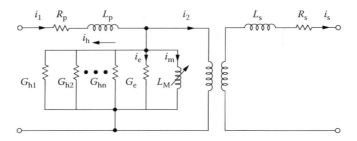

FIGURE C.14
Nonlinear model with superimposition of harmonic currents in the resistors.

For a sinusoidal voltage, this can be written as follows:

$$P_{fe} = K_h f^{1-s} E^s + k_e E^2 \quad (K_h \neq K_h \text{ and } K_e \neq K_e) \tag{C.34}$$

This defines two-conductance G_h for hysteresis loss and G_e for eddy current loss, given by the following equation:

$$G_h = K_h f^{1-s}, G_e = K_e \tag{C.35}$$

C.7 High Frequency Models

For the response of transformers to transients a very detailed model may include each winding turn and turn-to-turn inductances and capacitances [6]. Consider a disk-layer winding or pancake sections as shown in Figure C.15a. Each numbered rectangular block represents the cross section of a turn. The winding line terminal is at A and winding continues beyond E. Each section can be represented by a series of inductance elements with series and shunt capacitances as shown in Figure C.15b. Though the model looks complex,

Appendix C: Transformers and Reactors

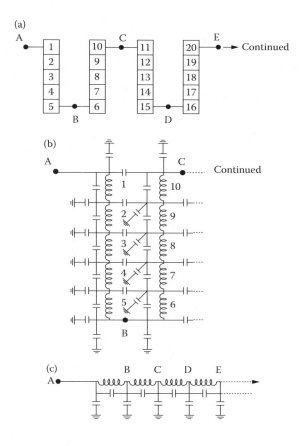

FIGURE C.15
(a) Winding turns in a pancake coil, (b) circuit of winding inductances and capacitances, and (c) simplified circuit model.

the mutual inductances are not shown, resistances are not represented, and no interturn capacitances are shown. This circuit will be formidable in terms of implementation. For most applications, representation of each turn is not justified and by successive lumping a much simpler model is obtained (Figure C.15c).

C.7.1 Surges Transferred through Transformers

The lightning and switching surges can be transferred through transformer couplings [7]. Winding capacitive couplings predominate at high frequencies. For protection of insulation of transformers primary and secondary surge arresters are necessary. Consider the circuit in Figure C.16. A 7.5-Mvar capacitor bank is switched at the 13.8-kV bus and the resulting switching over-voltages on the secondary of a 2.5-MVA, 13.8–0.48 kV transformer connected through 400 ft 500-KCMIL, 15-kV cable are simulated using EMTP. A transformer model with primary and secondary capacitances to ground, inter-winding capacitances and bushing capacitances is used. The results are shown in Figure C.17. This figure shows high-frequency components, due to multiple reflections in the connecting cable, and the peak secondary. The secondary voltage is 3000 volts. This is very high for a 480-V system. Secondary surge arresters and capacitors applied at transformer terminals will appreciably reduce this voltage and the high frequency of oscillations shown in this figure.

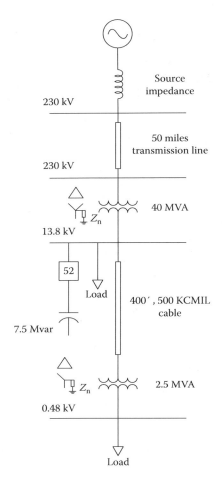

FIGURE C.16
Circuit for simulation of capacitor switching transients.

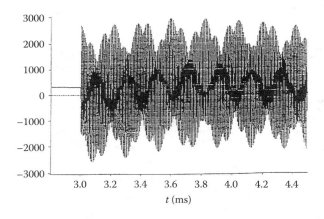

FIGURE C.17
EMTP simulation of transient voltage on 480-V secondary of 2.5 MVA transformer in Figure C.16.

Appendix C: Transformers and Reactors 617

The switching of transformers can give rise to voltage escalation inside the transformer windings. The windings have internal "ringing" frequencies and certain switching operations can excite these frequencies creating excessive intra-winding stresses. Sometimes the repeated restrikes in switching devices can also give rise to high rise transients. See Chapter 10 and also References [8–12].

C.8 Duality Models

Duality-based models can be used to represent transformers. These models are based on core topology and utilize the correspondence between electric and magnetic circuits, as expressed by the principle of duality. Voltage, current, and inductance in electrical circuits correspond to flux, MMF, and reluctance, respectively:

$$I = \frac{V}{R}$$

$$\Phi = \frac{\text{MMF}}{l/\mu\mu_0 a} = \frac{\text{MMF}}{S} \tag{C.36}$$

where l is the length of the magnetic path, a is the cross-sectional area, and S is the reluctance, which is analogous to resistance in an electrical circuit and determines the MMF necessary to produce a given magnetic flux. Permeance is the reverse of reluctance.

Figure C.18a shows electrical equivalent circuit of a three-winding core-type transformer portraying magnetic coupling in three-and five-limbed transformers [12]. Nonlinear inductances correspond to iron flux paths in the magnetic circuit, permitting each core limb to be modeled separately. Each L_k represents top and lower yokes and each L_b represents a wound limb; L_0 represents the flux path through the air, outside the core and around the windings. Finally, the ladder network between linear inductances L_0 and L_b represents winding leakages through air. Inductances L_h and L_y represent unequal flux linkages between turns due to finite winding radial build and these are small compared to L_0 and L_b. This model is simplified as shown in Figure C.18b. The various inductances are calculated from short-circuit tests.

Duality models can be used for low-frequency transient studies, such as short circuits, inrush currents, ferroresonance, and harmonics [13].

C.9 GIC Models

Geomagnetically induced currents (GICs) flow in earth surface due to solar magnetic disturbance (SMD) and these are typically of 0.001–0.1 Hz and can reach peak values of 200A. These can enter transformer windings trough grounded neutrals (Figure C.19a) and bias the transformer core to ½ cycle saturation. As a result, the transformer magnetizing current is greatly increased. Harmonics increase and these could cause reactive power consumption, capacitor overload and false operation of protective relays; etc.

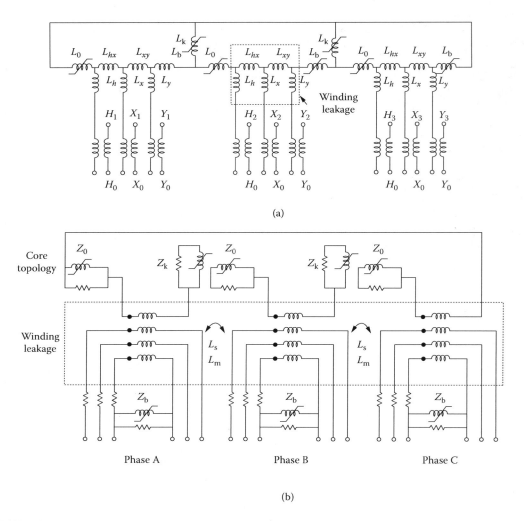

FIGURE C.18
(a) Duality based model of a core type three-winding transformer and (b) simplified circuit.

A GIC model is shown in Figure C.19b, as developed by authors of Reference [14]. This saturation magnetic circuit model of a single-phase shell form of transformer valid for GIC levels was developed based on 3D finite element analysis (FEM) results. This model is able to simulate not only the linear and knee region equations, but also the heavy saturated region and the so called "air-core" region; four major flux paths are included. All R elements represent reluctances in different branches. Subscripts c, a, and t stand for core, air, and tank, respectively, and 1, 2, 3, and 4 represent major branches of flux paths. Branch 1 represents sum of core and air fluxes within the excitation windings, branch 2 represents flux path in yoke, branch 3 represents sum of fluxes entering the side leg, part of which leaves the side leg and enters the tank. Branch 4 represents flux leaving the tank from the center leg. An iterative program is used to solve the circuit of Figure C.19b so that nonlinearity is considered, [14,15].

Appendix C: Transformers and Reactors

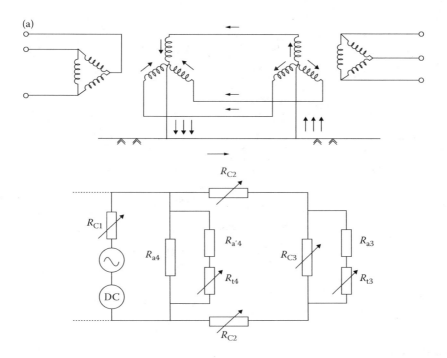

FIGURE C.19
(a) GIC entering the grounded neutrals of wye-connected transformers and (b) transformer model for GIC calculations.

C.10. Surge Voltage Distribution across Transformer Windings

The surge distribution across a transformer winding from the line to neutral terminal is nonlinear, and the line side turns are exposed to much steeper voltage gradient. For study of such surge distribution, we could use the model in Figure C.15c. After the initial incident of the surge, the first few capacitances to ground are charged, and the remainders are comparatively uncharged. The interchange of stored energy between capacitances through coil inductances generates complex oscillations at a variety of natural frequencies. Figure C.20 shows various patterns of surge distributions. The dotted curve indicated distribution of oscillatory peaks. When these become appreciably attenuated, and if the surge voltage is maintained, the distribution sinks to lower levels. A more detailed analyses can be performed by Fourier series of steep front wave, impressed on the end of a ladder network.

Earlier the end turns of the transformers were reinforced and strengthened to protect them from high insulation stresses. This did not prove to be effective in most cases. Currently metallic shields in strategic positions are placed adjacent to the coils. These shields are so placed and connected that the capacitance current flowing through them and the windings tends to compensate for the ground capacitance current, thereby making the current through the series capacitance more uniform. Another method is to adjust capacitance distribution depending upon how the windings are interconnected.

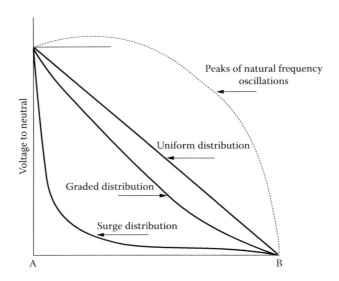

FIGURE C.20
Surge voltage distribution across transformer windings, various patterns (see text).

C.11 Ferroresonance

Ferroresonance can occur when a nonlinear inductance, like the iron-core reactance of a transformer is excited in parallel with capacitance through high impedance. Overvoltages up to three to four times the nominal voltage can occur. Ferroresonance occurs because of reactance of a distribution transformer X_m, and the capacitance from the bushings and underground cables form a resonant circuit. Highly distorted voltage waveforms are produced that include a 60 Hz component. Considerable harmonics are generated. We may postulate, based upon the current literature, that

- High overvoltages of the order of 4.5 per unit can be created.
- Resonance can occur over a wide range of X_c/X_m, possibly in the range [15,16]:

$$0.1 < X_c/X_m < 40 \tag{C.37}$$

- Resonance occurs only when the transformer is unloaded, or very lightly loaded. Transformers loaded to more than 10% of their rating are not susceptible to ferroresonance.

The capacitance of cables varies between 40 and 100 nF per 1000 ft, depending upon conductor size. However, the magnetizing reactance of a 35 kV transformer is several times higher than that of a 15 kV transformer; the ferroresonance can be more damaging at higher voltages. For delta connected transformers the ferroresonance can occur for less than 100 ft of cable. Therefore, the grounded wye-wye transformer connection has become the most popular in underground distribution system in North America. It is more resistant, though not totally immune to ferroresonance. During three-phase switching the

Appendix C: Transformers and Reactors

poles of the switch may not close simultaneously or a current limiting fuse in one or two lines may be open giving rise to ferroresonance.

Figure C.21a shows a basic circuit of ferroresonance. Consider that current limiting fuses in one or two line operate. X_c is the capacitance of the cables and transformer bushings to ground. Also the switch may not close simultaneously in all the three phases or while opening the phases may not open simultaneously. We can draw the equivalent circuits with one phase closed and also with two phases closed as shown in Figure C.21b and c, respectively.

Similar equivalent circuit can be drawn for wye–wye-ungrounded transformer.

The minimum capacitance to produce ferroresonance can be calculated from $X_c/X_m = 40$, say

$$C_{mres} = \frac{2.21 \times 10^{-7} MVA_{transf} I_m}{V_n^2} \quad \text{(C.38)}$$

where

I_m is the magnetizing current of the transformer as a percentage of the full load current. Typically magnetizing current of a distribution transformer is 1%–3% of the transformer full load current.

MVA_{transf} = Rating of transformer in MVA

C_{mres} = minimum capacitance for resonance in pF

V_n = line to netural voltage in KV

C.11.1 Parallel Ferroresonance

A less common phenomena is parallel resonance, not discussed here. Also see Reference [16].

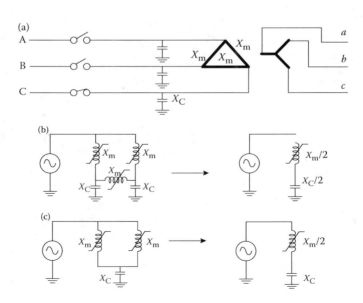

FIGURE C.21
(a) A circuit of possible ferroresonance. (b, c) Equivalent circuits derived with one-phase closed, two-phases closed.

622   *Appendix C: Transformers and Reactors*

C.12 Reactors

We have discussed the following applications of reactors in this series:

- Current-limiting reactors, mainly from the standpoint of limiting the short-circuit currents in a power system. These can be applied in a feeder, in a tie-line, in synchronizing bus arrangements, and as generator reactors.
- Shunt reactors for reactive power compensation.
- Reactors used in static var controllers, i.e., SVCs, TCRs, TSCs, and discharge reactors in a series capacitor, Volume 2.
- Harmonic filter reactors and inrush current-limiting reactors, Volume 3.
- Line reactors to limit notching effects and DC reactors for ripple current limitation in drive systems, Volume 3.
- Neutral grounding reactor.
- Series reactor for transmission line compensation.

Further applications are as follows:

- Smoothing reactors are used in series with an HVDC line or inserted into a DC circuit to reduce harmonics on the DC side, Volume 3. Filter reactors are installed on the AC side of converters. Radio interference (RI) and power-line carrier filter reactors are used to reduce high-frequency noise propagation, Volume 4.
- Reactors are installed in series in a medium-voltage feeder (high-voltage side of the furnace transformer) to improve efficiency, reduce electrode consumption, and limit short-circuit currents.
- An arc suppression reactor, called a Peterson coil, is a single-phase variable reactor that is connected between the neutral of a transformer and ground for the purpose of achieving a resonant grounding system, though such grounding systems are not in common use in the United States, but prevalent in Europe. The inductance is varied to cancel the capacitance current of the system for a single line-to-ground fault, Volume 1.
- Reactors are used in reduced-voltage motor starters to limit the starting inrush currents, Volume 2.
- Series reactors may be used in transmission systems to modify the power flow by changing the transfer impedance. Complexity of power grids results in power flow situations, where one area can be affected by switching, loading, and outage conditions occurring in other area. Strategic placement of reactors can increase power transfer capability and improve reliability, Volume 2.
- Shunt reactors are used in long high-voltage transmission lines, which generate a high amount of leading reactive power when lightly loaded. Conversely these absorb a large amount of reactive power when heavily loaded. This impacts the voltage profile. Shunt reactors absorb reactive power and lower the system voltage. The switching of shunt reactors gives rise to transients discussed in Chapter 10.

See References [17–19] for IEEE Standards on reactors.

Appendix C: Transformers and Reactors

The reactors are tested much alike transformers. Table C.1 shows the routine, design, and other tests for series reactors.

Mechanical and thermal short-circuit ratings are important. Series reactors will have a thermal and mechanical short-circuit current rating. The short-circuit rating is calculated by assuming a nominal system voltage of 105%. Higher voltages can impose higher short-circuit currents. The following expression can be used for the calculations:

$$X_R = V_s^2 \left(\frac{1}{MVA_A} - \frac{1}{MVA_B} \right) \tag{C.39}$$

where V_s is the system line-to-line voltage in kV

MVA_A = Three-phase symmetrical short-circuit level on the source side of the series reactor

MVA_B = Three-phase symmetrical fault level on the load side of the series reactor

The first cycle asymmetrical peak is given by

$$I = KI_{SC} \tag{C.40}$$

TABLE C.1

Routine, Design, and Other Tests for Dry Type Series Reactors

		Test Classification		
Test	**When Performed**	**Routine**	**Design**	**Other**
Resistance measurement	The DC resistance measurement shall be made on all units	X		
Impedance measurement	The impedance measurement shall be made on all units	X		
Total loss measurement	Total losses should be measured on all counts	X		
Temperature rise test	The test is performed on one unit out of a number of units of the same design		X	
Applied voltage test	The test shall be made on support insulators when specified			X
Radio interference voltage (RIV) test	The test id performed for nominal system voltage of 230 kV and above, when specified			X
Turn-to-turn test	This test is performed for nominal system voltage of 34.5 kV and below	X		
Lightning impulse test Nominal system voltage >34.5 kV Nominal system voltage <34.5 kV	The lightning impulse test should be performed on all units The lightning impulse test should be performed when specified	X		X
Switching impulse test	The switching impulse test shall be made on support insulators of series rated reactors rated 115 kV or above when specified			X
Chopped wave impulse test	The chopped wave impulse test shall be made on series units when specified			X
Audible sound test				X
Seismic verification test				X
Short-circuit test			X	

Source: IEEE C57.16-1996.

where I_{SC} is the rms symmetrical short-circuit current and K is a factor dependent on R/X of the reactor, which can be read form Table 7 of Reference [18] or calculated from the following expression:

$$K = \left[1 + (e^{-(\phi + \pi/2)R/X}) \sin \phi \right] \sqrt{2} \qquad (C.41)$$

where

$$\phi = \tan^{-1}(R/X) \text{ in radians}$$

For an R/X of 0.005, $K = 2.806$

The standard tolerance specified on the self impedance of the reactor, or on the minimum tap of three-phase reactor having more than one impedance rating shall not vary more than +7% or −3% from its guaranteed value. The self impedance on all other connections shall not vary more than +10% or −3%. For harmonic filter reactors it is often necessary to adhere to much closer tolerances, Volume 3.

The inter-turn test is recommended when applicable. The setup is shown in Figure C.22. It is performed by repeatedly charging a capacitor and discharging it through a sphere gap into the windings. This is more representative of a switching overvoltage. The test duration is for one min and the initial crest value of each discharge is to be $\sqrt{2}$ times the values specified in Table 5 of Reference [18]. Figure C.22 shows the test results, oscillogram in Figure C.23a shows that the reactor has passed turn-to-turn test, while oscillogram in Figure C.23b shows that it has failed the test. The use of oscillograms for failure test is based on change in ringing frequency and damping. The test consists of 7200 overvoltages of the required magnitude.

Following are specimen specifications of a 13.8 kV current limiting reactor suitable for indoor or outdoor installations:

- Dry type, coils vertically stacked one above the other. Middle coil reverse wound.
- 13 kA thermal 3 second withstand.
- 33.15 kA mechanical peak.
- 110 kV BIl across coil.
- 150 kV BIL across interphase insulators.

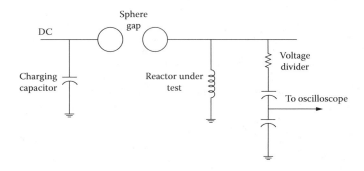

FIGURE C.22
Test circuit rig up for interturn tests on reactor windings. (From IEEE Standard C57.21, IEEE standard requirements terminology and test code for shunt reactors rated over 500 kVA, 1990)

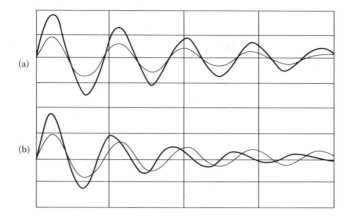

FIGURE C.23
(a) Test results of interturn tests on a reactor, the interturn test results are acceptable and (b) failure on interturn test.

- 110 kV BIL across base insulator.
- Q at 60 Hz = 95 ± 20%.
- Bus work braced for 130 kA peak for 1 s.
- Reactance = 0.4 Ω.
- Continuous current rating = 2000 A.

C.12.1 Duplex Reactor

We will discuss a duplex reactor, which can sometimes be useful and applied to limit short-circuit currents and at the same time improve steady-state performance as compared to a conventional reactor. It consists of two magnetically coupled coils per phase. The magnetic coupling, which is dependent on the geometric proximity of the coils, is responsible for desirable properties of a duplex reactor under short-circuit and load-flow conditions. The equivalent circuit is shown in Figure C.24a and b and its application in Figure C.24c. The mutual coupling between coils is

$$L_m = \sqrt{(L_{11} - L_1)(L_{22} - L_2)} = k\sqrt{L_{11}L_{22}} = kL \tag{C.42}$$

where k is the coefficient of coupling. Note that the inductance of sections 1–4 in the T equivalent circuit, for direction of current flow from source to loads, becomes negative ($= -kL$). Between terminals 2 and 4, and 3 and 4 it is L. The terminal 4 is fictitious and the source terminal is 1, while the load terminals are 2 and 3. Thus, the effective inductance between the source to a load terminal is $(1-k)L$, where L is the inductance of each winding and k is the coefficient of coupling. Effective reactance to load flow is reduced by a factor of k and voltage drops will also be reduced by the same factor. For a short-circuit on any of the load buses, the currents in one of the windings reverses and the effective inductance is $2L(1+k)$. The short-circuit currents will be more effectively limited [20]. The limitations are that k is dependent on the geometry of the coils and is almost independent of the current loading of the coils. Cancellation of the magnetic field under load flow occurs only

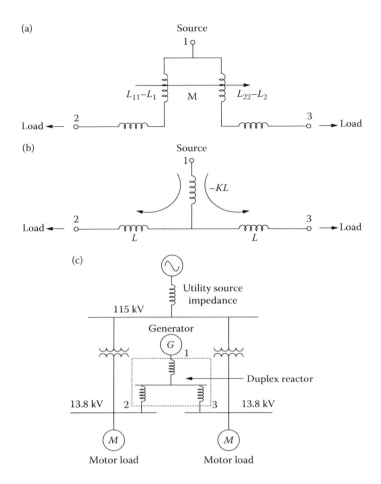

FIGURE C.24
(a, b) Equivalent circuits of duplex reactor and (c) a circuit showing application of a duplex reactor.

when the second winding is loaded. The advantages of a duplex reactor are well exploited if the loads are split equally on two buses (Figure C.24c). For a practical application, see Volume 2.

References

1. ANSI Standard C57.12.70. Terminal markings and connections for distribution and power transformers, 1978 (Revised 1992).
2. IEEE Standard C57.15. Standard Requirements, terminology and test code for step-voltage regulators, 1999.
3. CE Lin, JB Wei, CL Huang, CJ Huang. A new method for representation of hysteresis loops, *IEEE Trans Power Delivery* 4: 413–419, 1989.
4. Canadian/American EMTP User Group. ATP Rule Book. Portland, OR, 1987–1992.
5. JD Green, CA Gross. Non-linear modeling of transformers, *IEEE Trans Ind Appl* 24: 434–438, 1988.

Appendix C: Transformers and Reactors

6. WJ McNutt, TJ Blalock, RA Hinton. Response of transformer windings to system transient voltages, *IEEE Trans PAS* 9:457–467, 1974.
7. JC Das, Surge transference through transformers, *IEEE Ind Mag* 9(5): 24–32, 2003.
8. PTM Vaessen. Transformer model for high frequencies, *IEEE Trans Power Delivery* 3: 1761–1768, 1988.
9. T Adielson, A Carlson, HB Margolis, JA Halladay. Resonant overvoltages in EHV transformers-modeling and application, *IEEE Trans PAS* 100: 3563–3572, 1981.
10. IEEE Standard C37.010. IEEE application guide for high-voltage circuit breakers rated on symmetrical current basis, 1999.
11. JF Perkins. Evaluation of switching surge overvoltages on medium voltage power systems, *IEEE Trans PAS* 101, 1727–1734, 1982.
12. X Chen, SS Venkta. A three-phase three-winding core-type transformer model for low-frequency transient studies, *IEEE Trans Power Delivery* 12: 775–782, 1997.
13. A Narang, RH Brierley. Topology based magnetic model for steady state and transient studies for three-phase core type transformers, *IEEE Trans Power Syst* 9: 1337–1349, 1994.
14. S Lu, Y Liu, JDR Ree. Harmonics generated from a DC biased transformer, *IEEE Trans Power Delivery* 8: 725–731, 1993.
15. DR Smith, SR Swanson, JD Borst. Overvoltages with remotely switched cable fed grounded wye-wye transformers, *IEEE Trans Power Apparatus Syst* 94:1843–1853, 1975.
16. RH Hopkinson. Ferroresonance during single-phase switching of three-phase distribution transformer banks, *IEEE Trans Trans Power Apparatus Syst* 84: 289–293, 1965.
17. IEEE Standard C57.16. IEEE standard requirements terminology and test code for dry-type air-core series connected reactors, 1996.
18. IEEE Standard C57.21. IEEE standard requirements terminology and test code for shunt reactors rated over 500 kVA, 1990.
19. IEEE Standard 32. IEEE standard requirements, terminology and test procedures for neutral grounding devices, 1990.
20. JC Das, WF Robertson, J Twiss. Duplex reactor for large cogeneration distribution system—An old concept reinvestigated. *Proceedings of TAPPI Engineering Conference*, Nashville, TN, 637–648, 1991.

Appendix D: Solution to the Problems

D.1 Solution to the Problems Chapter 1

D.1.1 Problem 1.1

$$i(t) = \frac{e_C}{R} + C\frac{de_C}{dt} + \frac{1}{L}\int_{-\infty}^{t} e_c(t)dt$$

Taking derivative of the both sides

$$\frac{1}{C}\frac{di}{dt} = \frac{d^2e_C}{dt^2} + \frac{1}{RC}\frac{de_C}{dt} + \frac{1}{LC}e_C$$

It is a second-order system and can be characterized by two state variables. If we choose state variables as the inductor current and capacitor voltage, then

$$x_1 = e_C$$

$$x_2 = i_L$$

Then

$$\frac{de_C}{dt} = -\frac{1}{RC}e_C - \frac{1}{C}i_L + \frac{1}{C}i$$

Differentiating

$$\frac{di_L}{dt} = \frac{1}{L}e_C$$

Writing $i(t) = r(t)$

$$\begin{vmatrix} \dot{x}_1 \\ \dot{x}_2 \end{vmatrix} = \begin{vmatrix} -\dfrac{1}{RC} & -\dfrac{1}{C} \\ \dfrac{1}{L} & 0 \end{vmatrix} \begin{vmatrix} x_1 \\ x_2 \end{vmatrix} + \begin{vmatrix} \dfrac{1}{C} \\ 0 \end{vmatrix} r(t)$$

Note that the state variables describing the system are not unique. For example, we could choose flux linkage in the inductor as a variable.

629

D.1.2 Problem 1.2

$$y(t) = ax(0) + br(t)$$

$$y(t) = [r(t)]^2$$

The first relation represents a linear system, and it has the property of decomposition and is zero input and zero state linear. The second relation is not homogeneous and therefore, nonlinear.

D.1.3 Problem1.3

Note that in a well-coordinated system, the 1500 kVA transformers, their primary fuses, secondary main low-voltage breakers, or feeder breakers need not be considered. A fault in this part of the distribution system does not impact the reliability calculations at the required point. However, a fault in the primary disconnect switch, on the source side of the fuses will trip the main 13.8 kV breaker 1. Thus, the calculations are as shown in Table D1.1.

D.1.4 Problem 1.4

Only the following part of Equation 1.49 is of interest:

$$f_P r_P = \frac{\lambda_3 r_3 \lambda_4 r_4}{8760}$$

Substituting the values,

$$f_P r_P = \frac{1.3 \times 1.5 \times 0.67 \times 0.34}{8760} = 5.0708 \times 10^{-5}$$

TABLE D1.1

Reliability Indices in Problem 1.3

Equipment	λ	λ_r
13.8 kV Metal-clad breaker 1	0.001850	0.000925
Primary protection and controls	0.000600	0.003000
13.8 kV Cable 213.4 m	0.001652	0.026032
Cable terminations (4)	0.011840	0.008880
Disconnect switch (3)	0.005220	0.005220
2000 kVA Transformer	0.010800	1.430244
2.4 kV Breaker 2	0.001850	0.000925
Secondary protection and controls	0.000600	0.003000
2.4 kV Switchgear bus	0.009490	0.069182
2.4 kV Motor starter (4)	0.015300	0.061200
Total at point of use	0.059202	1.608608

Note: The primary system and utility tie to 13.8 kV bus are not shown.

Appendix D: Solution to the Problems　　　　　　　　　631

Note the efficacy and much higher reliability of circuits serving loads through paralleled systems.

D.1.5　Problem 1.5

This is not discussed in the text; a reader is supposed to investigate this problem.

Breakdown Maintenance

The maintenance efforts required to restore the equipment and systems to operating conditions after a failure or fault.

Predictive Maintenance

Predictive maintenance is the practice to conduct diagnostic tests and inspections during normal equipment operations in order to detect incipient weaknesses or impeding failures.

Preventive Maintenance

This is the practice of conducting routine inspections, tests, and servicing, so that impeding troubles can be detected and reduced or eliminated. There is always a debate: how much maintenance is required and at what cost benefits? The maintenance may require a process shutdown and the cost of the loss of production has to be balanced with cost reductions obtained through improved safety and performance. It depends upon the type of facility to be maintained—a shutdown for maintenance in an electronic manufacturing plant will be more expensive that in a small industrial facility. The effort starts with a survey of each equipment and then adopting some guidelines, for example, NFPA 70B and NETA MTS-1993.

Reliability-Centered Maintenance (RCM)

RCM is a systematic methodology that establishes initial preventive maintenance requirements or optimizes existing preventive maintenance requirements based on the consequences of equipment failure. RCM was first developed by the airline industry and has since been adopted by many other industries. The criticality and failure modes are analyzed and then the effective and preventive maintenance for each component is determined. All possible failure modes and preventive measures are evaluated for each component, thus establishing a program that is proven to result in improved facility reliability. It focuses on system components that are critical to maintain system functions.

Testing and repairs are integral parts of any maintenance activity. The repairs may be warranted immediately, as in the case of a breakdown, or the identified equipment can be earmarked for future upgrades and repairs depending upon its existing operational condition.

D.2　Solution to Problems Chapter 2

D.2.1　Problem 2.1

Overall efficiency $= 98.5 \times 98 \times 70 = 67.57\%$

　P = density of water $= 1000 \, \text{kg/m3}$

　Then from Equation 2.1

　$9.81 q h \eta = 9.81 \times 1000 \times 100 \times 0.6757 = 662 \times 10^3 \, \text{kW}$

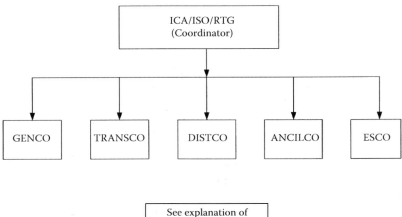

FIGURE D2.1
New structure of utility companies after deregulation.

D.2.2 Problem 2.2

The utility companies in the USA have a horizontal architecture. Thus, the organization chart can be drawn as shown in Figure D2.1.

D.2.3 Problem 2.3

The compressed air energy system consists of an air compressor, expansion turbine, M-G set, and an overhead or underground storage tank. When air is compressed from a pressure P_1 to P_2, then the energy stored is

$$E = \frac{n(P_2V_2 - P_1V_1)}{n-1}$$

where V_1 and V_2 are the volumes at P_1 and P_2.
This follows from the gas compression law:

$$PV^n = C$$

The isentropic value of n for air = 1.4, and under working conditions, it is approximately 1.3.
The temperature at the end of compression is given by

$$T_2 = T_1 \left(\frac{P_2}{P_1}\right)^{\frac{n-1}{n}}$$

Electrical power is generated by compressed air flow, under constant volume or constant pressure, through an expansion turbine. For constant pressure operation, a tank maintains a constant pressure by a weight on the tank cover.

Relatively, a compressed air storage system is more expensive than pumped storage systems or batteries.

Appendix D: Solution to the Problems 633

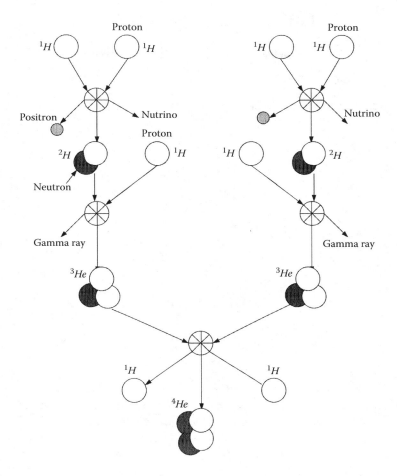

FIGURE D2.2
Proton cycle of fusion in stars.

D.2.4 Problem 2.4

A proton–proton reaction schematic is shown in Figure D2.2. A reader may search a web site for explanation of the process, not detailed here.

D.3 Solution to Problems Chapter 3

D.3.1 Problem 3.1

From the given data,

$$\eta_g = 0.98, \quad \eta_b = 0.95, \quad \eta_c = 0.97$$

$$\eta = \eta_g \eta_b \eta_c = 0.904$$

634   *Appendix D: Solution to the Problems*

From Equations 3.14 and 3.15 in the text,

$$P = c_P \left(\frac{1}{2} A \rho v^3 \right) W$$

Substituting

$$A = \frac{\pi}{4} D^2$$

where D is the rotor diameter in meters,

$$P = \eta \left(\frac{\pi}{8} c_P \rho D^2 v^3 \right)$$

Substituting the given values,

$$P = 0.904 \times \frac{\pi}{8} \times 0.49 \times 1.18 \times 110^2 \times 11^3 \times 10^{-6} \text{ MW}$$

$$= 3.305 \text{ MW}$$

D.3.2 Problem 3.2

See text. Based on the fundamental definitions of cells, module, panel, array, array sub-field, and array field, a schematic illustration is shown in Figure D.3.1.

D.4 Solution to Problems Chapter 4

D.4.1 Problem 4.1

The voltage at the source is $480\,(\sqrt{2})\sin(2\pi ft + 30°)$, its profile shown in Figure D.4.1. As we neglect the load currents in the short-circuit calculations, this is also the fault point voltage at the receiving end.

If a short circuit occurs when the voltage wave is crossing through zero amplitude on the x-axis, the direct current (DC) offset is maximum and if the short circuit occurs when the voltage wave peaks (60° from zero crossing in Figure D.4.1), the DC offset is zero. This statement is true only for a purely reactive circuit.

Note that in Equation 4.2 the angle $\phi = 90°$, that is, it is a purely reactive circuit. When resistance is present, the peak will occur *slightly before* the 1/2 cycle; this is further discussed in Chapter 10 of this volume.

D.4.2 Problem 4.2

In synchronous generators and dynamic loads like synchronous and induction motors, the assumption of constant inductance during short circuit is not valid. The trapped flux in

Appendix D: Solution to the Problems 635

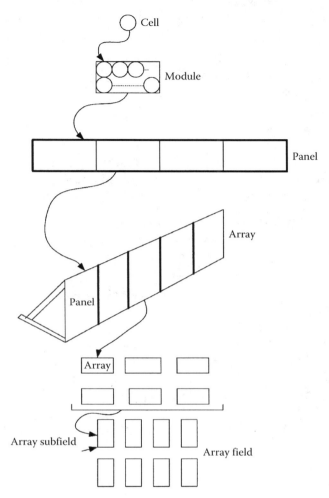

FIGURE D3.1
Configuration from cell to solar array field.

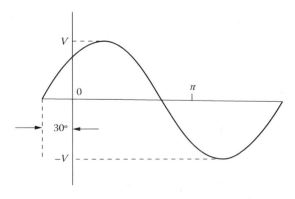

FIGURE D4.1
Waveform of the voltage for Problem P4.1.

Appendix D: Solution to the Problems

the rotating machines cannot change suddenly and decays depending upon machine constants, thus, giving rise to decaying alternating current (AC) component. (This is further discussed in Chapter 9.)

The decay of AC component will be most prominent close to a synchronous generator and will reduce with the added impedances in the circuit. For example, a generator connected through transformers, cables, or line impedances.

D.4.3 Problem 4.3

This question prompts a reader to familiarize with basic matrix concepts—study Appendix A: Matrix Methods (see Sections A.3 and A.3.1 for an answer). If a square matrix $\bar{A}$ of $n \times n$ has n independent eigenvectors, then a matrix $\bar{P}$ can be found, so that $\bar{P}^{-1} \bar{A} \bar{P}$ is a diagonal matrix. The transformation of $\bar{A}$ in to $\bar{P}^{-1} \bar{A} \bar{P}$ is called similarity transformation. *Diagonalization is a special case of similarity transformation.*

D.4.4 Problem 4.4

Following Section A.3.1 (Appendix A), the characteristic equation is

$$\begin{vmatrix} 6-\lambda & -2 & 2 \\ -2 & 3-\lambda & -1 \\ 2 & -2 & 3-\lambda \end{vmatrix} = 0$$

The eigenvalues are solution of λ in the above equation.

$$\lambda^3 - 12\lambda^2 + 35\lambda - 30 = 0$$

Factorizing,

$$\lambda^3 - 12\lambda^2 + 35\lambda - 30 = 0$$

$$(\lambda - 2)(\lambda^2 - 10\lambda + 15) = 0$$

$$(\lambda - 2)(\lambda - 1.838)(\lambda - 8.162) = 0$$

Therefore, $\lambda = 2, 8.162, 1.838$ are the eigenvalues.

D.4.5 Problem 4.5

The sequence impedances on 100 MVA base in per unit are shown in Table D4.1.

Here, we assume that zero sequence impedance of lines is twice that of positive sequence impedance.

The positive, negative, and zero sequence networks on the pattern of Example 4.3 are as shown in Figure D4.2a.

In order to avoid disappearance of a node, the positive and zero sequence networks can be drawn as shown in Figure D4.2b, reference Figure 4.15 of the text. The integrity of nodes is preserved in the computer simulations.

Appendix D: Solution to the Problems

TABLE D4.1

Sequence Impedance Data

Equipment	$Z+$	$Z-$	Z_0
G1 and G3	$j0.75$	$j0.90$	$j0.15$
G2	$j0.46$	$j0.57$	$j0.086$
G1 step-up transformer	$j0.53$	$j0.53$	$j0.53$
G2 step-up transformer	$j0.40$	$j0.40$	$j0.40$
G3 step-up transformer	$j0.60$	$j0.60$	$j0.60$
10 Ω 138 kV Lines	$j0.053$	$j0.053$	$j0.105$

D.4.6 Problem 4.6

For clarity, the impedances are connected as shown in Figure D4.3. This forms an unbalanced delta load. We can first find the currents in the delta load connection and the line currents and then convert these into sequence component currents.

The supply voltages are balanced, counterclockwise rotation

$$V_{ab} = 480 < 0°$$

$$V_{bc} = 480 < -120°$$

$$V_{ca} = 480 < 120°$$

Therefore,

$$I_{ab} = 48 < 0°$$

$$I_{bc} = 24 < -120°$$

$$I_{ca} = 24 < 120°$$

Then,

$$I_a = I_{ab} + I_{ac} = I_{ab} - I_{ca} = 48 < 0° - 24 < 120° = 54.15 < -22.58°$$

$$I_b = I_{bc} - I_{ab} = 24 < -120° - 48 < 0° = 54.14 < -157.43°$$

$$I_c = I_{ca} - I_{bc} = 24 < 120° - 24 < -120° = 41.58 < 90°$$

These currents can be converted to sequence component currents using Equation 1.33.

$$\begin{vmatrix} I_0 \\ I_1 \\ I_2 \end{vmatrix} = \frac{1}{3} \begin{vmatrix} 1 & 1 & 1 \\ 1 & a & a^2 \\ 1 & a^2 & a \end{vmatrix} \begin{vmatrix} 54.15 < -22.58° \\ 54.15 < -157.43° \\ 41.58 < 90° \end{vmatrix} = \begin{vmatrix} 0 \\ 43 - j25 \\ 7 + j4 \end{vmatrix}$$

Note that there cannot be any zero sequence currents in the lines.

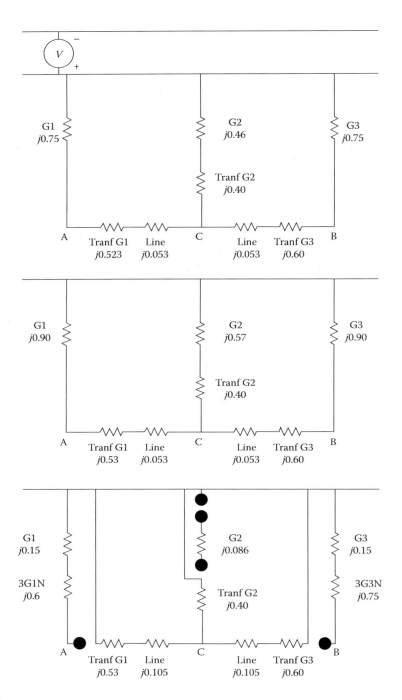

FIGURE D4.2
(a) Positive, negative, and zero sequence networks for circuit configuration of Problem 4.5 and (b) zero sequence network redrawn to avoid discontinuity.

(*Continued*)

Appendix D: Solution to the Problems

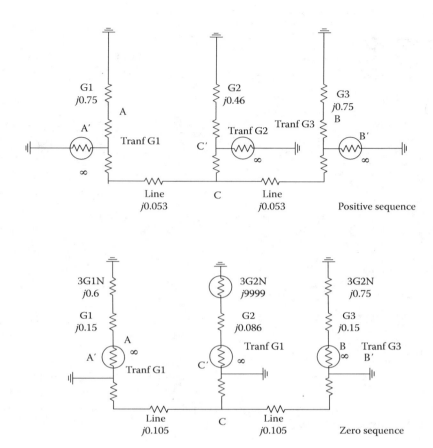

FIGURE D4.2 (CONTINUED)
(a) Positive, negative, and zero sequence networks for circuit configuration of Problem 4.5 and (b) zero sequence network redrawn to avoid discontinuity.

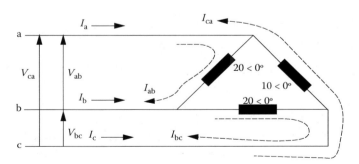

FIGURE D4.3
Unbalanced delta-connected load, Problem P4.6.

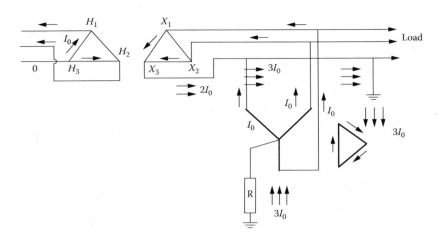

FIGURE D4.4
Distribution of zero sequence currents, Problem P4.7

D.4.7 Problem 4.7

This is shown in the Figure D4.4.

D.4.8 Problem 4.8

Note that due to load unbalance and ungrounded wye-connected load, the phase voltages will be unequal, as no unbalance current can return to the supply system neutral.

In this problem, the method of symmetrical components cannot be straightaway applied. At the neutral point of the load,

$$I_{an} + I_{bn} + I_{cn} = 0$$

Also it follows that

$$V_{ab} = V_{an} - V_{bn} = I_{an}Z_{an} - I_{bn}Z_{bn}$$

$$V_{bc} = V_{bn} - V_{cn} = I_{bn}Z_{bn} - I_{cn}Z_{cn}$$

Neither the voltages nor the currents are known in these equations. It can be shown with some manipulations that

$$V_{ao} = Z_{an}\left[\frac{V_{ab}(Z_{bn}+Z_{cn})+V_{bc}Z_{bn}}{Z_{an}(Z_{bn}+Z_{cn})+Z_{cn}Z_{bn}}\right]$$

where V_{ao} is the voltage of phase A to common neutral point.

The voltage vectors to neutral have been specified in Figure P4.2 as

$$V_{an} = 1 < 0°$$

$$V_{bn} = 1 < -120°$$

$$V_{cn} = 1 < 120°$$

Appendix D: Solution to the Problems

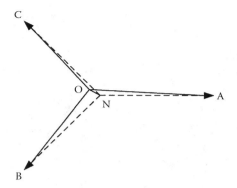

FIGURE D4.5
Displacement of neutral due to unbalance wye-connected loads, Problem 4.8.

Then the line-to-line voltages will be

$$V_{ab} = \sqrt{3} < 30°$$
$$V_{bc} = \sqrt{3} < -90°$$
$$V_{ca} = \sqrt{3} < 150°$$

Also

$$Z_{an} = 6.0 < 0° = Z_{bn}$$
$$Z_{cn} = 5.0 < 0°$$

Then from the equation above

$$V_{ao} = 1.032 - j0.054 = 1.0334 < -3°$$

Therefore, the residual voltage between O and N is

$$V_{on} = V_{oa} + V_{an} = -V_{a0} + V_{an} = -0.032 + j0.054 = 0.062 < 120.65°$$

The displacement is small because the load unbalance is small. This shift of neutral is illustrated in Figure D4.5.

D.4.9 Problem 4.9

The equations are simply

$$i = I_m \sin(\omega t + \theta - \phi) - I_m \sin(\theta - \phi) e^{-Rt/L}$$
$$i = I_m \sin(\omega t + \theta - \phi - 120°) - I_m \sin(\theta - \phi - 120°) e^{-Rt/L}$$
$$i = I_m \sin(\omega t + \theta - \phi + 120°) - I_m \sin(\theta - \phi + 120°) e^{-Rt/L}$$

D.4.10 Problem 4.10

Consider a load at a power factor of 0.7, 0.8, and 0.9. These are practical operating power factors of a load. The loads can be represented in a number of different models, which is further discussed in Volume 2. These power factors correspond to an X/R of 1.02, 0.75, and 0.48, respectively.

As we have seen in this chapter, the fault point X/R ratios in practical electrical systems are high, say in a 13.8 kV system close to a generator the X/R can be of the order of 60–80, depending upon the rating of the generator. Even in low-voltage systems, the X/R ratios at the fault point will be of the order of 4–10. Thus, by theorem of superimposition, the addition of loads to the short-circuit currents will reduce the X/R ratio overall and, therefore, the asymmetry.

D.4.11 Problem 4.11

The required distribution of flow of currents is shown in Figure D4.6.

D.4.12 Problem 4.12

Redraw the positive sequence network considering all sources connected to an infinite bus and the fault point F with intervening impedances according to the system configuration. This is shown in Figure D4.7.

The network can be reduced to single impedance. Wye–delta transformations as discussed in this chapter are applicable in this reduction. The negative sequence network can be similarly drawn. The zero sequence network to the fault point should consider the discontinuities. Generator G_3 is ungrounded and cannot contribute to the zero sequence currents. Generators G_1 and G_2 are grounded but cannot contribute any zero sequence currents due to transformer connections. Ground fault current at F can only return through the wye connected solidly grounded neutral of transformer T_1. Therefore, the zero sequence impedance is simply the zero sequence impedance of transformer T_1, which is given as $j0.10$.

D.4.13 Problem 4.13

Proof

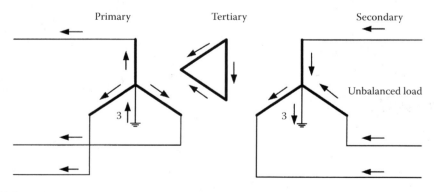

FIGURE D4.6
Flow of zero sequence currents in a three-winding transformer, with grounded wye neutrals and a tertiary delta, Problem P4.11.

Appendix D: Solution to the Problems

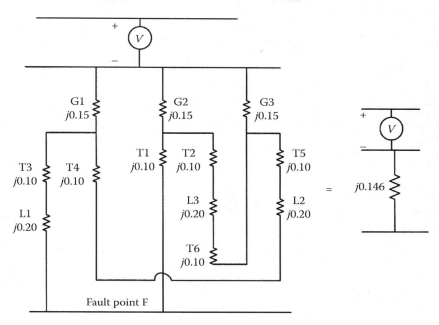

FIGURE D4.7
Positive sequence network and its reduction to a single equivalent impedance, Problem P4.12.

A differential equation of this single-phase circuit is written as

$$L\frac{di}{dt} + Ri = V\sin(\omega t + \theta)$$

Taking Laplace transform,

$$Ri(s) + Lsi(s) - LI(0) = V(\sin\omega t\cos\theta + \cos\omega t\sin\theta)$$

$$= V\left(\frac{\omega\cos\theta}{s^2+\omega^2} + \frac{s\sin\theta}{s^2+\omega^2}\right)$$

$$i(s) = \frac{V}{L}\frac{1}{s+(R/L)}\left[\frac{\omega\cos\theta}{s^2+\omega^2} + \frac{s\sin\theta}{s^2+\omega^2}\right]$$

This can be resolved into partial fractions.

$$\frac{1}{(s+\alpha)(s^2+\omega^2)} = \frac{1}{\alpha^2+\omega^2}\left[\frac{1}{s+\alpha} - \frac{s}{s^2+\omega^2} + \frac{\alpha}{s^2+\omega^2}\right]$$

where $\alpha = R/L$, therefore,

$$i(s) = \frac{V\omega\cos\theta}{L(\alpha^2 + \omega^2)} \left[\frac{1}{s + R/L} - \frac{s}{s^2 + \omega^2} + \frac{\alpha}{s^2 + \omega^2} \right]$$

$$+ \frac{V\sin\theta}{L(\alpha^2 + \omega^2)} \left[\frac{1}{s + R/L} - \frac{s}{s^2 + \omega^2} + \frac{\alpha}{s^2 + \omega^2} \right]$$

The inverse Laplace transform of function in Equation 6.4 is

$$L^{-1} \frac{1}{(s+\alpha)(s^2 + \omega^2)} = \frac{1}{(\alpha^2 + \omega^2)} \left[e^{-\alpha t} - \cos\omega t + \frac{\alpha}{\omega}\sin\omega t \right]$$

Therefore,

$$L^{-1} \frac{s}{(s+\alpha)(s^2 + \omega^2)} = \frac{1}{(\alpha^2 + \omega^2)} \left[-\alpha e^{-\alpha t} + \omega\sin\omega t + \cos\omega t \right]$$

Substituting these results in above equation for $i(s)$

$$i = \frac{V}{L(\alpha^2 + \omega^2)} \left[\omega\cos\theta \left(e^{-\alpha t} - \cos\omega t + \frac{\alpha}{\omega}\sin\omega t \right) + \sin\theta \left(\alpha\cos\omega t + \omega\sin\omega t - \alpha e^{-\alpha t} \right) \right]$$

$$= \frac{V}{L(\alpha^2 + \omega^2)} \left[(\omega\cos\theta - \alpha\sin\theta)e^{-\alpha t} - (\omega\cos\theta - \alpha\sin\theta)\cos\omega t + (\alpha\cos\theta + \omega\sin\theta)\sin\omega t \right]$$

Remembering that

$$\tan\phi = \omega L/R = \omega\alpha, \; \sin\phi = \omega/(\alpha^2 + \omega^2)^{1/2}, \; \cos\phi = \alpha/(\alpha^2 + \omega^2)^{1/2}$$

$$i = \frac{V}{\left(R^2 + \omega^2 L^2\right)^{1/2}} \left[\sin(\omega t + \theta - \phi) - \sin(\theta - \phi)e^{-\alpha t} \right]$$

A solution using differential equations is simpler.
The differential equation is

$$L\frac{di}{dt} + Ri = E_m \sin(\omega t + \theta)$$

This is a first-order nonhomogeneous differential equation.
The solution is given by

$$i = i_c + i_p$$

That is, sum of complementary solution and particular solution.
The auxiliary equation is

Appendix D: Solution to the Problems 645

$$L\frac{di}{dt} + Ri = 0$$

or

$$\frac{di}{i} + \frac{R}{L}di = 0$$

Integrating

$$\log i + \frac{R}{L}i = k$$

or

$$\log_e i = \log_e e^{(-R/L)t} + \log A$$

Taking antilog

$$i = Ae^{(-R/L)t}$$

A depends on initial conditions.
  Take a trial solution

$$i = C\cos(\omega t + \theta) = D\sin(\omega t + \theta)$$

If we take first and second differentials, then

$$C = -E_m\frac{\omega L}{R^2 + (\omega L)^2} = -E_m\sin\phi$$

$$D = E_m\frac{R}{R^2 + (\omega L)^2} = E_m\cos\phi$$

where

$$\phi = \tan^{-1}\frac{\omega L}{R}$$

Substituting

$$i = -\frac{E_m}{\sqrt{R^2 + \omega^2 L^2}}\sin\phi\cos(\omega t + \theta) + \frac{E_m}{\sqrt{R^2 + \omega^2 L^2}}\cos\phi\cos(\omega t + \theta)$$

$$= \frac{E_m}{\sqrt{R^2 + \omega^2 L^2}}\sin(\omega t + \theta - \phi)$$

Therefore, the complete solution is given as

$$i = \frac{E_m}{\sqrt{R^2 + \omega^2 L^2}} \sin(\omega t + \theta - \phi) + A e^{(-R/L)t}$$

A reader can prove the values of constant depending on the instant of switch closure as in the text.

D.5 Solution to Problems Chapter 5

D.5.1 Problem 5.1

From Figure D5.1, the sequence impedances calculated for a fault on the high side of the transformer T2 (point "O" in Figure 5.9) are as follows:

$$Z_1 = j0.26$$
$$Z_2 = j0.284$$
$$Z_0 = j0.139$$

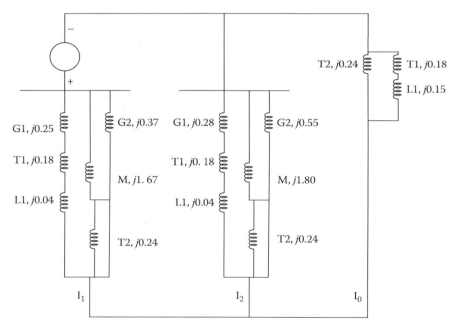

FIGURE D5.1
Sequence impedance diagram, double line-to-ground fault, to point O in Figure. 5.9, Problem P5.1.

Appendix D: Solution to the Problems 647

For a double line-to-ground fault and from Equation 5.18

$$I_1 = \frac{1}{j0.26 + \dfrac{j0.284 \times j0.139}{j0.284 + j0.139}} = -j2.830 \text{ pu}$$

$$V_b = V_c = 0$$

Also,

$$I_2 = \frac{-j0.139}{j0.284 + j0.139}(-j2.83) = j0.93 \text{ pu}$$

$$I_0 = -\frac{j0.284}{j0.284 + j0.139}(-j2.83) = j1.90 \text{ pu}$$

$$I_b = I_0 + aI_1 + a^2 I_2$$
$$= j1.90 - 0.5(I_1 + I_2) - j0.866(I_1 - I_2)$$
$$= -3.256 + j2.85$$

$$I_c = I_0 + a^2 I_1 + aI_2$$
$$= j1.90 - 0.5(I_1 + I_2) + j0.866(I_1 - I_2)$$
$$= 3.256 + j2.85$$

The total fault current is

$$I_b + I_c = j5.70$$

Check that total fault current is $3I_0 = 3 \times j1.90 = j5.70$, as calculated above.

Akin to Example 5.2, the sequence components of currents can be evaluated in each section of the network.

$$V_1 = 1.0 - (I_1 Z_1) = 1.0 - (-j2.83 \times j0.26) = 0.264 \text{ pu}$$

$$V_2 = 0 - I_2 Z_2 = 0 - (j0.93)(j0.284) = 0.264 \text{ pu}$$

$$V_0 = 0 - I_0 Z_0 = 0 - (j1.90)(j0.139) = 0.264 \text{ pu}$$

Therefore,

$$V_a = 0.79 \text{ pu} = (109 \text{ kV})$$

Positive sequence currents

Positive sequence current through T_1, $L_1 = -j1.513$ pu
Positive sequence current through T_2 and parallel combination of G_2 and $M = -j1.316$ pu

648   *Appendix D: Solution to the Problems*

This gives

$$I_{1,G2} = -j1.077 \, \text{pu}$$

$$I_{1,M} = -j0.239 \, \text{pu}$$

Negative sequence currents

Negative sequence current through T_1, $L_1 = j0.53$

Negative sequence current through T_2 and parallel combination of G_2 and $M = j0.40$

This gives

$$I_{2,G2} = j0.306 \, \text{pu}$$

$$I_{2,M} = j0.094 \, \text{pu}$$

Zero sequence currents

$$I_{0,G2} = 0$$

$$I_{0,M} = 0$$

Zero sequence current contributed by $T_{2=}j1.1 \, \text{pu}$

Zero sequence current flowing through L_1 and $T_1 = j0.80 \text{pu}$

Calculate generator G_2 currents

$$I_{aG2} = -j1.077 + j0.306 = -j0.771 \, \text{pu} = 3.23 \, \text{kA}$$

$$I_{bG2} = -0.5(-j1.077 + j0.306) - j0.866(-j1.077 - j0.306) = -1.1198 + j0.3855 = 4.95 \, \text{kA}$$

$$I_{cG2} = 1.1198 + j0.289 = 4.95 \, \text{kA}$$

The currents through the motor are as follows:

$$I_{aM} = -j0.239 + j0.094 = -j0.145 \, \text{pu} = 0.61 \, \text{kA}$$

$$I_{bM} = -0.5(-j0.239 + j0.094) - j0.866(-j0.239 - j0.094) = -0.288 + j0.0725 = 1.24 \, \text{kA}$$

$$I_{cM} = 0.288 + j0.0725 \, \text{pu} = 1.24 \, \text{KA}$$

Currents through line L_1 and T_1 are as follows:

$$I_{aL1} = j0.80 - j1.513 + j0.53 = -j0.183 \, \text{pu} = 0.077 \, \text{kA}$$

$$I_{bL1} = j0.80 - 0.5(-j1.513 + j0.53) - j0.866(-j1.513 - j0.53) = -1.77 + j1.292 = 0.917 \, \text{kA}$$

$$I_{cL1} = 1.77 + j1.292 = 0.917 \, \text{kA}$$

Calculate voltages at bus B

$$V_1 = 1 - (Z_{g2})(I_{1G2}) = 1 - (j0.37)(-j1.077) = 0.398 \, \text{pu}$$

$$V_2 = 0 - (j0.55)(j0.306) = 0.168 \, \text{pu}$$

$$V_0 = 0$$

Appendix D: Solution to the Problems 649

Therefore, the bus voltages are as follows:

$$V_a = 0.398 + 0.168 = 0.566 \, \text{pu} = 7.81 \, \text{kV}$$

$$V_b = -0.5(0.398 + 0.168) - j0.866(0.398 - 0.168) = -0.283 - j0.199 \, \text{pu} = 4.77 \, \text{kV}$$

$$V_c = -0.283 + j0.199 \, \text{pu} = 0.346 \, \text{pu} = 4.77 \, \text{kV}$$

Similarly, the voltages at other points can be calculated.

D5.2 Problem 5.2

Solve similar to solution to Problem 5.1 and Example 5.2.

D.5.3 Problem 5.3

We can apply theorem of superimposition in the short-circuit calculations as nonlinearity is ignored. Therefore, the 46 MVA generator contributes 12.03 kA to 13.8 kV bus (based on the saturated subtransient reactance of 16% on the generator 46 MVA base). As the three-phase fault current is to be limited to 28 kA, the contribution of the 40/60 MVA transformer through the utility source of 5200 MVA should not exceed 15.97 kA. One way of solution will be to try different values of the transformer percentage reactance and solve the problem by hit and trial. A better solution is obtained as follows:

28 kA at 13.8 kV, when converted to an equivalent impedance on 100 MVA base = 0.1494 pu.

Generator contribution of 12.03 kA, when converted to an equivalent impedance on 100 MVA base = 0.348 pu.

Let the equivalent impedance of the contribution from utility source and transformer be x.

Then:

$$\frac{0.348 \times x}{0.348 + x} = 0.1494$$

Solving $x = 0.262$ pu
Also,

$$x = Z_{pu}(\text{utility} + \text{transformer})$$

The utility source impedance can be expressed as 0.01923 pu.

$$x = 0.01923 + \text{Transformer} = 0.262$$

$$\text{Transformer} = 0.2428 \, \text{pu}$$

This is calculated on a 100 MVA base. Therefore, the transformer should have a percentage impedance of 9.71 on 40 MVA base.

Note that we have ignored resistances and X/R ratios. In practical applications, a vectorial calculation should be made to find out the impedance of the transformer.

650   *Appendix D: Solution to the Problems*

D.5.4 Problem 5.4

Based on Problem 5.3, and the transformer reactance as calculated, and a similar new generator added to the 13.8 kV bus, the short-circuit current at 13.8 kV bus is 40.03 kA rms symmetrical. It needs to be limited to 36 kA. In case we increase the subtransient reactance of the generators or the utility interconnecting transformer, the short-circuit current at 13.8 kV bus can be limited. But this will not be a practical solution. Consider the following:

- The impedances of the equipment cannot be arbitrarily increased. It may not be practical to manufacture such equipment or cost premiums must be paid to accommodate impedances beyond the normal manufacturing range. For example, the normal impedance of high-voltage transformers is dictated by the primary BIL level and the voltage of application.
- Figure P5.1 shows that two generators and a utility interconnection are all connected to the same bus. This is not a practical design, and a system configuration like this will not be used. A fault on the single bus will result in loss of all loads and all sources.
- As a simple system modification, the bus can be sectionalized and interconnected through a current-limiting reactor. The loads can be rearranged on each bus.
- Currently, 13.8 kV circuit breakers rated 63 kA are commercially available. Many times the existing underrated breakers can be retrofitted with higher interrupting rated circuit breakers.
- For the purpose of this problem, a reader can calculate two equal current-limiting reactors introduced in each generator to limit the short-circuit level at the bus to the required value of 36 kA. Note, however, that large generator reactors have an adverse effect on load flow (Volume 2).
- Adding reactors will have an adverse impact on load flow. The system configurations prima fascia should be designed to meet the requirements of short circuit and load flow. If reactors are introduced, it should be ensured that these do not cause unacceptable voltage dips during normal and contingency operations (see Volume 2).

D.5.5 Problem 5.5

This problem is similar to Problem 4.14 that the sequence networks must be drawn for the fault point and then reduced into single impedance. Again, we ignore resistance, which is not admissible in practical applications.

The positive, negative, and zero sequence impedances calculated at bus C are as follows:

$$Z^+ = j0.379$$

$$Z^- = j0.421$$

$$Z^0 = j0.182$$

Therefore, the three-phase short-circuit current is 1.128 < −90°kA, a low value indeed. The short-circuit currents in large 138 kV systems can even exceed 40 kA rms symmetrical. The low value is due to rather small generators and transformers in the example.

The single line-to-ground fault current using equations in the text is = 1.37 kA < −90°

Appendix D: Solution to the Problems 651

The presence of generators and rotating motor loads will increase the single line-to-ground fault current on the high side of a delta–wye-connected transformer, because the positive and negative sequence impedances will be reduced (see solution to Problem 5.8).

D.5.6 Problem 5.6

As we have already calculated the sequence impedances to the fault point, it is easy to calculate the double line-to-ground fault and also phase-to-phase fault.

The double line-to-ground fault current is $0.827 < -90°$ kA, and phase-to-phase fault is $0.905 < -90°$ kA.

Thus, in this example, in order of the short-circuit current magnitudes,

- Single line-to-ground fault = 1.37 kA
- Three-phase fault = 1.128 kA
- Two-phase fault = 0.905 kA
- Double line-to-ground fault = 0.827 kA

Therefore, it should not be concluded that in a power system the three-phase short-circuit current is *always* the highest, though normally it is so. In low-voltage industrial distribution systems, close to the solidly grounded secondary of a wye-connected three-phase transformer, a single line-to-ground fault current will be higher than the three-phase fault current, if the contributions from the rotating loads are ignored, that is, a transformer serving drive system loads.

It is easy to see that the relative magnitude of the short-circuit currents will depend upon the relative magnitudes of the positive, negative, and zero sequence components (see Figure 11.1).

D.5.7 Problem 5.7

Only positive sequence impedance network to the fault point is required to be constructed. The line impedances are on a 100 MVA base. The transformer, generator, and motor impedances are on their base ratings. The transmission line voltage is not specified. Assume it is 69 kV. This is more of an exercise of reducing the positive sequence network to a single element at the faulted bus. Figure D5.2 is constructed for the positive sequence network

The network reduction gives a positive sequence impedance of $j0.343$ at the fault point. Therefore, the three-phase symmetrical short-circuit current is 2.44 kA. Note that any transmission voltage could have been assumed, because the line impedances are specified in pu. Also in all previous examples, we have calculated the symmetrical currents and not the asymmetrical currents. A further observation of interest is that generator and motor impedances are considered time invariant. How the decay is accounted for in American National Standard Institute (ANSI)/Institute of Electrical and Electronics Engineers, Inc., USA (IEEE) calculations of short-circuit currents is discussed in Chapter 10.

D.5.8 Problem 5.8

With the transformer connection, the line-to-ground fault current in 115 kV system does not impact the 4.16 kV system. The delta windings act as a sink to the zero sequence currents. Therefore, as far as ground fault currents are concerned, the 115 kV and 4.16 kV

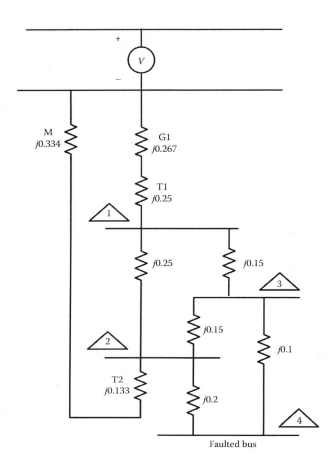

FIGURE D5.2
Positive sequence network for system configuration of Figure P5.2.

systems can be considered isolated. Yet the 8000 hp motor load contributes to the ground fault current on the 48 kV side indirectly by modifying the positive and negative sequence impedances. This is an important concept.

As the power factor and efficiency of the motor load is specified, 8000 hp = 7469 kVA. Using the same locked rotor impedance of 16.7% (Table 5.2), the motor impedance on 100 MVA base = $j2.236$ pu. The negative sequence impedance of rotating machines is not the same as the positive sequence impedance; here, we assume it to be the same.

Then, the ground fault current is 20.8 kA at the 4.16 kV bus. A reader can similarly demonstrate that on the 46 kV side the single line-to-ground fault current will slightly increase due to motor load (see Figure D5.3).

D.5.9 Problem 5.9

The neutrals of the wye–wye-connected transformer are isolated. The load served is between two phases on the secondary side of the transformer. As there is no connection to ground, the secondary load between two phases (32A) on the secondary will be reflected as load between corresponding phases on the wye-connected primary of the transformer. No current flows through the tertiary winding as there are no zero sequence currents.

Appendix D: Solution to the Problems

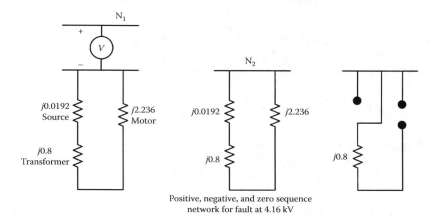

FIGURE D5.3
Positive, negative, and zero sequence network for a ground fault at 4.16 kV, Problem P5.8

D.5.10 Problem 5.10

In Example 5.2, the voltage at bus B on phases B and C is 1.034 pu. Therefore, COG=0.597 by definition.

The maximum voltage on primary side of transformer T2 is 1.17 pu. Therefore, COG is 0.675 by definition.

Also, Equation 5.37 for calculation of COG is

$$\mathrm{COG} = \frac{\sqrt{k^2 + k + 1}}{k + 2}$$

where $k = \dfrac{Z_o}{Z_1}$

We calculated $Z_0 = j0.2$ and $Z_1 = 0.212$ for a fault at bus 2.
Then $k = 0.9434$, and from Equation 2.37, COG=0.572.
Calculation of COG by calculating the bus voltages is more accurate.
For a double line to ground fault calculated in Problem 5.1, the COG is given by Equation 5.33.

$$\mathrm{COG} = \frac{\sqrt{3}k}{1 + 2k}$$

$$k = \frac{Z_0}{Z_1} = \frac{j1.90}{j2.83} = 0.671$$

$$\mathrm{COG} = 0.496$$

This can also be calculated directly from the voltage V_a at the fault point. By definition of COG, it is equal to 0.456.

654

D.5. 11 Problem 5.11

See Example 5.3, which shows calculation of relative impedances in a high-resistance grounded system. The positive and negative sequence impedances can be ignored because these are relatively small compared to the zero sequence impedance. This will not be the case for solidly grounded systems.

D.6 Solution to Problems Chapter 6

D.6.1 Problem 6.1

See Figure D6.1.

There are eight elements and four branches. The number of loops=4, number of tree branches=4

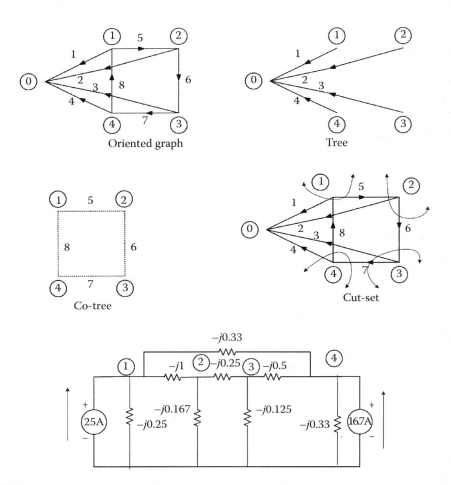

FIGURE D6.1
Oriented graph, tree, co-tree, and equivalent network, Problem P6.1

Appendix D: Solution to the Problems

By inspection the Y-bus matrix is,

$$\bar{Y} = \begin{vmatrix} -j1.583 & -j & 0 & j0.333 \\ j & -j1.417 & j0.25 & 0 \\ 0 & j0.25 & -j0.875 & j0.5 \\ j0.333 & 0 & j0.5 & -j1.166 \end{vmatrix}$$

To construct the matrix using Equation 6.30, construct a table, Example 6.4 from the graph in Figure D6.1

e/n	1	2	3	4	5	6	7	8
1	1				1			−1
2		1			−1	1		
3			1			−1	1	
4				1			−1	1

Then Matrix A is

$$\bar{A} = \begin{vmatrix} 1 & 0 & 0 & 0 & 1 & 0 & 0 & -1 \\ 0 & 1 & 0 & 0 & -1 & 1 & 0 & 0 \\ 0 & 0 & 1 & 0 & 0 & -1 & 1 & 0 \\ 0 & 0 & 0 & 1 & 0 & 0 & -1 & 1 \end{vmatrix}$$

and its transpose is

$$\bar{A}^t = \begin{vmatrix} 1 & 0 & 0 & 0 \\ 0 & 1 & 0 & 0 \\ 0 & 0 & 1 & 0 \\ 0 & 0 & 0 & 1 \\ 1 & -1 & 0 & 0 \\ 0 & 1 & -1 & 0 \\ 0 & 0 & 1 & -1 \\ -1 & 0 & 0 & 1 \end{vmatrix}$$

and primitive admittance matrix is

	1	2	3	4	5	6	7	8
	0–1	0–2	0–3	0–4	1–2	2–3	3–4	4–1
0–1	−j0.25							
0–2		−j0.167						
0–3			−j0.125					
0–4				−j0.333				
1–2					−j			
2–3						−j0.25		
3–4							−j0.5	
4–1								−j0.333

Then using Equation 6.30, the same bus admittance matrix is obtained as derived by inspection:

$$\bar{Y} = \bar{A}\bar{Y}_P\bar{A}^t = \begin{vmatrix} -j1.583 & -j & 0 & j0.333 \\ j & -j1.417 & j0.25 & 0 \\ 0 & j0.25 & -j0.875 & j0.5 \\ j0.333 & 0 & j0.5 & -j1.166 \end{vmatrix}$$

Formation of Loop impedance matrix

Assign loop currents as shown in Figure D6.2 and name the impedances. Then

$$I_a = I_1$$

$$I_b = I_1 - I_4$$

$$I_c = I_2 - I_4$$

$$I_d = I_3 - I_4$$

$$I_e = I_3$$

$$I_f = I_4$$

$$I_g = I_1 - I_2$$

$$I_h = I_2 - I_3$$

or

$$\begin{vmatrix} I_a \\ I_b \\ I_c \\ I_d \\ I_e \\ I_f \\ I_g \\ I_h \end{vmatrix} = \bar{C} \begin{vmatrix} I_1 \\ I_2 \\ I_3 \\ I_4 \end{vmatrix}$$

where

$$\bar{C} = \begin{vmatrix} 1 & 0 & 0 & 0 \\ 1 & 0 & 0 & -1 \\ 0 & 1 & 0 & -1 \\ 0 & 0 & 1 & -1 \\ 0 & 0 & 1 & 0 \\ 0 & 0 & 0 & 1 \\ 1 & -1 & 0 & 0 \\ 0 & 1 & -1 & 0 \end{vmatrix}$$

Appendix D: Solution to the Problems

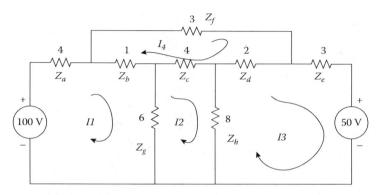

FIGURE D6.2
Numbering of the impedances and assignment of loop currents, Problem P6.1.

and Z primitive is

a	b	c	d	e	f	g	h
4							
	1						
		4					
			2				
				3			
					3		
						6	
							8

Then

$$\bar{Z}_B \bar{C} = \begin{vmatrix} 4 & 0 & 0 & 0 \\ 1 & 0 & 0 & -1 \\ 0 & 4 & 0 & -4 \\ 0 & 0 & 2 & -2 \\ 0 & 0 & 3 & 0 \\ 0 & 0 & 0 & 3 \\ 6 & -6 & 0 & 0 \\ 0 & 8 & -8 & 0 \end{vmatrix}$$

and

$$\bar{C}^t \bar{Z}_B \bar{C} = \begin{vmatrix} 11 & -6 & 0 & -1 \\ -6 & 18 & -8 & -4 \\ 0 & -8 & 13 & -2 \\ -1 & -4 & -2 & 10 \end{vmatrix}$$

Take an inverse of this matrix,

$$\left(\bar{C}^t \bar{Z}_B \bar{C}\right)^{-1} = \begin{vmatrix} 0.149 & 0.095 & 0.069 & 0.067 \\ 0.095 & 0.158 & 0.112 & 0.095 \\ 0.069 & 0.112 & 0.159 & 0.083 \\ 0.067 & 0.095 & 0.083 & 0.161 \end{vmatrix}$$

Calculate

$$\bar{C}\left(\bar{C}^t \bar{Z}_B \bar{C}\right)^{-1} = \begin{vmatrix} 0.149 & 0.095 & 0.069 & 0.067 \\ 0.082 & -0.00023 & -0.015 & -0.095 \\ 0.028 & 0.063 & 0.029 & -0.066 \\ 0.00207 & 0.017 & 0.075 & -0.078 \\ 0.069 & 0.112 & 0.159 & 0.083 \\ 0.067 & 0.095 & 0.083 & 0.161 \\ 0.054 & -0.063 & -0.043 & -0.029 \\ 0.026 & 0.046 & -0.047 & 0.012 \end{vmatrix}$$

Multiply the above matrix with transpose of matrix C:

$$\bar{C}\left(\bar{C}^t \bar{Z}_B \bar{C}\right)^{-1}\left(\bar{C}^t\right) = \begin{vmatrix} 0.149 & 0.082 & 0.028 & 0.00207 & 0.069 & 0.067 & 0.054 & 0.026 \\ 0.082 & 0.177 & 0.095 & 0.08 & -0.015 & -0.095 & 0.082 & 0.014 \\ 0.028 & 0.095 & 0.129 & 0.095 & 0.029 & -0.066 & -0.035 & 0.034 \\ 0.00207 & 0.08 & 0.095 & 0.153 & 0.075 & -0.078 & -0.015 & -0.058 \\ 0.069 & -0.015 & 0.029 & 0.075 & 0.159 & 0.083 & -0.043 & -0.047 \\ 0.067 & -0.095 & -0.066 & -0.078 & 0.083 & 0.161 & -0.029 & 0.012 \\ 0.054 & 0.082 & -0.035 & -0.015 & -0.043 & -0.029 & 0.117 & -0.02 \\ 0.026 & 0.014 & 0.034 & -0.058 & -0.047 & 0.012 & -0.02 & 0.093 \end{vmatrix}$$

Multiply the above calculated matrix with the voltage vector:

$$\bar{V}_B = \begin{vmatrix} 100 \\ 0 \\ 0 \\ 0 \\ 50 \\ 0 \\ 0 \\ 0 \end{vmatrix}$$

Then the currents in each element throughout the network are

Appendix D: Solution to the Problems

$$\begin{vmatrix} I_a \\ I_b \\ I_c \\ I_d \\ I_e \\ I_f \\ I_g \\ I_h \end{vmatrix} = \begin{vmatrix} 18.297 \\ 7.48 \\ 4.258 \\ 3.97 \\ 14.787 \\ 10.817 \\ 3.222 \\ 0.288 \end{vmatrix}$$

D.6.2 Problem 6.2

Only the calculation of Y-bus matrix with coupled branches is demonstrated. An oriented connected graph of the network is shown in Figure D6.3. Form the connection matrix,

e/n	1–2 (1)	1–3 (5)	4–3 (4)	2–5 (2)	5–4 (3)	4–1 (6)
1	1	−1				−1
2	−1			1		
3		1	−1			
4			1		−1	1
5				−1	1	

Primitive z with given branch couplings is given as

$$\begin{vmatrix} j0.5 & 0 & 0 & 0 & 0 & 0 \\ 0 & j0.4 & j0.2 & 0 & 0 & 0 \\ 0 & j0.2 & j0.6 & 0 & 0 & 0 \\ 0 & 0 & 0 & j0.5 & j0.3 & 0 \\ 0 & 0 & 0 & j0.3 & j0.7 & 0 \\ 0 & 0 & 0 & 0 & 0 & j0.3 \end{vmatrix}$$

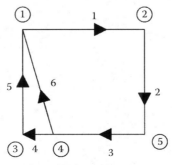

FIGURE D6.3 Oriented connected graph of network in Figure P6.2.

Then

$$
\begin{vmatrix} j0.4 & j0.2 \\ j0.2 & j0.6 \end{vmatrix}^{-1} = \begin{vmatrix} -j0.3 & j \\ j & -j2 \end{vmatrix}
$$

$$
\begin{vmatrix} j0.5 & j0.3 \\ j0.3 & j0.7 \end{vmatrix}^{-1} = \begin{vmatrix} -j2.69 & j1.154 \\ j1.154 & -j1.923 \end{vmatrix}
$$

Then the primitive y matrix is given as

$$
\bar{Y}_p = \begin{vmatrix} -j2 & 0 & 0 & 0 & 0 & 0 \\ 0 & -j0.3 & j & 0 & 0 & 0 \\ 0 & j & -j2 & 0 & 0 & 0 \\ 0 & 0 & 0 & -j2.69 & j1.154 & 0 \\ 0 & 0 & 0 & j1.154 & -j1.923 & 0 \\ 0 & 0 & 0 & 0 & 0 & -j3.33 \end{vmatrix}
$$

Also

$$
\bar{A} = \begin{vmatrix} 1 & -1 & 0 & 0 & 0 & -1 \\ -1 & 0 & 0 & 1 & 0 & 0 \\ 0 & 1 & -1 & 0 & 0 & 0 \\ 0 & 0 & 1 & 0 & -1 & 1 \\ 0 & 0 & 0 & -1 & 1 & 0 \end{vmatrix}
$$

and

$$
\bar{A}^t = \begin{vmatrix} 1 & -1 & 0 & 0 & 0 \\ -1 & 0 & 1 & 0 & 0 \\ 0 & 0 & -1 & 1 & 0 \\ 0 & 1 & 0 & 0 & -1 \\ 0 & 0 & 0 & -1 & 1 \\ -1 & 0 & 0 & 1 & 0 \end{vmatrix}
$$

Then the bus admittance matrix is

$$
\bar{A}\bar{Y}_p\bar{A}^t = \begin{vmatrix} -j5.63 & j2 & j1.31 & j2.33 & 0 \\ j2 & -j4.69 & 0 & -j1.154 & j3.844 \\ j1.3 & 0 & -j4.3 & j3 & 0 \\ j2.33 & -j1.154 & j3 & -j7.253 & j3.077 \\ 0 & j3.844 & 0 & j3.077 & -j6.921 \end{vmatrix}
$$

Appendix D: Solution to the Problems

D.6.3 Problem 6.3

A step-by-step procedure, referring to Figure D6.4, is illustrated.
 Step 1: Branches O1 and O2

$$\begin{vmatrix} 0.3 & 0 \\ 0 & 0.5 \end{vmatrix}$$

Figure P3.3. Network for Problems 3.3–3.8
 Step 2: Add link 1–2

$$Z_{e1} = Z_{11} - Z_{21} = 0.3$$

$$Z_{e2} = Z_{12} - Z_{22} = -0.5$$

$$Z_{ee} = Z_{12,12} + Z_{1e} - Z_{2e} = 0.4 + 0.3 + 0.5 = 1.2$$

$$\begin{vmatrix} 0.3 & 0 & 0.3 \\ 0 & 0.5 & -0.5 \\ 0.3 & -0.5 & 1.2 \end{vmatrix}$$

Eliminate last row and column

$$\begin{vmatrix} 0.225 & 0.125 \\ 0.125 & 0.292 \end{vmatrix}$$

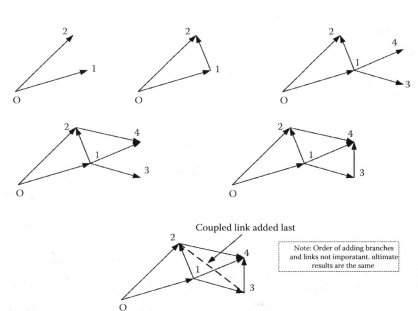

FIGURE D6.4
Step-by-step procedure for forming the impedance matrix, Problem P6.3.

Step 3: Add branch 1–3, $p=1, k=3$

$$Z_{31} = Z_{11} = 0.225$$

$$Z_{32} = Z_{12} = 0.125$$

$$Z_{33} = 0.225 + 0.2 = 0.425$$

$$\begin{vmatrix} 0.225 & 0.125 & 0.225 \\ 0.125 & 0.292 & 0.125 \\ 0.225 & 0.125 & 0.425 \end{vmatrix}$$

Step 4: Add branch 1–4, $p=1, k=4$

$$Z_{41} = Z_{11} = 0.225$$

$$Z_{42} = Z_{12} = 0.125$$

$$Z_{43} = Z_{13} = 0.225$$

$$Z_{44} = Z_{14,14} + Z_{41} = 0.225 + 0.1 = 0.325$$

$$\begin{vmatrix} 0.225 & 0.125 & 0.225 & 0.225 \\ 0.125 & 0.292 & 0.125 & 0.125 \\ 0.225 & 0.125 & 0.425 & 0.225 \\ 0.225 & 0.125 & 0.225 & 0.325 \end{vmatrix}$$

Step 5: Add link 2–4, $p=2, k=4$

$$Z_{e1} = Z_{21} - Z_{41} = 0.125 - 0.225 = -0.1$$

$$Z_{e2} = Z_{22} - Z_{42} = 0.292 - 0.125 = 0.167$$

$$Z_{e3} = Z_{23} - Z_{43} = 0.125 - 0.225 = -0.1$$

$$Z_{e4} = Z_{24} - Z_{44} = 0.125 - 0.325 = -0.2$$

$$Z_{ee} = Z_{24,24} + Z_{2e} - Z_{4e} = 0.467$$

$$\begin{vmatrix} 0.225 & 0.125 & 0.225 & 0.225 & -0.1 \\ 0.125 & 0.292 & 0.125 & 0.125 & 0.167 \\ 0.225 & 0.125 & 0.425 & 0.225 & -0.1 \\ 0.225 & 0.125 & 0.225 & 0.325 & -0.2 \\ -0.1 & 0.167 & -0.1 & -0.2 & 0.467 \end{vmatrix}$$

Appendix D: Solution to the Problems

Eliminate last row and last column

$$
\begin{vmatrix}
0.204 & 0.161 & 0.204 & 0.182 \\
0.161 & 0.232 & 0.161 & 0.197 \\
0.204 & 0.161 & 0.404 & 0.182 \\
0.182 & 0.197 & 0.182 & 0.239
\end{vmatrix}
$$

Step 6: Add link 3–4, $p=3$, $k=4$

$$Z_{e1} = Z_{31} - Z_{41} = 0.204 - 0.182 = 0.022$$

$$Z_{e2} = Z_{32} - Z_{42} = 0.161 - 0.197 = -0.036$$

$$Z_{e3} = Z_{33} - Z_{43} = 0.404 - 0.182 = 0.222$$

$$Z_{e4} = Z_{34} - Z_{44} = 0.182 - 0.239 = -0.057$$

$$Z_{ee} = Z_{34,34} + Z_{3e} - Z_{4e} = 0.2 + 0.222 + 0.057 = 0.479$$

Form augmented matrix

$$
\begin{vmatrix}
0.204 & 0.161 & 0.204 & 0.182 & 0.022 \\
0.161 & 0.232 & 0.161 & 0.197 & -0.036 \\
0.204 & 0.161 & 0.404 & 0.182 & 0.222 \\
0.182 & 0.197 & 0.182 & 0.239 & -0.057 \\
0.022 & -0.036 & 0.222 & -0.057 & 0.479
\end{vmatrix}
$$

Eliminate last row and last column

$$
\begin{vmatrix}
0.203 & 0.163 & 0.194 & 0.185 \\
0.163 & 0.229 & 0.178 & 0.193 \\
0.194 & 0.178 & 0.301 & 0.208 \\
0.185 & 0.193 & 0.208 & 0.232
\end{vmatrix}
$$

Step 7: Lastly add coupled link 2–3

$$
\text{Primitive matrix} =
\begin{vmatrix}
0.4 & -0.1 \\
-0.1 & 0.3
\end{vmatrix}
$$

$$
\begin{vmatrix}
0.4 & -0.1 \\
-0.1 & 0.3
\end{vmatrix}^{-1}
=
\begin{vmatrix}
2.727 & 0.909 \\
0.909 & 3.636
\end{vmatrix}
$$

$$Y_{pe,pe} = 3.636$$

$$Y_{pe,xy} = 0.909$$

$$Z_{e1} = Z_{21} - Z_{31} + \frac{Y_{pe,xy}(Z_{x1} - Z_{y1})}{Y_{pe,pe}}$$

$$= 0.163 - 0.194 + \frac{0.909(0.203 - 0.163)}{3.636} = -0.021$$

Similarly

$$Z_{e2} = 0.29 - 0.178 + \left(\frac{0.909}{3.636}\right)(0.163 - 0.229) = 0.0345$$

$$Z_{e3} = 0.178 - 0.301 + \left(\frac{0.909}{3.636}\right)(0.194 - 0.178) = -0.119$$

$$Z_{e4} = 0.193 - 0.208 + \left(\frac{0.909}{3.636}\right)(0.185 - 0.193) = -0.017$$

Also

$$Z_{ee} = Z_{2e} - Z_{3e} + \frac{1 + (0.909)(Z_{e1} - Z_{e2})}{3.636}$$

$$= 0.0345 + 0.119 + \frac{1 + (0.909)(-0.021 - 0.0345)}{3.636} = 0.415$$

Then the augmented matrix is

$$\begin{vmatrix} 0.203 & 0.163 & 0.194 & 0.185 & -0.021 \\ 0.163 & 0.229 & 0.178 & 0.193 & 0.0345 \\ 0.194 & 0.178 & 0.301 & 0.208 & -0.119 \\ 0.185 & 0.193 & 0.208 & 0.232 & -0.017 \\ -0.021 & 0.0345 & -0.119 & -0.017 & 0.415 \end{vmatrix}$$

Finally, eliminate the last row and last column

$$\begin{vmatrix} 0.202 & 0.165 & 0.188 & 0.184 \\ 0.165 & 0.226 & 0.188 & 0.194 \\ 0.188 & 0.188 & 0.267 & 0.203 \\ 0.184 & 0.194 & 0.203 & 0.231 \end{vmatrix}$$

D.6.4 Problem 6.4.

Here,

$$Z_1 = Z_2 = Z_0 = j0.226$$

Appendix D: Solution to the Problems 665

Therefore,

$$I_1 = \frac{1}{j0.226 + \dfrac{j0.226 \times j0.226}{j0.226 + j0.226}} = -j2.94 \text{ pu}$$

$$I_2 = I_0 = j1.47 \text{ pu}$$

Then

$$\begin{vmatrix} I_a \\ I_b \\ I_c \end{vmatrix} = \begin{vmatrix} 1 & 1 & 1 \\ 1 & a^2 & a \\ 1 & a & a^2 \end{vmatrix} \begin{vmatrix} j1.47 \\ -j2.94 \\ j1.47 \end{vmatrix} = \begin{vmatrix} 0 \\ -3.819 + j2.205 \\ 3.819 + j2.205 \end{vmatrix}$$

D.6.5 Problem 6.5

Calculate voltage at bus 2, the faulted bus

$$\begin{vmatrix} V_2^0 \\ V_2^1 \\ V_2^2 \end{vmatrix} = \begin{vmatrix} 0 \\ 1 \\ 1 \end{vmatrix} - \begin{vmatrix} Z_{2s}^0 & 0 & 0 \\ 0 & Z_{2s}^1 & 0 \\ 0 & 0 & Z_{2s}^2 \end{vmatrix} \begin{vmatrix} I_2^0 \\ I_2^1 \\ I_2^2 \end{vmatrix}$$

$$= \begin{vmatrix} 0 \\ 1 \\ 0 \end{vmatrix} - \begin{vmatrix} j0.226 & 0 & 0 \\ 0 & j0.226 & 0 \\ 0 & 0 & j0.226 \end{vmatrix} \begin{vmatrix} j1.47 \\ -j2.94 \\ j1.47 \end{vmatrix} = \begin{vmatrix} 0.332 \\ 0.332 \\ 0.332 \end{vmatrix}$$

Calculate line voltages

$$\begin{vmatrix} V_2^a \\ V_2^b \\ V_2^c \end{vmatrix} = \begin{vmatrix} 1 & 1 & 1 \\ 1 & a^2 & a \\ 1 & a & a^2 \end{vmatrix} \begin{vmatrix} 0.332 \\ 0.332 \\ 0..332 \end{vmatrix} = \begin{vmatrix} 0.996 \\ 0 \\ 0 \end{vmatrix}$$

The result is as expected, the voltages on the faulted phases is zero.
  Calculate the voltages at bus 1
  Voltages at bus 1

$$\begin{vmatrix} V_1^0 \\ V_1^1 \\ V_1^2 \end{vmatrix} = \begin{vmatrix} 0 \\ 1 \\ 1 \end{vmatrix} - \begin{vmatrix} Z_{21}^0 & 0 & 0 \\ 0 & Z_{21}^1 & 0 \\ 0 & 0 & Z_{21}^2 \end{vmatrix} \begin{vmatrix} I_2^0 \\ I_2^1 \\ I_2^2 \end{vmatrix}$$

$$= \begin{vmatrix} 0 \\ 1 \\ 0 \end{vmatrix} - \begin{vmatrix} j0.165 & 0 & 0 \\ 0 & j0.165 & 0 \\ 0 & 0 & j0.165 \end{vmatrix} \begin{vmatrix} j1.47 \\ -j2.94 \\ j1.47 \end{vmatrix} = \begin{vmatrix} 0.243 \\ 0.515 \\ 0.243 \end{vmatrix}$$

The line voltages are given as

$$
\begin{vmatrix} V_1^a \\ V_1^b \\ V_1^c \end{vmatrix} = \begin{vmatrix} 1 & 1 & 1 \\ 1 & a^2 & a \\ 1 & a & a^2 \end{vmatrix} \begin{vmatrix} 0.243 \\ 0.515 \\ 0.243 \end{vmatrix} = \begin{vmatrix} 1.001 \\ -0.136 - j0.236 \\ -0.136 + j0.236 \end{vmatrix}
$$

The voltages on other buses can be similarly calculated.

D.6.6 Problem 6.6

Voltages at buses 1 and 2 have been calculated in solution to Problem 6.5.
   Then, the current from bus 1 to 2 is given as

$$
\begin{vmatrix} I_{1,2}^0 \\ I_{1,2}^1 \\ I_{1,2}^2 \end{vmatrix} = \begin{vmatrix} 1/j0.1 & 0 & 0 \\ 0 & 1/j0.1 & 0 \\ 0 & 0 & 1/j0.1 \end{vmatrix} \begin{vmatrix} 0.243 - 0.332 \\ 0.515 - 0.332 \\ 0.243 - 0.332 \end{vmatrix} = \begin{vmatrix} j0.89 \\ -j1.83 \\ j0.89 \end{vmatrix}
$$

Then the line currents are given as

$$
\begin{vmatrix} I_{1-2}^a \\ I_{1-2}^b \\ I_{1-2}^c \end{vmatrix} = \begin{vmatrix} 1 & 1 & 1 \\ 1 & a^2 & a \\ 1 & a & a^2 \end{vmatrix} \begin{vmatrix} j0.89 \\ -j1.83 \\ j0.89 \end{vmatrix} = \begin{vmatrix} 0.005 \\ -0.136 - j0.236 \\ -0.136 + j0.236 \end{vmatrix}
$$

Similarly, the other currents can be calculated.

D.6.7 Problem 6.7

The link between buses 1 and 2 is intended to be removed in this problem. The required matrix following the procedure in Example 6.10 is given as

$$
\begin{vmatrix} 0.208 & 0.152 & 0.171 & 0.177 \\ 0.152 & 0.245 & 0.214 & 0.204 \\ 0.171 & 0.214 & 0.300 & 0.229 \\ 0.177 & 0.204 & 0.229 & 0.237 \end{vmatrix}
$$

D.6.8 Problem 6.8

$$
Z_1 = Z_2 = Z_0 = j0.237
$$

Then the single line-to-ground fault current is given as

$$
I_4^0 = I_4^1 = I_4^2 = \frac{1}{j0.237 + j0.237 + j0.237} = -j1.406\,\text{pu}
$$

Appendix D: Solution to the Problems   667

The line currents are given as

$$
\begin{vmatrix} I_4^a \\ I_4^b \\ I_4^c \end{vmatrix} = \begin{vmatrix} 1 & 1 & 1 \\ 1 & a^2 & a \\ 1 & a & a^2 \end{vmatrix} \begin{vmatrix} -j1.406 \\ -j1.406 \\ -j1.406 \end{vmatrix} = \begin{vmatrix} -j4.218 \\ 0 \\ 0 \end{vmatrix}
$$

The voltages at bus 4 are given as

$$
\begin{vmatrix} V_2^0 \\ V_2^1 \\ V_2^2 \end{vmatrix} = \begin{vmatrix} 0 \\ 1 \\ 0 \end{vmatrix} - \begin{vmatrix} j0.237 & 0 & 0 \\ 0 & j0.237 & 0 \\ 0 & 0 & j0.237 \end{vmatrix} \begin{vmatrix} -j1.406 \\ -j1.406 \\ -j1.406 \end{vmatrix} = \begin{vmatrix} -0.333 \\ 0.667 \\ -0.333 \end{vmatrix}
$$

Then the line voltages are given as

$$
\begin{vmatrix} V_2^a \\ V_2^b \\ V_2^c \end{vmatrix} = \begin{vmatrix} 1 & 1 & 1 \\ 1 & a^2 & a \\ 1 & a & a^2 \end{vmatrix} \begin{vmatrix} -0.333 \\ 0.667 \\ -0.333 \end{vmatrix} = \begin{vmatrix} 0 \\ -0.5 - j0.866 \\ -0.5 + j0.866 \end{vmatrix}
$$

The rest of the solution can proceed as in solution to Problems 6.5 and 6.6

D.7 Solution to Problems Chapter 7

D.7.1 Problem 7.1

Reignition: If the contact space of a circuit breaker breaks down within a period of a 1/4 cycle, the phenomenon is called reignition. At zero power factor, the maximum voltage is impressed across the gap that tends to reignite the arc in the hot arc medium. Reignitions do not give rise to overvoltages across the breaker contacts.

Restrikes: If the contact space of a circuit breaker breaks down after a period of a 1/4 cycle, the phenomenon is called restrikes. Restrikes can give high voltages across breaker contacts. The vacuum circuit breakers, in particular, can give rise to multiple reignitions and consequent escalation of voltages and part winding resonance in transformers.

Current chopping: Current chopping occurs due to instability of arc in low current region. It is a function of the breaker interrupting medium, breaker design as well as the circuit in which the breaker is connected. All circuit breakers, to an extent, exhibit current chopping phenomena. Earlier designs of vacuum circuit breakers had high chopping currents, and compared to other interruption technologies, the chopping currents are a problem in vacuum circuit breakers, due to arc instability close to the current zero. Much advancement in contact materials has lowered the values of chopping currents, but due to the nature of the interrupting medium and construction not much can be done to eliminate it.

668   *Appendix D: Solution to the Problems*

D.7.2 Problem 7.2

The delay line is dependent upon the ground capacitance of the input (source side) circuit (see Figure 7.6c and Equation 7.15).

D.7.3 Problem 7.3

Using Equation 7.29, the voltage $u_{2max} = 187\,kV$, which is approximately $= 1.36$ times the system voltage. Note that saturation is neglected. μ_m is conservatively $= 1$.

D.7.4 Problem 7.4

The source side oscillation frequency is zero. The load side frequency is 1.257 kHz (Equation 7.25). This means that the load side voltage rate of rise is 0.53 kV/μs. Table 8.3 gives a rating of 2 kV/μs for a 145 kV (maximum system voltage) class breaker. Therefore, this application is acceptable.

This rate of rise can be easily reduced by simply adding a small capacitor, if required. In transient recovery voltage (TRV) calculations, the system capacitance is of importance, and accurate modeling is a must. The source and load side capacitance of bus work, stray capacitance of transformers, cable models, accurate breaker models, and supply source models must be used—otherwise the results of calculations are misleading.

The resistance value to reduce the transient to zero can be calculated as 15.8 kΩ. The lower is the capacitance, the higher will be the value of the resistor.

D.7.5 Problem 7.5

See Figure 7.6 for a simplified circuit of a terminal fault. Here, for the given $L = 1.5$ mH and $C = 0.005$ μF, the frequency of oscillation from Equation 4.12 is 58.07 kHz. Note that no resistance is specified and therefore there is no damping.

Based on the inductance and the short-circuit current of 20 kA, the voltage is 11.31 kV. Therefore, E_m, the peak value of recovery voltage, phase-to-neutral $= \sqrt{2} \times 11.31 = 15.992$ kV The rate of rise of recovery voltage (RRRV) is given by

$$RRRV = \frac{E_m}{\sqrt{LC}} \sin \frac{t}{\sqrt{LC}}$$

The peak restriking voltage is given as

$$e = 2E_m \quad \text{at } t = \pi\sqrt{LC}$$

The maximum RRRV is given as

$$RRRV_{max} = \frac{E_m}{\sqrt{LC}} \quad \text{at } t = \frac{\pi}{2}\sqrt{LC}$$

Then from above equations, maximum RRRV occurs when

$$t = \sqrt{LC}\,\frac{\pi}{2} = \sqrt{1.5 \times 10^{-3} \times 0.005 \times 10^{-9}}\,\frac{\pi}{2} = 4.30\ \mu s$$

Appendix D: Solution to the Problems 669

and maximum RRRV is 5.813 kV/μs.

This is too high.

The example is a theoretical calculation. The TRV parameters for system source faults and generator source faults based on large generators connected through step-up transformers have been standardized in IEEE Standard C37.013 for High Voltage Generator Breakers. This specifies an inherent TRV rate of 1.6 kV/μs for generator ratings up to 100 MVA (see definition of ITRV). This assumes an ideal circuit breaker.

Practically, the calculation of TRV in real-world situations can be very complex. This requires accurate modeling of the system components and also the circuit breaker operation (see Cassie Mayr Theory briefly discussed in 7.2.1.3). Electromagnetic Transients Program (EMTP) and similar programs are valuable tools for the simulation. The accurate modeling of capacitances is important.

A transient model of the generator will be used in practical calculations with all the stray capacitances, stator capacitance to ground plus the capacitance of the surge capacitor. (See text in this chapter and also Chapter 8, which provides a brief description of hand calculation of TRV based on IEEE Standard C37.011, Annexure A.) The TRV profile depends upon the type of circuit.

A reader can recalculate the above parameters by increasing capacitance from 0.005 to 0.05 nF and examine the results, which will reduce maximum RRRV.

D.7.6 Problem 7.6

1. The TRV that occurs when interrupting asymmetrical current is less severe than those which occur when interrupting the rated symmetrical currents. This is so because the instantaneous value of the supply system voltage, when interrupting asymmetrical current is much less than the peak value. A reader can sketch out the asymmetrical current and voltage wave shapes.

2. This is a fundamentally incorrect statement. When a circuit breaker is called upon to interrupt a current much lower than its rated current, TRV will be higher. There is an IEEE standard for definite purpose circuit breaker for higher TRVs, see Chapter 8.

3. This is generally correct. After a current interruption, the hot arc channel is still conducting and may not be able to withstand a voltage stress of twice the system voltage. Note from Figure 7.31 that du/dt stress for capacitance current interruption is minimal. Thus, du/dt is not a factor for capacitor current interrupt.

4. See Figure 8.16. For a fault on the secondary side of the transformer, this is correct.

5. A breaker of higher interrupting rating does not have a higher TRV rating in the same voltage class. However, these can indirectly influence the TRV profile. For example, consider a breaker of 123 kV, interrupting rating of 40 kA and 63 kA. For a 100% terminal fault, the TRV profile is identical. Now consider a fault equal to 12.6 kA to be cleared. As a percentage of the interrupting rating for a 40 kA breaker, it is equal to 31.5% of the rated fault current, while for a 63 kA breaker, it is equal to 20% of the rated current. The 20% TRV profile will be higher than that at 31.5%. Thus, higher short-circuit ratings impact. Definite purpose TRV breakers (Chapter 8) or modifying the system, where TRV is being evaluated are other options, more valuable than increasing the short-circuit rating.

670 *Appendix D: Solution to the Problems*

D.7.7 Problem 7.7

Add a capacitor, see Equation 7.12. Also see solution to Problem 7.6.

D.7.8 Problem 7.8

See Figures 7.12 and 7.13. As discussed in the text some measures are as follows:

- Resistance switching
- Synchronous switching
- Multiple breaking gaps per pole, shunted by resistors or capacitors for voltage division
- Surge arresters, not discussed in the text, see work of Sluis in bibliography.

D.7.9 Problem 7.9

See Equation 7.61. The cable capacitance can form a resonant circuit with the locked rotor impedance of the motor, giving rise to restrikes. Surge arresters are used.

D.8 Solution to Problems Chapter 8

D.8.1 Problem 8.1

i. Maximum symmetrical interrupting capability = $41 \times 1.19 = 48.79$ kA rms. Tables in IEEE standard round it to 49 kA (see Table 8.2)

ii. Short-time rating for three seconds is 49 kA rms symmetrical

iii. Close and latch capability. In symmetrical amperes, the multiplying factor (MF) is 1.6 for K rated breakers, see text, thus, it is equal to 78 kA. In terms of peak, the MF is 2.7 for K rated breakers, see text, thus it is equal to 132.3 kA rounded to 132 kA peak in ANSI/IEEE standards.

iv. If this breaker is applied at 2.4 kV, the voltage ratio with respect to rated voltage is 1.98 > K factor of 1.19. Therefore, the breaker will have an interrupting rating of 49 kA rms symmetrical and close and latch rating of 132 kA peak.

If the K factor is 1, then the interrupting rating at 2.4 kV or at the rated maximum voltage remains at 40 kA. The close and latch rating for $K = 1$ rated breaker is $2.6 \times 40 = 104$ kA peak.

D.8.2 Problem 8.2

$K = 1.19$

At 4.16 kV, the interrupting rating = $(41 \times 4.76)/4.16 = 46.91$ kA rms symmetrical; at 2.4 kV, it is 49 kA rms symmetrical (see Problem 8.1).

A 5-cycle breaker has a contact parting time of 3-cycle. Then from Equations 8.4 and 8.5 and Figure 8.4, the DC component of the current to be interrupted is equal to 32.91%,

Appendix D: Solution to the Problems 671

which gives an asymmetrical rating of 1.1 times the symmetrical rating. That is, 51.60 kA rms asymmetrical at 4.16 kV and 53.9 kA at 2.4 kV.

$K=1$

The asymmetrical rating is $40 \times 1.1 = 44$ kA rms asymmetrical, both at 4.16. and 2.4 kV.

D.8.3 Problem 8.3

As the breaker is rated 15 kV, the rated permissible tripping delay is equal to 2 s. This is not shown in the tables reproduced from ANSI/IEEE standards in the text. The permissible delay is calculated from Equation 8.8. The left side of the equation implies the permissible delay is to be calculated in rms current.

Then

$$Y_{\text{permissible}} = 2 \left[\frac{1.3 \times 28}{22} \right]^2 = 5.47 \, \text{s}$$

D.8.4 Problem 8.4

With the specified change in the reactance and capacitance values, the following are the results of calculation:

- Peak inrush current when capacitor bank C1 is first switched: 5973 A peak
- Frequency: 555.14 Hz.
- Back-to-back switching current = 24.37 kA peak
- It's frequency = 5668 Hz.

First try a reactor of 70 µH, as in Example 8.3 is introduced; the back-to-back switching current reduces to 9.64 kA peak and its frequency to 2241 Hz.

This is still not acceptable, because a 15 kV and 40 kA short-circuit symmetrical interrupting current rated definite purpose breaker has an inrush current of 15 kA peak and its frequency should not exceed 2000 Hz. A reactor of 80 µH will reduce the inrush current and its frequency below 15 kA and 2000 Hz, respectively.

D.8.5 Problem 8.5

The oscillatory TRV occurs when a circuit breaker is called upon to clear a low-level three-phase ungrounded fault in system configuration as shown in Figure 8.16. This shows a reactor fault and a transformer secondary fault. A capacitor damps the TRV frequency (Chapter 7).

D.8.6 Problem 8.6

From Table 8.10 and for a 100% terminal fault,

$K_{\text{pp}} = 1.3$, $K_{\text{af}} = 1.4$, $u_1 = 98$, $t_1 = 49$, $u_c = 183$, $T_2 = 196$.

From this data, the TRV profile for a 100% terminal fault is plotted as shown in Figure D8.1.

From Figure 8.17,

$Ku_1 = 1$, $Kt_1 = 0.60$, $Ku_c = 1.07$, $Kt_2 = 0.5$. Assume these to be applicable for 50% fault.

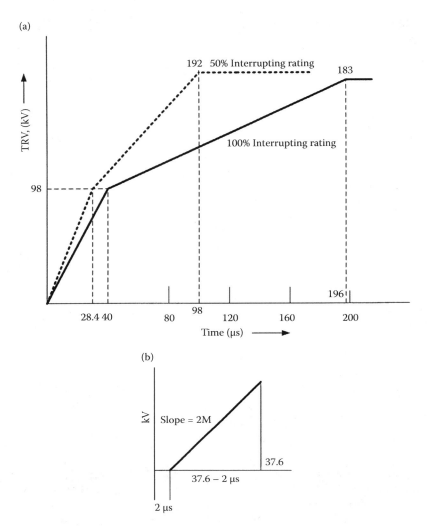

FIGURE D8.1
(a) Calculated TRV profile for 100% and 50% interrupting capacity and (b) short-line fault; Problem P8.6

Then from Equation 7.17,

$$u_c(T50) = u_c(T100)Ku_c = 192 \text{ kV}$$

With the other parameters and factors as given, the 50% terminal TRV profile is shown in dotted lines in Figure D8.1a.

Short-line fault
  $Z = 450, d = 1.8$
  Then the line side voltage from Equation 8.34 is 90.4 kV. This considers $M = 0.5$ and $E_{max} = 123$ kV.
  At breaker of interrupting current is equal to 20 kA rms symmetrical, from Equation 8.38, $R_L = 2.40$ kV/μs.

Appendix D: Solution to the Problems 673

This gives t_L from Equation 8.37 is equal to 37.6 µs.
The source side voltage = 35.6 kV.
The source side voltage will rise as shown in Figure D8.1b.
Then the total voltage from equation $e = e_L + e_s = 90.4 + 35.6 = 126$ kV.

D.8.7 Problem 8.7

Due to large bus capacitance, on account of closer spacing and clearances.

D.8.8 Problem 8.8

See Section 8.14.3.2. The short-time rating is of importance for selective protective device coordination. The low-voltage power circuit breakers (LVPCBs) have a short-time capability of 30 cycles and can withstand short-time duty cycle tests. The manufacturers publish curves for low-voltage trip programmers and the modern trend is that this 30 cycles time can be divided into 3–7 bands. Though the time gap between the bands is small and sometimes the bands may seem to overlap, coordination is achieved if a downstream breaker is set at a lower band and the upstream on the next higher short-time band, say 0.1 and 0.15 s. Insulated case circuit breaker (ICCB) may have a short-time withstand capability of 15 cycles and are provided with instantaneous overrides. A molded case circuit breaker (MCCB) does not have any short-time withstand capability.

Coordination between two instantaneous devices in series that see the same magnitude of short-circuit current is not generally attempted, though with careful analysis and selection of devices, it is a possibility (Volume 4).

D.8.9 Problem 8.9

From Table 8.12, for LVPCB, system short-circuit power factor of 15% (at which the breaker is tested), $X/R = 6.6$, MF = 1 for unfused breakers. This table gives a MF of 1.04 for an X/R of 8.27. At $X/R = 7$, it can be calculated from Equation 8.57 rather than interpolating and it is equal to 1.0116.

MFs for ICCBs and MCCBs are given in Table 8.14. For 50 kA, short-circuit current, $X/R = 6.59$, it is equal to 1.06 and for $X/R = 7.63$, it is equal to 1.09. Again, instead of interpolating, calculate from equations. Note that the test power factor is equal to 20%. This gives a factor of 1.073.

Appropriate MFs should be used while selecting low-voltage circuit breakers.

D.8.10 Problem 8.10

A reader can respond to this problem by reading through Section 8.15. The current-limiting fuses operate in 1/4 to 1/2 cycle in their current limiting zone, limiting the energy let-through (Figure 8.24). Further the prospective fault current is also limited. Considering a relayed breaker, the fault interrupting time, with instantaneous protection operating in 1 cycle will be 4-cycle (3-cycle breaker) and 6-cycle (5-cycle breaker). The UL listing of less inflammable liquid filled transformers for indoor applications is based upon the let-through energy capability of current limiting fuses. Expulsion-type fuses let the peak prospective current pass through and are not current limiting (see text).

D.8.11 Problem 8.11

The series ratings are discussed in Section 8.14.3.3. The series rating makes it possible to use a downstream device of lower short-circuit rating, protected by an upstream device of higher short-circuit rating (on the same switchboard or bus). This is by virtue of short-circuiting limiting characteristics of the upstream device. A wrong up-and-down method of application is illustrated in Figure 8.23. The results of this calculation are erroneous because the lower rated device may not be able to withstand the peak let-through illustrated in Figure 8.23. The series rated devices must be tested according to UL489. The manufacturers publish the tested combinations for their upstream and downstream devices.

While, it is possible to retain a lower short-circuit device in service by proper application of a higher rated current-limiting device, there are limitations in adopting this method as per NEC, discussed in Volume 4.

D.8.12 Problem 8.12

This is shown in Table D8.1 below based on IEEE standards. Note that all the ratings are not shown.

D.9 Solution to Problems Chapter 9

D.9.1 Problem 9.1

Using the data in Table 9.1, and following the procedure of Example 9.6, and constant excitation:

$$e'' = 1.077 \text{ pu}, \ e' = 1.097 \text{ pu}, \ i_d'' = 30.86 \text{ kA}, \ i_d' = 24.84 \text{ kA}, \ i_d = 2.41 \text{ kA}$$

Therefore,

$$i_{ac} = 6.02e^{-t/0.015} + 22.43e^{-t/0.597} + 2.41$$

$$i_{dc} = 47.34e^{-t/0.33}$$

The plotted profile is shown in Figure D9.1.

D.9.2 Problem 9.2

As a first step, the voltage at bus F needs to be ascertained. Consider that the voltage at bus G is maintained at 13.8<0°, then there is a voltage drop due to added resistance and reactance to point F. This is a load flow problem (Volume 2). If the reader calculates the current based upon assumed voltage of 13.8 kV at bus F and then calculates the voltage drop through the impedance, it will be acceptable for this problem.

Appendix D: Solution to the Problems

TABLE D8.1

Ratings of 13.8 kV Circuit Breakers for Various Applications

Breaker	Rated Short-Circuit, kA (rms)	Rated Short-Circuit, kA (rms)	Rated Continuous Current	TRV E2, kV peak	TRV T2, Ms	Dc Comp	Int. Time (ms)	Close and Latch, kA peak	Rated Cable Switching A, (rms)	Isolated Cap Bank A, (rms)	Back-to-back			
											Rated Cable Switching A, (rms)	Isolated Cap Bank A, (rms)	Inrush kA peak	Freq. (kHz)
GP	40	40, 3 s	2000	28	75	32%	83	104	250	250	0	0	0	0
DP-cap	40	40, 3 s	2000	28	75		83	104	1000	1000	1000	1000	15	2
DP-TRV Note 1	40	40, 3 s	2000	28 / 31.6 / 32.8	75 / 6.6 / 11		83	104	250	250	0	0	0	0
GB	40	40, 1 s Note 2	2000	Note 2		68.5% Note 2	83						0	0

Notes: K Factor=1, Rated voltage of the Circuit Breaker=15kV; TRV values specified at 30% (12 kA) and 7% (2.8 kA) of the rated short-circuit.

GP, general purpose; DP-cap, definite purpose, capacitor switching; DP-TRV, definite purpose, TRV applications.

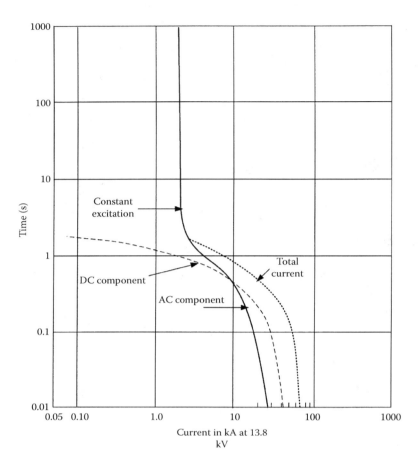

FIGURE D9.1
Decrement curves of generator, Problem P9.1.

More rigorously, the voltage calculation at bus F is an iterative process, also depending upon the load behavior on voltage dip. Volume 2 discusses it in detail. Considering a constant impedance load, the voltage at bus F will be 0.926 < −4.4° pu, and the current will be 4360A at 0.80 power factor.

Some texts will calculate the prefault voltages as follows:

$$e'' = V + I(\cos\phi - j\sin\phi)(jX'_d), \text{ for lagging current.}$$

This is like a voltage drop added to the terminal voltage. The steady-state phasor diagrams are constructed on this basis.

Alternatively, use Equation 9.15. The voltage behind subtransient reactance is

$$1 + (0.096)(0.6) = 1.0576 \text{ pu}$$

Note that no load flow calculations are involved in the above method. The factor $X'_d \sin\phi$ considers the effect of generator excitation, as on a sudden short circuit the excitation system will try to maintain the terminal voltage by beefing up the excitation current.

Appendix D: Solution to the Problems   677

Therefore, for a fault at G,

$$i_d'' = 1.0576/j0.096 = 46.09\,\text{kA rms, and maximum } i_{dc} = 65.17\,\text{kA}$$

For a three-phase fault at F,

$$i_d'' = \frac{1.0576}{(0.01+j0.10+j0.096)}\,\text{pu} = 22.54 < 87.1° = |22.54|\,\text{kA rms sym}$$

The maximum DC component is equal to 31.87 kA.

The solution for transient reactance X_d' can be similarly derived. For the synchronous reactance X_d, the voltage is assumed to be equal to the rated voltage (see Example 9.1).

D.9.3 Problem 9.3

The field current in Example 9.6 is given as

$$i_F = \frac{-i_d X_{ad}}{(X_{ad}+X_f)}(1-\cos\omega t)$$

From the given generator data,

$$X_{ad} = X_d - X_l = 1.121\,\text{pu}$$

$$X_f = \frac{X_{ad}(X_d'-X_l)}{(X_d-X_d')} = 0.0308\,\text{pu}$$

Therefore, for a fault at bus G

$$i_F = 34.5(1-\cos\omega t)\,\text{kA rms}$$

For a fault at bus F, the field current magnitude will reduce to 18.91 kA rms. The profile is shown in Figure D9.2. There is no decay as no resistances are considered.

Practically, the field current will be a function of the type of exciter, voltage regulator, and its response. ANSI/IEEE standards define a fast response excitation system as the one where the ceiling voltage is reached in 0.1 sec or less. Generator saturation also comes into play.

D.9.4 Problem 9.4

From Equation 9.4,

$$X_d'' = X_l + \left[\frac{1}{\dfrac{1}{X_{ad}}+\dfrac{1}{X_f}+\dfrac{1}{X_{kD}}}\right]$$

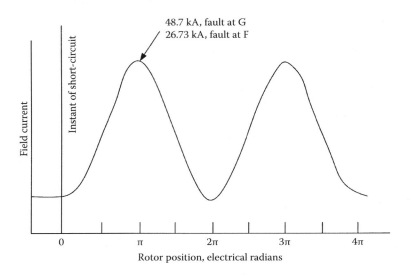

FIGURE D9.2
Field current profile with respect to rotor position, Problem P9.2.

In this equation, all parameters are known except X_{kD}, X_{ad}, and X_f are calculated in solution to Problem 6.3, $X_l = 0.0909$; all in per unit. Solving for $X_{kD} = 0.023$ pu.

Now from Equation 9.6,

$$T_d'' = \frac{1}{\omega r_D}\left[\frac{X_{ad}X_fX_l}{X_{ad}X_f + X_fX_l + X_{ad}X_l} + X_{kD}\right]$$

r_D is given as 0.00909 pu, and all the other parameters are known. Substitute, $T_d'' = 0.0134$ s. This is practically too low. The parameters in Problem 9.2 do not pertain to an actual generator, the parameters in Table 9.1 do.

From Equation 9.24,

$$T_d' = T_{do}'\left[\frac{X_d'}{X_d}\right]$$

T_{do}' is given as 4.8 s, Therefore, $T_d' = 0.48$ s.

These are the generator parameters. When a fault is fed through an external reactance, that is, fault at bus F, it is equivalent to increasing the armature leakage reactance to

$$X_l + X_e$$

where X_e is the external reactance = 0.1 per unit.
Then

$$T_{de}'' = T_d''\left[\frac{X_d'}{X_d''}\right]\left[\frac{X_d'' + X_e}{X_d' + X_e}\right] = 0.0141\,\text{s}$$

Appendix D: Solution to the Problems 679

Also

$$T'_{de} = T'_{do}\left[\frac{X'_d + X_e}{X_d + X_e}\right] = 0.809\,\text{s}$$

Here, we denote the time constants with added external reactance as T''_{de}, T'_{de}.

D.9.5 Problem 9.5

For a fault at G,

$$i_{ac} = \left[\frac{e''}{X''_d} - \frac{e'}{X'_d}\right]e^{-t/T''_d} + \left[\frac{e'}{X'_d} - \frac{e}{X_d}\right]e^{-t/T'_d} + \frac{e}{X_d}$$

$$= 9.05e^{-t/0.0134} + 33.59e^{-t/0.48} + 3.45$$

The fault currents at 0.05 and 0.10 s are 33.92 kA rms symmetrical and 30.72 kA rms symmetrical, respectively.

Fault at *F*

$$i_{ac} = \left[\frac{e''}{X''_d + X_e} - \frac{e'}{X'_d + X_e}\right]e^{-t/T''_{de}} + \left[\frac{e'}{X'_d + X_e} - \frac{e}{X_d + X_e}\right]e^{-t/T'_{de}} + \frac{e}{X_d + X_e}$$

$$= 2.18e^{-t/0.0141} + 17.1e^{-t/0.809} + 3.19$$

The fault currents at 0.05 and 0.1 s are 19.33 kA and 17.40 kA rms symmetrical, respectively.

D.9.6 Problem 9.6

We did not find the direct axis and quadrature axis voltages in solution to Problem 9.2. These can be found as follows:

(Refer to Example 9.5)

The load power factor $= 0.8$, angle $\phi = 36.8°$.

Load current $I_a = 4360\text{A} = 1.042$ pu on 100 MVA base.

$I_a X_q = 2.0$ pu. Resolving and referring to Figure 9.13.

$I_r = 0.8336$ pu and $I_x = 0.6252$ pu

Then in Figure 9.13,

$$(\delta - \beta) = \tan^{-1}\left(\frac{X_q I_r}{V_a + X_q I_r}\right) = 36.53°$$

Then

$$V_q = V_a\cos(\delta - \beta) = 0.8\,\text{pu}$$

$$V_d = -V_a\sin(\delta - \beta) = -0.6\,\text{pu}$$

Therefore,

$$V_a = \sqrt{0.8^2 + (-0.6)^2} = 1.0 \text{ pu as assumed (also see Equation 9.98, based on Park's transformation)}.$$

D.9.7 Problem 9.7

The phasor diagrams are as shown in Figure D9.3.

Note that the internal voltage is reduced with lagging power factor. The power can be approximately expressed as

$$P = \frac{VE}{X_d} \sin \delta$$

Thus, an overexcited motor is more stable. The reader can prove that for synchronous generator it is the leading power factor, which has a similar situation as lagging power factor for the synchronous motor.

D.9.8 Problem 9.8

This is not addressed in the text. The problem is meant for the readers how he will like to get deeper into the synchronous generator operation. The two phasor diagrams are shown in Figure D9.4.

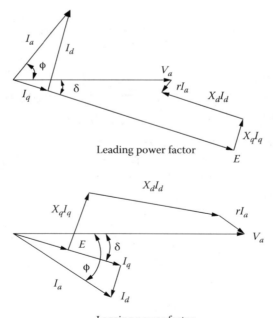

FIGURE D9.3
Phasor diagrams of a synchronous motor, operating at leading and lagging power factors, Problem P9.7.

Appendix D: Solution to the Problems

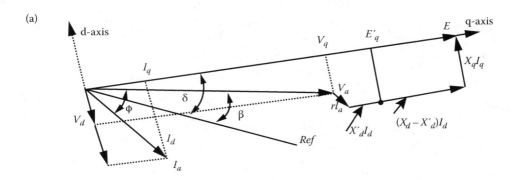

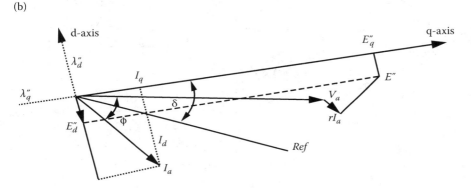

FIGURE D9.4
Phasor diagrams, Problem P9.8.

D.9.9 Problem 9.9

See Equations 9.61–9.72.

The simplification occurs because the columns of P' are the eigenvectors of L_{aa}, and therefore, similarity transformation yields a diagonal matrix of eigenvalues.

$$\overline{P} L_{aa} \overline{P}' = \begin{vmatrix} L_0 & 0 & 0 \\ 0 & L_d & 0 \\ 0 & 0 & L_q \end{vmatrix}$$

where

$$L_0 \approx L_s - 2M_s$$

$$L_d \approx L_s + M_s + \frac{3}{2} L_m$$

$$L_q \approx L_s + M_s - \frac{3}{2} L_m$$

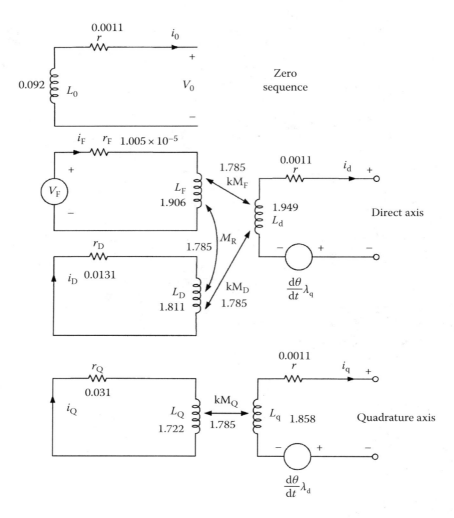

FIGURE D9.5
Solution to the Problem P9.10.

D.9.10 Problem 9.10

The values of all the parameters required have already been calculated in the text, except the resistance "*r*." The effective AC resistance can be calculated from Equation 10.21. The numerical values superimposed on Figure 9.10 are as shown in Figure D9.5.

D.9.11 Problem 9. 11

Based on the given power factor and efficiency of the motor, the motor kVA = 8356. As starting current is six times the full load current, this is equivalent to 16.7% reactance on a 100 MVA base. Therefore, $X_d = 1.998$ pu.

Appendix D: Solution to the Problems 683

$X_m = 3$ and $R_m = 130$.
From Equation 9.128, $X' = 1.748$ pu.
To calculate resistance, X/R ratio of the motor is required. ANSI standards publish typical ratios and for induction motors, refer to Figure 10.2. Assuming an $X/R = 40$, and equal stator and rotor reactances and resistances are as follows:

$$r_1 = r_2 = 0.025\,\mathrm{pu}$$

$$x_1 = x_2 = 0.874\,\mathrm{pu}$$

Then from Equations 9.130 to 9.134

$$T' = T_{dc} = 0.185\,\mathrm{s}$$

$$i_{ac} = 2392e^{-t/0.185}\,\mathrm{amperes}$$

$$i_{dc} = 5585e^{-t/0.185}\,\mathrm{amperes}$$

D.10 Solution to Problems Chapter 10

D.10.1 Problem 10.1

See Section 10.5.1. A 15 kV breaker, $K = 1$, interrupting rating of 25 kA can be selected. Though the rating is theoretically adequate for the intent of this problem, we do not select a circuit breaker, which is equal to its calculated interrupting rating. The next rating of 31.5, $K = 1$ rated breaker will be the correct choice (see Table 5.1).

For breakers rated $K > 1$, Table 8.2, a breaker of rated short-circuit current of 28 kA has to be selected. This has a maximum rated short-circuit current of 36 kA. At 13.8 kV, the interrupting rating will be 30.43 kA rms symmetrical.

Close and latch ratings are not specified in the problem. Sometimes these could be a limiting factor, that is, in order to meet the required close and latch capability a breaker of higher interrupting rating may be required.

D.10.2 Problem 10.2

See Section 10.5.2. A breaker can be selected without rigorous calculations; the selected breaker must have an interrupting rating of $(25/0.8) = 31.25$ kA rms symmetrical. Again referring to Tables 8.1 and 8.2, for $K = 1$ rated breakers; a breaker of standard interrupting rating of 31.5 is available. Again, it will be too close to the required interrupting duty. So, a breaker of 40 kA interrupting rating will be an appropriate choice.

D.10.3 Problem 10.3

For duties on the generator breaker, the currents from the generator source and the system source should be calculated. This calculation is best performed analytically as in

Example 10.2, and specimen example in IEEE standard. However, the problem does not specify enough generator data for the analytical calculation.

See Section 10.9.4 for double-ended faults. We will demonstrate the calculations for G_2 terminal fault. This also holds for G_1 terminal fault as the system is symmetrical. Similarly, the reader may calculate for 230 kV fault for breaker 52T1.

The three-winding transformer model can be derived using equations in Chapter 4. For a fault at terminals of G_2, the equivalent circuit is as shown in Figure D10.1

Generator source fault
The generator source fault is 30.99 kA at an angle of −89.48°. It is a local source current. Therefore, the MF for the interrupting duty from Figure 7.6 is approximately read as 1.16. That is a circuit breaker of 35.94 kA rating is required. If we select a general purpose breaker of 40 kA, $K=1$ from Table 8.1, it is not an appropriate selection.

This breaker will have a % DC interrupting rating of 32.91%. (Equation 8.4) But the generator breaker is required to have a DC interrupting rating of 68.5% (Figure 8.20). Moreover, it should be capable of interrupting 110% of DC current based upon the peak value of the symmetrical current. This cannot be calculated based upon the generator data specified in the problem.

The peak rating for generator source current can be calculated from Equations 10.1 and 10.2. This gives $t=0.49$ and peak=111, 56 kA. The 40 kA breaker from Equation 8.1 has a

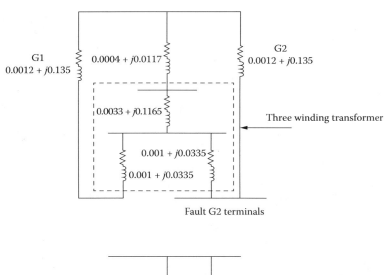

Fault G2 terminals

FIGURE D10.1
Equivalent circuit of network in Figure P10.1.

Appendix D: Solution to the Problems					685

peak rating of 104 kA. Thus, the interrupting rating of the breaker should be raised to meet the peak rating.

This shows that for generator source current the ratings should be carefully selected meeting the requirements of generator breakers. Figure D10.1 illustrates the equivalent circuit.

The source side current equivalent impedance is shown as $0.0026 + j0.1065$ in Figure D10.1. This gives a short circuit of 39.27 kA at an $X/R = 40.90$.

To calculate the breaker ratings, first calculate the X/R from separate R and X networks. This calculation gives an $X/R = 42$, which is slightly different from the value given by complex network reduction. To apply an MF NACD is required. Based upon the impedances in Figure D10.1, the utility and second generator contribution is remote. A remote MF from Figure 7.5 is 1.25. Therefore, the required interrupting duty is 49.09 kA rms symmetrical. The close and latch is calculated, which is equal to 103.34 kA peak. Therefore, a generator circuit breaker of 50 kA interrupting rating, $K = 1$ could be used—its calculated duty of 49.09 is too close to the rating of 50 kA interrupting. Better still, a breaker of 63 kA interrupting rating, Table 8.1 can be used.

The selection should be based on the higher of the two duties calculated above.

D.10.4 Problem 10.4

NACD ratio is defined in Equation 10.10

1. Figure 10.6, local source
2. Figure 10.6 (See 10.5.4, the reactor pu impedance$=0.21 < 1.5$ times generator sub-transient reactance)
3. Figure 10.6 (Generator impedance$=0.32$pu, transformer impedance$=0.25$pu)
4. No post calculation interrupting duty multiplier is applied
5. No post calculation interrupting duty multiplier

D.10.5 Problem 10.5

The calculations are demonstrated for fault at F1.

As a first step, convert all impedances to a 100 MVA (or a base of one's choosing, 100 MVA or 10 MVA being the most common), and construct a positive sequence diagram for a fault at F1. This is shown in Figure D10.2.

Two separate networks are required for calculations of first cycle and interrupting duty currents, as discussed in the text. Table 10.1 shows that impedance MFs are different for these two types of calculations. For analytical calculation, no MF is required.

The MF from Table 10.1 for turbo generators for interrupting or first cycle calculations is unity.

For motors, no efficiency and power factor data are given. The assumption that KVA is nearly equal to hp rating may be applied for a group of low-voltage motors and not for large medium-voltage motors. For 10,000 hp synchronous motor, assume a power factor of unity and efficiency of 96%; for 5000 hp induction motor, assume a power factor of 92% and efficiency of 93%; and for group of 8000 hp motors, assume a power factor of 0.9 and efficiency of 89%. Also all motors in group of 8000 hp motors are assumed >250 hp.

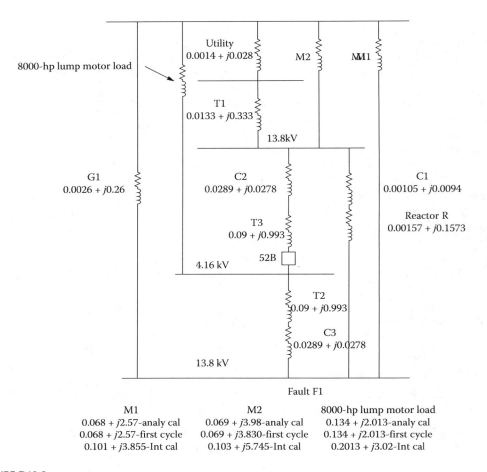

FIGURE D10.2
Positive sequence network for a fault at F1, Problem P10.5.

In medium-voltage applications, each motor is modeled individually and appropriate MF from Table 7.1 is applied. Based on above assumptions, the motor interrupting duty and first-cycle duty impedances are shown in Figure 7-S5.

Transformer T2 and T3 primary taps are set so as to provide a 5% secondary voltage boost. This means the taps are at 13.11 kV. Multiply the rated impedance by a factor of

$$\left(\frac{13.11}{13.8}\right)^2 = 0.902$$

From the given X/R ratio and impedance, resistance can be calculated from equation:

$$R = \frac{Z}{\sqrt{(X/R)^2 + 1}}$$

Note that the neutral grounding methods of transformers specified in Table P7.1 are not required for three-phase short-circuit calculations.

Appendix D: Solution to the Problems 687

The network shown in Figure D10.2 can be reduced for the interrupting and first-cycle calculations. This is time-consuming and laborious even for a simple system. The calculation results are as follows:

First cycle (close and latch)

$Z=0.0032+j0.15206$—from complex impedance reduction. This gives $E/Z=27.51$ kA.

X/R from separate R and X network$=71.69$

Therefore, first-cycle or close and latch peak current (Equations 10.1 and 10.2),

$$=\sqrt{2}\times 27.51\left(1+e^{-2\pi\tau/(X/R)}\right)=76.17\,\text{kA peak}$$

Interrupting duty

$Z=0.00321+j0.1580$—from complex impedance reduction. This gives 26.47 kA.

X/R from separate R and X networks$=73$

In order to calculate NACD, the utility contribution can be approximately calculated by ignoring resistance. From Figure P10.2, the utility source reactance to fault point is given as

$$0.361+\frac{2.04\times 0.1667}{2.04+0.1667}=0.515\,\text{pu}$$

This gives a current of 8.12 kA, thus NACD$=0.30$. From Figure 10.11, and considering 5-cycle breaker, the MF is 1.23. Therefore, for a 5-cyle, breaker interrupting duty current$=32.56$ kA rms symmetrical.

Analytical calculation

The generator source decrement curve can be calculated from the given generator data. Here, we ignore the generator preload and $e''=e'=e=13.8$ kV. Then

$$i_{\text{gen}}=4.88e^{-t/0.016}+10.29e^{-t/0.46}+0.92$$

As we consider 5-cycle breaker at the contact parting time of 50 ms, the generator contribution is 10.364 kA $<-89.4°$.

In the process of complex impedance reduction, the impedance of the rest of the system connected to faulted bus is given as

$0.013+j0.366$ pu, which gives 11.42 kA

Following example in IEEE standard 11, the X/R is 28.15.

Then

$$\tau=\frac{1}{\omega}\times\frac{X}{R}=0.075\,\text{s}$$

Therefore,

$$i_{sys}=11.42e^{-t/0.075}$$

At the contact parting time of 50 ms, this gives 5.86 $<-87.96°$

Then the AC component at contact parting time is equal to 16.224 kA.

The generator source DC component is given as

$$\sqrt{2}\times 16.9e^{-t/0.18}$$

688   *Appendix D: Solution to the Problems*

At contact parting time this gives 17.23 kA. The system DC component at contact parting time is given as

$$\sqrt{2} \times 11.42 e^{-t/0.075}$$

This gives 8.29 kA at contact parting time. Total DC component=25.52 kA. Then the interrupting duty is given as

$\sqrt{(16.224)^2 + (25.52)^2} = 30.24\,\text{kA}$. For comparison, we calculated 32.56 kA.

Analytical first cycle, close and latch calculations using ANSI/IEEE methods are not demonstrated in ANSI/IEEE standards, except that ANSI standard for generator circuit breakers demonstrates calculations of symmetrical and asymmetrical interrupting currents related to generator breakers.

It seems logical to calculate close and latch from the same equations, *approximately*, at 1/2 cycle. This gives

- Generator AC component=13.92 kA
- Generator DC component=21.70 kA
- System AC component=10.22 kA
- System DC component=14.453 kA

Therefore, first-cycle current=43.47 kA, *rms asymmetrical*. To calculate the peak proper, X/R should be considered. Note that in the above calculations, X/R of generator and other source currents have different X/Rs. If we consider the overall X/R approximately equal to 70, Table D10.1 from ANSI/IEEE standard C37.09, 2007 can be used.

Using a factor of 1.68, the peak is 43.47 × 1.68=73.03 kA. We calculated 76.17 kA using empirical ANSI/IEEE calculations.

Another calculation method can be used to calculate the peak for each of the component currents using Equation 10.1 and then add. This will be more like IEC procedure. If we do that,

TABLE D10.1

Asymmetrical Factors Based on X/R Ratio or Short-Circuit Power Factor

Short-Circuit Power Factor (%)	X/R Ratio	Ratio of rms Sym. Ampères To:	
		Max. Single-Phase Inst. Peak	Max. Single-Phase rms at 1/2 Cycle
0	∞	2.828	1.732
1	100.00	2.785	1.697
2	49.993	2.743	1.662
3	33.332	2.702	1.630
4	24.979	2.663	1.599
5	19.974	2.625	1.569
6	16.623	2.589	1.540
18	5.4649	2.231	1.278
40	2.2913	1.819	1.062
70	1.0202	1.517	1.002
100	0000	1.414	1.000

Appendix D: Solution to the Problems 689

Peak at $X/R=100$ for the generator component of 13.92kA=38.77 kA
Peak at $X/R=28.15$, system current of 10.22 kA=27.49 kA
Total=66.26 kA. Note the qualifications about 1/2 cycle peak in Section 10.1.2.

D.10.6 Problem 10.6

We will calculate X/R ratio from separate R and X networks and E/Z from complex impedance reduction. For duties on generator breaker, and from solution of Problem 10.5, the generator source current is higher equal to 16.09 kA $<-89.427°$ and therefore the duty can be based on it, ignoring the lower magnitude of rest of the system contribution. First, by empirical calculation, this is all a local contribution, and for interrupting duty, the MF from Figure 10.6 is 1.25. Then the required interrupting current is equal to 20.12 kA.

Also from Equation 10.1, the close and latch is equal to 44.8 kA peak.

A little reflection will show that this is not correct. ANSI/IEEE standard qualifies, "For duties above 80%, the derating method of 10.5.3 can be used provided that the X/R ratio does not exceed 45 at 60Hz (DC time constant no more than 120ms). For time constant beyond 120 ms consult the manufacturer."

The analytical calculation in solution to Problem 10.5 reveals that at the contact parting time:

AC component=10.364 kA and the DC component=17.23 kA. That means factor of asymmetry α=117.6%. According to ANSI/IEEE standard for generator breakers, a generator breaker must be able to handle 110% asymmetry at contact parting time. This standard recognizes that the asymmetry factor can be higher, but considers that arc fault resistance increases time constant T_a and forces current zero at contact parting time.

For interrupting duty at breaker 52T,

$Z=0.00533+j0.1773$. This gives 23.58 kA. X/R from separate X and R networks=22.86 kA.

The contribution from the utility source is 11.58 $< -87.7°$. Thus, for a fault on the source side, the breaker sees system contribution of 12.0 kA, and for a bus fault, it sees a contribution of 11.58 kA. These numbers are close to each other.

For a fault on the bus, all the contribution from the utility source is a remote contribution. Then the MF can be applied for an X/R of 25 from Figure 10.5. This is approximately 1.19, then the interrupting duty current to be considered is equal to 13.78 kA.

For the fault on the source side of the breaker, the generator contribution is a local contribution and there is no remote contribution. The generator impedance acts in series with parallel combination of reactor tie and impedances of C_1, T_2, T_3, and C_2 act in series. This gives approximately 9.50 kA local and a MF of 1.15 approximately can be read from Figure 10.5 This gives an interrupting duty current of 13.8 kA.

Practically, the lowest rated 13.8 kV circuit breaker has a short-circuit current of 18 kA.

D.10.7 Problem 10.7

This is simply an application of Equation 10.18. Substituting the specified data, I=25.26 kA. Note that the factor F_{ac} from Table 10.12=1.03.

D.10.8 Problem 10.8

Referring to Tables 8.12 and 8.14, it can be concluded that LVPCB is fused. MF for fused LVPCB is 1.07 for an X/R=6.6 and it is 1.12 for an X/R=8.27. For MCCB rated >20 kA, the

MF is 1.18 for $X/R=7$ and 1.20 for an $X/R=6$. Therefore, the X/R should be slightly > 7. From Equation 8.56, MF of 1.095 corresponds to an $X/R=7.56$.

D.10.9 Problem 10.9

For a fault at bus 3, the equivalent positive sequence network can be constructed similar to Figure 10.14 in the text. Then the separate R and X networks must be constructed. In each case, the circuits must be reduced to single equivalent Thévenin impedance at the fault point. These steps of calculations are not demonstrated. The final results of the calculation are shown in Table D10.2.

This does not imply that separate R and X calculations should be used for E/Z calculations.

D.10.10 Problem 10.10

LVPCB: Interrupting rating is expressed in kA rms symmetrical and short circuit withstand time rating is 30 cycles. However, a vendor may qualify this 30-cycle withstand rating with respect to the trip device. For example, the short-circuit rating for 30 cycles may be lower with short-time trip, as compared to the instantaneous trip settings. MF is applicable when X/R differs from test power factor.

ICCB: Interrupting rating is expressed in kA rms symmetrical and short-circuit withstand time rating is 15 cycles, generally. However, a vendor may qualify this 15-cycle with respect to the trip device. For example, the short-circuit rating for 15 cycles may be lower with short-time trip, as compared to the instantaneous trip. Instantaneous overrides are normally provided, that is, the short-time rating is not applicable for the full-specified interrupting rating. MF is applicable when X/R differs from test power factor.

MCCB: Interrupting rating is expressed in kA rms symmetrical. These do not have any short-time withstand ratings and are provided with instantaneous trips. Digital trip programmers are available. MF is applicable when X/R differs from test power factor. MCCB's can be current limiting type—the let-through characteristics are published by the manufacturers.

Power Cables: See discussions in the text.

ACSR Conductors: See discussions in the text.

Transformer primary switches and fused switches: The fuses can be current limiting or expulsion type. See text for their short-circuit ratings, expressed in kA rms symmetrical and asymmetrical. The switches should have compatible ratings with the fuses. These do not have short-circuit interrupting ratings but must withstand available peak currents and also have short-time withstand rating, till the fault is cleared by an upstream device. Coordination of switch and fuse short-circuit ratings is required.

Bus bars in electrical equipment: These must withstand the first-cycle or momentary short-circuit currents. Consider a medium-voltage switchgear lineup. The discussions

TABLE D10.2

Calculated X/R Ratios from Complex Impedance and Separate R and X Networks

Calculation	First Cycle, Close or Latch			Interrupting Duty		
	Z (pu)	X/R	I (kA)	Z (pu)	X/R	I (kA)
Complex	$0.0049+j0.1593$	41.05	$26.24 < -88.24°$	$0.0053+j0.1715$	41.16	$24.38 < -88.23°$
R and X	$0.00388+j0.1593$	41.05	$26.25 < -88.6°$	$0.00417+j0.1715$	41.16	$24.39 < -88.61°$

Appendix D: Solution to the Problems 691

in Section 10.9.4 are not applicable to the bus bars in the equipment. These must withstand all the short-circuit currents contributed by the load breakers, tie breakers, and the incoming breakers.

D.10.11 Problem 10.11

The high X/R ratios occur close to large synchronous generators—one impact is delayed current zeros, which make the generator source short-circuit current interruption difficult. Special circuit breakers may be needed—see Chapter 11 for an EMTP simulation and further discussions.

The impact of high X/R ratios results in high first-cycle close and latch duties as amply demonstrated in the text (see Equation 10.1).

D.10.12 Problem 10.12

Feeder breaker lightly loaded is the correct answer. A reader should be able to plot the component short-circuit currents at various locations specified in this problem and come to this conclusion.

D.10.13 Problem 10.13

The contributions are summarized in Table D10.3. A reader should verify the results.

D.10.14 Problem 10.14

To calculate 30-cycle current, drop all motor loads and use transient reactance of the generator. The results are shown in Table D10.4.

D.10.15 Problem 10.15

Similar to Example 10.4, the equivalent admittance diagram using *first-cycle impedance values* can be constructed as in Figure D10.3.

TABLE D10.3

Contributions of Short-Circuit Currents to 13.8 kV Buses

Bus	Contributions from	First Cycle (kA)	Interrupting (kA)
2	Motor M2	$1.092 < -88.97°$	$0.78 < -88.97°$
	From T3 through cable C2	$1.49 < -84.47°$	$1.284 < -84.43°$
	From reactor R	$10.28 < -89.4°$	$9.998 < -89.4°$
	Vectorial summation	$0.271 - j12.85$	$0.243 - j12.055$
	Equivalent impedance	$0.0164 + j0.078$	$0.0167 + j0.083$
4	From motor M1	$1.627 < -88.48°$	$1.085 < -88.5°$
	From T2 through Cable C3	$1.371 < -84.21°$	$1.16 < -84.05°$
	From reactor R	$8.425 < 88.39°$	$8.148 < 88.39°$
	Vectorial summation	$0.418 - j11.412$	$0.377 - j10.383$
	Equivalent impedance	$0.0032 + j0.088$	$0.0035 + j0.096$

TABLE D10.4
Calculated 30-Cycle Currents

Bus Identification	30-Cycle Currents (kA)
13.8 kV Bus connected to utility transformer T1	19.5 < −88.34
13.8 kV Bus with generator G1	19.92 < −88.74
4.16 kV Bus	19.33 < −88.92

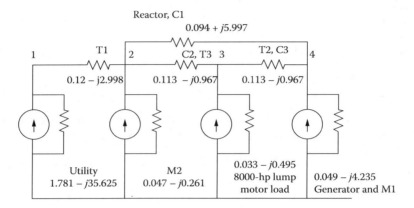

FIGURE D10.3
The equivalent admittance diagram, Problem P10.15.

Then the Y-bus matrix by examination is given as

$$\bar{Y}_{bus} = \begin{vmatrix} 1.901 - j38.623 & -0.12 + j2.998 & 0 & 0 \\ -0.12 + j2.998 & 0.3317 - j10.2331 & -0.113 + j0.967 & -0.094 + j5.997 \\ 0 & -0.113 + j0.967 & 0.259 - j2.429 & -0.113 + j0.967 \\ 0 & -0.094 + j5.997 & -0.113 + j0.967 & 0.225 - j11.199 \end{vmatrix}$$

The impedance matrix can be formed by inversion,

$$\bar{Z}_{bus} = \begin{vmatrix} 1.322 \times 10^{-3} + j0.0271 & 5.264 \times 10^{-4} + j0.013 & 2.384 \times 10^{-4} + j8.4 \times 10^{-3} & 2.846 \times 10^{-4} + j7.82 \times 10^{-3} \\ 5.264 \times 10^{-4} + j0.013 & 5.217 \times 10^{-3} + j0.171 & 2.079 \times 10^{-3} + j0.108 & 2.743 \times 10^{-3} + j0.101 \\ 2.384 \times 10^{-4} + j8.4 \times 10^{-3} & 2.079 \times 10^{-3} + j0.108 & 0.044 + j0.49 & 1.343 \times 10^{-3} + j0.101 \\ 2.846 \times 10^{-4} + j7.82 \times 10^{-3} & 2.743 \times 10^{-3} + j0.101 & 1.343 \times 10^{-3} + j0.101 & 3.182 \times 10^{-3} + j0.152 \end{vmatrix}$$

This gives an impedance of 0.003182+j0.1520 for the first-cycle fault at bus 4. We calculated Z=0.0032+j0.15206—from complex impedance reduction. This verifies the calculation. Interrupting duty admittance and bus impedance matrices can be similarly constructed.

Mathcad has been used for the manipulation of complex numbers and matrix inversion and solution of problems.

Appendix D: Solution to the Problems　693

D.11 Problems Chapter 11

D.11.1 Problem 11.1

For a bus-connected generator, from Equations 11.13 and 11.14,

$$K_G = \frac{U_n}{U_{rG}}\left(\frac{C_{\max}}{1+X_d^{''}\sin\phi_{rG}}\right) = \frac{1.1}{1+0.11\times0.6} = 1.032$$

$$Z_{GK} = K_G(R_G + jX_d^{''}) = 1.032(0.07\times0.11 + j0.11) = 0.0079 + j0.11352$$

Here, we consider $U_n = U_{rG}$.
For a fault on the high side,

$$K_S = \left(\frac{C_{\max}}{1+\left|X_d^{''}-X_T\right|\sin\phi_{rG}}\right) = \frac{1.1}{1+0.01\times0.6} = 1.0934$$

Note that transformer and generator are both 60 MVA, and the specified impedances are at 60 MVA base. The results of calculations are also on 60 MVA base.
　The modified power station impedance is

$$Z_S = K_s(Z_G + Z_T) = 1.0934(0.0077 + j0.11 + 0.0286 + j0.9996) = 0.0396 + j1.213$$

This is in pu on 60 MVA base.

D.11.2 Problem 11.2

138 kV fault
　Utility contribution,

$$I_{kQ}^{''} = \frac{1.1}{0.00081 + j0.0154} = 0.0374 - j7.142\ (100\,\text{MVA base, pu}) = 29.88\ \text{kA}$$

From solution to Problem 11.1,

$$I_{k,\text{PSU}}^{''} = \frac{1.1}{0.066 + j2.022} = 0.016 - j0.494 = 2.06\ \text{kA}$$

Thus, $I_k^{''} = 0.0869 - j8.645$ pu, 100 MVA base $= 8.6454 < -88.12° = |36.17|$ kA.

$$i_p = i_{p,\text{PSU}} + i_{pQ}$$

Power station $R/X = 0.033$. Then, χ_{PSU} from Equation 11.4 is equal to 1.91. Utility system $R/X = 0.0526$, $\chi_{PQ} = 1.857$

694   *Appendix D: Solution to the Problems*

Then,

$$i_p = 1.857 \times \sqrt{2} \times 29.88 + 1.91 \times \sqrt{2} \times 2.06 = 84.01 \text{ kA peak}$$

$$i_b = I_{b,\text{PSU}} + I_{bQ} = I_{b,\text{PSU}} + I_k''$$

For a contact parting time of 0.03s, $\mu=0.85$ (note that equations give calculations for $t_{\min}=0.02$ and 0.05 s). It is, therefore, necessary to interpolate. Also, the ration I_{kG}'' / I_{rG} must be based on the same voltage base. This gives $I_{kG}'' / I_{rG}=8.20$.

$$I_{b,\text{PSU}} = 0.85 \times 2.06 = 1.751$$

Therefore, breaking current=31.63 kA rms symmetrical.
   DC current utility source, Equation 11.39 is given as

$$\sqrt{2} \times 29.88 \times e^{-2\pi ft/R/X} = 23.30 \text{ kA at 0.03 s. Similarly component from PSU} = 2.0 \text{ kA.}$$

Then, the asymmetrical breaking current is equal to

$$\sqrt{(31.63)^2 + (25.30)^2} = 40.50 \text{ kA}$$

Fault at 13.8 kV
   From solution to Problem 11.1,

$$Z_{G,\text{PSU}} = 0.0092 + j0.1892 \text{ pu, 100 MVA base:}$$

Therefore,

$$I_{kG}'' = 0.256 - j5.273 \text{ pu} = 22.09 \text{ kA}$$

$$K_{T,S} = \frac{C_{\max}}{1 - x_T \sin \phi_{rG}} = \frac{1.1}{1 - 0.1 \times 0.6} = 1.17$$

$$I_{kT}'' = \frac{1.1}{1.17(0.0048 + j0.1667) + 0.00081 + j0.01538} = 0.212 - j6.036 \text{ pu} = 25.27 \text{ kA}$$

$I_{kT}''=0.468-j11.309$ pu$=47.35$ kA.
   Factor $\chi(= \kappa)$ for generator 1.863, and for $I_{kT}'' =1.90$, based on R/X
   Therefore,

$$i_p = \sqrt{2} \times 1.863 \times 22.09 + \sqrt{2} \times 1.90 \times 25.26 = 126 \text{ kA peak.}$$

$$I_{bG} = \mu I_{kG}''$$

$$I_{kG}'' / I_{rG} = 8.8$$

Appendix D: Solution to the Problems 695

Therefore, for $0.05\,\text{s}$ $t_{\min}$, $\mu = 0.746$.

Then $I_{bG} = 16.48$ kA, and total breaking current $= 16.48 + 25.27 = 41.75$ kA.

For calculation of DC component, we do not use the generator resistance given by IEC equations (see qualifications in IEC standards).

From Figure 7.4, an $X/R = 55$ is used.

This gives a DC component of 22.17 kA at $t_{\min} = 0.05\,\text{s}$ from the generator. For the transformer component, $X/R = 28.47$. This gives a DC component of 18.43 kA. Total DC component $= 40.60$ kA. Therefore, asymmetrical breaking current $= 58.23$ kA.

D11.3 Problem 11.3

Contribution from asynchronous motors

From the given data,

$$I_{LR} / I_{rM} = 6, \quad I''_k / I_{rM} = 6.6, \quad m = 8.5\,MW, \quad R_M / X_M = 0.10 \ (\text{Equation 11.20}),$$

$$\kappa_M(\chi) = 1.746, \quad \mu(t_{\min} = 0.05s) = 0.7804, \quad q\,(\text{Equation 8.36}) = 1.047$$

Then,

$$I''_{kM} = \frac{1.1}{0.167 + j1.67} = 0.065 - j0.652 = |0.6552|\,\text{pu} = 9.46 \text{ kA}$$

$$i_{pM} = 9.46 \times \sqrt{2} \times 1.746 = 23.35\,\text{kA}$$

$$i_{bsym} = 9.46 \times 0.7804 \times 1.047 = 7.30\,\text{kA}$$

$$i_{DC} = 9.46 \times \sqrt{2} \times e^{-2\pi ftR/x} = 2.031\,\text{kA}$$

Synchronous motors

$$K_M = \frac{c_{\max}}{1 + X''_d \sin \phi_{rM}} = 1.01$$

$$R_M = 0.07 X''_d$$

Therefore,

$$I''_{kM} = \frac{1.1}{1.01(0.21 + j3.00)} = 0.026 - j0.365 = |5.28|\,\text{kA}$$

$$I''_{kM} / I_{rM} = 7.02$$

$$\kappa_M = 1.814, \quad \mu = 0.772 \text{ at } t_{\min} = 0.05\,\text{s}$$

696 *Appendix D: Solution to the Problems*

Therefore,

$$i_{pM} = \sqrt{2} \times 1.814 \times 5.28 = 13.541 \text{ kA}$$

$$i_{bsym} = 0.772 \times 5.28 = 4.076 \text{ kA}$$

$$i_{DC} = \sqrt{2} \times 5.28 \times e^{-2\pi f t R/X} = 1.996 \text{ kA}$$

The two similar transformers operating in parallel through a common utility transformer, and for a fault at F, do not constitute a meshed network. The equivalent impedance to fault point is

Transformer $= 0.0212 + j0.42447$
Cable $= 0.0155 + j0.0258$
Utility source $= 0.00486 + j0.0465$

All in pu on 100 MVA base.

Therefore, impedance to the fault point is equal to $0.01326 + j0.2715$ pu

$$I''_{k,U} = \frac{1.1}{0.01326 + j0.2715} = 0.197 - j4.042 = 4.047 < -87.2o = |56.17| \text{ kA}$$

$$\kappa_{US} = 1.87$$

$$i_{p,US} = \sqrt{2} \times 1.87 \times 56.17 = 148.54 \text{ kA}$$

$$i_{b,US} = I''_{k,US} = 56.17 \text{ kA}$$

$$i_{DC,US} = \sqrt{2} \times 56.17 \times e^{-2\pi f t R/X} = 31.67 \text{ kA}$$

These currents are calculated at 4.16 kV. The motor contributions are calculated at motor voltage of 4 kV. These can be multiplied by a factor of 0.96 (the motor currents at a higher rated voltage will reduce). Then by simple addition,

$$i_{pT} = 148.54 + 13.02 + 22.41 = 183.97 \text{ kA}$$

$$i_{bT} = 56.17 + 3.92 + 7.01 = 67.10 \text{ kA}$$

$$i_{DC,T} = 31.67 + 1.92 + 1.95 = 35.54 \text{ kA}$$

The asymmetrical breaking current is 75.93 kA. No circuit breaker meeting these duty requirements will be commercially available. One simple remedy will be not to parallel the transformers and provide a bus section breaker.

D.11.4 Problem 11.4

We will demonstrate short-circuit calculations at $F3$.

First calculate the system impedances for a fault at $F2$.

Appendix D: Solution to the Problems 697

As in previous problems, based upon the data in Table P11.1 and calculation in pu based on 100 MVA base, we have,

Utility source: $Z_Q = 0.006796 + j0.114869$

200 MVA transformer, $Z_{t,PSU} = 0.0014238 + j0.049979$

230 MVA generator, $R_G = 0.05X_d''$

$$K_{G,PSU} = \frac{1.1}{1 + 0.16 \times 0.5267} = 1.0145$$

Then $Z_{G,PSU} = 1.0145(0.00345 + j0.069) = 0.0035 + j0.070$

$$K_{TS} = \frac{1.1}{1 - 0.1 \times 0.5267} = 1.161$$

Then the transformer and generator equivalent in parallel is given as

$$1.161(0.001428 + j0.049979) \text{ in parallel } (0.0035 + j0.070) = 0.0012 + j0.032$$

The medium-voltage motor contributions to the faulted 4.16 kV bus is shown in Table D11.1 and for the unfaulted 4.16 kV bus in Table D11.2.

This gives a total: $I_{kM,MV}'' = 12.21$, $i_{pM,MV} = 29.53$, $i_{bM,MV} = 10.38$, $i_{DC,MV} = 1.98$ (s4–1).

The combined impedance of the medium-voltage motors is given as $1.4345 + j9.55$ and $2.505 + j16.70$ and $2.233 + j14.89$ per unit motor impedances in parallel. These three

TABLE D11.1

Medium-Voltage Motors, Faulted 4.16 kV Bus

Motor details →	M1 5000-hp, 6-pole Sync., 0.8 pf	M2 2500-hp, 8-pole Induction	M3 2500-hp, 2-pole Induction	M4 2500-hp, 12-pole Sync., 0.8 pf
Efficiency(assume)	0.95	0.90	0.92	0.95
PF (assume)		0.85	0.90	
P_{rM} (MW)	4.91	2.438	2.252	2.453
S_{rM} (MVA)	6.135	2.868	2.502	3.066
I_{rM} (A)	708.4	351.9	325.1	354.2
X_M	15%	16.7%	16.7%	15%
m	1.636	0.61	2.252	0.41
R_M/X_M	0.10	0.15	0.10	0.15
κ_M	1.75	1.65	1.75	1.65
I_{kM}'' / I_{rM}	7.3	6.6	6.6	7.3
M (at $t_{min} = 0.05$s)	0.767	0.7804	0.7804	0.767
q	-	0.73	0.887	-
I_{kM}'' (kA)	5.17	2.32	2.15	2.57
i_{pM} (kA, peak)	12.80	5.41	5.32	6.0
i_{bM}(kA, rms sym)	5.60	1.32	1.49	1.97
i_{DC} (at $t_{min} = 0.05$s)	1.11	0.19	0.46	0.22

TABLE D11.2
For Medium-Voltage Motors Connected to Unfaulted 4.16 kV Bus

Motor details →	M5 2000-hp, 4-pole Sync., 1.0 pf	M6 1000-hp, 2-pole Induction	M7 1000-hp, 6-pole Induction
Efficiency(assume)	0.95	0.92	0.90
PF (assume)		0.90	0.86
P_{rM} (MW)	1.571	0.901	0.964
S_{rM} (MVA)	1.571	1.0	1.121
X_M	15%	16.7%	16.7%
m	0.785	0.45	0.321
R_M/X_M	0.15	0.15	0.15
X_M (pu, 100 MVA, base)	9.55	16.70	14.89
R_M	1.434	2.505	2.233

impedances in parallel give $0.648 + j4.315$ Group of low-voltage motors MG_1 on the faulted bus feeds through a 1 MVA transformer.

There are two motors of 100 hp and six motors of 40 hp. Total connected motor hp $= 440$. We can consider an average power factor of 0.8 and 1 hp rating is approximately equal to 1 kVA. Again considering $X_M = 16.7\%$ (because even for low-voltage motors a locked rotor current is equal to six times the full-load current is a valid assumption) and $R_M / X_M = 0.42$ according to IEC Equation 8.20, we have an impedance of $15.941 + j37.954$ pu on a 100 MVA base.

To this, add the impedance of 1 MVA transformer that with given parameters equals to $0.985 + j5.665$. Thus, the two impedances in series are $16.926 + j43.619$ pu.

Then,

$$I''_{K,LV,\text{fault-side}} = \frac{1.1}{16.926 + j43.619} = 0.0085 - j0.022$$

This is equal to 0.327 kA $< -68.87°$

This is small enough to be even ignored, but continuing with this calculation, $\kappa = 1.327$. Then $i_p = 0.614$. kA, and $i_b = 0.327$ kA. Add this to calculation in above

$$I''_{k,T} = 12.537, \quad i_{pT} = 30.14, \quad i_{bT} = 10.71, \quad i_{DC,T} = 1.98$$

The effect of the LV motor group on the unfaulted side MG_2 is similarly considered. This group has four motors of 150 hp and eight motors of 75 hp. Total $= 1200$ hp, nearly equal to 1200 kVA. On the same basis as before, the motor group equivalent impedance is given as $5.845 + j13.92$ pu, 100 MVA base. Also see Table D11.2.

To this, add the impedance of 1.5 MVA transformer ($= 0.506 + j3.8$). This gives $6.351 + j17.72$. This low-voltage motor impedance through 1.5 MVA transformer acts in parallel with the medium-voltage motor impedance calculated above. This parallel combination gives $0.656 + j3.492$.

Appendix D: Solution to the Problems

The three-winding transformer impedances are calculated based on Chapter 1. This gives,

$$Z_H = j0.025$$

$$Z_{L1} = Z_{L2} = j0.375 \text{ pu on 100 MVA base}$$

The X/R ratios are not specified for three-winding transformer, but we can assume it is equal to 15.

Then the equivalent circuit based on the calculations so far reduces to that shown in Figure D11.1.

Therefore, the equivalent impedance of the system to the fault point (the motor contributions to the faulted bus are already calculated and should not be considered in this impedance calculation),

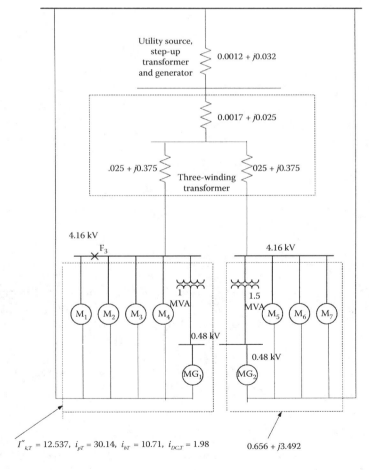

FIGURE D11.1
Equivalent circuit for fault at F3, Problem P11.4.

700   *Appendix D: Solution to the Problems*

{(0.656 + j3.492) + (0.025 + j0.375)} in parallel with {(0.0017 + j0.025) + (0.0012 + j0.032)} in series with 0.025 + j0.375

This gives 0.00296 + j0.056 + 0.025 + j0.375 = 0.028 + j0.431

Therefore,

$$I''_{K,sys} = \frac{1.1}{0.028 + j0.431} = 0.165 - j2.541\,\text{pu}$$

This gives 35.33 <−86.28°

This is also the breaking current.

$$\chi = 1.82$$

Multiply this by factor 1.15, this gives 2.09. However, the maximum is limited to 2.0.

Therefore, $i_{p,\,sys}$=99.92 kA peak.

The DC at t_{min}=0.05s is

$$\sqrt{2} \times 35.33 e^{-2\pi ftR/X} = 14.69\,kA$$

The total short-circuit currents are, therefore,

$$I''_k = 47.87 \text{ kA}$$

$$i_p = 130.06 \text{ kA peak}$$

$$i_{b,sym} = 46.04 \text{ kA}$$

$$i_{DC} = 16.67 \text{ kA}$$

$$i_{b,asym} = 48.96 \text{ kA}$$

D.11.5 Problem 11.5

While this can be done using the same method as in the solution of Problem 11.4, except that motor and generator resistances can be taken from Tables 10.2 and 10.4. Practically, this should never be done. A user cannot alter the methodology and calculation procedure laid down in ANSI/IEEE or IEC standards. The results of such a calculation will be totally invalid. A reader must recognize that the calculations must be carried out in total according to one standard or the other. ANSI/IEEE and IEC standards approach asymmetry from different prospective (see Section 11.6).

D.11.6 Problem 11.6

Note that the EMTP simulation uses a more elaborate model of three dampers. Rotor windings f (field) and k_d (damper) in direct axis designated 1 and 2, and g and k_q in the q-axis, again denoted by windings 1 and 2.

Appendix D: Solution to the Problems 701

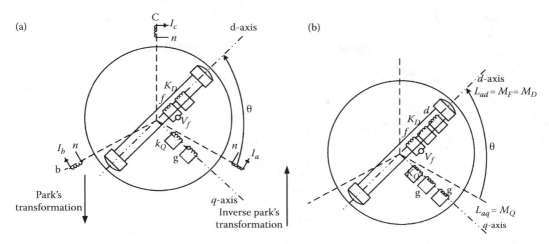

FIGURE D11.2
Diagrammatical representation of Park's transformation.

The following matrix equation is reproduced from ATP Rule book,

$$\begin{vmatrix} \lambda_d \\ \lambda_q \\ \lambda_0 \\ \lambda_f \\ \lambda_g \\ \lambda_{kd} \\ \lambda_{kq} \end{vmatrix} = \begin{vmatrix} L_d & 0 & 0 & L_{af} & 0 & L_{akd} & 0 \\ 0 & L_q & 0 & 0 & L_{aq} & 0 & L_{akq} \\ 0 & 0 & L_0 & 0 & 0 & 0 & 0 \\ L_{af} & 0 & 0 & L_f & 0 & L_{kkd} & 0 \\ 0 & L_{ag} & 0 & 0 & L_g & 0 & L_{gkq} \\ L_{akd} & 0 & 0 & L_{fkd} & 0 & L_{kd} & 0 \\ 0 & L_{akq} & 0 & 0 & L_{gkq} & 0 & L_{kq} \end{vmatrix} \times \begin{vmatrix} i_d \\ i_q \\ i_0 \\ i_f \\ i_g \\ i_{kd} \\ i_{kq} \end{vmatrix}$$

Compare this with Equation 9.72 in Chapter 9. The correspondence between Tables 11.12 and 11.13 cannot be established using equations in Chapter 6. If two dampers are considered in direct axis, this adds another row and column to the matrix (8×8).

Note that parameters L_d, L_q, L_o are direct conversion from Table 8.12, unit values to ohms in Table 8.13 (see Figure D11.2).

D.11.7 Problem 11.7

We will verify the IEC calculations

$$K_{G,SO} = \frac{1}{1+p_G} \frac{c_{max}}{1+X_d'' \sin\phi_{rG}}$$

702   *Appendix D: Solution to the Problems*

Generation voltage is equal to system voltage, and substituting $K_{G.SO}=1.1$, $R_{Gf}=0.05$ times $X_d''=0.0032$ pu on generator base of 234 MVA,

$$I_{kG}'' = \frac{1.1}{1.1(R_{Gf} + X_d'')} = \frac{1.1}{1.1(0.0032 + j0.0641)} = 0.779 - j15.586 \text{ pu on } 100 \text{ MVA base} = |49.97| \text{kA}$$

The factor χ from equation based on $X/R=20$ is 1.863. Therefore, $i_{pG}=131.60$ kA.

For calculation of breaking current μ from Equation 11.22 is 0.779. This gives breaking current$=38.50$ kA rms symmetrical.

For calculation of DC component, IEC recommends that vendor's data should be used and Equation 8.16 is not to be applied. Considering $X/R=125$ from Table 8 through 12, the DC component is given as

$$\sqrt{2} \times 49.97 e^{-2\pi ftR/X} = 60.73 \, kA \text{ at minimum time delay } t_{\min} = 0.05 \, \text{s}$$

Asymmetry factor$=112\%$.

D11.8 Problem 11.8

The AC decrement current profile based upon given data is

$$6.39e^{-t/0.015} + 15.11e^{-t/0.638} + 1.88$$

This gives a current of 16.08 kA at contact parting time of 0.05s.

The DC component is 29.97 kA.

Therefore, asymmetry factor$=131.83\%$.

This shows that higher asymmetry at contact operating time can occur in bus-connected industrial generators also.

D.12 Solution to Problems Chapter 12

D.12.1 Problem 12.1

First make a layout plan. Based on the cell data furnished, the layout in Figure 12.5 is acceptable, per tier. There will be two such tiers, one above the other, each tier housing 60 cells. The cell width is 6.8 inches, 30 cells per row gives 17 feet. Thus, 18 feet length shown in Figure 9.5 is adequate. As the cell length is 15.5 inches, 2 feet center line spacing is also acceptable.

Battery resistance from Equation 12.11 is equal to 18.6 mΩ. Consider battery connectors that are rectangular have a total length of $18 \times 4 = 72$ ft (in two tiers). Then from the given data of 0.0321 mΩ/ft, the connector resistance$=2.31$ mΩ.

The external battery resistance is 2 mΩ. Then the total resistance in battery circuit is 22.91 mΩ.

Then the maximum short-circuit current is $240 / (22.91 \times 10^{-3}) = 10.47$ kA.

Calculate the inductance

Of the rectangular connectors is given by Equation 9.13:

Appendix D: Solution to the Problems

$$L = \frac{4\pi \times 10^{-7}}{\pi}\left[\frac{3}{2} + \ln\frac{24}{1+0.5}\right] = 1.708\ \mu H/m,\ \text{total for 36 ft (in two tiers)} = 10.97\ \mu H$$

Similarly from the same equation, the inductance of battery cells=0.743 µH/m, total=8.14 µH for 120 cells in two tiers.
   Inductance of the cable=15 µH. Thus, total inductance=34.11 µH.
   Then the rate of rise is

$$\frac{240}{34.11 \times 10^{-6}} = 7.03 \times 10^{6}\ A/s$$

and the time constant is

$$\frac{34.11 \times 10^{-6}}{22.91 \times 10^{-3}} = 1.49\ ms$$

IEC Calculation
   From Equation 12.14,

$$i_{pB} = \frac{1.05 \times 240 \times 10^{3}}{0.9(18.6) + 2 + 2.31} = 11.97\ kA$$

From Equation 12.15,

$$\frac{1}{\delta} = \frac{2}{\dfrac{22.91 \times 10^{-3}}{34.11 \times 10^{-6}} + \dfrac{1}{30 \times 10^{-3}}} = 2.83\ ms$$

Then from Figure 9.6, time to peak t_{pB}=7 ms, and time constant t_{1B}=1.3 ms.
   The quasi steady-state current given by Equation 9.16 is

$$I_{kB} = \frac{0.95 \times 240 \times 10^{3}}{[0.9(18.6) + 2 + 2.31] + 0.1(18.6)} = 9.95\ kA$$

The profile calculated according to IEC method is plotted in Figure D12.1.

D.12.2 Problem 12.2

Calculation based on Example 12.3

$$I_{a}' = \frac{I_{a}}{r_{d}'} = \frac{178}{0.07/1.29} = 3.28\ kA$$

Note that transient reactance is given in ohms.
   From Figure 9.9, for 50 hp motor L_{a}'=2.5 mH. The rate of rise is 92 kA/s. The time constant at second step in Figure 9.8 is 11.90 ms.

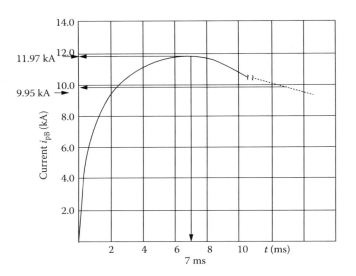

FIGURE D12.1
Short-circuit current profile of the battery, Problem P12.1.

The calculation shows that 2/3rd of 3.28 kA = 2.186 kA will be achieved in 23.76 ms. Then the time constant of 11.90 ms is applicable for the balance 1.094 kA.

IEC Calculation as in Example 9.3

It must be understood that the calculations shown above as per Example 9.3 are not comparable with the calculations according to IEC. IEC considers the impact of the field excitation and load characteristics. Again to be valid, a host of motor, excitation, and load data are required, which should be accurate (see Equation from 12.27 to 12.37); and the complete curves and procedures as detailed in IEC, which are not reproduced in the text.

The quasi steady-state current is given by Equation 12.29,

$$I_{kM} = \frac{L_F}{L_{0F}} \frac{U_{rM} - I_{rM} R_M}{R_{MB}} = 0.3 \left[\frac{240 - (0.07)(178)}{0.07} \right] = 975 \text{ A}$$

Here, we consider a motor resistance $R_M = 0.07$ ohms, though transient resistance is not the same as motor armature resistance. Also $R_M = R_{MB}$, as there is no external resistance in the motor circuit.

$$\tau_M = \frac{2.5 \times 10^{-3}}{0.07} = 35.7 \text{ ms}$$

Peak current is given by

$$i_{pM} = \kappa_M \left(\frac{U_{rM} - I_{rM} R_M}{R_{MB}} \right)$$

We need to ascertain κ_M

Appendix D: Solution to the Problems 705

The mechanical time constant is given by

$$\tau_{mec} = \frac{2\pi J n_0 R_{MB} I_{rM}}{M_r U_{rM}} = \frac{2\pi \times 2 \times 690 \times 0.07 \times 178}{52.68 \times 240} = 8.54 \text{ ms}$$

Here, $\tau_{mec} < 10\tau_F$. Figure 12.10 cannot be applied. A reference to IEC is required. This requires calculation of $1/\delta$, and ω_0

$$1/\delta = 21/\tau_M = 71.4$$

$$\omega_0 = \sqrt{\frac{1}{\tau_{mec} T_M}\left(1 - \frac{I_{rM} R_M}{U_{rM}}\right)} = \sqrt{\frac{1}{8.54 \times 35.7}\left(1 - \frac{178 \times 0.07}{240}\right)} = 17.9$$

These values when entered in the curve in IEC give $\kappa_M = 0.6$. This gives a peak current of 1950 A.

Again from curves in IEC, the time to peak is equal to 60 ms.

Now consider that resistance is 0.0904 instead of 0.07 ohms. Then $\tau_M = 27.6\text{ms}, \tau_{mec} = 11.03, \tau_M > 10\tau_F, \kappa_M = 1, I_{pM} = 2.48\text{kA}$.

Then from Figure 12.10, time to peak $= t_{pM} = \kappa_{1M}\tau_M = 4 \times 27.6 = 110.4.4\text{ ms}$. The profile is shown in Figure D12.2.

D.12.3 Problem 12.3

Based on the given data

The utility source and transformer impedance in ohms referred to 230-V side (Figure 12.13).

$$R_N + jX_N = 0.00183 + j0.00916 \ \Omega$$

$$R_{DBR} = 0.001\Omega$$

$$L_{DBR} = 3\mu H$$

$$\frac{R_N}{X_N} = 0.2$$

$$\frac{R_{DBR}}{R_N} = 0.546$$

Then from Equation 12.48,

$$\lambda_D = \sqrt{\frac{1 + 0.04}{1 + (0.04)(1 + 0.667 \times 0.564)^2}} = 0.985$$

From Equation 12.47,

$$i_{kD} = (0.985)\frac{3\sqrt{2}}{\pi}\left(\frac{1.05 \times 480}{\sqrt{3} \times 0.00935}\right)\left(\frac{230}{480}\right) = 19.84 \text{ kA}$$

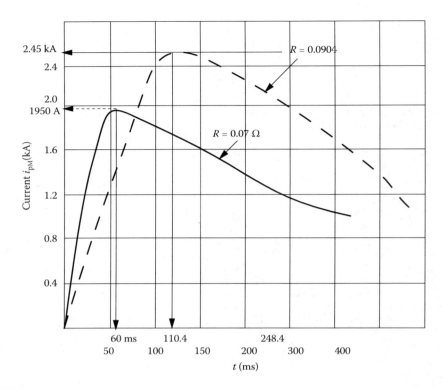

FIGURE D12.2
Short-circuit current profile of DC motor, Problem P12.2.

To calculate the peak current, ratios

$$\frac{R_N}{X_N}\left(1+\frac{2}{3}\frac{R_{DBR}}{R_N}\right)=0.273$$

$$\frac{L_{DBR}}{L_N}=0.123$$

From Equation 12.52, $\phi_D = 74.73°$
From Equation 12.51, κ_D

$$\kappa_D = 1+\frac{2}{\pi}e^{-\left(\frac{2}{3}+\phi_D\right)\cot\phi_D}\sin\phi_D\left(\frac{\pi}{2}-\arctan\frac{L_{DBR}}{L_N}\right)=1.383$$

Then, the peak short-circuit current is

$$i_{pD} = \kappa_D i_{kD} = 27.44\,\text{kA}$$

The time to peak is given by Equation 12.53,

Appendix D: Solution to the Problems 707

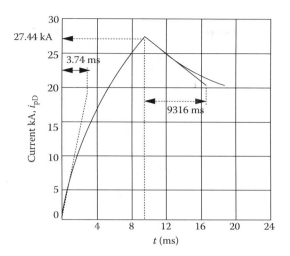

FIGURE D12.3
Short-circuit current profile of the rectifier, Problem 12.3.

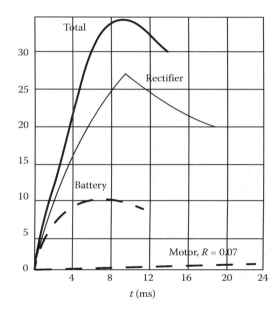

FIGURE D12.4
Summation of short-circuit current profiles of battery, motor, and rectifier, Problem 12.4.

$$t_{pD} = (3 \times 1.383 + 6)\,\text{ms} = 10.15\ \text{ms and the rise time constant is}$$

$$\tau_{1D} = \left[2 + (1.383 - 0.9)(2.5 + 9 \times 0.123)\right] = 3.74\,\text{ms}$$

The IEC equations are for 50 Hz. For 60 Hz, the peak will occur around 8.46 ms.

The decay time constant is given by Equation 12.54,

$$\tau_{2D} = \frac{2}{(0.2)(0.6+0.9(0.546))} = 9.16 \text{ ms}$$

The short-circuit current profile is plotted in Figure D12.3.

D.12.4 Problem 12.4

The summation of the partial currents is shown in Figure D12.4. Again the short-circuit current from rectifier predominates, it reaches the peak value in approximately 1/2 cycle.

The DC circuit breaker should have a peak rating of 40 kA and short-circuit rating of 40 kA

Index

Note: Page numbers followed by "*f*" and "*t*" refer to figures and tables, respectively.

A

Acetylene, 296
Additive polarity, 600–601
Adjoint matrix, 559
Adjustable speed drives (ASDs), 70
Aluminum conductor steel reinforced (ACSR)
    conductors, 456
American Electric Power (AEP), 85, 304
Amplitude constant, 342
Amplitude factor, 270, 480
Ancillary services company (ANCILCO), 39
Arc-back short circuits, 414
Arc interruption theories, 261
    Cassie-Mayr theory, 262–263
    Cassie's theory, 261–262
    Mayr's theory, 262
Arc suppression reactor, 622
Armature time constant, 375, 438, 461, 531
As Low an exposure As is Reasonably
    Achievable (ALARA), 47
Asymmetrical breaking current, 498, 502,
    506, 508, 512
Asymmetrical current waveform, 314
Asymmetrical ratings, 316, 366, 480
    contact parting time, 316–317
Asynchronous medium-voltage
    motors, 510
Asynchronous motors, 491, 492, 495
    ANSI fault current calculations
        from, 504*t*
    IEC fault current calculations
        from, 505*t*
Automatic Generation Control (AGC), 12, 36
Autoregressive moving average (ARMA)
    model, 109
Autotransformers, 604–608
Auxiliary transformer (AT), 456, 486*f*, 487
Availability, 16
    defined, 16
    exponential distribution, 17–18
Available Transfer Capability (ATC), 91
Azimuth-altitude dual axis trackers
    (AADAT), 120

B

Back-to-back switching, 328, 329, 331–334, 334*f*
Backward substitution, 580–583
Basic insulation levels (BILs), 301, 304, 306, 365
Basic loop, 224
Battery short-circuit model, 525
Beacon Power, 67
Becquerel, Edmond, 120
Betz limit, 98
Bifactorization method, 582–583
Binary cycle geothermal plant, 49, 50*f*
Biofuels, 50–51
Breakdown maintenance, 631
Breaker duty calculations, 437
Breaking capability, 479–480
Breaking current, 481, 491–492, 494
    asymmetrical, 498, 502, 506, 508, 512
    calculations of, of asynchronous motors,
        496–497
    symmetrical, 485, 503, 506, 508
Breaking current factor, 502
Breeder reactors, 47
Bright Source Energy, Inc., 118
Bubble theory, 296
Bulk oil circuit breakers, 296
Bus admittance matrix, 214–218, 226
    incidence matrix, 228–232
    primitive network, 226–228
    Y-matrix, node elimination in, 232–233
Bus bracings, 444, 452–455
Bushing current transformer (BCT) secondary
    voltage, 335
Bus impedance matrix, 219, 220*f*
    algorithms for construction of, 233
        adding a link, 236–238
        adding a tree branch to an existing node,
            234–236
        coupled branch, removal of, 238–246
        uncoupled branch, changing impedance
            of, 238
        uncoupled branch, removal of, 238
    incidence matrix, 228–232
    from open-circuit testing, 220–221

709

Bus impedance matrix (*cont.*)
  primitive network, 226–228
  short-circuit calculations with, 246
    double line-to-ground fault, 247–256
    line-to-ground fault, 246
    line-to-line fault, 246
    Y-matrix, node elimination in, 232–233
Bus impedance method, 256
Bus incidence matrix, 228

C

Cables
  HTS cables, 68
  HVDC cable systems, 88
  power cables, 455–456
  switching of, 331–335
Calculations, short-circuit (ANSI methods), 421
  accounting for short-circuit current
    decay, 423
    low-voltage motors, 424–425
  breaker duty calculations, 437
  calculation procedure, 439, 440
    adjustments for transformer taps and
      ratios, 443–444
    circuit breakers with sources on either
      side, 441–443
    dynamic simulation, 441
    hand calculations, 441
    necessity of gathering accurate data,
      439–440
    switching devices without short-circuit
      interruption ratings, 443
  comparison with ANSI and IEE calculation
    procedures, 497–499
  examples of, 444
    bus bracings, 452–455
    equivalent impedance, deriving, 464–469
    4.16 kV circuit breakers and motor
      starters, 452, 453*t*, 454*t*
    generator breaker, selection of, 463–464
    generator source asymmetrical
      current, 461
    generator source symmetrical short-
      circuit current, 460
    K-rated 15 kV breakers, 448–452
    low-voltage circuit breakers, 452
    overhead line conductors, 456–460
    power cables, 455–456
    required closing latching capabilities,
      462–463
    short-circuit duties, calculation of,
      444–448

system source asymmetrical short-circuit
  current, 462
system source symmetrical short-circuit
  current, 461–462
transformer primary switches and fused
  switches, 452
fault current calculations, 500
fault current limiters, 470
  superconducting fault current limiters,
    473–474
generator source asymmetry, 437–439
methods of calculation, 427
  E/X method for AC and DC decrement
    adjustments, 427–428
  fault fed from local sources, 430–435
  fault fed from remote sources, 428–430
  simplified method, 427
  weighted multiplying factors, 435
network reduction, 435–436
  E/X or *E/Z* calculation, 436–437
rotating machine model, 425–426
thirty-cycle short-circuit currents, 469–470
type and severity of system short circuits,
  426
types of, 421
  assumptions, 422
  maximum peak current, 422
Calculations, short-circuit (IEC standards), 479
  ANSI/IEE calculation procedures, 497–499
  comparison of calculations using IEEE/IEC
    standards and EMTP simulations, 516
  differences in IEC and ANSI standards, 479
    breaking capability, 479–480
    breaking current, 481
    highest short-circuit currents, 482
    initial symmetrical short-circuit current,
      480
    peak making current, 481
    rated making capacity, 480
    rated opening time and break time, 480
    rated restriking voltage, 480
    steady-state current, 481–482
  electromagnetic transients program
    simulation of generator terminal short
    circuit, 513–516
  far-from generator faults, 483
    meshed networks, 487–489
    nonmeshed sources, 485–487
  influence of motors, 495–496
    asynchronous motors, calculations of
      breaking currents of, 496–497
    low-voltage motor groups, 496
    static converter fed drives, 497

Index 711

near-to-generator faults, 489
    generators and unit transformers of
        power station units, 490–491
    generators directly connected to systems,
        489–490
    motors, 491
    short-circuit currents fed from one
        generator, 491–493
    short-circuit currents in meshed
        networks, 494–495
    short-circuit currents in nonmeshed
        networks, 493–494
    prefault voltage, 483
Calculations, short-circuit, 6, 26–28, 29, 131, 132,
        213, 421, 422, 424, 425, 441, 444–448,
        479, 513
    with bus impedance matrix, 246
        double line-to-ground fault, 247–256
        line-to-ground fault, 246
        line-to-line fault, 246
    current–time profiles, 530*f*
    practical, 417
Capacitance currents, 197–198, 199–200, 326,
        327*t*, 330*t*
    interruption of, 276–278, 276*f*
Capacitance switching, 326
Capacitive and inductive circuits
    transient recovery voltage (TRV) in, 278–279
Capacitors, 276–278
    charged capacitor, short-circuit of, 543–544
    contribution to short-circuit currents,
        413–414, 413*f*
    defined, 2
    power capacitors, 75
    shunt capacitors, 35, 62, 76, 328, 336, 413
    static converters contribution to short-circuit
        currents, 414–417
    surge capacitor, 197, 302, 306, 326, 328
    synchronous capacitor switching, 283–284
Carbon emissions, reducing, 52–53
Carbon-neutral fuels, 51
Cascade/Choleski form, 583–585
Cassie-Mayr theory, 262–263
Cassie's theory, 261–262
Cayley–Hamilton theorem, 561–562
Cellulosic ethanol, 50
Cement plants, 70
Characteristic equations of the matrix, 139
Characteristic matrix, 561
Characteristic roots, 561
Characteristic vectors, 562
Charged capacitor, short-circuit of, 543–544
Chernobyl disaster, 48

Circuit breakers, 73, 269, 272, 273, 275, 277, 313, 479
    asymmetrical ratings, 316
        contact parting time, 316–317
    chopping numbers, 285, 285*t*
    current-zero breaker, 263
    DC circuit breakers, 545–548
    definite-purpose, 326
    failure modes of, 22*t*, 293–295, 293*f*, 294*f*
    4.16 kV circuit breakers, 452
    general-purpose, 313, 326
    generator, 353–357, 354*f*, 437, 438–439
    high-voltage, 358
    high-voltage DC circuit breakers, 550–552
    indoor oil-less, 317, 318*t*, 319*t*
    line closing switching surge factor, 335
        switching of transformers, 336–337
    low-voltage, 358, 452, 454*t*
        insulated case circuit breakers (ICCBs), 359
        low-voltage power circuit breakers
        (LVPCBs), 359–363
        molded case circuit breakers, 358–359
    maximum peak current, 321–322
    out-of-phase switching current rating, 337
    peak recovery voltage across, 287
    permissible tripping delay, 322
    prestrikes in, 279–280
    rated 100 kV and above, 338–342
    rated below 100 kV, 338
    resistors, 289
    rheostatic, 259
    service capability duty requirements and
        reclosing capability, 322
        transient stability on fast reclosing,
        323–326
    shunt capacitance switching, 326
        switching of cables, 331–335
    solid-state, 306–308
    with sources on either side, 441–443
    stresses in, 295
    timing diagram, 320–321, 321*f*
    total and symmetrical current basis, 314–315
    Transient Recovery Voltage (TRV), 337
        adopting IEC TRV profiles in IEEE
        standards, 345–350
        circuit breakers rated 100 kV and above,
        338–342
        circuit breakers rated below 100 kV, 338
        definite-purpose TRV breakers, 350
        initial TRV, 345
        oscillatory TRV, 344
        short-line faults, 342–344
        TRV calculation techniques, 350–353
    voltage range factor, 317–320

Circuit breakers, interrupting mediums in, 295
  operating mechanisms, 299–300
    SF6 circuit breakers, 296
      electronegativity of SF_6, 297–299
    vacuum interruption, 300
      current chopping and multiple ignitions, 301–302
      unloaded dry-type transformers, switching of, 302–304
Circuit equations of unit machines, 386–390
Clarke component transformation, 138, 143, 146–149
Closing resistors, 283, 291
Coal-based plants, 52
Coefficient of grounding (COG), 195, 203–204
Cofactor, 559
Combined heat and power plants (CHP), 92
Community Choice Aggregation, 51
Commutative law, 558
Concentrated solar power (CSP) plants, 115, 116, 117f
  solar energy collectors, 116
    parabolic dish concentrators, 117–118, 118f
    solar tower, 118, 119f
  trackers, 118
    photovoltaic trackers, 119–120
Concentrator photovoltaic (CPV) trackers, 119, 120
Condensing power plants, 92
Congestion management, 89, 93
Connected graph, 223–224
Consistent equations, 574
Constituency relations, 3
Consuming function, 611
Contact parting time, 263, 314, 315, 316–317, 353, 430, 432, 433, 434, 438, 460, 461, 516
Contingency management, 87
"Controlling the blade pitch," 99
Cooperatively owned utilities, 36
Coplanar vector, 137
Core losses, nonlinearity in, 612–614
Corp residues, 50–51
Cotree, 224, 225f
Council for superconductivity (CSC), 474
Coupled branch, removal of, 238–246
Coupled primitive impedance matrix, 244
Critical damping resistor, 291
Critical resistance, evaluation of, 291
Crout's transformation, 576–578
Cube law, 97–99
Cumulative distribution function (CDF), 16
Current chopping, 667
Current-limiting fuses, 363f, 364–365, 364f, 365f, 366, 443

Current-zero breaker, 261, 263–264, 264f
Cut-set method, 18–19, 23f, 224
Cylindrical rotor generators, 492, 492f

D

Damper leakage reactance, 407
Decoupling a three-phase symmetrical system, 141–145
Decoupling a three-phase unsymmetrical system, 145–146
Decrement factor of a decaying exponential current, 134
Definite-purpose circuit breakers, 313, 326, 333
Deionization, 261
Delta–delta system, 158f, 201f
Delta–delta transformer, 158
Delta–wye transformer, 155–157, 157f, 168f, 169f, 180
Demand response (DR), 59
Demand side management (DSM), 38
Deregulation of power industry, 38
  Distribution Company (DISTCO), 39
  Generation Company (GENCO), 38
  Transmission Company (TRANSCO), 39
Diagonalization technique, 139
Diagonal matrix, 563
Dielectric failure, of circuit breakers, 293, 294f
Digital computer-based calculation, 434, 439
Direct axis subtransient reactance, 381
Direct coupled induction generator, 105
Direct current (DC) systems, 521
  calculation procedures, 523
    IEC calculation procedure, 523–525
    matrix methods, 525
  charged capacitor, short-circuit of, 543–544
  DC arc characteristics for different lengths, 551
  DC circuit breakers, 545–548
  DC motors and generators, short-circuit of, 531–537
  DC rated fuses, 548
  DC short-circuit current sources, 521–523
  DC value, 422
  high-voltage DC circuit breakers, 550–552
  lead acid battery, short-circuit of, 525–531
  rectifier, short-circuit of, 537–543
  semi-conductor devices, protection of, 548–550
  short-circuit current–time curve for DC motor/generator, 532
  total short-circuit current, 544–545
Dish Sterling system, 118

Index 713

Distortion component, 610
Distributed generation, microgrids and, 59–62
Distributed parameter system, 11–12
Distributed Resources (DR) technologies, 123
  utility connections of, *see* Utility connections
    of distributed resources
Distribution Company (DISTCO), 39
Distribution systems, 13, 21, 23*f*, 39, 59, 71, 71*f*,
    76*f*, 77, 209*f*
  network/grid system, 73, 74*f*
  parallel/loop system, 73, 74*f*
  primary, 75–76
  radial system, 72–73, 75*f*
Double line-to-ground fault, 132, 177–179, 203,
    247–256
  in a three-phase system, 178*f*
  equivalent circuit, 178*f*
Doubling effect, 133
Doubly fed induction generator (DFIG), 106–107,
    108, 110
Doubly fed induction motor (DFIM), 112
Drive train, 94–95
Dry steam plant, 49
DSTATCOM, 110
D–T–Li cycle, 47
Duality models, 617
Dual-voltage rated circuit breaker, 482
Duplex reactor, 625–626
Duplicate utility sources, system with, 25*f*
Dynamic system, 2

E

Eastern Interconnection, 37
Eddy current loss, 598
"Effectively grounded" systems, 195, 346*t*, 347*t*,
    351, 352
Eigenvalues, 138–139, 142, 561
Eigenvectors, 138–139, 562
Einstein, Albert, 120
Electrical Generating System Association, 15
Electrical power system (EPS), 1, 123
  dynamic system, 2
  electrical standards and codes, 14–15
  linear system, 5
    property of decomposition, 6
  lumped and distributed parameters,
    11–12, 11*f*
  nonlinear system
    linearizing, 6–9
    property of decomposition, 6
  optimization, 12
  planning and design of, 12–14

power system studies, 28–29
power system studies software, 29
reliability analyses, 15
  availability, 15–18
  data for reliability evaluations, 18
  methods of evaluation, 18–22
  reliability and safety, 22–25
state variables, 3–5
static system, 2
system modeling, extent of, 25
  harmonic analysis, 28
  load flow calculations, 28
  short-circuit calculations, 26–28
time duration of transient phenomena in, 1
time-invariant systems, 9–11
units, system of, 30
Electric Power Research Institute (EPRI)
    projects, 307
The Electrification Council (TEC), 14
Electromagnetic force (EMF), 372, 377, 387, 388,
    390
ElectroMagnetic Transients Program (EMTP),
    306, 334, 380, 441, 513, 612, 613–614
Electromotive force (EMF), 180, 182, 423, 595
Elementary row operations, 568–569
Elements of the set, 555
Energy Independence and Security Act (EISA),
    58
Energy Service Companies (ESCOs), 39
Energy storage, 58, 63, 65–66*t*
  flywheel storage, 64–67
  superconductivity, 67
    electrical systems, applications in, 68
Environment Protection Agency (EPA), 53
Equal area criterion of stability, 325
Equivalent circuit for short-circuit current
    calculations of rectifier, 540
Equivalent circuits during fault, 380–383
Equivalent impedance, 207, 423, 462
  deriving, 464–469
Equivalent machine model, for short-circuit
    calculations, 426*f*
Equivalent surge impedance, 352
Ethanol, cellulosic, 50
Excitation current, 610
Exempt wholesale generators (EWGs), 38
E/X method for AC and DC decrement
    adjustments, 427–428, 437
E/X or *E/Z* calculation, 436–437
Exponential cosine envelope, 339
Exponential distribution, PDF for, 17–18
Extended transformer models, 610–614
Extra high voltage (EHV) networks, 87, 282, 314

F

Failure mode and effect analysis (FMEA), 19
Far-from generator faults, 483
  meshed networks, 487–489
    equivalent frequency method, 488–489
    ratio R/X or X/R, at short-circuit location, 487–488
    uniform ratio R/X or X/R ratio method, 487
  nonmeshed sources, 485–487
"Far-from-generator" short-circuit, 483
Fast breeder reactors (FBR), 47
Fault current limiters (FCL), 68, 470–473, 471f, 472f
  superconducting fault current limiters, 473–474
Fault decrement curve, 383–386, 386f, 460
Fault point X/R ratio, 427
Federal Energy Regulatory Commission (FERC), 38, 39, 103
Federally owned utilities, 36
Feeder circuit breakers, 442, 451t
Ferroresonance, 194, 620–621
Field emission, 300
Field leakage reactance, 407
Filter reactors, 622
First-cycle peak, 499
First generation cells, 122
Fission process, 46, 46f, 47
Flash steam plant, 49
Flexible AC transmission system (FACTS), 35
Flow graphs, 588–591
Flywheel energy storage systems (FESs), 64, 67, 67f
Flywheels, 67
Flywheel storage, 64–67
Foot-pound-second (FPS) system, 30
Forced oscillation, 329
Forced outage overlapping schedule outage, 20
Forward substitution, 580–583
Fossil fuels, 40, 44
  transportation of, 52
Free oscillation, 329
Fuel cells, 51–52
Fundamental frequency component, 610
Fuses, 363
  current-limiting fuses, 363f, 364–365, 364f, 365f, 366, 443
  DC rated fuses, 548
  high-voltage fuses, 365–366
  interrupting ratings, 366–367
  low-voltage fuses, 365
Fusion, nuclear, 47–48

G

Gas turbines, 59, 92
Gates, Bill, 48
Gaussian elimination, 578–579
General-purpose circuit breaker, 313, 326
General Service Administration (GSA), 53
Generation Company (GENCO), 38
Generator breaker, 353–357, 439, 441, 461
  selection of, 463–464
    analytical calculation, 468–469
    simplified calculations, 466–467
Generator fault currents, 509, 509t
Generator flux, 109
Generator resistance, 489, 505
Generators and unit transformers of power station units, 490–491
Generator-saturated subtransient reactance, 187
Generators circuit breakers, 439
Generators directly connected to systems, 489–490
Generator source asymmetrical current, 461
Generator source asymmetry, 437–439
Generator source short-circuit current, 354, 355, 509
Generator source symmetrical short-circuit current, 460
Generator step-up (GSU) transformers, 103, 356
Generator terminal short circuit
  electromagnetic transients program simulation of, 513
Generator voltage, 460, 489
Geomagnetically induced currents (GICs) models, 617–619
Geothermal energy, 49
Geothermal plants, 49–50, 56t
Graph approach
  bus admittance and impedance matrices by, 226
    incidence matrix from graph concepts, 228–232
    node elimination in Y-matrix, 232–233
    primitive network, 226–228
Graph theory, 223–226
Green energy, 41
  local green energy systems, 51
  nuclear power as, 48–49
Greenhouse gases (GHS)
  limits on, 59
  reductions, 53
Grid controlled converter, 550

Index 715

Grid/network system, 73
Grounding, 123
    coefficient of, 203–204
    system grounding, *see* System grounding methods
Grounding transformer, 200, 201*f*, 202
GTOs, 306, 307*f*, 548

H

Harmonic analysis, 1, 28
Harmonic emission, 70
Harmonics, 35, 70, 112–113
Hazard function, 16, 18
Hazard rate, 16
Helium-4 ion, 47
Hermitian matrix, 557
High- and low-pressure arcs, 261
Highest short-circuit currents, 481, 482
High frequency models, 614–617
High-level wastes (HLW), 44
High-resistance grounded (HRG) systems, 197
    limitations of, 199–202
High-speed reclosing, 321, 323
High-temperature superconductor (HTS), 67–68
High-voltage circuit breakers, 282, 292*f*, 294*f*, 295*f*, 314
    specifications of, 358
High-voltage, direct current (HVDC) transmission, 80, 87–89, 552
    HVDC configurations and operating modes, 80–82
    HVDC light, 80
High-voltage DC circuit breakers, 550–552
High-voltage fuses, 365–366, 366*t*
Homogeneity, 5–6
Horizontal axial flow arrangement, 45*f*
Horizontal single axis trackers (HSAT), 120
Horizontal single axis trackers with tilted modules (HTSAT), 120
HV lines, overvoltages on energizing, 280
    overvoltage control, 282–283
    shunt reactors, 284
        oscillation modes, 287–288
    synchronous capacitor switching, 283–284
    synchronous operation, 283
Hydroelectric plants, 41–43
    pumped storage systems, 43
Hydrogenation, 92
Hydropower, 41–42
Hysteresis loop, model of, 611
Hysteresis loss, 598

I

Idempotent matrix, 558
Identity matrix, 557
IEC standard, 479
IEEE Industry Applications Society and Power Engineering Society, 14
Impedance matrices, 141–142
    bus admittance and, 226
        incidence matrix from graph concepts, 228–232
        node elimination in Y-matrix, 232–233
        primitive network, 226–228
    loop admittance and, 221
        selection of loop equations, 223
Incidence matrix from graph concepts, 228–232
Inconsistent equations, 574
Individual power systems, 35
Induction generators, 103–105
    connected to grid through full size converter, 105–106
    direct coupled, 105
    doubly fed, 106–107
Induction machine, 103, 104*f*, 410–411
    direct coupled, 105
    stator of, 106
    transformation of, 410*f*
Induction motors, 103, 105, 409–413
    hp rating, 428
Inductors, defined, 2
Industrial bus-connected generators, 197
Industrial systems, 69–70, 193, 357*f*, 408
Inherent availability (Ai), 16
Inherent transient recovery voltage (ITRV), 269
Initial short-circuit current, 491, 494, 498, 501, 505, 507, 508, 511, 512
    calculation of, 484, 484*f*, 488*f*, 495*f*
Initial symmetrical short-circuit current, 480, 489, 494, 496, 497
Injection well, 49
Input-output-state relations for a system, 4
Institute of Electrical and Electronics Engineering (IEEE), 313
Insulated case circuit breakers (ICCBs), 359–360, 361, 452
Insulated gate bipolar transistor (IGBT) topology, 80, 111
Integrated Resource Planning (IRP), 38
International Council for Large Electrical Power Systems (CIGRE), 1
International Electrotechnical Commission (IEC) standard, 182, 313, 362, 421, 521
International grids, 35

716   *Index*

Interrupting ratings, 320, 362, 366–367
Interruption of shortcircuit currents, 259
    arc interruption theories, 261
        Cassie-Mayr theory, 262–263
        Cassie's theory, 261–262
        Mayr's theory, 262
    capacitance currents, interruption of, 276–278
    circuit breakers
        failure modes of, 293–295
        stresses in, 295
    circuit breakers, classification of, 295
        operating mechanisms, 299–300
        SF_6 circuit breakers, 296–299
        vacuum interruption, 300–304
    current-zero breaker, 263–264
    low inductive currents, interruption of, 273
        virtual current chopping, 275–276
    out-of-phase closing, 288–289
    overvoltages on energizing HV lines, 280
        overvoltage control, 282–283
        shunt reactors, 284–288
        synchronous capacitor switching, 283–284
        synchronous operation, 283
    part winding resonance in transformers, 304
        snubber circuits, 306
    prestrikes in circuit breakers, 279–280
    resistance switching, 289–292
    rheostatic breaker, 259–261, 260*f*
    short-line fault, 271–272
    solid-state circuit breakers (SSBs), 306–308
    terminal fault, 269
        four-parameter method, 269–270
        two-parameter representation, 270–271
    transient recovery voltage (TRV), 264
        in capacitive and inductive circuits,
            278–279
        first pole to clear factor, 266–269
Inverse of a matrix, 568–574
    by calculating the adjoint and determinant
        of the matrix, 568
    by elementary row operations, 568–569
    by partitioning, 570–574

K

K factor-rated breakers, 317
Kirchoff's current law, 224
Kirchoff's voltage law, 222

L

Laplace transform, 399, 404, 405, 551
Large power stations of the world, 53–57

LDU product form, 583–585
Lead acid battery, short-circuit of, 525–531
Leakage reactance, 372
Linear independence and dependence of
    vectors, 565
Linear network graphs, 223
Linear system, 5, 6
    decomposition, property of, 6
Line closing switching surge factor, 335, 336*t*
    transformers, switching of, 336–337
Line drop compensator, 610
Line-to-ground fault, 132, 158*f*, 173–175, 174*f*,
    186, 193, 246, 431*f*
    double, 177–179, 178*f*, 203, 247
    single, 187, 189*f*, 203, 269, 325, 326*f*, 361, 427
Line-to-line fault, 175–177, 246
    in a three-phase system, 176*f*
    equivalent circuit, 176*f*
Line-to-neutral voltage, 153, 180
Links/link branches, 224
Load-commutated inverter (LCI), 70
Load flow calculations, 28
Load management systems, 77
Load shedding, 93
Load side oscillations, 286, 287
Loop admittance matrix, 221, 222
    selection of loop equations, 223
Loop impedance matrix, 222, 226
    selection of loop equations, 223
Low inductive currents, interruption of,
    273, 274*f*
    virtual current chopping, 275–276
Low-level wastes (LLW), 44
Low-pressure arc, 261
Low-resistance grounded system, 197
Low-voltage circuit breakers, 358, 452, 454*t*
    insulated case circuit breakers
        (ICCBs), 359
    low-voltage power circuit breakers
        (LVPCBs), 359–363
    molded case circuit breakers, 358–359
    short-circuit duties on, 454*t*
Low-voltage fuses, 365
Low-voltage motor groups, 496
Low-voltage motors, 424–425, 511
Low-voltage power circuit breakers (LVPCBs),
    359, 452
    series connected ratings, 362–363
    short-time ratings, 361–362
    single-pole interrupting capability, 361
L-type fuse, time current characteristics of, 549*f*
Lumped and distributed parameters, 11–12, 11*f*
Lumped parameter system, 11–12

Index

717

M

Magma, 49

Magnetomotive forces (MMFs), 159, 373, 377, 601

Maintenance downtime (Mdt), 16

Man-machine interface (MMI) errors, 22

Matrices, 557–560

Matrix methods
bifactorization method, 582–583
Cayley–Hamilton theorem, 561–562
characteristic roots, 561
characteristic vectors, 562
Crout's transformation, 576–578
diagonalization of a matrix, 563–564
eigenvalues, 561
eigenvectors, 562
forward and backward substitution method, 580–583
Gaussian elimination, 578–579
inverse of a matrix, 568–574
LDU product form, 583–585
linear independence and dependence of vectors, 565
matrices, 557–560
quadratic form expressed as product of matrices, 565–566
scalar and vector functions derivatives, 566–567
sets, 555–556
similarity transformation, 563–564, 636
solution of large simultaneous equations, 574–576
vectors, 556
vector spaces, 565

Matrix methods for network solutions, 213
bus admittance and impedance matrices by graph approach, 226
incidence matrix, 228–232
node elimination in Y-matrix, 232–233
primitive network, 226–228
bus admittance matrix, 214–218
bus impedance matrix, 219
from open-circuit testing, 220–221
bus impedance matrix, algorithms for construction of, 233
adding a link, 236–238
adding a tree branch to an existing node, 234–236
changing impedance of an uncoupled branch, 238
removal of a coupled branch, 238–246
removal of an uncoupled branch, 238

bus impedance matrix, short-circuit calculations with, 246
double line-to-ground fault, 247–256
line-to-ground fault, 246
line-to-line fault, 246
graph theory, 223–226
large network equations, solution of, 256–258
loop admittance and impedance matrices, 221
selection of loop equations, 223
network models, 213–214

Maximum peak current, 321–322, 422

Maximum Transfer Capability (MTC), 91–92

Mayr's theory, 262

Mean time between failures (MTBF), 22

Medium-voltage turbine generators, 491

Megavolt amp (MVA) base, 187, 428

Meshed networks, 487–489
short-circuit currents in, 494–495, 495f

Metallic return transfer breaker (MRTB), 81, 552

Metal-oxide surge arresters, 287, 306

Meter-kilogram-second-ampere (mksa) system, 30

Microgrids and distributed generation, 59–62, 60f

Microturbines, 60

Modern electrical power systems, 35, 36f
biofuels and carbon-neutral fuels, 50–51
carbon emissions, reducing, 52–53
classification, 35
North American Power Systems Interconnections, 37–38
utility companies in the USA, 36–37
deregulation of power industry, 38
Distribution Company (DISTCO), 39
Generation Company (GENCO), 38
Transmission Company (TRANSCO), 39
distribution systems, 71, 71f
network or grid system, 73, 74f
parallel or loop system, 73, 74f
primary, 75–76
radial system, 72–73, 75f
energy storage, 63, 64–65t
flywheel storage, 64–67
superconductivity, 67–68
fuel cells, 51–52
future load growth, 77
general concept of, 35, 36f
geothermal plants, 49–50
green energy, 41
HVDC transmission, 80, 82f
HVDC configurations and operating modes, 80–82

Modern electrical power systems (*cont.*)
HVDC light, 80
hydroelectric plants, 41–43
industrial systems, 69–70
large power stations of the world, 53–57
local green energy systems, 51
microgrids and distributed generation, 59–62, 60f
nuclear power, 43
breeder reactors, 47
as green energy, 48–49
nuclear fusion, 47–48
around the globe, 48
pumped storage hydroelectric plants, 43
renewable energy, 40–41
smart grid, 57
legislative measures, 58
technologies driving, 58–59
solar and wind energy, 50
sustainable energy, 40
transmission systems, 68–69
underground versus OH systems, 77
spot network, 78–79
Molded case circuit breakers (MCCB), 358–359, 361–362, 452
Molten carbon fuel cell (MCFC), 51
Motor circuit protectors (MCPs), 359
Motor-locked rotor reactance, 411
Multivoltage-level distribution system, for short-circuit calculations, 444, 447f

N

National Electrical Code (NEC), 14
National Electrical Manufacturer's Association (NEMA), 14, 443
National Electric Safety Code (NESC), 15
National Energy Policy Act (NEPA), 38
National grids, 35, 213
National Institute of Occupational Safety and Health (NIOSH), 14
National Reliability Council, 38
National Renewable Energy Laboratory, 51
Near-to-generator faults, 489
generators and unit transformers of power station units, 490–491
generators directly connected to systems, 489–490
meshed networks, short-circuit currents in, 494–495
motors, 491
nonmeshed networks, short-circuit currents in, 493–494

short-circuit currents fed from one generator, 491–493
breaking current, 491–492
steady-state current, 492–493
Negative sequence components, phase shift for, 183–186
Negative sequence reactance, 374
Network for node elimination, 588–589
Network/grid system, 73, 74f
Network models, 213–214
Network protectors, 76, 78–79, 124
Network reduction, 435–436
E/X or E/Z calculation, 436–437
Neutrons, 45
number of neutrons released per capture of a neutron, 46
The New Oil Order, 39
Nilpotent matrix, 558
Niobium–Titanium (Ni–Ti) alloy, 67
Node elimination in Y-matrix, 232–233
Node/vertex of a graph, 224
Noncurrent-limiting fuses, 365
Nonlinearity, in core losses, 612–614
Nonlinear resistors, 2–3
defined, 2
Nonlinear system, 5, 6
decomposition, property of, 6
linearizing, 6–9
Nonmeshed networks, short-circuit currents in, 493–494
Nonsingular matrix, 565
Non-Sustained Disruptive Discharge (NSDD), 329
Nonutilities, 37
Normal forms, 560
North American Power Systems Interconnections, 37–38
North American Reliability Council (NERC), 38, 91
Norton's current equivalent, 214
Notice of proposed rulemaking (NOPR), 38
Nuclear energy production, 48, 48f
Nuclear fuel, 43, 44
Nuclear fusion, 47–48
Nuclear power, 43
breeder reactors, 47
as green energy, 48–49
nuclear fusion, 47–48
around the globe, 48
Nuclear power plants, 92
Nuclear Regulatory Commission (NRC) regulation for LLW, 44
Nuclear waste, 44, 48

Index

719

O

Occupational safety and Health Administration (OSHA), 14

Omnibus Trade and Competitiveness Act 1988, 30

One-minus-cosine curve, 338

On-load tap changers, 490

Open-circuit direct axis transient time constant, 384

Open-circuit driving point impedance, 221

Open-circuit testing, bus impedance matrix from, 220–221

Open-circuit test, of two-winding transformer model, 599

Open-circuit time constant, 375, 384

Open-circuit transfer impedance, 221

Open-circuit transient time constant, 411

Open conductor faults, 204
  one-conductor open, 204–208
  two-conductor open fault, 204

Opening resistors, 291

Operational availability (Ao), 16

Optimal ordering, 587
  schemes, 591–594

Optimal Power Flow (OPF), 91

Optimization, in power system, 12

Oriented graph, 223

Oscillating neutral, 157

Oscillation modes, 287–288

Oscillatory TRV, 344, 353
  exponential (overdamped) TRV, 344

Out-of-phase closing, 288–289

Out-of-phase switching capability, 355, 463

Out-of-phase switching current rating, 337

Overhead line conductors, 456–460

Overhead versus underground systems, 77
  spot network, 78–79

Overvoltage factor, defined, 280

Overvoltages on energizing HV lines, 280
  overvoltage control, 282–283
  shunt reactors, 284
    oscillation modes, 287–288
  synchronous capacitor switching, 283–284
  synchronous operation, 283

P

Pacific Gas and Light Company, 118

Parabolic dish concentrators, 117–118, 118f

Parallel ferroresonance, 621

Parallel operation of transformers, 601–604

Parallel short-circuit currents, short-circuit currents and, 486f

Parallel/loop system, 73, 74f

Park's transformation, 390
  reactance matrix
    of synchronous machine, 390–393
    transformation of, 393–395

Park's voltage equation, 395–397

Partial short-circuit currents, 487, 491, 496, 507, 510t, 511t, 545f

Partitioned matrices, 570–574

Part winding resonance in transformers, 304
  snubber circuits, 306

Peak making current, 355, 481

Peak short-circuit current, 484, 485, 490, 491

Periodic matrix, 558

Permissible tripping delay, 322

Peterson coil, 622

Phase currents, asymmetries in
  in three-phase short circuit, 136f

Phase shift
  for negative sequence components, 183–186
  in three-phase transformers
    for negative sequence components, 183–186
    transformer connections, 180
    in winding connections, 180–183
  in winding connections, 180–183

Phase-to-ground peak voltage, 286

Phase-to-phase mutual inductances, 392

Phosphoric acid fuel cell (PAFC), 51, 52f

Photoelectric effect, 120

Photovoltaic (PV) cells, direct conversion of solar energy through, 120
  cells modules panels and systems, 120
    PV array, 121
    PV array subfield, 121
    PV module, 120
    PV panel, 121

Photovoltaic trackers, 119–120

Pitch angle reference, 114

Point of common coupling (PCC), 62, 108–109

Polar aligned single axis trackers (PSAT), 120

Polarity (additive and subtractive), 600–601

Polymer electrolyte fuel cell (PEFC), 51

Potier reactance, 374, 374f

Power cables, 455–456

Power conditioning unit (PCU), 123

Power electronics, 111–112

Power-line carrier filter reactor, 622

Power station units (PSUs), 497
  generators and unit transformers of, 490–491

Power systems, *see* Modern electrical power systems

Power system studies, 13, 27f, 28–29
  software, 29

720   *Index*

Predictive maintenance, 631
Prefault voltage, 483
Prevention through design (PtD), 15
Preventive maintenance, 631
Prima facie, 138
Primary distribution system, 71, 75–76
Primitive network, 226–228, 227f
Probability density function (PDF), 16
    for exponential distribution, 17–18
Publicly owned utilities, 36
Pulse width modulation (PWM), 111, 112
Pumped storage hydroelectric plants, 43

Q

Quadratic form expressed as product of
    matrices, 565–566
Quadrature axis reactances, 373
Quadrature axis subtransient open-circuit time
    constant, 384
Quasi steady-state short-circuit current, 529,
    530, 534, 541

R

Radial distribution system, 23f, 72f
Radial system, 72–73, 73f, 76, 458
Radio interference (RI) reactor, 622
Rank of matrix, 560
Rated asymmetrical breaking capability, 479
Rated line closing switching surge factors, 282,
    335, 336t
Rated making capacity, 480
Rated opening time and break time, 480
Rated restriking voltage, 480
Rated symmetrical breaking current, 479
Rated transient inrush current, 328
Rate of rise of dielectric strength (RRDS),
    283, 284f
Rate of rise of recovery voltage (RRRV),
    266–267, 267f, 268f, 269, 291f, 293, 295,
    309, 344, 352
Reactance network, 427, 435, 440, 458
Reactances, saturation of, 375
Reactive power and wind turbine controls,
    107–111
Reactors, 622–626
Reclosing capability factor, 323
Rectifier, 111
    short-circuit of, 537–543, 537f
Reduced incidence matrix, 228
Reference power source, 336
Regional grids, 35

Reignition, 667
Reignition overvoltage oscillation, 287, 288f
Reliability analyses, 15, 20t
    availability, 16
        exponential distribution, 17–18
    methods of evaluation, 18–22
    reliability and safety, 22–25
    reliability evaluations, data for, 18
    of simple radial system, 24t
Reliability-centered maintenance (RCM), 631
Reliability evaluations, data for, 18
Renewable energy, 40–41, 51
Renewable energy sources (RES), 57, 85, 87
Renewable methanol (RM), 51
Renewable portfolio standards (RPS), 59
Resistance grounding, 195, 196–197
    coefficient of grounding (COG), 203–204
    high-resistance grounded (HRG)
        systems, 197
    limitations of, 199–202
Resistance network, 427, 458
Resistance of system components for short-
    circuit calculations, 428t
Resistance switching, 279, 283, 289–292, 358
Resistor, 2, 3, 11, 267, 290, 291, 294
Resonant delta, 607
Resonant frequencies, 96, 305, 306
Resonant overvoltages, 305
Restrikes, 667
Rheostatic breaker, 259–261, 260f
Rise of recovery voltage (RRRV), 668
Root mean square (RMS), 422
    value, of symmetrical sinusoidal wave, 314
Rotating machine model, 425–426
Rotor blades, 96–97

S

Safety instrumented systems (SIS), 22–24
Safety integrity levels (SIL), 22–24
Safety lifecycle, 22–23
Salient pole generators, 135, 491
Salient-pole machines, 493
Saturated synchronous reactance, 373, 481, 493
Saturation, 610
Sawtooth TRV waveform, 342
Scalar functions, derivatives of, 566–567
Scott connected transformer, 607–608
Secondary selective system, 22, 26f
Second generation cells, 122
Second-order minimum cut set, 20
Self-blast technology, 300
Self-correcting system, 22

Index

Self-impedance, 226
Self-monitoring system, 22
Semi-conductor devices, protection of, 548–550
Sequence impedance of network components, 153
  static load, 163
  synchronous machines, 163–168
  transformers, 155
    delta–delta transformer, 158
    delta–wye/wye–delta transformer, 155–157
    three-winding transformers, 159–163
    wye–wye transformer, 157–158
    zigzag transformer, 158–159
Sequence network
  computer models of, 168–169
  construction of, 153–155
  interconnections, 174, 176f, 178f
  negative, 154f, 166f, 167, 167f, 168, 240f, 241
  positive, 153, 154, 154f, 166f, 167, 167f, 169, 169f, 173, 240f, 241
  zero, 154f, 155, 166f, 167f, 168, 169
Series faults, 204
Series reactors, 622, 623
Service capability duty requirements and reclosing capability, 322
  transient stability on fast reclosing, 323–326
Sets, 555–556
SF_6 circuit breakers, 272, 296
  electronegativity of SF_6, 297–299
Short-circuit calculations, *see* Calculations, short-circuit
  ANSI methods, *see* Calculations, short-circuit (ANSI methods)
  IEC standards, *see* Calculations, short-circuit (IEC standards)
Short-circuit currents, nature of, 132–135
Short-circuit direct axis transient time constant, 384
Short-circuit test, of two-winding transformer model, 600
Short-circuit transient time constant, 411, 460
Short-line faults, 268, 269, 271–272, 342–344
Shunt capacitance switching, 326
  cables, switching of, 331–335
Shunt capacitors, 35, 62, 76, 328, 336, 413
Shunt reactors, 284, 285t, 286f, 622
  oscillation modes, 287–288
Silicon-controlled rectifiers (SCRs), 306, 548
Similarity transformation, 563–564, 636
Similar matrices, 141
Singular matrix, 565

Small-scale superconducting magnetic energy storage (SEMS) systems, 68
Smart grid, 57
  legislative measures, 58
  technologies driving, 58–59
Smoothing reactors, 622
Snubber circuits, 306
Software programs, 29
Solar and wind energy, 50, 63
Solar cells, classification of, 121–123
Solar energy collectors, 116
  parabolic dish concentrators, 117–118, 118f
  solar tower, 118, 119f
Solar energy, direct conversion of, 120
  cells modules panels and systems, 120
    PV array, 121
    PV array subfield, 121
    PV module, 120
    PV panel, 121
Solar power, 35, 85, 115–116; *see also* Concentrated solar power (CSP) plants
Solidly grounded systems, 173, 195–196
Solid oxide fuel cell (SOFC), 51
Solid-state circuit breakers (SSBs), 306–308
Sparsity and optimal ordering, 587–594
Specific rated capacity (SRC), 96
Spectral matrix, 140
Spot network, 78–79, 79f, 124
Square matrix, 557
Standard transmission line, 336
State variable model of time-invariant system, 10
State variables, 2, 3–5, 5f, 11, 92
Static converter fed drives, 497
Static converters contribution, to short-circuit currents, 414–417
Static exciters, 491
Static load, 163
Static power converter devices, 497
Static system, 2
Static VAR compensators, 35
Stator-to-rotor mutual inductances, 392
Steady-state analysis of power systems, 1
Steady-state current, 132, 329, 481–482, 492–493
  calculations
    ANSI method, 513
    IEC method, 513
Steady-state resistance, 532
Step voltage regulators, 608–610
Sterling engine, 118
Stresses in circuit breakers, 295
Subgraph, 224
Subtractive polarity, 600–601

722   *Index*

Subtransient reactance, 372, 374, 375, 424, 439
Subtransient short-circuit time constant, 375
Subtransmission systems, 36, 90
Sulfur fluorides, 299
Sun tracking, parabolic trough with, 117f
Superconducting fault current limiters (SFCL), 473–474, 473f, 474f
Superconductivity, 67
   electrical systems, applications in, 68
Superposition, 5–6
Suppression peak overvoltage, 286–287
Surge capacitors, 197, 306
Surge voltage distribution across transformer windings, 619–620
Sustainable energy, 40
Sustained fault current factor versus rectifier terminal voltage, 538
Switching devices without short-circuit interruption ratings, 443
Switching overvoltages, 282, 296, 335, 358
Switching resistors, 282, 291
Symmetrical breaking current, 264, 484, 485, 498, 502, 503, 506, 508, 512
Symmetrical components, 29, 131, 135–138, 149, 169, 204, 208, 266
   characteristics of, 150–153
   transformation, 138, 139, 149, 152
      decoupling a three-phase symmetrical system, 141–145
      decoupling a three-phase unsymmetrical system, 145–146
      power invariance in symmetrical component transformation, 146
      similarity transformation, 139–141
Symmetrical interrupting capacity, 317
Symmetrical short-circuit breaking current, 481, 491, 494
Symmetrical three-phase short-circuit current, 482
Sympathetic inrush, 337
Synchronous capacitor switching, 283–284
Synchronous compensators, 491, 495
Synchronous condensers, 35, 68, 408
Synchronous generator impedances, sequence components of, 165f
Synchronous generators, 101, 107, 135, 356, 371, 408, 429f, 469, 491
Synchronous machine, 105, 163–168, 323, 371, 441, 481, 489, 494
   behavior on short circuit, 375
      equivalent circuits during fault, 380–383
      fault decrement curve, 383–386
   calculation procedure and examples, 399

      manufacturer's data, 406–408
   circuit equations of unit machines, 386–390
   circuit model of, 397–398
   induction motors, 409–413
   Park's voltage equation, 395–397
   Park transformation, 390
      reactance matrix of a synchronous machine, 390–393
      transformation of reactance matrix, 393–395
   practical short-circuit calculations, 417
   reactance matrix of, 390–393
   reactances of, 372
      leakage reactance, 372
      negative sequence reactance, 374
      Potier reactance, 374, 374f
      quadrature axis reactances, 373
      subtransient reactance, 372
      synchronous reactance, 372–373
      transient reactance, 372
      zero sequence reactance, 374
   saturation of reactances, 375
   short-circuit currents, capacitor contribution to, 413–414
   short circuit of synchronous motors and condensers, 408–409
   static converters contribution to short-circuit currents, 414–417
   stator resistance of, 490
   time constants of, 375
      armature time constant, 375
      open-circuit time constant, 375
      subtransient short-circuit time constant, 375
      transient short-circuit time constant, 375
   V-curves of, 408, 409f
Synchronous motors, 321, 325, 408, 429f, 481, 491, 495, 502
   ANSI fault current calculations from, 500
   IEC fault current calculations from, 501
Synchronous operation, 283
Synchronous reactance, 372–373, 493
System grounding methods, 193, 194f, 195
   resistance grounding, 196–197
      coefficient of grounding (COG), 203–204
      high-resistance grounded (HRG) systems, 197–202
   solidly grounded systems, 195–196
System modeling, extent of, 25
   harmonic analysis, 28
   load flow calculations, 28
   short-circuit calculations, 26–28
System protection, 15

Index

723

System reliability assessment and evaluation methods, 15
System source asymmetrical short-circuit current, 462
System source symmetrical short-circuit current, 461–462

T

Taylor's series, 8–9
    linearizing a nonlinear system with, 9*f*
Teaser transformer, 607
Terminal antiresonance, 304–305
Terminal fault, 267, 269, 293, 372, 377, 412, 425
    four-parameter method, 269–270
    two-parameter representation, 270–271
Terminal resonance, 304
Texas Interconnection, 38
Thermal failure, of circuit breakers, 293
Thévenin branch equivalent, 214
Thévenin sequence voltages, 154
Thévenin theorem, 25
Third generation cells, 122
Thirty-cycle short-circuit currents, 469–470
Three-phase fault, 102, 173, 179–180, 266, 456
Three-phase modified decrement curve, 434
Three-phase power system, 426
Three-phase short circuit, 371, 378, 385, 486, 507, 509
    asymmetries in phase currents in, 136*f*
Three-phase symmetrical system, decoupling, 141–145
Three-phase-to-ground faults, 350–351
    interruption of, 351*f*
Three-phase transformers
    core form of, 156*f*
    phase shift in
        negative sequence components, phase shift for, 183–186
        transformer connections, 180
        winding connections, phase shifts in, 180–183
Three-phase ungrounded faults, 350, 426
Three-phase unsymmetrical system, decoupling, 145–146
Three-winding transformers, 159–163
    equivalent positive, negative, and zero sequence circuits of, 162*t*
    wye equivalent circuit of, 161*f*
Tie-set, 224
Tilted single axis trackers (TSAT), 120
Time constants of synchronous machines, 375
    armature time constant, 375

open-circuit time constant, 375
    subtransient short-circuit time constant, 375
    transient short-circuit time constant, 375
Time duration of transient phenomena in power systems, 1
Time-invariant systems, 9–11
Time-variant system, 9
Tip speed ratio (TSR), 97, 114
Tip-tilt dual axis trackers (TTDAT), 120
Title 10 of Code of Federal Regulations (10CFR), 47
Topographical graph or map of the network, 223
Total fault current profile, 545*f*
Total short-circuit current, 314, 353, 442, 544–545
Towers, 94, 96
Townsend coefficient, 294, 297
Trackers, 118
    photovoltaic, 119–120
Transformer polarity and terminal connections, 600–601
Transformers, 76, 78, 110, 155
    autotransformers, 604–608
    connections, 180
    delta–delta transformer, 158
    delta–wye/wye–delta transformer, 155–157
    duality models, 617
    extended models
        EMTP models, 612, 613–614
        hysteresis loop, model of, 611
        nonlinearity in core losses, 612–614
    ferroresonance, 620–621
    GIC models, 617–619
    impedance, 444
    with non-base voltage, 445–446
    parallel operation of, 601–604
    part winding resonance in, 304
        snubber circuits, 306
    polarity and terminal connections, 600–601
    primary switches and fused switches, 452, 454*t*
    and reactors, 622–626
    self-cooled MVA rating, 429
    step voltage regulators, 608–610
    surges transferred through, 615–617
    surge voltage distribution across transformer windings, 619–620
    switching of, 336–337
    three-winding transformers, 159–163
    two-winding transformer model, 595–600
    wye–wye transformer, 157–158
    zigzag transformer, 158–159

724   *Index*

Transformer taps and ratios, adjustments for, 443–444
Transient reactance, 372, 411
Transient recovery voltage (TRV), 264, 265, 266–267, 267*f*, 313, 337, 552
  adopting IEC TRV profiles in IEEE standards, 345–350
  calculation techniques, 350–353
  in capacitive and inductive circuits, 278–279
  circuit breakers rated 100 kV and above, 338–342
  circuit breakers rated below 100 kV, 338
  definite-purpose TRV breakers, 350
  first pole to clear factor, 266–269
  initial, 345
  oscillatory, 344
    exponential (overdamped) TRV, 344
  short-line faults, 342–344
Transient short-circuit time constant, 375, 382
Transient stability on fast reclosing, 323–326
Translation operator, effect of, 10*f*
Transmission Company (TRANSCO), 39
Transmission/distribution system network, 213
Transmission system operators (TSO), 93
Transmission systems, 37, 68–69, 87, 356, 408
Transmission Transfer Capability (TTC), 91, 91*f*, 92
Tree branches, 224
Tree-link concept, 224
Triggered current limiter (TCL), 470, 471
Tripping delay, 263, 316
  permissible, 322
Tritium, 47, 48
Two-winding transformers, 159, 160*t*, 180, 185, 595–600
24-bus system, 35, 37*f*

U

Ultra-high voltage (UHV), 68, 314
Unbalanced voltage vectors, 137, 152*f*
Uncoupled branch
  changing impedance of, 238
  removal of, 238
Underground versus overhead (OH) systems, 77
  spot network, 78–79
Underwriters Laboratories (UL), 14
Uninterruptible Power Supplies (UPS), 67, 414, 550
Unit inductance, 532
Unit machines, circuit equations of, 386–390
Unit matrix, 557

Units, system of, 30
Unloaded dry-type transformers, switching of, 302–304
Unsymmetrical fault calculations, 173, 186–193
  double line-to-ground fault, 177–179
  line-to-ground fault, 173–175
  line-to-line fault, 175–177
  open conductor faults, 204
    one-conductor open, 204–208
    two-conductor open fault, 204
  phase shift in three-phase transformers
    for negative sequence components, 183–186
    transformer connections, 180
    in winding connections, 180–183
  system grounding, 193
    resistance grounding, 196–204
    solidly grounded systems, 195–196
  three-phase fault, 179–180
  unsymmetrical fault calculations, 186–193
Unsymmetrical system, three-phase decoupling, 145–146
Upper triangular matrix, 557
Uranium, 40, 43, 45
US National Institute of Occupational Safety and Health (NIOSH), 14
Utility companies in the USA, 36–37
Utility connections of distributed resources, 123
  abnormal frequencies, 125
  area faults, 125
  distribution secondary spot networks, 124
  electromagnetic interference (EMI) interference, 124
  grounding, 123
  harmonics, 125–126
  inadvertent energization, 124
  isolation device, 124
  metering, 124
  paralleling device, 125
  reconnection, 125
  surge withstand, 125
  synchronizing, 123
  voltage control, 123

V

Vacuum interruption, 300, 302*f*
  current chopping and multiple ignitions, 301–302
  unloaded dry-type transformers, switching of, 302–304
V-curves of synchronous machines, 408, 409*f*
Vector functions, derivatives of, 566–567

Index

Vectors, 556
Vector spaces, 565
Verband der Elektrotechnik (VDE) standard, 421
Vermont Yankee 604 MW plant, 48
Vertical axis trackers (VSAT), 120
Vertical Francis unit, 44f
Virtual current chopping, 275–276
Voltage range factor, 317–320
Voltage source converters (VSC), 88, 106

W

Water hammer, 43
Wave propagation, 2, 11
Western Electricity Coordinating Council (WECC), 102, 107
Western Interconnection, 38
Wind energy conversion, 93, 95f
  drive train, 94–95
  rotor blades, 96–97
  towers, 95
Wind generation task force (WGTF), 102
Wind generators, 102, 103
  direct coupled induction generator, 105
  doubly fed induction generator, 106–107
  grid connections of, 106f
  induction generator connected to grid through full size converter, 105–106
  induction generators, 103–105
  synchronous generators, 107
Winding connections, phase shifts in, 180–183
Winding resonance response, 305
WindPACT drive train project, 95
Wind power, 29, 41, 49, 59, 88f, 92
Wind power generation, 85, 86f
  characteristics of, 87

congestion management, 93
  Maximum Transfer Capability (MTC), 91–92
  power reserves and regulation, 92–93
computer modeling, 113
  wind turbine controller, 113–115
cube law, 97–99
harmonics, 112–113
operation, 99
  behavior under faults and low-voltage ride through, 102–103
  speed control, 101–102
power electronics, 111–112
prospective of wind generation in the USA, 85–86
reactive power and wind turbine controls, 107–111
Wind turbine controller, 113–115, 115f
Wind turbine control zones, 114f
Wye–delta transformer, 155–157, 168f
Wye equivalent circuit of three-winding transformer, 161f
Wye–wye transformer, 157–158

Y

Y-matrix, node elimination in, 232–233

Z

Zero-input response, 6
Zero sequence eigenvector, 143
Zero sequence matrix, 253–256
Zero sequence reactance, 195, 352, 374
Zero-state response, 6
Zigzag transformer, 158–159, 193